Canadian Bulletin of Fisheries and Aquatic Sciences 226

# Perspectives on Canadian Marine Fisheries Management

*Edited by*

**L. S. Parsons and W. H. Lear**

*Department of Fisheries and Oceans*
*Biological Sciences Directorate*
*200 Kent Street*
*Ottawa, Ontario, Canada K1A 0E6*

NATIONAL RESEARCH COUNCIL OF CANADA
AND
DEPARTMENT OF FISHERIES AND OCEANS
Ottawa 1993

ISBN 0-660-15003-4
ISSN 0706-6503
NRCC No. 36176

**Canadian Cataloguing In Publication Data**

Main entry under title:
Perspectives on Canadian marine fisheries management

(Canadian Bulletin of Fisheries and Aquatic Sciences; no. 226)
Includes abstracts in French.
Includes bibliographical references.
ISBN 0-660-15003-4
DSS Cat. No: Fs 94-226E

1. Fishery management — Canada. 2. Fisheries — Canada. 3. Fisheries policy — Canada. I. Parsons L.S. II. Lear W.H. (William Henry). III. Canada. Dept. of Fisheries and Oceans. IV. Series.

SH328.P47 1993 333.95'6'0971 C93-099546-5

This publication is available from:

Subscription Office, Research Journals, National Research Council of Canada, Building M-55, Montreal Road, Ottawa, Ontario, Canada K1A 0R6

Enquiries: Tel.: 613-993-9084 FAX No.: 613-952-7656

Remittances should be made payable to the Receiver General for Canada, credit National Research Council of Canada.

---



---

Printed on recycled paper.

[*The views expressed are those of the author and not necessarily those of the Department of Fisheries and Oceans or the National Research Council of Canada.*]

This publication has been peer reviewed.

Correct citation for this publication:

Parsons, L. S., and W. H. Lear [ed.]. 1993. Perspectives on Canadian marine fisheries management. Can. Bull. Fish. Aquat. Sci. 226: 446 p.

# Contents

# Abstract

Parsons, L. S., and W. H. Lear [eds.] 1993. Perspectives on Canadian marine fisheries management. Can. Bull. Fish. Aquat. Sci. 226: 446 p.

The diversity and complexity of Canada's marine fisheries present major challenges to their successful management. Alternating periods of boom and bust confound fisheries managers' efforts to achieve stability. This volume describes management approaches in different fisheries arising from this diversity. Five chapters on Atlantic species depict the boom and bust phenomenon in the Atlantic fisheries and the contrasting approaches to finfish and invertebrate management. The important role of the marine environment in determining resource productivity is also emphasized. Arctic fisheries are small, widely scattered and diverse, are difficult to manage and rely on self-compliance by resource users. In the Pacific, despite the enormous complexities of managing five species and hundreds of stocks, managers have achieved considerable success in maintaining, rebuilding and enhancing salmon populations in recent years. The management process for Pacific herring illustrates the benefits of involving fishermen closely in the assessment process. Government financial interventions to sustain the fisheries through various recurring crises are described. The role of scientific advice in the management process is examined. There have been many studies of Canada's marine fisheries; two early 1980's studies — the Kirby and Pearse reports — illustrate the influence of process in determining the impact of such studies on public policy-making. Finally, the potential and limitations of aquaculture to supplement or replace some of the traditional wild fisheries is examined.

---

La diversité et la complexité des pêches en eau salée effectuées dans les eaux canadiennes présentent d'importants défis sur le plan de la gestion. L'alternance de forte expansion et de récession dans cette industrie compliquent les efforts des gestionnaires des pêches en vue d'atteindre un certain niveau de stabilité. On décrit dans le présent document des approches de gestion de diverses pêches axées sur cette diversité. Cinq chapitres consacrés aux espèces de l'Atlantique décrivent ce phénomène de l'alternance de forte expansion et de récession, et les approches opposées de gestion de la pêche de poissons et d'invertébrés. On met aussi l'accent sur l'important rôle que joue l'environnement marin dans le niveau de productivité de la ressource. Étant donné que les pêches dans l'Arctique sont de faible ampleur, isolées et diverses, leur gestion est difficile et repose sur l'auto-contrôle des usagers de la ressource. Malgré le niveau élevé de complexité de la gestion de cinq espèces et de centaines de stocks de saumons dans le Pacifique, les gestionnaires ont connu un grand succès dans le rétablissement et la mise en valeur des salmonidés au cours des dernières années. La gestion du hareng du Pacifique illustre bien les avantages de l'engagement des pêcheurs dans le processus d'évaluation. On décrit les subventions fédérales qui ont été versées afin de renflouer cette pêche dans ses diverses périodes de creux. On examine aussi le rôle des avis scientifiques dans le processus de gestion. Des études menées sur les pêches canadiennes en mer, les Rapports Kirby et Pearse au début des années '80 illustrent l'influence du processus dans la détermination de l'impact de telles études sur la formulation de politiques officielles. En dernier lieu, on examine le potentiel et les limites de l'aquiculture pour ce qui est de compléter ou de remplacer certaines des pêches traditionnelles.

## Acknowledgements

We would like to thank about 30 reviewers who made constructive suggestions for improving the various manuscripts. Some of the reviewers are acknowledged by individual authors but others preferred to remain anonymous. We would especially like to thank Dr. Phil Symons, the consulting editor, for his invaluable assistance and advice. Thanks go to Ronnie Sanford, who typed several of the manuscripts and made revisions to others.

## Introduction

This volume is a companion to L.S. Parsons' *Management of Marine Fisheries in Canada* (1993). Parsons describes and evaluates the impact of the major changes in the management of Canada's marine fisheries in recent decades. He concludes that, despite the benefits which flowed from the 1977 extension of fisheries jurisdiction to 200 miles, Canada's marine fisheries continued to be plagued by instability because of various problems and constraints. These include:

1. natural resource variability, often environmentally determined;
2. the common-property nature of fisheries resources and resultant overcapacity;
3. market fluctuations;
4. heavy dependence on the fisheries in isolated coastal communities;
5. recurrent conflict among competing users; and
6. conflicting objectives for fisheries management.

Despite abundant resources, various combinations of these factors have contributed to recurrent boom-and-bust patterns in many marine fisheries. Parsons suggests that extended jurisdiction and various post-extension initiatives have not solved the problems of the fisheries sector. He identifies an urgent need to bring harvesting and processing capacity into balance with sustainable resource levels in many fisheries.

In the first chapter of this volume, Gough provides an historical sketch of fisheries management in Canada. He outlines the early attempts at management and demonstrates that, although it may appear that the solutions to modern problems are recent innovations, the ideas behind them have been around for a long time. The overfishing by foreign fleets and the depletion of many stocks provided the catalyst for major changes in management approaches in the 1960's and 1970's. Social, political, economic and biological factors demanded action to prevent a major catastrophe. The convergence of these forces plus the combination of timing and political will precipitated a revolution in fisheries management.

The next five chapters depict the diversity of the Atlantic fisheries and the complexity of their management. The five species/stocks chosen — Northern cod, Atlantic lobster, Atlantic herring, Atlantic salmon and Gulf of St. Lawrence snow crab — exemplify the tremendous challenges of managing the Atlantic fisheries. Approaches to the management of each of these five species or stocks evolved differently in response to a different array of management problems. These case studies from the Canadian Atlantic demonstrate that there is no simple cookbook solution to the management of marine natural resources. Successful management is contingent upon reasonably reliable scientific estimates of biomass, recruitment and fishing mortality rates. This presents scientists with a formidable task in the face of changing fishing technology, fishing patterns and generally increasing efficiency of fishing effort even when overall entry to the fishery is limited. The upsurge in lobster productivity through the 1980's and the sudden cataclysmic decline in the northern cod stock at the beginning of the 1990's illustrate the important role of the marine environment in determining the productivity of living aquatic resources.

The difficulties encountered recently in managing the northern cod resource underscore the uncertainties inherent in fisheries management. They exemplify the great difficulty in long-term planning when dealing with a naturally variable resource. Attempts to achieve stability in this situation have been frustrating. The stochastic nature of fish population abundance does not lend itself readily to linear models. The effects of large-scale environmental perturbations such as those which affected northern cod and other Atlantic groundfish in the early 1990's can confound attempts to manage marine fish resources. One lesson which can be distilled from the northern cod management experience is the necessity for a cautious, conservation-oriented approach whereby the stock is allowed to grow and the harvest is controlled at a low fishing mortality, in conjunction with other targets such as minimum spawning biomass constraints.

Conflicts over who gets to harvest the resource complicate the management process as illustrated by the cases of northern cod, Atlantic salmon and herring. The extreme difficulty of managing a natural resource in the absence of an historical scientific data base is exemplified by the case histories for northern cod and snow crab in particular, and, to a lesser extent, for the other Atlantic species.

International conflicts also complicate the management of fisheries within the Canadian zone. Applebaum, in his chapter on the straddling stocks problem, uses the example of the Northwest Atlantic fisheries to demonstrate how extending fisheries jurisdiction to 200 miles provided the key to conservation of most of the world's fish stocks. The 200 mile limit, however, left some stocks or portions of stocks exposed to fishing pressures from international fleets on the high seas beyond 200 miles. The fact that some stocks straddle the boundary between coastal-state jurisdiction and the high seas created a jurisdictional gap. In the Canadian case, this has undermined the ability of the coastal state to effectively manage these straddling stocks. International organizations such as NAFO are limited in their effectiveness since they are dependent on consensus among their members and on the willingness of other countries to respect management decisions made by either the international organization or the coastal state.

Clarke, in the chapter on Arctic marine fisheries, illustrates the integral and major role that fisheries play in the lives of northern Canadians. These fisheries are small and widely scattered and their development and management is made difficult by a lack of knowledge of the biology of the fishery resources and limited economic potential. For most of these fisheries, there is heavy reliance on self-compliance by the resource users. Fisheries management in the Arctic is evolving in the direction of formal cooperative management arrangements in which resource users share responsibility for fisheries management with government. One of the greatest challenges in the Arctic is to ensure that industrial activities do not affect the fishery resources adversely through deterioration of fishery habitats.

The section on the Pacific fisheries illustrates further the enormous challenge of managing fisheries which display such biological, economic and social complexities. Healey describes the management of the fisheries based upon five species of Pacific salmon, all of which differ widely in their ecology and productivity and which originate from many rivers comprising hundreds of different populations. The combination of the numbers of species and populations with the inherent uncertainty of survival rates and ensuing recruitment makes for mind-boggling complexity when devising management plans. Complexity is compounded by conflicts among the variety of user groups which harvest the Pacific salmon, particularly the commercial, recreational and native fisheries. Migration across international boundaries further confounds the management process. Despite these difficulties, Pacific salmon managers have been successful in maintaining most populations. Variability in recruitment, however, will continue to be a significant factor in influencing the relative stability of catches. Some recent evidence indicates that large-scale climatic factors may have influenced the upsurge in salmon stocks throughout the north Pacific during the 1980's.

Stocker describes the management of Pacific herring in British Columbia and the susceptibility of these stocks to large fluctuations in recruitment and to overfishing. The herring resource exhibits considerable natural variability in productivity. Variability in the markets for roe herring lends a further element of uncertainty to this fishery. Stocker describes how the credibility of the scientific assessment process for these stocks has been enhanced by involving fishermen closely in this process.

Fundamental to successful fisheries is the maintenance of suitable habitat to enable the successful reproduction, rearing and growth of fish species. This is especially true for freshwater habitats for salmonids. These are particularly vulnerable to human impact, because it is here that the adults spawn, the eggs incubate and the fry and smolts rear. The success of a year-class of a Pacific salmon species is heavily influenced by conditions during the freshwater phase of its existence, although survival rates in the marine environment are also important. Waldichuk emphasizes that there are no universally applicable rules or formulae for fish habitat management. Most of the decisions taken to protect fish habitats are based on experience, inference and intuitive judgement since the models relating habitat variables to production of fish have been usually site-specific and stock-specific. Thus the fishery manager is under great pressure to make prudent decisions and be a trail blazer when industrial or other activities threaten fish habitat.

Marine fisheries support more than a thousand coastal communities in Canada and for many fishermen and plant workers provide the ticket of entry to the social safety net. During the decades since the Second World War, in response to major swings in the fishery, a variety of federal programs have been initiated to assist the fishing industry to address financial problems. These problems have stemmed from fluctuations in the resource, the market or from general downturns in the economy. Substantial sums of money have been expended to subsidize the fishery and attempt to stabilize it during these downturns. Crowley, McEachern and Jasperse examine the various federal government initiatives to financially assist the Canadian fishing industry from 1945 to 1990. They conclude that, although assistance objectives changed over time, they were seldom in harmony with fisheries management objectives. Federal and provincial assistance programs often fostered increased fishing and processing capacity and undermined attempts to improve efficiency in the industry. Despite this, governments were repeatedly called upon to intervene with financial assistance during industry downturns. More recently, the Northern Cod Adjustment and Recovery Program introduced in 1992, in addition to providing income support to fishermen during the moratorium, was intended to foster structural adjustment in this segment of the industry. In December 1992, in response to further downturns in the Atlantic cod fisheries, Fisheries Minister John Crosbie called for a fundamental restructuring of the Atlantic groundfish fishery leading to reduced harvesting and processing capacity.

Underpinning the fisheries management process is the need for reliable scientific information on the status of, and prospects for, fish stocks. Doubleday discusses several key issues concerning the reliability and consistency of scientific advice for fisheries management. He defines the reliability of scientific advice in terms of regularly meeting the expectations of fishermen and the fisheries managers who receive it. He concludes that, in many situations, consistency of scientific estimates and advice over time is as important as accuracy. He notes that expectations of fishermen and the general public may be much higher than the requirements for conservation. Scientific estimates may be sufficiently reliable for conservation purposes even if the accuracy is low. A much higher accuracy is required if the aim of fisheries management is to maintain high constant catches by means of catch quotas. Doubleday identifies a number of factors which can reduce the need for high standards of consistency and accuracy of catch forecasts.

Fisheries science is linked to fisheries policy. On many occasions, government has commissioned special inquiries or task forces to examine the factors contributing to problems in the marine fisheries. There have been over 100 reports of task forces, special inquiries and royal commissions investigating Canada's marine fisheries over the past century. In Chapter 14, Shaping Fisheries Policy, Parsons examines the impact of two of the most significant recent reports — the Kirby and Pearse reports of the early 1980's — on Canadian marine fisheries policy. He analyses why most of the recommendations of the Kirby Report were accepted by government while Pearse's major recommendations were rejected amidst a storm of opposition. He links the differing impact of the two reports on fisheries policy to the differing composition, nature and scope of these inquiries and the process followed by each.

The final chapter discusses a dynamic new aspect of Canada's fisheries. Aiken describes the dramatic growth in Canadian aquaculture during the 1980's from a hobby activity to a producer of high value seafood capable of generating more than a hundred million dollars for the Canadian economy. The "landed" value of aquaculture increased almost tenfold from 1986 ($27 million) to 1992 (about $250 million). It is predicted that by the year 2000 aquaculture's contribution to landed value will increase to 25% or more. Aiken concludes that diversification will be the key to the continuing success of aquaculture in Canada. Rapid expansion has lead to public resistance, and resulted in multi-use conflicts. These conflicts need to be handled with great sensitivity and skill by the aquaculture industry and all levels of government. The interesting question is the extent to which aquaculture can supplement or replace some of the traditional wild fisheries. It is already replacing the commercial fishery for Atlantic salmon.

Many of the papers in this volume illustrate the boom-and-bust pattern characteristic of many of Canada's marine fisheries and the factors which contribute to that pattern. They also indicate that, because of the diversity of the fishery resources and the fisheries which exploit them, the situation varies considerably from fishery to fishery. While some fisheries are experiencing difficulty, e.g. Atlantic groundfish, others are experiencing relative prosperity at the present time, e.g. Atlantic lobsters and Pacific salmon. Solutions to fisheries problems must take account of this diversity.

L. S. Parsons and W. H. Lear
*December, 1992*

# A Historical Sketch of Fisheries Management in Canada

Joseph Gough

*Department of Fisheries and Oceans*
*Scotia-Fundy Region, Communications Branch*
*P.O. Box 550, Halifax, Nova Scotia, Canada*

Gough, J. 1993. A historical sketch of fisheries management in Canada, p. 5–53. *In* L. S. Parsons and W. H. Lear [eds.] Perspectives on Canadian marine fisheries management. Can. Bull. Fish. Aquat. Sci. 226.

The Canadian fishing industry developed differently in different regions, and so did fishery management. After Confederation, demography, canning technology, and the nature of salmon migrations gave the Pacific industry a more consolidated and dynamic character than the Atlantic industry. The Pacific fishery generally yielded better incomes for fishermen.

Pacific management was more far-reaching than on the Atlantic, and often more successful. With a more concentrated fleet and a more vulnerable resource, the industry could readily see dangers and accept regulation. Social factors such as more alternate employment also favored active management.

From the 1970's, however, Atlantic management became if anything more interventionist than on the Pacific. Domestic and foreign overfishing with new technology had made the resource more visibly vulnerable. Fishery dependence had decreased. The fisheries department and industry were more conscious of limits to growth and of new opportunities.

Federal management became more assertive than ever on both coasts. Major changes in the 1970s and 1980s included licence limitation, quotas, subdivision of quotas, extension of fisheries jurisdiction to 200 miles, the Pacific Salmonid Enhancement Program, and an elaborate advisory committee system. Although these initiatives produced positive results, Canada's fisheries are still subject to major fluctuations.

---

L'industrie canadienne de la pêche a évolué différemment dans différentes régions. Il en va de même pour la gestion des pêches. Après la Confédération, les tendances démographiques, les techniques de conserverie et les migrations mêmes du saumon ont conféré à l'industrie du Pacifique un caractère plus uni et dynamique que celui de l'industrie de l'Atlantique. Le revenu des pêcheurs du Pacifique était habituellement supérieur à celui de leurs homologues de l'Atlantique.

La gestion des pêches dans le Pacifique était de portée beaucoup plus grande que dans l'Atlantique et donnait souvent de meilleurs résultats. La flotte du Pacifique étant plus concentrée, et les ressources plus vulnérables, l'industrie était plus apte à reconnaître les dangers et à accepter la réglementation. Des facteurs sociaux, notamment un plus grand nombre d'emplois de rechange, ont aussi favorisé cette gestion active.

À compter des années 70, toutefois, la gestion des pêches dans l'Atlantique est devenue plus interventionniste que celle dans le Pacifique. La surpêche au pays et à l'étranger grâce aux nouvelles technologies a rendu les ressources visiblement plus vulnérables. La dépendance de l'industrie de la pêche a diminué. Le ministère des pêches et l'industrie est devenu plus conscient des limites de croissance et des nouvelles occasions.

La gestion par le gouvernement fédéral est devenue plus autoritaire que jamais dans les deux régions. Des changements importants sont survenus dans les années 70 et 80 : restriction du nombre de licences, imposition de contingents, extension des territoires de pêche à 200 milles, Programme de mise en valeur du saumon du Pacifique et création d'un réseau élaboré de comités consultatifs. Bien que ces mesures aient donné d'excellents résultats, l'industrie canadienne de la pêche demeure très sujette à d'importantes fluctuations.

## Introduction

This paper outlines briefly the early growth and management of major Canadian fisheries, and compares the different course of development in different regions. Management is discussed mainly in the following terms:

- Understanding the fisheries;
- Conserving and protecting the fisheries;
- Sharing the fisheries (i.e., who gets the fish);
- Developing the fishing industry;
- The fishery as a system; and
- Linkages of the fishery to the surrounding society.

Much of the factual material in this compressed and truncated sketch derives from Gough (1988 and 1991) and a longer historical work is awaiting publication. This sketch proceeds mainly in chronological fashion. For each period, it briefly outlines the industry's development and accompanying management, and presents some general observations. The conclusion focuses on some perennial management questions, and comments on the many changes since 1968.

## Pre-European Native Fisheries

Across what is now Canada, Indians and Inuit used a large variety of fishing methods. The west coast Indians are the best documented. To many of their groups, fish were all-important. They speared, netted, trapped, trolled, and longlined fish. A post-Confederation report by fisheries inspector Thomas Mowat described a special hook for black cod (*Anoploma fimbria*) that when snaring the fish released a signal to the surface. Natives would dry or freeze fish for storage.

On the west coast, the importance of salmon manifested itself in many ways. The Indians were familiar with every stage of the salmon's life, from egg to mature adult. They knew how to transplant fish from stream to stream. Catching was often a community endeavor. At least in some instances, the native regulatory customs were strict indeed. According to Roderick Haig-Brown, among the Bella Coola on the Pacific coast the penalty for fouling salmon streams during spawning runs could be death (Haig-Brown 1974).

Then as now, the perennial question was "who gets the fish?" There were instances of Indians claiming fishing sites as their own. After Confederation, officials of the Canadian fisheries service found certain Indians claiming particular spots on the Fraser. In 1888, the chief on the Nass River told fisheries inspector McNab that he owned all the fish, the white fishermen would have to get licences from him, and he would keep half the money collected (Annual Report, Department of Marine and Fisheries, 1888).

Community sharing, however, seems to have typified the fisheries. The great anthropologist Diamond Jenness (Jenness 1960) said that:

> "The community weir, like the buffalo or caribou pound, was a powerful factor in welding the different families of Indians into a single, social unit. Every man contributed his labour to the building and maintenance of the weir or pound, and every man was entitled to his share of the booty. ... At the weirs each man retained whatever fish he caught, but allowed no family to remain in want."

On the west coast in particular, the fishery figured in ceremonies, status, and communal endeavors. It was important physically and spiritually. When Alexander Mackenzie crossed the continent by land and went down from the mountains along the Bella Coola river, the Indians he met considered their large fish weir too important to let him view it closely (Jenness 1960).

Two things strike a modern observer about the pre-European fishery: the Indians' inventiveness in fishing gear, and the intertwining of the West Coast fishery with community and religion. What were the results of their approach to the fishery?

When the Europeans came, the waters were rich with fish. Although the Indians had large weirs and the ability to destroy salmon runs, the early Europeans reported no instances where runs were destroyed. As regards fish per person, however, there was often poverty amid plenty. Fish migrations were seasonal, and could vary from year to year. There were no means of large-scale transport and storage. Instead, the tribes had to migrate from place to place in search of food.

The Indian fishery declined as the Europeans advanced. To quote Jenness again:

> "In British Columbia the canneries that sprang up at the mouths of the Columbia, Fraser, and Skeena rivers depleted the salmon on which the Indians had depended for their daily food. ... Throughout the whole country, indeed, there was a serious diminution in the food resources of the tribes that depended on fishing and hunting. ... No longer was each tribe a self-contained and self-supporting unit, but ... all alike found themselves inextricably enmeshed in the economic system forced upon them from without." (Jenness 1960).

## Eastern North American Fisheries (1500–1750)

### Early Colonial Fisheries

Fish and whales attracted Europeans to the northwest Atlantic in the 16th century. Distant-water fishers included the English, French, Spanish, Portuguese, and Basques. Fishermen arrived in the spring and stayed until about September. They fished directly from the boats using hooks and lines. The plentiful, easy-to-catch cod (*Gadus morhua*) was the most valuable commodity. Dried or salted, it could be transported long distances and kept for many months.

The early European fishery was a major marine movement. In the 1600's, men in the fleets from France and England at times numbered well over 20 000 (cf Innis 1940). They soon affected some resources. The Basques, among the earliest fishermen, largely abandoned the whale fishery in the 1600's. By the latter 18th century, the walrus fishery in the Gulf of St. Lawrence had practically disappeared (Dawson 1905).

By the late 1500's, the French and English were the chief rivals. The French fished partly from shore stations in North America, where they could use small boats and bring the fish ashore for drying. But having abundant salt, many French vessels came only as far as the Grand Banks. In this "green fishery," vessels would heavily salt the fish without drying them, and head straight back for France. French fishermen thus could get the fish to market faster than the English and could return to the banks more than once in a season. But the green fishery yielded a shorter-lived product more suited to home use than distant travel.

The English fishery in Newfoundland, operated primarily by "West Countrymen" from Bristol and other sites, concentrated in semi-permanent fishing stations in protected harbors on the island's southeast coast. The captain of the first ship to arrive at a harbor became the fishing admiral and governed the station. Fish were caught close to shore from small boats brought from England. The day's catch was unloaded directly onto a "stage" (wharf and buildings), where the fish were cleaned, split, and lightly salted. They were then dried on "flakes" (open tables that allowed maximum circulation of air). Essentially the same processing system prevailed into the 20th century.

The shore-based "dry fishery" fostered English settlement in Newfoundland, laid the basis for territorial claims, and aided the growth of England's trading empire. The dry fishery produced a "hard-cure" cod suitable for trade to distant markets. Some English vessels began to bring passengers who fished from small "bye-boats," and sometimes settled. In time, European vessels began to take aboard fish cured by settlers, or "planters." Some vessels became traders only: they brought goods and traded them for fish to take away. The planters' fish thus competed against the fish produced by fishing ships themselves. Accordingly, the operators of the fishing ships fought a long battle against settlement in Newfoundland.

In the other English base, the Puritans of Massachusetts Bay were under no such control by the West Country. They built their own fishery "empire." New Englanders developed the fishing schooner, and used it especially for cod and for whaling. They soon began to fish off Nova Scotia. The New Englanders used the cod fishery to build a thriving colony. Massachusetts symbols attest to the fishery's importance: automobile licence plates used to bear a codfish design, and the State House in Boston still has its "sacred cod."

After the Treaty of Utrecht in 1713, mainland Nova Scotia became British, and Yankee settlers began moving in.

As the British fishery developed in Nova Scotia, cod dominated. Settlers typically fished in small, open boats from shore using single lines with hooks.

Everywhere on the Atlantic, the English interests depended on the shore. The French were more likely to salt down cod at sea, and accordingly built up less strength on shore. Placentia, Newfoundland served as French headquarters from 1662 until the Treaty of Utrecht in 1713. France then lost her claim to Newfoundland and mainland Nova Scotia. The French fishery became dispersed, with fishermen using Cape Breton, the Gulf of St. Lawrence, and temporary fishing stations on Newfoundland's north and west coast.

Big cod fish from the trap at Battle Harbour, Labrador. The larger fish measured five feet five inches, and weighed 60 pounds. Antecdotal evidence often suggests very large fish and great abundance in earlier fisheries. From R.E. Holloway's 1910 book, *Through Newfoundland with the Camera.* (Photograph courtesy of the National Archives of Canada.)

**Early Management and Colonial Fishery Policy**

Until the 19th century, there was little of fishery management as we now think of it — such matters as size limits and seasons and assessing fish populations. There were precursors — for example, inspection laws in Scandinavia, and scattered ad hoc laws on river conservation. But there was little of substance, especially for sea fisheries.

Yet, there was one key fisheries question that existed then as now: "who gets the fish?" That question was at the shadowy background of imperial conflicts. In their state policy, England and France paid strong attention to the fishery. The Tudors passed several laws to promote the sea fishery. And France by the 1700's was subsidizing its fishing fleet.

The fishery encouraged the growth of the British, French, and Dutch empires, partly because of its value as a "nursery for seamen," but also because fishing, shipbuilding, shipping, and trading reinforced one another economically. Other countries also saw the fishery's value. Indeed, Innis (1940) noted that Denmark in 1580 imposed a licence fee on English vessels fishing at Iceland. This accelerated England's move to the northwest Atlantic fisheries.

In New France, the government made large grants of fishing rights to Nicolas Denys and others. But the French faced the same situation on the coast as in their far-flung and visionary claims to the interior heartland of the continent. They lacked the North American manpower to live up to their visions.

While French designs for settlement had weak and scattered success, English settlement grew on its own. Although the English dry fishery fostered coastal occupation, that country made no systematic effort to settle Newfoundland. England did make some early grants. Examples include those of Gorges off New England and of John Guy at Newfoundland. Guy put forth a series of fishery rules, mainly regulating the shore practices of the fishing industry. But these regulations faded away as his early colony faded.

The West Countrymen, as they were known, and their fishing ships became dominant. The Bristol concerns and others saw Newfoundland as a big fishing base. For them, settlers would only create competition. Various laws held back the settlers. The Western Charter of 1634 gave the fishing admirals strong powers; and they saw settlers as the enemy.

But Newfoundlanders managed to colonize despite the law, and spread along the coast. The cod-fishing technology — the stages and flakes — required large amounts of space and natural materials. A small number of families could fill up a cove. This and the search for fish kept people from settling in large numbers at any one point. Instead they would move on to another cove. As settlement took hold more and more, the law recognized settlers' rights — for example, to hold property — only step by grudging step.

**Some Observations on Early Policy**

Although the fishery figured in the state policy, government paid little attention to managing the fishery itself. In terms of understanding as we now think of it, with researchers pondering everything from the precise biology of fish to satellite monitoring of fleets, there was little. Nor was there much conservation and protection. Such regulations as existed tended to deal with the curing and shipping of fish.

On the question of who gets the fish, the French made more effort to organize property and settlement, used grants in seigneuries, made larger-scale grants like that of Denys, and also made grants of fishing rights. But the French lacked the power to enforce them over time.

The English on the other hand tried to discourage settlement, at least in Newfoundland, yet ended up with an occupied coastline. The English did allow some ownership of fishing rights on rivers, like those of Skeffington in Newfoundland, for it was natural for fishing rights to go along with river occupation. But for most of the period, politics and greed put hindrances in the way of settlers, since the West Countrymen wanted all the fish for their ships. In open waters, the English leaned towards free and open fishing. For example, Newfoundland's Governor Palliser in 1765 did away with grants along the Labrador given under the French regime, so as to "lay the fisheries open to all His Majesty's subjects from Britain and ... keep it a free British fishery...." (Innis 1940).

The idea of fisheries "development" through new methods, species, and products has preoccupied fishery managers in Canada for much of this century. In the early European period, the emphasis was different. England and France did encourage the fishing fleet; but less in a technological way than as an aspect of state policy relating to food supplies, trade, and defence. Indeed, the fishing-shipping trade was part of a larger English system — the Navigation Acts and their protective rules about shipping and trading. The Navigation Acts prevailed from the 1600's till the mid-1800's, and among other things helped to shelter the growing Canadian colonies.

As for linkages to the larger society, the English fishery both fostered the growth of empire and drew benefits from it. The English were more emigration-minded than the French. The dry fishery accelerated the English tendency to settle in parts of the New World. Fast-growing settlements in turn aided the English fishery and empire itself.

What were the effects of colonial management, or the lack of it, on the fishery itself? There still seemed to be plenty of fish, although there were no doubt natural cycles of abundance. There was little thought that humankind could affect the number of fish in the sea. In terms of economic vigor, the fishing ships from England and France went through their ups and downs. The English West Country fleet at Newfoundland fell from 275 vessels before the Civil War to 100 after. Relatively small during the wars at the beginning of the 1700's, the Northwest Atlantic fleet rebuilt to 300 and more after the Seven Years' War.

In New England's temperate climate, fishing and whaling combined with farming and the beginning of small manufactures to create a growing prosperity. In Newfoundland, with starkly different conditions, there were few if any reports of fishery-based prosperity. In 1714, an observer noted that Newfoundland had about 500 families but "their condition ... is more to be pitied than that of slaves and negroes." (Innis 1940).

Yet something must have pulled them to settle there and fish on their own. The historian Prowse (1895) speaks of "hard-working humble settlers from the West of England; oppressed by the harsh laws of the Stuarts, and persecuted by the western adventurers, they clung with sturdy tenacity to the land they had made their home. ... They lived in the midst of a rude plenty, game and fish being simply in unlimited supply."

## The Period 1750 to 1867 (Confederation)

### New England, Maritimes, and Quebec

#### *The Fishery*

New England Yankees had begun settling in Nova Scotia after 1713. Their settlement picked up speed

This detail from an old map of North America shows the making of saltfish in Newfoundland around 1720. (Source: National Archives of Canada.)

after the Acadian expulsion in 1755, and the fall of Quebec. The Treaty of Paris in 1763 opened all of New France to settlers from New England and the British Isles. In Cape Breton and the Gulf of St. Lawrence, family firms from Jersey in the Channel Islands became prominent, aided by their ability in French.

The American Revolution beginning in 1776 brought a new wave of settlers. Loyalists poured into Canada, bringing with them a certain suspicion and resentment of America. New England, formerly Britain's imperial stronghold, became a sometimes friendly enemy. Britain became by turns hostile and placating to the Americans.

Especially from the second half of the 18th century, a great schooner fleet developed in the Atlantic fishery, with New England leading the growth. These fore-and-aft rigged vessels (evolving into vessels like the Bluenose featured on the Canadian dime) searched for cod, halibut (*Hippoglossus hippoglossus*) and haddock (*Melanogrammus aeglefinus*) on the Atlantic banks. New England builders also developed the marvellous flat-bottomed dory, which handled well on the water and pulled easily up a beach. Later, schooners began to carry them for fishing. They could be nestled in stacks on a schooner's deck.

As New Englanders began to fish down some of their home stocks, their hunger grew for the fish of British North America. They sent large fleets to the provinces. John Quincy Adams, then Secretary of State and later President of the United States, wrote in 1822 that "the portion of the fisheries to which we are entitled, even within the British territorial jurisdiction, is of great importance to this Union. To New England it is the most valuable of earthly possessions." And again:

> "The shores, the creeks, the inlets of the Bay of Fundy, the Bay of Chaleurs, and the Gulf of St. Lawrence, the Straits of Bellisle and the coast of Labrador, appear to have been designed by the God of Nature as the great ovarium of fish; -the inexhaustible repository of this species of food, not only for the supply of the American, but of the European continent. At the proper season, to catch them in endless abundance, little more of effort is needed than to bait the hook and pull the line, and occasionally even this is not necessary. In clear weather, near the shores, myriads are visible and the strand is at times almost literally paved with them." (Departmental International Compilation 1898).

After the Revolution, an initial agreement on fishing rights fell apart, in the turmoil of the Napoleonic wars and the War of 1812. The British-American Convention of 1818 tried to resolve the situation. By its terms, New England fishermen could generally use British North American waters only for shelter and repair, and to purchase wood and water. Otherwise, at least in southern BNA waters, they had to stay outside the three-mile limit.

But there was a major fishing privilege. Americans could fish within three miles at the Magdalen Islands, southwest and western Newfoundland, and Labrador east of about Natashquan. In these same areas, they could also dry fish ashore in unsettled bays, or with permission from inhabitants. Thus the Convention perpetuated American fisheries, and became a chronic source of disputes about specific limits on the Americans.

The New England schooner fishery for cod grew ever stronger during the period, and stretched all along the coast to Labrador. In the years 1790–1810, with no British-American wars, it was estimated that an average 584 vessels would go to the banks off New England and Nova Scotia, 648 to Chaleur Bay and Labrador (Innis 1940). The invention of the mackerel jig accelerated the U.S. fishery for bait, especially in the Gulf of St. Lawrence.

Imperial conflicts after the American Revolution gave the Maritimes a new opportunity for growth, as Britain turned to North America for raw materials. From the American Revolution in 1776 until after the War of 1812, British relations with the United States were frequently hostile. Napoleon in his heyday blocked the usual British access to timber from the Baltic. As Britain bought timber and fish from new entrepreneurs, the mutually reinforcing industries of lumbering, ship-building, shipping, trading, and fishing brought vigor to the Atlantic economy. Even today the period is considered a golden age. But the fishery grew more as part of a self-reinforcing marine economy than as a strong industry on its own.

In the Gulf of St. Lawrence, large fish-curing and schooner-owning firms, particularly the Robin interests, held much of the fishery in an iron grip. Moses H. Perley's remarkable reports in 1849 and 1850 on the New Brunswick fisheries mentioned the "Jersey Houses" on the north and east of New Brunswick keeping local fishermen in some areas in "poverty, and [a] state of bondage" (Perley 1852). He wrote of inefficiency and backwardness, and of fishermen being "even in a worse position than southern slaves...." (Perley 1852). There were few amenities. Schools were ruled out. "They have no need of education," said Philippe Robin in 1836. "If they were educated would they be any cleverer as fishermen?" (Innis 1940).

On the Bay of Fundy shore, Perley described an industry which was getting along fairly well, with independent fishermen. Indeed, southwestern Nova Scotia and New Brunswick's Bay of Fundy area were beginning to lead the Atlantic fishery. This region had a good lumber and trading base, a long ice-free fishing season, an abundance of fish, a good mix of species, a relatively short coastline allowing some concentration of enterprise, nearby markets, and nearby alternative employment in the United States.

Even in Nova Scotia, however, there seem to have been few rich fishermen. A member of the provincial legislature said in 1827 that: "I have never yet known among my constituents one solitary instance of a man getting beforehand by fishing and fishing only; those who own vessels or parts of vessels did not earn them by fishing; they earned them in better times, by sailing coastwise, etc. ..." The fishermen, "although they compose a large proportion of the population are literally in a state of bondage...." (Innis 1940). Overall in the Maritimes and Quebec, many or most fishermen appear to have been poor.

Most fishermen operated small shore boats rather than schooners, and many, especially in the southern areas, alternated between fishing and the shipping trade. The British North American fishery trailed the Americans in adopting new technology. From the 1850's, Americans began using the longline (originally "bultow" and also known as "trawl"), adopting it from the French. The longline method used long "groundlines" anchored near or on the bottom, with many short sections of hook and line fastened onto them. Instead of a single hook on a line, now fishermen could use hundreds of hooks.

The longline gave great impetus to the fishery from "dory schooners," with dories fishing from a mother vessel, each dory setting lines. By the 1870's, with faster schooners, growing markets, and frequent inshore catch fluctuations and failures, the Americans were turning more and more to the offshore or "bank" fishery. When Kipling published *Captains Courageous* in 1897, about the Gloucester dory schooners on the Grand Banks, the longlining technique for New Englanders was still only about half a century old, and the stress on offshore banks was also relatively new.

Fishermen were still making frequent use of the beach seine, to encircle fish close to shore. But in the 19th century, steam engines allowed American fishermen to use large purse seines. Operated in open water, the purse seine surrounded surface-schooling fish with a net hanging down from a line of corks. The fishermen then tightened a purse line running through metal rings at the bottom of the net, to enclose the fish in a floating trap.

Although New England led in most things, British North America was first in one. In Lower Canada (Quebec), Richard Nettle — later the superintendent of fisheries for Lower Canada — in 1857 became the first person to breed salmon in North America. At the same time, some private parties also wanted to revitalize salmon fisheries, by getting control of rivers to provide better management. Interest in both aquaculture and leases became significant.

### *Management*

The beginnings of modern problems and preoccupations appeared in this period.

Aware of the usefulness of fisheries to the general economy, the French and New Englanders subsidized fleets during the period. The French were particularly generous. The Maritimes and Quebec also in the early 1800's got into the subsidy game. In Nova Scotia, the noted diarist Simeon Perkins wrote in 1811 about sending a schooner out to fish for an extra period of time, in order to qualify for the subsidy (Innis 1940).

Various laws in the Maritimes and Quebec tried to promote fish quality, for example, an 1828 inspection act in Nova Scotia. But these had no great effect. Perley in his travels gave a long explanation to a group of Grand Manan fishermen about the benefits of taking better care of product quality. The fishermen made a response that seemed to fall on deaf ears for the next century: "Unless the system was general, it was useless for anyone person to cure his fish better than his neighbour, as he would obtain no better prices...." (Perley 1852).

Another form of what we would now call "development" took place in New Brunswick. After Perley's report, the province sponsored fishery societies, in an attempt to advance the industry. For example, the society on Campobello Island received money to offer premiums for the best quality product shown at the newly-established Campobello Fish Fair (which lasted into the 20th century and revived briefly in the 1970's).

By the end of the 1700's, Nova Scotia was trying to patrol the fisheries. George Leonard in 1786 became superintendent of trade and fisheries (Innis 1940). Later his authority extended to all the Maritimes and Newfoundland. Leonard tried to get Americans to obey Canadian rules, and use local labor to dry their fish. The vessel the *Earl of Moira* helped in enforcement. But to keep foreign vessels outside the three-mile limit, the colonial governments relied primarily on the Admiralty, calling frequently for British assistance.

From 1818 to 1851 the British colonies seized 51 American vessels, destroying 25 of them (United States Tariff Commission 1944 and Departmental International Compilation, 1898). By mid-century, several provinces were trying to mount patrols in concert (Innis 1940). Fishery protection foreshadowed Canadian Confederation and sovereignty.

After 1854, the situation became calmer and friendlier. Fostered by the fishery conflicts, a Reciprocity treaty with the U.S. allowed fishermen from each jurisdiction to fish within the other's territorial waters; and it provided for free trade in fish and other natural products. But renewed British-American tensions ended Reciprocity in 1866. This coincided with some economic distress among Nova Scotian and Newfoundland fishermen.

By that time, alarms were rising about inshore resources, as they had risen earlier in New England. Perley's great outline of the New Brunswick and Bay of Fundy fisheries at mid-century spoke of overfishing and local depletion. He lamented the pollution from sawdust, and other problems in salmon streams. In Massachusetts, the growing scarcity of inshore stocks had already prompted fishermen to turn offshore to Georges and the eastern banks.

The Maritime provinces had already taken some conservation action, chiefly for salmon rivers. New Brunswick had regulations as early as 1784. But these and later ones had little effect. "Drift netting was illegal, but still carried out, and fishways were required, but few were ever built...." After Perley's report, New Brunswick enacted new regulations setting closed seasons and weekend closures, requiring fishways, forbidding pollution of streams forbidden, and so on. But these, like earlier laws, had little effect (Dunfield 1985).

Elsewhere as well, conservation laws had typically dealt largely with river fisheries, often on a local, ad hoc basis, and without much apparent result. As early as 1763, Nova Scotia gave the right to regulate the river fisheries to the respective county courts (Dunfield 1985). At mid-century, following Perley's report, there was renewed interest in conservation in Nova Scotia, but again, to no great effect. Contemporary opinion often found the Maritime provisions useless, and sometimes laughable (Innis 1940; Prince 1921; Dunfield 1985). Writing after Confederation, Peter Mitchell, the first Minister of Marine and Fisheries after Confederation, said that the fisheries of the Maritimes were poorly managed and nearly exhausted.

It was Mitchell who put forward Canada's 1868 Fisheries Act. Earlier laws in the Maritimes had been mainly ad hoc and local. The new Fisheries Act created sweeping powers. The impetus came not from the fisheries of the broad Atlantic, but mainly from the rivers of Upper and Lower Canada. The smaller the body of water, the more visible the fishery problem. Atlantic salmon, so visibly vulnerable when they run the gauntlet of river pollution and human predation, helped give the Fisheries Act its great strength.

In the period before Confederation, there was obvious concern with fishery depletion generally, in the Maritimes and the Canadas. There was also interest in leases of salmon privileges and other fisheries. In New Brunswick, Perley's report stressed the need for leases. Interest was especially apparent in Lower Canada, where riparian rights were already common from the old seigneurial system. For example, in 1852 some entrepreneurs wanted the exclusive privilege for the porpoise (beluga) fishery at the mouth of the Saguenay. Others wanted clear holdings in salmon rivers, so they could take better care of the rivers and get better returns. But old proprietary fisheries clouded the picture as to who held what.

Richard Nettle — the same man who pioneered aquaculture — in the early 1850's carried out investigations of the Lower Canada fishery. The Crown Lands department was already familiar with forest leases. In 1857, Crown Lands Commissioner Joseph Couchon put forward legislation that established the government's control over rivers and its right to grant leases and licences. The same act, which applied to the joint jurisdictions of Upper and Lower Canada, set up a Fisheries Branch. Nettle became Superintendent of Fisheries for Lower Canada (Dunfield 1985).

Another act in 1859 reasserted the licensing power, and introduced broad powers for "better management." In 1865, a significant act integrated licences and leases, broad management powers, and anti-pollution powers, including a key phrase forbidding "deleterious substances." These powers and the general assertive approach for "better management" were the pillars of the post-Confederation Fisheries Act of 1868. In particular, the licensing provisions of the Fisheries Act seem intended to a large degree to bring order to the salmon fishery, and to protect private interest and private incentive, including incentive for conservation. In other words, the Fisheries Act aimed to promote what we might now call quasi-property rights.

There were earlier instances of licences being used or suggested in the sea fisheries, as a means to keep law and order. But the Annual Reports of the Department of Marine and Fisheries in 1873 and 1874 suggest an overriding concern with the private-incentive aspect of licences. W. F. Whitcher was writing as Commissioner of Fisheries. (Whitcher may have helped draft the 1868 Fisheries Act; for earlier, as head of the Fisheries Branch for Upper and Lower Canada, he wrote a memorandum explaining the forerunner Act of 1859 to the Commission of Crown Lands who was to put it forward (Whitcher 1865). He mentioned the removal of salmon nets from the Restigouche and Moisie rivers, and how it had increased the fish.

> "At both places it is now clearly proved that immoderate netting is a serious hindrance to the restoration of the salmon fishery, and a positive disadvantage to the fishermen themselves. It also is quite as clearly established that a moderate quantity of nets, judiciously situated, render at once a far more profitable return to the owners and admit of maintaining a permanent stock of mature salmon. This fact has a peculiar bearing on the regulation of the salmon fishery. The occupancy of salmon stands under formal titles enables the occupiers to economise both their own capital and labor and the public property in salmon. Where the fishery is carried on in a desultory and improvident manner, under such incitements to excess as are created by contentious rivalry and the prospect of mere temporary gain, it is extremely difficult to control fishing operations within reasonable bounds. But, on the other hand, where occupants can rely on the permanence of their holdings, and enjoy in successive years the benefit of their own moderation in each preceding season, the Department finds very little difficulty in controlling the pursuit."

Whitcher later suggested that the:

> "...system of leasing and licensing fishery privileges under the Fisheries Act, already introduced in the Provinces of Nova Scotia and New Brunswick, be now further extended conformably with the practice existing in the Provinces of Ontario and Quebec.
>
> "In these Provinces the system has been brought gradually into operation since the year 1856. It is confined almost exclusively to salmon and sea-trout fishery in Quebec, and to white fish and salmon trout fishery in Ontario. There is still open a large field for its extension, without encroaching on the deep sea fisheries for cod, halibut, mackerel, herring and other scale fishes.
>
> "At the date of Confederation a similar principle existed in Nova Scotia and New Brunswick, but was limited in its application to very few instances. ...
>
> "The **Fisheries Act** evidently contemplates the system of granting titles for fishing privileges as a basis of administration. Certain of its provisions are predicated on the supposition that leasing and licensing would become general, providing always for necessary exceptions as to legal titles, prior occupancy and preferential claims.
>
> "It is unnecessary, after several years of its beneficial operation, even though but partially carried out, to explain at length its advantages. Primarily, it systematizes the fishing business, and it also induces private expenditure both in guarding and improving the streams, which outlay would otherwise require to be defrayed from public funds. Secondarily, it promotes investment of capital, and gives permanence and security to fishing industries, enhancing the value of fishing privileges to both individual fishermen and the public, which hitherto had but a fitful existence and were fast becoming altogether unproductive. Revenue is only an incident and not a main object."

Whitcher now wanted the government to use licences and leases more aggressively in the Maritimes, although not yet in the sea fisheries:

> "... Attention should be directed to carrying out this system with every regard for the obvious desirability of enlisting the sympathies of the public and promoting the truest interests of the fishermen. There should be a thorough examination into each case; and the greatest possible care and precautions should be observed in order to avoid doing violence to the prejudices, or injury to the position and interests of persons affected thereby. Scrupulous regard will require to be paid to priority of occupation and recognized user. A careful distinction must be observed between the deep-sea and inland and the estuary and river fishings. These latter should alone, in my humble opinion, be subject (for the present at least) to the system of occupation under lease or license.
>
> "The undersigned considers it undesirable to anticipate the production of direct revenue from fishery rentals, the rates of which are for the most part nominal. Any system of regulation and economic use of fishing privileges under titles may be more profitable adapted as an auxiliary to protection of inland fisheries, and to enhance their productive value. It is not improbable, however, that in due course of

time sufficient funds may be derived to render the service self-sustaining."

The second great pillar of the Fisheries Act — the provision against deleterious substances in waters frequented by fish — arose partly from local concerns about mill dams and other sources of pollution, and seemingly in part from contemporary thinking in Europe, where there was new concern with pollution. Canadians in authority made reference to European legislation.

The third pillar of the Canadian fisheries regime — the generally assertive approach by government in regulating for the general good — was clear in the United Canada's Act of 1859. It carried a provision to let "the Governor in Council ... make all and every such regulation or regulations as may be found necessary and expedient for the better management and regulation of the Fisheries...."

The licensing and pollution powers and the general assertive approach, passing from the fisheries legislation of Upper and Lower Canada into Canada's law of 1868, made the Fisheries Act very strong. Fisheries expert Alfred Needler has referred to it as "the best Fisheries Act in the world" (pers. comm.).

### Newfoundland

#### *The Fishery*

Despite the obstacles put up by Parliament and the West Countrymen, British settlers gradually spread along the south and northeast coasts. The British government gradually recognized reality and allowed settlement. Still, the long opposition to settlement slowed the social development of Newfoundland. The colony received Responsible Government in 1855, without having had much previous experience in government.

What some termed the "fishocracy" of Newfoundland took shape during this period. Merchant firms became very powerful. They dominated the saltfish trade, which dominated almost everything else in Newfoundland. The truck system took hold: fishermen operated largely on credit from merchants. They got outfitted on credit, got food and goods on credit, and settled up at season's end according to the merchant's books. Small merchant firms in the outports could advance only limited credit. The main financing came from the larger houses based in St. John's. These firms also had fish-buying and curing stations along the coast.

The fishing ships from England gradually gave way to the Newfoundland fleet. Following the Americans and Maritimers, Newfoundlanders adopted schooners and longlines. There was little bank fishing. Schooners and open boats worked along the shore. Conception Bay was important in the schooner fishery. In the summer and fall fishing season, "floaters" moved from place to place especially along the northeast and Labrador coast. "Stationers" lived in shacks or buildings at shore bases. On the Labrador, the few settlers were "liveyers" (live-heres).

After 1713, the French had abandoned their south coast base at Placentia. All of Newfoundland was clearly English, but fishing rights complicated the picture. The Treaty of Paris in 1763 gave the French title to Saint-Pierre and Miquelon, and let French vessels from there or metropolitan France retain a fishery on the French Shore, which later became the west coast and the far northeast coast of Newfoundland. France sent large subsidized fleets across the ocean. Disputes emerged about the extent and meaning of French fishing rights.

American fishing vessels also coursed the waters off Newfoundland. Indeed, the American Shore and the French Shore partly overlapped. But the American fleet to some degree provided a market. The Gloucester fleet and other American vessels, as well as those from the Maritime provinces and from France, purchased bait from the Newfoundland shore fishery. This was the earliest form of what we now call "over-the-side sales" — raw fish transferred at sea from a domestic fishing vessel to a foreign vessel. As well, the Americans in winter bought frozen herring, and took it back to New England for both food and bait. This was probably the first commercial sale of frozen fish in North America.

The Americans, although traditionally welcome traders, irritated Newfoundlanders by the strength of their presence and their competition in the fishery.

#### *Management*

In Newfoundland as elsewhere, there appears to have been little attention to management in terms of understanding the fish, or of conservation and protection. But as part of state policy, government got involved in questions of sharing, development, and the fishery as a system.

Newfoundland's former Governor Palliser, back in England as a Member of Parliament, in 1775 secured passage of a law — "Palliser's Act" — promoting development of the British fishery at Newfoundland, in part because "the fisheries carried on by His Majesty's subjects ... have been found to be the best nurseries for able and experienced seamen, always ready to man the Royal Navy when occasions require." Providing subsidies and tariff exemptions, the Act helped the British fishing ships at the expense of Newfoundland and mainland vessels. Other rules at the time restricted fishing vessels from New England, which was about to break out in revolution partly because of just such restrictions on commercialism.

A modification to Palliser's Act in 1786 required larger mesh sizes, of 4 inches, to let small fish escape. This was one of the earliest purely conservation regulations in the groundfish fishery. Palliser's Act gave new life to the fishing ships and the bank fishery (Innis 1940).

Newfoundland began to assert her own rights in the fishery. Shortly after Newfoundland got Responsible Government in 1855, the British in 1857 attempted to settle a dispute with the French by allowing them exclusive fishing rights on parts of the coast. Newfoundlanders raged in protest; Britain promised to consult better in future; and the affair became known as Newfoundland's "Magna Carta."

### General Observations on Fisheries Development and Management 1750–1867

Throughout this period in British North America, the overriding fact was still the intermixing of fisheries with state policy. But concerns about depletion were growing in the river and shore fisheries. In the Canadas, the growing worries about the fishery, combined with a tradition of leases, prompted a fisheries act with strong licensing and anti-pollution powers.

The elements of management as we think of them today — understanding, conservation, sharing, developing, making the industry systematic — were beginning to take shape. Perley's remarkable report — probably a better description than any since written on any province's fisheries — demonstrated the growth of understanding and was itself a major contribution to knowledge. Rules about conservation increased in the river fisheries; and concern about international protection showed up in the sea fishery.

As for the question of who gets the fish: proprietary practices and instincts helped put the licensing power into the Fisheries Act. In the shore fisheries, local regulations could favor local residents. For example, a 19th century regulation for the Peggy's Cove, N.S. area allowed only residents of the area to use beach seines.

The subsidies (or "bounties" as they were sometimes called) by the Americans, French, and British constituted attempts at development. Also, Perley was beginning to make explicit some idea of the fishery as a system, from ocean to market place.

The fishery was beginning to do better where the rest of the economy was stronger: in the southern Maritimes, and in New England. The fishery was linked to sovereignty, with the Maritime provinces first calling on Britain to protect their zone, and in the mid-1800's patrolling on their own. By that time, Britain was placing less importance on the fishery. The Industrial Revolution was well under way. The British were abandoning the Navigation Acts and turning to free trade.

What were the results of such management as existed? In a nutshell, the first half of the 19th century saw more management, or at least attempts at it, and less fish. The authorities were doing more things, but doing them poorly for the most part. By 1867, salmon stocks had declined in many rivers. Anecdotal evidence — such things as herring spawn piled two feet deep for miles on the beaches — suggests that river and sea fisheries formerly had great abundance. In some cases these may have passed out of memory by the end of the 19th century.

Subsidies had become increasingly evident — the sign of a weaker industry. Contemporary writings suggest that fishermen were most often poor; the farther north, the poorer. Observers often associate the fishery with freedom, and rightly so. It is one of the great satisfying occupations. But by 1867, one finds instances of distress, and the beginning of frustrated reports, like parts of Perley's, about how good the fishery could be if only those involved followed proper conservation and other practices.

## Confederation to World War I

### Maritimes and Quebec

#### *The Fishery*

In the late 19th century, the growing urban and industrial economy brought changes to the Atlantic coast. Steel and steam vessels with greater reliability, safety, and size began to displace wooden trading vessels. Markets for coastal timber became less voracious. Timber stands themselves were becoming less abundant.

In the fishery, Americans began to make use of steam-powered vessels and winches. After the turn of the century, some larger operators in Nova Scotia began to use steam-powered trawlers. By that time, the gas engine was starting to revolutionize the shore fishery, bringing mobility and captainship to more small-boat fishermen.

On the North Shore of the Gulf of St. Lawrence, fisheries official William Whiteley in the 1870's invented the cod trap. This device, which had its precursors in other waters, spread like wildfire in the Newfoundland and Labrador fishery, often replacing seines and hook and line. Whitely himself later became concerned about the growth of cod-trap fishing, and helped devise regulations for it. This was an early example of the fisheries service being caught between its twin gods of conservation and development.

As the Americans turned from the Maritimes, Newfoundland, and Labrador shore fishery to the Grand Banks and other banks, Canadians gradually followed. Among Canadians, Lunenburgers led the way. The British in 1753 had imported a group of Protestant Germans and Swiss to Lunenburg County, Nova Scotia, to help counterbalance the remaining Acadian population. Little seemed to differentiate Lunenburg physically from other Nova Scotian ports. It appears to have been hard-working habits and co-operative ventures that let Lunenburg take the lead in the Canadian schooner fishery by the latter 1800's.

The Americans meanwhile were pioneering in the fresh-fish fishery. In the late 1800's, their fast schooners started landing fish fresh from the nearest banks, including LaHave Bank off Nova Scotia. Nova Scotians by the end of the century also made a modest start in the fresh-fish trade. Freezing became common in the bait fishery, usually by means of a brine of salt and ice (like the mixture in old hand-cranked ice-cream freezers). It was still rare in the food-fish industry. Mechanical/electrical refrigeration was however becoming more common in other industries, and the fishing industry would eventually follow.

Tristram Halliday at Saint John, N.B. in 1839 started the first commercial canning of salmon in North America (Lyons 1969). Americans soon took over the lead, putting up canned products of many kinds, especially lobster. They started up some lobster factories in Canada, and then Canadians began setting up their own — hundreds of them by the turn of the century.

Canadians also followed Americans in the herring industry. Maritimers had established a pickled and smoked herring industry, using small boats. In the post-Confederation period, Americans began to use larger vessels with mechanized purse seines to catch herring and mackerel off Canada. This alarmed the Canadian authorities, who feared depletion.

The Americans in eastern Maine also began to can sardine herring. The supply came from weirs — large fish traps using twine mounted on stakes driven into the bottom. This picturesque technique had spread early in the 1800's from Nova Scotia's Digby area to southern New Brunswick and eastern Maine. Canadian weirs began feeding the new American sardine factories. In time, Canadians themselves set up sardine canneries.

As parts of the fishery became more "industrial," the larger economy was moving the same way. The American mid-west and west were opening up. Quebec and Ontario were growing apace, and then the Canadian west. After Confederation in 1867, Canadian politicians generally wanted to turn from a commercial-marine economy to a continental economy. Maritime leaders tried to take advantage of new opportunities in railways and manufacturing, and made little effort to promote the self-reinforcing lumbering-fishing-exporting marine economy. By World War I, only fishing remained as a major coastal employer.

### *Management*

#### *Early enforcement*

Before Confederation, some provinces had established fishery overseers; but the system was often laughable, with some officials being unaware of their appointment for as long as two years. Now the British North America Act gave the federal government jurisdiction over fisheries. Under Peter Mitchell, Minister of Marine and Fisheries, and Deputy Minister William Smith, the notable W. F. Whitcher as Commissioner of Fisheries set up a workable though far from perfect system. By 1909–10 the fisheries Outside Service numbered about 1 200 persons, including 680 seasonal guardians and 255 men on the 13-vessel patrol fleet.

Meanwhile, England was losing interest in northwest Atlantic fisheries, and in Canada itself. The last of the English fishing ships had left the Labrador. Free-trading, industrial England was looking to manufacturing for growth. Many "Little Englanders" had no interest in the colonies, and foresaw Canada becoming part of the United States.

The United States wanted to continue fishing off the provinces, even after Reciprocity ended in 1866. In response, Canada passed a strong law to protect coastal fisheries. The aggressive Peter Mitchell, Minister of Marine and Fisheries, commissioned a patrol fleet which captured and condemned 15 American vessels. Washington grew angry, and London grew alarmed.

The fisheries conflict and Canadian assertiveness helped bring about the 1871 Treaty of Washington. Sir John A. Macdonald helped negotiate the treaty. The Treaty provided for renewed American fishing privileges, which helped the New Englanders; but it also provided for reciprocity (free trade) in fishery products, which helped the Canadians. In effect, the Treaty of Washington constituted recognition and respect, and confirmed Canadian sovereignty (Innis 1940; Departmental International Compilation 1898; Brebner 1945).

After the Treaty of Washington, however, New Englanders resented "unfair" Canadian competition under reciprocity, and also resented the $5.5 million ("the Halifax Award") Canada had received in return for U.S. fishing privileges. This led to abrogation of fisheries reciprocity in 1885. Conflict erupted anew. The

Early days in the New England lobster fishery.

patrol fleet, which had shrunk to one vessel, took on new life, and aggressive patrolling resumed. Emotions at times were high. One historian recounts an incident in which Canadian authorities at Prince Edward Island, accusing American seamen of illegally obtaining food supplies, administered emetics to make them disgorge their potatoes (Brebner 1945). The conflict ended with a "modus vivendi" allowing Americans to purchase licences which gave them port privileges.

The Department of Marine and Fisheries strengthened Canada's rule in the north. After Canada took over Britain's Arctic claims, the department became responsible for Arctic sovereignty. Fishery officials including Andrew Gordon, William Wakeham, and Joseph Elzear-Bernier were key figures in confirming Canada's Arctic sovereignty.

The fishery patrol fleet was also the forerunner of the Canadian navy. When the idea of creating a navy, rather than contributing to an imperial force, won the day, a key element was the already-existing fisheries patrol service. It was the nucleus of the navy. The patrol fleet had the first modern warship built in Canada, the *Vigilant*. It was Marine and Fisheries Minister Brodeur who introduced the Canadian naval bill in Parliament, and in 1910 became the first Minister of the Naval Service, while initially retaining the Department of Marine and Fisheries.

*A Knowledge Base for Management*

Fishery management based on biological knowledge was coming closer, as understanding grew. After some earlier university work on fisheries science, research accelerated in 1899 with the building of a floating laboratory that spent time at various Atlantic ports. Permanent stations followed in St. Andrews, N.B., and Nanaimo, B.C. Another station operated for some years at Go Home Bay on Georgian Bay. More stations were to follow, including the technological labs in Halifax and Prince Rupert; after World War II, the Arctic Biological Station in Montreal and the Freshwater Institute in Winnipeg; and some smaller stations.

The Biological Board soon became independent of the Fisheries Service. University scientists dominated,

Fishing codfish on board the *Diana* at Port Burwell, N.W.T., 1905, on a Canadian Government Expedition. The CGS *Arctic* is in the background. (Photograph courtesy of the National Archives of Canada.)

doing summer research with students mainly on the natural history of fish. Relations with the fisheries service were hesitant and sometimes aloof. The fisheries service itself contributed to knowledge by collecting statistics, from very early days onward. In the 1890's it set up an Intelligence Bureau to report on such matters as where bait could be obtained.

Fears of overfishing and pollution increased rapidly in the latter 1800's, as depletion became apparent in many estuarine and inshore fisheries. Those fears spurred a series of royal commissions, which broadened understanding of how the fishery itself worked. E. E. Prince became Commissioner of Fisheries in 1893, and dominated fishery management until World War I. The key figure in setting up the Biological Board, Prince also spearheaded many of the royal commissions. Like Perley's report, Prince's articles and royal commissions still repay the reading. But then as today, there was inadequate provision for spreading such knowledge.

*Regulations and licences*

In this post-Confederation period, Whitcher and the early officials, but more especially the many royal commissions after about 1890, set the basic regulations and practices which dominated most fisheries until after World War II. Indeed, the same regulatory system still prevails, though it is now often buttressed by licences and quotas.

Typical regulations dealt with:

- the minimum size of fish, to give them a chance to grow up. Size limits were especially common in the inland fisheries. Mesh size regulations also became common.
- the amount of fishing effort, with controls on the amount of gear and the time it was in the water. Typical regulations set closed seasons, controlled the length of nets, and stipulated the spacing between traps and nets.
- the types of gear, with restrictions sometimes applying for particular areas only.

As well, fishways became more common in the rivers. Licences became common; but the number of licences was limited in only a few fisheries. For the most part, the royal commissions and fisheries service were restricting what fishermen could do, without trying to control their numbers.

A complete summary of regulations as of 1911 appears in Gough (1991). The number of regulations had increased tremendously since Confederation, with rules now existing for a great many aspects of fishing.

"Depletion" became a common term. Although nobody had defined what overfishing was, fears of depletion led to some strict regulations. Department officials and fishermen blamed the purse-seine, used largely by Americans, for depleting mackerel. In 1891 the Department banned purse-seines. The ban lasted until the 1930's. This was the first widespread ban of a major fishing method, other than such destructive methods as explosives.

The early fisheries service was also sensitive to the need to protect fish habitat. For example, Whitcher in the 1870's warned officials to be strict on mill dam rubbish.

Then as now, there were doubts about enforcement and obedience to laws. But then as now, the rules must have had some effect, or why would industry often lobby strongly against them? A report to the U.S. Commissioner of Fish and Fisheries in the 1870's said that: "The great reason for failure in the effect of Fishery laws has not been their character, but the fact that they were not enforced. This has been the almost universal history of the laws *except* in Canada, Scandinavia, and portions of Russia. It has been notably so throughout the United States." (Hind 1877) After Confederation, the Canadian goverment was trying in earnest to mend matters.

As noted, among the rain of new regulations in the period was the widespread imposition of licences. Licences seem first to have served mainly to keep some sort of law and order in the fishery, by identifying participants and establishing that the fishery was a privilege under the Crown. The great exception to that pattern was Whitcher's use of leases on salmon rivers. This aimed at both conservation and orderly sharing, with a reward for private initiative. But court decisions in the 1880's and 1890's weakened the federal licensing power in inland waters, where Whitcher had seen their main application.

Despite this check to the licensing approach, by World War I licences prevailed in many commercial fisheries. And even if they were applied less rigorously than today, still they showed in some instances potential power and flexibility. In certain cases, e.g. in the whale and salmon fisheries, licences were conditional on approval of plant operations. Also, licence-holders sometimes paid a fee based on gross earnings (as in the whale fishery) or on the amount of gear used (three cents a fathom in the New Brunswick salmon fishery).

Even though licences served mainly for law and order and conservation, they sometimes dealt with the question of allocation, or who gets what share of the fish. For instance, the cod trap licences on the Quebec North Shore gave residents first claim to location. Another regulation required that anyone using a beach seine at Peggy's Cove, N.S. must live within 5 miles of the village.

Although E. E. Prince wrote of the department's licensing strictness in early days, I have found little evidence of widespread application of "limited entry" on the Atlantic. Presumably he was thinking primarily of the river leases. When he tried in 1913 to get limited entry for lobster, after an earlier false start in 1906, he failed. Prince wrote: "So long as the taking of lobsters on Canadian shores is a free fishery, so long will it be difficult to carry out the preservative measures that are desirable." The lack of such measures would eventually exhaust the fishery (Department of Marine and Fisheries, 1913).

### *Other Management Practices*

In the half-century after Confederation, the Department of Marine and Fisheries developed an extensive hatchery program. It was a giant attempt at development. Although fishery authorities claimed excellent results, by the mid-1930's the program's success was minimal, and most hatcheries were closed, especially on the Pacific.

Using the word "development" broadly, there were some other notable attempts. For instance, Whitcher tried and failed to get legislation requiring compulsory inspection of products. There followed various and largely futile attempts at a voluntary system, which rarely worked. In Prince's time, the government built three sizeable dogfish plants, in a failed attempt to develop a fishery for this still underutilized species. Work was also undertaken on better ways of catching and curing herring.

The department began subsidizing the train transport of fish, which seemed to help the fish trade. The idea was to break into the continental rather than just the coastal market. The department provided bait freezers for groups of fishermen, the idea being both to provide bait and to encourage fishermen's organization.

To some degree, the early regulators were trying to make the fishery work as an integrated system. Their measures went well beyond conservation into questions of sharing; beyond the vexing question of the number of fish caught, into the equally vexing question of who gets them. Some of the development efforts took into account the whole chain from water to market. Also, Whitcher made clear that he saw private holdings as a way to systematize the fishery.

Even so, regulators for the most part were responding to specific issues, without worrying overmuch about the totality of the industry as a system. They

Women tending fish flakes in Newfoundland, 1903. (Photograph courtesy of National Archives of Canada.)

groped towards the right action without much elaboration or definition.

## Newfoundland 1867–1914

### *The Fishery*

In terms of volume, Newfoundland's traditional salt-cod fishery reached a plateau in this period. Newfoundland fishermen developed a bank fishery, but only gradually. The Labrador fishery along the shore remained very important. As well, Newfoundlanders turned to longlines and cod traps. There was almost no development of trawlers. With little real growth in the industry, some of the old-line merchant companies retrenched to buying and selling operations, rather than running their own fishing stations.

In Newfoundland's seal fishery, steamers were replacing schooners by the 1860's. The lobster fishery entered a boom by the turn of the century, with hundreds of small canning operations. But cod was always king. Compared with the Maritimers, Newfoundlanders caught far smaller quantities of lobster, herring, haddock, and pollock (Gough 1991).

The fishery by the French, Americans, and Canadians was still strong, on the Treaty Shores and more and more on the Newfoundland banks. Even when foreigners increased their bank fishing, they continued to fish inshore for bait, and to buy bait from Newfoundlanders.

### *Management*

Disputes with foreign fleets consumed a lot of Newfoundland's political energy. The heavily subsidized French fleet used Newfoundland bait to catch cod just off Newfoundland's shore. Newfoundland passed Bait Acts to control the sale of bait; but this seemed to have no major impact on French fishing. A complex lobster dispute on the French Shore also engaged the government. After the French-English Entente Cordiale at the beginning of the century, however, the lobster conflict faded away. Newfoundland settlement of the French Shore speeded up.

As it had against the French, Newfoundland tried the bait-control strategy against Canada. This led to another complex dispute with inconclusive results. In the strained atmosphere, the Canadians thwarted a broader Newfoundland trade initiative. They prevailed on Britain to veto a tentative Reciprocity agreement ("the Bond-Blaine treaty") between Newfoundland and the United States.

The third friendly-enemy fleet at Newfoundland was that of the Americans. They still fished the Treaty Shore, and also bought fish. Newfoundland tried to bring the Americans under control of her own regulations, such as the Newfoundland law prohibiting purse-seines. In "the Fortune Bay incident," a crowd of Newfoundlanders drove American fishermen away from their inshore fishing. This caused international tension. An international tribunal finally resolved the regulatory dispute in 1910, ruling that Newfoundland could apply reasonable regulations to Americans on the Treaty Shore.

In 1898, Newfoundland set up a Department of Marine and Fisheries; but there was little regulation or enforcement, and no science. Like Canada, Newfoundland set up hatcheries for cod, lobster, and other species, only to gradually lose faith in them. For a time the government paid lobster fishermen for berried lobsters, which the government would impound and later return to sea. But this prompted fishermen to catch berried lobsters on purpose. There was no scientific evaluation of such attempts.

The Newfoundland authorities promoted a whale fishery, and set up a management regime that Canadians considered a model. It provided for limited entry (i.e., a limited number of licences), but put no controls on the number of whales taken. The fishery soon dwindled, and most plants closed.

Newfoundland in 1898 set a minimum price for barrelled herring. This appears to have been the first such price-stabilization measure in what is now Canada.

## Freshwater 1867-1914

### *The Fishery*

On the Great Lakes between Confederation and World War I, rapid growth took place, especially in the gillnet fishery. The chief species were trout, whitefish, herring, pickerel, and pike. Near the burgeoning cities, fresh fish were important from the early days of the fishery. On the upper lakes, the American-controlled Booth interests became important.

The fishery also grew fast on the Prairies, especially near Winnipeg. The early lake fishery was dominated by companies which rented small boats to fishermen, often Indians.

### *Management*

By the end of the century, the federal government lost clear control in freshwater fisheries. At Confederation, the old proprietary rights in river fisheries had transmuted themselves to give strong licensing powers to the Fisheries Act. Now those old rights resurfaced, this time to take power away from the Fisheries Act. In an 1882 court case, the Queen versus Robertson, the court judgment upheld the claims of a private holder to a river fishery in New Brunswick. The licences and leases, in which W. F. Whitcher put such faith, became inoperable in non-navigable waters.

The Queen versus Robertson and subsequent court decisions created an awkward semi-standoff regarding management of freshwater fisheries. Provincial management came to prevail in Ontario and Quebec, although it remained under the ultimate authority of the federal Fisheries Act.

The freshwater decisions may have worked against the use of licences elsewhere. Whitcher himself had viewed licences and leases mostly in terms of river fisheries, "at least for the present." Although they might later extend to sea fisheries, for now the rivers were the great example: he saw licences and leases as fostering conservation by appealing to the self-interest of the licence-holders. By taking away federal licensing powers from the fisheries where licences and leases were most prevalent, the court decision must have been a set-back for the licensing mentality of Dominion fishery managers. Yet the idea of control through licences persisted, especially in British Columbia, where licences became important in the salmon fishery.

After Ontario took over management, the Great Lakes fishing industry fished down one desirable species after another, letting less valuable ones take over. Provincial authorities seemed less fearful of overfishing than the federal managers. Even before the provincial takeover, however, environmental changes resulting from urban and industrial growth killed off the Atlantic salmon in Lake Ontario.

Federal authorities directly managed prairie province fisheries until 1930. Early fears of overfishing in Lake Winnipeg led to early regulation. By World War I, on Lake Winnipeg there were rules to:

— curtail the use of powered tugs;
— favor residents in granting licences; and,
— set quotas for the summer fishery — probably the first systematic use of quotas.

As well, there were attempts to curtail the influence of American buyers.

Here as elsewhere, E. E. Prince and his fellow royal commissioners were showing their concern with the whole system of the fishery, but without any widely accepted definition of first principles, and sometimes without the means to put their concepts into thorough and effective practice.

## Pacific 1867–1914

### *The Fishery*

In this period, the salmon fishery grew in headlong fashion. Earlier on the Pacific coast, salted and dried fish

Professor E.E. Prince (*left*), ex-chief Justice McGuire, and Dr. Euston Sisley at Big Quill Lake for the 1910 Alberta and Saskatchewan Fisheries Commission.

had been used by Indians, fur traders and miners. The first salmon cannery started up in 1870. By 1901, there were 73 of them (Lyons 1969). The canneries needed large numbers of workers, and the migration patterns of salmon and B.C. geography itself helped concentrate fishermen and plant workers at specific landing sites. This was a major contrast to the frequently isolated and fragmented Atlantic fishery.

Railway transport to market soon aided the growing Pacific salmon industry and the halibut fishery. Increased bait requirements helped encourage the herring fishery. The fur seal fishery also became important. It was conducted on the high seas by Indians in canoes launched from schooners, in a daredevil fishery that impressed observers.

By World War I, many organizations represented fishermen and processors. With more concentrated groups of fishermen and plant workers than on the Atlantic, organization came easier. The usual spur was price bargaining and collective selling.

### *Management*

Federal managers in British Columbia took an assertive approach from the beginning. Many wanted to avoid the fishery problems already apparent in the United States. There were early difficulties with the Americans over conservation. Although U.S. authorities had seen major losses of salmon in their own rivers, they repeatedly refused to enter agreements with Canada to conserve the Fraser River stocks, fished by both nations.

The "Bering Sea dispute" over Pacific fur seals became a major issue, rivalling the Atlantic "Fisheries Question" for complexity and longwindedness. The Americans resented harvesting by Canadians and others of seals that whelped on U.S. territory. The dispute included seizures of vessels, for example the Canadian seizure in 1890 of the U.S. vessel *Davy Crockett* (presumably a blow to American pride) (Annual Report, Department of Marine and Fisheries, 1890). The Americans attempted to extend control over high-seas harvesting. An international conference finally settled the matter in 1911. The Americans and Russians were to do all the harvesting on land at their rookeries, and Canada and Japan were to get a share of the proceeds.

For the salmon and other fisheries, the Canadian authorities set strict rules. They had in mind "the lessons of the Sacramento and Columbia." They restricted purse seines (although they later weakened this rule, since it gave too much advantage to the Americans, who were allowed to use seines). They restricted net mesh size, net length, and fish size. They forbade "fish wheels," a deadly effective automatic fishing device common in the U.S. They used closed times and closed areas.

The federal authorities in British Columbia also instituted "limited entry" (that is, restricting the number of fishing licences) in 1889. This continued sporadically until after World War I. The federal authorities also at times limited the number of canneries. The fishing industry itself took a strong interest in management questions, and made its own efforts to regulate the number of boats and plants (Gough 1991).

Besides using limited entry for conservation itself, by controlling the fishing effort, the fisheries service used licences to control who got the fish. Already by World War I, they had different rules for whites, orientals, and the native fishery (natives were restricted to fishing for food only).

The fisheries service was assertive in other ways, for example by requiring fish plants to meet standards as a condition of their licences. From a summary of regulations in 1911 (Gough 1991): "Salmon canneries and curing establishments must have a license. Fee: $50. These licenses are conditional on the favourable report of the Fishery Inspector respecting the sanitary condition of the establishment."

## Progress in Fisheries Development and Management, 1867 to World War I

The above sketch foreshadows certain trends that lasted until after World War II. Although the Fisheries

Fraser River salmon fleet, 1900 (*top*). Fifteen hundred fishing boats at twilight at Steveston, near Vancouver, British Columbia (*bottom*). (Photographs courtesy of the Naitonal Archives of Canada.)

Act was the same everywhere, management by now differed significantly in different areas, and the industry itself differed strikingly.

*In the Maritimes and Quebec*, the industry was varied, fragmented, and often poor. Yet, many parties did fairly well, especially in the southern areas; and many in the industry had a gritty streak that resented interference.

The regulators had by now touched at least tangentially on most methods of management. One way or another, in one Maritime and Quebec area or another, they had dealt with the questions of who could fish (including the principles of limited entry, and preference for those closest to the resource), what they could fish, when, where, and how. Although they never defined quotas in the Maritimes, they did use other restrictions to limit the catch of fish, thus dealing by implication with the question of how much fish should be caught. (Setting exact quotas would have been difficult without more biological knowledge; although the department did set quotas for Lake Winnipeg by the time of World War I).

Also, regulators had on occasion set fees and landings taxes. Whitcher even speculated on making the fisheries service self-sustaining. (This idea later became known as "resource rent," a buzzword of the 1970's and 80's).

Despite touching on so many things, and despite the power of the Fisheries Act, resource regulation in the Maritimes was ad hoc and less than thorough-going. The university-based biologists were working on questions of their own, without much relation to the fisheries service. The department often bewailed events, as in the constantly lamented declines of the oyster fishery, without being able to change their course. E. E. Prince tried and failed to limit the number of lobster licences, at the very time when the B.C. industry itself was helping to control entry into the fishery.

It seemed that British Columbia was dealing more firmly with problems of growth, the Maritimes more weakly with problems of decline. The Maritime provinces were becoming less vigorous generally. They had been a centre for an international marine economy; now they were on the fringe of a continental economy. This coastal fringe had few industrial centres of its own. The fishery had helped scatter Maritimers and Atlantic Quebeckers in a thin layer all along the coast, as they tried to get close to the fish and find room for their flakes. The scattered nature of settlement worked against the "critical mass" of population, information, organization, and infrastructure required for new developments generally, and in the fishery specifically.

Department official J. J. Cowie, later a Deputy Minister, in the 1909–1910 Annual Report said that the industry's habitual preoccupation with salt fish, rather than fresh fish as in the U.S., had held back development. Only recently had anyone put smoked fillets onto the Halifax market. Retailing was poor. Despite the new subsidy, transportation remained a problem. Canadian fish products needed promotion. He advised bringing in an inspection program and revamping the inadequate Fisheries Intelligence Bureau. And he pointed out that "the aggregate value of the fisheries of the four eastern provinces has almost stood still for the last twenty-five years." (Cowie 1910).

The Commission of Conservation in 1915 reported remarks by D. J. Byrne of the new Canadian Fisheries Association. Byrne and others, including Prince and other department officials, pointed to the lack of education and organization among fishermen, and to problems of enforcement, of quality, of the transportation system, of marketing and the retail markup, of attitude, and of no public opinion backing up the department (Anon, 1916).

Regulators had touched on development, and making the fishery work as a system. The underlying assumption seemed to be that management could concern itself with anything it cared to for the betterment of the industry — thus the wide range of regulations. But tempering that was a lack of explicit goals and a reluctance to interfere unduly.

Regulating Atlantic fishermen was and is no easy matter. For one thing, the occupation stirs elemental feelings related to pride, strength, the struggle for survival, and the fisherman's and the community's identity. The small communities create strong relationships, like living in a stage play. Fishing towns and villages mark people growing up, so that many natives are never happy elsewhere.

The hundreds of isolated villages in remote coves, with all their self-knowledge and hardihood, often had the defects of their virtues. Fishermen might know their own village to the heart, the whole human picture of everything that can happen to a person. But there were great gaps in knowledge of the outside world, and a frequent suspicion. Picture the fisherman who has lived through fifty years of local struggle and local satisfaction; and picture the fishery officer obliged to tell him that Ottawa has a better idea of how he should do things. The Maritime situation bred a certain reluctance to change. And regulation, although it went wide to touch many matters, never went deep enough to meddle with such fundamental questions as licence limitation.

The Maritime fragmentation by area and fishery, the frequent isolation, the dependence on the fishery for subsistence, and the relative lack of capital, communications, and other mobilizing forces, held back both development and management. Neither industry members nor government could readily gather the forces for positive change. Although fishing and processing always had their built-in pleasures, they were far from a dynamic, leading-edge occupation. The Maritimes industry was becoming a less than proud one, seemingly more of a problem than an opportunity.

*In Newfoundland*, the fishery was more homogenous, and was generally poor. Although lagging behind the Maritimes in science and management, Newfoundland was already showing an aptitude for bold measures such as the Bait Acts and price support. Fishing was everybody's business; and everybody was in the same fishery, for cod. Compared with the Maritimes, it was sometimes easier for everyone to think alike, and take action.

*In Ontario and the prairies*, the industry seemed to do best near the cities, and elsewhere to resemble the worse-off parts of the Maritimes. As the lake fishery on the prairies expanded into more and more remote areas, it would become the subject of many concerns and of various inquiries by government, which never seemed to bring the dreamt-of cure. There was to be a long history of mainly complaints until the 1970's, when important changes took place.

*In British Columbia*, broadly speaking, the industry had more money, better boats, better organization, and more voice in management. And management itself did better at taking hold of situations and trying to mold them to the general satisfaction.

Indeed, the British Columbia fishery grew up with management from the beginning. The visibility and vulnerability of salmon created more conservation consciousness than on the Atlantic coast. Early losses of salmon runs such as those on the Adams River and at Quesnel heightened this consciousness. The concentration of processing plants in the Fraser created early concerns with offal and pollution. The canning technology, with easily countable cans, aided the keeping of statistics. At an early stage, fishery officials began keeping records of returns to rivers. There was more education and organization than on the Atlantic coast, and less isolation and fragmentation.

By the end of the forty years from 1870 to 1910, B.C. had already built a better, more prosperous fishery than that on the Atlantic, and one with a higher consciousness of management.There was more money per person in the B.C. industry, and it was more organized and more involved in management.

Already the B.C. industry lived less with the threat of poverty and hard times than the east, and more with the problem of over-crowding the rivermouths and destroying the resource. But the industry and managers were already finding ways of responding to this constant challenge. In B.C., fishing appeared not that far behind other occupations, or even equal.

## Between the Wars

### Maritimes and Quebec

#### *The Fishery*

The Maritime and Quebec fishing industry boomed during World War I. The war interrupted fish supplies to Europe, and made manpower scarce. With fewer people fishing and demand high, the Atlantic fishery came into unprecedented prosperity.

After the war, the industry slipped back as others advanced. In the 1920's, Americans and Europeans using draggers and trawlers began to outstrip the Canadians and Newfoundlanders in catching methods. New England fishermen and processors developed the filleting and quick-freezing processes. Now they were able to sell packaged fresh or frozen fillets to a wider market. The Birdseye company began using plate freezers to put out retail packs.

Canada and Newfoundland still relied on saltfish. Losing its European markets to the resurgent overseas fleets, Newfoundland started putting more fish into West Indies markets formerly served by Maritimers. Groundfish prices in the Maritimes and Newfoundland dropped. More people turned to other fisheries, and prices went down there too. The saltfish crisis ricocheted through the industry. The Great Depression started early in the Atlantic fisheries, and brought misery to overcrowded fisheries (Bates 1944). In Nova Scotia, this led to new attempts at organization and to strikes in the fishing, processing, and fish-handling sectors.

A well-documented instance of over-crowding was the lobster fishery of Prince Edward Island. Gordon (1952) pointed out that from 1929 to 1939, the number of fishermen on Prince Edward Island increased by more than 50%. But the lobster catch, already at its limit, began to fall. "...By the fundamental nature of the case, an increase in the numbers of fishermen could only reduce the average yield per man. ... The increase in fishermen aggravated still further a situation that was already bad due to falling catch and falling prices."

Gordon noted that extreme cases had earlier led to limits on the number of licences in particular fisheries.

> "On the one hand, it is a serious interference with the freedom of an individual to work and to choose his occupation.... On the other side, one must raise this question: if the size of a particular cake is fixed, is it not both

proper and just to prevent too many from sharing it?"

Again:

"What was happening during the depression years was that additional men were trying to gain a living from the fishery of the Island; they were spending their savings and incurring heavy debts in order to provide themselves with boats and gear; the older fishermen were increasing their gear fleets in an effort to sustain their incomes in the face of falling prices. Yet, despite this heavy increase in effort and investment the quantity of the fish catch as a whole did not rise, the lobster catch actually fell to a substantial degree and the dollar receipts of fishermen declined. A smaller gross income was being shared by a larger number of fishermen running more boats and fishing larger fleets of gear. It is difficult to estimate the effect of all these factors together, but a conservative guess would be that the average fisherman's net income declined by at least fifty per cent between 1929 and 1937. The pathetic feature of this situation was that the fishermen of Prince Edward Island were adding to the economic difficulties which the depression had thrust upon them. It was an experience that had the same tragic character as one of those fires in a community hall or a theatre which, after the flames have been put down, discloses that few were burnt to death but many lost their lives in the rush for the exits."

Shortly after this report, Gordon began publishing his fundamental work on common property in the fisheries.

***Management***

The fisheries service in the years following the wartime prosperity showed little vigor or innovation. It seemed to take the attitude, "leave well enough alone." The Biological Board showed more initiative, in technological developments as well as science. Its development work, however, produced little in the way of industry change.

The transport subsidies ended. Development attempts stagnated. In 1922 the federal government allowed Quebec to manage its own fixed-gear fisheries. In 1930 it turned over freshwater fishery management to the Prairie provinces.

When prosperity faded in the latter 1920's, federal fisheries authorities reacted like the government at large, mainly in a conservative rather than a developmental way. In a chief example, while the Europeans and New Englanders were rushing to use trawlers, Canada put restrictions in their way. Earlier, in 1908, the trawler *Wren* had begun working out of Canso. When inshore fishermen opposed it, the government forbade steam trawlers within the three-mile limit. Line fishermen continued to oppose trawlers, and despite the lack of scientific evidence, politicians declared that they destroyed marine life. The government brought in more restrictions, forcing them to fish out beyond 12 miles (Gordon 1951). Still the trawler fleet increased in the 1920's to more than twenty vessels. Lunenburgers soon became important in the fleet.

The 1927–28 Royal Commission on fisheries of the Maritimes and Magdalen Islands heard the heart-felt fears of inshore fishermen that trawlers would ruin their markets. New and stern restrictions followed: special licences for trawlers, and a trawler tax. By the time of World War II, there were only three trawlers left operating.

In small fishing villages, there is a great tendency to see anything new as the enemy — especially if there is no chance to get in on it. Independent fishermen were fast enough to adopt gas engines, but trawlers were beyond their means,so they opposed them.

The restriction on trawlers, combined with decreased markets and investment capital, extended an existing technological lag, and slowed development for many years. Improvements did take place in the existing fleet: for example, use of diesel engines and radios. The Lunenburg schooner fleet in particular began winter-fishing for the fresh-fish market. But schooners were a lesser technology for reliable supply. The trawler restriction slowed the growth of a fresh and fresh-frozen fishery and of a year-round fishery.

In 1930, as a result of the 1927–28 Royal Commission, the government separated the Department of Fisheries from the Department of Marine. Science remained separate, with the Biological Board working on its own studies. W. A. Found, now the top person in the Fisheries Service, tried to set up its own science branch; but the Biological Board checked him (Annual Report, Department of Naval Services, 1919 and Johnstone 1977). Scientific knowledge gradually increased. Alfred Needler and others began mapping the boundaries of the main deep-sea stocks.

There was also an impressive new attempt to figure out the industry as a system. A remarkable report in 1932 by the Cockfield Brown consulting firm (Cockfield Brown 1932) found failings at every level.

"The disjointed condition of the Canadian fishing industry ... is a heritage, in part, of the

> geographical dispersion of resources, and in part of the complete lack of intercommunication in the days when each of its branches was first established. It is not a new problem, but has become steadily more acute over a period of many years. ... Return for effort and investment in every phase tends towards the vanishing point, and new capital cannot be directed into it for necessary improvements and additions to equipment, plant and organization."

The report noted about the end product:

> "The purchase of fish is not accompanied with such confidence as in the case of the purchase of meat. This is due to the fact that fish have hitherto reached the consumer's table in an unattractive condition, resulting in a large proportion of the consumer's market being apathetic towards the use of fish."

Quality problems were common, e.g. bones, poor flavor, bad odor, lack of freshness. The inspection system was unsatisfactory, and the industry itself wanted improvements. The report said that the fragmented and uncoordinated industry needed basic improvements in quality and marketing, and needed some control of overexpansion.

Conservation policy should restrict seasons, places open to fishing, and licences, said the report, but otherwise let people fish as they wanted. The industry needed more use of trawlers, refrigeration, and quick freezing, and it needed better inspection. There should be systematic collecting of market information. Promotion was almost entirely lacking; practically the only form of advertising was to list fish prices. Bit by bit, there should be reforms, integration, consolidation, until the industry worked as a system.

Except for two dramatic results of the 1927–28 Royal Commission — the trawler ban and a movement to aid fishermen's co-operatives — no major and fundamental effort came forth from the department between the wars. The Cockfield Brown report had little impact. There seemed to be little will to take bold initiatives.

More might have happened had the scientists been closer to the operations side. Both science and management suffered from the divorce between them. Except by fishery officials themselves in a rough and ready way, there were few true experiments in management. Occasionally the Biological Board ventured widely. Generally, however, it was unclear to the Board scientists how much the department wanted or needed them.

In the day-to-day work of the department, reform of the civil service after World War I brought some improvements. The frugal W. A. Found ran a tight service, and instilled a conscientious attitude. A lot of work was still concentrated on the river and estuary fisheries. Sea fishermen were generally left freer, although more rules were being introduced for the lobster fishery.

There was no allocation system per se. The "squeaky wheel" principle played a part. The trawler ban favored the smaller fishermen, who had the biggest voice with politicians. It is notable, as Gordon points out, that no major competitors banned trawlers (Gordon 1951). Writing in 1951 on the trawler question, he said:

> "...Canadian fishermen are not the foremost users of the Atlantic Banks and the natural superiority of the Canadian ports ... has not resulted in a continuous growth of the Canadian industry relative to the other nations whose vessels use the Atlantic Banks. ... At practically all levels - production, processing, marketing — the Canadian industry lags behind in implementing the developments in science and technology which have been made during the past century."

The Biological Board, still separate from the department, undertook more development work in this period. In a remarkable science and development effort, Alfred Needler of the Biological Board in the 1930's developed new techniques for oyster farming, and encouraged many parties (often farmers) into it. Although successful, the effort lost impetus when War II changed the economic landscape. The Biological Board also experimented with freezing fish at its new technological lab in Halifax, and increased its attention to such matters as fish transport.

Broadly speaking, the department itself did little in the way of Atlantic development. It made some attempt to promote fish products; and fish inspection was improved, with compulsory inspection for some products, and more encouragement of good quality in production. However, as Gordon (1951) observed: "The upside-down economy of the pre-war decade was one in which restriction of economic innovation was more in line with current thought and practice than was encouragement."

Attempts at change tended to come from outside the department. In the desperation of the Depression, the government relaxed the purse-seine ban. Fishermen at Campobello Island, N.B., accustomed to using purse-seines inside weirs, wanted to use them in open water to strengthen their fishery. They made representations to the Price Spreads Commission under H. H. Stevens, and the sardine-packing firm Connors Brothers supported the idea. The government changed the rules. This paved the way for growth in the purse-seine fleet after the war (Gough 1971).

As a major result of the 1927–28 Royal Commission, the department engaged Moses Coady to promote fishermen's co-operatives. Scores of co-operatives sprang up, mainly along the Gulf of St. Lawrence. Later many combined into the United Maritime Fishermen's Co-operative and the Quebec United Fishermen. Both of these organizations lasted to the 1980's.

The federal government on occasion during the Depression offered special relief payments to fishermen, many of whom were desperate. In Nova Scotia, a special arrangement channelled some of the federal money into loans for boats and gear. This arrangement was a forerunner of provincial Loan Boards that sprang up in the Maritimes after the war.

The Dominion government also, in the latter 1930's, set up a Salt Fish Board to regulate and subsidize exports. But the question of subsidies rapidly became irrelevant, as another World War started to drive up demand.

### Newfoundland Between the Wars

#### *The Fishery*

Newfoundland was even slower to adopt new technology than the Maritimes. It had only one trawler in the latter 1920's. Like the Gulf of St. Lawrence fleet, that of Newfoundland made great use of old vessels sold off by the Nova Scotians. Newfoundland exporters, often selling on consignment, had little power in the marketplace. There was no significant move to the fresh-frozen trade. The island was still the captive of cod, and the salt cod trade was weakening. Meanwhile the lobster industry plummeted from overfishing, and went through a three-year closure in the 1920's.

#### *Management*

There was one major attempt at reforming the fishery in Newfoundland, and it failed. A remarkable fishermen's movement had begun even before World War I. After the war, William Coaker built the Fishermen's Protective Union (FPU) into a powerful industrial force. The union became active politically; and Coaker became a powerful Minister. Immediately after the war, he tried to consolidate marketing, by controlling the flow and price of salt cod. Major exporters bucked him, and the attempt failed. The FPU itself faded from prominence in the 1930's (Smallwood 1927; McDonald 1971 and 1980).

After Coaker's valiant attempt, nobody else attempted fundamental change. Newfoundland faced the same problems that the Cockfield Brown report outlined for Atlantic Canada, only more so. By the 1930's, Newfoundland had sunk deeply into the Depression. Bankruptcy followed. In 1934, three hundred years after the Western Charter empowered the fishing admirals over the would-be settlers, a Commission of Government took over — rule by British civil servants.

A 1937 Commission of Enquiry into the Newfoundland fisheries noted that "the Department of Marine and Fisheries was not created until the year 1898, and does not seem to have been able to make itself of any great importance to the Fisheries. It did not compile any statistics, of which today there is an almost complete absence.... There is even a doubt as to the laws and local regulations in effect for the control of fishing." (Newfoundland 1937).

The Newfoundland government in 1930 had set up a research station, but fishery inspectors had made no use of it. The Commission noted with approval, however, that under Responsible Government the authorities set up bait stations at central points, to supply hook and line fishermen. The Commission of Government extended the system, which by 1937 consisted of eight cold storage plants.

By the end of the 1930's, the Commission government was taking a number of interventionist measures, including a grouping of exporters. To judge by a broadcast speech in 1944 by commissioner P. D. H. Dunn, the government was looking towards stronger measures to strengthen, streamline, and centralize the fishing industry (Dunn 1944). But Confederation was to intervene.

### The Freshwater Fisheries Between the Wars

The Great Lakes fishery continued largely as before, most prosperous in the lower lakes. The pattern continued of fishing down desirable species, with less desirable ones filling the ecological niches.

In the Prairie provinces, overfishing and overcrowding, lack of organization, and weak marketing created an unstable, low-income fishery. Complaints were rife, and there were dark accusations of undue market influence by American buyers.

A royal commission on Manitoba fisheries in 1933 (Manitoba 1933) called for a thorough overhaul of the fisheries, including lower catch quotas, stricter control of fishing licences, and a "clearing house" to co-ordinate the market. Nothing notable happened.

In the fishery itself, the Manitoba government in the 1930's began experimenting with individual boat quotas. In 1939 for Lake Winnipeg, the government combined these with limitation of licences — thus preceding by three decades the first such attempt in the sea fisheries. The government took into account seniority and other factors when granting licences — for example, the number of dependent children the fisherman had. However, the number of licences and amount of quota both rose in the early 1940's. Management in the prairie fisheries was still fumbling with controls, and was still weak (Judson 1961).

## The Pacific Fisheries Between the Wars

### *The Fishery*

British Columbia fishermen weathered the Depression better than their Atlantic counterparts. The salmon industry, with roughly 100 plants at the beginning of the century, began to consolidate in the late 1920's, as it did again in the 1950's and the late 1970's.

The pilchard (*Sardinops sagex*; also known as California sardine) industry grew rapidly from the late 1920's. It fed the purse seine fleet and the "reduction" fishery, which reduced fish flesh and bones into fertilizer or fish meal. The pilchard fishery boomed in the 1930's but failed in the 1940's, when the resource declined.

B.C. fishermen continued to organize themselves more than did Atlantic fishermen, and their organizations gained influence. The Prince Rupert Fishermen's Co-operative Association took hold in the 1930's and become one of the world's most successful fishermen's co-operatives. It still dominates the northern B.C. fishery.

B.C. fishermen in the 1930's developed the gillnet drum and in the 1940's, the drum seine.

### *Management*

Internationally, the fisheries service had helped strengthen Canadian sovereignty after Confederation, and did so again in the 1920's. W. A. Found helped negotiate the 1923 halibut treaty with the U.S., the first treaty Canada negotiated and signed independently of Britain. The resulting International Pacific Halibut Commission set up a research arm, and began using quotas to regulate the fishery. Thus began the first sea quotas in Canada's fishery.

Efforts to regulate Pacific salmon internationally had started earlier, but took longer. After years of aborted attempts at a salmon agreement, Canada and the United States finally signed one in 1937. This came about after sport-fishing groups in the U.S. realized it was in their own interest. Like the Halibut Commission, the Salmon Commission had its own research arm. The first major efforts went to overcoming the results of the 1913 Hell's Gate slide, with research and fishways on the Fraser. Again, there was joint regulation of openings and closures, and allocation of catch shares by country.

In British Columbia, fishery officers had early begun doing river reports and monitoring salmon and spawnings. As time went by, they began making forecasts of future runs of salmon, with tolerable results. Only in the 1960's and 1970's did scientists begin making similar forecasts.

Albert Tester of the Biological Board investigated herring (*Clupea harengus pallasi*) stocks, and worked with the Fisheries Service to set quotas. This remained standard practice until the early 1960's, when a new wave of scientists decided controls were more or less unnecessary. As quotas ended and fishing efficiency increased, a drastic decline of herring followed.

While controls on fishing waxed, controls on the number of fishermen waned. After World War I, the pressure from returning veterans wanting jobs led the fisheries service to abandon strict control of licensing, at least for whites. Various restrictions remained on Japanese fishermen. With the disappearance of limited entry, there vanished also some of the rules about northern waters, including the restriction on motorboats. As well, following a court case in 1927, the department lost its power to control the licensing of canneries.

The fishing industry lobbied for something new, to control the massing up and destructive power of fleets. The result was a system whereby salmon boats transferred into and out of areas. When the number of vessels in an area exceeded a certain number — even if only one additional vessel came in — the Fishery Officer would shorten the fishing week, making everybody tie up an extra day.

The fishing industry also encouraged the fisheries department to set up an inspection laboratory for canned salmon. Opened in 1936, this was to be the first of several inspection laboratories across Canada. The B.C. industry, quicker to seek inspection to safeguard its reputation, was also quicker to promote its products than the Atlantic industry.

Meanwhile, an earlier attempt at development fell into disarray. After the department and others raised questions about the usefulness of salmon hatcheries, the Biological Board carried out a thorough and widescale investigation, and found their contribution to the fishery to be negligible. The department largely abandoned salmon culture in B.C. for the time. The Atlantic coast continued using hatcheries for stocking rivers (Johnstone 1977).

### Fisheries Development and Management Between the Wars: Discussion

In the Maritimes between the wars, the department was regulating in many ways, but seemed to have no great conviction about improving the fishery. There was no concerted attempt to make the different elements of the industry mesh better. Innovations came mostly from outside the department, through royal commissions or political pressures. Nothing seemed to work that well. As usual, the farther south in the Maritimes, the better off the fishermen; but in the Depression there was some desperation everywhere.

The Maritimes industry by the mid-1930's appeared troublesome and backward to most observers. The Cockfield Brown (1932) report took a close look at the whole system, and found failings everywhere. The fishing industry was in even worse shape than other industries, just as the Maritimes and Atlantic Quebec were in worse shape than other parts of the country. As the Halifax Herald wrote, looking for a villain: "Is it, as rumor declares, a matter of interlocking directorates, of mutual shareholding, of definite collusion in certain quarters to keep prices down...? ... Is there a devil fish somewhere in the industry, with slimy tentacles reaching into every cove and harbor in Nova Scotia, sucking the very life blood out of men, women and children?" (Halifax *Herald*, 7 January 1937)

Yet, as always in the fishery, the situation was complex. The Maritime fisherman usually owned his own house, and had the sometimes claustrophobic comforts of his own community. Maybe in some cases he felt he had more than the salaried city man. On the other hand, Maritimer after desperate Maritimer from fishing communities was migrating to the cities, looking for work.

Newfoundland was worse. The 1937 Commission of Enquiry painted the fishery in these words:

> "Not being able or willing to purchase his requirements for the fishery for cash, the fisherman is compelled to seek someone who is ready to let him have them on credit. This may be a merchant supplier in St. John's, or in, or near, the settlement or town where he resides. ... Having secured his supplies, the fisherman may fail in making payment for them at the end of the season, and, if his merchant continue to supply him in succeeding years and misfortune should still prevail, he may become so heavily in debt that he is unable to find a merchant who will give him an outfit. ... A continuation of bad years results in the depreciation of the vessels and the property and eventually the fisherman is helpless and hopeless. That is the condition of the majority of the fishermen in Newfoundland today." (Newfoundland 1937).

The prairie authorities were talking of new measures such as marketing agencies, and trying new measures such as individual boat quotas, without thoroughly carrying them through.

On the Pacific, however, managers and industry were swifter to adapt to circumstances, and were tackling management issues head-on.

In understanding and in sharing information, the Pacific industry was ahead of the prairies and Atlantic. In conservation and protection, there was less of a game of cops and robbers, hide and seek, than on the Atlantic. It was easier to see the consequences if the fishery officer failed to set or enforce fishing boundaries at the river mouth, and this created a certain support for the department. When limited entry in the salmon fishery fell apart, the fishing industry itself sought new methods to control fishing power. The Halibut Commission pioneered in international quotas. The fishing industry supported inspection labs for quality, and everyone supported Fraser River restoration.

Although the B.C. industry had its problems, it seemed to benefit from its linkages with the surrounding society. As the industry went through successive consolidations, fishermen — although they would roam the whole coast — were based more and more in areas such as Prince Rupert, Nanaimo, and the Vancouver environs. Their industry had a more concentrated, even urban character, with easier flow of information and access to capital.

## World War II to Canada's Centennial (1967)

### Maritimes, Quebec, and Newfoundland

#### *The Fishery*

World War II and the growth of applied science brought new technology for navigating and fish-finding: radio, radar, loran, sonar. It brought new technology for fish-handling: better engines including diesel, better ropes and nets, haulers, hydraulics; and bigger boats, including steel and later fibreglass vessels. The war also brought increased prosperity and productivity through a high demand for fish. The fishing industry and government began to emphasize development.

The vital Atlantic groundfish sector, the fishing industry's main employer on sea and shore, went through a revolution. Refrigeration — the "cold chain" in transportation and storage facilities, stores, and homes — boosted demand for frozen fish. New money and new markets prompted a rapid changeover from saltfish for most of the industry. The remaining saltfish sector got into repeated financial trouble, and needed repeated subsidies.

The government lifted the trawler ban, and encouraged building of draggers and trawlers, including midwater trawlers. The industry made more use of offshore species such as flatfish and redfish. In the groundfish industry, vertically-integrated companies (fishing, processing, and marketing) began to dominate, developing a powerful fleet of 150 or more large trawlers. By the end of the period, they were extending partial dominance into the lobster and scallop industries.

Following the industry-pioneered build-up of the Bay of Fundy herring-seine fishery, federal and especially provincial governments jumped aboard, subsidizing additional boats and plants into the industry. In the late 1960's, several vessels from the closed-down B.C. herring fishery came to the Atlantic. Overcapacity led to overfishing, which led to a 1970's crisis.

New fisheries started up for offshore scallops; for inshore scallops in new places; for shrimp; and for crabs. The Irish Moss fishery also started up during the war. Federal and provincial government explorations and demonstrations of techniques helped the new fisheries get going.

### *Management and Development*

After the war, Deputy Minister Stewart Bates sought a stronger industry for the Maritimes and Quebec, with operations better integrated from sea to market place. Bates reshaped the Department of Fisheries accordingly, creating the modern department. Management became more active in terms of all the elements noted at the outset: understanding, conservation and protection, sharing out the fish, development, and making the fishery work better as a system. Some efforts succeeded better than others. Development was the dominant trend, more than the other "twin god" of conservation.

The Fisheries Research Board (1937 successor of the Biological Board) became stronger, with more full-time staff, and worked more and more closely with the department. Work took a more practical turn, with Board scientists finding new fish stocks and introducing new equipment. Knowledge of population dynamics began to increase in the late 1950's and in the 1960's. However, few scientists worried about overfishing. The renowned A. G. H. Huntsman had written in 1943 that "'depletion,' except as an uneconomic condition which may exist when fishing starts, is little more than a bogie to frighten the credulous" (Huntsman, 1943). Even in the 1960's, scientists were slow to take alarm.

Bates himself had already amassed a great deal of fisheries knowledge. While still a professor of commerce at Dalhousie University, he wrote a 1944 *Report on the Atlantic Sea-Fishery* for Nova Scotia's Royal Commission on Provincial Development and Rehabilitation. He described a largely poor and backward industry. For fishermen after expenses in 1939, "the net income remaining for living was inadequate, no matter how it is estimated."

Bates wanted "a fuller use of the fishery resources" of the Maritimes, and wrote that "this revolution can be achieved only by new methods that will increase its efficiency, and its productivity per man.... On this matter the industry itself is unlikely to voice any definite policy, both fishermen and firms being individualistic in outlook, and being inclined to regard any imposed action that requires a change in their traditional ways as 'interference' or 'regimentation'". He called for improvements at every level, and a bold plan of expansion. Figure 1 from Bates (1944) exemplifies his thinking. The implication is that the more money you put into the fishery, the more you will get out of it, with no end in sight.

In conservation and protection, the department buckled down after the war, with a special training program for a few score fishery officers, mostly veterans. The "Class of '47" became important in management. Gradually the department got a few patrol boats of substantial size.

Canadian conflicts with overseas fleets in this century were minimal until after World War II. In the 1950's, development of the factory freezer trawler gave new impetus to foreign fishing in the northwest Atlantic. The International Commission for the Northwest Atlantic Fisheries, established in 1949, gathered comprehensive data on fish abundance and location and established mild restrictions. However, it lacked the power of enforcement, and the political will to take effective measures. The governing idea in ICNAF and in Canada was Maximum Sustainable Yield (see Parsons 1993a).

In 1945, new restrictions confined lobster boats to their own districts. In the period 1966–68, spurred by representations from the north shore of Prince Edward Island, the department limited the number of traps per boat, and also, a century after the Fisheries Act provided the power and more than half a century after Prince called for it, put limits on the number of licences. For New Brunswick salmon, too, licences became limited

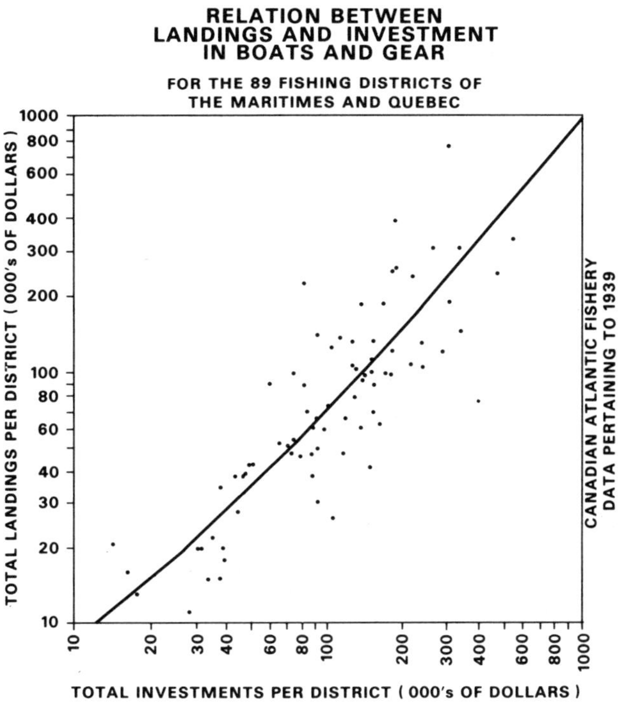


FIG. 1. This graph redrawn from Bates (1944) shows landings rising with investment, with no suggestion of the downturn that occurs as returns on investment diminish. It exemplifies the optimistic, developmental attitude that prevailed in the Department from post-World War II until the late 1960's.

in number in the 1960's. This seems to have been less from any theoretical rationale than from desperation to control the fishery.

The department began closing various bays and harbors to roaming trawlers or purse seiners. But these measures were sometimes a consequence of its own efforts to encourage more mobile fisheries. Development prevailed as the overriding idea until the late 1960's. It influenced the thinking of scientists and economists as well as operations people. Few voices warned about possible overexpansion and its consequences.

The Industrial Development Service and the Fisheries Research Board helped to find new stocks of fish and encourage modern fishing methods and gear, for example, mechanical haulers. The first boat-building subsidies, probably reflecting Bates's ideas, favored co-ops and groups and what we would now call midshore fishermen (independent but with sizeable boats able to fish well away from home, for days or weeks). Political influence got the subsidies extended to smaller boats; and a shipbuilding subsidy from another department helped the large trawlers. The federal government also created fishing vessel improvement loans and vessel-insurance programs for fishermen. The provinces set up loan boards, offering advantageous interest rates to fishermen building boats. Provincial and federal aid also helped build new fish plants (see Crowley et al. 1993).

Before the war, a chief problem had been too little capacity to catch offshore groundfish. By the late 1960's, a chief problem was too great a capacity. The Canadian fleet of vessels 50 tons and over multiplied its capacity fivefold from 1959 to 1974; but the groundfish catch actually fell as both domestic and foreign fleets added pressure (Department of the Environment, 1976). By the early 1970's, it was clear that overexpansion was helping to deplete stocks and drive down average incomes.

Although "more fish" was the main idea, Bates and the department tried to strengthen the rest of the system. From the early 1950's there was new, much stronger work on inspection; although complaints continued about the quality of Canadian fish as compared to that from Scandinavia. To lend stability to the fishery, the federal government created the Fisheries Prices Support Board in 1947, and in the 1950's extended unemployment insurance to self-employed fishermen. The provinces set up fisheries training schools, largely dealing with navigation and practical matters; there was little or no imparting of fishery management principles.

An economist himself, Bates created the first strong economics division of the department. There was a greater attempt to apply economics to particular fisheries, as in the collaborative work of Rutherford and Wilder on lobsters. Bates encouraged marketing analysis, and strengthened public information work. He helped encourage fish processors to set up the Fisheries Council of Canada, a stronger successor to the earlier Canadian Fisheries Association. British Columbia already had an active cannery association, which became part of the FCC. The Fisheries Council became very influential in regulation. But Maritime fishermen remained generally unorganized.

Bates had wanted an overall strengthening of the fisheries system, with attention to volume, value, and stability. In practice, volume got by far the most attention.

The post-war changes in the Maritimes and Quebec filtered into Newfoundland after the 1949 union. The department gradually transplanted the Canadian science and management system to the new province, and did away with the Commission government's collective marketing of saltfish. Newfoundland politicians exerted their own influence, and were instrumental in extending boat-building subsides to smaller vessels and creating unemployment insurance for self-employed fishermen.

Premier Smallwood in the 1950's set up a small resettlement program. Outports voting 100% to pack up and move could get help to do so. In the 1960's, federal funding enlarged the resettlement program, and the required percentage of approval dropped to 80%. Smallwood wanted to strengthen the overall system, with good ports, fewer "raggedy-ass" (to use a Smallwoodism) outports, and a return to the collective marketing of saltfish that the Commission Government created before Confederation. He did move the fishery in that direction, but at great cost. Many fishery observers in Newfoundland consider the resettlement program an essentially good idea, carried out poorly. The resettlement caused a major emotional upheaval.

### Freshwater and Arctic

In the freshwater fishery, technological change was less marked. Although frozen-fish production became more common, the fresh-fish trade remained strong. Tugs on the Great Lakes and skiffs on the prairies remained the prevalent types of fishing vessels.

Long before the introduction of limited entry and individual quotas in the sea fisheries, as early as 1939 the Manitoba government had combined limited entry with individual catch quotas. But application was lax. There was, however, one effort at development where the instigators kept pushing and pushing until they got a substantial measure of success. Saskatchewan's CCF government enacted a series of interventionist measures, eventually creating a system of producers' co-operatives (Judson 1961).

### Pacific

#### *The Fishery*

In the Pacific industry, vessel ownership by processors decreased. The B.C. fleet became more independent and the salmon-canning industry more consolidated. Fishing power became ever stronger, as salmon seiners, trollers, and gillnetters all improved techniques. The herring seine fleet expanded, and techniques including the power block, the drum seine, and "pitlamping" (attracting the fish with mercury lights) brought huge catches.

#### *Management*

In the herring fishery, as in many fisheries since, the growth of fishing power masked a decline in stocks. Scientists did not become concerned about depletion of the stocks until the 1960's (see Iles 1993). The theory had grown up that commercial fishing would become uneconomic before there was any real danger; as the fishery declined, the stock would bounce back. However, herring school up even when scarce, enabling large catches that give an illusion of abundance. Under heavy fishing, herring stocks went into a decline that by the end of the 1960's became obvious to all.

Salmon, always returning to the rivers, were easier to count. Knowledge of salmon biology and populations grew steadily. W. E. Ricker made great contributions to the theory of population dynamics. The department people and the FRB more and more filled out the picture of salmon behavior in the rivers and at sea.

Major changes took place in conservation and protection. The department gave up the increasingly cumbersome system of area licensing, with fishermen transferring in and out. How then did they cope with an ever-stronger fleet, grouping up in ever-greater numbers? By means of a superb corps of fishery officers, under the redoubtable A. J. Whitmore.

Whitmore instilled a great esprit de corps, and set up a streamlined system of management, relying on up-to-the-moment field reports and the day-to-day judgment of well-motivated, well-trained officers. Alfred Needler later remarked in the 1980's that "the reason we have so many Pacific salmon today is because of Joe Whitmore" (Alfred Needler, pers. comm.).

With changes in salmon management came new efforts in restoration and development. The International Pacific Salmon Commission finally built the Hell's Gate fishway, and took an ever stronger role in managing Fraser stocks. The Fish Culture Branch of the department grew stronger. In the 1950's and 60's this branch did pioneering work on spawning channels and pursued hatchery technology, which was resurging after the earlier hatchery closures. The branch also fought the ever-increasing encroachments by the urban-industrial environment. It helped to stimulate environmental consciousness in British Columbia, and to stave off the threat of a new system of power dams on B.C. rivers. Indeed, this Fish Culture Branch was in effect the embryo of the federal Department of Environment, created later by the B.C.-based Minister Jack Davis.

The fishing industry itself was a force for change. Canners had long been interested in controlling entry into the fishery, as shown by their own attempts early in the century, and their representations later (Campbell, 1973). In the late 1940's, the United Fishermen and Allied Workers Union pressed for a licensing system to protect the fishermen who had invested their money and lives in the fishery against newcomers and moonlighters. The union's interest helped prompt the federal government to commission an early 1960's study by Professor Sol Sinclair, on a future licensing system

(Campbell, 1973 and Sinclair, 1960). The idea of limited entry was gaining ground on the West Coast.

Homer Stevens was a dominant figure in the UFAWU from the late 1940's until the latter 1970's. He helped make the union's influence felt in many aspects of management.

The department and FRB never devoted as much work to technical development on the West Coast as on the East. Still, the west coast produced some of the department's best work on development, as officials fostered the use of fish pumps for handling fish, and refrigerated sea water for preserving quality.

**Fisheries Development and Management from World War II to the Centennial: Discussion**

The Atlantic fishery was paralleling the general progress of society after the war, although at a lower level. The Atlantic economy still lagged behind the rest of the country, and often enough, the Atlantic fishery lagged behind other parts of the Atlantic economy. Still there was headway in the industry. Fishermen had more money. In the Maritimes, though less so in Newfoundland, fishermen were typically moving up from open boats to decked vessels with a pilot-house. Although British Columbia still led the sea fisheries in incomes and in value of production per capita, the Maritimes were gaining ground.

While gaining in cash, many Maritime fishing villages were losing part of their character, as boats grew bigger and cars and television became common. Broadly speaking, however, the higher cash flow made up for it. For example, more families could hope that their children would go to college. The Atlantic coast was gaining a bit more security. Fishermen were gaining ability to pursue fish rather than wait for migrations; and had not yet begun overfishing in a clearly demonstrable way. Outside the fishery, they gained security also from government social policy initiatives such as the family allowance and unemployment insurance, and from a gradual increase in alternative employment, often stemming directly or indirectly from government efforts.

As with hatcheries almost a century earlier, the idea of development now swayed people who rarely stopped to evaluate the results. Anyone could get into the fishery, to race for the harvest with bigger and better boats. By the late 1960's, technology was outrunning biology. Increases in fishing power were failing to bring parallel increases in the catch of groundfish.

Bates' economics division in the 1950's made considerable use of H. Scott Gordon. Along with others in the division, Gordon had begun looking at basic questions, well before overfishing became dramatically apparent in the late 1960's. In his seminal 1954 paper, "The Economic Theory of a Common Property Resource: The Fishery," Gordon pointed out that what counted was the amount of fish per person. Though a great deal was known about biology, little was known about the economic characteristics of the fishing industry. The idea of an inexhaustible sea had dominated until late in the 19th century, and still had advocates (Gordon, 1954).

Gordon quoted the Russian Baranoff (a pioneer of population theory) to the effect that mankind like any other predator reduced the stock of fish, and the more fishing, the less fish would remain in the water. New predation at an increased level would drive a fish population back to a new equilibrium, at a lower level of abundance. "Fisheries management" (a relatively new term) had focussed on attaining the largest sustainable catch, but neglected "the inputs of other factors of production which are used up in fishing and must be accounted for as costs. ... In fact, the very conception of a *net economic yield* has scarcely made any appearance at all."

Gordon cited the work of the international halibut commission. Its influence on the increase in catches was debatable. More important, there was no clear evidence it had made halibut fishermen more prosperous. Fishing costs had risen:

> "Since the method of control was to halt fishing when the limit had been reached, this created a great incentive on the part of each fisherman to get the fish before his competitors. During the last twenty years, fishermen have invested in more, larger, and faster boats in a competitive race for fish. In 1933 the fishing season was more than six months long. In 1952 it took just twenty-six days to catch the legal limit in the area from Willapa Harbor to Cape Spencer, and sixty days in the Alaska region. What has been happening is a rise in the average cost of fishing effort, allowing no gap between average production and average cost to appear, and hence no rent."

Gordon referred also to the case of Atlantic lobster, and found analogies in the destruction of other common-property resources, for example the buffalo. For hunting and trapping, "even in the most civilized communities, conservation authorities have discovered that a bag-limit *per man* is necessary if complete destruction is to be avoided." More than a decade before Hardin (1968) made popular the phrase "tragedy of the commons," Gordon referred to the problem of common pastures in the medieval manor economy, and the development of elaborate rules for "stinting" the common: "limitations on the number of animals, hours of pasturing, etc." Gordon concluded:

> "There appears, then, to be some truth in the conservative dictum that everybody's property is nobody's property. Wealth that is free for all is valued by none because he who is foolhardy enough to wait for its proper time of use will only find that it has been taken by another. ... The fish in the sea are valueless to the fisherman, because there is no assurance that they will be there for him tomorrow if they are left behind today."

By merely setting physical limits to landings, without regard to fishing costs, biologists had overlooked essential elements of the problem. In contrast to Bates' graph implying (although Bates never put it that way) an ever-increasing revenue, Gordon used graphs showing how profits decline as fishing increases; and that the best profits occurred at a fishing level well below the Maximum Sustainable Yield.

Although Gordon's basic idea had occurred earlier — that fish and profits per man were the main consideration, and that open entry to the fishery drove down both abundance and average incomes — his was probably the clearest expression yet. Gordon's thinking gradually became common in fishery management circles — but only after years had passed. For a fuller discussion of attempts to deal with the common property problem in Canada's marine fisheries, see Parsons (1993a).

When Gordon wrote his 1954 paper, other fishery economists in Canada and elsewhere had already begun thinking in a similar vein. Canadians led the way in spreading the new thinking and eventually in its practical application (W. C. MacKenzie, 1992, pers. comm.). Yet, even in the late 1960's, long reports by the Department of Fisheries were projecting ever-greater growth, without much worrying about side effects on abundance and average incomes.

The department approach on the Atlantic was to a degree unintegrated. The conservation and protection arm of the department was gradually increasing its regulations — sometimes creating new restrictions even where other parts of the department were subsidizing new technologies for growth.

In British Columbia, the problems of growth and development were more familiar from the early days of the fishery. Now as the social and industrial pressures on B.C. fisheries grew more acute, the industry and the fishery managers were quicker to rise to the challenge. Perhaps the same visibility of the salmon which made problems apparent gave them belief in the solutions. When you see the salmon swarming up the river, you know you have done something right.

The Pacific industry and the Fisheries Service shared more of a common consciousness than on the Atlantic. This served well in the battle against proposed hydro dams on British Columbia salmon rivers. Industry pressure in British Columbia was also a key factor in driving the department towards a renewal of licence limitation — today a key element of management almost everywhere. Indeed, the B.C. industry, especially the United Fishermen and Allied Workers Union, may have done as much as H. Scott Gordon to make limited entry a cornerstone of the modern system.

As management grew ever more complex, the character of managers became more important. Joe Whitmore seems to have exemplified the ideal manager. Three characteristics seem to mark successful fishery managers.

(1) Knowledge. The good manager has to gain knowledge from long experience, or by intensive study and research. Whitmore had very long experience, in Ottawa and B.C.

(2) Toughness. Whitmore was said to be able to face down a group of angry fishermen as if they were schoolboys. As the senior Richie Nelson of B.C. Packers remarked, "When Whitmore was around, fights were fun." Homer Stevens respected Whitmore as he did few other fishery managers.

(3) Responsiveness. To take an example, the renowned Jimmy Sewid, a leading commercial fisherman and a leader among Pacific Coast Indians, recounted the day he was working behind his house at Alert Bay. He turned around and there was Whitmore, who had dropped in for a chat, as he used to do all along the coast. When Sewid brought up a local problem with regulation, Whitmore solved it on the spot.

Whitmore brought along behind him a corps of fine managers such as C. R. Levelton. Although Whitmore retired in 1963 and is now deceased, his legacy of dedication and operational efficiency still influences the management of B.C. fisheries.

## Into the Second Century

### Maritimes, Quebec, and Newfoundland

In earlier sections, this article briefly summarized changes in the industry, then discussed regional management. After 1967, industry trends and management became so intertwined that it is simpler to discuss them together.

In most places during the last 25 years, the keynote has been the struggle to control overexpansion, and to make the industry work as a more streamlined, orderly system.

The first great crisis came with overexpansion by foreign fleets which decimated certain Atlantic stocks. Canadians in the 1960's and early 1970's raced to keep up with the foreigners, and themselves overexpanded on sea and shore. The federal government finally took H. Scott Gordon's thinking to heart, and tried to curtail Canadian fleet expansion with controls on the number and size of boats. This long struggle had a good deal of success; but even though the number of boats became more stable, fishing power still grew, with better and costlier boats and gear and electronics. The method of "limited entry" controlled the number of pre-existing licences, but did nothing else to control the number of fishermen. A groundfish licence that once employed one person could come to employ several. The number of fishermen, plants, and plant workers increased. Overexpansion hampered Canadian plants, and helped bring the larger concerns into recurrent financial crises. Long after federal fisheries joined the battle to control expansion, some provincial governments kept promoting it.

Federal managers did advance the industry during this period, and formed a better though imperfect partnership with industry in management. Most major fisheries became more stable and valuable, based on a higher level of harvesting. This came about only after years of turmoil, and the price of the gains is constant vigilance and a complex system of regulations. Today a drastic decline in groundfish catches is forcing a major re-evaluation likely to lead to a further tightening of controls.

In 1964 Canada, under the new Territorial Sea and Fishing Zones Act, had extended fisheries control nine miles beyond the territorial sea, and laid down the basis for determining it from straight baselines. But this and ICNAF controls had little effect on foreign fishing, or for that matter on Canadian expansion.

The year 1968 saw record landings. On the Atlantic, the groundfish industry is the most widespread fishery and the major employer. It often affects other parts of the industry. Despite good landings, for market reasons the groundfish industry needed government help in 1968.

In subsequent years groundfish landings began to drop. By the 1970's it was clear that overexpansion by foreign fleets had decimated certain Atlantic stocks. Although Canadians too had overexpanded in some instances, the foreigners were the main culprits. As it had been a century earlier, fisheries sovereignty became a major public and governmental concern, peaking in the mid 1970's just before the 200-mile limit.

In 1968 the B.C.-based Minister Jack Davis took over the new Department of Fisheries and Forestry. Canada asserted the need for new international controls, and got the sometimes alarmed attention of European nations fishing in the northwest Atlantic. ICNAF in the late 1960's and early 1970's began setting overall quotas (Total Allowable Catches) for certain stocks of fish with haddock on Georges Bank being the first.

In 1970, Canada's Arctic Waters' Pollution Prevention Act heightened international attention to the Law of the Sea and to Canada's concerns. In 1971, the three-mile territorial sea and nine-mile fishing zone were replaced by a twelve-mile territorial sea. Also in 1971, Canada established special fishing zones in the Gulf of St. Lawrence and Bay of Fundy on the Atlantic coast, and in Queen Charlotte Sound, Hecate Strait, and Dixon Entrance on the Pacific coast (Department of External Affairs, 1977). A 1972 agreement with France, phasing out distant-water fisheries from the Gulf of St. Lawrence, foreshadowed extended jurisdiction. Agreements with other nations followed, to phase out foreign fishing from enclosed waters.

By that time, ICNAF had begun subdividing herring quotas among member states. These were the first multinational quota allocations in the world. National quotas of groundfish followed, with Canada pressing for coastal states to have first claim on all the fish they wanted.

In 1970 the government created the Department of Environment, taking in the Fisheries and Marine Service and many other agencies, including marine science agencies. Federal fisheries officials, many of whom had believed in and helped create the new department, soon began a fight to get clear of the new and cumbersome bureaucracy and re-create a separate department for fisheries.

For market reasons, 1973 was a record year for landed value. But the groundfish industry again needed help in 1975–78, when depleted fish stocks and high fuel and other costs faced the industry. A subsidy went to both fishermen and plants. This and other authorized assistance came to some $200 million.

In 1973, Davis followed up precursor actions in the Atlantic salmon and lobster fisheries, and in the B.C. salmon fishery, by limiting entry into all major Atlantic fisheries. The industry accepted the move quite readily. But as noted, controls on the number of people registering as fishermen and on fishing power were loose.

Foreign overfishing, new examples of domestic overfishing, for example in herring and Gulf of St. Lawrence redfish, and the chronic troubles of the

Atlantic industry helped crystallize a new departmental attitude. The fisheries service with strong industry support wanted to control the foreigners, and with less industry support wanted to control the fishery at home.

Roméo LeBlanc in 1974 had become Minister of State for Fisheries. In 1975, to the shock of certain foreign fishing nations, LeBlanc announced that Canada was banning the Soviet Union's fishing fleet from using Canadian ports, and he threatened the same for the Spanish and Portuguese. From 1975, the department negotiated several bilateral agreements with foreign fishing nations, in which they effectively pledged co-operation with Canada's future extension of jurisdiction. In these bilaterals and in Law of the Sea negotiations, Canadians advanced their own cause and furthered international acceptance of 200-mile limits generally. For a fuller discussion of events leading to Canada's extension of fisheries jurisdiction to 200 miles, see Parsons (1993a).

In 1976 the department issued *Policy for Canada's Commercial Fisheries* (Department of the Environment, 1976). In it, W. C. MacKenzie, F. J. Doucet, K. C. Lucas, and others crystallized contemporary thinking. To quote the summary:

> "Canada's fishing industry has a history of crises, many antedating the growth of foreign fishing that helped deplete fish stocks off our coats. Although the forthcoming extension of Canadian jurisdiction over fisheries will help our industry, it is as much a challenge as an opportunity. ... The state of the natural resource is only one of the factors affecting the success of the fishing industry. The fortunes of the fishing industry depend also on markets, costs of production, and the industry's structural strength. The commercial fisheries of Canada often have been unstable, self-weakening, prone to crises, and providing low and insecure incomes for participants."

The document pointed out the typical pattern:

> "The fisheries are common property, and each fisherman tries to carve out the biggest possible share of the catch. This naturally leads to a race for the biggest and best vessels."

And again:

> "A brief period of prosperity while fish are plentiful attracts additional fishing craft and processing facilities. In the consequent over-expansion, fish stocks become smaller and profit margins shrink or disappear. Enterprises fail, or survive as cripples in a debilitated industry. The end result is severe social and economic distress in fishing communities...."

The idea was to reform the whole system. The document stated the government's intentions at various levels. In partial summary, it proposed:

— In resource management, to get the 200-mile limit; enhance certain stocks; manage on an ecological basis for the best interests of Canadian society; allocate access to resources "in the short-run on the basis of a satisfactory trade-off between economic efficiency and the dependency of the fleets involved."

— In fishing, to apply entry control everywhere; coordinate the deployment of mobile fishing fleets; provide for withdrawal of excessive catching capacity and for the best possible mix of fleet units; abolish wasteful and destructive gear and fish-handling practices.

— In fish processing, "facilitate price differentiation according to quality for fish landed"; provide that landings go to the most profitable end use; concentrate assistance programs on improving rather than building new plants; make more use of unwanted by-catches.

— In product marketing, promote consolidation of marketing; encourage inter-firm developmental and promotional programs; improve market intelligence.

— Regarding fishermen and their communities, develop programs to counteract instability of revenues; provide for relief of chronically income-deficient fishermen; foster the acquisition of professional status by commercial fishermen; help communities adjust to change; integrate fishery programs with other regional development programs; ensure "the fullest possible involvement of all the people concerned ... in the decision-making process associated with fishery management and development."

The Policy Document said that "the strategies adopted reflect a fundamental redirection.... Fishing has been regulated in the interest of the fish. In the future it is to be regulated in the interest of the people who depend on the fishing industry."

The Policy Document marked a change in attitude. Federal fisheries did make major changes at all levels of the industry, though most of them came where jurisdiction was clearest, in fishing itself. The present management system, largely reshaped in the 1970's and early 1980's, reflects the Policy Document.

The federal government extended Canadian fishing limits on the Atlantic and Pacific coasts to 200 nautical miles (roughly equal to 230 statute miles or 370 kilometres) on 1 January 1977. In March, it extended the same authority to the Arctic coast.

Both Department of National Defence and Canadian Coast Guard vessels helped with early patrols. The fisheries department's own growing fleet carried the main burden. Offshore patrol days roughly doubled beginning in 1976, to about 2 000 sea days. The offshore patrol fleet gained some modern vessels, and air surveillance became an important tool.

In an understanding reached at the Law of the Sea negotiations, Canada and other countries with 200-mile limits took on an obligation to share "surplus" resources with other states. They were to use all the fish they needed and share the rest, within conservation limits they established.

In practice, as Canada's fishing industry showed more and more interest in "underutilized species" and formerly surplus stocks, the amount allocated to foreign countries has steadily shrunk. Today there is almost no foreign fishing for species of traditional high interest to Canada, such as cod, haddock, pollock, herring, and shellfish. For certain other species of relatively little interest to the Canadian industry, Canada licenses several dozen foreign vessels yearly, and places fisheries observers aboard each one of them, at foreign expense. Observer coverage has also become common on the larger Canadian vessels.

Certain international frictions lingered after the 200-mile limit, although some have now been resolved. In extending their fisheries zones, the United States and Canada made overlapping claims on Georges Bank, in the Beaufort Sea, and on the Pacific. The Georges Bank claim was resolved through a 1984 arbitration by the International Court of Justice. This arbitration gave Canada most of its claim for half of the Bank, and this area included some of the richest sections. Canadian management has guarded the health of the stocks. Scallops are much more plentiful on the Canadian side of the line. To a degree that surprised scientists, even the "straddling stocks" of finfish seem more abundant on the Canadian side (For more details, see Parsons 1993a).

Just as in the 1800's, New England fishermen tend to look with a certain appetite to the Canadian resources. Despite Canadian patrols on Georges, American vessels have sometimes slipped across. Canadian authorities have convicted more than 20 American vessels of fishing on the Canadian side. The biggest penalty amounted to a quarter of a million dollars.

In 1987, Canadian fisheries patrol vessels began carrying deck guns for the first time in decades. In October 1988 a Halifax-based patrol vessel, the *Chebucto*, fired warning shots across (but well ahead of) the bow of an American fishing vessel, the *Donna Lynn*. In December 1989, when the Department of National Defence was helping patrol Georges Bank during a strike of civilian ships' crews, a destroyer fired warning shots ahead of the U.S. fishing vessel *Concordia*, and a minor collision took place.

These incidents caused concern on both sides of the border. Negotiators for the two countries sought a more satisfactory way of dealing with the problem. In 1991 the Canada-United States Reciprocal Enforcement Agreement came into force, through which the U.S. undertook to impose penalties of up to $100 000 U.S. for American vessels transgressing Canadian fisheries regulations, and provided for the use of Canadian evidence by the American authorities. Court cases have now gone forward under the new arrangements. As well, the U.S. Coast Guard has begun mounting its own patrols on Georges Bank, to keep American vessels within bounds.

Besides the United States, another old competitor complicated the picture after the 200-mile limit. France maintained a certain claim to fishing rights in Canada, stemming from the historic French fishery and from the 1972 Canada-France agreement. Like the United States, France made an overlapping claim of jurisdiction, based on the French islands of St. Pierre and Miquelon off southern Newfoundland.

In 1992, an international arbitration awarded France a small zone contiguous to the islands, and a ten-mile-wide corridor stretching 200 miles towards the open sea. As for French fishing in Canadian waters under old treaty provisions, final arrangements remain to be worked out.

Canadian officials consider the St. Pierre and Miquelon arbitration another in a series of Canadian successes in fisheries diplomacy over the last 20 years. Typically, the Department of Fisheries and Oceans and the Department of External Affairs have worked in tandem, with fisheries ministers and officials providing an assertive edge to negotiations, and External providing back-up expertise and connections.

In the burst of euphoria immediately following the 200-mile limit in 1977, Newfoundland, Nova Scotia, New Brunswick, and Quebec all put forward plans for major expansions, to be funded by the federal government. Fisheries Minister LeBlanc paid little attention, and tried to keep a strict control on licences and quotas. The department suffered the outrage of editorial writers in Nova Scotia, and advertising campaigns by some larger companies.

Yet LeBlanc seemed to grow more popular. He had identified himself with inshore fishermen and the little man. Although fish were scarce, he still had the subsidy to sweeten the atmosphere while stocks recovered. In retrospect, the strict controls by the department, allowing only a slow increase in catches, helped Canada start off the new regime with good management.

The department had excellent officials who rose to the occasion. Some had gained experience ever since the Class of '47. Others had advanced rapidly through ability. They made the strongest attempt yet to get science and management working in tandem. The Department of Fisheries and Oceans, with responsibilities now including marine science, came into being in 1979.

The 1976 Policy Document had for the first time tried explicitly to combine scientific, economic, and social knowledge with policy at all levels. The federal fisheries bureaucracy itself began to form a clearer collective idea about its goals, especially the control of licences, the control of quotas, and the setting up of a better consultative system.

The new and thorough approach to management brought changes at many levels. Enforcement gained strength for the domestic fleet as well as foreigners. Understanding advanced in several areas. Biological science work roughly doubled. Oceanographic work also increased in the 1960's and 70's.

Although for decades reports had talked loosely of overfishing and depletion, only now did officials begin to carry clear definitions in their minds. Today we think mainly of two kinds of overfishing. First is "growth overfishing": harvesting at such a rate that the average fish never reaches the best size for yield. This kind of overfishing does not generally threaten disappearance. The stock just settles into a new equilibrium, yielding less weight than it should, like a forest habitually cropped before the trees reach their best size. Second is "recruitment overfishing," which interferes with reproduction, and threatens a stock's ability to sustain itself. (See also Parsons 1993a and Doubleday 1993.) In either case, a shorthand definition of overfishing is that by fishing more, fishermen are catching less; and that by fishing less they would over time catch more.

From the 1970's, the idea of Maximum Sustainable Yield, which had in practice fostered overfishing, gave way to the idea of optimum yield. Quotas and other measures began aiming for a level of harvesting which would ensure good profits and good catch rates. Exceeding this level is a kind of economic overfishing. It is a relatively short time since these ideas about overfishing and proper yields reached their present level of knowledge and acceptance, which is still far from universal (see Parsons 1993).

Careful rebuilding of finfish stocks took place in the first decade following the 200-mile limit. Both catches and markets grew stronger, particularly in 1979. Then the groundfish industry, the focus of crisis in 1974–75, again needed help in 1983–84. The larger processing-fishing operations, just before a severe recession, had "overexpanded beyond their capacity to manage," as Michael Kirby, leader of a government task force, once phrased it. They got major financial aid. Total expenditures, some recoverable, on their behalf came to several hundred million dollars. The four largest companies operating large trawlers, through a government-aided "restructuring," amalgamated into two — National Sea Products and Fishery Products International. These companies consolidated some operations and, with the advent of "Enterprise Allocations," reduced the number of trawlers. Independent fishermen got no financial aid. (For more detail on the Atlantic fisheries restructuring initiatives, see Parsons 1993a.)

With the restructuring came an important attempt at reform. In February 1983, the government published the report of the *Task Force on Atlantic Fisheries*, better known as the Kirby Report (Kirby 1982). This recommended many small reforms and certain large ones. The latter included better control of industry capacity, with an individual quota system for much of the fleet, and concerted and government-orchestrated improvements in both quality and marketing.

None of the large reforms came to pass exactly as envisaged. Yet the Kirby Report, like many special inquiries before and since (Pearse Report, Haché Report, Harris Report, Dunne Report, etc.) seemed to clear the air and let people get their problems off their chest. As well, the Kirby and Pearse reports helped strengthen the still-new idea of individual quotas for fishing enterprises. (For a discussion of the impact of the Kirby and Pearse inquiries, see Parsons 1993b.)

Perhaps the greatest change of the last two decades came in the system of sharing Canada's fisheries. In the 1970's and 1980's, limited entry — which affects both conservation and sharing — became the rule. Fishermen were categorized as full-time and part-time, and full-timers received the most benefits — for example, they might get first chance at a new fishery for underutilized species.

Fishermen rarely opposed limited entry on principle, apparently since they saw benefits in security. Complaints about application were however frequent, from those who wanted better licences or more licences, and who claimed a history in the fishery which they were unable to prove. Others outside the fishery sometimes opposed limited entry: either those who wanted entry to the fishery, or in some cases, academics and romantics who saw the new controls as an imposition on traditional freedoms and a way of life.

In 1977, the department began sub-dividing groundfish quotas to vessels of different size and in different areas, during different periods of time. More and more subdivisions took place, usually with historical percentages of the catch serving as the starting point.

The imposition of licences and quotas sharpened some of the old conflicts about "who gets the fish." Particularly in the groundfish fishery, the old struggles among different fleets or gear types trying to outrace one another on the fishing grounds changed partly into a new type of contest: industry representatives trying to outargue one another during government-industry consultations about quotas. In fact, the percentage of quotas allocated to various sectors remained fairly stable over the period. However, even relatively small changes are of large importance to someone. Fiery arguments preceding the annual Atlantic Groundfish Management Plan became a tradition.

In the groundfish and herring fleets, the department began keeping small and medium vessels within their own broad area. A 65-foot southwest Nova Scotia dragger, for instance, would be unable to start fishing in the Gulf of St. Lawrence.

In 1976, the Bay of Fundy purse-seine fleet started using individual boat quotas. These spread to the larger vessels in the groundfish fleet, and to larger-vessel fisheries for crab, scallop, lobster, and shrimp. Meanwhile, the Bay of Fundy herring quotas became transferrable, i.e., holders with the department's permission can buy and sell them. The groundfish fishery is now moving in that direction.

Various documents spelled out the rationale for individual quotas. They would create less incentive to overinvest in an over-powered boat to compete in a race that could lead to overfishing. Fishermen would cut costs, pace their fishing, and aim for top quality products. In practice, individual quotas often dampened "the race for the fish," but seldom did away with it entirely. The big selling point for individual quotas (or Enterprise Allocations) among fishermen seems to be security.

Today individual quotas have become highly controversial. Usually they pertain to larger boats with more fishing power. Other gear types accuse IQ-holders of dumping less valuable species in order to get maximum value from their quota, thus worsening conservation problems. To their adherents, individual transferrable quotas, as well as company-based "Enterprise Allocations," offer a painless method of fleet-reduction as some operators buy out others. Opponents charge that the best-financed will eventually monopolize the fishery, changing its character, and taking undue advantage from a public resource.

The nature of development work changed. The department sharply reduced its work on fishing gear and "volume," but did new work on such matters as cost-cutting, selective gear, and underutilized species. Promotion of fish products, for example through National Fish and Seafood Month and advertising campaigns, increased in the 1970's and early 1980's, but then practically vanished. By then, the market itself had strengthened for diet and health reasons, and seemingly because of a growing consciousness of how good fish and seafood could be. Under the spur of competition and of quality-improvement campaigns by the department, the industry improved product quality. Use of boxes and containers became common.

In another development that constitutes a complex story of its own, salmon aquaculture became a resounding success in southern New Brunswick, increasing in the space of ten years to a landed value of more than $80 million in 1990. Elsewhere as well on the Atlantic, aquaculture became a new employer. For a fuller discussion of the growth of aquaculture in Canada, see Aiken (1993).

In an important social change, again laid out in the 1976 Policy Document, fishermen became eligible to collect seasonal unemployment insurance payments for longer periods of time. The idea was to supplement the incomes of the worst-off fishermen. Like many fishery schemes, this had side effects. Relatively easy access to UI helped increase the number of registered fishermen from about 40 000 in the late 1970's to about 60 000 in the late 1980's. Some of the additional entrants made only a few hundred dollars fishing, and collected far more from Unemployment Insurance benefits.

A fundamental change in the 1970's and 1980's was the growth of fisheries advisory committees, chaired by the department but incorporating individual fishermen and their representatives. By the 1980's these advised on management for all major fisheries. More than a hundred committees now existed for various fisheries at local, regional, and Atlantic-wide levels. Fishermen have sometimes doubted the effectiveness of their voice through these committees. DFO officials, however, often considered them to have almost a veto power.

Handicapping the advisory committees was the lack of information, organization, and education among fishermen. Some important new organizations did appear, especially in Newfoundland. The department especially in the later 1970's encouraged organization of fishermen. The fisheries media and advisory committees spread more information. Provincial governments continued their fisheries training schools and often gave courses in local communities on practicalities of the trade, especially navigation. But it was all too little. Too few fishermen ever got fully acquainted with principles of biology, management, quality, and marketing. The lack of a common understanding, at the same time the department was bringing in more and more regulations, caused endless misunderstandings.

Even before the 1976 *Policy on Canada's Commercial Fisheries*, a new marketing agency came into being to resolve the chronic problems in marketing saltfish — the historic mainstay of Newfoundland fisheries, but now a weak and stagnating sector. In 1970 the federal government with Newfoundland's strong backing created the Canadian Saltfish Corporation, giving it a monopoly on marketing salt fish from Newfoundland and parts of Quebec. The number of saltfish producers decreased to less than two dozen, and a new stability entered the industry. Creation of the crown corporation, however, never sat well with the private industry.

The department was now making multi-level interventions. Perhaps the best example of the new approach was the Bay of Fundy herring fishery, mentioned above. Canadian and international vessels had fished down the stock. In 1976, with fishermen's fortunes at a low ebb, fishermen and the fisheries department (LeBlanc, Fernand J. Doucet, Derrick Iles, and others) worked out an individual boat quota scheme, a first for the Atlantic. The department outlawed the high-volume fish meal fishery that had helped create overfishing. Fishermen and processors switched into the slower-paced, higher-value food fishery. To ease the transition, the department allowed over-the-side sales directly to foreign vessels at a higher price. This was another first for modern times. The department also fostered formation of a marketing co-operative, which raised the fishermen's influence on prices. As always, some problems remained. Overall, however, the combined measures brought more security and higher incomes to this fishery (see Iles 1993).

The Bay of Fundy project changed the fishery's system at a stroke. Elsewhere as well, the department tried to make the fishery work more systematically. Despite a great outcry from boat-owners, the department tried with some success to limit increases in the size of boats. Officials feared Canadian boats would otherwise grow to do more damage than the foreigners they had displaced.

Although the controls dampened growth, fishing power grew anyway. By the later 1980's, overcapacity had become the dominant problem for fishery managers. They typically pointed to groundfish fishing in southwest Nova Scotia as a problem area: not because of low incomes, because these were among the highest on the Atlantic, but largely because of the problem of controlling this fleet, which was helping to deplete groundfish stocks.

Until this period, the health of the fishery in particular regions usually parallelled the health of the general economy in the region, but at a lower level. With the post-1967 changes, especially licence limitation, the fishery could now to some degree operate as a self-contained system, separate from the larger society. In certain fisheries and certain places, it became a leader. The crab-fishing captain on the Acadian Peninsula, the trawler captain in a Newfoundland outport, the herring seiner in the Bay of Fundy, or the dragger captain in southwest Nova Scotia, could have an income far above the norm for his community.

In 1987, an extraordinary year of high prices and good landings, the Atlantic fishery was well into another boom. Some voices warned that it depended greatly on the value of the Canadian dollar, and the current strength of markets. Groundfish quotas were declining. By 1989, another downturn was upon the groundfish industry. Again the problem was centred in the larger companies and plants, including some which were the main support of one-industry towns.

In 1990, the government approved the 5-year, $584 million Atlantic Fisheries Adjustment Program (AFAP). A variety of measures came into play under three government departments, to help some plants survive, to help plant workers adjust to the loss of employment, and to help communities find other sources of income. Outside the large plants, the situation was less drastic than in the 1974–75 and 1981–84 downturns, and financial assistance was less. There were no direct industry subsidies. Some of the AFAP money appeared destined for special programs to strengthen regular government operations.

Until 1992, independent fishermen appeared to be surviving the latter downturn as they did the 1982–83 one, with no special aid. This seemed to speak well for fleet and fishery management measures of the last twenty years. In both of the last two downturns, problems had occurred mainly in major groundfish plants rather than the fleet, and plants are to a large degree outside federal management.

But in the early 1990's, problems became evident with the "northern cod stock," the huge cod population east and north of Newfoundland. Although scientists remain uncertain as to the weight of various factors in the decline, there was a sudden, unexpected decline in the stock abundance probably due largely to environmental factors exacerbated by fishing pressure. The biomass of spawning cod dropped so low that the government took the unprecedented and sweeping step of banning the northern cod fishery for two years. The Minister of

Fisheries and Oceans, the Hon. John Crosbie, announced special aid measures for fishermen and plant workers (For more details, see Lear and Parsons 1993).

In 1993, other groundfish stocks suffered quota cuts. All told, the groundfish decline has taken work from more than 20 000 people. Many major plants that were seemingly on the leading edge of the industry have closed, either temporarily or permanently.

Other fisheries have done better, for example lobster landings tripled in the 1980's. But the groundfish crisis has caused new questioning of fisheries management; not so much with basic intentions or methods, such as limited entry and quota management, as with the way they were applied. Despite licenses and quotas, the number of people and the fishing pressure grew anyway. Scientific ability to forecast stock abundance proved to be much more limited than many people had hoped. Nature itself appeared to have chastened those who would predict it.

In spite of the 200-mile limit, the foreign fishery still affects northern cod and other stocks. At the Law of the Sea negotiations in the 1970's, Canada tried to get international recognition of the special problem posed by areas where a nation's continental shelf — the main fish-producing area — stretches out beyond 200-miles. Negotiators failed to win international support. The Northwest Atlantic Fisheries Organization, a successor body to ICNAF managing certain stocks at the offshore edge of Canada's zone, has no means to cope with nations that wilfully overfish. The consequences have now become clear. European fishing on the "Nose" and "Tail" of the Grand Banks — areas beyond the 200-mile line — has helped deplete stocks of cod and flatfish, with tragic effects for Newfoundland communities.

Canada has pursued this issue forcefully. At the United Nations Conference on Environment and Development in 1992, Canada won agreement for an international conference on the issue.

In the domestic fleet, as fishing power grew despite controls, fishermen themselves in the 1980's began worrying more about overfishing and illegal discarding and dumping of fish. Many wanted higher penalties for fishery offences, and maximum fines increased in the early 1990's from $5 000 to $500 000. Habitat-protection laws became stronger, and maximum fines rose to $1 million.

All told, the last two decades saw an unprecedented burst of government activity in the fisheries. By about the mid-1980's government had become so entangled with the industry, albeit with mostly good results, that Ministers and regulators began to worry and draw back. The department got out of boat subsidies; got out of fisheries improvement loans; distanced itself from market promotion; and gave over to a board of private citizens some of the say regarding fishermen appealing for changed or reinstated licences. In the early 1990's the department began looking for more ways to simplify its role in the Atlantic fishery and involve the industry more in management.

**Freshwater and Arctic**

The Ontario-managed Great Lakes fishery in the 1980's moved ahead with individual boat quotas, with some success. On the Prairies, the federal government in 1969 created the Freshwater Fish Marketing Corporation (FFMC), bringing to fruition the recommendations of the 1930's. The FFMC has a monopoly on selling fish in interprovincial and export trade in the Prairie Provinces, the Northwest Territories, and parts of northwestern Ontario. To some degree, the FFMC uses the higher prices from fresh-fish markets in the south to benefit fishermen in more remote areas. To date, the corporation seems to be making money and doing a good job for fishermen, who in the more remote regions nevertheless have frequent complaints about prices and markets.

**Pacific**

The herring fishery has been a frequent forerunner in both progress and problems. With new technology and fishing power in the 1960's, Pacific herring catches soared and stocks dwindled. By the time the department realized how much trouble the herring fishery was in, extreme measures were necessary. The Department of Fisheries banned the British Columbia herring fishery from 1967 to 1972. When it reopened, a new and rich Japanese market for herring roe helped to fuel investment (see Stocker 1993).

The herring collapse helped make clear the dangers of too much fishing power. Following the Sinclair Report, the pressure from the UFAWU, and the gradual forming of a consensus among leading officials, fisheries minister Jack Davis in 1969 enacted a grand scheme for the salmon and herring fleet, to limit the number of licences and buy back part of the fleet. This was probably the biggest such attempt to that time in the western world, and the "Davis Plan" influenced Canadian and international thinking.

The Davis Plan seemed to have good initial results. Price increases in salmon, good harvests, the start-up of the roe herring fishery, and the sense of security brought new confidence — and new investment. The Davis Plan failed to halt the growth in fishing power. Some fishermen bought up the licences of others, and through "pyramiding," got authority to build vessels of somewhat larger size, but many times more fishing power. The fleet turned into a high-technology, ultra-modern fisheries navy. (For more details see Parsons (1993a) and Healey (1993).

During the early 1980's recession, many thought the British Columbia fleet needed drastic surgery. The Commission on Pacific Fisheries Policy under university professor Peter Pearse in 1982 recommended drastic fleet reduction, and a system where fishermen would bid for fishing licences and pay taxes on their landings. The idea was to remove excess capacity, obviate windfall profits, and dampen the boom-and-bust pattern (Pearse, 1982).

Fishermen rebelled against what some saw as a cruel mechanical approach. No such action came about. Meanwhile, good catches and prices reduced the urgency. Today the Pacific salmon and herring fleet continues to operate with massive overcapacity, but with most parties seemingly happy. Like the Kirby Report, the Pearse Report helped define the issues and educate all concerned. Also like the Kirby Report, the Pearse Report led to minor changes but no major ones. (For more details see Parsons 1993b).

On shore, the 1980's saw another of the waves of consolidation that have taken place every 25 years or so in British Columbia. Some major processors cut back or vanished, leaving B.C. Packers more and more dominant.

Meanwhile, a very large development was moving forward. Federal fisheries with support from the commercial and recreational fishing industries and the province of British Columbia launched the Pacific Salmonid Enhancement Program, or SEP. This major effort involved hatcheries, spawning channels, lake fertilization, new research, and many "public participation" projects to improve salmon streams and rivers. SEP began in 1977 with $150 million in federal funding and $7.5 million in provincial funding. The program aims to restore salmon abundance to historic levels, and at least some good results are now appearing. Private aquaculture also grew apace in the late 1980's and early 1990's (see Aiken 1993).

**Overall Achievements Since 1968**

After 1967, the department started reshaping the whole fisheries system, especially on the Atlantic, traditionally the great problem area. The end result turned out well, by and large. But the groundfish industry on parts of the Atlantic coast now has grave problems, stemming from high fishing pressure and environmental factors. It remains to be seen how deeply this will affect the industry, and how much it stemmed from management as distinct from causes beyond national control. In 1991 at least, it was still possible to say that fishermen in general had access to more fish than in the pre-extension period; as well as more money and a greater voice in management.

When compared internationally, Canadian management since implementation of the 200-mile limit stands up very well (Pinhorn and Halliday 1990; Anthony 1990; and Parsons 1993a). In New England after the 200-mile limit, the domestic fleet increased sharply, with few controls on fishing. The result was that ten years after the new limit, New England draggers and trawlers were catching 26% less groundfish, while increasing the number of boats 43%.

Canadian results until the 1990's were far better. Groundfish catches increased dramatically over levels of the mid-1970's, although some stocks and catches declined in the late 1980's, partly as a result of reduced resource productivity. Herring abundance and catches increased. Lobster catches in the Scotia-Fundy Region almost quadrupled in the 1980's. The scallop fishery has done well. The snow crab fishery first burgeoned, then in the 1980's contracted, but in the early 1990's was recovering. New shrimp fisheries developed.

By 1986, per capita production figures in the Maritimes and Quebec were finally even with those of British Columbia, though Newfoundland was still far behind, (Appendix I) partly as a result of a lower-value mix of species. In short, the department after 1967 made a special and unprecedented effort, and it paid off.

However, the departmental effort created an enormous number of new regulations. It brought new expectations, which suffered when groundfish catches began to decline from an early-1980's peak. It gradually created a new breed of committee fishermen, spending half their time at meetings, as well as a new class of representatives who dressed like fishermen and talked on their behalf but were not. The new regulations brought new frustrations. Yet, in an industry with as many layers of feeling as a Russian novel, some also began to take new pride in their new voice in setting regulations.

Typically, Atlantic fishermen with any length of time in the fishery, especially those in more southern waters, will grudgingly admit that the fishing business overall is better for them today than it was thirty years ago.

Compared with the Atlantic, British Columbia at first led in innovations. Then, following the major changes of the salmon buyback and licensing scheme, the Salmonid Enhancement Program, and the wave of plant consolidations, the B.C. industry seemed to quiet down. The participants make generally good incomes and seem to have a satisfactory life. Yet all depends on a visibly vulnerable resource, with a highly-capable fleet balanced against a wide array of regulations. Without the strict controls by federal fisheries, the B.C. fleet could quickly wipe out salmon runs.

The Canadian fishery still contains great variety. It has fishermen with incomes in the hundreds of dollars, the thousands of dollars, and even the hundreds of thousands of dollars. At the best of times, some fishery somewhere is in a downturn. Broadly speaking, however, the changes in the period 1967 to the present have created overall a more stable, more valuable, better-working industry.

## Discussion and Conclusions

Until the late 1960's, fishery managers and industry members in B.C. were generally ahead of the Atlantic participants. This seemed to stem from the influence of the larger society — more prosperous and more concentrated — and from the nature of salmon themselves. Because the salmon schooled up visibly at focal points, the industry developed more of a critical mass and management got more thorough attention than on the Atlantic.

Officials and industry members managed to keep the resource fairly healthy; indeed, they helped restore major salmon runs. The B.C. industry until recently produced more money per capita than the Atlantic. The Pacific coast always had more articulate and organized fishermen with higher incomes (and generally speaking, easier fishing conditions). Even now, if you ask a B.C. fisherman how he is making out, you are likely to hear that notwithstanding various problems, he's doing all right. Ask an Atlantic fisherman and you are likely to hear complaints.

Is there a more accurate way to gauge the welfare of a fishery? What is the real good? Is it the size of stocks, the amount of fish caught, the number of jobs in the industry, the per capita income, the industry's efficiency purely in terms of dollars spent and earned, or the survival of fishing communities and their quality of life? After several decades of intensive fisheries management, the answers remain elusive.

Although no one has found a magic yardstick to measure the health of a fishery, the 1976 Policy Document defined many desirable goals, and the 1982 Task Force on Atlantic Fisheries put them in a nutshell. The "Kirby Report" said that objective number one for the Atlantic industry should be economic viability; within that, maximum employment; and within that, Canadian operation and control of the industry. Officials sometimes quote these goals, and nobody seems to have disavowed them.

On a day-to-day basis, managers in the late 1980's seemed to be seeking stability as much as anything else, and to cling above all to the key tenet of limited entry, to conserve stocks and protect incomes. Parsons (1993a) has recently questioned the stability objective in light of the boom and bust syndrome evident in many fisheries. Almost all managers seem to fear repeating the pattern in which the first sign of success attracts new entrants and new investment that ultimately becomes harmful. As in the B.C. salmon fishery of the 1880's, the lobster fishery of the 1930's, or the herring and groundfish fisheries of the late 1960's, new entrants can soon fish down the stock and divide the shrinking catch into smaller and smaller shares for more and often more costly boats. Limited entry gives the fishermen a better chance at a decent income and a stable future. The recent fisheries minister Bernard Valcourt put it well: "Do we want to feed two people or starve five?"

Although in the 1970's officials typically spoke of limiting the number of participants, today they stress "over-capacity" as the enemy of fishery success: not just the number of boats, but also their fishing power. The new goal is to match capacity to the available resource. However, this goal gets only limited support from fishermen, who generally support controls on the number of licences, but always seem to want better boats. Atlantic fishermen sometimes argue that a better boat is a safety measure, and that limited entry, quotas, and other regulations are sufficient to protect fish stocks. They occasionally point to the example of British Columbia, where the fleet has such overcapacity that large sectors can fish only one day a week, yet most fishermen seem to do well.

The prosperity of the B.C. salmon, Fundy herring, and southwest Nova Scotia dragger fleets suggest that fishermen can live with a controlled overcapacity, provided the rest of the equation favors viability. However, overcapacity does tend towards overfishing and lost profits. To keep an overpowered fleet under control takes great effort by fishery managers.

On the Atlantic, the double struggle against growth in the number of vessels and their fishing power has helped the fleet survive the ups and downs since the early 1970's. The processing industry has had few restrictions on number or capacity, and this contributed to problems in the 1980's, when the federal government spent millions of dollars on assistance to the larger enterprises. Fishermen survived the downturns with no such large-scale direct assistance, until the recent collapse of the northern cod fishery.

If controlling the number and size of boats can help the general welfare of fishermen, what about the type of gear? For example, would conservation, fish quality, employment, and incomes benefit if the dragger fleet gave way to longlining? In recent decades, the

The fisheries service becomes totally engrossing to many of its officials; probably because the challenges never stop. William Wakeham (*left*) was an early example, leaving aside his medical career as he got caught up in fisheries. He also demonstrated the wide reach of fisheries matters. Wakeham helped establish Canadian sovereignty in the Arctic; and in his time the fisheries protection fleet became the nucleus of the Canadian navy. The fisheries vessel *Vigilant* (*below*) was "the first modern warship to be built in Canada." Wakeham also exemplified the department's persistent concern with social matters, as shown in his account of an early resettlement. On the north shore of the St. Lawrence in 1886, there was distress because of ..."the almost complete failure of the cod fishery between Esquimaux Point and Blanc Sablons. The distress was greatest at and about Natashquan, and it was wisely decided to aid all who were willing to do so, to leave the coast. With this object, your Department sent the Dominion steamer Napoleon to Natashquan, and she carried to Quebec about one hundred and sixty persons; at Quebec these people were taken charge of by the Local Government, and removed to the county of Beauce, where arrangements had been made to settle them. It is to be hoped that they will succeed in making a better living as farmers, than they did as fishermen." (Wakeham photograph courtesy of Charles Friend.)

department has seldom fully argued out such questions. It did restrict the use of factory freezer trawlers. But for the most part, ever since the subdivision of Atlantic quotas began in the 1970's, the department gave allocations to the different gear types based mainly on their historical percentages of the catch. To quote Dr. Arthur May, deeply involved in the 1970's changes: "We froze everything and held on for dear life" (May, 1983).

In evaluating fisheries and their management, one falls back on what seems obvious: the amount of fish, the amount of money per person, and the opinions of observers who visit fishing places and ask people how they are doing. On both counts, B.C. was well ahead of the Atlantic until at least the 1960's. Why did B.C. lead for so long?

The British Columbia managers and industry for many decades had a clearer consciousness of management goals and more willingness to act thoroughly than their Atlantic counterparts. This stemmed from:

— the visibility of the salmon and their struggle for survival, a great educating force;
— the lateness of immigration, after Europeans had learned more about running their own affairs (this contributed for example to the higher prevalence of fishermen's organizations);
— the higher level of education and urbanization among fishermen;
— the easier flow of information in a more concentrated and more organized industry, which insisted on being heard.

As in the political commonwealth of democracy, so in the common property of the fishery: management works best with an educated public. B.C. had more information and organization. On the Atlantic coast over many decades, and even today, there has been untold loss of money, despair, frustration, and heartbreak because of ignorance. Part of it is ignorance or low consciousness of basic principles, such as how overfishing operates — the harder you work, the less you catch. Part of it is ignorance of current department rules, policies, and attitudes. The department has never made a large, concerted, and effective effort at imparting knowledge. This situation causes endless difficulties for all parties; not least for fishery managers, who find themselves giving the same long explanations to one party after another, with little discernible impact.

The concentrated nature of the B.C. fishery fostered organization. On the Atlantic, organizations formed mostly to deal with the government, often in a kind of last-resort reaction against regulations or to lobby for help. In B.C., they mostly started up for price bargaining or collective selling, but also took a strong interest in fishery management.

B.C. managers and industry members generally had more of an attitude that productive change was possible. The Hell's Gate fishway was an example — anyone could visualize the possibilities of restoring lost runs. Similarly, the Salmonid Enhancement Program in its origins inspired everybody.

Since the mid-1970's, however, fisheries management on the Atlantic has been if anything more innovative than on the Pacific, with good results. Both coasts moved forward, but especially the Atlantic. Despite the continuing day-to-day problems that will always afflict fisheries, both statistics and the opinions of observers show major progress throughout Canadian fisheries. There remain areas, especially in Newfoundland and the northern prairies, where geography and demography are so stacked against fishermen that progress is difficult. But by and large, the fishery of the 1990's is more prosperous, stable, and valuable that than of the 1960's, and more responsibly run by a better government-industry partnership.

What explains the recent progress? In management methods, there is little if anything new in principle. Individual transferable quotas may be a partial exception; but the same thinking appeared in Whitcher's words about the leasing system and the Fisheries Act. Although the jury is still out, individual transferable quotas do promise a major change. That change will still depend on the old human factors of knowledge, organization, and attitude.

If there has been no magic new idea in fisheries management, whence the recent successes? What happened was that the department in the late 1960's and the 1970's faced unprecedented challenges from domestic and foreign overfishing and overcapacity, and made an unprecedented effort to meet those challenges. It gathered the courage of its convictions, and made new use of old ideas. And the industry itself, buffeted by strong forces, showed a new willingness to accept changes.

The key factors seem to be these:

— The thinking exemplified by H. Scott Gordon was percolating through the bureaucracy, notably in the person of W. C. MacKenzie. More and more knowledge was accumulating on fish stocks, thus defining the limits to growth.
— Fishermen themselves were beginning to call for limited entry, at least in the form of keeping out "moonlighters," and were more willing to accept controls on effort. The United Fishermen and Allied Workers Union, and lobster fishermen in P.E.I., both made their voices heard.

— The foreign fishing invasion, the beginnings of Canadian overfishing on the Atlantic, and the herring collapse and expanding fleet on the Pacific, all made problems visible and dramatic. The managers now knew more and tried harder. They had to do something. These circumstances fostered actions that MacKenzie and others had long contemplated. Under the cloak of the foreign over-fishing crisis, managers and federal ministers brought in the long-term changes they wanted for Canadians.

— With licences and quotas, the department's sphere of control greatly increased. These new tools could affect not only conservation and sharing, but such things as fish quality and cost-benefit ratios.

— Underneath it all, the Atlantic was becoming more like B.C., with a wider consciousness, more organization, somewhat more of a "critical mass" in the industry, and less dependence on the fishery. The variety of available jobs increased, and so did the availability of unemployment insurance. One might make bold to say that, where the fishery had been the employer of last resort, the unemployment insurance system became the non-employer of last resort. The worst-off people could usually get enough weeks of work somewhere to "get their stamps." Now the fishery could stand apart as a separate system, and this made it easier to change.

— Finally, there were a lot of the right people in the right place at the right time — including Ministers ready to make decisions, and a corps of experienced and capable officials.

The post-1967 changes in management left almost nothing untouched. It was a revolution. When all is said and done, the new management regime has had remarkable success (and has averted what could have been remarkable disasters, given the great improvements in fishing technology). The increases in per capita revenues make that clear. So do comparisons with other countries. Failures since the system came into place, such as unduly heavy fishing pressure or some stocks, seem to have stemmed less from the intentions or methods of management than from insufficient thoroughness in living up to the principles of limited entry, quota management, and the involvement of fishermen in management. But to the degree such new approaches have succeeded, the industry today is better equipped to readjust itself to meet changing circumstances. Most fishermen who remember the small boats and small incomes of the 1940's and 50's, the foreign destruction of fish stocks in the late 1960's, and their old lack of voice in management, will admit that things are better today, notwithstanding the Atlantic groundfish crisis of the early 1990's.


## Acknowledgments

For their help in various ways on this paper, I thank two key figures in recent fisheries history: W. C. MacKenzie and Fernand J. Doucet. I also thank R. W. Dunfield of DFO's Scotia-Fundy Region, author of *The Atlantic Salmon in the History of North America*.


## References


Aiken, D. 1993. Aquaculture in Canada, p. 411–446. *In* L.S. Parsons and W.H. Lear [ed.] Perspectives on Canadian marine fisheries management. Can. Bull. Fish. Aquat. Sci. 226.

Annual Reports, Department of Fisheries, 1930–1968.

Annual Reports, Department of Marine and Fisheries, Fisheries Branch, 1868–1914 and 1921–1930.

Annual Reports, Department of Naval Services, Fisheries Branch, 1914–1920.

Anon. 1916. Commission of Conservation, 1915. Toronto, Ont.

Anthony, V.C. 1990. The New England groundfish fishery after 10 years under the Magnuson Fisheries Conservation and Management Act. N. Am. J. Fish. Manage. 10: 175–184.

Bates, S. 1944. Report on the Canadian Atlantic sea-fishery. Kings Printer, Halifax, N.S. 179 p.

Brebner, J.B. 1945. North Atlantic Triangle. New Haven, CT.

Campbell, B.A. 1973. An historical review of developments in salmon licensing up until 1968. Internal paper, Fisheries and Marine Service, Department of the Environment.

Cockfield Brown & Co. Ltd. 1932. Summary of Report on the Marketing of Canadian Fish and Fish Products to the Hon. E. N. Rhodes, Minister of Fisheries. Ottawa, Ont.

Cowie, J.J. 1910. The non-progression of the Atlantic fisheries of Canada. Annual Report, Department Marine and Fisheries, 1909–10.

Crowley, R.W., B. MacEachern, and R. Jasperse. 1993. A review of federal assistance to the Canadian fishing industry 1945–1990, p. 339–367. *In* L.S. Parsons and W.H. Lear [eds.] Perspectives on Canadian marine fisheries management. Can. Bull. Fish. Aquat. Sci. 226.

Dawson, S.E. 1905. The St. Lawrence Basin. London.

Departmental International Compilation. 1898. Memorandum upon the Questions at issue between Great Britain and the United States in Relation to the North American Fisheries and other information on the Nineteenth Century "Fisheries Question". Department of Fisheries and Oceans Library, Ottawa.

DEPARTMENT OF THE ENVIRONMENT. 1976. Policy for Canada's commercial fisheries. Fisheries and Marine Service, Ottawa, Ont.

DEPARTMENT OF EXTERNAL AFFAIRS. 1977. Canada and the Law of the Sea: Resource Information. May 18, 1977.

DEPARTMENT OF MARINE AND FISHERIES. 1913. Dominion Shell-Fish Fishery Commission. Ottawa.

DOUBLEDAY, W.G. 1993. Reliability of scientific advice on fishery management measures, p. 369–383. *In* L.S. Parsons and W.H. Lear [ed.] Perspectives on Canadian marine fisheries management. Can. Bull. Fish. Aquat. Sci. 226.

DUNFIELD, R.W. 1985. The Atlantic salmon in the history of North America. Can. Spec. Publ. Fish. Aquat. Sci. 80: 181 p.

DUNN, P.D.H. 1944. Fisheries re-organization in Newfoundland. Radio Address 21, January 1944 (transcript). Memorial University of Newfoundland, St. John's, Nfld.

GORDON, H.S. 1951. The trawler question. Dalhousie Rev. 31(2).

GORDON, H.S. 1952. The fishing industry of Prince Edward Island. Report for Department of Fisheri Ottawa, Ont.

GORDON, H.S. 1954. The economic theory of a common-property resource: the fishery. J. Polit. Econ. 62: 124–142.

GOUGH, J. 1971. Frankenstein's Seine. Canadian Fisherman, October and December, 1971.

1988. Fisheries history, p. 781–784. In The Canadian Encyclopedia (2nd Edition), Hurtig Publishers Limited, Edmonton, Alta.

1991. Fisheries management in Canada, 1880–1910. Can. Manuscr. Rep. Fish. Aquat. Sci. 2105: iv + 96 p.

HAIG-BROWN, R. 1974. The salmon. Fisheries and Marine Service. Department of the Environment, Ottawa.

HARDIN, G. 1968. The tragedy of the Commons. Science 162: 1243–1248.

HEALEY, M.C. 1993. The management of Pacific salmon fisheries in British Columbia, p. 243–266. *In* L.S Parsons and W.H. Lear [eds.] Perspectives on Canadian marine fisheries management. Can. Bull. Fish. Aquat. Sci. 226.

HIND, H.Y. 1877. The effect of the fishery clauses of the Treaty of Washington on the fisheries and fishermen of British North America. 2 parts. Study prepared for the Fishery Commission, Halifax.

HUNTSMAN, A.G.H. 1943. Fisheries research in Canada. Science 98(2536): p. 121.

ILES, T.D. 1993. The management of the Canadian Atlantic herring fisheries, p. 123–150. *In* L.S. Parsons and W.H. Lear [ed.] Perspectives on Canadian marine fisheries management. Can. Bull. Fish. Aquat. Sci. 226.

INNIS, H.A. 1940. The cod fisheries: the history of an international economy. New Haven, CT. 520 p.

JENNESS, D. 1960. The Indians of Canada. Fifth edition. National Museum, Ottawa, Ont.

JOHNSTONE, K. 1977. The aquatic explorers: a history of the Fisheries Research Board of Canada. University of Toronto Press, Toronto, Ont. 342 p.

JUDSON, T.A. 1961. The freshwater commercial fishing industry of Western Canada. Ph.D. thesis, University of Toronto, Toronto, Ont.

KIRBY, M.J.L. [Chairman]. 1982. Navigating troubled waters. A new policy for the Atlantic fisheries. Report of the Task Force on Atlantic Fisheries. Supply and Services, Ottawa, 379 p.

LEAR, W.H., AND L.S. PARSONS. 1993. History and management of the northern cod fishery, p. 55–89. *In* L.S. Parsons and W.H. Lear [ed.] Perspectives on Canadian marine fisheries management. Can. Bull. Fish. Aquat. Sci. 226.

LYONS, C. 1969. Salmon — Our heritage. Mitchell Press, Vancouver, B.C.

MANITOBA. 1933. Report of the Commissioners appointed to Investigate the Fishing Industry of Manitoba. Winnipeg, Man.

MAY, A.W. 1983. Notes for an Address by Dr. A.W. May, Deputy Minister of Fisheries and Oceans to the Fisheries Council of Canada, St. John's, Newfoundland. May 2, 1983. SP-HQ-083-006E. 16 p.

MCDONALD, I. 1971. W.F. Coaker and the Fishermen's Protective Union in Newfoundland Politics 1908–1925. Ph.D. thesis, University of London, 1971.

MACDONALD, I. 1980. W.F. Coaker and the Balance of Power Strategy: The Fishermen's Protective Union in Newfoundland Politics. In Hiller and Neary [ed.] Newfoundland in the Nineteenth and Twentieth Centuries: Essays and Interpretation. Toronto, Ont.

NEWFOUNDLAND. 1937. Report of the Commission of Enquiry Investigating the Sea Fisheries of Newfoundland and Labrador Other than the Seal Fishery. St. John's, Nfld.

PARSONS, L.S. 1993a. Management of marine fisheries in Canada. Can. Bull. Fish. Aquat. Sci. 225. 784 p.

PARSONS, L.S. 1993b. Shaping fisheries policy: The Kirby and Pearse Inquiries — Process, Prescription and Impact. p. 385–409. *In* L.S. Parsons and W.H. Lear (eds.) Perspectives on Canadian marine fisheries management. Can. Bull. Fish. Aquat. Sci. 226.

PEARSE, P.H. (Commissioner). 1982. Turning the tide: A new policy for Canada's Pacific fisheries. Final Report. Vancouver, B.C. 292 p.

PERLEY, M.H. 1852. Report of the sea and river fisheries of New Brunswick. J. Simpson, Printer, Fredericton, N.B.

PINHORN, A.T., AND R.G. HALLIDAY. 1990. Canadian versus International Regulation of Northwest Atlantic Fisheries: Management practices, fishery yields, and resource trends, 1960–1986. N. Am. J. Fish. Manage. 10: 154–174.

PRINCE, E.E. 1921. Fifty years of fishery administration in Canada. Trans. Am. Fish. Soc. 50.

PROWSE, D.W. 1895. A History of Newfoundland from the English, Colonial, and Foreign Records. Facsimile edition. Mika Studio, Belleville, Ont. 1972.

SINCLAIR, S. 1960. Licence limitation — British Columbia. A method of economic fisheries management, p. 13–38. In The salmon fishery. Department of Fisheries and Oceans, Ottawa, Ont.

STOCKER, M. 1993. Recent management of the British Columbia herring industry, p. 267–293. *In* L.S. Parsons and W.H. Lear [ed.] Perspectives on Canadian marine fisheries management. Can. Bull. Fish. Aquat. Sci. 226.

SMALLWOOD, J. 1927 Coaker of Newfoundland. London.

UNITED STATES TARIFF COMMISSION. 1944. Treaties affecting the northeastern fisheries. Washington.

WHITCHER, W.F. 1865. Memo to the Commissioner of Crown Lands, the 1865 Act and related materials. Department of Fisheries and Oceans Library, Ottawa, Ont.

## APPENDIX I
## A Statistical Snapshot

Below are two sets of key statistics a half-century apart: 1936, in the depth of the Depression, and 1986, after nearly a decade of the 200-mile limit and the modern management system.

The per capita production figures shown are a rough guide only, obtained by dividing the number of persons into the value of production. The large numbers of part-time and casual workers distort the picture. Although their absolute accuracy is questionable, per capita production figures are a useful yardstick for comparing regions and trends. In this century they show the Great Lakes fishery as a consistently high per capita producer; the Newfoundland and prairie fisheries generally lagging all others; British Columbia generally leading the sea fisheries; and the Maritimes and Quebec fishery catching up with British Columbia in recent years.

### 1936: Maritimes and Quebec

Dollar production per person dropped sharply in the Depression. The mid-1930's statistics also showed a drop in shore employment, reflecting a drop in curing as more fish went into green-salted and fresh product.

| Craft | $ landed (sea) | # fishermen (sea) | $/man (sea) |
|---|---|---|---|
| 29 960 | $9.5 M | 46 978 | $202 |

| # fishermen | # in industry | $ marketed | $/person |
|---|---|---|---|
| 49 437 | 58 079 | $16.4 M | $282 |

Note the two sets of per capita figures for production value: one gained by dividing the number of fishermen into the landed value, and one gained by dividing the total number in the industry into the marketed value.

### 1935: Newfoundland

It is convenient to use 1935 statistics from the 1937 Commission of Enquiry. No landed value per se is given. A comparison of export value per person with Maritime production per person shows Newfoundland well behind.

| # vessels | # craft o/a | # men fishing | $ export | $ /fisherman |
|---|---|---|---|---|
| 1 317 | 30 295 | 34 458 | $4.0 M | $116 |

*The bank or deep sea fishery:*

| # vessels | # men | Quintals | Av. vessel | Av. man |
|---|---|---|---|---|
| 74 | 1 305 | 115 965 | 1 567 | 88 |

### 1936: Ontario

| # craft | # fishermen | $ marketed | $marketed/fisherman |
|---|---|---|---|
| 2 398 | 4 280 | $3.2 M | $748 |

### 1936: Manitoba, Saskatchewan, Alberta, Yukon

By now the Prairie lag behind the Great Lakes was clear. In the Depression, per capita production on the Prairies dropped much more than in Ontario or the Maritimes and Quebec. But it still came out to a higher figure than in the Maritimes and Quebec.

| # craft | # fishermen | $ marketed | $marketed/fisherman |
|---|---|---|---|
| 1 243 | 6 625 | $2.4 | $360 |

**1936: Pacific**

B.C. suffered with the Depression, but there was still a lot more money circulating per person than on the Atlantic.

| # craft | $ landed (sea) | # fishermen (sea) | $/man (sea) |
|---|---|---|---|
| 8 716 | $7.5 M | 11 393 | $658 |

| # in industry | $ marketed | $/person |
|---|---|---|
| 17 989 | 17.2 M | $956 |

**1986: Maritimes and Quebec**

These figures are from the *Annual Statistical Review* of the Department of Fisheries and Oceans.

By the rough yardstick of average landed value per fisherman, the Maritimes and Quebec in 1986 had caught up with British Columbia.

| # craft | lb. landed | $ landed | # fishermen |
|---|---|---|---|
| 13 089 | 1 608 M | $650.5 M | 33 338 |

| $/fisherman | # in industry | $ marketed | $/person |
|---|---|---|---|
| $20 112 | 47 391 | $1 534.3 M | $32 375 |

**1986: Newfoundland**

| # craft | lb. landed | $ landed | # fishermen |
|---|---|---|---|
| 15 792 | 1 136 M | $209.6 M | 27 075 |

| $/fisherman | # in industry | $ marketed | $/person |
|---|---|---|---|
| $7 741 | 37 360 | $732.0 M | $19 593 |

**1986: Ontario**

| # craft | lb. landed | $ landed | # fishermen |
|---|---|---|---|
| 1 023 (1985) | 53.7 M | $46.3 M | 1 837 |

(Note: in the line above, the number of fishing craft in Ontario comes from the 1985 *Annual Statistical Review* of the Department of Fisheries and Oceans; the other figures are for 1986.)

| $/fisherman | # in industry | $ marketed | $/person |
|---|---|---|---|
| $25 204 | n/a | $92.7 M | n/a |

**1986: Manitoba, Saskatchewan, N.W.T., Yukon**

| # craft | lb. landed | $ landed | # fishermen |
|---|---|---|---|
| 2 212 | 42.0 M | $27.8 M | n/a |

| $/fisherman | # in industry | $ marketed | $/person |
|---|---|---|---|
| $4 752 | n/a | $45.6 M | n/a |

Note: although there is no breakout, the *Annual Statistical Review, 1955–1976, vol. 9* published in Ottawa, 1977 gives processing employment for 1974 in Ontario, the Prairies, and the Territories as 1 125.

**1986: British Columbia**

| # craft | lb. landed | $ landed | # fishermen |
|---|---|---|---|
| 6 135 | 497.8 M | $402.0 M | 20 033 |

| $/fisherman | # in industry | $ marketed | $/person |
|---|---|---|---|
| $20 066 | 23 821 | $766.6 M | $32 182 |

# History and Management of the Fishery for Northern Cod in NAFO Divisions 2J, 3K and 3L

W. H. Lear and L. S. Parsons

*Department of Fisheries and Oceans, Biological Sciences Directorate, 200 Kent Street, Ottawa, Ontario, Canada K1A 0E6*

Lear, W. H., and L. S. Parsons. 1993. History and management of the fishery for northern cod in NAFO Divisions 2J, 3K and 3L, p. 55–89. *In* L. S. Parsons and W. H. Lear [eds.] Perspectives on Canadian marine fisheries management. Can. Bull. Fish. Aquat. Sci. 226.

The northern cod stock off Labrador and eastern Newfoundland has provided the basis for the economic life of much of the province for several centuries. Catches from this stock have fluctuated dramatically over several centuries. Offshore fishing beginning in the late 1950's and continuing into the 1970's depleted the enormous resource of older, larger cod in the late 1960's and early 1970's. TACs introduced by ICNAF in the early 1970's failed to arrest the decline. Following its extension of fisheries jurisdiction to 200 miles in 1977, Canada managed the northern cod stock on a more conservative basis, with the objective of rebuilding the stock rapidly. The biomass of cod aged 3 years and older (age 3+) increased from 450 000 tons in 1976 to 1.2 million tons in 1984 but subsequently declined to about 800 000 tons by 1990. There was a sudden, unexpected decline in the spawning stock during 1991 to a level at, or near, the lowest observed. The decline of the stock in the early 1990's precipitated a reduction in TAC from 266 000 tons in 1988 to an initial conservation cap of 120 000 tons in 1992 with concurrent economic and social hardships through plant closings and layoffs. The abrupt decline in 1991 led to a 2-year moratorium beginning in July 1992. The rebuilding of this stock is imperative for the future well-being of communities that depend upon this resource for their livelihood and their continued existence. This chapter describes the history and management of this stock from colonial times to the present.

---

Pendant plusieurs siècles, la vie économique d'une grande partie de Terre-Neuve a reposé sur l'exploitation du stock de morue du nord occupant les eaux du large du Labrador et de la côte est de cette province. Les prises faites à partir de ce stock ont fluctué de façon extrêmement marquée au cours des siècles. La pêche hauturière qui a débuté à la fin des années 1950 et qui s'est poursuivie jusqu'aux années 1970 avait épuisé les ressources énormes en morues plus âgées et plus grosses à la fin des années 1960 et au début des années 1970. Les TPA imposés par la CIPANO au début des années 1970 n'ont pas permis de stopper le déclin. Suite au prolongement à 200 milles de sa zone de compétence en matière de pêche, réalisé en 1977, le Canada a été en mesure de gérer le stock de morue du nord de façon plus prudente et dans l'optique d'un rétablissant rapide du stock. La biomasse des morues âgées de trois ans et plus (âge 3+) a augmenté pour passer de 450 000 tonnes en 1976 à 1,2 million de tonnes en 1984, mais elle a ensuite diminué pour atteindre 800 000 tonnes environs en 1990. En 1991, le stock de géniteurs a chuté, de façon soudaine et imprévue, au niveau le plus faible pratiquement jamais enregistré. Le déclin du stock du début des années 1990 a donné lieu à une baisse du TPA qui, de 266 000 tonnes en 1988, a été fixé à 120 000 tonnes en 1992. Il s'agissait d'un maximum visant à assurer la conservation de la ressource qui a été la cause de difficultés économiques prenant la forme de fermetures d'usines et de mises à pied. La chute soudaine de 1991 a donné lieu en juillet 1992 à l'imposition d'un moratoire de deux ans. La rétablissement de ce stock constitue un impératif pour des collectivitiés dont la vie économique et l'existence même dépendent de son exploitation. On trouve dans ce chapitre l'historique et une description de la gestion de ce stock de la période coloniale à nos jours.

## Introduction

The cod stock off Labrador and eastern Newfoundland (northern cod stock) provided the economic basis for the settlement of eastern Newfoundland and Labrador (NAFO Divisions 2J, 3K and 3L (Fig. 1). The early English fishery was a migratory one based mainly in southeastern Newfoundland while the French fished mainly along the northeast coast. The landings by Newfoundland-based fishermen exhibited wide annual fluctuations (Fig. 2).

The modern decline in the inshore fishery from 1960 to 1974 was caused by overexploitation of the offshore components of the stock which was first brought under catch quota control in 1973 in an attempt to stop their decline. With the introduction of the Canadian 200-mile fisheries limit in 1977, there was optimism that the northern cod stock would rebuild. The first quotas under international management were set too high but were eventually set at a more conservative level ($F_{0.1}$). It was agreed that the inshore sector would be given priority in allocation during the rebuilding phase. During the 1980's there was fierce competition among the Atlantic provinces and among fleet sectors over access to northern cod as the offshore catches increased and the inshore catches declined. In 1989 Canadian scientists produced a dramatic reassessment of stock status and projected that the total allowable catch (TAC) corresponding to $F_{0.1}$ would be 125 000 tons, down from a previous recommended level of 266 000 tons. An independent inquiry confirmed the CAFSAC advice and recommended lowering the TACs. Subsequent advice from CAFSAC on biomass levels led to the setting of an interim ceiling of 120 000 tons early in 1992 and a 2-year moratorium beginning in July 1992.

The purpose of this chapter is to trace the development of the northern cod fishery and describe the scientific basis and the evolution of the management regime for northern cod from the rapidly expanding fishery in the 1960's to the overexploitation and near collapse of the stocks in the early 1970's. The stock was partially rebuilt after 1977 to a peak in the mid 1980's when high exploitation rates again led to a reduction in the stock biomass. During 1991, there was a sudden, drastic and unexpected decline in the spawning stock to a very low level, probably due to ecological factors. This led to a 2-year moratorium on fishing for northern cod, commencing in July 1992.

## The Northern Cod Stock Complex

### Stock Boundaries

The Labrador-East Newfoundland stock of cod extends from Ungava Bay and northern Labrador (NAFO Divisions 2G and 2H) to the northern Grand Bank and the Avalon Peninsula (Templeman 1981). This, the largest cod stock in the northwest Atlantic, extends over a distance of 900 miles from north to south (Fig. 1).

Although it has not been possible thus far to subdivide this large stock complex into smaller units, two separate portions have been identified separately for management purposes:

1. Northern Labrador Portion (NAFO Divisions 2G and 2H)
2. Southern Labrador-East Newfoundland Portion (NAFO Divisions 2J, 3K and 3L). This is the portion commonly referred to as the "northern cod stock".

The cod of northern Labrador (NAFO Divisions 2G and 2H) were considered separately for management purposes from the rest of the Labrador-East Newfoundland stock complex because the effect of the fishery on this part of the complex had been more severe than in the south (Pinhorn 1976). The abundance of cod in this area was severely reduced during 1965–69 when the average annual catch was 68 000 tons. There was lower than average recruitment of cod in this area during 1967–69.

The evidence for separation of the Labrador-East Newfoundland Stock complex from the neighbouring

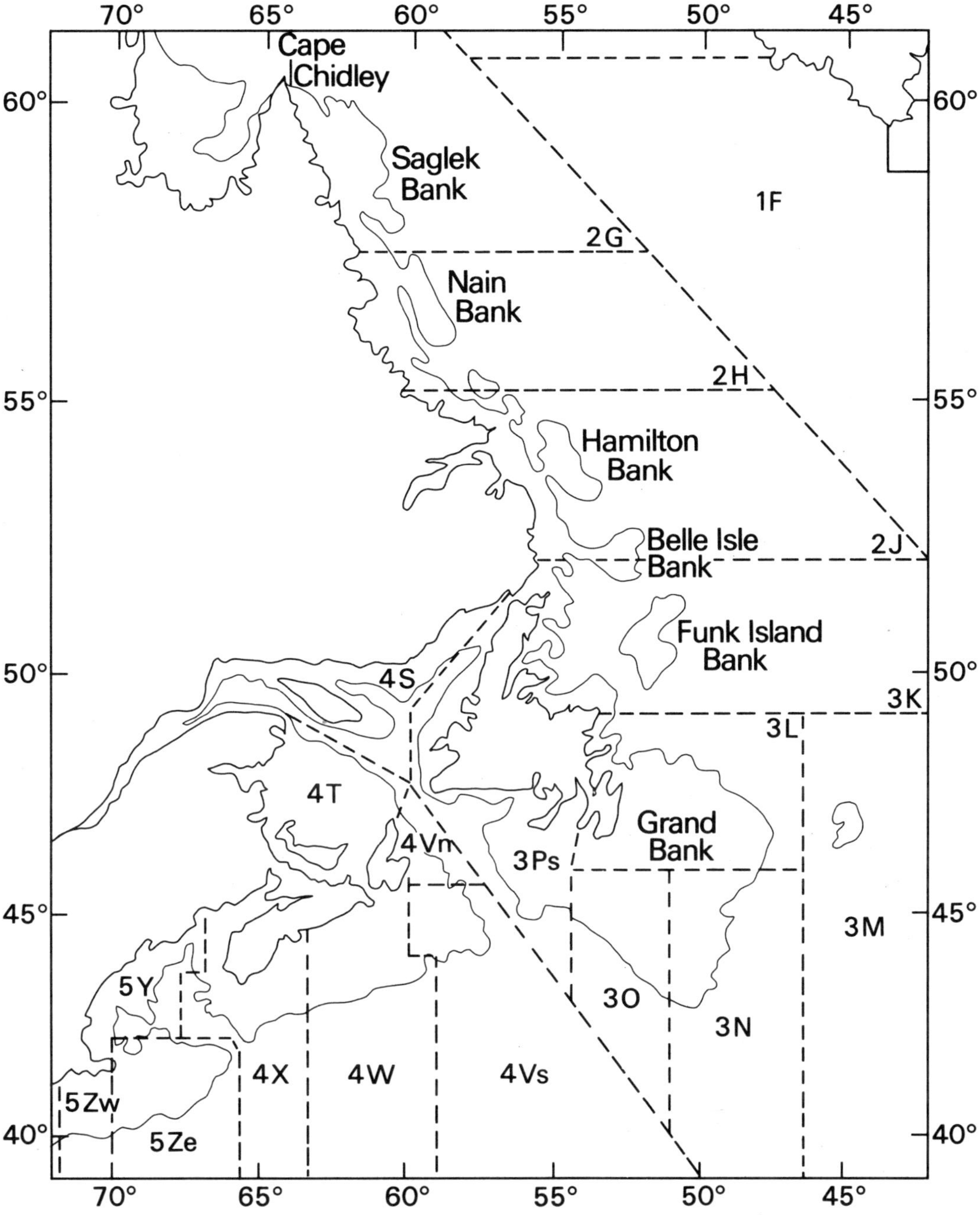


FIG. 1. The northwest Atlantic showing the location of the northern cod stock in NAFO Divisions 2J, 3K, 3L and its northern component in Divisions 2G and 2H.

stocks on the Grand Bank and in the northern Gulf of St. Lawrence included vertebral averages, growth rates, migration studies based upon tagging experiments, spawning times and the areas occupied by the cod during their spawning and feeding concentrations and their ensuing migrations (Lear 1986). Inferences were also drawn from the distribution of cod containing the nematode *Terranova (Porrocaecum) decipiens* and *Anisakis* sp. (Templeman et al. 1957) and the copepod *Lernaeocera branchialis* (Templeman and Fleming 1963; Templeman et al. 1976) and from biochemical methods (Jamieson 1975; Cross and Payne 1978).

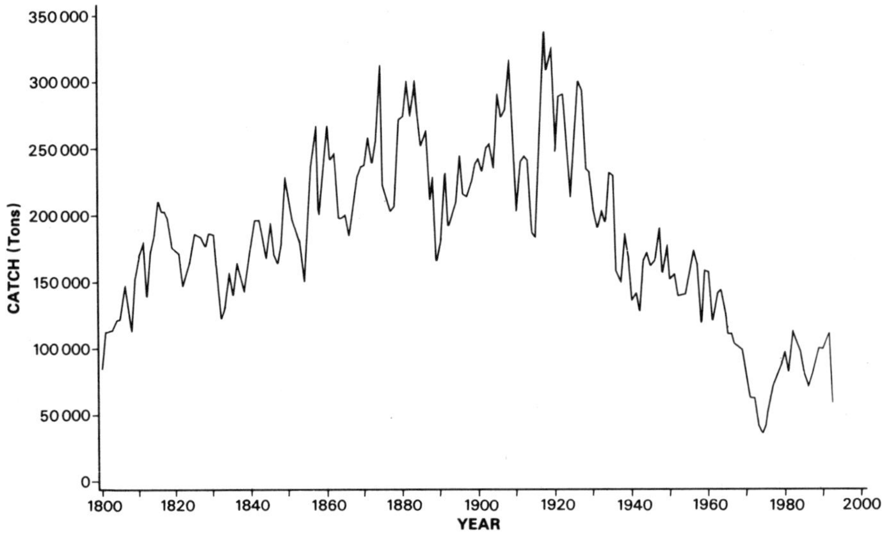


FIG. 2. Inshore catches of northern cod in Newfoundland and Labrador during 1800–1991.

Within this stock complex there are separate populations associated with offshore banks and with coastal shelves. The cod from these populations intermingle especially during the summer feeding period. There are also marked differences in characteristics of the cod from north to south, most notably growth rate, condition factors, spawning times, and size and age at sexual maturity.

Differences in growth rate exist throughout the area with the growth rate increasing from north to south (May 1961; Fleming 1960; May et al. 1965).

There are considerable differences in the condition of the cod when they arrive in the inshore waters. The cod of the Labrador-East Newfoundland stock at the southern end of the range are in excellent condition with fat creamy-white livers when they arrive in June and this condition declines northward until off Labrador the cod arrive on the coast as thin fish with thin brown livers, depleted of fat reserves (Templeman 1962).

Templeman (1962) postulated that each shelf region, e.g. the Bonavista Shelf, Fogo Shelf and the St. Anthony Shelf, projecting seaward with deep water on each side, had a basic stock of its own and that overfishing on these local stocks could cause local depletion (Templeman 1959; Fleming 1959). Tagging studies during 1954–55 and 1962–66 (Templeman 1974, 1979) produced continued evidence of homing of local populations of cod to original inshore tagging locations along the coast from the Avalon Peninsula to northern Labrador.

The size at first sexual maturity in the cod of the Labrador area appears to be small, that on the east coat of Newfoundland and on the northern and western edges of the Grand Bank is intermediate, and that on the southern part of the Grand Bank is the largest of all (Templeman 1953). Age of cod at first sexual maturity does not show such great differences as does size. Fleming (1960) found that more cod were sexually mature at the smaller sizes in Labrador than in any other region.

Spawning occurs chiefly from April to early June throughout most of the area occupied by the Labrador-Newfoundland stock, at least from southern Labrador and Hamilton Bank southward (Templeman 1962).

Templeman (1979, 1981) reviewed earlier studies on spawning times and locations. The spawning times of cod from northern Labrador (2G and 2H) were inferred from the assumed drift of eggs and larvae found off northern and southern Labrador (Serebryakov 1967). Off Labrador (Subarea 2) cod were presumed to spawn mainly in March–April in deep water on the continental slopes, with most of the remainder spawning in May. There is a small amount of spawning, as late as June, in the colder water channels and slopes closer to the coast.

In Division 3K off northeast Newfoundland, at least as far south as Funk Island Bank and vicinity, cod spawning begins in March but occurs mainly in April to early May with some spawning continuing to June.

In Division 3L at the northern slope of the Grand Bank and in the Avalon Channel and on the northwestern slope of the Grand Bank, cod spawn mainly in April–June. Most of the spawning is usually over by June. In the northern part of 3L, however, the mature cod typically lie shallower in winter than in Subarea 2 and Division 3K and spawning may be delayed in years when unfavourably low temperatures extend deeper than usual.

Based on tagging results, Lear (1984) produced evidence of a consistent annual pattern of migration to inshore waters during summer and to offshore areas during winter for each group of cod tagged along the outer continental shelf. The Hamilton Bank component contributes to the southern Labrador and northeast Newfoundland coastal fisheries mainly from Notre Dame Bay northward. The Belle Isle Bank component migrates during summer mainly to southern Labrador, Strait of Belle Isle entrance and northeastern Newfoundland as far south as Notre Dame Bay. The pattern of movement is similar to that of the Hamilton Bank component except that a greater proportion migrate to the Strait of Belle Isle. Cod on the northern and northeastern slopes of Funk Island Bank migrate during summer to eastern and southeastern Newfoundland, with smaller proportions going to southern Labrador and the Strait of Belle Isle than from the taggings on Belle Isle Bank. Cod from the southwestern slope of Funk Island Bank contribute mainly to the summer inshore fishery of Notre Dame Bay and Bonavista Bay and, to a lesser extent, to the fishery in Trinity Bay, Conception Bay and the eastern part of the Avalon Peninsula. Cod which overwinter on the northern Grand Bank migrate southwards across the bank to the Virgin Rocks and the Eastern slope of the bank. This component contributes mainly to the inshore fishery from Trinity Bay southward to St. Mary's Bay, with little effect on the fishery north of Cape Bonavista.

Each component appears to contribute to the inshore fishery in specific, although wide, overlapping areas. Excessive exploitation on any one component of the stock complex could have adverse effects on the inshore fishery in the areas frequented by the summer migrants.

Lear (1984) suggested that the offshore effort should be spread evenly over the three divisions so that the inshore fishery in summer and autumn would not be affected detrimentally by excessive exploitation of a particular component of the stock complex.

The cod of NAFO Divisions 2J and 3K also enter the Strait of Belle Isle and intermingle with cod of the northern Gulf (Lear 1986). The highest contribution to the Strait of Belle Isle is from Belle Isle Bank, around 8–9% on average annually. Cod tagged in 2J and 3K have been recaptured as far south as Port-au-Choix on the Newfoundland side and as far south as Blanc Sablon on the Labrador side of the Strait of Belle Isle.

### Inshore Migration

The inshore migration of northern cod, as reflected in the inshore cod fishery of Labrador and eastern Newfoundland, has a long history of variability both in terms of annual landings and in local abundance. Lear *et al* (1986) summarized several hypotheses that had been proposed to explain variability in the inshore cod fishery. These were as follows:

a) The inshore migration of cod in late June and July is a feeding migration in which cod actively pursue mature migrating capelin towards the coast. Inshore catch would be positively correlated with mature capelin biomass.
b) The inshore migration of cod is negatively correlated with the biomass of immature capelin which are mainly offshore and which would provide an adequate food supply and reduce the necessity to migrate inshore.
c) The proportions of cod moving across the thermal barrier in the cold intermediate layer (CIL) of the Labrador Current decrease as the thickness of the core increases or the temperature decreases.
d) The inshore cod catch (especially in cod traps and other shallow-water gear) is influenced by the surface temperatures near the coast in the vicinity of cod traps and other fixed shallow-water gear.

Thus, even when there is an increasing offshore biomass of cod, there may not be an increase in inshore catches in a particular year but a generalized increase with wide fluctuations in catches in the inshore fishery.

### The Cod-Capelin Interaction

The inshore, summer-feeding migration of cod to the coastal region is linked to the inshore migration of mature capelin which spawn on the beaches and in the shallow water of coastal Labrador and eastern Newfoundland. From mid-June to early August cod feed almost exclusively on capelin in the inshore area, whereas from mid-August to late November they feed mainly on bottom invertebrates, especially crustacea (Templeman 1965). Lilly and Fleming (1981) reported that capelin represented 99% by weight of thc stomach contents of cod from traps at St. John's and Bonavista in July 1968–69. Offshore, cod in 3L feed on both large and juvenile capelin, especially during spring (Popova 1962; Templeman 1965 and Lilly 1987).

Generally, cod must be about 35 cm in length to feed on adult capelin, but cod as small as 20 cm can feed on juvenile capelin. Cod 40–60 cm in length tend to feed most heavily on capelin but capelin are also found in the stomachs of even the largest cod (Lilly 1987).

Lilly (1987) reviewed studies of the interactions between cod and capelin in waters off Labrador and eastern Newfoundland. Although the broad patterns of the feeding of cod on capelin have been described, he concluded that little is still known about the dynamics of the interaction. Several tentative conclusions were drawn from a short series of cod stomach examinations in Division 2J, 3K in autumn (Lilly 1987). The rate of predation by cod on capelin varied with capelin abundance. At times of low capelin abundance, cod did not compensate by preying more intensively on other prey. This led to the hypothesis that the interannual variability in capelin abundance would be reflected in changes in cod vital rates, such as growth rate, fecundity, age (or size) at maturity and mortality. These questions are being addressed by current investigations into cod-capelin interactions.

## Historical Development of the Northern Cod Fishery

### The Fishery: 1500–1954

The discovery of the rich codfish resource off the coast of Newfoundland precipitated a rapid expansion of the fishing industry at the beginning of the sixteenth century. The Portuguese began fishing in Newfoundland in 1501 and the French by 1504. The French and Spanish Basques were also fishing for cod in Newfoundland in the early 1500's (NORDCO 1981). Not until the second half of the sixteenth century did the English cod fishery at Newfoundland begin to expand significantly. This occurred especially in the 1580's after the destruction of the Spanish and Portuguese fisheries at Newfoundland, which provided England with golden opportunities to supply the southern European markets (Rowe 1980).

Also about this time the population of England started to increase more rapidly than it had previously and the growth of the English Navy and England's military expeditions in Ireland and elsewhere created a large and continuous demand for food in the form of fish (Rowe 1980). The lightly salted, hard, sun dried product which the British favoured, required sun and wind and also demanded a land-based operation. This led to settlements devoted to the curing of cod destined for the English trade and was mainly responsible for the pattern of settlements in southeastern Newfoundland from Cape Freels in the north to Cape Pine in the south (Matthews 1973). The English settled in this area because of the proximity to the cod resource and also because they were generally precluded from the area north of Bonavista Bay by the Spanish and from Cape St. John to Cape Bauld by the French. Thus the availability of the cod resource and the constraints of the curing method predicated the pattern of settlement.

Fishing at first was carried out by migratory fishermen from Spain, Portugal, France and England, who sailed to Newfoundland in the spring and returned in the fall with cargoes of salt fish. The English fishermen built up their stages, flakes, warehouses and living quarters and returned to them annually (Ryan 1986). As these holdings increased in size and became valuable, especially from the vantage point of location, caretakers were left behind over the winter to build new premises, rebuild old ones and to have everything ready for an early start when the fishing ships, arrived from England. These caretakers, plus deserters from the fishing ships, became the first European residents (Ryan 1986). This migratory fishery reached its peak in 1788 and declined during the remainder of the century. By the end of the Napoleonic War it had been fully eliminated and replaced by a resident fishery. During the eighteenth century there was a northward expansion of the English fishery along the northeast Newfoundland coast.

The settling of Fogo and Twillingate appears to have taken place between 1728 and 1732 (Head 1976). This northward expansion was in part necessitated by a collapse in yield in southeastern Newfoundland that occurred in the mid eighteenth century. The catch rates declined from about 400 quintals or more per boat in every major region to 150 quintals along the southern shore in 1754, as little as 50 quintals in St. John's in 1755 and 150 quintals in Conception Bay in 1753 (Fig. 3) (Head 1976) (A quintal was 112 pounds of split, salted, dry cod). It is presumed that extended heavy exploitation of the inshore cod may have contributed to falling catches since the stocks did not appear to recover until the fishing pressures were reduced. The fishery on the offshore banks offered some relief from the low inshore yields. The northern cod fishery was supplemented by other natural resources. Atlantic salmon, seals and, to some extent furs added a good deal to the income from cod especially during the great decline in inshore cod yield after the 1740's (Head 1976).

During the Napoleonic War, many fishermen from Newfoundland's southeast coast, particularly Conception Bay, became engaged in a migratory cod fishery on the North Shore in the area between Quirpon and Cape St. John, a part of what was the French Shore. The Newfoundland fishermen occupied the fishing areas vacated by the French at the outbreak of the war. The men and ships participating in this fishery were mainly the same ones engaged in the spring seal fishery (Ryan 1986). Out of this North Shore fishery there grew the

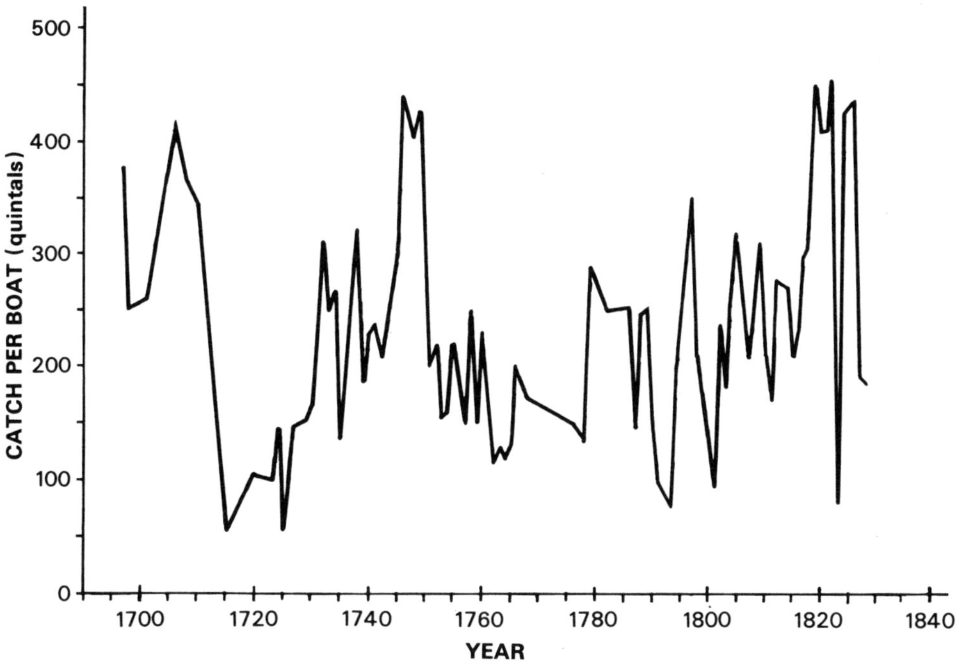


FIG. 3. Catch Rates (catch/boat) by inhabitants boats in Conception Bay during 1697–1828.

Labrador fishery as vessels would occasionally continue on to the Labrador coast to secure a catch if the fishery along the North Shore was poor.

By the late 1820's the Labrador fishery was firmly established. It consisted of two types of fisheries in addition to the small resident fishery. Migratory fishermen from the island of Newfoundland were known as:

1. *Stationers* who established themselves on shore at a "room" and who caught and cured their fish in one place;
2. *Floaters* who lived on board their schooners and moved around to the various fishing grounds and generally salted their catch on board and brought it back to the island to be dried.

At its peak the stationer fishery generally operated along the southern coast of Labrador and did not extend north of Cape Harrison. The floaters extended their fishery farther north than the stationers until they were fishing as far north as Cape Chidley. The Labrador fishery, during its peak and in conjunction with the seal fishery, provided the Conception Bay-St. John's area and to a lesser extent, Trinity Bay and points north with a basis for greater population concentration than could be expected from the shore fishery alone (Ryan 1986).

The traditional methods of the inshore fishery that prevailed in the eighteenth century generally continued on into the nineteenth century. An exception was the introduction of the cod trap, invented in 1866 by Captain W.H. Whiteley in the Strait of Belle Isle. The use of the cod trap had become widespread by the last decade of the nineteenth century because of its applicability to the shore fishery and to the Labrador fishery and may have partly accounted for the increased production of cod during the late nineteenth and early twentieth century (Ryan 1986). The number of traps increased from 2 588 in 1891 to 4 182 in 1901 (Innis 1940). It is interesting to note that this was approximately the number of traps in use in 1990 (about 4 000). Also during this time the use of long lines from which hundreds of baited hooks were suspended (commonly called bultows) began to be adopted mainly by the banks fishery. Their use was prohibited on many inshore fishing grounds (Anon. 1943).

Also during the 1860's steam power was introduced into the seal fishery and as a result into the supplementary Labrador fishery. This did not change the catching efficiency of the inshore fishery but it enabled larger cargoes of fish to be carried to market than ever before and caused market gluts (Ryan 1986). At the beginning of the twentieth century, important changes began to take place in the Atlantic fisheries although these changes were slower to take hold in

Newfoundland. The gasoline engine came into use in the small-boat inshore fishery in the first decade of the century (Gordon et al. 1956).

The use of draggers or otter-trawlers was introduced in Atlantic coast waters in 1908. The first trawler began commercial operations in Newfoundland in 1935. This technological innovation met powerful opposition from inshore line-fishermen who saw it as a threat to their livelihood. The dragger was a highly efficient instrument of production. It was also expensive and outside the financial means of independent fishermen who perceived that it would undermine their sales outlets. As a result, dragging operations were severely restricted for many years. The intent was to retain the largest possible labour force in the fishing industry. It was thought that the introduction of highly efficient fishing units would reduce employment. The ultimate result of the dragger restriction was to retard the growth of productivity and contribute to low individual incomes in the fishery as a result of the large number of people employed.

During the early years of the twentieth century the development of transportation and of refrigerated fish handling facilities, as well as the growth of urban centres in Canada, had begun to create an outlet for fresh and frozen fish products. Establishment of a refrigerated railway freight service for fish products from both coasts to points in Central Canada was supported by a federal government subsidy from 1909 to 1919. Diversion to fresh and frozen products especially affected the output of salted fish. During World War I, because of the drop in European supplies, there was increased production of salted dried codfish to fill the demand in Europe. After the war the Europeans restored and expanded their production and improved their cure. This resulted in a shrinking of the European market for Newfoundland salt codfish and Newfoundland turned to the West Indies, which before this had been served by Nova Scotian exporters (Gordon et al. 1956). The ensuing struggle beginning in the mid 1920's resulted in the forcing down of the price of fish throughout the whole Atlantic coast region. Between 1926 and 1939 it fell on the average more than 50%. As a result the income of fishermen in Newfoundland was reduced to an extremely low level. It had a devastating effect on the island's economy and was one of the reasons for the collapse of Responsible Government in 1933.

During the 1920's the fresh (chilled and frozen) fish trade expanded. This was made possible by two innovations both originating in New England:

1. Filleting at source of production.
2. Quick-freezing.

Filleting at source reduced the transportation costs per unit of edible product and enabled utilization of waste material for by-products. It also resulted in a revolution in packaging and handling of fish products, making possible brand differentiation and the creation of other display values.

Quick-freezing permitted greater standardization of quality and added greatly to the carrying properties of perishable foodstuffs through time and space. It made it possible to avoid or at least minimize fluctuations in supply on the one hand and to extend markets geographically on the other (Gordon et al. 1956).

Newfoundland lagged far behind in these developments in the exploitation of the northern cod fishery. It was not until World War II that the demand for fishery products for both food and industrial purposes was vastly strengthened. The salted cod trade of the Atlantic coast was revitalized and in Newfoundland a filleting and freezing industry was established. After World War II marketing of the Newfoundland production of salted codfish was undertaken by the International Emergency Food Council (IEFC). Inspection of the fish to assure quality, size and grade was compulsory and was carried out by the Newfoundland Fisheries Board and by the Newfoundland Associated Fish Exporters Limited (NAFEL). Price levels were fixed for each grade of fish. The realized values were four times that of the depression years of 1930–33. The IEFC was discontinued in 1947 and the dollar shortage removed the financial support that had bolstered the industry in the immediate post-war period. This resulted in great instability in the industry. As a result of Confederation with Canada in 1949, Newfoundland became part of the high-priced dollar area. Its fishing industry demanded higher prices for a low-grade product, thus contributing to the decline of the floater fishery for cod on the Labrador coast (Black 1960).

The adverse marketing conditions in 1950 and the carry-over of salt codfish from the previous year into 1950 as a result of international restrictions on the free flow of salt fish led to a fall in prices for the Labrador grade product. Even though annual subsidies were established, it was too late to avert a disastrous fishing season. The net result was that the shoreman's margin of profit was almost eliminated. There was virtually no incentive for the shoreman to take part in the fishery. The result was a sharp reduction in the number of floaters operating on the Labrador coast to 29 vessels in 1951, to 25 vessels in 1952 and to 5 vessels in 1954. Black (1960) observed: "The cumulative effect of unfavourable marketing conditions upon an industry that was submarginal and uneconomic in its operations resulted in the cessation of the floater fishery in 1954, thereby bringing to an end an industry that had been in operation for almost two centuries".

**Fluctuations in Inshore Cod Catches**

The inshore cod fishery of Labrador and eastern Newfoundland has a long history of variability in terms of annual catches and local abundance. Catch rates in the inshore fishery at Newfoundland have shown very wide fluctuations over the years as demonstrated in the catch per boat by residents of Conception Bay during 1697 to 1828 (Fig. 3). Inshore catches of northern cod during the period 1802 to 1991 fluctuated greatly (Fig. 2). From 1800 to about 1880 there was a general increase in catches from about 100 000 tons in the early 1800's to about 200 000–300 000 tons in the 1870's. Catches during 1880–1920 fluctuated but were generally in the range of 200 000–300 000 tons. Beginning in 1920, the inshore catches of northern cod declined precipitously from more than 300 000 tons to 35 000 tons in 1974. From 1974 onward inshore catches increased to about 113 000 tons in 1982 but declined again during the mid-1980's to about 70 000 tons–80 000 tons. Catches during 1988–90 were about 100 000 tons. The catch declined again to 60 000 tons in 1991.

Ryan (1986) lists two reasons why the Newfoundland cod fishery failed to expand to any worthwhile extent, namely: periodic catch failures and inferior curing, both of which hindered the production and sale of fish. The failure of the cod to appear in the coastal area and to be available to the traditional gear types translated into lost market opportunities and losses to fishermen and processors. The production of inferior quality fish led to lower prices, which in turn contributed to a lack of motivation to improve the cure. Thus the fishing industry found itself in a situation where market prices were spiralling downwards. Everybody was aware during the century 1850–1950 of the necessity for upgrading the quality and for some level of standardization of the cull of fish. However, no attempt was made to standardize the cull during this period although some exporters tried to improve their marketing practices (Ryan 1986).

**Build-up of the Foreign Fishery**

As previously mentioned, in addition to the fishery by Newfoundlanders, the northern cod stock had been fished by France for centuries especially along that portion of the French North Shore from Cape St. John to Cape Bauld in NAFO Division 3K. Portugal had fished in Newfoundland for centuries although predominantly on the Grand Bank in NAFO Division 3L. Spain had participated in the fishery for cod north of Bonavista Bay during the sixteenth century but their fishery had diminished considerably by the first decade of the seventeenth century. Spain began otter trawling in 1927 and by 1953 around 23 Spanish otter trawlers were exploiting Newfoundland cod stocks.

The fishery by France, Portugal and Spain on northern cod off Labrador (NAFO Subarea 2) is a modern phenomenon. In 1952–53 Portugal, Spain and France began catching significant quantities and several other countries small quantities of cod in Subarea 2 (Templeman 1966).

During the 1960's other European otter trawlers carried out very intensive fisheries on the pre-spawning, spawning and post-spawning concentrations of cod on the Southern Labrador Shelf. This fishery was mainly on the eastern and southeastern slopes of Hamilton Bank. This new offshore fishery affected the inshore fisheries of Labrador and northern Newfoundland by reducing both quantities and sizes of cod (Templeman 1966). Catch per unit of effort increased in the offshore fishery but declined markedly in the Labrador inshore fishery, apparently as a result of the increased offshore catches during 1959–66 (May 1967).

In NAFO Division 3K off eastern Newfoundland total catches increased from about 79 000 tons during 1955–58 to about 114 000 tons during 1961–64, mainly as a result of otter trawling in deep water during summer. The inshore catch declined correspondingly and the inshore proportion of the total declined from 73% in 1955–58 to 36% in 1961–64.

In 3L there was an increase in total annual catches from 153 000 tons during 1955–58 to 170 000 tons during 1961–64 while the inshore proportion of the catch declined from 53 to 43% (Templeman 1966).

The catch from the northern cod stock complex peaked in 1968 at 810 000 tons. The main countries involved in the cod fishery during 1968 in NAFO Divisions 2J, 3K and 3L were as follows:

| **Country** | **% of Total Catch (810 000 tons)** |
|---|---|
| Portugal | 19.5 |
| Spain | 16.7 |
| Canada | 15.7 |
| USSR | 15.0 |
| France | 10.4 |
| Poland | 8.9 |
| West Germany | 6.1 |
| United Kingdom | 4.2 |
| Other countries | 3.5 |
| **Total** | 100.0 |

The catch dropped slightly to 753 000 tons in 1969 and then declined precipitously to 214 000 tons by 1976. This was less than the 1976 Total Allowable Catch (TAC) of 300 000 tons. The fishery from 1973 (the first year when quotas were introduced) until 1976 failed to take the TAC. This shortfall, plus the steep decline in the catch rate index and sharply declining inshore catches, indicated that the stock was declining at an alarming rate (Fig. 4).

In addition to the fishery for northern cod in NAFO Divisions 2J, 3K and 3L, there was also a fishery on the northern portion of the stock complex in NAFO Divisions 2G and 2H. During the 1950's catches were low and were mainly by inshore fishermen. In 1960 about 8 000 tons of cod were caught in 2GH mainly by France, Spain, Portugal and Canada in that order (Fig. 5). Catches remained low until 1965 when they reached about 50 000 tons and increased to 94 000 tons in 1966 mainly by West Germany, Poland, Portugal, Spain, Soviet Union and France. Inshore fishermen caught only about 1% of the total in 1966. Catches rapidly declined to about 12 000 tons in 1971 and to 297 tons in 1973. Since 1976 catches have remained low, always less than 6 000 tons except for 1982 when about 13 500 tons were caught. TACs for this stock have remained at 20 000 tons since their introduction in 1974, largely because of the lack of any analytical assessment of cod in this area. For 1993 the TAC was reduced to 1 000 tons.

## Early Attempts to Manage the Fishery

Prior to 1950 there were attempts to limit effort and to avoid gear conflicts in the fishery for northern cod. These dealt mainly with mesh sizes in cod nets and cod traps, prohibitions of cod trawls (line trawls) on specific grounds at various times of the year and the designation of cod trap berths by local cod trap committees (Anon. 1943).

When the ICNAF Convention entered into force in 1950, the Commission was given the mandate "to make possible the maintenance of a maximum sustained catch from the (ICNAF) fisheries." This maximum was not defined. The most common interpretation was that the objective was to obtain maximum yield in weight of fish, separately from each stock of fish (Templeman and Gulland 1965). The early ICNAF regulations dealt exclusively with the regulations of mesh size and construction features of otter trawls (Pinhorn and Halliday 1990). The expansion of international fishing effort in the Northwest Atlantic signalled very strongly that direct control of the amount of

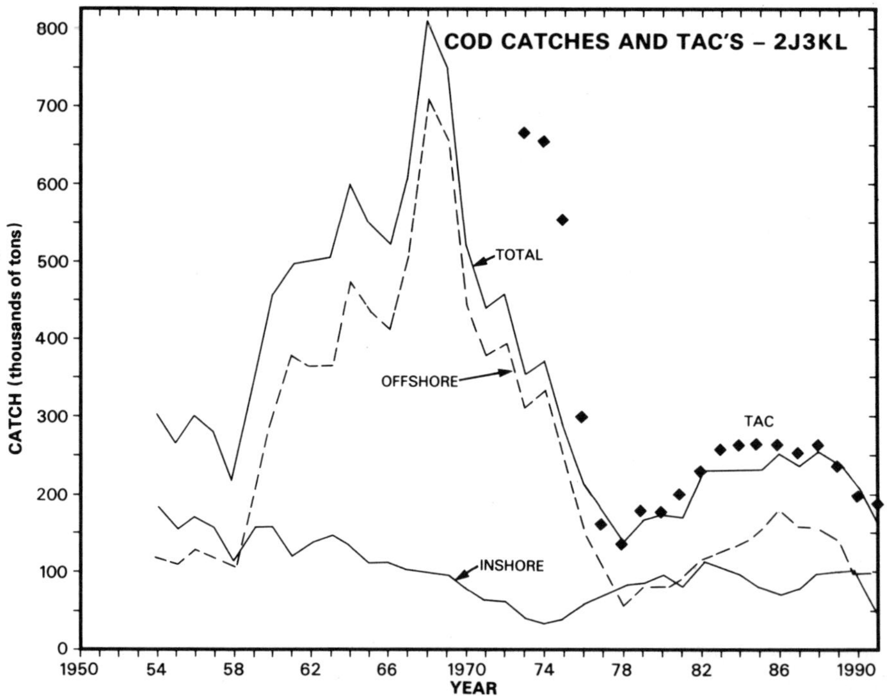


FIG. 4. Cod catches (1954–1991) and TACs (1973–1991) in NAFO Divisions 2J, 3K and 3L, the southern component of the northern cod stock complex.

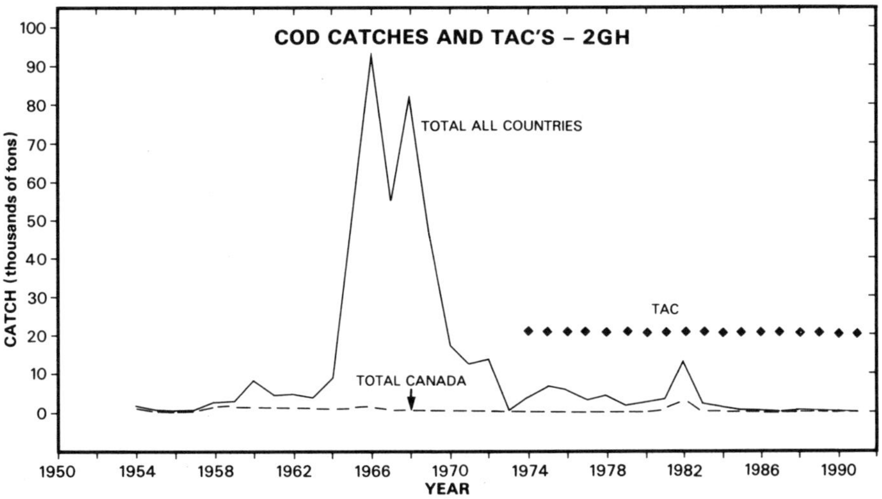


FIG. 5. Cod catches (1954–1991) and TACs (1974–1991) in NAFO Divisions 2G and 2H, the northern component of the northern cod stock complex.

fishing was required if fisheries were to be exploited at the maximum sustainable level. As early as 1965 Templeman and Gulland (1965) stated that further expansion of the fishing effort would produce little or no increase in catch. An increase in mesh size, aimed at protecting small fish, would allow only a temporary small increase which would be negated by further expansion of fishing. They advocated that the least difficult method to achieve direct control of the amount of fishing was by means of catch quotas. They suggested that there be separate quotas for each stock of fish and that preferably the quotas should be allocated to each section of the industry. See Parsons (1993) for a fuller discussion of the evolution of the international management regime.

The first TAC for the 2J3KL cod stock complex was agreed to at the 1972 ICNAF meeting. It allowed for a catch of 575 000 tons in the Convention area plus an estimated catch of 90 000 tons within the Canadian 3-mile territorial limit for a TAC in 1973 of 665 000 tons.

For the northern portion of the stock complex in Divisions 2G and 2H, the member countries of ICNAF agreed in 1973 that the TAC for this area in 1974 be set at 20 000 tons even though the Maximum Sustainable Yield (MSY) was estimated to be 30 000 tons (Pinhorn 1976). This decision was taken since the abundance of cod in 2GH had been severely reduced during 1965–69 when the annual catch was 68 000 tons. Also, the numbers of newly recruited cod in this area in 1967 to 1969 appeared to be lower than average and consequently would yield a smaller than average catch when they were exploited by the fishery.

The TAC on this portion of the stock complex has remained at 20 000 tons from 1974 to 1992. With the exception of a catch of 13 500 tons in 1982 (68% of the 20 000 tons TAC) the catch has never been in excess of 35% of the TAC since TACs were introduced in 1974, and the percentage of the TAC taken has varied from less than 1% to 35%. This is possibly attributable to the low abundance of cod in this area and to the inaccessibility during the winter when cod are concentrated on the offshore banks and during the summer when the cod are dispersed in their migratory summer distribution. For 1993 the TAC was reduced to 1 000 tons.

For the 2J3KL portion of the northern cod stock complex the percentages of the TAC caught in 1974, 1975 and 1976 were 53, 57 and 52%, respectively. These early TACs were not effective in curbing the overexploitation because the TACs were set at too high a level. Meanwhile catches in the inshore fishery had declined to a historically low level of 35 000 tons in 1974.

## Pressures for Extension of Fisheries Jurisdiction to Save the Canadian Fishery

With the drastic decline in the fish catches in Atlantic Canada during the early 1970's, there was public pressure on the government to deal with the problem. There was a concerted effort by fish processors, the fishermen and the public whose livelihoods depended on the fisheries to speak with one voice concerning the severity of the problem. On September 3, 1971 the Save Our Fisheries Association (SOFA) was launched under the leadership of Gus Etchegary. On October 8, 1971, SOFA presented a brief to the federal government at a meeting in Ottawa. This brief contended that it was the responsibility of the federal government to take whatever steps necessary to save the thousands of jobs that depended on the fishery. SOFA demanded that stronger measures be taken to protect and preserve the fish stocks in the northwest Atlantic. The traditional Canadian approach through ICNAF and the 1973 Law of the Sea Conference was considered insufficient to protect the welfare of those eastern Canadians who depended on the fishery for a livelihood. SOFA presented 11 recommendations, many of which directly affected the northern cod. The main thrust was that Canada seek full management control over the marine resources of the Continental Shelf by the end of 1973 (*Evening Telegram*, Oct. 12, 1971).

At the 1972 ICNAF Meeting a quota of 575 000 tons in the ICNAF Convention Area was agreed for 1973. At the 1973 ICNAF Meeting a quota of 606 700 tons in the Convention Area was established for 1974. This, with an estimated catch of 50 000 tons within the Canadian 3-mile territorial limit, allowed for a total catch of 656 000 tons in 1974 (Pinhorn 1976). The 1975 TAC was reduced to 554 000 tons and this was further reduced to 300 000 tons for 1976 and even further reduced to 160 000 tons and 135 000 tons in 1977 and 1978 respectively as the stock size declined and it was realized that the former TACs had been set too high to allow the stock to recover. ICNAF decided to set TACs at the $F_{0.1}$ level for 1977 in the context of a Canadian decision to extend fisheries jurisdiction to 200 miles effective January 1977 (May et al. 1981). The adoption of the $F_{0.1}$ strategy appeared to be an attractive economic measure in that theory suggested that only about two-thirds of the fishing effort is required to catch the $F_{0.1}$ TAC, which in the long term should produce 85–90% of the yield-per-recruit at $F_{max}$. The $F_{0.1}$ strategy was considered a conservation measure in that it would provide a buffer against stock assessment error and enforcement deficiencies and would generate a larger spawning biomass to increase the probability of good year-classes occurring (Pinhorn and Halliday 1990). The more conservative TACs during the mid and late-1970's appeared to have arrested the stock decline. The biomass of cod age 3 years or older (3+) increased from a low of 450 000 tons in 1976 to about 1.2 million tons in 1984 (CAFSAC 1989). This rebuilding was facilitated mainly by the conservative management measures adopted by Canada following extended jurisdiction on January 1, 1977.

### "Rebuilding Phase"

The euphoria of the 200-mile limit generated expectations of an ever increasing abundance of northern cod. Public discussion began on how Canadians were going to catch and process all this cod. At the Fogo Island Fisheries Conference, January, 1979, Dr. A.W. May, then the Assistant Deputy Minister for Atlantic Fisheries of Department of Fisheries and Oceans (DFO) told the participants that Canada would be wise to take a slow approach to fleet expansion (Anon. 1979). At that time there were enough boats in the traditional fisheries to fully harvest the northern cod resource. The priority was to rebuild the resource and then implement an orderly fisherman-oriented expansion in the fleet, building from the coast out.

On June 28, 1979, the Minister of Fisheries and Oceans, the Honourable James McGrath, announced that the Department of Fisheries and Oceans would sponsor a government-industry seminar on northern cod. Projections by DFO scientists indicated that the TAC for the stock, which was 180 000 tons in 1979, could be as high as 350 000 tons by 1985 presenting significant opportunities for expansion of the Canadian fishery. These projections were made on the assumption that the number of new fish entering the commercial stock (recruitment) and hence the commercial fishery at age 4 each year would be the average of that observed over recent years including the strong year-classes of the 1960's. In 1979, for example, 60% of the projected catch for 1985 was dependent on the accuracy of this assumption. It was stressed very strongly that, if recruitment in future was lower than average, then the stock would rebuild more slowly than projected and projected catches would be lower (Pinhorn 1979).

At the Northern Cod Seminar in August 1979 Minister McGrath reiterated that the Department recognized that the northern cod were the "staff of life" to the people of Northeast Newfoundland and Labrador (DFO 1979a). He stated that the policy of the government was that the inshore fishermen had first call on this resource: "For these people, for the people working in processing plants, for their families and the communities in which they live, the right and the ability to reap this harvest is indispensable, because they have a special, a very special relationship to these stocks."

The intent was to derive the greatest possible benefit and to provide for greater stability of supply, a longer processing season and a better way of life in coastal communities, which was within the context of protection for the inshore fishermen. A need for an offshore fishery to complement the inshore fishery was also recognized.

In a paper on the Utilization of Northern Cod, Doucet (1979) concluded:

"1. The current Newfoundland east coast inshore fleet, including the operation of cod traps, and the Atlantic Coast Fleet of stern trawlers of 500 tons and over, if ice reinforced, can catch the projected 350 000 MT of northern cod in 1985. Indeed, I believe that with the planned replacements the fleet will be able to catch more.
2. In terms of benefits and costs the most advantageous vessel is the 55 foot long-liner. The least advantageous are the factory and freezer trawlers.
3. The cheapest method of catching fish is the cod trap — less than half that of any other type of gear.
4. Open access will not only dissipate the (economic) rent but lead to eventual bankruptcy for all its participants (referring to an open unregulated fishery on a common property resource)".

Shortly after the Northern Cod Seminar DFO released a discussion paper (DFO 1979b) entitled "Toward a Policy for the Utilization of Northern Cod".

The main policy directions suggested were as follows:

"1. The first and over-riding priority in allocations is to the inshore fishery. The consensus from the seminar participants was that two-thirds of the TAC of northern cod should be set aside as an allowance for the inshore fishery.
2. There must be some control on the number of units in the inshore sector.
3. An amount not to exceed 10% of the TAC of northern cod or alternatively an amount of 20 to 30 thousand tons should be set aside for negotiating, through NAFO, a larger allocation to Canada of fish beyond the 200 mile zone.
4. Replacements of existing wetfish trawlers by similar vessels would be allowed to fish on the same basis as the vessels they replaced.
5. Replacements of existing wetfish trawlers by vessels capable of freezing at sea (providing that they come within the 125% guideline) would receive a separate quota within the offshore allocation.
6. With respect to the possibility of access to northern cod for additional offshore vessels, this matter remains under review pending further Provincial and Industry consultations.
7. Consideration should be given to placing an upper size limit on offshore vessels, irrespective of technology used, e.g. 175–200 feet, on the assumption that vessels of this size can be equipped to freeze and/or process at sea, and to prevent excessive concentration of offshore catching capacity."

In response to the federal proposals, the Province of Newfoundland set out its own position in a discussion paper on November 5, 1979. In this they reiterated the Newfoundland view articulated at the Northern Cod Seminar, namely:

"1. Northern Cod must be reserved to inshore and middle distance effort to the extent it can be harvested by that effort.
2. Where, within the total allowable catch a surplus to inshore effort can be clearly shown to exist, it must be reserved to offshore effort landing into Newfoundland ports for distribution to processing plants which now operate on a seasonal basis" (Newfoundland 1979b).

Newfoundland stated that the inshore fleet could and should take up to 85% of the northern cod catch and that any residual volume should be taken by Newfoundland-based offshore effort to supply resource-short plants in Newfoundland. The Province could see no role for freezer trawlers in the harvesting of northern cod. While the replacement of existing vessels with larger vessels had considerable merit overall, the replacement of existing wetfish trawlers by larger freezer trawlers was not acceptable. The Province also emphasized that its ultimate fishery policy objective was for constitutional change to give Newfoundland a greater share of jurisdiction in fisheries management. See Parsons (1993).

As a result of the Northern Cod Seminar, the federal government adopted as policy that the TAC be set at a level below $F_{0.1}$. The level that was adopted for 1980 was F = 0.165 (180 000 tons) compared to $F_{0.1} = 0.20$. The TAC continued to be set at this level of fishing mortality until 1983 when, in the light of scientific advice which would have seen the TAC at F = 0.165 reduced, it was decided to set the TAC at the calculated $F_{0.1}$ level for 1984. For 1986–1988 the TACs for this stock were set below the $F_{0.1}$ levels as calculated at that time.

Prior to the Northern Cod Seminar, in 1976 and 1977 a drastic decline in redfish stocks in the Gulf of St. Lawrence led to the displacement of the larger Nova Scotia and Newfoundland trawlers out of the Gulf. The 1977 Atlantic Groundfish Plan attempted to divert some of the offshore trawler effort toward northern cod. There were incentives to encourage offshore trawlers to fish northern cod in the winter months. Under the Groundfish Vessel Dislocation Program, and the Northern Fishery Incentive Program, approximately $4.5 million was provided by the federal government to Newfoundland and Maritime-based trawlers in the form of incentives for fishing in northern areas, for fishing redfish and turbot and assistance for fishing unutilized quotas in traditional areas. For more detail, see Parsons (1993). As catch rates in the winter offshore cod fishery increased, the effort focused very sharply on northern cod. The Canadian offshore catch increased from 1 000 tons in 1975 to 45 000 tons in 1980 with Newfoundland-based offshore vessels harvesting 82% of the total and the Maritimes-based vessels 18%. By 1982 the Canadian offshore trawler fleet was harvesting about 95 000 tons. The Canadian offshore fleet had adapted quickly to fishing the northern cod stock during the winter in ice-covered areas despite earlier resistance by the Canadian fishing industry to attempts to divert fishing effort to the northern area.

The development of this northern cod offshore fishery sparked interprovincial rivalry between Nova Scotia and Newfoundland. In December 1979 the Newfoundland Government embarked on a major interprovincial battle with Nova Scotia over allocations of northern cod. Newfoundland attempted to force all offshore vessels fishing this stock to land their catch in Newfoundland for processing. Premier Peckford summoned the executives of National Sea Products and H.B. Nickerson Ltd. to his office and told them that if they landed northern cod caught in "Newfoundland" waters at mainland plants, the Government would reserve the right to "consider them ineligible for any provincial assistance programs or to apply for any further processing licenses" (Newfoundland 1979a). Until this time Nova Scotia had supported Newfoundland's demand for constitutional realignment of jurisdictional responsibilities respecting fisheries. The Newfoundland ultimatum to the Nova Scotia-based companies was firmly resisted by Nova Scotia which changed its position on the question of provincial jurisdiction over fisheries (Parsons 1993). On September 4, 1980 a new Nova Scotia Fisheries Minister responded to Newfoundland's claim to northern cod stating that it had been "a warning as to how they could conduct themselves if the provinces got together to decide allocations of various species. Newfoundland's demands, when they haven't the authority to make demands, has been a great instructor as to what they would demand if they had the authority" (*Halifax Chronicle Herald*, September 4, 1980).

The 1979 TAC for northern cod was set at 180 000 tons of which 100 000 tons was set aside as an *allowance* (unregulated portion of the TAC) for the inshore fishery. In 1980, again with a TAC of 180 000 tons, the inshore allowance was increased to 110 000 tons. For 1981 the TAC was increased to 200 000 tons of which 120 000 tons were set aside as a "quota" (regulated portion of the TAC) for the inshore fishery. The concept of a quota for the inshore fishery was short-lived. In 1982 an *allowance* of 120 000 tons was set aside for the inshore fishery. The Federal Government decided to maintain this approach in future because the inshore fishery had fallen significantly short of its 120 000 tons quota in 1981. It appeared preferable to manage the inshore allowance as an average over several years. The rationale was that this would allow the inshore fishery to take its full potential with existing effort levels in years of high cod abundance inshore while ensuring that the offshore fishery was not unduly constrained by high and possibly erroneous estimates of potential inshore catch. This resulted in a 55:45 percentage inshore/offshore split of the northern cod TAC in 1982, a reduction from the 60% allocated to the inshore during 1980–81. Newfoundland never did obtain its 85:15 percentage split in favour of the inshore fishery. During later years (1984–86, 1988) when the TAC was 266 000 tons, the inshore allowance remained at 115 000 tons (43% of the total TAC).

During the early 1960's the total number of inshore fishermen in NAFO Divisions 2J, 3K and 3L increased from about 12 500 in 1960 to about 15 600 in 1964. Subsequently as the stocks and inshore catch rates declined, the number of fishermen declined very sharply to about 8 200 in 1974 (Fig. 6). The numbers in Labrador declined somewhat but the declines were greatest in 3K and 3L. After 1974 and especially after the proclamation of the 200-mile limit the number of inshore fishermen in 2J, 3K and 3L increased dramatically to a peak of about 24 000 in 1980. With the levelling off of inshore catches and a subsequent catch decline the numbers of fishermen declined to about 16 000 in 1984 and subsequently stabilized at about 18 000 inshore fishermen. The rate of change since 1984 has been about the same for 3K and 3L. A noticeable change has occurred since the 1960's in that prior to 1972 3L had significantly more fishermen than 3K. This situation has been reversed and since 1984 there have been slightly more fishermen in 3K than 3L (8 400 in 3K and 7 800 in 3L). The numbers for Labrador (2J) remained fairly stable during 1980–89 (1 253 to 1 547) representing a relatively small proportion of the total number of inshore fishermen.

From the mid 1970's onward several initiatives were introduced to limit participation in the inshore cod fishery. In 1976 a freeze was introduced on the licensing of small otter trawlers less than 65 feet. In 1979 a moratorium was introduced on all vessels greater than 35 feet in length. In 1980 the categorization of inshore fishermen as full-time and part-time was initiated

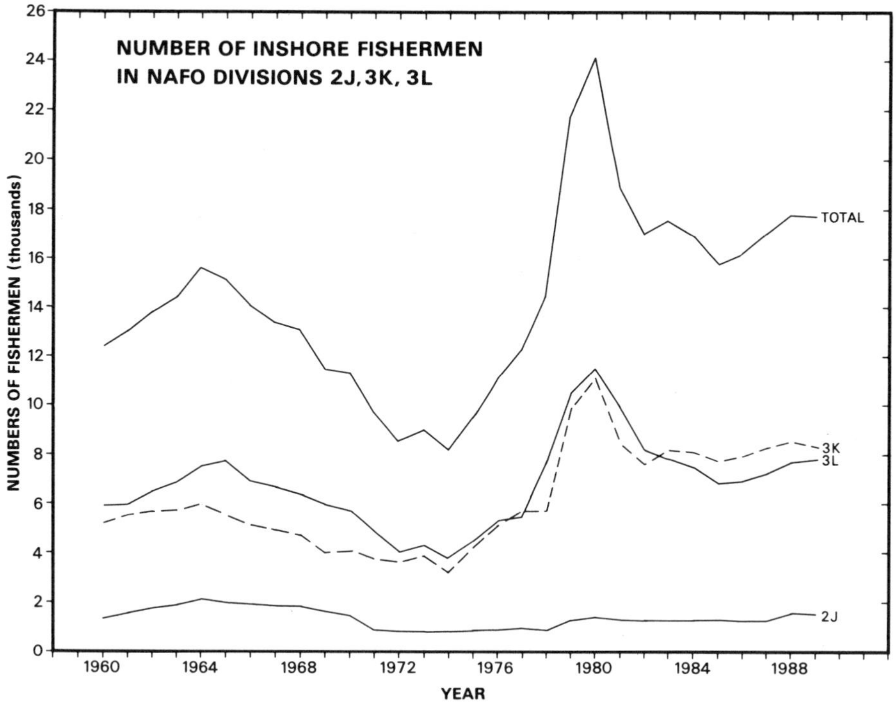


FIG. 6. Numbers of inshore fishermen in NAFO Division 2J, 3K and 3L (1960–1989).

(Parsons 1993). This initial categorization of fishermen did not limit effort. The later introduction of groundfish fixed gear licences for full-time fishermen limited effort to some degree. Under this policy a full time fisherman could register a vessel up to 34'11" and attain a groundfish fixed gear licence. A fisherman designated as part-time could only register a vessel up to 22' and fish using handline/jigger only. However, an individual originally categorized as a full-time fisherman and downgraded to a part-time could retain his groundfish fixed gear licence.

When categorization was introduced in 1980 there were about 7 000 full-time fishermen and this number increased to about 8 200 in 1981. From 1981 onward the numbers of full-time inshore fishermen in NAFO Divisions 2J, 3K and 3L remained surprisingly constant at about 8 000 fishermen (Fig. 7). The numbers of part-time fishermen declined from 17 200 in 1980, when a large number of licences were issued for part-timers for squid, to 10 600 in 1981. From 1982 to 1989 the numbers of part-timers fluctuated from a low of 8 000 in 1985 to a high of 9 800 in 1988.

## Role of Foreigners in Development of the Northern Cod Fishery

With Canada's extension of jurisdiction to 200 miles, it was generally assumed that stock rebuilding would increase significantly the inshore catches of northern cod. This would in turn exert pressure on the seasonal plants in eastern Newfoundland to process all the cod during the glut in the cod trap catch when it was anticipated the catch would exceed the local processing capacity. To address this potential problem, several programs were devised involving foreign vessels in the short term to absorb the surplus production while Canada developed the capacity to fully utilize the resource. These programs were as follows:

1. Over-the-side Sales (direct sales).
2. Over-the-wharf Sales (direct sales).
3. Resource-short Plant Program.

### Over-the-side Sales

Over-the-side sales started on the east coast of Canada in 1976 for herring. Fishermen were permitted to sell unprocessed herring directly to foreign processing vessels at a price higher than was available from Canadian buyers. There have been significant changes in the over-the-side sales policy in the post-extension era (Parsons 1993). Generally "over-the-side" sales of northern cod were permitted in situations where fishermen were unable to sell their catch, sometimes due to plant closures, but in most cases during a

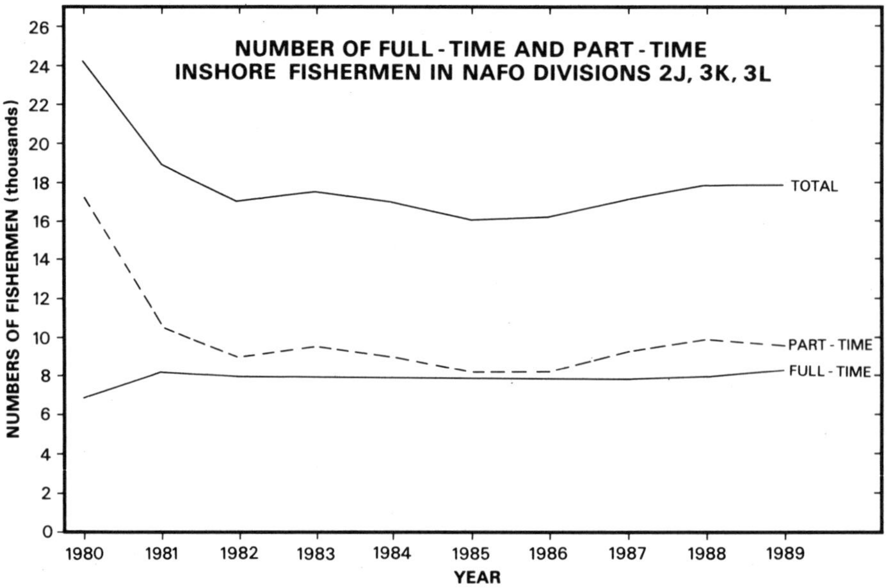


FIG. 7. Numbers of full-time and part-time inshore fishermen in NAFO Division 2J, 3K and 3L (1980–1989).

period of glut in the cod trap catch around eastern Newfoundland and Labrador.

### Over-the-wharf Sales

Over-the-wharf sales consisted of the sale of semi-processed cod (split frozen or lightly salted) by processors to a foreign vessel tied up at a wharf near the processing plant. Over-the-wharf sales involved only Portuguese vessels.

The scale of direct sales (over-the-side and over-the-wharf) of cod in Newfoundland and Labrador during 1979 to 1984 is illustrated below. There were no direct sales of cod from Newfoundland and Labrador prior to 1979.

| Year | Over-the-side Sales (tons) | Over-the-wharf Sales (tons) |
|---|---|---|
| 1979 | 565 | 0 |
| 1980 | 5 504 | 0 |
| 1981 | 4 429 | 8 525 |
| 1982 | 7 018 | 10 230 |
| 1983 | 6 248 | 4 222 |
| 1984 | 7 203 | 3 587 |
| Total | 30 967 | 26 564 |

Overall, these programs had only a marginal impact on development of the inshore fishery.

### Resource-short Plant Program

The concept that later evolved into the Resource-short Plant Program developed in the 1976–78 period. It was based upon two assumptions and one conclusion:

**Assumption (1)** That inshore catches of northern cod were bound to rise considerably after extended jurisdiction and that there would be a need for an increase in seasonal processing capacity on the East Coast of Newfoundland.

**Assumption (2)** That more cod could be caught as a whole if more fish were allocated to the offshore sector as the biomass increased. However, the Canadian trawler fleet was not expected to be able to harvest the offshore quota. (This assumption later proved false). Therefore, irrespective of how the northern cod was allocated between inshore and offshore fleets, there would be a surplus that would have to be allocated to foreigners.

**Conclusion** If it was inevitable that foreigners were to harvest this surplus, why not direct these foreign catches to the seasonal plants on the East Coast of

Newfoundland to extend their operating season?

Based initially upon this type of reasoning, programs aimed at extending the operating season of "inshore" plants have been in existence since 1977. In the 1977–78 period these were known as "cooperative" or "joint-ventures". In 1979 it was termed "northern cod program" and in 1980–82 it was called the "Inshore Plant Delivery Program." In 1983 it became formalized as the Resource-short Plant Program.

In 1977 individual Canadian companies were granted authority to enter into contracts with foreign partners on an ad hoc basis for the delivery of fish to their respective plants for further processing. The processing requirements were not very stringent.

In 1978 plant operators complained of a critical shortage of raw material during the off-season but no program was established. In 1979, 12 000 tons of northern cod (2J3KL) were set aside for this Program. There was considerable controversy respecting the allocation of the 12 000 tons among provinces. Fisheries Minister Romeo LeBlanc allocated 9 000 tons to Newfoundland, 2 500 tons to the Maritimes and 500 tons to Québec. About 7 700 tons of the total allocation was harvested by two foreign companies and delivered to plants in Newfoundland and the Maritime Provinces.

In response to the short-fall in landings caused by the 1980 Newfoundland lockout/strike and the significant effect this had on inshore processing plants and workers, a special program entitled "Newfoundland Inshore Fish Plant Delivery Program" was approved in September 1980. The program authorized the harvesting of 10 000 tons of northern cod to be caught by foreign and/or Canadian vessels for delivery to inshore seasonal plants within 2J3KL. Deliveries were permitted to processing plants that had groundfish freezing capacity, with the requirement that the fish would be processed to at least the fillet stage. Overall, the program landed about 4 500 tons.

In 1981 the Department responded to poor inshore landings in Divisions 2J3KL in Newfoundland by authorizing three Northern Cod Programs. Initially, a program was authorized to harvest 10 000 tons of cod from 2J3KL to be caught by Canadian vessels. When it became apparent that plants could not obtain vessels to catch this allocation, a second program of 10 000 tons was made available to be caught by foreign vessels. Deliveries were permitted to inshore seasonal plants located in 2J3KL that had freezing capacity, with the requirement that 60% of the fish must be processed to at least the fillet stage. About 4 100 tons were landed under this program. A third program, a special allocation of 10 000 tons was also approved to provide funds for inshore fishermen's benefits. Under this program about 3 200 tons of northern cod were landed.

As a result of the pressure for the continuation of some such program, the department reserved 5 250 tons (from the 2J3KL cod TAC of 230 000 tons) for such programs in 1982. This 5 250 tons was allocated as an extension of the 1981 program commencing on January 1, 1982. The rationale for allowing the 5 250 tons to be taken in the January-April period rather than in the fall was that the northern cod catch rates are traditionally much higher in the early part of the year. As well, those involved preferred to continue the 1981 program while vessels were still under contract, thus eliminating the extra cost and inconveniences of a second phase of the program in late 1982. A total of about 3 700 tons of cod was caught from the overall allocation of 5 250 tons. During the 1980–82 period these programs were confined to Newfoundland.

In 1983 as a result of the Report of the Task Force on Atlantic Fisheries, 10 000 tons of northern cod were allocated for a Resource-short Plant Program (RSPP). That year negotiations on harvesting these allocations were never finalized and there were no landings against the 10 000 tons allocation in 1983.

On December 8, 1983, Fisheries and Oceans Minister Pierre De Bané officially announced the implementation of a long-term Resource-short Plant Program. Under the Program, a specific amount of offshore-caught fish would be allocated for delivery to seasonal resource-short fish processing plants in the five eastern provinces during the off-peak fishing season. An allocation of 10 000 tons of 2J3KL cod and an additional 1 000 tons of 2GH cod were set aside for the program. Seventy-five percent of the northern cod to be caught under this program was designated to be landed at plants on the northeast coast of Newfoundland immediately adjacent to this stock. The remainder was to be made available to plants in the other four regions including the west coast of Newfoundland which lay in the Gulf Region. Of the total allocation of 10 000 tons of 2J3KL cod, 5 401 tons were taken by Newfoundland and 2 254 tons by Gulf, Scotia-Fundy and Québec Regions for a total of 7 665 tons (77% of the total). None of the 1 000 tons allocation of 2GH cod was caught.

Allocations of 10 000 tons of 2J3KL cod were accorded to the RSPP during each of the years 1984, 1985 and 1986, of which 75% were for Newfoundland plants and 25% for Scotia-Fundy, Gulf and Québec plants. All of the cod landings had to be made by Canadian vessels. This did not materialize as anticipated. The continued use of foreign vessels in the RSPP was a major issue until 1989 when Newfound Resources Limited, a consortium of the eligible Resource-short

Plants in the Newfoundland Region, purchased a freezer trawler, the POLAR STORM, to catch and deliver fish to the Resource Short Plants.

To maximize employment benefits, the Department required that all cod be landed and that 100% be brought to the fillet stage by each processor (DFO 1988a). As a result of a reduced TAC in 1987 the RSPP allocation was reduced to 9 300 tons but was increased to 15 350 tons in 1988 when the TAC was increased from 256 000 tons to 266 000 tons. This was reduced to 12 200 tons in 1989 when the TAC was reduced to 235 000 tons. In 1990, with a further reduction in the TAC to 197 000 tons, the RSPP allocation was reduced from 12 200 tons to 5 000 tons.

This reduction in the allocation to the RSPP was a poignant counterpoint to the projections of the Task Force on Atlantic Fisheries which had envisaged a major program of up to 50 000 tons by 1987 to extend the operating season of inshore plants. This did not materialize. Processors were generally positive toward the 1984 RSPP and its contribution toward the viability of the inshore plants (Davis 1986). The 1990 reduction in the program to 5 000 tons severely reduced the level of benefit since this allocation was being spread among 32 eligible plants (DFO 1988a). Fisheries and Oceans Minister Valcourt eliminated the allocation of northern cod to the RSPP in December 1990.

**Allocations for Market Access**

At the time of extended jurisdiction, Canada decided to allocate stocks of fish surplus to Canadian capacity to foreign countries in exchange for commensurate benefits. The Canadian catch was expected to increase as stocks rebuilt. There was some concern that Canada might not be able to market this increased catch. In exchange for allocations to foreign countries, Canada sought market advantages for its fishery products. A commensurate benefits approach was outlined to ICNAF member countries in 1978.

In 1978, 25 000 tons of northern cod were declared surplus to Canadian requirements. Seventy-five hundred (7500) tons of this were allocated to foreign countries which had signed bilateral agreements with Canada. The remaining 17 500 tons were considered available for commensurate benefits. Of this total, 8 000 tons were allocated to Portugal in view of its traditional cod fishery and its expressed willingness to expand markets for salt cod from Canada. In 1979 Canada shifted its policy emphasis from commensurate benefits to a principle of "a satisfactory trading relationship". The surplus was to be allocated on a proportional basis, taking into account past catches prior to 1979, to states which had signed bilateral agreements with Canada, which cooperated in the conservation of transboundary stocks and whose fish-trading relationship with Canada was deemed satisfactory.

In 1980 Canada offered to enter into long-term agreements with the EEC, Spain and Portugal in which specific fish allocations e.g. northern cod, in the Canadian zone would be made available in exchange for specific market commitments for Canadian fish products. After protracted negotiations, this resulted in a Long Term Agreement (LTA) with the EEC. Spain and Portugal did not enter into LTAs. Spain embraced a policy of fishing freely stocks beyond 200 miles. Portugal opted for year-to-year agreements with Canada. The question of northern cod allocations was central to the LTA with the EEC, the ad hoc arrangements with Portugal and the fish war with Spain. For details, see Parsons (1993).

By 1981 it was clear that the 25 000 tons of northern cod set aside for foreign allocations were no longer surplus to Canadian requirements. The Canadian offshore fleet had demonstrated its ability to take more than its quota. Thus the foreign allocations were constraining the Canadian harvesting effort. Also the LTA was not producing the expected benefits. As a result opposition to the allocation of non-surplus northern cod developed among various components of the Canadian industry.

Following a Seminar on International Fisheries Policy in 1982 (Oak Island I) and recommendations of the Task Force on Atlantic Fisheries, the policy regarding allocation of non-surplus fish for market access was changed. The Task Force recommendation that access to non-surplus allocations be strictly limited in total, be allocated unilaterally on the basis of good performance and be offered to selected nations "after the fact" rather than in return for a promise of access to markets, was adopted.

At a Second Oak Island Seminar in 1985, it was recognized that the allocation of non-surplus cod to the EEC under the Long-Term Agreement and to Portugal annually had not been beneficial to Canada. Anticipated market benefits with the EEC had failed to materialize. Spain was continuing to overfish beyond 200 miles. With the imminent accession of Spain and Portugal to the EEC in 1986, the EEC was about to embrace the policy of obstructing NAFO efforts to deal with the foreign overfishing problem. The Oak Island II Seminar recommended that instead of rewarding "good behaviour" outside 200 miles, the policy emphasis should be shifted to punishing "bad behaviour" by cutting off access to surplus allocations inside the zone. The attempt to improve market access in foreign countries through the use of fish allocations was abandoned with the exception of certain agreements with Eastern Bloc countries.

## The Task Force on Atlantic Fisheries and Northern Cod

On January 8, 1982 the federal government announced the formation of a Task Force on the Atlantic Fisheries, chaired by Michael Kirby. The Task Force had two mandates:

1. To advise the government on the long-term policies it should implement to promote a healthy fishery in the future.
2. To advise the government on how to deal with the specific requests which came in late 1981 for financial assistance, from the Lake Group Ltd. and H.B. Nickerson and Sons Ltd.

The Kirby Task Force Report, "Navigating Troubled Waters" released in February 1983, contained an entire chapter on the "Utilization of the Northern Cod Stock". The Task Force based its recommendations on what proved to be rather optimistic projections of the future growth of the northern cod stock. The TAC of northern cod was forecasted to increase from 230 000 tons in 1982 to at least 400 000 tons by 1987. The Task Force recommended that the 1987 allocation of the Canadian quota of northern cod should be approximately as follows:

| | 1982 | 1987 |
|---|---|---|
| 1. Inshore allowance (<65 ft.) | 120 000 tons | 145 000 tons |
| 2. Existing trawler fleet (>100 ft.) | 87 250 | 145 000 |
| 3. Resource-short plants | 5 250 | 50 000 |
| 4. Other fixed and mobile gear | 2 500 | 40 000 |
| | 215 000 tons | 380 000 tons |

The 1982 allocations are shown for comparison.

The Task Force recommended that the allowance to existing inshore vessels not be increased proportionally to the growth in the TAC. It considered that adding 90 000 tons to the inshore allowance over the next 5 years would add about 60 000 tons to the summer peak catches and the processing plants would not be able to handle the delivery and glut situations would result. It concluded that a proportional increase in the inshore allowance would be inconsistent with the economic viability objective recommended by the Task Force (Kirby 1982).

The Task Force recommended that a greater share of the increase go to the existing trawler operators. It also recommended that a significant amount be set aside for resource-short plants. In view of the seasonality of the inshore fishery and, as a result, the extreme under-utilization of many large inshore freezing plants it was suggested that the wisest course was to supply a portion of growth in the northern cod catch to these plants to improve their prospect of viability. The Task Force recommended an allocation of 50 000 tons of northern cod to be delivered to the resource-short plants by 1987. This formed the basis for the initiation of the Resource-short Plant Program on a formal basis (see previous section).

Based on the assumption that the extreme seasonality and low annual productivity of the inshore fishery on the east coast of Newfoundland and Labrador was fundamentally a problem of technology, the Task Force placed a high priority on the development of a fleet of large "Scandinavian type" longliners that would have the feasibility of fishing for most of the year. The offshore longliner was considered to be a significant technological option because there appeared to be no other intermediate gear type between the smaller inshore vessels (<65 feet) and a large stern trawler that was capable of fishing northern cod. These longliners were to form what was later to be known as the Middle Distance Fleet. An allocation of 40 000 tons was proposed for such a fleet by 1987.

The long term objective was the complete Canadianization of the northern cod fishery but in the interim it was suggested that chartering of foreign vessels be permitted to supply cod under the Resource-short Plant Program and to experiment with the Scandinavian-type longliner technology.

The Task Force recommendations on northern cod were accepted by the federal government and shaped allocations policy for the next several years. The expectations were frustrated by the failure of the stock to grow at the projected rate and achieve the projected level of 400 000 tons by 1987. Nevertheless, allocations were set aside for the Resource-short Plant Program and the development of a Scandinavian longline fleet (Table 1).

No specific allocations were set aside for the Scandinavian longline fleet until 1986 when 1200 tons were allocated during April 1986 after the management plan was announced. In 1987, 1 050 tons were allocated. This allocation was increased to 4 270 tons in 1988, was reduced proportionally to 3 393 tons in 1989 and was eliminated entirely in 1990 when the Canadian TAC was reduced to 197 000 tons. As a result six longliners owned by the province of Newfoundland were put up for sale in January 1990.

Meanwhile, the Canadianization of the northern cod fishery progressed. The foreign allocation was reduced from 20 000 tons in 1983 to 9 500 tons in 1987. In 1989

TABLE 1. Allocations of 2J3KL Cod during 1983–1991. Foreign Allocations were held in reserve in some cases. They may not have all been allocated to countries but reassigned to the Canadian quota.

| Year | TAC | Foreign Alloc. | Canadian Quota | Vessels <65' Fixed gear | Vessels <65' Mobile gear | Vessels 65'–100' Fixed gear | Vessels 65'–100' Mobile gear | Vessels >100' | RSPP | Fixed Gear>100' |
|---|---|---|---|---|---|---|---|---|---|---|
| 1983 | 260 000 | 20 000 | 240 000 | 115 000 | 5 000 | 3 000 | 1 000 | 106 000 | 10 000 | — |
| 1984 | 266 000 | 20 000 | 246 000 | 115 000 | 5 000 | 3 000 | 1 000 | 112 000 | 10 000 | — |
| 1985 | 266 000 | 16 300 | 249 700 | 115 000 | 5 000 | 3 000 | 1 000 | 115 700 | 10 000 | — |
| 1986 | 266 000 | 16 300 | 249 700 | 115 000 | 5 000 | 3 000 | 1 000 | 115 700 | 10 000 | 1 200 |
| 1987 | 256 000 | 9 500 | 246 500 | 115 000 | 5 000 | 3 300 | 700 | 112 150 | 9 300 | 1 050 |
| 1988 | 266 000 | — | 266 000 | 115 000 | 5 000 | 4 180 | 1 000 | 121 200 | 15 350 | 4 270 |
| 1989a | 266 000 | — | 266 000 | 115 000 | 5 000 | 4 180 | 1 000 | 121 200 | 15 350 | 4 270 |
| 1989b | 237 950 | 2 950 | 235 000 | 115 000 | 3 975 | 3 566 | 316 | 96 553 | 12 200 | 3 390 |
| 1990 | 199 262 | 2 262 | 197 000 | 111 007 | 3 993 | 2 545 | 610 | 73 845 | 5 000 | — |
| 1991 | 190 000 | 2 140 | 187 860 | 111 000 | 4 000 | 2 408 | 577 | 69 875 | — | — |

1989a — Proposed based on TAC of 266 000t.
1989b — Allocations after TAC announced at 235 000 t.

and 1990 Canada set aside 2 950 tons and 2 262 tons respectively of northern cod as part of its commitment to France under the interim agreement on fish quotas pending arbitration of the Canada-France boundary claims in 3Ps (Parsons 1993).

### Inshore/Offshore Shift

At the Northern Cod Seminar in August 1979, Newfoundland put forward the position that the northern cod resource should be reserved for the inshore and middle distance effort and where an offshore surplus could be shown to exist, it should be allocated for landing in Newfoundland ports. In response to the September 1979 federal Discussion Paper, Premier Peckford in a Discussion Paper on Bilateral Issues submitted to Fisheries Minister James McGrath in November 1979, reiterated the Newfoundland position.

The 1979 Newfoundland position on the harvesting of northern cod was "that the inshore fishery can and should take up to 85% of the northern cod, with any residual volume being taken by Newfoundland-based offshore effort to supply resource-short plants in this province." The stated rationale was to utilize the entire northern cod catch to provide the maximum economic and social benefits first of all to the Newfoundland communities adjacent to the resource and anything surplus to their needs was to go to other plants in Newfoundland. Only if there was shown to be a surplus to Newfoundland requirements should northern cod be allocated to other provinces.

Coincident with the debate on the allocation of northern cod and concurrent jurisdiction in fisheries matters was the debate on the use of factory freezer trawler technology for the harvesting of northern cod. In 1979 the Nova Scotia Fish Packers Association in a brief to the Northern Cod Seminar argued that the choice of appropriate technology should be left to the private sector (Nova Scotia Fish Packers Association 1979). National Sea Products, supported by the Province of Nova Scotia, requested permission to acquire a factory freezer trawler.

Although the Northern Cod Seminar did not approve of utilizing factory freezer trawlers to harvest northern cod, there was no consensus as evidenced by the following statement:

> "Although there was general agreement that neither factory nor freezer trawlers were required to catch the northern cod, there was a strong feeling expressed by some participants that it was unwise to consider the harvesting of northern cod except in the context of the total Atlantic fishery. Because some species such as squid, offshore mackerel, silver hake, etc., have to be frozen at sea, some participants expressed the view that if northern cod is required to ensure the viability of such vessels then an allocation should be made for this purpose" (DFO 1979a).

In September 1979 a Department seminar on squid concluded that freezing-at-sea technology was necessary if Canada was to take advantage of squid and other non-traditional species.

A 1979 DFO discussion paper proposed that replacements of existing wetfish trawlers by vessels capable of freezing at sea be permitted (providing that they fell within the existing 125% length restriction) (DFO 1979b). Filleting of traditional species at sea would not be permitted. An upper size limit of 200 feet was also suggested.

In November 1979 the federal government announced its policy setting out guidelines for the licensing of freezer and/or factory freezer trawlers. Four licences were made available for 1980. These vessels could be no larger than 200 feet in length, with foreign charters permitted for 1980 only. Vessels were permitted to carry the filleting equipment necessary for processing of non-traditional species but would not be permitted to fillet any traditional groundfish species. Thus northern cod was still required to be brought to shore for filleting to provide on-shore processing employment. Three of the four licences were revoked after one year as the licensees demonstrated no serious attempt to acquire vessels. Only one Canadian vessel with capacity to freeze fish at sea was acquired as a result of this initiative. The major proponent of factory freezer trawlers, National Sea Products, did not succeed in acquiring one of the four non-traditional species licences.

National Sea Products resumed its fight to obtain a factory freezer trawler licence in 1985. It requested a licence with:

— no restriction on the amount of the vessel's catch to be processed at sea;
— conditions limited to those normally applied to wetfish trawlers;
— unrestricted groundfish utilization; and
— unrestricted non-traditional species utilization.

The company proposed to process at sea an annual average catch of 15 000 tons (less than 8% of the company's total enterprise allocation). The company also indicated its willingness to surrender currently held licences for vessels with an annual capacity of 15 000 tons. The operating plan called for reliance upon traditional groundfish species, with half the annual production to be cod of which northern cod was a large component. There was a great deal of discussion and debate including an August 1985 Discussion Paper on Factory Freezer Trawlers, meetings of the Federal-Provincial Atlantic Fisheries Committee and the Atlantic Council of Fisheries Ministers. The matter was also discussed with the Atlantic Groundfish Advisory Committee in September 1985 and by the newly established Atlantic Regional Council in August 1985 and October 1985.

The introduction of FFT technology was opposed by fishermen's groups, particularly the Newfoundland Fishermen, Food and Allied Workers Union (NFFAWU). The province of Newfoundland strongly opposed FFT technology and in September 1985 released a discussion paper entitled *Appropriate Offshore Fish Harvesting Technology: An Assessment of the Detrimental Effects of the Use of Factory Freezer Trawlers*. The Newfoundland government claimed that by virtue of clause 12 of the Canada/Newfoundland Restructuring Agreement (1983), the federal government was bound to continue the existing policy of prohibiting the filleting of traditional groundfish at sea (Newfoundland 1985).

On November 7, 1985 National Sea Products was granted a licence to operate one factory freezer trawler in the Atlantic Groundfish fishery for a 5-year introductory trial period during which time the operational and socioeconomic impact would be assessed. On November 8, 1985 the Acting Minister of Fisheries and Oceans Erik Nielsen announced that a maximum of three FFT licences would be granted to selected Atlantic coast companies for a 5-year introductory period. These were reserved for:

1. National Sea Products
2. Fishery Products International
3. A company or consortium from the remaining offshore groundfish companies.

One of the conditions of the licences required that no more than 6 000 tons of a company's northern cod quota could be harvested by a factory freezer trawler in any one year.

National Sea Products acquired the factory freezer trawler and commenced fishing it in 1986. The two other licences were not utilized. The end result was that factory freezer trawler technology was introduced to the Canadian offshore fishery and in particular the harvesting of northern cod.

Over time the inshore/offshore allocation ratio shifted in favour of the offshore. In 1980 out of a Canadian quota of 155 000 tons, the "inshore allowance" was set at 110 000 tons, while the offshore fleet was allocated 45 000 tons. For 1981 the Canadian quota was 200 000 tons of which 62 500 tons were allocated to the offshore fleet and 120 000 tons were allocated to the inshore fleet. By 1984, when the Canadian quota was set at 246 000 tons, the inshore allowance remained at 115 000 tons while the larger offshore vessels received 112 000 tons. Thus, apart from the allocations to the Resource-short Plant Program, all the growth from the increased northern cod TACs was accruing to the offshore fishery while the inshore catches remained static and during the mid-1980's declined. The inshore fishery experienced declining catches which fell short of the inshore allowance in spite of a reported increasing cod biomass. The decreasing inshore share of the catch led to a perception that it was incapable of catching its allowance. Later revelations demonstrated that its inability to catch the allowance was more a function of biomass or availability than of any lack of catching capacity in the inshore sector.

## Enterprise Allocations (Offshore)

In the 1981 Groundfish Plan, which allocated 62 500 tons of northern cod to the offshore fleet, it was also stipulated that when 47 500 tons had been taken the directed fishery would close. The remainder of the offshore quota would be used to support a flexible by-catch in the flatfish fisheries. Unusually light ice conditions in the winter of 1981 resulted in a "race for the fish". It became apparent that the offshore quota would be caught by the end of February. Because of a lack of consensus among industry representatives on how to spread out the catch, no action was taken to close the fishery before the quota was reached. The effects of this 1981 "race for the fish" produced a consensus that a new approach was required to avoid the destructive aspects of the common property competition for the greatest share of the available resource. This sparked the introduction on a trial basis of enterprise allocations for the 1982 offshore groundfish fishery. Late in 1983 a 5-year trial Enterprise Allocation Plan was introduced for the offshore fleet (see Parsons 1993). In addition to the five (later two) large offshore companies, the smaller independent companies, the Independent Offshore Group (IOG) were included in the Enterprise Allocation Plan for the 1984–88 period. Separate allocations were set aside for trawlers over 100 feet owned and/or operated by members of the IOG. Initially the IOG allocation was a pool and fished on a competitive basis. Twelve percent of the Canadian offshore quota for northern cod was set aside for the IOG for each of the 5 years.

## Foreign Overfishing on the Nose of the Grand Bank

When the 200-mile fishing limit was declared effective on January 1, 1977, most stocks of fish were contained within Canada's fisheries zone. There were however a number of important stocks which occurred partially outside the 200-mile zone. These straddling stocks later proved to be a thorn in the side of Canadian attempts to successfully manage these stocks (Sullivan 1989). The northern cod (2J3KL) is one of those stocks. A small portion of the 2J3KL cod stock migrates outside the Canadian zone to the Nose of the Grand Bank (Division 3L) in some years. Since 1977 there had been some overfishing of allocations in the earlier years by some countries (Table 2) but generally there were few problems with the exception of Spain and Portugal until about 1985 (Table 3). In the winter and spring of 1985 unusual ice and temperature conditions prevailed in the area and a portion of the 2J3KL stock concentrated on the eastern part of the northern Grand Bank commonly called the "Nose" of the Grand Bank. After fishing its EEC (LTA) allocations in the Canadian zone, the Federal Republic of Germany (FRG) commenced fishing in Division 3L outside the zone. The result was that it overfished its allocation by about 15 000 tons. As a result of bilateral negotiations, the FRG agreed to cease fishing. Canada, at the 1985 annual meeting of NAFO sought to impose a restraint on a directed fishery for 3L cod outside 200 miles (NAFO 1985). Canada indicated that it would restrict its fishermen from a directed fishery in 3L outside the 200-mile zone and that the TAC of 266 000 tons for 2J3KL cod was "fully subscribed" in the Canadian zone in accordance with the scientific advice. The Canadian proposal was accepted with the delegates of EEC, Portugal and Spain declaring their objection to the moratorium on cod in 3L.

Spain, which had been ignoring NAFO quota restrictions since the early 1980's (Table 3), continued to overfish despite the NAFO moratorium. In 1982 it caught 1449 tons in 3L. This escalated to a reported catch of 21 849 tons from 3L in 1986 with continued high catches in 1987 and 1988.

Portugal, which had been given a separate allocation of 2J3KL cod by Canada up to 1985, had also consistently overfished its allocation every year since 1979, although in most years the magnitude of the overrun was not great. Beginning in 1986, the first year for which no allocation was granted, Portugal reported a 2J3KL cod catch of 30 463 tons, of which 29 903 tons were taken in 3L (Table 3). Reported catches in 1987 and 1988 were lower but nevertheless substantial at 13 620 tons and 8 615 tons respectively.

From 1985 through 1990 Canada fought vigorously in NAFO to curtail overfishing of northern cod in 3L beyond 200 miles. It was successful in having the NAFO moratorium extended but this did not limit EEC fishing activities. During 1988 to 1990 the matter was elevated to ministerial level discussion between Canada and the EEC. Following the decline in the northern cod stock and the imposition by Canada of the 2-year moratorium in 1992, within the Canadian zone, the EEC finally agreed, in September 1992, to a NAFO moratorium on fishing cod in 3L beyond 200 miles for 1993. In December 1992 Canada and the EEC reached a memorandum of understanding which, if ratified, would end their 7-year "fish war." See Parsons (1993).

## The Darkening Cloud

Inshore catches of northern cod increased from about 35 000 tons in 1974 to about 96 000 tons in 1980

TABLE 2. Foreign allocations and catches of 2J3KL cod (tons) 1977–1991 by country as reported to NAFO. A = Allocation (T); C = Catch (T).

| COUNTRY | 1977 | | 1978 | | 1979 | | 1980 | | 1981 | | 1982 | | 1983 | | 1984 | |
|---|---|---|---|---|---|---|---|---|---|---|---|---|---|---|---|---|
| | A | C | A | C | A | C | A | C | A | C | A | C | A | C | A | C |
| BULGARIA | 600 | 0 | 0 | 0 | 0 | 0 | 0 | 0 | 0 | 0 | 0 | 0 | 0 | 0 | 0 | 0 |
| CUBA | 1,810 | 18 | 700 | 28 | 1,400 | 1,031 | 700 | 12 | 0 | 0 | 0 | 0 | 0 | 0 | 0 | 0 |
| E.E.C. | 0 | — | 0 | — | 0 | — | 5,500 | — | 0 | — | 8,000 | — | 9,500 | — | 9,500 | — |
| FAROES | 1,690 | 1,196 | 660 | 782 | 700 | 768 | 700 | 1,829 | 700 | 1,812 | 700 | 692 | 1,000 | 989 | 0 | 0 |
| FRANCE | 5,630 | 5,385 | 2,190 | 2,511 | 1,390 | 1,311 | 880 | 696 | 0 | 0 | 1,300 | 1,102 | 1,545 | 0 | 1,545 | 1,205 |
| F.R.G. | 8,030 | 14,661 | 3,120 | 3,103 | 1,980 | 1,894 | 4,320 | 4,469 | 0 | 0 | 6,000 | 5,477 | 7,125 | 1,793 | 7,125 | 7,505 |
| ITALY | 600 | 33 | 200 | 0 | 0 | 0 | 0 | 0 | 0 | 0 | 0 | 0 | 0 | 0 | 0 | 0 |
| U.K. | 1,330 | 1,641 | 520 | 462 | 330 | 458 | 300 | 449 | 0 | 0 | 700 | 851 | 830 | 0 | 830 | 765 |
| G.D.R. | 4,970 | 4,300 | 1,950 | 1,990 | 1,500 | 10,012 | 700 | 509 | 0 | 23 | 0 | 12 | 0 | 29 | 0 | 77 |
| NORWAY | 1,610 | 194 | 630 | 410 | 700 | 1,094 | 700 | 963 | 700 | 799 | 1,000 | 1,070 | 800 | 2,107 | 0 | 996 |
| POLAND | 7,430 | 7,429 | 2,890 | 2,874 | 3,000 | 4,256 | 700 | 733 | 0 | 196 | 0 | 11 | 0 | 792 | 0 | 112 |
| PORTUGAL | 21,100 | 18,695 | 8,210 | 8,027 | 8,000 | 8,283 | 7,000 | 10,127 | 5,000 | 9,685 | 5,700 | 8,690 | 6,500 | 6,962 | 6,700 | 9,749 |
| ROMANIA | 700 | 24 | 250 | 3 | 0 | 0 | 0 | 0 | 0 | 0 | 0 | 0 | 0 | 0 | 0 | 0 |
| SPAIN | 18,760 | 20,725 | 9,330 | 8,609 | 5,000 | 5,996 | 7,000 | 7,014 | 5,000 | 8,636 | 0 | 1,451 | 0 | 5,062 | 0 | 8,159 |
| U.S.S.R. | 18,880 | 18,850 | 7,330 | 7,349 | 1,000 | 1,009 | 700 | 297 | 0 | 114 | 0 | 452 | 0 | 159 | 0 | 776 |
| JAPAN | 0 | 5 | 0 | 34 | 0 | 0 | 0 | 1,132 | 0 | 2,412 | 0 | 2,459 | 0 | 0 | 0 | 470 |
| OTHERS | 0 | 3 | 0 | 0 | 0 | 8 | 0 | 0 | 0 | 0 | 0 | 1 | 0 | 0 | 0 | 0 |
| TOTAL | 93,140 | 93,159 | 38,000 | 36,182 | 25,000 | 36,120 | 20,700 | 28,230 | 11,400 | 23,677 | 15,400 | 22,268 | 17,800 | 17,893 | 16,200 | 29,814 |

| COUNTRY | 1985 | | 1986 | | 1987 | | 1988 | | 1989 | | 1990 | | 1991 | |
|---|---|---|---|---|---|---|---|---|---|---|---|---|---|---|
| | A | C | A | C | A | C | A | C | A | C | A | C | A | C |
| BULGARIA | 0 | 0 | 0 | 0 | 0 | 0 | 0 | 0 | 0 | 0 | 0 | 0 | 0 | 0 |
| CUBA | 0 | 0 | 0 | 0 | 0 | 0 | 0 | 0 | 0 | 0 | 0 | 0 | 0 | 0 |
| E.E.C. | 9,500 | — | 9,500 | — | 9,500 | — | 0 | — | 0 | — | 0 | — | 0 | — |
| FAROES | 0 | 0 | 0 | 0 | 0 | 401 | 0 | 0 | 0 | 0 | 0 | 963 | 0 | 83 |
| FRANCE | 1,545 | 880 | 1,545 | 1,539 | 1,545 | 1,785 | 0 | 0 | 2,950 | 3,099 | 2,262 | 2,315 | 2,140 | 2,197 |
| F.R.G. | 7,125 | 22,684 | 7,125 | 7,313 | 7,125 | 7,463 | 0 | 515 | 0 | 484 | 0 | 1,533 | 0 | 6,459 |
| ITALY | 0 | 0 | 0 | 0 | 0 | 0 | 0 | 0 | 0 | 0 | 0 | 0 | 0 | 0 |
| U.K. | 830 | 0 | 830 | 821 | 830 | 822 | 0 | 0 | 0 | 84 | 0 | 0 | 0 | 24 |
| G.D.R. | 0 | 45 | 0 | 8 | 0 | 32 | 0 | 29 | 0 | 33 | 0 | 0 | 0 | 0 |
| NORWAY | 0 | 935 | 0 | 5,274 | 0 | 0 | 0 | 0 | 0 | 0 | 0 | 0 | 0 | 0 |
| POLAND | 0 | 15 | 0 | 2 | 0 | 1 | 0 | 6 | 0 | 0 | 0 | 0 | 0 | 0 |
| PORTUGAL | 6,700 | 8,813 | 0 | 30,463 | 0 | 13,620 | 0 | 8,615 | 0 | 23,047 | 0 | 12,443 | 0 | 9,459 |
| ROMANIA | 0 | 0 | 0 | 0 | 0 | 0 | 0 | 0 | 0 | 0 | 0 | 0 | 0 | 0 |
| SPAIN | 0 | 10,638 | 0 | 21,849 | 0 | 11,702 | 0 | 17,429 | 0 | 11,979 | 0 | 9,782 | 0 | 6,893 |
| U.S.S.R. | 0 | 125 | 0 | 146 | 0 | 134 | 0 | 68 | 0 | 95 | 0 | 135 | 0 | 3 |
| JAPAN | 0 | 64 | 0 | 171 | 0 | 113 | 0 | 137 | 0 | 10 | 0 | 10 | 0 | 4 |
| OTHERS | 0 | 0 | 0 | 0 | 0 | 2 | 0 | 8 | 0 | 3 | 0 | 623 | 0 | 5 |
| TOTAL | 16,200 | 44,199 | 9,500 | 67,586 | 9,500 | 36,075 | 0 | 26,807 | 2,950 | 38,802 | 2,262 | 27,804 | 2,140 | 25,127 |

EEC Allocation was distributed to France, Federal Republic of Germany and United Kingdom during 1980 and 1982–1987.
1989–91 Figures are provisional.
Catches for France during 1989, 1990 and 1991 are from CAFSAC Res. Docs. 91/53 and 92/75.

TABLE 3. Allocations (t), total catches (t) in 2J3KL and the 3L catches (t) of northern cod for selected countries, 1977–1991 as reported to NAFO.

| | F.R.G. | | | France | | | Norway | | | Portugal | | | Spain | | |
|---|---|---|---|---|---|---|---|---|---|---|---|---|---|---|---|
| Year | Alloc. | 2J3KL | 3L | Alloc. | 2J3KL | 3L | Alloc. | 2J3KL | 3L | Alloc. | 2J3KL | 3L | Alloc. | 2J3KL | 3L |
| 1977 | 8 030 | 14 661 | 0 | 5 630 | 5 385 | 1 757 | 1 610 | 194 | 0 | 21 100 | 18 695 | 7 956 | 18 760 | 20 725 | 15 419 |
| 1978 | 3 120 | 3 103 | 43 | 2 190 | 2 511 | 2 158 | 630 | 410 | 206 | 8 210 | 8 027 | 2 916 | 9 330 | 8 609 | 5 634 |
| 1979 | 1 980 | 1 894 | 1 | 1 390 | 1 311 | 513 | 700 | 1 094 | 52 | 8 000 | 8 283 | 3 389 | 5 000 | 5 996 | 5 996 |
| 1980 | 4 320 | 4 469 | 0 | 880 | 696 | 690 | 700 | 963 | 4 | 7 000 | 10 127 | 5 621 | 7 000 | 7 014 | 5 755 |
| 1981 | 0 | 0 | 0 | 0 | 0 | 0 | 700 | 799 | 0 | 5 000 | 9 685 | 3 429 | 5 000 | 8 636 | 8 636 |
| 1982 | 6 000 | 5 477 | 0 | 1 300 | 1 102 | 314 | 1 000 | 1 070 | 147 | 5 700 | 8 690 | 5 588 | 0 | 1 451 | 1 449 |
| 1983 | 7 125 | 1 793 | 0 | 1 545 | 0 | 0 | 800 | 2 107 | 1 240 | 6 500 | 6 962 | 4 521 | 0 | 5 062 | 5 062 |
| 1984 | 7 125 | 7 505 | 82 | 1 545 | 1 205 | 313 | 0 | 996 | 654 | 6 700 | 9 749 | 6 339 | 0 | 8 159 | 8 159 |
| 1985 | 7 125 | 22 684 | 15 466 | 1 545 | 880 | 0 | 0 | 935 | 0 | 6 700 | 8 813 | 5 035 | 0 | 10 638 | 10 638 |
| 1986 | 7 125 | 7 313 | 0 | 1 545 | 1 539 | 0 | 0 | 5 274 | 2 187 | 0 | 30 463 | 29 903 | 0 | 21 849 | 21 849 |
| 1987 | 7 125 | 7 463 | 0 | 1 545 | 1 785 | 231 | 0 | 0 | 0 | 0 | 13 620 | 13 620 | 0 | 11 702 | 11 702 |
| 1988 | 0 | 515 | 515 | 0 | 0 | 0 | 0 | 0 | 0 | 0 | 8 615 | 8 615 | 0 | 17 429 | 17 429 |
| 1989* | 0 | 484 | 484 | 2 950 | 3 099 | 979 | 0 | 0 | 0 | 0 | 23 047 | 23 047 | 0 | 11 979 | 11 979 |
| 1990* | 0 | 1 533 | 1 533 | 2 262 | 2 315 | 743 | 0 | 0 | 0 | 0 | 12 443 | 12 443 | 0 | 9 782 | 9 782 |
| 1991* | 0 | 6 459 | 6 459 | 2 140 | 2 197 | 1 104 | 0 | 0 | 0 | 0 | 9 459 | 9 459 | 0 | 6 893 | 6 893 |

Catches for France during 1989, 1990 and 1991 are from CAFSAC Res. Docs. 91/53 and 92/75.
*Provisional figures.

(Baird and Bishop 1989), declined to 80 000 tons in 1981 and peaked at 113 000 tons in 1982. After 1982 catches showed a steady decline to 72 000 tons in 1986. The declining inshore catches, occurring at a time when TACs were being increased from 180 000 tons in 1980 to 266 000 tons in 1984, were the object of considerable debate and deep concern to the inshore fishing sector. The offshore fishery had increased its catch from 79 000 tons in 1980 to 179 000 tons in 1986 (an increase of 100 000 tons or about 127%). Inshore fishermen particularly were sceptical that the cod stocks were increasing at the rate at which scientific assessments indicated. That the stock was not increasing at the rate projected earlier was evidenced by a levelling off of TACs at 260 000–266 000 tons during 1983–1986. In 1987 the TAC was lowered to 256 000 tons but raised again to 266 000 tons in 1988.

Strong opposition developed to the escalating offshore catches. This opposition coalesced and resulted in the formation of the Newfoundland Inshore Fisheries Association (NIFA) in August 1986 (Anon. 1987). NIFA's 800 members included fishermen, fish processors, plant workers, business people, regional development associations, concerned citizens and other interest groups who had an interest in the inshore fishery. NIFA sponsored a report on the 2J3KL cod stock which questioned various aspects of the CAFSAC assessments (Keats et al. 1986).

There was a very intense public debate and concern was growing on the part of the industry (FANL 1986). A series of questions were posed to CAFSAC by Departmental managers in October, 1986. CAFSAC, in November 1986, provided responses to these specific questions. The most recent assessment (June 1986 at NAFO) (Baird and Bishop 1986) had indicated that the age 4+ annual mean biomass had increased by a factor of 3 (386 to 1228 thousand tons) while the age 7+ spawning biomass increased by a factor of 7 (86 to 582 thousand tons) from 1977 to 1985. However, it was noted that the retrospective $F_{0.1}$ catches based upon revised estimates of annual population size were considerably lower than the $F_{0.1}$ catches advised at the time of the actual assessments (CAFSAC 1986). In other words, the catch levels advised for this and a number of other stocks had tended to be rather optimistic.

Several factors were identified as possibly contributing to the low catches in 1985 and 1986:

> "The decline in 1985 in inshore catches was likely due mainly to severe environmental conditions since the biomass of the stock has apparently not declined in 1985 or 1986. There was also an indication that effort in the gillnet fishery declined in 1985 and thus

the catch by this sector was reduced. Capelin of the very abundant 1983 year-class were mostly immature in 1985 and their presence offshore may have reduced the intensity of the inshore migration of cod.

"In 1986, environmental conditions were close to normal. Capelin of the 1983 year-class were 3 years old in 1986 and would normally be almost entirely mature. In fact, fewer than 40% were mature, and remaining in the offshore area, would again form a substantial source of food. The inshore migration may have thus been reduced. In 1986 also, there was a substantial increase in the number of capelin licences and some increase in the number of crab licences as well. Increased participation in these fisheries probably diverted fishing effort from the fixed gear cod fishery."

During 1984–86 the catches by the offshore sector had tended to occur in the two southern Divisions. Tagging studies on winter concentrations had shown that cod tended to return to distinct areas to spawn and the concentration of effort in one or more years was potentially liable to cause over-exploitation of localized stock components (Lear 1984).

To protect against such possible overexploitation of localized spawning components, CAFSAC advised that catches offshore be taken equally in all three Divisions. Spawning components of cod in Division 3K appeared to contribute more substantially to inshore catches than spawning components in the other two Divisions. CAFSAC therefore advised, that no more than one third of the offshore catch in any one year be taken in Division 3K. This recommendation was adopted for the 1987 offshore fishery.

## TGNIF-1987

The prognosis for northern cod arising out of the 1987 CAFSAC assessment was optimistic. The catch in 1988 that would approximate the $F_{0.1}$ level was estimated to be 293 000 tons on the assumption that the 1987 TAC (256 000 tons) was caught. This upward revision of the estimated size of the stock was largely due to the increase in estimated sizes of the 1980–82 year-classes (CAFSAC 1987). In view of the low inshore landings in 1986 (at 72 000 tons, the lowest since 1977), inshore fishermen became extremely concerned that the TAC might be increased in spite of declining inshore catches. During the summer of 1987 inshore catches were poor.

On August 4, 1987, Fisheries Minister Tom Siddon, announced the establishment of a Task Group to examine the question of the decline in catches since 1982 in the inshore cod fishery in eastern Newfoundland and Labrador (DFO 1987a). The independent task group was composed of five prominent scientists led by Dr. Dayton L. Alverson, former Director of the Northwest and Alaska Fisheries Centre, Seattle.

The Task Group on the Newfoundland Inshore Fishery was requested to investigate the nature of and reasons for the recent decline in inshore catches. The Task Group was also requested to advise whether the observed decline was more than could be expected as a natural fluctuation, whether the causal factors were controllable by fishery management measures, and if so, what measures should be adopted to reverse the decline. The results of the TGNIF report were released on November 30, 1987 (DFO 1987b).

The report attributed the decline in recent inshore cod catches in the northern cod stock to a combination of several factors:

1. changes in availability resulting from predator-prey and/or environmental relationships, coupled with a slower rebuilding of the major stocks;
2. uneven distribution of fishing on stocks or components of offshore stocks migrating to inshore fishing grounds;
3. potential over-fishing in the southern areas by the inshore fishermen of separate inshore stocks of cod;
4. redeployment of effort to other target species;
5. possible effects of fishing on recruitment; and
6. a slower growth rate of individual cod.

The conclusions of the Task Group and CAFSAC were quite similar (Fig. 8) with respect to the current stock size and the causes for the decline in the inshore fishery since 1982 (DFO 1988b). The Task Group estimated a 5-fold increase in stock size from 1976–86 whereas CAFSAC scientists had estimated a 5.5 fold increase over the same period (DFO 1987c). These studies indicated that the northern cod was continuing to increase but perhaps not as fast as projected several years earlier. The Task Group recommended a cautious approach to managing the stock to protect the inshore fishery. The 1988 TAC was set at 266 000 tons, which was 27 000 tons below the calculated $F_{0.1}$ level of 293 000 tons.

The Task Group noted the beneficial effect of the initiative in 1987 to distribute fishing more evenly on the major stock components and to increase observer coverage on offshore trawlers (Task Group on Newfoundland Inshore Fisheries 1987). One new possibility introduced in the Task Group report was the potential over-fishing of inshore stock components by inshore fishermen in southeastern Newfoundland.

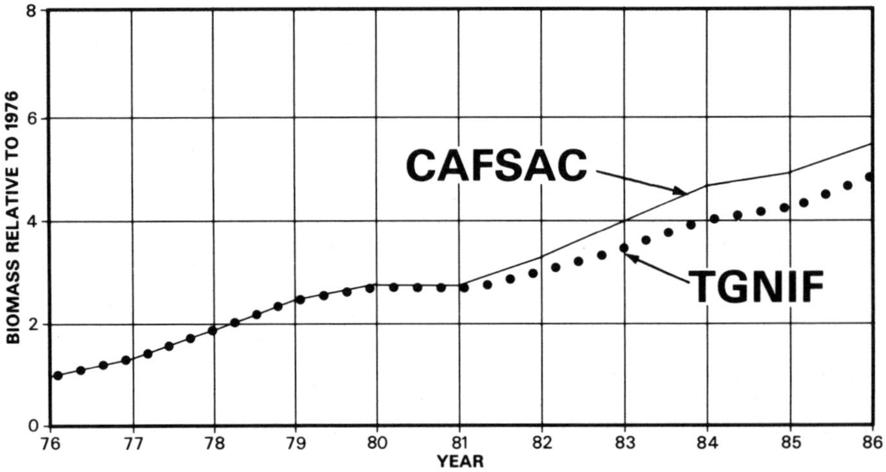


FIG. 8. The conclusions of the Task Group on the Newfoundland Inshore Fishery (TGNIF) and of Canadian Scientists (CAFSAC) about the growth and size of the northern cod stock based upon assessments performed in 1987.

The total offshore catch for 1988 remained the same as it did for 1987 at about 156 000 tons while the inshore catch increased from about 79 000 tons in 1987 to about 102 000 tons in 1988 (Baird and Bishop 1989). Things appeared to bode well for the northern cod stock.

### A Dramatic Change in the Scientific Advice

Assessments prior to 1989 had indicated a northern cod stock that was increasing from the low levels of the 1970's. CAFSAC in 1987 (CAFSAC 1987) had estimated the catch in 1988 that would approximate the $F_{0.1}$ level to be 293 000 tons, up from the 266 000 tons for 1987 advised by NAFO Scientific Council and CAFSAC (CAFSAC 1986). TGNIF generally agreed with this CAFSAC advice but estimated fishing mortalities somewhat higher than CAFSAC. It concluded that the stock was increasing, but at a slower rate than that indicated by CAFSAC (Task Group on Newfoundland Inshore Fisheries 1987). CAFSAC postponed the assessment of 2J3KL cod in the spring of 1988 in order to address, as fully as possible, all recommendations regarding the assessment process that were made by the TGNIF. In the fall of 1988 the assessment of the stock was postponed still further.

In January 1989, CAFSAC reassessed the status of cod in Divisions 2J3KL (CAFSAC 1989). The resulting assessment was a bombshell that shattered the expectations that had been generated within the fishing industry in Newfoundland and to a great degree within Atlantic Canada since the extension of jurisdiction in 1977. The scientific perspective of this stock abruptly changed from one of optimism (an $F_{0.1}$ catch of 293 000 tons for 1988) to a drastically lower $F_{0.1}$ catch of 125 000 tons for 1989. CAFSAC concluded that previous assessments had underestimated fishing mortalities and that mortalities were about double those estimated earlier. The catch of 258 000 tons in 1988, for example, was now considered to have exerted a relatively high fishing mortality ($F$=0.44), more than double the $F_{0.1}$ level.

The drastic change in the advice was due to the use of a much higher level for the estimate of fishing mortality in the latest fishing year. The addition of data for 2 more years (1987–88) indicated that the 1986 research vessel survey index was unrealistically high and also that the reduction in the commercial catch rate in 1986, which had previously been thought to be a "blip" in the data, was real. The higher fishing mortality in the new assessment was supported by the consistency in the research survey data of the overall pattern of decline in the numbers of fish in each year-class as they became older and were harvested or died due to natural causes. Retrospective analysis had indicated levels of fishing mortality in the early 1980's (1981–84) of around 0.4, regardless of the assumed mortality in the most recent years.

Coupled with the foregoing factors was a lower level of recruitment in the most recent years. The period of relatively strong year-classes (1978–82) appeared to have ended and the 1983 and 1984 year-classes were among the smallest on record.

Recruitment (at age 3) to this stock appeared to have declined precipitously from 880–1195 million for the 1961–1963 year-classes to 137–140 million for the 1970–1971 year-classes (Fig. 9). These earlier

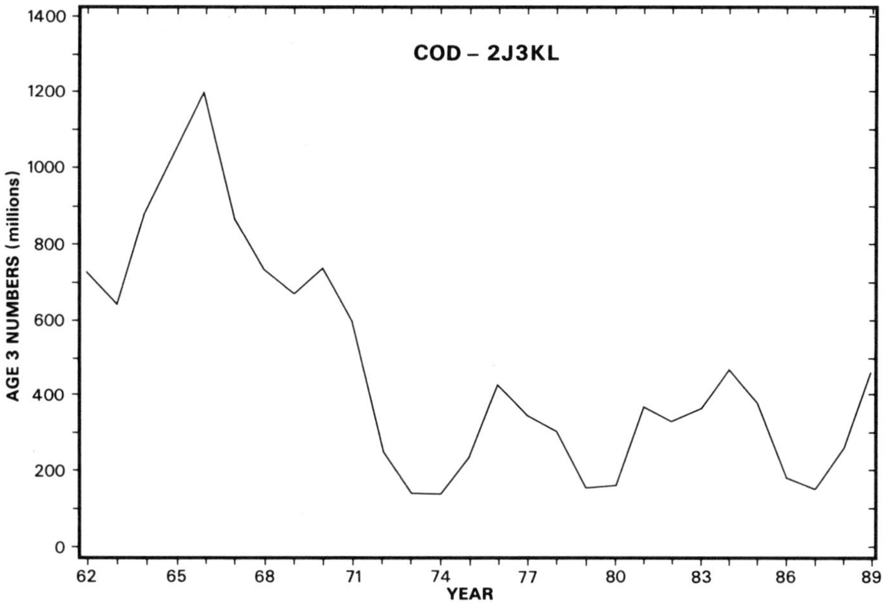


FIG. 9. Recruitment (at age 3) to the northern cod stock (1962–1989) (Baird et al. 1990).

estimates are based upon foreign catch data, which are suspect. It is possible that recruitment during this period had been overestimated. Recruitment improved somewhat for the 1973–1975 year-classes (303–418 million) but was poor for the 1976 and 1977 year-classes. A series of fairly good year-classes followed during 1978–1982 (340–539 million). The 1983 and 1984 year-classes were considered, on the basis of survey data and commercial catches to date, to have numbered around 250 million at age 3, among the smallest on record and prospects for large year-classes in the immediate future did not appear good at the time of the 1989 assessment (CAFSAC 1989).

On February 8, 1989 Fisheries Minister Siddon announced a revised provisional TAC for northern cod (2J3KL) of 235 000 tons, a reduction of 31 000 tons from the 266 000 tons provisional TAC set on December 30, 1988, when the Atlantic Groundfish Management Plan was announced. The Minister also announced his intent to establish an independent northern cod review panel (DFO 1989a).

On February 12, 1989 the Minister announced that the traditional inshore sector was to retain its allowance of 115 000 tons with the entire 31 000 tons reduction in the TAC to come from the other fleet sectors, primarily the offshore fleet. The Minister also announced the terms of reference and the membership of the independent panel to review the scientific advice (DFO 1989b).

The chairman of the Northern Cod Review Panel was Dr. Leslie Harris, President and Vice-Chancellor of Memorial University of Newfoundland.

On May 15, 1989 an interim Report of the Independent Review of the State of the Northern Cod stock was submitted to Minister Siddon.

The Panel concluded that the most recent CAFSAC advice was in the right domain. It believed that a fishing mortality in the region of 0.45 was correct but that it could lie somewhere in the range of 0.35 to 0.55. The Panel members were inclined to think that the mortality was more likely to be at the higher end of the range than the lower (Harris 1989).

The Panel recommended that for 1990 the fishing mortality should be reduced to a value half way between the current level of 0.45 and the $F_{0.1}$ level of 0.20, i.e. to approximately 0.32 which implied a TAC of 190 000 tons for 1990. The consequences of fishing at this level should be closely monitored and there should be further reductions in subsequent years if there was no evidence of spawning stock recovery.

On January 2, 1990 Minister Siddon announced that the Canadian Northern Cod quota would be reduced from 235 000 tons in 1989 to 197 000 tons in 1990 (DFO 1990a).

The socioeconomic impacts of these reductions in the northern cod quota were far-reaching. On January 5, 1990 Fishery Products International announced that as a result of reductions in its northern cod quota and further quota reductions for 3NO cod and 3LNO flatfish, the company would be catching about 15 000 tons less fish in 1990 than in 1989. It announced that 13 trawlers were to be retired from the fishery and trawlers were to be reallocated and realigned. The processing plants at

Gaultois, Grand Bank and Trepassey were to be closed in 1990. They were kept open in 1991 with funding from the provincial government and were to be permanently closed at the end of their operating period in 1991. This would result in a projected loss of about 1300 jobs involving both plant workers and trawlermen.

Quota reductions in 1988–1990 had resulted in National Sea Products closing major fish plants in Nova Scotia and Newfoundland, with a resultant loss of employment involving more than 4000 workers.

On March 30, 1990 Bernard Valcourt, the new Minister of Fisheries and Oceans, released the final report of the Independent Review Panel on Northern Cod. The Panel made a total of 29 recommendations regarding northern cod under several broad headings (Harris 1990). The Minister accepted the basic principles of the report as well as its major recommendations aimed at rebuilding fish stocks through intensified scientific research, and conservation management measures as well as direct communications with fishermen and the fishing industry.

Under 'Management Actions', the Panel recommended an immediate reduction of fishing mortality to 0.30 and as early as possible to 0.20, a limit on fishing mortality during the spawning season, a reduction of the harvest of young cod by various gears and of the by-catch of young fish, and distributing fishing effort in divisions 2J, 3K and 3L relative to the distribution of the exploitable biomass. At the international level, it recommended that Canada seek to manage all fish stocks indigenous to the Canadian Continental Shelf and extending beyond the 200 mile zone and reexamine policies regarding foreign fisheries with the intention of eliminating any catches of cod. It also recommended that the stock allocations take into account the principles of contiguity and the "vital needs" of communities.

The Panel recommended several initiatives for scientific research including:

- the development of a CPUE index independent of research vessel and large trawler CPUE data,
- expanded efforts to understand the integrity and interrelationships of spawning aggregations,
- examination of the stock-recruitment relationship,
- analysis of by-catch of cod in other fisheries not targeted on cod,
- increased research vessel sampling,
- conduct of a seal survey,
- the integration of information from other biological and ocean disciplines into the assessment process,
- the study of seal feeding patterns and consumption rates,
- the determination of interactions among cod, capelin and seals and,
- expansion of the collection of catch and effort data especially from the inshore fisheries.

The Panel advocated the improvement and expansion of relevant technologies in the annual assessment process and in management activities and the use of advanced technology for surveillance purposes. The Panel also recommended that the Government of Canada carefully reexamine its biological, ecological and socioeconomic goals in respect of the fisheries.

Several recommendations were presented emphasizing the need for DFO to review its management approaches, expand the observer program and increase surveillance. The Panel recommended a joint management board involving the Governments of Canada and Newfoundland for information sharing, classifying objectives and coordination of policy directions. The Panel urged that penalties for infringement of regulations be made more onerous. Development of an educational program and improved lines of communication with clients, and a regular appraisal of research activities and the assessment process were also proposed.

The Harris Report confirmed the CAFSAC assessment (Baird and Bishop 1989) that the total population (ages 3 and older) and total biomass had been declining since 1984. The spawning biomass of northern cod appeared to have increased from 1984 to 1988. The Panel agreed with the CAFSAC assessment that the fully recruited fishing mortalities since 1982 had been about 0.45. It agreed that the more recent trend in recruitment was definitely downward.

### Atlantic Fisheries Adjustment Program

As a result of the social and economic dislocation caused by the reduction in the quotas of northern cod and other Atlantic coast quota reductions, and overcapacity problems in other areas of the Atlantic groundfish fishery, the federal government announced an Atlantic Fisheries Adjustment Program on May 7, 1990. This program consisted of three elements aimed at ensuring a viable fishery in the long-term for Atlantic Canadians, while supporting individuals and communities in the fishery to adjust to the realities of declining fish stocks and plant closures.

These three elements were:

— Rebuilding the Fish Stocks ($150 million)
— Adjusting to Current Realities ($130 million)
— Economic Diversification ($146 million)

The Rebuilding the Fish Stocks element included expanded research on northern cod, other cod stocks off southwest Newfoundland and in the Gulf of St. Lawrence, Scotia-Fundy groundfish and Gulf snow crab. There was to be a greater focus on the ecology of cod in relation to its ocean environment and the impact of trawling on spawning grounds and nursery areas. Greater involvement of fishermen in DFO scientific research was emphasized. Research on harp and grey seals was to be increased. Surveillance and Enforcement activities were to be expanded to ensure compliance with fisheries regulations. Increased penalties to deter violators were to be pursued through Amendments to the Fisheries Act (Government of Canada 1990).

At the same time the government responded to the Harris Report. The government accepted all but 3 of the 29 recommendations. The three recommendations not accepted were:

- a further reduction in the 1990 TAC. This had been rejected on March 30, 1990 by Fisheries Minister Valcourt because of the additional hardships it would produce in 1990, following TAC reductions previously announced.
- a new joint management board involving Canada and Newfoundland. This was rejected on the grounds that existing mechanisms already provided for the requisite cooperation.
- unilateral action by Canada to acquire management rights over straddling stocks beyond the 200 mile limit. This recommendation was rejected on the grounds that it was incompatible with the Law of the Sea Convention.

On June 7, 1990, Minister Valcourt requested the Director General of the Newfoundland Region, Mr. E.B. Dunne, to lead a task force of officials to carry out consultations with interested groups. They were to develop recommendations on how best to implement recommendations of the Harris Report dealing with fisheries conservation, management and harvesting, surveillance and enforcement, communications, and the licensing of inshore fishermen and vessels. The Report of the Task Force released on October 24, 1990 recommended a number of goals as a minimum starting point for further discussions with the industry (Dunne 1990). Regarding the level of TAC the Task Force concluded that a decision was required whether the short-term objective was: (1) to continue the status quo and keep the biomass constant, (2) to start rebuilding by lowering catches, or (3) to rely solely on future recruitment to improve stock status.

The Task Force concluded that, as a minimum, age 3+ biomass should be rebuilt to 1 000 000 tons by 1994 and 1 300 000 tons by 2000 while the spawning stock biomass should be rebuilt to 450 000 tons by 1994 and to 650 000 tons by 2000. This would necessitate some reduction in current catch levels. The Task Force recommended that multi-year TACs be used to reach $F_{0.1}$ and that multi-year plans be continued thereafter to maintain stable catches and reduce pressure for major upward revisions. The ultimate goal was a larger stock with more older fish.

### Multi-Year Groundfish Plan

On December 14, 1990 Fisheries and Oceans Minister Valcourt announced a new Multi-Year Groundfish Plan (DFO 1990b). He indicated that the new multi-year approach to managing the groundfish fishery was being introduced to allow for better planning by Atlantic Canada's groundfish industry and allow Canadian exporters and their customers to develop long-term planning as fish stocks were being rebuilt. The TAC for northern cod was set at 190 000 tons for 1991, 185 000 tons for 1992 and 180 000 tons for 1993. The inshore allowance was to remain constant at 115 000 tons for the 3-year period. The allocation of northern cod to the Resource-short Plant Program was eliminated. A period of relative stability in TACs was proposed in the expectation that the stock would rebuild. This rebuilding was expected to occur very slowly since the fishing mortalities estimated at about 0.45 or higher were far in excess of the $F_{0.1}$ value of 0.25 and were approaching the $F_{max}$ value of 0.52 (Baird et al. 1990). Hopes for the stock rebuilding were based upon a forecast that the 1986 year-class would be above average strength.

In 1991 CAFSAC concluded that the most recent assessment on 2J3KL cod indicated that fully recruited (ages 7–9) fishing mortality in 1990 was about 0.52 (Baird et al. 1991). The age 3+ biomass at the beginning of 1990 was estimated to be about one million tons. About 47% of the biomass and over 70% of the population numbers in 1990 consisted of the 1986 and 1987 year-classes.

### Shattered Expectations

Early in 1992 hopes for a quick rebuilding were dashed when CAFSAC advised that the biomass of cod aged 3 and older had declined to 800 000 tons, mostly because of the disappearance of older fish (CAFSAC 1992a). This indicated that under the 3-year plan fishing mortality would remain high. The spawning biomass for 1991 was estimated at 130 000 tons, amongst the lowest observed. CAFSAC expressed concern about the low estimate of the adult stock biomass. It advised that 1992 catches for the first half of the year should be restricted to the lowest level possible, to about 50% of catches during the first half of 1991.

In response, Fisheries Minister John Crosbie slashed the Canadian northern cod TAC to 120 000 tons. The offshore quota was reduced from approximately 72 000 tons in 1991 to 20 000 tons for the first six months of 1992. The implications of this reduction sent shock waves through the Atlantic fishing industry.

These shock waves intensified when new scientific advice was received in June and July 1992. Because of EC failure to curtail its fishery for cod on the Nose of the Grand Bank, Canada requested the NAFO Scientific Council to review the status of this stock. The Scientific Council at a Special Session, in June 1992, confirmed that the biomass of northern cod had declined sharply during 1991. It concluded that the stock was at a very low level. The age 3+ biomass at the beginning of 1991, between 520 000 and 640 000 tons, and the 7+ biomass (approximately the spawning stock biomass–SSB), between 72 000 and 110 000 tons, were at, or approaching, the lowest levels ever observed for this stock. It expressed concern about the low level of the SSB and recommended that fishing mortality should be reduced in 1992 from the level of recent years. While calculating a range of possible $F_{0.1}$ catches under various assumptions and using different mathematical models, the Council suggested it would be wise to consider the $F_{0.1}$ catch for 1992 to be at the lower value of 50 000 tons (NAFO 1992).

Following receipt of the NAFO Scientific Council Report, Fisheries Minister Crosbie called upon all countries to observe the NAFO moratorium on northern cod outside the Canadian 200-mile zone. The European Community responded by announcing that its vessels would cease fishing for northern cod, at least until the fall of 1992.

At the beginning of July 1992, CAFSAC released its final advice on the northern cod stock for 1992. CAFSAC concluded that the abundance of cod had declined drastically in the previous 12–18 months. It estimated the age 3+ biomass at the beginning of January 1992 to be between 527 000 and 696 000 tons and the age 7+ biomass to be between 48 000 and 108 000 tons, about 10% of the long-term average. All indications were that the stock had recently declined abruptly and was at, or near, its lowest observed abundance. CAFSAC advised the 1992 catches be kept to the lowest level possible (CAFSAC 1992b).

The exact cause of the sudden, drastic and unexpected decline during 1991 was not clear. Water temperatures off Labrador and northeastern Newfoundland were considerably below normal in 1991. The ice coverage was greater than normal and persisted for a record length of time, with ice present inshore into what would normally have been summer. Low ocean temperatures persisted throughout summer and early autumn and the cold intermediate layer of water less than 0° C of the Labrador Current was at, or near, its greatest size (CAFSAC 1992b).

The entire stock was essentially composed of the 1986 and 1987 year-classes (5 and 6 year-olds in 1992). The following four year-classes appeared to be weak. Fisheries in the next few years would depend heavily on the 1986 and 1987 year-classes and would exploit them before they made their full contribution to the spawning stock. CAFSAC recommended that all efforts be made "to enhance the possibilities of stock rebuilding."

The Federal Government reacted swiftly to the new scientific advice. On July 2, 1992, Fisheries Minister John Crosbie imposed a 2-year moratorium on the northern cod fishery to continue until the spring of 1994. The rationale was that, since 1990, a devastating decline in the northern cod stock, due primarily to ecological factors, had reduced the biomass by half and the spawning biomass by three-quarters. A 2-year moratorium appeared to offer the only chance for the spawning biomass to recover quickly to its long-term average, permitting resumption of an inshore fishery in the spring of 1994 (DFO 1992a). The moratorium affected 19 000 people directly dependent on this fishery. Although such an action would have been unthinable previously, the moratorium was generally considered necessary by industry participants, including many inshore fishermen.

Inshore fishermen, however, condemned the Federal Government's initial emergency assistance payments of $225 a week to fishermen and plant workers affected by the moratorium. On July 17, 1992, Minister Crosbie announced an increase in compensation and a comprehensive adjustment program for northern cod fishermen and plant workers. New income replacement benefits, to a maximum of $406 per week with a minimum of $225 per week, took effect on August 1, 1992. The income replacement payments would be based, in most cases, on individuals' average unemployment insurance benefits over the previous 3 years (DFO 1992b).

Between August 1 and December 31, 1992, fishermen and plant workers would be asked to choose among a variety of alternative skills development programs. These included training for work outside the fishery as well as professionalization and certification within the fishery. Those who opted for these programs would continue to receive the full income replacement payments up to the maximum of $406 per week throughout the northern cod moratorium to the spring of 1994. Eligible individuals who chose not to enter one of these programs would revert to basic payments of $225 a week for the remainder of the moratorium. Older fishermen could choose early retirement through extension of an existing program for plant workers and trawlermen between the ages of 50 and 65.

The program also included a plan to retire licences from fishermen choosing to leave the fishery. This option would be available to the 4 000 active and 2 000 inactive groundfish licence holders in the northern cod fishery. Minister Crosbie indicated that, when the moratorium ended, priority of access to the resource would be given to full-time, professional fishermen.

With respect to the processing sector, Minister Crosbie stated that it was up to the Newfoundland Government, which has jurisdiction over processing plants, to decide the level of processing capacity and the distribution of fish plants. However, he indicated that the Federal Government "believes that capacity should be matched to thc limits of the resource and be distributed along the coast in a way that provides reasonable access by fishermen for the sale of their catches."

The stated goals of the Northern Cod Recovery Plan were as follows:

1. Protection and rebuilding of the northern cod stock;
2. A sustainable fishery for that resource once the moratorium ended;
3. A professional work force of fishermen and plant workers deriving adequate incomes from harvesting and processing northern cod;
4. Processing and harvesting capacity matched to the limits of the resource; and
5. Processing plants distributed along the coast to provide reasonable access to fishermen to sell what they harvest.

In response to questions from the media, Minister Crosbie indicated that he viewed the northern cod crisis as "an opportunity as well as a tragedy." He also affirmed his belief in the importance of the inshore fishery "which has priority in the fishery scheme of things." He also stated that the historic share for the inshore sector must be maintained.

According to Minister Crosbie, the cost of the income replacement and adjustment program would be well in excess of $500 million. Various sources estimated the likely cost at $750–800 million.

The social and economic implications of the 2-year moratorium were profound for the Newfoundland economy. Few would have believed that the bright expectations of a prosperous future raised by the extension of fisheries jurisdiction to 200 miles in 1977 could have been shattered to such an extent. The people and communities dependent on northern cod had weathered many previous crises, but none of the magnitude they now faced. Few doubted that the fishery in the future would be substantially different from that of recent decades.

## Conclusions

The degree of dependence of Newfoundland on the northern cod resource is of such magnitude that the wise management of this resource is imperative for the future well-being of Newfoundland communities. The overexploitation of the northern cod stocks in the 1960's and the resultant decline in the biomass and the inshore catches in the 1970's caused great hardship in fishing communities in eastern Newfoundland and Labrador. The imposition of TACs in 1973 did nothing to alleviate the stock decline as the TACs were set in excess of the maximum sustainable yield. Only when TACs were reduced to a more conservative level during 1977 and 1978 was the stock decline arrested. The 3+ biomass increased from about 500 000 tons in 1976 to about 1.2 million tons in 1984 when it levelled off and declined again to about 850 000 tons in 1989. This biomass level falls far short of that capable of sustaining inshore catches similar to historical levels of 200 000–300 000 tons.

The sudden, drastic and unexpected decline in the abundance of northern cod during 1991 reduced the total biomass at the beginning of 1992 by half and the spawning biomass by three-quarters to a very low level of 48 000–108 000 tons, at or near the lowest level ever observed (Fig. 10). This very low level of spawning stock biomass (only 10% of the long-term average between 1962 and 1991) generated conservation concerns about the future of the stock. The Federal Government reacted swiftly by introducing a 2-year moratorium on fishing for northern cod, beginning in July 1992. The hope was that the spawning stock could be rebuilt rapidly to the long-term average by protecting the 1986 and 1987 year-classes from exploitation for 2 years. This, the most dramatic conservation action in Canadian marine fisheries management history, represented a stark contrast to the expectations generated following Canada's extended jurisdiction in 1977 and, particularly the Task Force on Atlantic Fisheries projections of 1983 which anticipated catches of 380 000 tons by 1987.

The scientific reassessment of stock status in 1989 came as a considerable shock to the fisheries system. Yet the variances in the earlier estimates of fishing mortality, biomass and TACs and the retrospective estimates of these parameters were not surprising in view of the inadequacies of the data base and the enormous variation around research vessel estimates, arising from environmental factors and the behaviour of fish. The

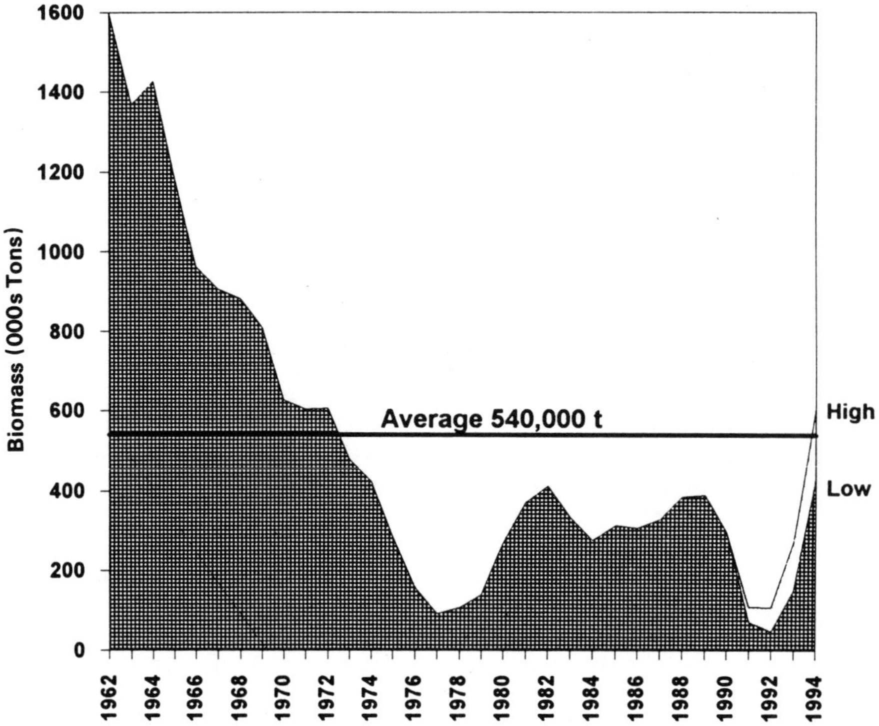


FIG. 10. Trends in the spawning stock biomass of northern cod from 1962 to 1992 with projections of the impact of the 2-year moratorium on spawning biomass in 1993 and 1994 (DFO 1992a).

sheer enormity of the task of doing anything resembling an accurate assessment is highlighted by the lack of a research vessel survey in NAFO Divisions 2J and 3K prior to 1977. Thus assessment scientists were hamstrung by a lack of a research vessel index and had to rely on offshore catch rates with all the inherent possibilities of bias arising from the development of a new Canadian offshore fishery in 2J and 3K from 1977 onward.

The problems in accurately estimating stock abundance are not unique to northern cod. The history of the management of fish stocks throughout the world is replete with examples of variance in estimates of equal proportion (see Parsons 1993). In the context of attempts to rebuild stocks or even of maintaining stocks at status quo levels, the international record contains many dismal examples of fishing mortalities increasing or remaining at high levels while the biomass declines. Had Canada been managing for $F_{max}$ instead of the more conservative $F_{0.1}$ level of fishing mortality, the margin of error may have been greater.

The difficulties encountered in managing the northern cod resource underscore the uncertainties inherent in fisheries management. They exemplify the great difficulty in long-term planning when dealing with a naturally variable resource. Attempts to achieve economic stability in this situation have been frustrating. The stochastic nature of fish population abundance does not lend itself readily to linear models. The effects of large-scale environmental perturbations such as that which affected northern cod in 1991 can confound attempts to manage marine fish resources.

Perhaps one lesson which can be distilled from the experience of northern cod management is the necessity for a cautious, conservation-oriented approach whereby the stock is allowed to grow and the harvest is controlled at a low fishing mortality (e.g., $F_{0.1}$) level, in conjunction with other targets such as minimum spawning biomass constraints. The consistent pattern of natural stock variability in Canada and worldwide should serve as a grim reminder that we cannot will stability into a naturally chaotic system by imposing arbitrary regulatory mechanisms. If this resource is to serve the best interests of those who depend upon it for their livelihood, there has to be a shift in our perception of how we utilize natural resources. Expectations for the level of catch which can be sustained by the northern cod resource will have to be lowered. This will necessitate a reduction in the number of fishermen and plant workers dependent on this resource. The shape of the future fishery is unclear.

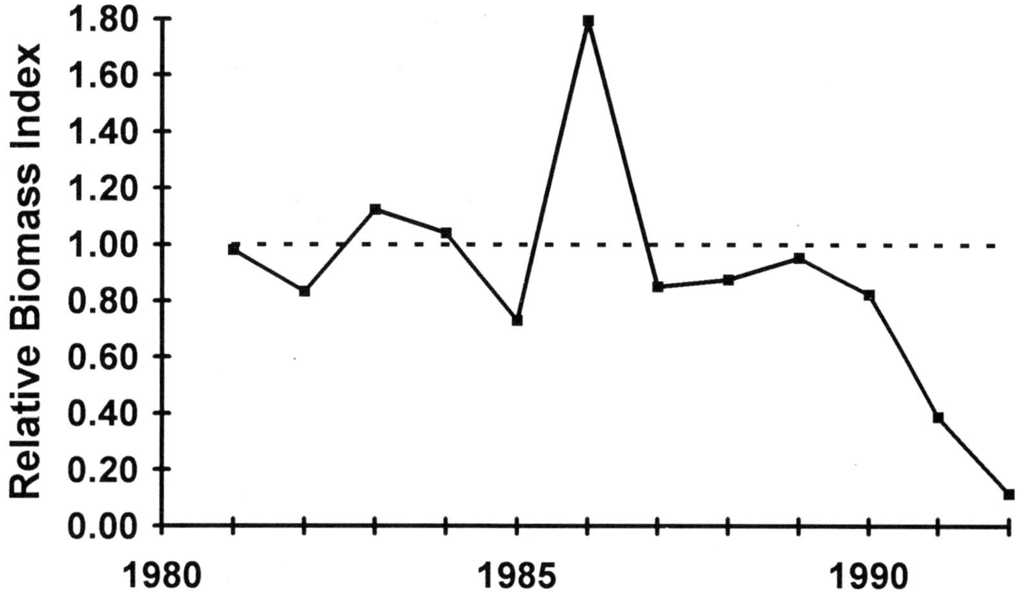


FIG. 11. Relative biomass index for northern cod based upon research vessel surveys during 1981 to 1992. (----) represents the average from 1981 to 1990.

Indeed, at the time this book was going to press, the most recent survey information showed no signs of improvement. Survey indices had indicated a continuing decline in resource abundance during 1992 (Fig. 11).

In July 1993, DFO scientists, in a report on the Status of Groundfish Stocks in the Canadian Northwest Atlantic, concluded that the stock was at its lowest level of abundance in the 20th Century. The spawning biomass was estimated to have declined to 22,000 tons and incoming year-classes appeared to be weak.

Given the current state of the stock, and considering that fish normally mature at about age 7, recovery of the spawning stock appears unlikely before the end of the 20th Century. The implications of this for the communities and people of Newfoundland and Labrador who have depended on this resource for centuries will be catastrophic.

### Acknowledgements

We thank A.T. Pinhorn and W.J. Bruce for their helpful suggestions for improving the manuscript. Art Read prepared the illustrations and Ronnie Sanford typed various drafts of the manuscript.

### References

ANON. 1943. Codfishery regulations. Approved 5th February 1943. Published by authority King's Printer, St. John's, Nfld. 52 p.

1979. The challenge has changed. Fogo Island Fisheries Conference, Fogo Island, Newfoundland, January 16, 17, and 18, 1979. Report prepared by Extension Service, Memorial University of Newfoundland. 73 p.

1987. The Newfoundland Inshore Fisheries Association Looks Ahead. Decks Awash 16(No. 2): 67–69.

BAIRD, J.W., AND C.A. BISHOP. 1986. Assessment of the cod stock in NAFO Divisions 2J&3KL. NAFO SCR Doc. 86/47: 50 p.

1989. Assessment of the cod stock in NAFO Divisions 2J and 3KL. CAFSAC Res. Doc. 89/6. 102 p.

BAIRD, J.W., C.A. BISHOP, AND W.A. BRODIE. 1990. The assessment of the cod stock in NAFO Divisions 2J, 3K, and 3L. CAFSAC Res. Doc, 90/18: 89 p.

BAIRD, J.W., C.A. BISHOP, AND E.L. MURPHY. 1991. An assessment of the cod stock in NAFO Divisions 2J3KL. CAFSAC Res. Doc. 91/53: 84 p.

BLACK, W.A. 1960. The Labrador floater codfishery. Ann. Assoc. Am. Geogr. 50(3): 267–295.

CAFSAC. 1986. Advice on the status and management of the cod stock in NAFO Divisions 2J, 3K and 3L. CAFSAC Adv. Doc. 86/25: 32 p.

1987. Advice on the management of cod in Divisions 2J3KL. CAFSAC Adv. Doc. 87/17: 12 p.

1989. Advice for 1989 on the management of cod in Divisions 2J3KL. CAFSAC Adv. Doc. 89/1: 14 p.

1992a. Advice on the management of cod in Divisions 2J3KL. CAFSAC Res. Doc. 92/2. 3pp.

1992b. Advice on the management of groundfish stocks. CAFSAC Adv. Doc. 92/7.

CROSS, T.F., AND R.H. PAYNE. 1978. Geographic variation in Atlantic cod, *Gadus morhua*, off eastern North America: a biochemical systematics approach. J. Fish. Res. Board Can. 35: 117–123.

DAVIS, J.E. 1986. Economic analysis of the 1984 Resource Short Plant Program, Newfoundland Region. Econ. Branch, Newfoundland Region. 29 p. + app.

DFO. 1979a. Final Report, Northern Cod Seminar, Glynmill Inn, Corner Brook, Newfoundland, August 28–30, 1979. Prepared by Conference Secretariat, Department of Fisheries and Oceans, Ottawa. 214 p.

1979b. Toward a policy for the utilization of northern cod: a discussion paper. Dept. of Fisheries and Oceans, Sept. 28, 1979. 30 p.

1987a. Task Group on the Newfoundland Inshore Fishery. DFO News Release, August 4, 1987. NR-HQ-87-86E. 4 p.

1987b. Minister releases results on Task Group Report. DFO News Release, November 30, 1987. NR-HQ-87-105E. 3 p.

1987c. Minister comments on Task Group on the Newfoundland Fisheries. DFO News Release, December 30, 1987. NR-HQ-87-110E. 2 p.

1988a. Resource Short Plant Program (RSPP) Newfoundland Region — 1988. Fish and Habitat Management, Resource Management Division, Newfoundland Region. 20 p.

1988b. The Science of Cod. Fo'c'sle 8(No. 2): 29 p.

1989a. Northern Cod TAC for 1989 Announced. DFO News Release, February 8, 1989. NR-HQ-89-005E. 3 p.

1989b. Further Measures Regarding Northern Cod Announced. DFO News Release, February 12, 1989. NR-HQ-89-006E. 6 p.

1990a. "Fisheries — A National Priority" says Siddon. DFO News Release, January 2, 1990. NR-N-90-01E. 4 p.

1990b. Minister Valcourt Says New Multi-Year Groundfish Plan Strikes Balance for Sustainable Atlantic Fishery. DFO News Release, December 14, 1990. NR-HQ-90-50E. 3 p.

1992a. Crosbie Announces first steps in Northern Cod (2J3KL) Recovery Plan. News Release, July 2, 1992. NR-HQ-58E.

1992b. Notes for a Statement by the Honourable John Crosbie, Minister of Fisheries and Oceans on Income Support and Related Measures for fishermen and plant workers affected by the northern (2J3KL) cod moratorium — July 17, 1992, St. John's, Newfoundland.

DOUCET, F.J. 1979. The utilization of northern cod. Prepared by F.J. Doucet for Fisheries and Oceans Canada. Northern Cod Seminar. August 28–30, 1979. Annex c.6: 161–206.

DUNNE, E.B. 1990. Report of the implementation task force on northern cod. Fisheries and Oceans. 104 p.

FANL. 1986. Statement concerning the inshore fisheries of Newfoundland and Labrador. Fisheries Association of Newfoundland and Labrador, August 8, 1986. St. John's, Nfld. 9 p.

FLEMING, A.M. 1959. The longline fishery at Bonavista, Newfoundland, 1952–58. ICNAF Meeting Doc. 59/30 Ser. No. 629: 3 p.

1960. Age, growth and sexual maturity of cod (*Gadus morhua* L.) in the Newfoundland area. J. Fish. Res. Board Can. 17: 775–809.

GORDON, W.L., O. LUSSIER, A.E. GRAUER, A. STEWART, R. GUSHUE, AND D.V. LEPAN. 1956. The commercial fisheries of Canada. Royal Commission on Canada's Economic Prospects. Prepared by the Department of Fisheries of Canada and the Fisheries Research Board. 193 p.

GOVERNMENT OF CANADA. 1990. Atlantic Fisheries Adjustment Program. News Release, May 7, 1990. 4 p. + appendices.

HARRIS, L. 1989. Independent Review of the State of the Northern Cod Stock: interim report prepared for the Hon. Thomas E. Siddon, May 15, 1989. 67 p.

1990. Independent Review of the State of the Northern Cod Stock. Final Report prepared for the Honourable Tom Siddon, Minister of Fisheries and Oceans. 154 p.

HEAD, G.C. 1976. Eighteenth Century Newfoundland: a geographer's perspective. McClelland and Stewart Ltd. 296 p.

INNIS, H.A. 1940. The cod fisheries: the history of an international economy. Yale University Press, New Haven, CT. 520 p.

JAMIESON, A. 1975. Enzyme types of Atlantic cod stocks on the North American banks. Isoenzyme IV, Genetics and Evolution, Academic Press, San Francisco, CA. 491–515.

KEATS, D., D.H. STEELE, AND J.M. GREEN. 1986. A review of the recent status of the northern cod stock (NAFO Div. 2J3KL) and the declining inshore fishery. A report to the Newfoundland Inshore Fisheries Association on scientific problems in the northern cod controversy. 43 p.

KIRBY, M.L.J. 1982. Navigating troubled waters. A New Policy for the Atlantic Fisheries. 379 p.

LEAR, W.H. 1984. Discrimination of the stock complex of Atlantic cod (*Gadus morhua*) off southern Labrador and eastern Newfoundland, as inferred from tagging studies. J. Northw. Atl. Fish. Sci. 5: 143–159.

1986. A further discussion of the cod stock complex of Atlantic cod (*Gadus morhua*) in NAFO Div. 2J, 3K and 3L. NAFO SCR Doc. 86/118 Ser. No. N1245: 18 p.

LEAR, W.H., J.W. BAIRD, J.C. RICE, J.E. CARSCADDEN, G.R. LILLY, AND S.W. AKENHEAD. 1986. An examination of factors affecting catch in the inshore cod fishery of Labrador and eastern Newfoundland. Can. Tech. Rep. Fish. Aquat. Sci. 1469: iv + 71 p.

LILLY, G.R. 1987. Interactions between Atlantic cod (*Gadus morhua*) and capelin (*Mallotus villosus*) off Labrador and eastern Newfoundland. Can. Tech. Rep. Fish. Aquat. Sci. 1567: vii+37 p.

LILLY G.R., AND A.M. FLEMING. 1981. Size relationships in predation by Atlantic cod, *Gadus morhua*, on capelin, *Mallotus villosus*, and sand lance, *Ammodytes dubius*, in the Newfoundland area. NAFO Sci. Coun. Studies: 41–45.

MATTHEWS, K. 1973. Lectures on the history of Newfoundland 1500–1830. Memorial University of Newfoundland, Maritime History Group.

MAY, A.W. 1961. Cod investigations in the Labrador area, 1960. Fisheries Research Board of Canada, Biological Station, St. John's, Nfld. Annual Report for 1960-61. Appendix No. 1: 1–4 p.

1967. Effect of offshore fishing on the inshore Labrador cod fishery. ICNAF Res. Bull. 4: 67–75.

MAY, A.W., R.G. HALLIDAY, R. WELLS, AND E. DUNNE. 1981. Management of Canadian cod stocks. NAFO SCR Doc. 81/141. Ser. No. N449: 13 p.

MAY, A.W., A.T. PINHORN, R. WELLS, AND A.M. FLEMING. 1965. Cod growth and temperature in the Newfoundland area. ICNAF Spec. Publ. No. 6: 545–555.

NAFO. 1985. Seventh Annual Meeting — September 1985. Report of the fisheries commission. NAFO/FC Doc. 85/8 (Revised) Ser. No. N1098: 14 p.

1992. Report of the Special Meeting of the Scientific Council, June 1–4, 1992. NAFO SCS Doc. 92/20.

NEWFOUNDLAND. 1979a. Statement by the Hon. A. Brian Peckford, December 4, 1979.

1979b. Discussion Paper on Fisheries Issues. Presented by Premier A.B. Peckford to The Hon. J. McGrath, Minister of Fisheries and Oceans. November 5, 1979.

1985. Appropriate offshore fish harvesting technology: an assessment of the detrimental effects of the use of factory freezer trawlers. Government of Newfoundland and Labrador discussion paper.

1988. Northern cod under attack: The Newfoundland and Labrador perspective on the allocation of northern cod. Government of Newfoundland and Labrador. 11 p.

NORDCO. 1981. It were well to live mainly off fish. The place of the northern cod in Newfoundland's development. Prepared for the Government of Newfoundland and Labrador. Jesperson Printing Ltd. 211 p.

NOVA SCOTIA FISH PACKERS ASSOCIATION. 1979. Freezing at sea. A Canadian opportunity.

PARSONS, L.S. 1993. Management of marine fisheries in Canada. Can. Bull. Fish. Aquat. Sci. 225. 784 p.

PINHORN, A.T. [ed.]. 1976. Living marine resources of Newfoundland-Labrador: status and potential. Bull. Fish. Res. Board. Can. 194: 64 p.

1979. The northern cod resource. Northern Cod Seminar, August 28–30, 1979. Corner Brook Newfoundland. Annex C.1: 53–72.

PINHORN, A.T., and R.G. HALLIDAY. 1990. Canadian versus international regulation of Northwest Atlantic fisheries: management practices, fishery yields and resource trends, 1960–1986. N.Am. J. Fish. Manage. 10: 154–174.

POPOVA, O.A. 1962. Some data on the feeding of cod in the Newfoundland area of the Northwest Atlantic, p. 228–248. Y.Y. Marti [ed] Soviet Fisheries investigations in the Northwest Atlantic, VNIRO-PINRO, Moskva. (Transl. from Russian for U.S. Dep. int. Nat. Sci. Found., Washington, D.C., by Israel Prog. Sci. Transl. 1963).

ROWE, F.W. 1980. A history of Newfoundland and Labrador. McGraw-Hill Ryerson Ltd. 563 p.

RYAN, S. 1986. Fish out of water. The Newfoundland saltfish trade, 1814–1914. Breakwater Books Ltd. 320 p.

SEREBRYKOV, V.P. 1967. Cod reproduction in the Northwest Atlantic. Trudy PINRO. (Fish. Res. Bd. Canada Transl. Ser. No. 1133, 1968) 20: 205–242.

SULLIVAN, K.M. 1989. Conflict in the management of a northwest Atlantic transboundary cod stock. Mar. Policy 13(2): 118–136.

TASK GROUP ON NEWFOUNDLAND INSHORE FISHERIES. 1987. A study of trends of cod stocks off Newfoundland and factors influencing their abundance and availability to the inshore fishery. A report to the Hon. Tom Siddon, Minister of Fisheries, Canada. 101 p. + app.

TEMPLEMAN, W. 1953. Knowledge of divisions of stocks of cod, haddock, redfish, and American plaice in Subareas 3 and 2 of the Northwest Atlantic

Convention Area. ICNAF Annu. Proc. 3: 62–66.
1959. Canadian research report, 1958 A. Subareas 2 and 3. ICNAF Annu. Proc. 9: 20–26.
1962. Divisions of cod stocks in the Northwest Atlantic. ICNAF Redbook 1962, part III: 79–129.
1965. Canadian research report, 1964. A. Subareas 2 and 3. ICNAF Redbook 1965, part II: 3–20.
1966. Marine resources of Newfoundland. Bull. Fish. Res. Board Can. 154: 170 p.
1974. Migrations and intermingling of Atlantic cod (*Gadus morhua*) stocks of the Newfoundland Area. J. Fish. Res. Board Can. 31: 1073–1092.
1979. Migrations and intermingling of Atlantic cod (*Gadus morhua*) stocks of the Newfoundland and adjacent areas from tagging in 1962–66. ICNAF Res. Bull. No. 14: 5–50.
1981. Vertebral numbers in Atlantic cod (*Gadus morhua*) of the Newfoundland and adjacent areas, 1947–71, and their use for delineating cod stocks. J. Northern Atl. Fish. Sci. 2: 21–45.

TEMPLEMAN, W., AND A.M. FLEMING. 1963. Distribution of *Lernaeocera branchialis* (L) on cod as an indicator of cod movements in the Newfoundland area. ICNAF Spec. Publ. No. 4: 318–322.

TEMPLEMAN, W., AND J.A. GULLAND. 1965. Review of possible conservation actions for the ICNAF area. ICNAF Annu. Proc. 15: 47–56.

TEMPLEMAN, W., V.M. HODDER, AND A.M. FLEMING. 1976. Infection of lumpfish (*Cyclopterus lumpus*) with larvae and of Atlantic cod (*Gadus morhua*) with adults of the copepod *Lernaeocera branchialis*, in and adjacent to the Newfoundland area, and inferences there from on inshore-offshore migrations of cod. J. Fish. Res. Board Can. 33: 711–731.

TEMPLEMAN, W., H.J. SQUIRES, AND A.M. FLEMING. 1957. Nematodes in the fillets of cod and other fishes in Newfoundland and neighbouring areas. J. Fish. Res. Board Can. 14: 831–897.

# The Canadian Lobster Fishery and its Management, with Emphasis on the Scotian Shelf and the Gulf of Maine

J.D. Pringle and D.L. Burke

*Department of Fisheries and Oceans*
*P.O. Box 550, Halifax, Nova Scotia, Canada B3J 2S7*

Pringle, J. D., and D. L. Burke. 1993. The Canadian lobster fishery and its management, with emphasis on the Scotia Shelf and the Gulf of Maine, p. 91–122. *In* L. S. Parsons and W. H. Lear [eds.] Perspectives on Canadian marine fisheries management. Can. Bull. Fish. Aquat. Sci. 226.

Lobster is the backbone of the eastern Canadian inshore fishery, now fished by 11,573 licensed vessels and 32,000 fishermen. The cost of lobster management was estimated to be about 5% of the 1988 value. The development of biological advice and management plans involve a peer review system. Management controls now include minimum legal size, non-possession of egg-bearing females, limited entry, lobster fishing areas, open seasons and trap limits. The small (3% of total landings) offshore fishery initiated in 1971 is managed by total allowable catch (TAC) and enterprise allocation (EA). Inshore fishery management plans vary with both markets and the impact of local environmental conditions on lobster abundance and fishing conditions. Biological and economic advice recommends increases in minimum legal size in most management units to reduce yield overharvesting, to optimize yield per recruit and to increase egg production. Canadian landed volume, between 1981 and 1990, increased 115%; ex-vessel value had increased by 317% between 1981 and 1987, but 1990 landed value was only 249% above the 1981 level. A pulse in recruitment was the main cause for enhanced landings; factors responsible for the recruitment pulse are under investigation. The current challenge is to better understand the factors controlling both lobster abundance and the fishery's economics; more urgent is the need to develop a management structure that will permit the development of sound advice by personnel who have industry's confidence.

---

Le homard, échine de la pêche côtière dans l'est du Canada, est exploité par 32 000 pêcheurs titulaires d'un permis oeuvrant à partir de 11 573 bateaux de pêche. On situe le coût de la gestion du homard à environ 5 % de la valeur de la ressource en 1988. L'élaboration de conseils scientifiques et de plans de gestion comprend un système d'examen par des pairs. La réglementation repose actuellement sur une taille légale minimum, la possession interdite de femelles oeuvées, une limite du nombre de participants à la pêche, des arrondissements de pêche, des saisons de pêche et des limites du nombre de casiers exploités. La pêche hauturière de faible envergure, soit 3 % des débarquements totaux, entreprise en 1971 est gérée selon un système de total des prises admissibles (TPA) et d'allocation aux entreprises (AE). Les plans de gestion de la pêche côtière varient selon les marchés et l'impact des conditions environnementales locales sur l'abondance de la ressource et les conditions de pêche. Les avis biologiques et économiques formulés recommandent une augmentation de la taille légale minimum dans la plupart des unités de gestion afin de réduire la surexploitation au détriment du rendement, d'optimaliser le rendement par recrue et d'accroître la production d'oeufs. Entre 1981 et 1990, le volume de homard débarqué au Canada a vu une augmentation de

115 %; même si, de 1981 à 1987, la valeur de la ressource après le débarquement ait augmenté de 317 %, la valeur au débarquement en 1990 n'était que de 249 % plus élevée qu'en 1981. Une pointe du recrutement, dont la cause est à l'étude, est en grande partie responsable de l'augmentation des débarquements. Le défi qui se présente actuellement est de mieux comprendre les facteurs sous-tendant l'abondance du homard et l'économie de la pêche. Il est par contre plus urgent d'élaborer une structure de gestion qui permettra aux personnes à qui l'industrie fait confiance de formuler des avis judicieux.

## Introduction

The American lobster (*Homarus americanus* Milne-Edward) occurs in commercial concentrations (Fig. 1) from southern Labrador to New Jersey (Campbell 1989). The Canadian lobster is common property under the jurisdiction of the federal government. Regulations were promulgated as early as 1878, and have been modified during the intervening period to protect and enhance the resource (De Wolfe 1974).

Canadian lobster yields, nevertheless, dropped to near record lows in the late 1960's (Pringle et al. 1983). A task force was struck in 1974 to investigate the fishery (Anon. 1975). The study recognized the diverse nature of the Canadian lobster fishery, specifically its structure, its importance to local economics and its

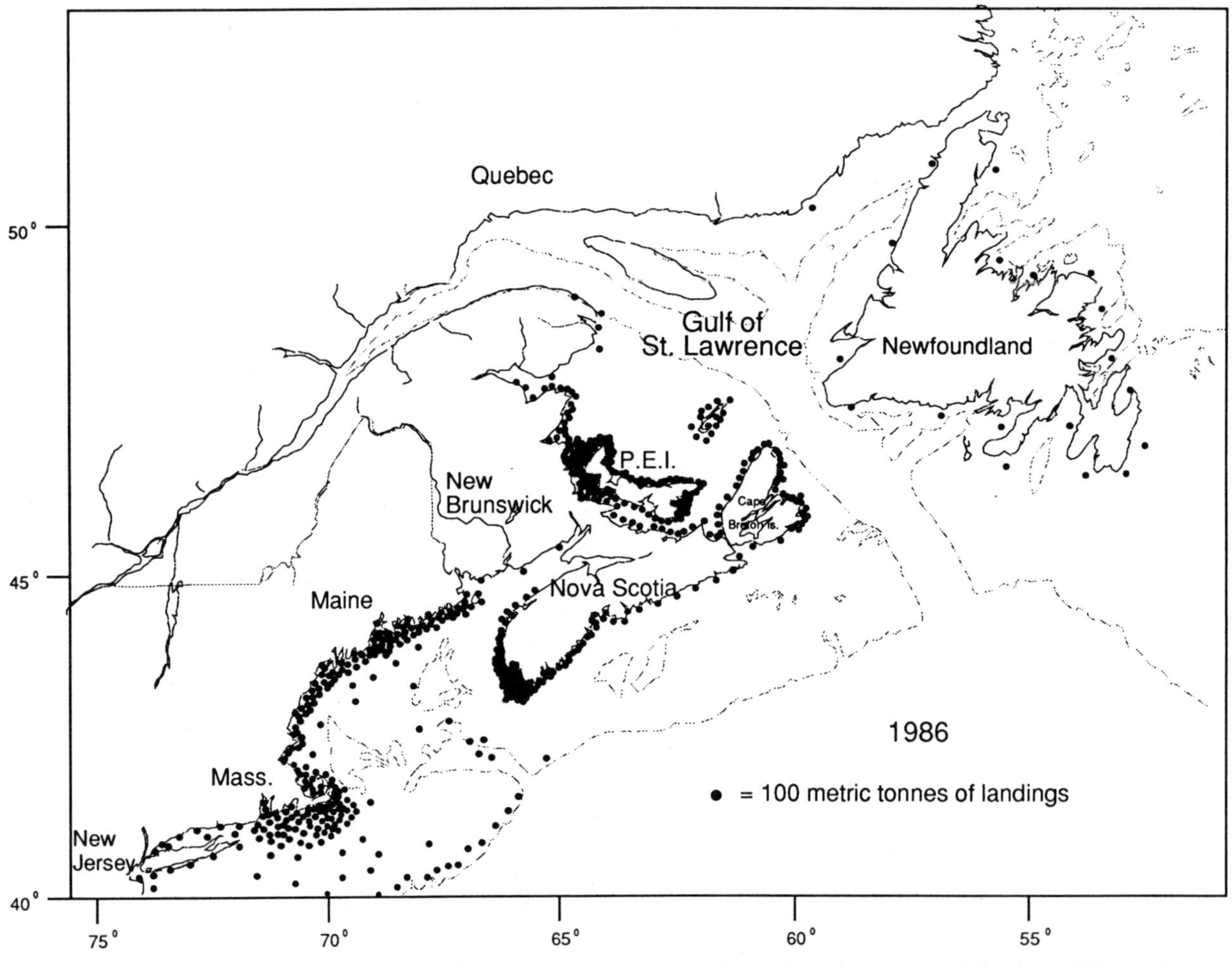


FIG. 1. The distribution of 1986 annual landings, as 100 t allocations, throughout the commercial range of *H. americanus* (Source: A. Campbell, DFO unpubl. data).

monetary returns to fishermen. Recommendations were made in a wide range of areas from trap design through biological research to market enhancement. A later overview (Pringle et al. 1983) noted that many of the recommendations had not been implemented; this failure to act may have spawned the problems which led to the most recent review (Anon. 1989).

The challenges that face the lobster fishery are many. The information provided here will allow one to understand the source of the problems facing the fishery and the mechanisms and strategies available to permit the fishery to move to a higher level of longterm economic return.

## Lobster Biology and Ecology

### Background and Life Cycle

The American lobster grows, following the molt of its chitinous exoskeleton, by expanding prior to the hardening of the new shell (Aiken 1980). Lobsters usually mate within hours of the female molt, although females do breed in the hard-shell state (Dunham and Skinner-Jacobs 1978). Relative size between sexes may be important to breeding success as small males tend not to mate with large females (Dr. D. Cowan, Woods Hole Oceanographic Institute, Woods Hole, Massachusetts; pers. comm.). Sperm, attached by the male to a receptacle on the female, fertilize the eggs upon their extrusion the following summer. They incubate on the tail's underside for 10 to 12 months (Aiken and Waddy 1986). Consequently, sexually mature females of the size commonly observed in most inshore fisheries (small, about 0.5 kg), molt and produce eggs in alternate years. While larger females greater than 1 kg may molt only once every 3–5 years, they have the ability to maintain viable sperm for more than one year (Waddy and Aiken 1986). The number of eggs produced increases with the female's size (Table 1).

Eggs hatch into planktonic, free-swimming larvae between July and October (Fig. 2). Development time of the three larval stages to the postlarva (first juvenile stage) (Templeman 1936) decreases with increasing water temperature (Table 2). Late summer larvae may not have sufficient time to develop into post larvae before temperatures decline below minimum for larval development (Scarratt 1964; Caddy 1979; Harding et al. 1983; MacKenzie 1988). During cold-water years this may be a significant cause of mortality (Harding et al. 1983).

Postlarvae seek a suitable bottom type (stones, shells, etc. see Pottie and Elner 1982), settle and are cryptic until about 35 mm CL (carapace length) (Lawton 1987). Lobsters less than 45 mm CL are rarely seen in traps, and in laboratory experiments moved only a few metres beyond their shelters (Lawton 1987).

TABLE 1. Mean number of eggs produced by female lobsters of different sizes. (Source: D. Moore, DFO unpubl. data).

| Carapace Length (mm) | Number of Eggs |
|---|---|
| 80 mm | 7,600 |
| 90 mm | 11,600 |
| 100 mm | 16,000 |
| 140 mm | 49,500 |
| 160 mm | 77,400 |

TABLE 2. Approximate number of days required to pass through each larval stage at different temperatures (MacKenzie 1988).

| Temperature (°C) | Larval stage | | | | |
|---|---|---|---|---|---|
| | 1 | 2 | 3 | 4 | Total |
| 8 | 20 | 26 | 42 | | |
| 10 | 14 | 15 | 25 | 49 | 103 |
| 12 | 10 | 11 | 15 | 32 | 68 |
| 14 | 7 | 8 | 11 | 25 | 51 |
| 16 | 5 | 6 | 9 | 22 | 42 |
| 18 | 3 | 5 | 7 | 18 | 33 |
| 20 | 2 | 3 | 6 | 10 | 21 |

Lobsters reach the minimum legal size of 81 mm CL in cool Canadian waters at age 6 to 8 yr (Wilder 1953; Aiken 1980), though time to recruitment may be longer in colder water such as the Bay of Fundy. Most female lobsters reach minimum legal size prior to sexual maturity. The size of 50% female maturity ranges from 75 mm in eastern Cape Breton [Lobster Fishing Area (LFA) 27-see Fig. 3], to over 105 mm in parts of the Bay of Fundy. Males mature at smaller sizes than females (Aiken and Waddy 1980).

### Distribution and Ecology

Lobsters occur in commercial numbers along much of eastern Canada's coast south of Labrador (Fig. 1) at depths generally less than 35 m. Yields vary greatly from area to area (Fig. 1). Commercial yields per bottom area fished along the open coast of eastern and southeastern Nova Scotia ranged between 82 kg/km$^2$ in LFA 32 to 1,249 kg/km2 in LFA 27 (Pringle and Duggan 1985). Few direct measurements of lobster density are available. Often, where density is low, such as along Nova Scotia's eastern shore, the availability of lobster habitat is not limiting (Moore et al. 1986).

Lobsters occur on a variety of combined bottom types from mud/silt through sand/rock to bedrock/rock

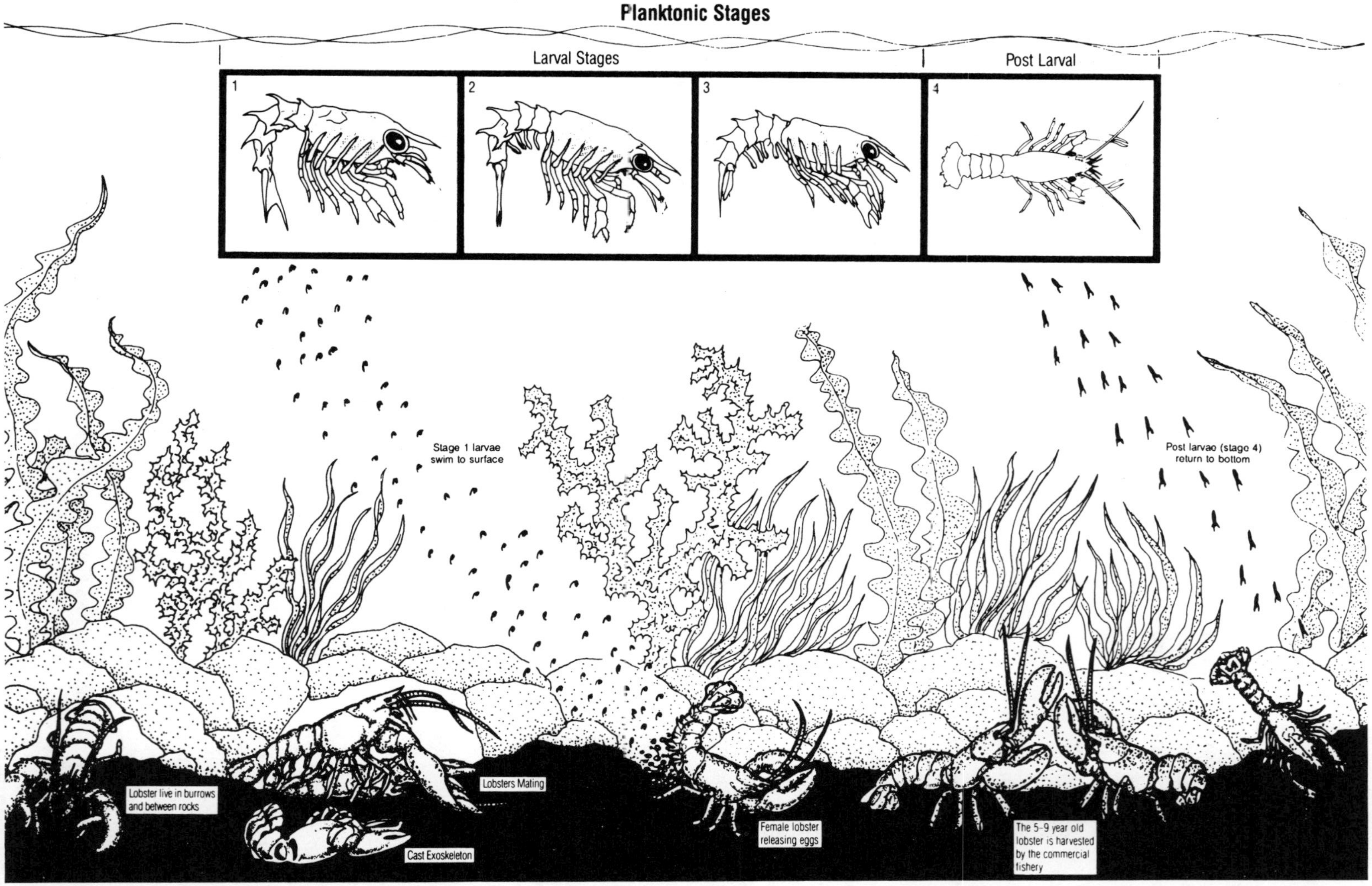


FIG. 2. A sketch illustrating the lobster life cycle including the postures assumed when mating and upon release of larvae (Source: R. Duggan, DFO unpubl. data).

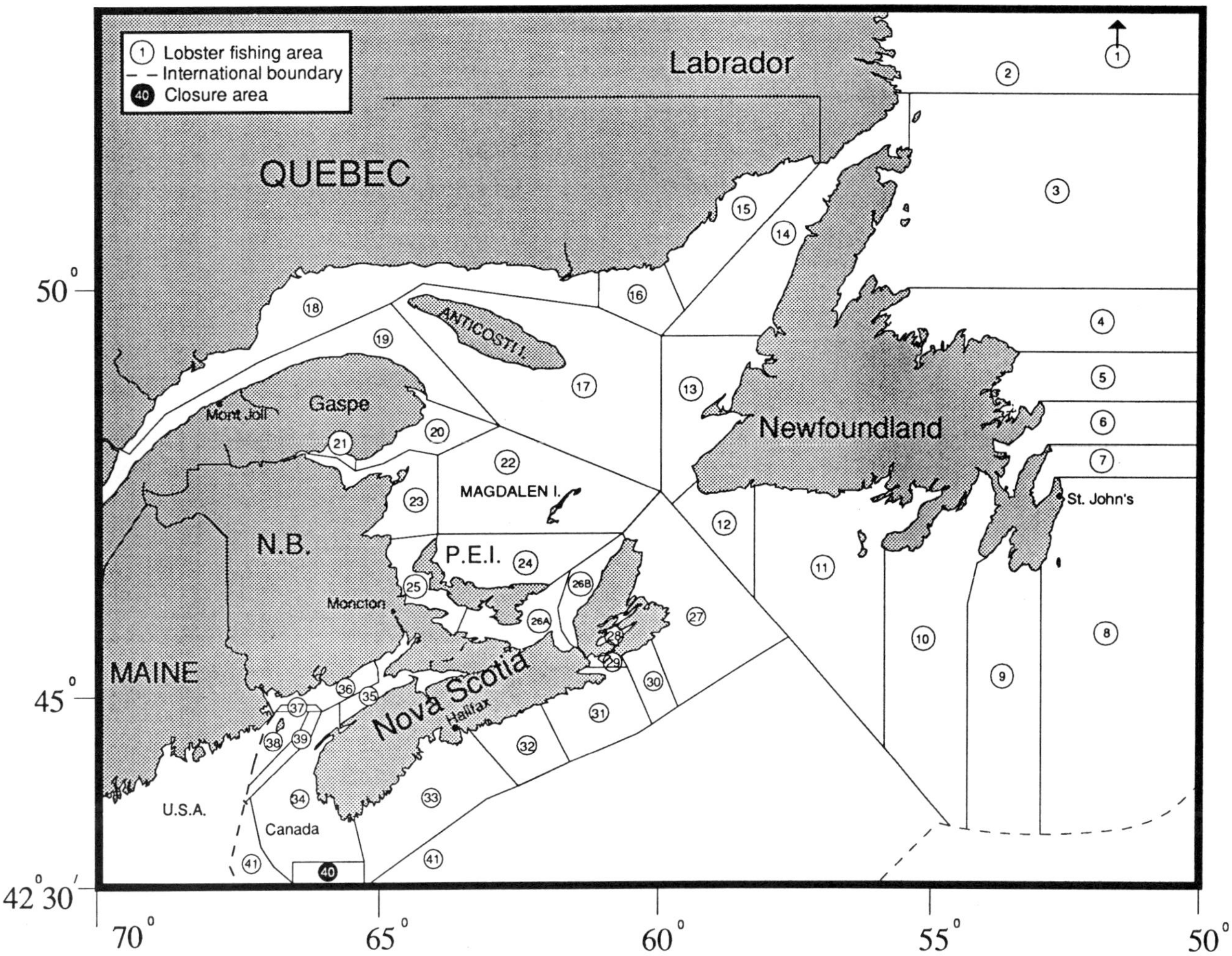


FIG. 3. Lobster Fishing Areas (LFA) for Canada's Atlantic fishing zone (Source: R. Duggan, DFO, unpublished data).

(Pottie and Elner 1982). The most common near-shore habitat is rock and boulders overlying sand (Miller 1985). Frequently, in depths less than 15 m, lobsters are found among large seaweeds (Jarvis 1989), but research (Miller 1985) has found little evidence to support the popular concept that lobsters require seaweeds.

Until recently, little was known about the distribution of juveniles. Studies near Shelburne, Nova Scotia (A. Campbell, Dept. of Fisheries and Oceans, Nanaimo, B.C. V9R 5K6, unpubl. data), and off Iles de la Madeleine, Quebec (Hudon 1987), indicate high abundances on shallow rocky bottoms, with low densities in sanded eel grass communities.

Lobsters along the outer Scotian Shelf and upper slope and in the Gulf of Maine's deep basins (Fig. 1 and 3) are commercially fished by the offshore fleet at depths to 450 m, and have been observed to depths of 700 m (Pezzack and Duggan 1985). Densities here are generally less than 0.1% of average coastal areas, although certain offshore canyons have densities up to 5.0% of coastal areas (Cooper and Uzmann 1980). Offshore lobsters occur in canyon wall burrows, and more generally in shallow, dish-shaped depressions, on open sand or gravel bottom (Cooper and Uzmann 1980).

Marked differences in mean lobster size can occur between closely situated grounds. For example, trap-caught lobsters off Grand Manan Island's (LFA 38, see Fig. 3) north side are significantly larger than off the south side (Campbell and Duggan 1980). These differences may be due to both a more recent exploitation of deep water, overwintering sites and to different habitats. Similarly, the relatively large size of offshore lobsters may be due only to a more recent exploitation history (Pezzack and Duggan 1987). The small mean size of coastal lobsters is thought to be largely the result of heavy fishing pressure. Lobsters averaged 0.9–1.3 kg in the late 1800's (Venning 1910), similar to current sizes captured offshore. Offshore stocks subjected to 35 years heavy fishing in southern New England now have average lobster sizes similar to adjacent coastal grounds (Cooper and Uzmann 1980).

### Movement of Juveniles and Adults

Distance travelled by lobsters larger than 60 mm has been estimated by releasing tagged lobsters and recording location and time of recapture. During the 1940's and 1950's, 27,500 tagged lobsters were released in three nearly contiguous lobster grounds off northeastern Nova Scotia; 15,000 were recaptured, with 2,000 at large over a year. No movement was observed among the contiguous fishing grounds within LFA 30, a distance of only 6.4 km, and no recaptures were made more than a few kilometres from release sites (Wilder 1974). More recently, animals tagged with sphyrion tags (Scarratt and Elson 1965) off southern and eastern Nova Scotia (LFA's 31–33) tended to move <12 km over a 1–6 year period (Miller et al. 1989).

Gulf of Maine tagging studies indicate mature lobsters move significantly farther than immature ones, and that long-distance movement (greater than 100 km) allows some interchange between the Bay of Fundy, Gulf of Maine, and the adjoining continental shelf (Campbell and Stasko 1985, 1986; Campbell 1986). Mature lobsters make seasonal migrations to shallow, warmer water of either the offshore banks (Pezzack and Duggan 1986) or coastal regions during summer/fall, and into deeper water during winter-spring. Higher temperatures in these locations facilitate molting, growth (Cooper and Uzmann 1980) and egg development (Campbell 1986). As well, larvae in warm, shallow, coastal or bank waters may have a survival advantage through reduced exposure in the water column prior to moving to the cryptic, bottom-dwelling stage (Huntsman 1923; Caddy 1979). The seasonal migration concept explains some of the long and short-distance movements observed (Campbell and Stasko 1986). Although certain lobsters appear to return to the same location year after year (Pezzack and Duggan 1986), not all do. Some mixing of mature lobsters probably occurs along the continental shelf area with time.

A pelagic larval stage led to the hypothesis that larvae are subject to considerable transport (Harding et al. 1983), drifting up to 160 km along the eastern shore of Nova Scotia before settlement (Dadswell 1979). Larvae hatched on the offshore, Gulf of Maine banks (Fig. 1) could settle on the coasts of Nova Scotia or Maine (Stasko 1980; Harding and Trites 1988). However, observations of lobster larval behaviour (Harding et al. 1987) indicate they occur less frequently in surface residual currents than previously thought. Thus, before distances moved can be calculated, better data on subsurface currents are required.

### Growth

Growth rate is a function of both molt frequency and size increase at molt. Lobsters reach legal size after 15 to 20 molts (Templeman 1948; Aiken 1980). Around legal size (70–90 mm CL) they generally molt once annually, increasing in length by about 15%, and in weight by about 45% (Miller et al. 1989). While percentage increase in lobster size at molt remains relatively constant through life, mean annual growth rate slows after sexual maturity because these animals, particularly the females, molt less frequently.

The maximum sizes recorded for male and female lobsters are 260 mm CL (19.1 kg) and 216 mm CL (7.4 kg), respectively (Wolff 1978). Lobster longevity is not known due to our inability to age them.

### Natural and Fishing Mortality

Lobster mortality rates are difficult to obtain because of a discontinuous growth and no demographic data. Natural mortality is thought to be highest during egg and larval stages, and lowest following sexual maturity. Mean egg mortality estimates at incubation range from 36% (Perkins 1971) to 54% (Campbell and Brattey 1985). Larval mortality estimates range from 97.5 to 99.5% (Scarratt 1964, 1973; Harding et al. 1982). Annual, natural mortality rate of adult lobsters varies from 2 to 30% (Thomas 1973), though the higher estimates are considered less reliable (Fogarty and Idoine 1986). Values of 5% annual mortality for non-molting animals and between 5% and 10% for molting animals are commonly accepted (Ennis and Akenhead 1978; Fogarty and Idoine 1986). Fishing appears to be the dominant control of densities of recruit-sized, lobsters. Exploitation rates vary from about 50% along Atlantic Nova Scotia (Miller et al. 1987) to 90% in parts of Newfoundland (Ennis 1981).

### Food

Larval lobsters have been raised in the laboratory on live copepods (small crustaceans) and frozen adult brine shrimp, but little is known of their natural diet (Phillips and Sastry 1980). Juveniles and adults locate food primarily by chemosensory means (Phillips et al. 1980). They eat live prey such as crabs, mussels, sea urchins, marine worms, periwinkles, sea stars, fish and seaweeds and scavenge dead animal matter (Ennis 1973; Scarratt 1980; Elner and Campbell 1987). Feeding activity increases directly with water temperature and immediately following molting. Feeding is believed to stop when water temperatures drop below approximately 3°C (Paloheimo 1963).

### Stock-Recruitment Relationships

A weak correlation was found between lobster larval abundance and subsequent recruit abundance about 6 years later (Scarratt 1973). Otherwise, lobster stock-recruit relationships are not known (Fogarty and Idoine 1986). Data on spawner numbers in successive years does not exist. Recent log book data, however, suggest that the percentage of trap-caught berried females varies markedly along coastal eastern Nova Scotia (J.D. Pringle and R.E. Duggan unpubl. data). In recent years high landings (Fig. 4) were obtained from offspring hatched when catches were at record lows. This indicates spawner number is not the sole factor limiting lobster recruitment.

## Brief History of the Fishery

### Maritimes [New Brunswick, Prince Edward Island, and Nova Scotia]

The first Maritime cannery, constructed in 1845, was situated on Portage Island, New Brunswick. Yarmouth, in the late 1840's, was the site of the first Nova Scotia cannery. Canneries were developed on Prince Edward Island and Newfoundland about 25 years later (DeWolfe 1974). The first large cannery was constructed by U.S.A. investors who were supplementing their declining catches. Growth of the industry was rapid; by 1873, 67 canneries (Table 3) preserved 2,208,710 one kg cans. Forty (93%) of the Nova Scotia canneries were located from the south shore coast to

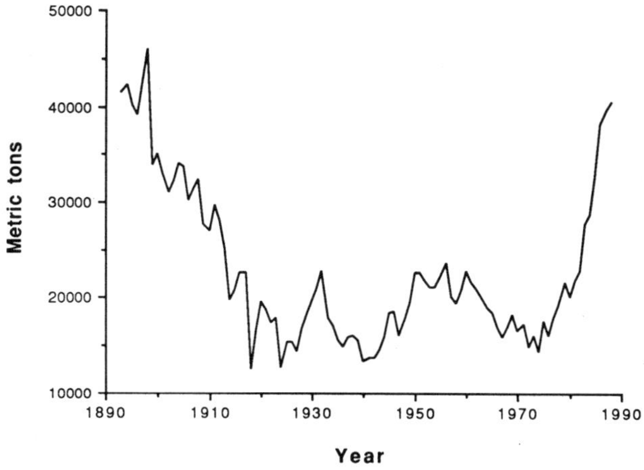


FIG. 4. Annual Canadian lobster landing's from 1892 to 1988 (D. Pezzack, DFO, Scotia-Fundy Region).

Cape Breton Island; New Brunswick had 24 canneries. Lobsters (543 t) were first shipped live to the Boston, Massachusetts market in 1881, which complemented the production of 7 950 t of canned lobster. Distance from Boston set the limit on the development of the Canadian live lobster trade at this time.

There were more than 331 canneries in the Maritimes by 1890; 96 on mainland Nova Scotia (Anon. 1890): about one-third of the western Nova Scotia catch was now shipped live to the United States (Robinson 1979). Canadian lobster landings peaked in the late 1880's (Fig. 4), with demand increasing in concert with a decline in U.S.A. catches. Price per lobster increased 10-fold (Prince 1899). Canned Canadian lobster demand in 1897 was spurred by cannery closings in Maine, whereafter this fishery was restricted to lobsters large enough for live market trade (hereafter referred to as "markets"). The number of Canadian canneries more than doubled (760) by 1900, the peak year (Table 3). However, live lobster shipments from eastern Nova Scotia were enhanced by improved railroads; by 1925, 30% of the catch was exported live.

TABLE 3. Number of lobster canneries per province for selected years between 1873 and 1978 (Robinson 1980).

| Year | Nova Scotia | New Brunswick | Prince Edward Island | Total |
|---|---|---|---|---|
| 1873 | 43 | 24 | — | 67 |
| 1890 | 133 | 100+ | 98 | 331+ |
| 1892 | 182 | 185 | 212 | 579 |
| 1900 | 277 | 237 | 246 | 760 |
| 1910 | 214 | 185 | 187 | 586 |
| 1925 | 139 | 144 | 148 | 431 |
| 1939 | — | — | — | 191 |
| 1944 | 39 | 49 | 49 | 137 |
| 1978 | 2 | 11 | 10 | 23 |

Declining canned lobster meat prices in the late 1920's, due in part to competition on the European market from canned Japanese crab meat (80% of pre-World War II lobster cannery production was sold overseas) and increasing demand for live lobsters modified marketing. Market size limits were imposed throughout western Nova Scotia in 1933 (Sutherland 1937). A subsidized collection service by vessel was inaugurated in 1930 along Nova Scotia's eastern shore to facilitate live shipments.

A small section of Cape Breton Island, currently LFA 30, observed a market size limit in 1940. Its present boundaries were created in 1947. A program to phase out the "cannery" fishery in LFAs 27 and 31/32, by gradually increasing the minimum legal size (MLS), was started in the late 1950's. It was completed in LFA 31/32 in 1961, but was halted in LFA 27 at 70.0 mm in 1957. Currently, only fishermen in LFAs 26 and 27 fish canner lobsters in Nova Scotia.

The basic structure of the Nova Scotia fishery was established by the turn of the century. The use of motor driven vessels, and decreasing nearshore catches, led certain western Nova Scotia fishermen to move out to waters 36–90 m deep (Prince 1899). Somewhat later, Halkett (1916) noted fishermen in southern LFA's 33 and 34 fished about 30 km from shore (still inshore by today's definition). This required modern equipment. The introduction of the parlour trap early in the century was a significant development. Halkett (1916) claimed their efficiency and gear, including a mechanical "trap hoist", were unmatched elsewhere. By contrast, fishermen interviewed in the 1970's claimed trap haulers were not introduced into LFA 31 until after World War II (Robinson 1980).

The fishery, by the 1890's, had been underway for about 40 years in some areas. The average weight of canners in the catch was about 0.45 kg, whereas markets averaged 0.68 kg (Anon. 1890). The fishery was no longer based on a stock of large lobsters. Even with a catch of about 37 195 t in 1898, Prince (1899) observed signs of what would later be termed yield overfishing.

Landings were declining by 1898, but the worst was yet to come (Fig. 4 and 5). The fishery off eastern Nova Scotia declined from about 5 900 to 1 800 t in the early 1920's. The landings in western Nova Scotia similarly declined, but rebounded in the early 1940's, about 7 years after the application of a market size limit. Landings were relatively constant from the mid 1940's to the early 1980's.

In the early 1980's, the Department of Fisheries and Oceans (DFO) divided the Maritimes into two Regions, Scotia-Fundy (S-F) and Gulf (Gulf of St. Lawrence). Both Nova Scotia and New Brunswick are represented in each Region giving each both cold (S-F) and warm (Gulf) water lobster production areas. The ecological differences between the two Regions have created quite different harvest strategies leading to different marketing problems. The colder waters of S-F, with the exception of LFA 27 in northern Nova Scotia, dictate a larger MLS than that of the Gulf. Lobsters in waters of the large production states of Maine and Massachusetts have similar environmental conditions, thus the similar MLS. In addition, water transport to U.S.A. markets was more viable for S-F landings, and the presence of ice-free ports in southwestern Nova Scotia (the most productive Canadian fishery) gave potential for year-round landings. These characteristics permitted the S-F industry to market the more profitable live lobsters during periods when demand was high.

By contrast, the Gulf lobster industry processed most of its catch because MLS was below that of the

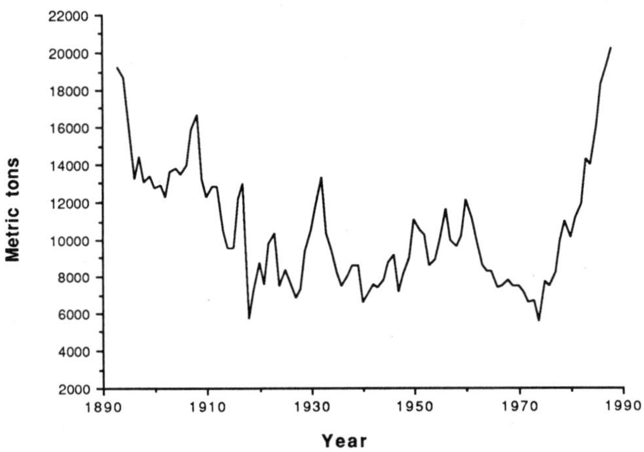


FIG. 5. Annual lobster landings for the southern Gulf of St. Lawrence (LFA 23–26B) for the years 1892–1988.

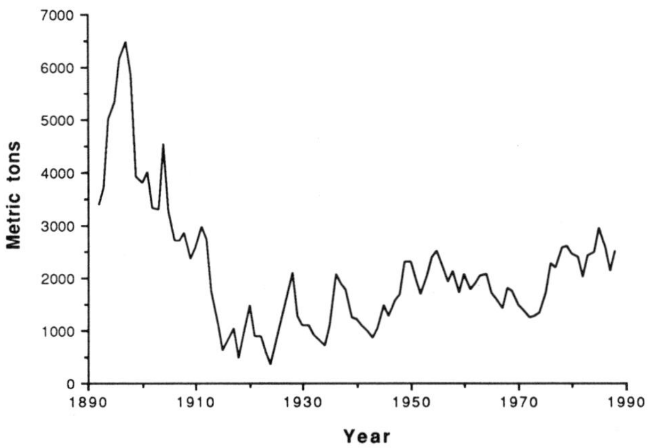


FIG. 6. Annual lobster landings for Newfoundland for the years 1892–1988 (G. Ennis, DFO, Newfoundland Region).

U.S.A. Furthermore, a 5-month winter ice cover dictated seasons be open during periods of high supply. Distance prevented water transport of product.

### Newfoundland

The Newfoundland lobster fishery began in 1874, with around 68 t landed; landings peaked at 8 000 t in 1889. Subsequently there has been a number of sharp fluctuations superimposed on a distinct downward trend (Fig. 6). Reduced abundances forced closure of the fishery between 1925 and 1927 inclusive. During these early years the catch was processed in small canning operations around the coast. Regulations, largely unenforced, defined a MLS and protected berried females.

A certain portion of the catch was shipped live to the U.S.A. market in the early 1930's, where size limit and berried female regulations were strictly enforced. By the early 1950's, most of the catch was shipped to the U.S.A. and has since remained a live-market industry. Size limit and berried female regulations were enforced in the 1940's, which may have caused the enhanced landing in 1949.

### Québec

The Québec lobster fishery is divided into Iles de la Madeleine (LFA 22) and Gaspé-Bonaventure areas (LFA's 15 to 21). Commercial lobster fishing started in Québec (Bergeron 1967) in LFA 20 in 1871. Bonaventure and Gaspé catches increased quickly to peak at 1 100 t in 1880. Two dramatic decreases occurred in 1900 and 1913. The latter was followed by a steady decline which bottomed out in 1939 at 63 t. A gradual increase in catches occurred from 1939 to 1975.

Effort, measured as trap number fished, tended to follow landing patterns, though accurate numbers are unavailable. Despite apparent increases in numbers of fishermen between 1950 and 1971, it is difficult to conclude a proportional increase in effort, as many rarely fished, or only used a few traps (unpublished data from the Bureau de la statistique de Québec 1950–1971). The lobster fishery developed in concert with legislation, which required the use of at least 100 traps (Belzile 1981). Part-time fishermen were slowly eliminated. The fishery became more professional between 1975 and 1981. Trap number per vessel approached the legal maximum (250 traps); fishing effort increased by 150% (Belzile 1981) and landings rose from 204 to 538 t.

Lobster fishing began in 1875 on Iles de la Madeleine, 5–6 years after it began in Chaleur Bay (Lavoie 1877). The fishery appeared to develop more slowly here than in Gaspesie. Catches peaked at 2 174 t in 1905 and declined between 1914 and 1942, reaching a minimum of 719 t (Fig. 7).

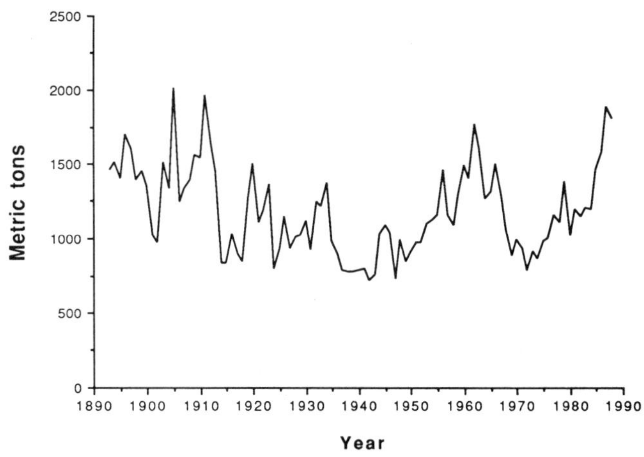


FIG. 7. Annual lobster landings for the Iles de la Madeleine for the years 1892–1988.

## Resource Management

### Structure

An important factor in the quality of resource management is the structure of the resource management agency (Pringle 1986a). The 1975 task force noted that no individual in the management hierarchy was responsible for the lobster fishery (Anon. 1975). Consequently, in the early 1980's a lobster management structure was set up after a plan outlined by Pringle et al. (1983). The basic unit is the Lobster Fishing Area Advisory Committees (LFAAC). Industry is represented by elected fishermen and processors. DFO members include a biologist, an economist, the local fishery officer, the Regional Senior Advisor for Invertebrates and a chairperson (usually the DFO Area Manager). Each committee meets a minimum of once yearly to advise the Regional Directors-General on all issues impacting the lobster fishery. LFA management plans are modified only after discussions with the respective committees.

Each LFAAC elects one fisherman to the Regional Lobster Advisory Committee (RLAC), chaired by a senior DFO advisor. The remainder of the committee resembles that of the LFAAC. The RLAC meets at least twice yearly to advise on Regional or Atlantic Zonal matters.

The committee biologist provides each LFAAC with an annual "stock" assessment (see next section for details). Until 1992, the biological advice must first have been peer reviewed by the Canadian Atlantic Fisheries Scientific Advisory Committee (CAFSAC).

The economist provides information on fishery and market performance. Data come from commercial landings throughout the Atlantic. Special studies provide information on costs, earnings, income and industry capacity. Statistics Canada data are used to analyze markets, prices, exports, general economics and trends, and community dynamics.

Biological and economic advice, along with advice from the advisory committee process and other levels of government are combined to develop each LFA's annual management plan and to modify regulations as required. Advice can impact seasons, trap limits and other effort related measures. A licensing unit administers the annual issuance of licenses, their transfers and conducts licensing appeal hearings. Fishery officers promote, monitor and ensure compliance with Acts, Regulations, policies, management practices and specific fishing plans applicable to lobster.

## Management Controls

A unique feature of the Canadian lobster fishery is the division of the coast into Lobster Fishing Areas (Fig. 3). Each LFA has a management plan tailored to its characteristics. The LFA's vary both in length of coastline, from about 110 to 2,000 km, and in level of lobster production (Pringle and Duggan 1985). Sizes and numbers of LFAs are dynamic, having been shaped by issues over time.

Effort is controlled by limited entry licensing (Table 4), by limiting trap number and by limiting maximum number of days fished within an open season. Limited entry plays a significant role in controlling the economics and exploitation rate within each LFA. A licence is issued annually to each fishermen who held one the previous season. Licences are privileges, not property, nevertheless a market has developed for eligible licenses by permitting reassignment to eligible fishermen. The price attained by the previous holder, upon licence transfer, reflects potential licensee earned income. A licensed fishermen can operate a vessel of unlimited size (except LFA 34) within the boundaries of the LFA.

The maximum legal trap number per fisherman varies amongst LFA's; varying with overall lobster production, fishermen density and historic circumstance. The number tends to be higher than deemed optimal (Anon 1989). Trap limits are controlled by issuance of trap tags, equal in number to the licensee's trap limit. A trap, to be legally set, must be tagged.

The number of days per open season varies amongst LFAs as does the occurrence of the open season. The latter is a crude attempt to maintain a constant supply of good quality lobsters through most of the year.

TABLE 4. The number of lobster licenses per Atlantic Zone Region and per LFA for the Gulf and Scotia-Fundy Regions for 1988 (Anon. 1989 and DFO Statistics).

| Region | LFA | Number of Licenses | | |
|---|---|---|---|---|
| | | CLASS A | CLASS B | Total |
| Gulf | 12 | 64 | 12 | 76 |
| | 13 | 482 | 92 | 574 |
| | 14 | 611 | 48 | 659 |
| | 23 | 713 | 60 | 773 |
| | 24 | 641 | 6 | 647 |
| | 25 | 865 | 25 | 890 |
| | 26A | 754 | 20 | 774 |
| | 26B N.S. | 244 | 17 | 261 |
| | 27 N.S. | 47 | 0 | 47 |
| Regional Total | | 4224 | 280 | 4704 |
| Scotia-Fundy | 27 | 459 | 41 | 500 |
| | 28 | 16 | 2 | 18 |
| | 29 | 59 | 20 | 79 |
| | 30 | 19 | 1 | 20 |
| | 31 | 140 | 11 | 151 |
| | 32 | 152 | 16 | 168 |
| | 33 | 658 | 117 | 775 |
| | 34 | 975 | 7 | 982 |
| | 35 | 92 | 3 | 95 |
| | 36 | 186 | 10 | 195 |
| | 38 | 135 | 1 | 136 |
| | 41 | 8 | | 8 |
| Regional Total | | 2899 | 280 | 3128 |
| Newfoundland | | | | 3196 |
| Quebec | | | | 545 |
| Atlantic Zone | | | | 11,573 |

The only regulation aimed at optimization of yield-per-recruit is MLS. This regulation along with prohibition of berried female possession, is designed to prevent recruitment overfishing. There are four minimum legal CLs; 63.5, 70.0, 76.0 and 81.0 mm which apply in different LFA's. The lower MLS's occur in warm water areas. MLS change over the past 75 years has been triggered for the most part by market pressures.

## Management Costs

Resource management costs are rarely available and have not been calculated for the total eastern Canadian lobster fishery. DFO's accounting structure does not permit the easy separation of lobster management costs from its myriad of programs. However, data are available for the Scotia-Fundy lobster fishery (Anon. 1989).

Scotia-Fundy Region's 1987 budget for fisheries management was $38.7 million, 34% of the Region's budget. Lobster management costs were $8.5 million, which included $2.4 million, $0.3 million, $1.0 million and $4.8 million, respectively, for biological research, economic research, conservation and allocation, and surveillance and enforcement. The total management cost was 5.9% of the total landed value ($144 million) of the inshore/offshore lobster fishery. Depreciation of capital assets was not included in the analysis.

The government does not have a rent recovery policy to offset these costs. A nominal fee of $30.00 is charged for licences.

## Resource Science and Stock Assessment

### *Resource Science*

Despite lobster research having been carried out in the northwest Atlantic since the late 1800's, there is yet a paucity of data in areas of lobster biology pertinent to resource management. Indeed, the lack of significant change in lobster management in eastern Canada over 40 years, despite major fishery problems, is in part due to the shallowness of our understanding of lobster ecology and natural history. For example, the lobster fishery in LFAs 29 to 33 (Fig. 8 and 9) collapsed in the late 1960's (Robinson 1980). Four disparate hypotheses were suggested by four groups of departmental scientists to explain the collapse. Dadswell (1979) suggested the damming of Canso Strait (the Canso Causeway, completed in 1955, connects Cape Breton Island and mainland Nova Scotia-see Fig. 1) interrupted larval recruitment to Chedabucto Bay from the southern Gulf of St. Lawrence. Harding et al. (1983) supported Dadswell's (1979) interpretation, but went on to suggest that oceanic temperatures along the eastern Scotian Shelf were generally too cold to permit significant larval maturation except in warm water years. Occurrence of anomalous warm water years was then suggested as the causal mechanism for the periodic, short-term recruitment peaks in lobster (Fig. 8). By contrast, Robinson (1979) argued recruitment overfishing was responsible for lobster catch decline. Mann and Breen (1972) agreed the catch decline was due to recruitment overfishing, but linked reduction in adult lobster density to release of predation pressure from the keystone species, *Strongylocentrotus droebachiensis*, the green sea urchin. This was their explanation for the explosion in urchin densities, which led to macrophyte overgrazing. Mann (1977), then predicted this to be a non-cyclic phenomenon; lobster densities would remain low because juvenile lobsters required macrophytes for habitat.

A similar number of hypotheses have been constructed around inshore and offshore interaction of Gulf of Maine lobster stocks. Industry was puzzled at the lack of consensus amongst scientists, which was in part responsible for the poor reception DFO received from Maritime fishermen in the early 1980's when attempting to present support for the implementation of CAFSAC advice. Inshore fishermen claimed lack of consensus demonstrated a lack of understanding of lobster ecology. They refused to support MLS increases until three conditions were filled: first, they wanted more knowledge of adult lobster movement; second, they

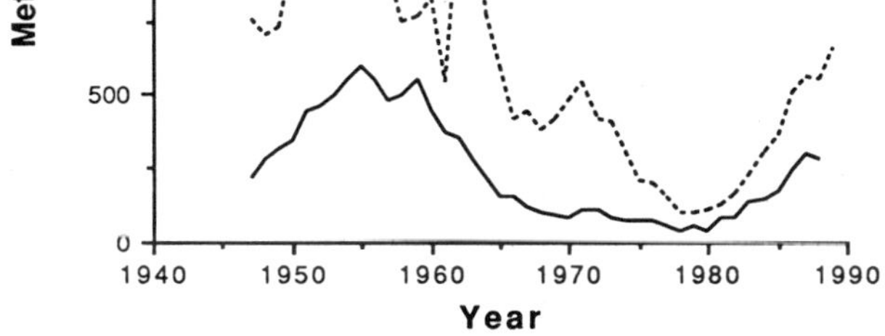


FIG. 8. Annual lobster landings for LFA 28–30 (combined) and LFA 31–32 (combined) between 1947 and 1988.

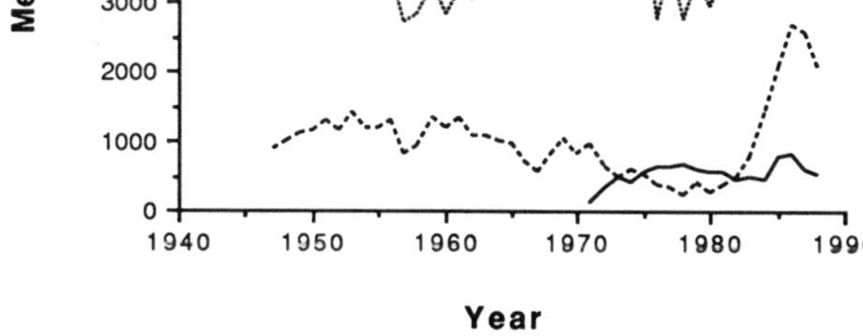


FIG. 9. Annual lobster landings for LFA 33, 34 & 40 between 1947 and 1988.

insisted DFO attain a better understanding of recruitment source; and thirdly, that the offshore lobster fishery be closed (inshore fishermen from the Scotian Shelf and northern Gulf of Maine felt offshore stocks were the broodstock of inshore stocks).

The direction for future lobster research over the next few years was obvious. No management action would gain fishermen's consent without an improved understanding of lobster natural history and population biology.

*Resource Assessment*

Each Atlantic Zone Region (Québec, Newfoundland, The Gulf and Scotia-Fundy) employs a team, consisting of at least one scientist and technician, which provides lobster biological advice. Scotia-Fundy Region is Fisheries and Oceans' "center of expertise" for lobster resource science. Here, a team of about 10 individuals was assembled in 1982. Each resource scientist was assigned both a stock to assess plus a mandate for resource science. Rationale for the dual role was, that mutual trust between fishermen and science personnel is required within the Canadian fisheries management structure to permit good science to become good resource management (Pringle 1985; Miller 1990a). To attain this trust, the scientist must be visible within the fishing communities, conducting that research most important to improving the stock assessment (Miller 1990a). There is a greater likelihood of this happening if the scientist has both stock assessment and resource science responsibilities.

The approach to follow was debated; the traditional fisheries approach, employing fishery statistical data and little biology within yield models (Gulland 1971) was rejected. In support of this decision, it was noted by Dow (1980) that little had come of this approach [see models by Dow et al. (1975) and Flowers and Saila (1972)] mainly because the basic biological knowledge was insufficient (Dow 1980). Indeed the latter concluded that *H. americanus* stocks were in trouble because biological knowledge was inadequate for effective management. Science personnel in Scotia-Fundy Region thus decided to emphasize biological research with emphasis on larval and juvenile natural history and ecology, reproductive ecology and trap selectivity (Pringle and Campbell 1988).

Fisheries catch and effort data, collected by non-science DFO personnel for other major fisheries, are not available for the lobster fishery. To gather these data a logbook system involving volunteer fishermen was initiated; three to four fishermen in each selected port record daily trap hauls and catch. Information on frequency of both undersized lobsters and berried females are also recorded by some volunteers.

Additionally, each team visits selected ports within all LFAs at season's beginning and end. The catches of two to four fisherman are haphazardly sampled. 500 lobsters are sexed and measured each sampling and the total sample weighed. Exploitation rate (A) and annual mortality (Z) are calculated respectively as:

$$A = 1 - (M2/M1) \text{ and}$$
$$Z = \log_e (M2/M1)$$

where $M1 = M1'/tm1$ and $M2 = M2'/tm2$. $M1'$ and $M2'$ are the number of lobsters in the catch in the first and second molt classes of market size (81–93 and 94–107 mm CL, respectively), and tm1 and tm2 are the average times (measured in years) spent in each molt class. Estimates of natural mortality vary, but the commonly assumed value is 0.1 (Thomas 1973; Campbell 1985).

Yield-per-recruit (Y/R) (calculated separately for males and females, then averaged) in grams and egg number per female recruit, when required, are calculated using Caddy's (1979) program with suggested changes by Campbell (1985). See Miller et al. (1987) for Y/R analyses for S-F stocks.

Specific knowledge on effort distribution has been unavailable. Requirements for finer resolution prompted indirect methods of assessment. Investigators in the Gulf of St. Lawrence (Gulf Region) developed an aerial visual scanning methodology (Conan and Maynard 1983), while aerial color photography (Pringle and Duggan 1983) and interviews with individual fishermen (Pringle and Duggan 1984) were tried elsewhere. A cost evaluation of these techniques showed the photography and interview techniques were the most and least costly, respectively (Pringle et al. 1986). Sharp et al. (1989) used the photography technique to plot fishing effort distribution on LFA 34's middle grounds, grounds that were not traditionally fished and may hold the major broodstock for this LFA.

## Management Objectives

Prior to the late 1960's, fishery regulations focused on the animal harvested, thus lobster regulation dealt with MLS, and protection of broodstock. As number of fishermen increased and modern fishing techniques enhanced fishing power, there was deemed a need to regulate effort and catch levels.

The Lobster Task Force of 1974 (Anon. 1975) sought means to attain a viable, self-sustaining fishery, managed in response to local needs with input from industry which would have greater access, than in the past, to biological, economical and technological knowledge. Later, Kirby (1982) called for creation of an economically viable industry with maximum employment, subject to the provision of economically reasonable returns to participants, and that the industry remain Canadian based. These objectives guide fisheries management to the present.

## Characteristics of a Modern Fishery

### Harvesting

The lobster fishery in Atlantic Canada is primarily an inshore, small boat fishery (Fig. 10) with landings concentrated in the southern Gulf of St. Lawrence (Gulf Region) and Scotian Shelf (Scotia-Fundy Region). The fishery occurs along the entire coast, in some LFA's up to 90 km from shore, though in most areas the lobster grounds extend only a few kilometres from the coast (Pringle and Duggan 1984). A small offshore sector exists in the northern Gulf of Maine, beyond 90 km from shore, in LFA 41 (Fig. 3) and is concentrated on Georges and Browns Bank. It began in 1971 with six vessels. Since 1985, the offshore fishery has been managed with both a total allowable catch (TAC) and an enterprise allocation (EA) scheme. In recent years, the offshore fishery accounted for 1–3% of Canadian landings.

Following the introduction of limited licensing in 1967, the industry entered a period of low landings and prices (Fig. 4 and 5). A Lobster Task Force (Anon. 1975) recommended that licenses be reduced both to decrease effort and to increase incomes. To assist this process, licenses were categorized into "A" [full time] and "B" [part time] based on economic dependence on the fishery. The "A" licence holders were able to fish the maximum trap limit permitted in the LFA and the licenses were made "transferable" or assignable. To improve efficiency, two "A" licenses can be combined temporarily as a partnership. Partnerships can deploy 150% of the trap limit. By contrast, "B" licence holders fish a reduced number of traps and the licence retires with the fisherman.

FIG. 10. A typical inshore lobster fishing vessel, about 12 m in length, with traps on stern (Source: DFO S-F Region files).

To further increase the earning power of individual fishermen, government sponsored a buy-back program which retired 1,569 licenses between 1978 and 1982 (unpublished DFO data). Initially, the buy-back program appeared to have little significant impact on incomes. Most licenses purchased were from marginal fishermen who were having little impact on the fishery; for example, the least productive 20% of the fishermen accounted for less than 1% of total landings. However, landings underwent consistent yearly increases throughout the 1980's, increasing from 20 949 t in 1981 to 44 963 t in 1990. A concomitant increase in lobster prices drove the value up by more than threefold between 1981 ($89 million) to a peak in 1987 ($282.8 million). Lobster fishermen's incomes underwent significant yearly increases during this period. Many of the licenses that were bought back would have been activated or transferred during these good times, and more fishermen would have participated in the fishery. Input costs for vessels and labour would thus have increased to produce the same output, resulting in lower net incomes.

Lobsters today are the backbone of the inshore fishery in Nova Scotia, New Brunswick, Prince Edward Island, and parts of Québec. Most of the 11,573 licenses are active for the fishing season. The average boat is operated by an owner-captain and an average crew of 1.5 to 2, providing employment for about 32,000. In most LFA's lobsters provide the base for a mixed fishery supplemented with groundfish, herring and other species. The degree of lobsters' contribution to income varies widely between Regions and is dependent on the productivity of the lobster grounds relative to that of other species. In some LFA's of the Scotia-Fundy Region, lobsters represent over 80% of a lobster fishermen's total landed value, while in others this may be as low as 60% (Anon 1989). Gulf Region fishermen rely on lobsters for about 76% of their total income.

Commercial fishing is done exclusively by baited traps, set in one of three configurations (Fig. 11). The traps are made of wood or wire mesh and are generally of one or two basic designs (Fig. 12 A and B). For further information on lobster trap design and fishing power see Krouse (1989). Miller (1990b) provides a thorough review of the selectivity of lobster traps. Most fishermen use Cape-Island-style, open deck boats between 10 and 13 m in length, and increasingly, fibreglass boats replace wooden ones. Several technological advancements have emerged such as diesel engines, hydraulic trap haulers and navigation equipment. In Scotia-Fundy for example, about 40% of the fleet was equipped with Loran C in 1987 (Table 5) while only 20% had such equipment a decade previously.

These improvements to the way lobsters are harvested have caused significant increases in costs (Anon. 1989). Cost increases were sustainable while the landing values were rapidly increasing prior to 1987 (Table 6). Since then, the value of landings has decreased substantially. Lobster fishermen are now finding it harder to finance the new equipment they purchased during the prosperous mid-1980's. Although net incomes vary widely between LFAs, they are falling.

## Processing — The On-Shore Sector

The lobster industry's on-shore sector receives and sorts lobsters at more than 1,000 ports, for either the live or processed market. Recent studies of concentration levels indicate that Scotia-Fundy's top 10 buyers accounted for 34% of purchases; the next 68 accounted for 56%; and the 102 smallest buyers accounted for 10% (Anon. 1989). The top 10 Gulf Region buyers account for

Table 5. Vessel/gear characteristics for the Scotia-Fundy fleet in 1987 (Anon. 1989).

| | | | % of Vessels | | | | |
|---|---|---|---|---|---|---|---|
| LFA | Average Vessel Length (m) | Average Crew Size (including skipper) | Wood Hulls | Diesel Engines | Loran | Hydraulic Haulers | Wire Traps (1) |
| 27 | 9.3 | 2.4 | 66.2 | 22.5 | 39.7 | 90.0 | 21.9 |
| 28, 29, 30 | 8.2 | 2.1 | 77.8 | 16.7 | 13.3 | 50.0 | 17.4 |
| 31 | 8.3 | 1.6 | 87.5 | 10.0 | 39.5 | 72.5 | 11.2 |
| 32 | 8.5 | 1.6 | 87.2 | 17.9 | 28.2 | 63.2 | 31.1 |
| 33 | 9.9 | 2.4 | 52.6 | 42.2 | 54.5 | 83.2 | 48.0 |
| 34 | 11.1 | 2.9 | 23.8 | 58.7 | 76.7 | 83.3 | 36.8 |
| 35 | 10.2 | 1.9 | 75.0 | 33.3 | 41.7 | 50.0 | 21.7 |
| 36 | 10.7 | 2.1 | 65.0 | 30.0 | 45.0 | 90.0 | 36.7 |
| 38 | 11.7 | 2.8 | 78.6 | 57.1 | 85.7 | 85.7 | 59.4 |

Note: The reported use of wire traps may be overstated. The percentage noted likely includes traps that are wire and part wood.

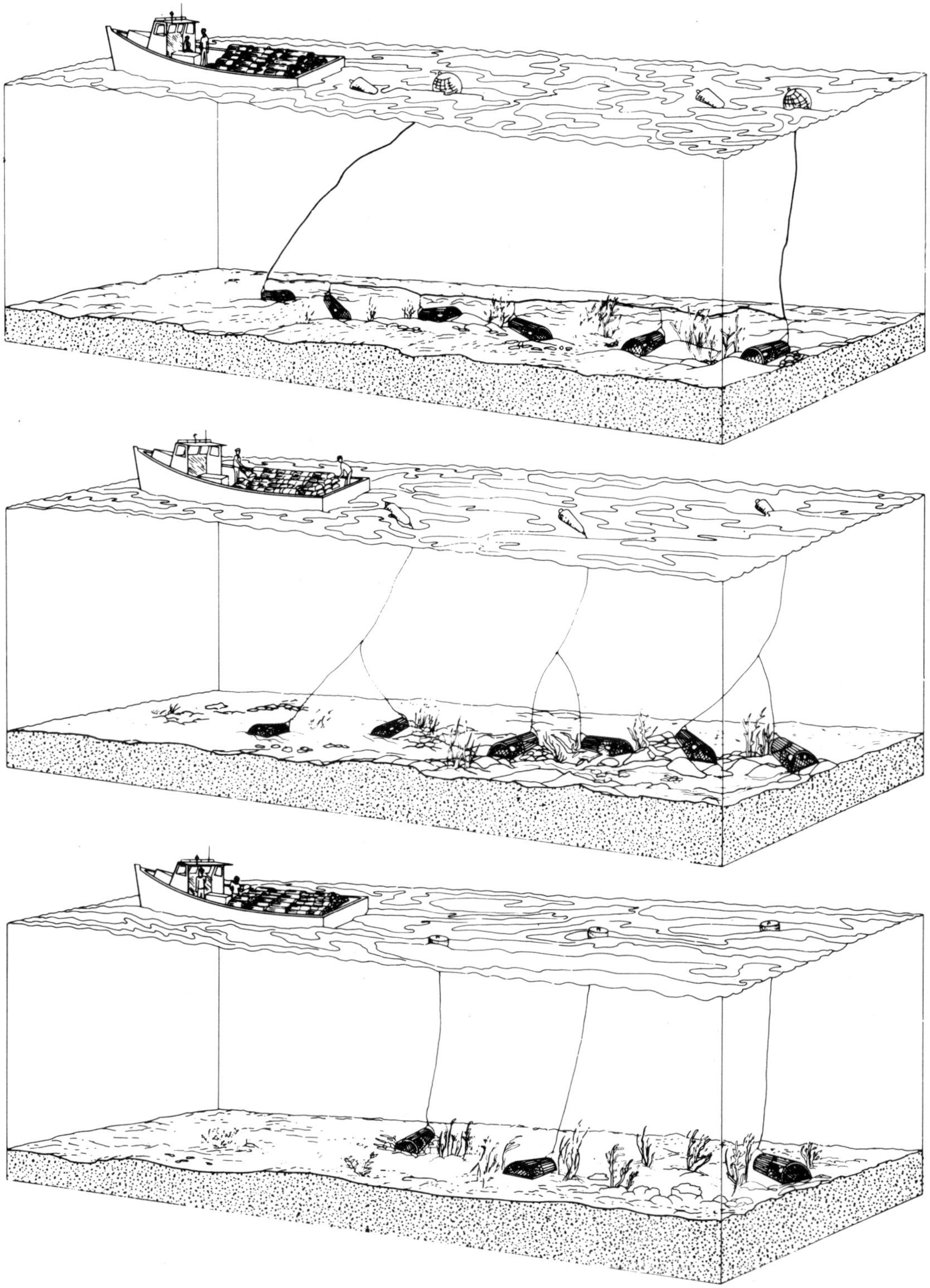

FIG. 11. A sketch showing the three basic methods of rigging lobster gear: upper panel — up to 12 traps hooked sequentially to a main line which is anchored at either end and buoyed above; middle panel — traps hooked together in pairs to a common buoy; and, lower diagram — one trap per surface buoy (Source: G. Sharp, DFO, unpubl. data).

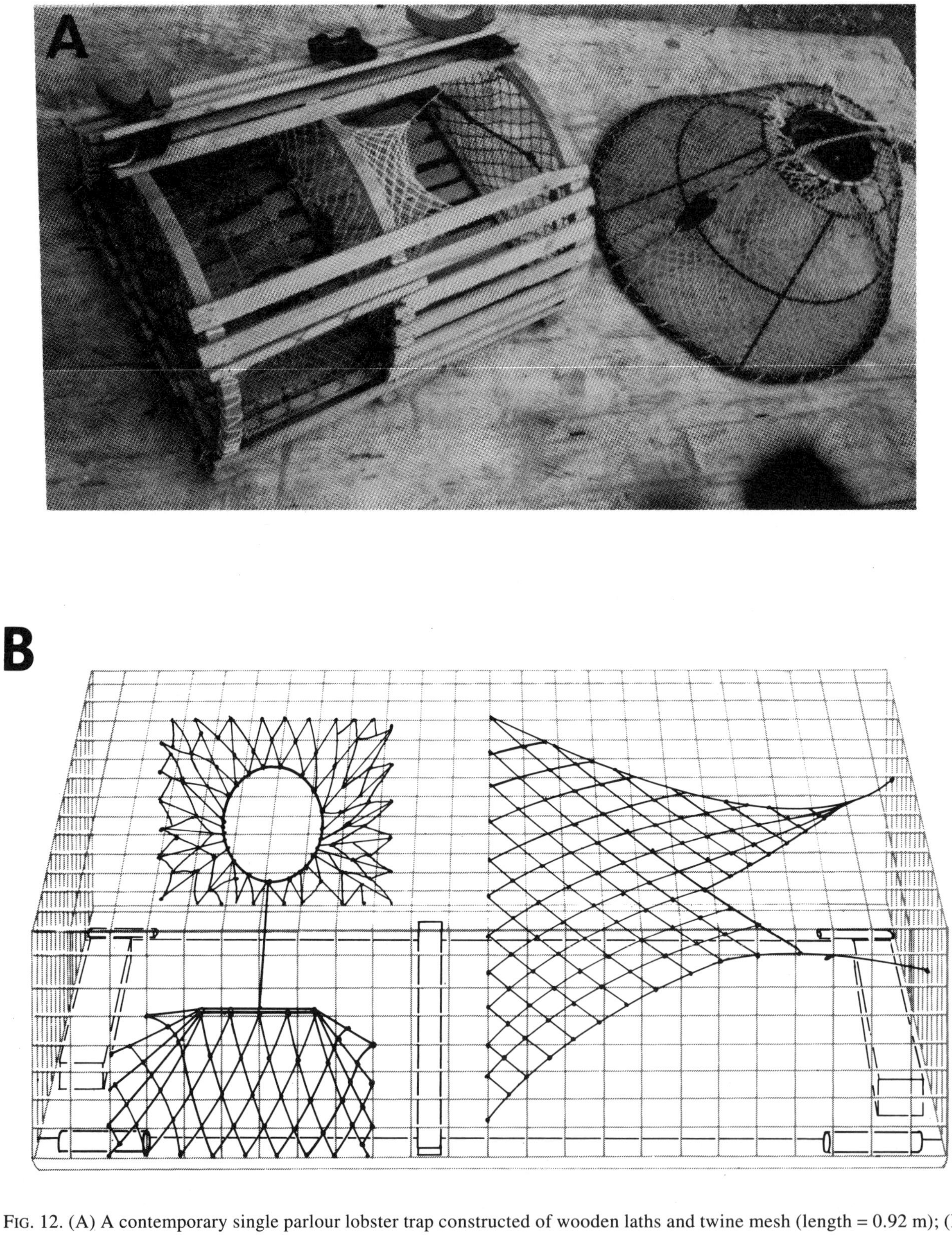


FIG. 12. (A) A contemporary single parlour lobster trap constructed of wooden laths and twine mesh (length = 0.92 m); (B) A sketch of a modern wire mesh, single parlour lobster trap (length = 1.2 m).

TABLE 6. The value, per LFA, of lobsters to Scotia-Fundy fishermen for the years 1978 and 1987 (Anon. 1989).

| | 1978 | | 1987 | |
|---|---|---|---|---|
| LFA | ($000) | % | ($000) | % |
| 27 | 2 530 | 10.0 | 18 908 | 13.6 |
| 28, 29,30 | 157 | 0.6 | 2 474 | 1.8 |
| 31 | 199 | 0.8 | 2 341 | 1.7 |
| 32 | 253 | 1.0 | 2 576 | 1.9 |
| 33 | 1 546 | 6.1 | 25 755 | 18.5 |
| 34 | 17 537 | 69.2 | 77 065 | 55.5 |
| 35 | 753 | 3.0 | 3 403 | 2.4 |
| 36 | 1 053 | 4.2 | 3 167 | 2.3 |
| 38 | 1 328 | 5.2 | 3 267 | 2.4 |
| Total | 25 356 | | 138 956 | |

30% of purchases. Companies tend to buy inter-Regionally and estimates suggest the Scotia-Fundy/Gulf top 10 companies account for 40% of all purchases (Gardner 1991). However, since the price of lobsters is derived mostly from demand in the highly competitive U.S.A. market, and there are close to 700 buyers in Atlantic Canada with no substantial barriers to entry at this level, competition is adequate to assure that no one agent in this sector can influence price.

About 90% of lobsters from Newfoundland and Scotia-Fundy, 80% of lobsters from Québec, and 30% from Gulf are marketed live. Live lobsters must be marketed through a network of "pounds and tanks". There are times when lobsters are best held, since Canadian landings do not coincide totally with peak live market demand (Pringle et al. 1983). Industry developed five types of holding facilities. Fishermen and buyers often store lobsters for short periods in moored crates called "cars". For medium term storage, crated lobsters will be held in either "inside pounds", which are buildings with cement tanks of circulating sea water, or natural sea enclosures called "tidal pounds". The latter are found mainly in the cold Bay of Fundy waters of Southern New Brunswick, which has traditionally served as an assembly area for Canadian lobsters destined for U.S.A. markets. "Tank houses", where lobsters are held without crates in small tanks of circulating sea water, and "dry-land pounds", where lobsters are stored individually in separate trays through which cold sea water is continuously circulated, are best for long-term storage. Two dry land pounds have been built in Atlantic Canada since 1982. Lobsters in these pounds have been graded by sex, size, molt stage and health when placed in individual trays. The system holds lobsters for long periods at low mortality rates; this approach best permits the wholesaler to better meet market demand year round.

Lobsters are best stored at temperatures below 4°C and are vulnerable to disease and stress when crowded. Healthy lobsters can be kept out of water for up to 36 hours, which permits shipment by truck or air to markets throughout North America, Europe and Japan. They are returned to saltwater tanks until sold, usually as a luxury item to restaurants, some supermarkets and fishmongers. The holding technology is relatively specialized and the costs, including the losses due to mortality, must be factored into retail prices (Fig. 13).

Lobsters, especially those from the Gulf Region, are processed into a number of products. Some of these are meat-in-shell products such as "popsicle packs" or tails. The price of these products reflect the shell weight and the processing cost. Shucked meat can be put into hermetically sealed cans, but most is frozen either in blister or cold-packs (frozen cans). These non-shell products have a higher price per pound. The price of shucked products reflects the 25% yield from live lobster and the processing costs (Fig. 14).

## Markets

The American and the European lobster (*Homarus gamarus*), the latter found along the European coast from Norway to Morocco, are the only crustaceans which can legally be called "lobster" in most western countries. They are a premium seafood, and Canada is the only net exporting nation with close to 90% of its catch exported. The U.S.A., Europe and Japan are the biggest importers, with the U.S.A. importing about 75% of Canadian landings. The U.S.A. imported close to 19 000 t of Canadian lobster products in 1988, the peak year of Canadian exports, of which over 14 000 t were live. Europe and Japan imported about 8 000 and 2 000 t of lobster products, respectively. While the volume of live lobster imported by the U.S.A. continues to increase, imports to both Europe and Japan are increasing more rapidly (Fig. 15).

Canadian export of all lobster products grew steadily through the 1980's until 1989, when all three major foreign markets imported significantly less product. U.S.A. demand fell for all forms, while Europeans imported less processed, but more live lobster since 1985. Japanese imports increased by a factor of 20 between 1981 and 1988, but fell by 30% between 1988 and 1989.

Suggested reasons for the reduced demand for Canadian lobster are numerous. First, a recent U.S.A. trade restriction precluded about 8% of the Canadian live catch (that proportion which falls between 81.0 and 82.5 mm CL). Second, Canadian exporters are facing increased competition in their traditional export markets from American exporters. For example, U.S.A. exports of lobster recently increased by 89%,

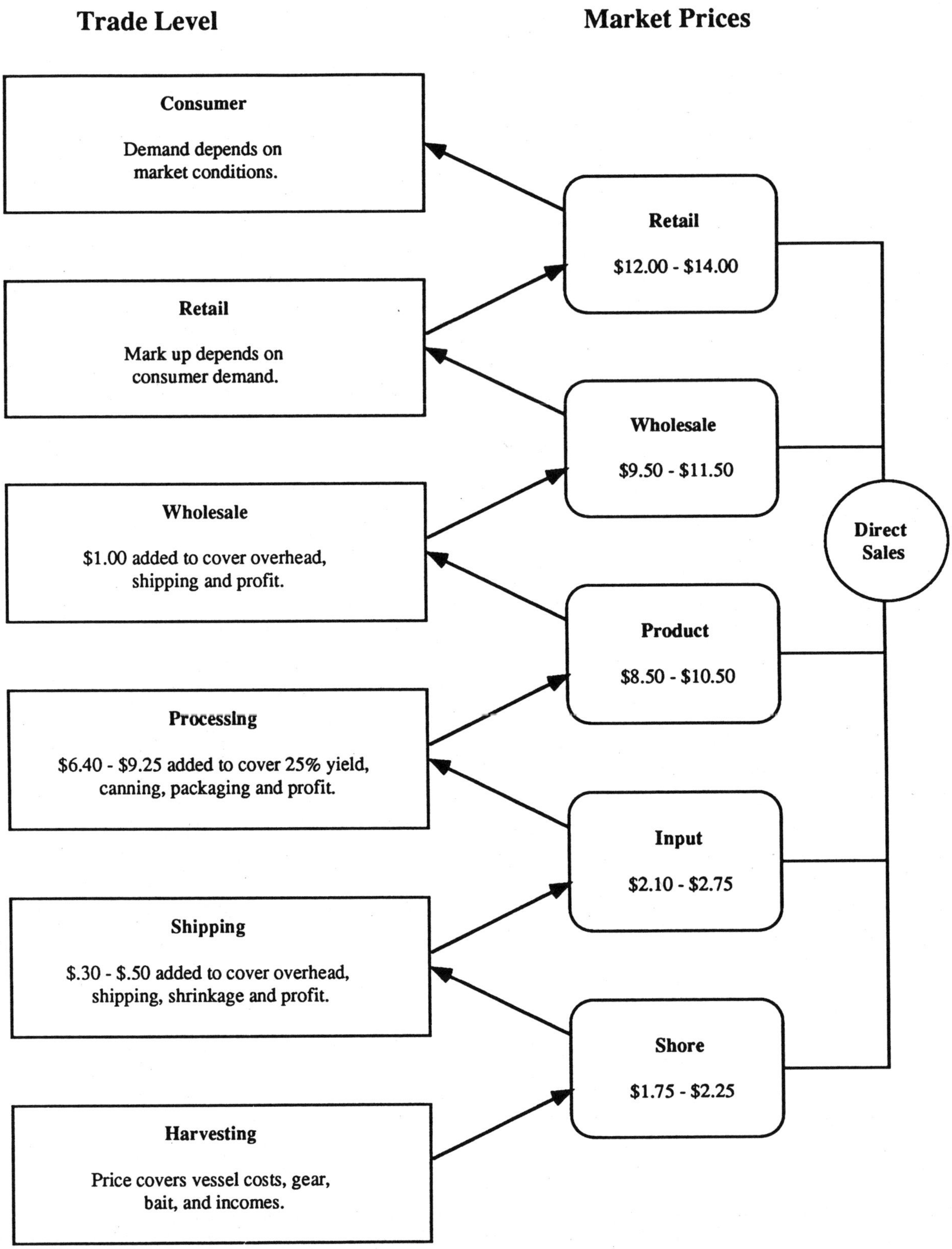


FIG. 13. Schematic of price formation in North American lobster markets (Gardner Pinfold Consulting Economists Ltd., Halifax, N.S., Canada).

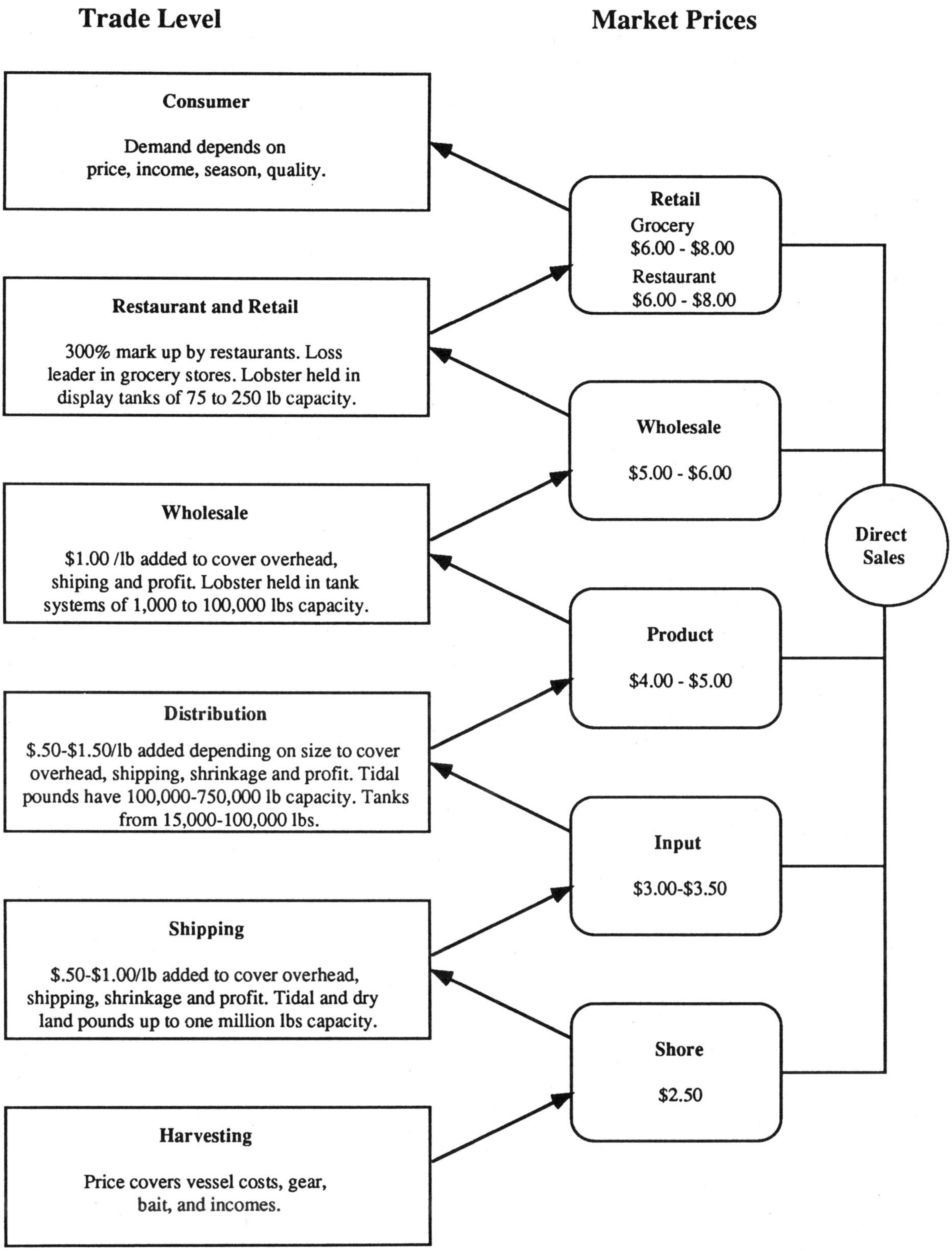


FIG. 14. Schematic of price formation for a 321 g (11.3 oz.) can of frozen lobster meat. (Gardner Pinfold Consulting Economists Ltd., Halifax, N.S., Canada).

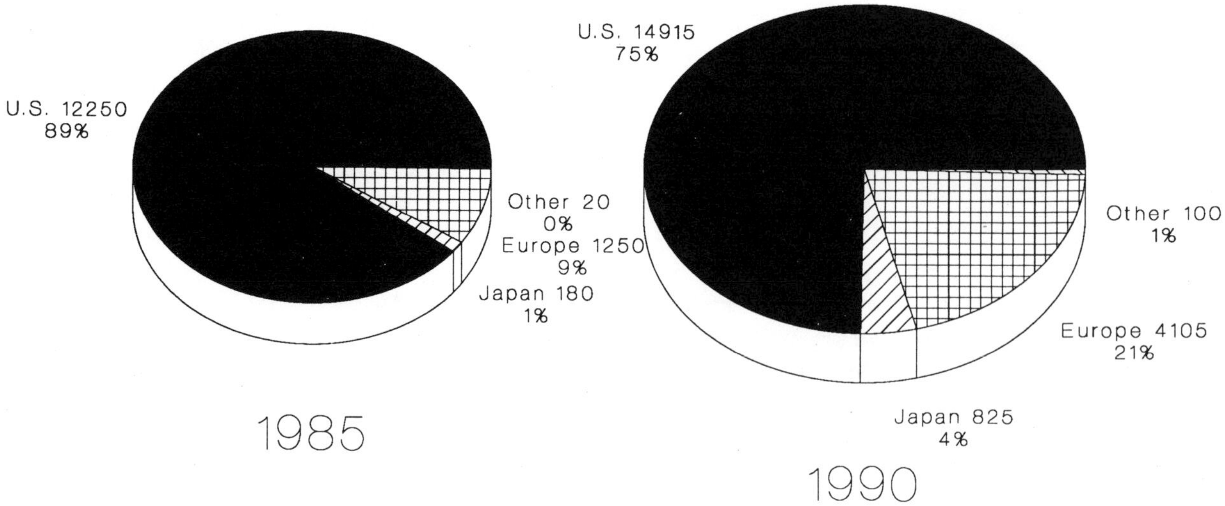


FIG. 15. Canadian live lobster exports by market (U.S.A., Europe, Japan and others), 1985 and 1990. Quantities in metric tons. Source: Statistics Canada.

principally in Europe and Japan. Finally, the Canadian dollar appreciation relative to the U.S.A. dollar, since 1987, makes U.S.A. exports relatively less expensive on the world market.

**Price Trends**

Canadian lobster prices are derived from world demand. This was demonstrated between 1980 and 1987, when lobster ex-vessel price increased 7.7% per annum, while landings also increased (Fig. 4). However, this relationship reversed when price fell by 18% between 1987 and 1990 and supply continued to increase. Although Scotia-Fundy landings attained higher prices (due to the region's higher proportion of market size landings), all Regions have followed similar pricing trends.

While long-term lobster price appears demand driven, supply exerts influence on lobster price over the short-term. The emergence of intra-annual price fluctuations depicts a sensitivity of price to supply. U.S.A. live lobster supply is lowest in February, rises gradually through April, more rapidly in May and remains high until October, declines in November, increases in December and then falls. Canadian lobster seasons tend to coincide with periods of U.S.A. low-landings, thereby achieving some price stability at premiums levels. Canadian exporters attempt to stabilize price and maximize revenues by timing pound releases to coincide with periods of low-supply. Product is released when exporters discern appropriate market conditions. Misreading the market can create under or over supply, causing short-term price swings.

The final price in the U.S.A. market tends to influence all levels of prices in Canada. Canadian lobsters exported live to the U.S.A. are graded by weight into five grades at the wholesale market {chicks (1 lb), eights (1 1/8 lb), quarters (1 1/4 lb), halves (1 1/2 lb), and selects (2 lb)}. Higher prices per kilogram are paid for the larger grades of lobster; the difference between quarters and halves was as much as $2.20 U.S.A. in 1987, but by the end of 1990 this difference diminished to about $0.35 (Fig. 16). In Canada, lobster is sorted into three grades: canners, markets and jumbos. "Canners" are excluded from the U.S.A. live market. "Markets" include the first four U.S.A. grades, and "jumbos" are selects. In Canada, "markets", which include all lobsters between 1 and 3 lb, command the highest price per pound, while "jumbos" (lobsters greater than 3 lb) command a slightly lower price (Fig. 17). "Canners" can sometimes command only 70% of the "market's" price.

Market information appears adequate for pricing parity to be reached. Exporters and importers are knowledgeable of current market prices and the size distribution of lobsters shipped. New England wholesale prices, Nova Scotia export prices, and Scotia-Fundy shore prices fluctuate in concert (Fig. 18).

In Canada's other major export markets, Europe and Japan, a higher premium is paid for lobsters (Fig. 19). The Japanese have traditionally paid the highest premium, but in 1989, Japanese consumers paid only $1.50 more per pound than U.S. consumers.

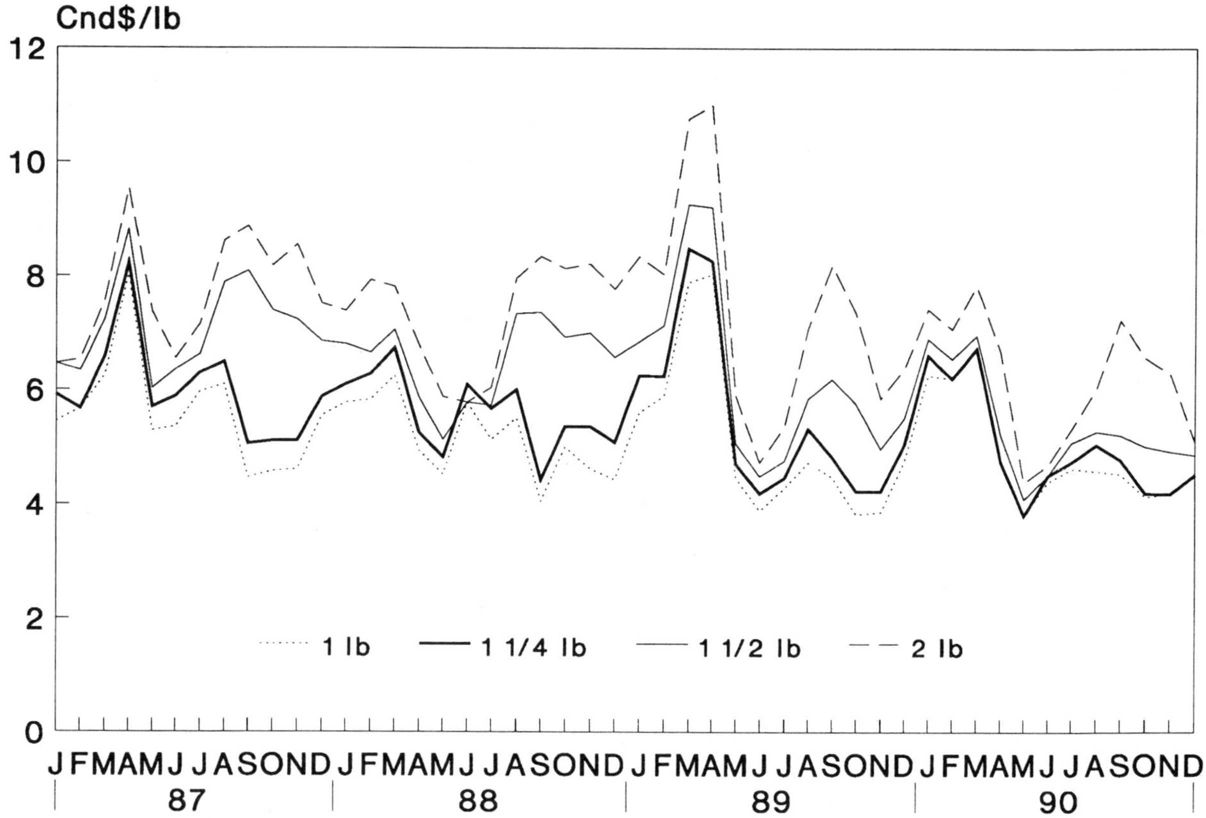


FIG. 16. New England wholesale lobster prices by grade, 1987 to 1990, current Canadian dollars. Source: Seafood-Price Current.

**Stock Status**

The term "stock" is defined as "the part of a fish population which is under consideration from the point of view of actual or potential utilization" (Ricker 1975). A population on the other hand is all the members of one species that inhabit an area sufficiently small to enable them to interbreed freely (Pielou 1976). Thus "stock" is a resource management term, whereas "population" has biological connotations. Studies to discern lobster population structure using morphometrics (Saila and Flowers 1969), and biochemical markers (Hedgecock et al. 1977) have been inconclusive (Pezzack and Pringle 1986).

Using indirect techniques, Harding et al. (1983) and Campbell and Mohn (1983) attempted lobster population discrimination for eastern Canada by analyzing 90 years of landing data. Independently both groups suggested three populations; the Gulf of St. Lawrence, Nova Scotia's east and southern coasts, and southwestern Nova Scotia and the Bay of Fundy. Given the lack of unequivocal evidence for population structure, the term stock will be employed in the remainder of the present study.

The calculated exploitation rate of all Canadian lobster stocks is high (Ennis 1986; Miller et al. 1987). Thus, if lobsters decline in abundance, recruitment overfishing is frequently hypothesized, as legal minimum size is substantially below that at which most lobsters become sexually mature (Robinson 1979; Anthony and Caddy 1980; Ennis 1986). As well, levels of fecundity increase with lobster size (Aiken and Waddy 1980) thus, most fecund females are small in size and with a concomitant small clutch.

*Southern Gulf of St. Lawrence*

Annual mortality and recruits is around 80 to 90% (D.G. Robinson, Lobster biologist, Clearwater Fine Foods, Halifax, Nova Scotia, unpubl. data). The harvest has been sustained for the central portion of eastern New Brunswick (portions of LFA 23 and 25), the northern coast of Prince Edward Island (LFA 24), and the southeastern corner of the Gulf, including the north shore of

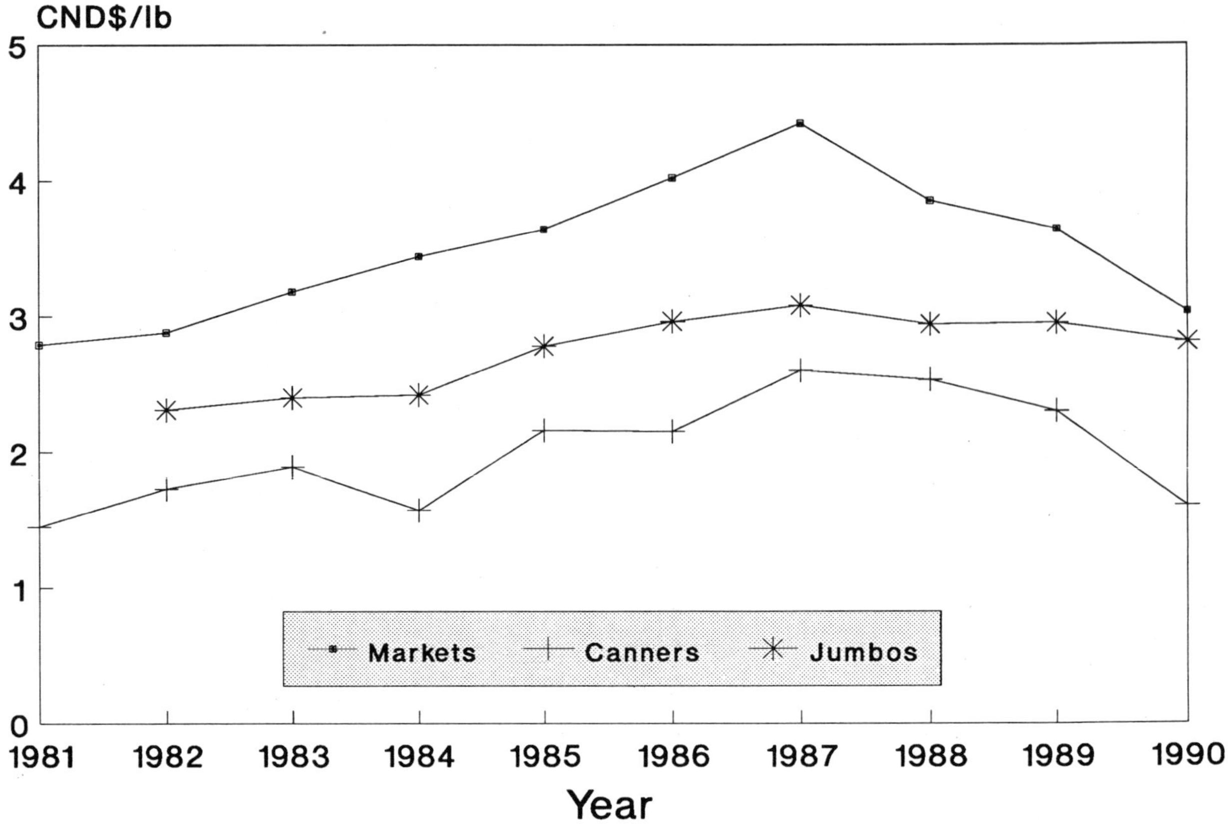

FIG. 17. Scotia-Fundy average annual ex-vessel live lobster price by grade, 1981–1990. Current dollars. Source: DFO.

Cape Breton County (LFA 26B). Stock collapse occurred in central Northumberland Strait (eastern portion of LFA 25 and the western portion of LFA 26A) between the late 1960's and early 1980's, with landings at about 5% of historic maximum catches. Landings in northern New Brunswick and the Iles de la Madeleine have undergone long-term cyclical trends of varying magnitudes.

Although the evidence is not conclusive (Wilder 1963; Munroe and Therriault 1981), high summer water temperatures in the southern Gulf may permit two molts per year in newly recruited lobsters. Mature females would extrude eggs rather than molt the second time. Legal minimum size (LMS) is at the small end of the reproductively mature size range, thus only a few females are functionally mature upon first molt into legal size (molting occurs in July/August). Egg extrusion occurs following molt and these "berried" females (those incubating eggs) are protected in the spring fishery. The timing of the fishing season likely plays a crucial role in sustaining these stocks.

Recruits in the fall fishery (open season, August 10 to October 10) of LFA 25, have only just molted and quality is poor. Many females are captured prior to egg extrusion. Recruitment overfishing has been suggested as the cause of the collapsed Northumberland Strait fishery, where source of recruitment may be northern LFA 25 (D.G. Robinson, pers. comm.). This hypothesis, however, does not explain sustained landings in the latter area. Available data are inadequate to explain the fishery collapse in central Northumberland Strait (portions of LFAs 25 and 26A); hypotheses can only be formulated and tested. It has been suggested that adult migration played a role in a recent recruitment pulse (G. Conan, Lobster Research Scientist, DFO, Moncton, New Brunswick, pers. comm.).

*The Gaspé and Iles de la Madeleine*

In the cooler waters of the Gulf of St. Lawrence (LFA 20 and 22), it has been hypothesized that egg production per recruit is 5–10 times less than in the warmer waters, because there is either only one molt or one egg extrusion per reproductively mature female per year. The cyclic landing patterns may have resulted from recruitment limited by egg production (D.G. Robinson, pers.

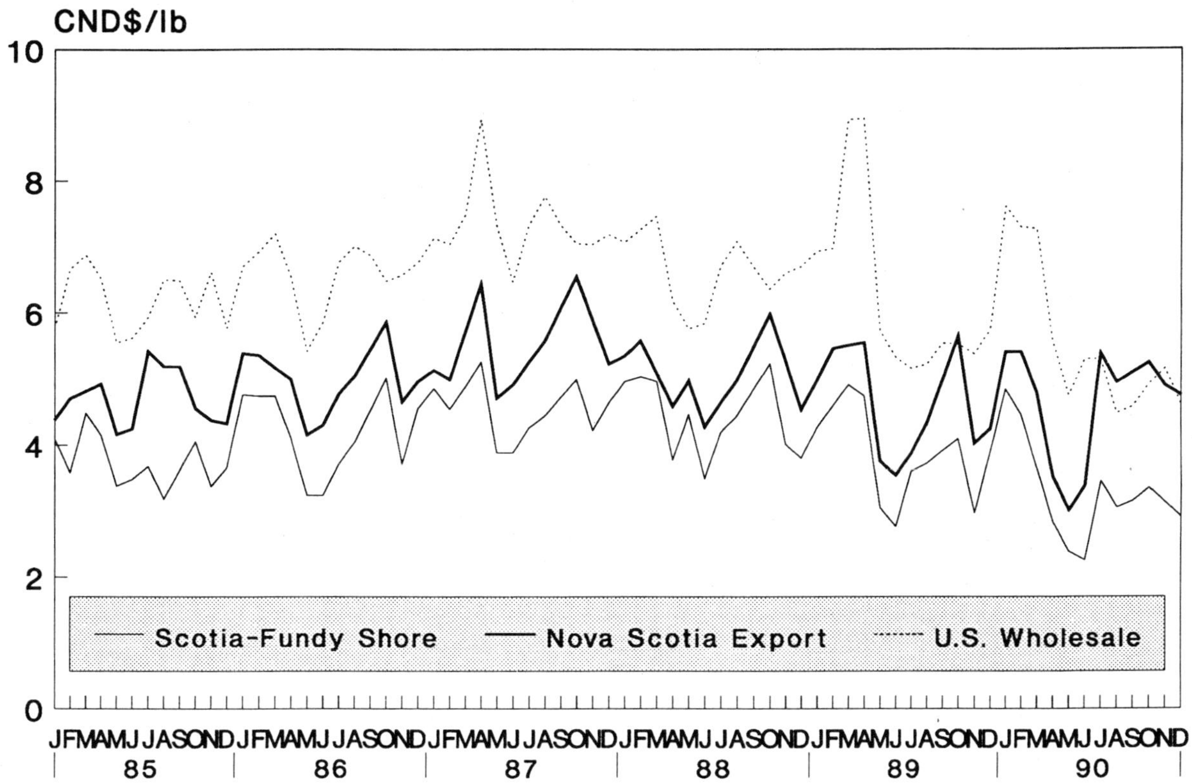


FIG. 18. Monthly live lobster prices, Scotia-Fundy Shore, Nova Scotia export and U.S.A. wholesale. Source: DFO and Seafood-Price Current.

comm.). Fluctuating environmental parameters, linked to St. Lawrence River outflow have been implicated in the cycles of lobster abundance (Sutcliffe 1973; Sinclair et al. 1988).

*Sydney Bight (LFA 27)*

The mean size and size at maturity of animals in the catch from LFA 27 increase markedly from west to east. This reflects the influence of waters from the warm Gulf of St. Lawrence (west) and the cooler Scotian Shelf (east). The warm waters will enhance larval survival and are reflected in the intermediate minimum legal size of 70 mm CL. Landings were relatively stable between the mid 1940's and 1982. Annual catches have since increased at a rapid rate, to reach a record high in 1989 (Fig. 20). Landing stability may be due to larval retention caused by a gyre in the Sydney Bight (B. Petrie, Physical Oceanographer, BIO, DFO, Dartmouth Nova Scotia, pers. comm.).

*Atlantic Coast of Nova Scotia (LFA 29–33)*

The harvest between 1960 and 1985 was not consistently low, but was deemed collapsed in the 1970's at 5% of peak landings (Fig. 8 and 9). Given colder water temperatures than the southern Gulf, growth rates for all life history stages will be lower (Robinson 1980); conditions for larval survival have been suggested as poor (Harding et al. 1983). Annual landings since 1982 have increased at a moderate rate in the eastern LFA's (29–32); in the west (LFA 33), however, gains were much more rapid.

Assessments suggest recruitment was sustained during the 1940's when an estimated 5,000 eggs per recruit were produced (D.G. Robinson, unpubl. data). Calculated production of about 10,000 to 15,000 eggs per recruit in the early 1980's suggested the stock could be in recovery mode (Campbell and Robinson 1983). The recovery rate would have doubled or tripled current landings per generation length (8–10 years). It is likely the enhanced recruit biomass could have been quickly harvested by the existing latent fishing power within the fishery, prior to any significant increase in larval production (Pringle and Duggan 1984). Management has, however, kept exploitation rates down to between 40 and 60% (Miller et al. 1987). This may over a period of years, permit the rebuilding of the broodstock.

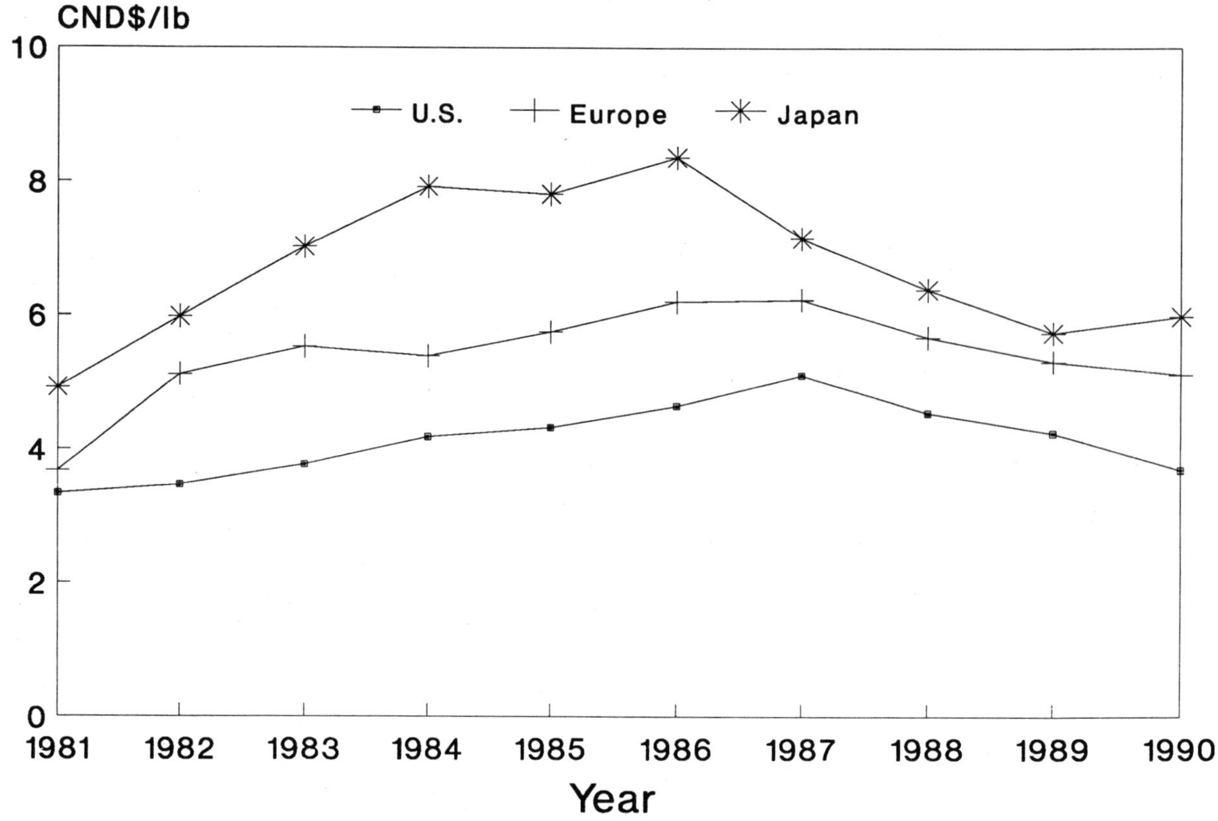


FIG. 19. Average annual export price, Canadian live lobster, 1981–1990. Prices adjusted for exchange rates. Source: Statistics Canada.

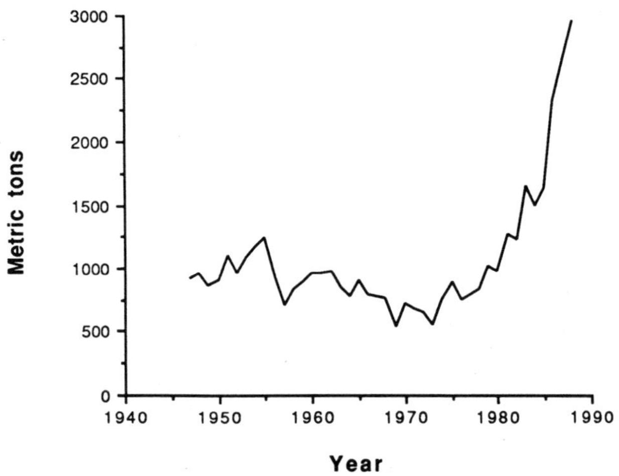


FIG. 20. Annual lobster landings for LFA 27 between 1947 and 1988.

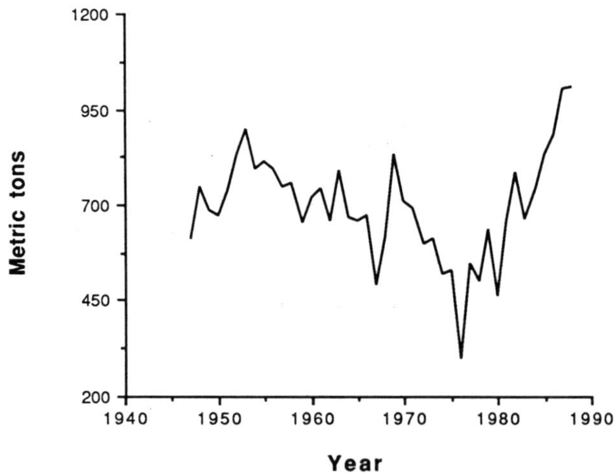


FIG. 21. Annual lobster landings for LFA 35, 36 & 38 (combined) between 1947 and 1988.

*Southwestern Nova Scotia*

Management techniques are quite different between the two Canadian stocks (inshore and offshore) in the northern Gulf of Maine (Pezzack and Pringle 1986). Little information is available on the distribution of fishing effort (Sharp and Duggan 1985) over the inshore stock, which confounds attempts at interpreting source of landings. Despite cold waters, production has been high and stable since the 1940's and since 1983, landings increased by 166% (Fig. 9), effort increased by 81% (Campbell 1989) and exploitation rates are about 85% (Miller et al. 1987). Abundance increase is due in part to a pulse in juvenile recruits and a greater exploitation of large lobsters (possible inshore broodstock) in the middle grounds, a former refuge area between 45 and 90 km from shore (Anthony and Caddy 1980; Miller et al. 1987). The impact of this increased effort on inshore abundance will only be discerned in the mid 1990's, a generation after enhanced exploitation (Campbell 1989). A recent review of the northern Gulf of Maine stocks suggests the inshore fishery is self sustaining (Pezzack et al. 1992).

*Bay of Fundy*

LFA's 35, 36, and 38 are wholly within the Bay of Fundy. The water is cold, suggesting low larval survival rates. Lobster mean size at capture in the Bay's upper reaches is large, similar to that found off northern Grand Manan Island. Berried females tend to concentrate in summer on the western side of the Bay (Campbell and Pezzack 1986); these females may be the source of recruitment for the Bay's stocks. However, residual currents from the northern Gulf of Maine enter the Bay on the east side and exit on the west side (Bumpus and Lauzier 1965), suggesting that if lobster larvae are merely current drifters, then recruitment could be from the offshore banks (Harding et al. 1983).

Interannual variation in yields has been high during the last 40 years (Fig. 21). Annual catches reached a minimum in 1976, increasing steadily thereafter. Much of the annual landings were from upper Bay grounds, although the most productive grounds are at the Bay's mouth. Current landing levels have not been equalled since the 1950's. Bay of Fundy annual exploitation rates were estimated at 70% (Miller et al. 1987).

*Newfoundland*

Stock assessments clearly indicate exploitation rates are much higher than those that would maximize Y/R at current MLS (Ennis 1978; 1980). The possibility of recruitment failure is real given high exploitation rates (Ennis 1981). Moreover, though the current MLS is well above the smallest size at which females lay eggs in most LFA's (Ennis 1981), in certain Newfoundland waters, egg bearing females as small as 65 mm CL have been observed. Landings comparable to recent levels were taken in 1954 and 1955 (Fig. 6) with less effort, and therefore presumably with lower exploitation rates. This did not prevent a 20 year decline in abundance likely caused by less-than-favourable environmental conditions for larval and juvenile growth and survival combined with recruitment overfishing (Ennis 1986).

Increased effort in recent years has caused low, or marginally productive areas to be fished; these "refugia" (sensu Anthony and Caddy 1980) were not previously, or only lightly, fished. Lobsters here grew to large sizes; egg production per female would be high. The contribution of these areas to overall egg production might have been substantial (G.P. Ennis, Fisheries scientist, DFO, St. John's, Newfoundland; pers. comm.).

Analysis of data from local field work demonstrates a 91% exploitation rate or higher (Ennis 1981). Slight variations in annual recruitment will result in substantial interannual fluctuations in landings. Natural fluctuations in lobster abundance will occur no matter the management plan, but excessive exploitation rates combined with a juvenile MLS no doubt exacerbate the phenomenon. Under present management, one can expect a downward trend similar to, but probably steeper, than that which occurred between 1955 and 1972 (Ennis, pers. comm.).

## The Recruitment Pulse of the 1980's

Canadian lobster landings, since 1979, have increased at an unprecedented rate, attaining levels not previously seen this century (Fig. 4). The increase in fishing effort of the 1980's (Miller et al. 1987) appeared not to be the direct cause of the increase in landings, (Pringle and Duggan 1985). Improved fishing success in most LFA's was due to an increase in young recruits (Miller et al. 1987). The recruitment pulse occurred in the Gulf of St. Lawrence (Fig. 5), perhaps the inner Scotian Shelf (Fig. 8), and in southwest Nova Scotia (Fig. 9) thus crossing the boundaries of the three "populations" of Harding et al. (1983) and Campbell and Mohn (1983). It occurred as far south as Massachusetts (Pezzack 1992). Even the collapsed stocks within Northumberland Strait and off Nova Scotia's open Atlantic coast show sign of some recovery. Hypotheses to explain these stock collapses (Mann and Breen 1972, Dadswell 1979 and Robinson 1979) must now be reconsidered.

The cause of the recruitment pulse likely occurred in the early 1970's, 5–9 years (the approximate time interval from larval settlement to size at first capture, depending on water temperature) from its initial appearance. A recruitment pulse covering such a broad geographic area, and encompassing two significantly different resource management schemes (Canadian and

U.S.A.), suggests that an oceanic climate more favourable to larval and/or juvenile lobster survival than in the immediate past, played a significant role; likely a larger role than resource management.

Pringle (1986b) noted a coincidence in urchin mortality levels, unstable lobster landings and finfish distribution patterns along the Scotian Shelf. He hypothesized a link to Gulf of St. Lawrence outflow to the Scotian Shelf. The seed for this concept was the tight correlation between Iles de la Madeleine lobster landings and St. Lawrence River discharge (Sutcliffe 1973). This discharge influences Gulf of St. Lawrence outflow. The link is strengthened when it was noted the relatively small (only 4% of the Canadian landings) Iles de la Madeleine annual lobster landings tracked the Canadian landings ($r^2 = 0.71$) between 1940 and 1988. Thus, large scale, lobster recruit success in the Gulf of St. Lawrence, along Nova Scotia's eastern and southern shores and in Massachusetts appears to have been strongly influenced by a common, but unknown factor(s).

## Challenges to Future Resource Management

Some years ago Alverson and Paulik (1973) grouped the problems facing fishery resource managers (resource management was defined as "everything done to maintain or improve fisheries resources and their utilization"). Six of their seven groupings encompass most issues facing the managers of Canada's lobster fishery.

### A Changing World: Enhanced Fishing Technology

Alverson and Paulik (1973) propose that the gap between the ability to catch fish and the ability to manage is broadening. Fishermen are constantly improving their ability to catch fish. Indeed modernization of the lobster fishery has moved at a rapid pace, particularly in the major producing areas of the southern Gulf of St. Lawrence and the northern Gulf of Maine. For example, bigger, faster, modern and more dependable vessels moved into the middle grounds in LFA 34 in the early 1980's. It was believed to be a refuge area (Anthony and Caddy 1980) and possibly the major brood stock area for the inshore fishery. Fishermen now fish with trawls of up to 15 traps (Fig. 11). Directional equipment and radar permit the rapid location of gear. Fishermen tend to stay on these distant grounds for a few days with 50 to 100% more crew; trap soak time is much less than 24 hours. To worsen the situation, there has been considerable abuse of the trap limit in these distant waters. The impact on nearshore production should be apparent in the mid 1990's.

### Nature of Objectives

"Nature of objectives" includes the problems of differing objectives both between the resource management agency and industry and amongst the many disciplines within the agency. Fishing is an ancient occupation, but only recently has there been a rapid change in the understanding of the dynamics between fishermen and the resource. This has complicated the process of defining management objectives. For example, should exploitation levels offer protection against overfishing or maximize yield in bio-economic terms? Regarding participation, should the objective be to maximize the distribution of wealth, or should wealth creation override number of participants?

Fisheries resource management is a young and complex discipline. No single person or discipline is all knowing. It is thus important that an inter-disciplinary team approach be developed within the management agency, and that objectives be set following comprehensive discussions. The loss in time and in human and fiscal resources is significant when the wrong path is chosen; more significant is the adverse impact on agency credibility with industry, which can impede resource management for years.

Setting objectives is a process which should involve fishermen, resource managers and interested Canadians. The objectives must be ratified or endorsed through the political process. The challenge is to set clear objectives that will provide the framework for management action and the benchmark against which progress can be measured.

### Management Institutions

Historians of resource management have concluded that institutional structure is important (McEvoy 1982). Most resource management agencies yield to political pressure (Alverson and Paulik 1973). Setting the stage within the lobster fishery for the implementation of a major change in a resource management plan, such as CL increase, may take years. Having set the stage, a poor decision by a senior bureaucrat could set lobster resource management back years. Government institutions within a democracy are inevitably responsive to political direction. Within Canada, mechanisms are currently being developed to separate policy decision-making from day-to-day operational decisions on licensing and allocation matters.

### Allocation of Resources

There are two dimensions to the issue of allocation of any renewable resource; first, when should a particular year class be harvested and second, who is given the right to harvest? For lobsters, minimum legal size dictates when a cohort will recruit to the fishery. This regulatory feature has varied little over time. Strategies to increase or decrease the MLS (within safe conservation practices) to optimize returns in response to supply and demand could be considered. The value of the fishery, by taking advantage of market demands, could theoretically be increased by delaying or advancing the time when a cohort recruits.

The question of right-of-access to this limited entry fishery has been hotly debated during the period of enhanced abundance of the 1980's. The contentious issues that require resolution are at what point can additional effort be allocated, what form should the allocation take, and what groups should decide who should participate? There were non-lobster fishermen who wanted to participate in this lucrative fishery; there were lobster fishermen who wanted to increase fishing effort. Existing license holders naturally resisted any efforts to arbitrarily increase the number of licenses since new entrants would erode their earning potential.

Other approaches to lobster allocation are being discussed. The lobster fishery has matured under a "rights based" management scheme (Rettig 1989) with restrictive licensing and without catch limits. The fishery has enjoyed considerable success through the 1980's with high landings and high earnings to fishermen. The fishery has a small standard unit of effort, the trap-day, which could underlie a refined capacity-based management regime. There are few fisheries with such a relatively homogenous input unit. Most of the ongoing experiments with EA's or individual quotas, IQ's, are based on individually assigned catch quotas. A credible trap-based unit of capacity would allow for the further development of some of the six characteristics identified by Scott (1989) as properties of rights based fisheries: duration, flexibility, exclusivity, quality of title, transferability, and divisibility in a capacity regime. A system based on tradeable trap fishing rights would provide many of the advantages of the output based schemes without the need for TAC's.

## Adequacy of Scientific and Economic Advice

Lobster science has been carried out for a century or more, resulting in a wealth of knowledge on many aspects of lobster biology. Fundamental information is lacking in key areas of larval, juvenile and reproductive ecology and in the understanding of population structure. Given that we are dealing with complex ecological systems there will always be deficiencies in our understanding. Often this paucity in knowledge has been used to stifle sensible changes to lobster management plans.

Economic and marketing advice are generated from within complex systems as well. Complete confidence in predicted impact cannot always be given. Moves to enhance lobster yield-per-recruit and reduction in trap number have been stifled, in part, for want of more thorough economic and marketing analyses. S-F fishermen have requested a better understanding of the relationship between demand, supply and price (Anon. 1989).

Given the nature of the fishery and market systems, there is a need for scientists and economists to work closely with, and gain the confidence of industry. Where solid relationships were developed, progress has occurred. For instance there is now a consensus to increase CL in LFA 26B, and we witnessed a demand from the fisherman for escape vents in LFA 33. It also requires that economic, marketing and resource science projects be priorized, based on their potential to allow management plans to progress.

## Enforcement

Canadian lobster surveillance and enforcement tends to receive about 44% of the lobster resource management budget (Anon. 1989), a percentage much higher than in U.S.A. waters, a fact readily apparent during deliberation of the U.S.A/Canada free-trade panel. Overtrapping, the most widespread violation, causes fishermen to lose an estimated $1,700,000 in profits for each 10% of traps fished in excess of the legal number (Anon. 1989). Abuse of MLS and poaching occur in isolated areas. Fishermen's associations have successfully lobbied for higher penalties, and in conjunction with DFO have begun a "Fishermen's Watch" program. The potential for increasing compliance through the use of stiffer penalties to violaters is significant.

## Summary

The commercial lobster fishery has enjoyed a long and sustained history in most of Atlantic Canada in spite of high exploitation rates. Management measures have been introduced which appear to have protected many of the stocks against recruitment overfishing. Licensing and effort controls have had a positive impact on fishing costs and hence net earnings. As a result lobster has become the mainstay of the inshore, small boat fishery.

The industry, beyond the Gulf of St. Lawrence, went through a major transition around mid-century when it converted from a canner to a live market fishery. An appropriate mix of seasons permitted Canada to nicely take advantage of a lucrative American market. The minimum legal size of 81 mm was adopted in the major lobster producing states and in the cold water fisheries of Canada. Thus it was a size chosen, not because it was the most appropriate for the biology of the lobster in Canadian waters, but because of current market demands. To this day Canada has a canner fishery in its warm water areas (Gulf of St. Lawrence) and a market fishery elsewhere.

Conditions prevailed which tended to thwart change, particularly in the Gulf. First, the warm water conditions promoted faster growth and earlier maturation, which permitted exploitation at smaller size. Secondly, winter ice coverage restricted seasons to the April/November period. The consequences of this are that much of the Gulf catch occurs during the period of the American peak catch. Also, 25% of the catch is "soft shell", which precludes live sales. These conditions along with recent American trade restrictions have encouraged the persistence of the less-than-satisfactory "canner" size for 80% of the southern Gulf catch, which is about 40% of the total Canadian catch.

Despite concerns through the mid to late 1970's of the potential for recruitment overfishing, yields moved to near record highs through the 1980's due in large part to a pulse in juvenile recruitment. Since systematic changes in lobster management had not occurred, the increased lobster abundance is thought to be due to improved, but unknown, environmental conditions. If the conditions are cyclical, there is no guarantee the current abundance levels can be sustained.

The dramatic increase in supply created a significant holding and marketing challenge to industry, which was successfully met. Air transport was used in a large way to tap underdeveloped or new markets. All of this luxury product was absorbed to 1987 inclusive, and without erosion of real price. Seafood markets softened in 1988, and persist to the present. The impact on the processed lobster product has been acute.

Biological and econometric advice for many years has been that the present Canadian fishery is severely yield overfished. Science has encouraged the adoption for most stocks, of a minimum legal size, that is greater by one molt length (about 12 mm), than the current size. Industry has rejected the proposal in part because of the uncertainty of both the scientific and marketing data, and in part because of a potential short-term loss in income. Like the science data, recent marketing data indicate a positive correlation between size (to a size three times the present minimum legal size) and price. The analyses suggest that increases in minimum legal size could lead to a 50% to a 100% increase in value at dock side. It is impossible to predict optimal sizes (they would vary amongst stocks), but the direction of change can be identified through both biological and economic research. There are a number of stocks, for example LFA 27, where the benefits would be greater than elsewhere, for example in LFA 34. The American industry, after years of prodding from scientists and economists, increased its minimum legal size from 81 to 83.5 mm. Canadian fishermen chose not to follow suit. The percentage of the Canadian catch that is in the canner class increases in concert with the gradual upward move of the American size. The impact in the Canadian industry is most severe in the Gulf fishery where a greater percentage of the catch is canner size. This accompanied by enhanced surveillance and enforcement of present trade regulations, new, more restrictive American regulations, and increased yields has created a glut of canner sized lobsters. Canada lodged a complaint under the Free Trade Agreement; the dispute panel declared the trade restriction fair. A move to slowly increase southern Gulf of St. Lawrence MLS's has begun.


### Acknowledgements

DFO's Scotia-Fundy Region, over the past few years, has expended considerable resources on attaining a better understanding of all aspects of the lobster fishery. The work of co-workers R. Barbara, R. Miller and D. Pezzack and former co-worker, R. Robinson, contributed significantly to the economics, marketing, biology and stock assessment aspects of this paper. The latter two workers plus R. Duggan, G. Ennis and G. Sharp permitted the citation of CAFSAC Research Documents. Figures 1 and 11, and 12 were contributed by D. Pezzack and G. Sharp, respectively. The original manuscript was much improved by incorporating the recommendations of G. Ennis and an anonymous reviewer. Technical assistance was provided by V. Clayton, R. Duggan and B. Field.


### References


AIKEN, D.E. 1980. Molting and growth, p. 91–163. In J.S. Cobb and B.F.Phillips [ed.]. The biology and management of lobsters. Vol. 1, Physiology and Behaviour, Academic Press, Toronto, Ont.

AIKEN, D.E., AND S.L. WADDY. 1980. Maturity and reproduction in the American lobster. Can. Tech. Rep. Fish. Aquat. Sci. 932: 60–69.

1986. Environmental influence on recruitment of the American lobster, *Homarus americanus*: A perspective. Can. J. Fish. Aquat. Sci. 43: 2258–2270.

ALVERSON, D.L., AND G.J. PAULIK. 1973. Objectives and problems of managing living resources. J. Fish. Res. Bd. 30: 1936–1947.

ANONYMOUS. 1890. Annual Report of the Department of Marine Fish. Appendix 8: 27–136.

1975. Lobster Fishery Task Force, Final Report, March 1975. Fisheries and Oceans Canada, Halifax, Nova Scotia. 180 p.

1989. The Scotia-Fundy Lobster Fishery Phase One: Issues and considerations. Fisheries and Oceans Canada, Halifax, Nova Scotia. 106 p.

ANTHONY, V.C., AND J.F. CADDY. 1980. Proceedings of the Canada-U.S. Workshop on status of assessment science for N.W. Atlantic lobster (*Homarus americanus*) stocks (St. Andrews, New Brunswick, Oct. 24–26, 1978). Can. Tech. Rep. Fish. Aquat. Sci. 932: 186 p.

BELZILE, L. 1981. La pêche commerciale du homard en Gaspésie de 1975 à 1981. Office de planification et de développement du Québec — Rimouski, Québec.

BERGERON, J. 1967. La pêche commerciale du homard (*Homarus americanus* Milne-Edwards) au Quèbec des origines à nos jours. Sta. Biol. mar. Grande-Rivière, Cah. Inf. 42: 47 p.

BUMPUS, D.F., AND L.M. LAUZIER. 1965. Surface circulation off the continental shelf of eastern North America between Newfoundland and Florida, p. 92. Folio 7. In W. Webster [ed.] Serial Atlas of the Marine Environment. Am. Geogr. Soc. New York, NY.

CADDY, J.F. 1979. The influence of variations in the seasonal temperature regime on survival of larval stages of the American lobster (*Homarus americanus*) in the southern Gulf of St. Lawrence. Rapp. P.-v. Reun. Cons. Int. Explor. Mer. 175: 204–216.

CAMPBELL, A. 1985. Application of a yield and egg-per-recruit model to the lobster fishery in the Bay of Fundy. N. Am. J. Fish. Manage. 5: 91–104.

1986. Migratory movements of ovigerous lobsters, *Homarus americanus*, tagged off Grand Manan, eastern Canada. Can. J. Fish. Aquat. Sci. 43: 2197–2205.

1989. The lobster fishery of southwestern Nova Scotia and the Bay of Fundy, p.141–158. *In* J.F. Caddy [ed.]. Marine invertebrate fisheries: their assessment and management. Wiley Interscience, Toronto, Ont.

CAMPBELL, A., AND J. BRATTEY. 1985. Egg loss from the American lobster, *Homarus americanus*, in relation to nemertean, *Pseudocarcinonemertes homari*, infestation. Can. J. Fish. Aquat. Sci. 43: 772–780.

CAMPBELL, A., AND D.R. DUGGAN. 1980. Review of the Grand Manan lobster fishery with an analysis of recent catch and effort trends. Can. Tech. Rep. Fish. Aquat. Sci. 997: 20 p.

CAMPBELL, A., AND R.K. MOHN. 1983. Definition of American lobster stocks for the Canadian Maritimes by analysis of fishery-landing trends. Trans. Am. Fish. Soc. 112: 744–759.

CAMPBELL, A., AND D.G. ROBINSON. 1983. Reproductive potential of three American lobster (*Homarus americanus*) stocks in the Canadian Maritimes. Can. J. Fish. Aquat. Sci. 40: 1958–1967.

CAMPBELL, A., AND A.B. STASKO. 1985. Movements of tagged American lobsters, *Homarus americanus*, off southwestern Nova Scotia. Can. J. Fish. Aquat. Sci. 42: 229–238.

1986. Movement of lobster (*Homarus americanus*), tagged in the Bay of Fundy, Canada. Mar. Biol. 92: 393–404.

CONAN, G., AND D.R. MAYNARD. 1983. Aerial survey of spatial distribution of effort in lobster fishery of southern Gulf of St. Lawrence. Int. Counc. Explor. Sea, C.M.1983/K.13: 13 p.

COOPER, R.A., AND J.R. UZMANN. 1980. Ecology of adult and juvenile *Homarus*. p. 97–142. In J.S. Cobb and B.F. Phillips [ed.] The biology and management of lobsters. Vol. 2, Academic Press, New York, NY.

DADSWELL, M.J. 1979. A review of the decline in lobster (*Homarus americanus*) landings in Chedabucto Bay between 1956 and 1977 with an hypothesis for a possible effect by Canso Causeway on the recruitment mechanism of eastern Nova Scotia lobsters stocks. Can. Fish. Mar. Serv. Tech. Rep. 834 (Part 3): 113–144.

DEWOLF, A.C. 1974. The lobster fishery of the Maritime provinces: economic effects of regulations. Bull. Fish. Res. Board 187: 59 p.

DOW, R.L. 1980. The clawed lobster fisheries, p. 265–316. *In* J.S. Cobb and B.F. Phillips [ed.]. The biology and management of lobsters. Vol. 2, Ecology and Management. Academic Press, Toronto, Ont.

DOW, R.L., F.W. BELL, AND D.M. HARRIMAN. 1975. Bioeconomic relationships for the marine lobster fishery with consideration of alternative management schemes. NOAA Tech. Rep., NMSF SSRF 683.

DUNHAM, P.J., AND D. SKINNER-JACOBS. 1978. Intermolt mating in the lobster (*Homarus americanus*). Mar. Behav. Physiol. 5: 209–214.

ELNER, R.W., AND A. CAMPBELL. 1987. Natural diets of lobster *Homarus americanus* from barren ground and microalgal habitats off southwestern Nova Scotia, Canada. Mar. Ecol. Prog. Series 37: 131–140.

ENNIS, G.P. 1973. Food, feeding and condition of lobster, *Homarus americanus*, throughout the seasonal cycle in Bonavista Bay, Newfoundland. J. Fish. Res. Board Can. 30: 1905–1909.

1978. Growth curves for Newfoundland lobsters from data on molt increment and proportion molting. Can. Atl. Fish. Sci. Adv. Comm. Res. Doc. 78/29: 7 p.

1980. Further yield per recruit analysis for Newfoundland lobsters. Can. Atl. Fish. Sci. Adv. Comm. Res. Doc. 80/39: 9 p.

1981. An examination of the effect of size and sex of animals tagged on estimates of exploitation rate in a Newfoundland population of lobsters (*Homarus americanus*). Can. Atl. Fish. Sci. Adv. Comm. Res. Doc. 81/64: 6 p.

1986. Stock definition, recruitment variability, and larval recruitment process in the American lobster, *Homarus americanus* a review. Can. J. Fish. Aquat. Sci. 43: 2072–2084.

ENNIS, G.P., AND S.A. AKENHEAD. 1978. A model and computer program used to assess yield per recruit in Newfoundland lobster stocks. Can. Atl. Fish. Sci. Adv. Comm. Res. Doc. 78/30: 13 p.

FLOWERS, J.M., AND S.B. SAILA. 1972. An analysis of temperature effects on the inshore lobster fishery. J. Fish. Res. Board Can. 29: 1221–1225.

FOGARTY, M.J., AND J.S. IDOINE. 1986. Population assessment of the American lobster on the outer continental shelf. Woods Hole Lab. Ref. Doc. No. 86–01.

GARDNER, M. 1991. Lobster pricing analyses. Econ. Comm. Analyses Rept. 91, Fisheries and Oceans Canada, Ottawa, Ont. 34 p.

GULLAND, J.A. 1971. Ecological aspects of fishery research. Adv. Ecol. Res. 7: 115–170.

HALKETT, A. 1916. In Annual Report of the Department of Marine Fish. Natural History Report. Appendix 16: 349–355.

HARDING, G.C., K.R. DRINKWATER, AND W.P. VASS. 1983. Factors influencing the size of American lobster (*Homarus americanus*) stocks along the Atlantic coast of Nova Scotia, Gulf of St. Lawrence, and the Gulf of Maine: a new synthesis. Can. J. Fish. Aquat. Sci. 40: 168–184.

HARDING, G.C., J.D. PRINGLE, W.P. VASS, S. PEARRE, JR., AND S.J. SMITH. 1987. Vertical distribution and daily movements of larval *Homarus americanus* over Browns Bank, Nova Scotia. Mar. Ecol. Prog. Ser. 41: 29–41.

HARDING, G.C., AND R.W. TRITES. 1988. Dispersal of *Homarus americanus* larvae in the Gulf of Maine from Browns Bank. Can. J. Fish. Aquat. Sci. 45: 416–425.

HARDING, G.C., W.P. VASS, AND K.R. DRINKWATER. 1982. Aspects of larval lobster (*Homarus americanus*) ecology in St. Georges Bay, Nova Scotia. Can. J. Fish. Aquat. Sci. 39: 1117–1129.

HEDGECOCK, D.K., NELSON, J. SIMMS, AND R. SHLESER. 1977. Genetic similarity of American and European species of lobster *Homarus*. Biol. Bull. 152: 41–50.

HUDON, C. 1987. Ecology and growth of post larval and juvenile lobster, *Homarus americanus*, off Iles de la Madeleine (Québec). J. Fish. Aquat. Sci. 44: 1855–1869.

HUNTSMAN, A.G.V. 1923. Natural lobster breeding. Bull. Biol. Board Can. 5: 1–11.

JARVIS, C. 1989. Movement patterns of late-stage ovigerous female lobsters (*Homarus americanus* Milne Edwards) at Jeddore, Nova Scotia. M.Sc. thesis, Dalhousie University, Halifax, Nova Scotia. 148 p.

KIRBY, M.J.L. 1982. Navigating Troubled Waters. A New Policy for the Atlantic Fisheries. Report of the Task Force on Atlantic Fisheries. Fisheries and Oceans Canada, Ottawa, Ont. 379 p.

KROUSE, J.S. 1989. Performance selectivity of trap fisheries for crustaceans. p. 307–325. In J.F. Caddy [ed.] Marine Invertebrate Fisheries their assessment and management. Wiley Interscience, Toronto, Ont.

LAVOIE, N. 1877. Report of the cruise of the government steamer *Lady Head* in the protection of the fisheries of the Gulf and River St.-Lawrence during the season of 1876, under command of Napoleon Lavoie, Esq. fishery officer. Report of the Commissioner of Fisheries for the year ending 31 December 1876 (1877). Appendix 3: 44–147.

LAWTON, P. 1987. Diel activity and foraging behaviour of juvenile American lobsters, *Homarus americanus*. Can. J. Fish. Aquat. Sci. 44: 1195–1205.

MACKENZIE, B. 1988. Assessment of temperature effects on interrelationships between stage durations, mortality and growth in laboratory-reared *Homarus americanus* Milne Edwards larvae. J. Exp. Mar. Biol. Ecol. 116: 87–98.

MANN, K.H. 1977. Destruction of kelp beds by sea urchins: a cyclical phenomenon or irreversible degradation. Helgoländer. wiss. Meeresunters. 30: 455–467.

MANN, K.H., AND P.A. BREEN. 1972. The relation between lobster abundance, sea urchins, and kelp beds. J. Fish. Res. Board Can. 29: 603–605.

MCEVOY, A.F. 1982. Scientific research and the twentieth-century fishing industry. Calif. Coop. Oceanic Fish. Invest. Rep. XXIII: 48–55.

MILLER, R.J. 1985. Seaweeds, sea urchins, and lobsters: a reappraisal. Can. J. Fish. Aquat. Sci. 42: 2061–2072.

1990a. Properties of a well managed fishery. Fisheries 15: 7–12.

1990b. Effectiveness of crab and lobster traps. Can. J. Fish. Aquat. Sci. 47: 1228–1251.

MILLER, R.J., R.E. DUGGAN, D.G. ROBINSON, AND Z. ZHENG 1989. Growth and movement of *Homarus americanus* on the outer coast of Nova Scotia. Can. Tech. Rep. Fish. Aquat. Sci. 1716: 17 p.

MILLER, R.J., D.S. MOORE, AND J.D. PRINGLE. 1987. Overview of the inshore lobster resources in the Scotia-Fundy Region. Can. Atl. Fish. Sci. Adv. Comm. Res. Doc. 87/85: 20 p.

MOORE, D.S., R.J. MILLER, AND L.D. MEADE. 1986. Survey of shallow benthic habitat: eastern shore and Cape Breton, Nova Scotia. Can. Tech. Rep. Fish. Aquat. Sci. 1546: 49 p.

MUNROE, J., AND J.C. THERRIAULT. 1981. Abundance distribution, probilité et fréquencie de miue de la population de homards des lagunes des Iles de la Madeleine. Rapp. tech. can. sci. halieut. aquat. 1034: 35 p.

PALOHEIMO, J.E. 1963. Estimation of catchabilities and population sizes of lobsters. J. Fish. Res. Board Can. 20: 59–88.

PERKINS, H.C. 1971. Egg loss during incubation from offshore southern lobster (Decapoda:Homaridae). Fish. Bull. U.S.A. 69: 451–453.

Pezzack, D.S. 1983. Distributions of American lobster (*Homarus americanus*) in the midshore and offshore regions of S.W. Nova Scotia during October 1980, July 1981, and October 1981 trap Surveys. Can. Atl. Fish. Sci. Advis. Comm. Res. Doc. 83/67: 27 p.

1992. A review of lobster (*Homarus americanus*) landing trends in the Northwest Atlantic 1947–1987. J. N. Atl. Fish. Sci. 14: 115–127.

PEZZACK, D.S., AND D.R. DUGGAN. 1985. The Canadian offshore lobster fishery 1971–1984: catch history, stock condition and management options. Can. Atl. Fish. Sci. Advis. Comm Res. Doc. 85/89: 26 p.

1986. Migration and homing tendencies of offshore lobsters (*Homarus americanus*) on the Scotian Shelf. Can. J. Fish. Aquat. Sci. 43: 2206–2211.

1987. Canadian offshore lobster fishery, 1985–86, and assessment of the potential for future increases in catch. Can. Atl. Fish. Sci. Adv. Comm. Res. Doc. 87/79: 25 p.

PEZZACK, D.S., AND J.D. PRINGLE. 1986. Gulf of Maine Area lobster management areas, and suggestions on stock structure. CAFSAC Res. Doc. 86/85: 17 p.

PEZZACK, D.S., M.J. TREMBLAY, C. HUDON, AND R.J. MILLER. 1992. The inshore-offshore lobster issue in southwestern Nova Scotia. Can. Manuscr. Rep. Fish. Aquat. Sci. 2165: 198 p.

PHILLIPS, B.F., J.S. COBB, AND R.W. GEORGE. 1980. General biology, p. 1–82. *In* J.S. Cobb and B.F. Phillips [ed.]. The biology and management of lobsters. Vol. 1., Academic Press, New York, NY.

PHILLIPS, B.F., AND A.N. SASTRY. 1980. Larval ecology, p. 1–57. *In* J.S. Cobb and B.F. Phillips [ed.] The biology and management of lobsters. Vol. 2., Academic Press, New York, NY.

PIELOU, E.C. 1976. Population and community ecology, principles and methods. Gordon and Breach Science Publishers, New York, NY. 424 p.

POTTIE, R.A., AND R.W. ELNER. 1982. Substrate preference behaviour of juvenile American lobsters, *Homarus americanus*, in gravel and silt-clay sediments. Can. J. Fish. Aquat. Sci. 39: 928–932.

PRINCE, E.E. 1899. Report of the Canadian Lobster Commission, 1898. *In* 31st Ann. Rep. Dept. Mar. Fish., 1899, Supplement 1, 41 p.

PRINGLE, J.D. 1985. The human factor in fishery resource management. Can. J. Fish. Aquat. Sci. 42: 389–392.

1986a. Structure of certain North American government fishery agencies and effective resource management. Ocean Manage. 10: 11–20.

1986b. A review of urchin/macroalgal associations with a new synthesis for nearshore, eastern Canadian waters. Monogr. Biol. 4: 191–218.

PRINGLE, J.D., AND A. CAMPBELL. 1988. Lobster research — back to the basics. Science Review 1987. Government of Canada ISBN 0-662-16944-1: 18–21.

PRINGLE, J.D., AND R.E. DUGGAN. 1983. A remote sensing technique for quantifying lobster fishing effort. Can. Tech. Rep. Fish. Aquat. Sci. 1217: 16 p.

1984. Latent lobster fishing effort along Nova Scotia's Atlantic coast. Can. Atl. Fish. Sci. Adv. Comm. Res. Doc. 84/56: 21 p.

1985. An estimate of yields for oceanic Nova Scotia lobster grounds. Can. Atl. Fish. Sci. Adv. Comm. Res. Doc. 85/102: 16 p.

PRINGLE, J.D., R.E. DUGGAN, AND G.J. SHARP. 1986. A cost evaluation of techniques designed to assess canadian lobster fishing effort. Fish. Res. 4: 283–295.

PRINGLE, J.D., D.G. ROBINSON, G.P. ENNIS, AND P. DUBÉ. 1983. An overview of the management of the lobster fishery in Atlantic Canada. Can. Manuscr. Rep. Fish. Aquat. Sci. 1701: 103 p.

RETTIG, R.B. 1989. Is fishing management at a turning point? Reflections on the evolution of rights base fishing, p. 47-64. *In* P.A. Neher, R. Arnason, and N. Mollett [ed.]. Rights based fishing. Kluwer Academic Publishers, Boston, MA.

RICKER, W.E. 1975. Computation and interpretation of biological statistics of fish populations. Bull. Fish. Res. Board Can. 191: 382 p.

ROBINSON, D.G. 1979. Consideration of the lobster (*Homarus americanus*) recruitment overfishing hypothesis, with special reference to the Canso Causeway. Fish. Mar. Serv. Tech. Rep. 834 (Part 3): 77–99.

1980. History of the lobster fishery on the eastern shore of Nova Scotia, p. 8–23. *In* J.D. Pringle, G.J. Sharp and J.F. Caddy [ed.]. Proceedings of the workshop on the relationship between sea urchin grazing and commercial plant/animal harvesting. Can. Tech. Rept. Fish. Aquat. Sci. 954.

SAILA, S.B., AND J.M. FLOWERS. 1969. Geographic morphometric variation in the american lobster. Syst. Zool. 18: 330–338.

SCARRATT, D.J. 1964. Abundance and distribution of lobster larvae (*Homarus americanus*) in Northumberland Strait. J. Fish. Res. Board Can. 21: 661–680.

1973. Abundance, survival, and vertical and diurnaldistribution of lobster larvae in Northumberland Strait, 1962–63, and their relationship with commercial stocks. J. Fish. Res. Board Can. 30: 1819–1824.

1980. The food of lobsters, p. 144–169. *In* J.D. Pringle, G. Sharp and J. Caddy [ed.] Proceedings of the workshop on the relationship between sea urchin grazing and commercial plant/animal harvesting. Can. Tech. Rep. Fish. Aquat. Sci. 954.

SCARRATT, D.J., AND P.F. ELSON. 1965. Preliminary trials of a tag for salmon and lobsters. J. Fish. Res. Board Can. 22: 421–423.

SCOTT, A.D. 1989. Conceptual origins of rights based fishing, p. 11–38. *In* P.A. Neher, R. Arnason, and N. Mollett [ed.]. Rights based fishing. Kluwer Academic Publishers, Boston, Massachusetts.

SHARP, G.S., AND R.E. DUGGAN. 1985. An aerial survey of near-shore and mid-shore lobster fishing effort in southwestern Nova Scotia, spring and fall 1983–1984. Can. Manuscr. Rep. Fish Aquat. Sci. 1847: 37 p.

SHARP, G.S., J.D. PRINGLE, AND R.E. DUGGAN. 1989. Assessing fishing effort by remote sensing in the Scotia-Fundy Region of Fisheries and Oceans Canada. 12th Can. Symp. Remote Sens. 4: 2056–2060.

SINCLAIR, M., J.T. ANDERSON, M. CHADWICK, J. GAGNÉ, W.D. MCKONE, J.C. RICE, AND D. WARE. 1988. Report from the national workshop on recruitment. Can. Tech. Rep. Fish. Aquat. Sci. 1626: 254 p.

STASKO, A.B. 1980. Lobster larval surveys in Canada, p. 157–165. *In* V.C. Anthony and J.F. Caddy [ed. Proceedings of the Canada — U.S. workshop on status of assessment science for N.W. Atlantic lobster (*Homarus americanus*) stocks, St. Andrews, New Brunswick, Oct. 24–26, 1978. Can. Tech. Rep. Fish. Aquat. Sci. 932.

SUTCLIFFE, W.H., JR. 1973. Correlations between seasonal river discharge and local landings of American lobster (*Homarus americanus*) and Atlantic halibut (*Hippoglossus hippoglossus*) in the Gulf of St. Lawrence. J. Fish. Res. Board Can. 30: 856–859.

SUTHERLAND, D.H. 1937. Annual reports of chief supervisors of fisheries, Appendix I. Ann. Rep. Dept. Fish. 30.

TEMPLEMAN, W. 1936. The influence of temperatures, salinity, light, and food conditions on the survival and growth of the larvae of the lobster (*Homarus americanus*). J. Biol. Board Can. 2: 485–497.

1948. Growth per molt in the American lobster. Bull. Newfoundland Gov. Lab. 18: 26–48.

THOMAS, J.C. 1973. An analysis of the commercial lobster (*Homarus americanus*) fishery along the coast of Maine, August 1966 through December 1970. U.S. Dep. Commer. NOAA Tech. Rep. NMFS SSRF 667: 57 p.

VENNING, R.N. 1910. The marine and fisheries committee and lobster industry. Ann. Rep. Dept. Mar. Fish., Spec. Appended Report II, Canada Annual Reports, Marine and Fisheries, Ottawa, Ontario, Sess. Paper 22.

WADDY, S.L., AND D.E. AIKEN. 1986. Multiple fertilization and consecutive spawning in large American lobsters, *Homarus americanus*. Can. J. Fish. Aquat. Sci. 43: 2291–2294.

WILDER, D.G. 1953. The growth rate of the American lobster (*Homarus americanus*). J. Fish. Res. Board Can. 10: 371–412.

WILDER, D.G. 1963. Movement, growth and survival of marked and tagged lobsters liberated in Egmont Bay, Prince Edward Island. J. Fish. Res. Board Can. 20: 305–318.

1974. Inshore and offshore lobster stocks. Fish. Res. Board Can. Manuscr. Rep. 1293: 14 p.

WOLFF, T. 1978. Maximum size of lobsters (*Homarus*) (Decapoda, Nephropidae). Crustaceana 34: 1–14.

# The Management of the Canadian Atlantic Herring Fisheries

T. D. Iles

*Department of Fisheries and Oceans,*
*Biological Station, St. Andrews, New Brunswick, Canada E0G 2X0*

Iles, T. D. 1993. The management of the Canadian Atlantic herring fisheries, p. 123–150. *In* L. S. Parsons and W. H. Lear [eds.] Perspectives on Canadian marine fisheries management. Can. Bull. Fish. Aquat. Sci. 226.

The management history of east coast Canadian herring is briefly reviewed. Management-related research for herring was a relatively recent development in eastern Canada, stimulated by a marked increase in exploitation rates in the 1960s both nationally and internationally.

Internationally, Canada sought control of exploitation through the establishment of Total Allowable Catches and national allocation by the International Commission for the Northwest Atlantic Fisheries. The foreign fisheries for herring diminished following extension of fisheries jurisdiction by Canada and the U.S. in 1977.

Canada developed new management approaches for the herring resources within its 200-mile fisheries zone. This management system and trends in the fisheries in the various management areas are described. Two case histories are examined. One illustrates the difficulties in assessing resource abundance in the absence of adequate historical records. The other describes a cooperative attempt, involving many interest groups, to resolve complex management issues under the new system of national jurisdiction. The importance of market factors in determining appropriate management strategies is emphasized.

---

L'histoire de la gestion du hareng de la côte est du Canada est brièvement passée en revue. La recherche sur la gestion du hareng était relativement récente dans l'est du Canada; elle fut cependant stimulée par l'augmentation de l'exploitation dans les années soixante, à l'échelle nationale et internationale.

À l'échelle internationale, le Canada a cherché à contrôler l'exploitation en instaurant un Total des prises admissibles et une allocation nationale par la Commission internationale des pêches de l'Atlantique Nord-Ouest. La pêche du hareng étrangère a diminué par suite de l'extension de la juridiction des pêches par le Canada et les États-Unis en 1977.

Le Canada a mis au point de nouvelles méthodes de gestion du hareng dans sa zone de pêche de 200 milles. Ces méthodes, ainsi que les tendances dans divers domaines de gestion des pêcheries, sont décrites. Deux cas typiques sont examinés. Le premier illustre les difficultés à évaluer l'abondance des ressources en l'absence de dossiers fiables. Le deuxième décrit les efforts concertés de nombreux groupes d'intérêt pour tenter de résoudre des problèmes de gestion relevant de la nouvelle juridiction nationale. On souligne l'importance des facteurs du marché dans le choix de stratégies de gestion appropriées.

## Introduction

Herring was fished in coastal Canadian and American waters by the original inhabitants of the continent and, since European colonization, large, economically important fisheries both nearshore and offshore have existed for more than 200 years (Perley 1852; McFarland 1911), but the current management regime can be said to have originated as late as the mid- to late 1960's. Modern management developed in response to major changes in both the nature and degree of exploitation by the fisheries in the early 1960's that put greatly increased pressure on the herring resource in the northwest Atlantic area as a whole. This increased exploitation was the result of the diversion of effort from herring fisheries in other areas that had collapsed under extreme fishing pressure (Iles and Tibbo 1970). One such fishery, on Pacific herring, was Canadian; the others in European waters, exploited the Atlantic herring and involved a large number of European countries.

By the late 1960's, economically important traditional Canadian and US fisheries off the Atlantic coast were threatened, and this during a period when Canada, in particular, was developing an increasingly active policy of jurisdictional extension (Anon. 1971). The resolution of what became a critical situation by the turn of the decade involved the balancing of national, bilateral (with the US), and international interests, and became an integral part of Canada's progress towards the extension of jurisdiction which was realized in 1977. Herring resources in the Canadian area of jurisdiction by then were exploited exclusively by Canadians but, by the late 1970's, management problems that had been resolved at the international level re-emerged at the national level — a process that had been termed "internalization" (Gulland 1974). These problems arose largely in relation to the allocation of shares of the Canadian resource among individual conflicting interests involving both the fisheries and processing sections of industry and individual provinces with access to Canadian waters.

The management process for Atlantic herring has evolved significantly since it became an issue in the late 1960's and documentation of the early period is relatively poor. This chapter, therefore, deals with the important management developments of the last quarter century, as well as outlining the current management structure.

## Herring Biology

The general biology and life history of Atlantic herring has been well reviewed (Parrish and Saville 1965). Certain of its biological characteristics contribute to its economic importance, and influence its management.

Herring is an ecologically dominant species, and the single most abundant fish species in the temperate Atlantic (Parrish and Saville 1965). This abundance is the natural result of its ability to feed on the most readily available fish food, plankton, throughout life. It is also gregarious at all life history stages and the often huge and dense schools of feeding or spawning herring have been the target of fisheries throughout historical time. Its relatively small size of some 30 cm or so, maximum, is a convenient size for a meal, and helps to account for its value as a food fish.

In northern temperate regions, the primary production season in the sea is short and, as a result, zooplankton feeding on this production become abundant only over a few months. Herring have evolved the ability to store enough food material, both fat and protein, during its short feeding and growth season to allow for later gonad development and to sustain it over the winter period of fast (Iles 1964, 1984; Iles and Wood 1965). As a result, there is a large seasonal range of biological "condition" in herring, that is matched by a wide range of marketable products that can be processed. Herring is the raw material for a diverse range of food products, more so than most other fish species and the seasonal matching of fish to market can be an important management factor.

Like nearly all bony fishes, herring produces large numbers of small eggs, about 1 mm in diameter, that develop into pelagic larvae feeding on various stages of zooplankton. Herring is unusual amongst important marine species in that it lays its eggs on the bottom in shallow waters, usually near the shore (MacKenzie 1964; Tibbo et al. 1963) but sometimes on offshore banks. On Georges Bank, in the Gulf of Maine where the largest concentrations of spawning herring have been recorded, a single egg bed represented some 50 thousand tons of spawning fish (Caddy and Iles 1973).

Herring is found on both sides of the Atlantic but does not have a continuous distribution (Haegele and Schweigert 1985). In some areas, such as the Bay of Fundy or the Gulf of St. Lawrence, there are a number of spawning groups each with its own particular spawning locality, substrate preference and spawning period, typically lasting only a few weeks and occurring at the same time each year. These individual spawning groups represent what are considered to be individual "stocks," that are considered to be largely, or even completely, biologically distinct. Individual stocks in the northeast Atlantic can be found spawning in every month of the year (Sinclair and Tremblay 1984) but, in the northwest Atlantic, the range is from about May to November (Haegele and Schweigert 1985).

The biological status of herring stocks has been debated for a century and still is very much in question

(Iles and Sinclair 1982). The ecological discreteness and separate identity is indicated by the documented existence of discrete fisheries over many hundreds of years in some instances (Parrish and Saville 1965). This is the basis for the use of the standard assessment techniques that assume that recruits to a stock are solely the progeny of the same stock, and this assumption underlies the study of stock-recruitment relationships (Ricker 1954; Iles 1973). This "ecological" discreteness is supported by the economic collapse of some stocks, following heavy fishing, without it being immediately replaced by other healthy stocks even those well within the known cruising range of the adults, as established by tagging experiments. The Georges Bank stock in the northwest Atlantic is a good example of this (Stephenson and Kornfield 1990) as is the Dogger Bank spawning group in the central North Sea (Anon. 1987).

However, the difficulty of demonstrating systematic differences between stocks by morphometric methods (Stephenson and Gordon 1991) or genetic analysis (Smith and Jamieson 1986), even when the stocks are sampled at the spawning stage, can be interpreted as evidence that they are not discrete genetically.

The absolute size of individual stocks, or stock complexes, varies enormously, by a factor of over 100,000 (Iles and Sinclair 1982). Nonetheless, the smallest stock presents the same research and logistical difficulties to management as does the largest.

Individuals from different spawning groups are known to mix at other stages of the life history and other times of the year. Exploitation in a single area can, therefore, affect a number of different stocks simultaneously. To complicate the issue, different stocks vary in seasonal spawning time and, in some cases, as in the Baltic Sea, in Icelandic waters and in the Gulf of St. Lawrence, more than one stock may occupy the same spawning area but at different spawning seasons (Haegele and Schweigert 1985). The allocation of catches from fisheries or areas where stock mixing is known or suspected to occur is often impossible because of the degree of morphometric similarity. These difficulties are compounded by inadequate or misleading catch information and lack of comprehensive biological sampling, and these can combine to make it impossible for a sound biological basis for management.

## The Herring Fisheries

### The Historical Fisheries

The herring fisheries of the northeast Atlantic helped shape the course of European history for a thousand years. Herring was responsible for the rise of the Hanseatic League in the 12th century and, later, for the development of both Dutch and British sea power (Samuel 1918). Indeed, to a significant extent the conquest of the sea can be traced to the rise of the herring fisheries and, as late as the beginning of the 20th century, its economic value exceeded that of any other fish (Whitehead 1985).

In the northwest Atlantic herring has not played as dominant a role. Inshore herring fisheries based on brush "weirs", the fore-runner of today's "sardine" industry, were used long before Columbus but the early importance of herring after European colonization of North America was largely as a convenient source of bait for the much more important cod fishery (Innis 1954). As early as 1662, a French settlement at Placentia in Newfoundland took an abundant amount of herring for bait to support a cod industry that supplied food for the slaves of the French West Indies (Innis 1954). In the 18th century the success of the American Revolution was followed by the expansion of American fisheries into distant waters on the Grand Banks and the Gulf of St. Lawrence. American fishermen were exploiting the Magdalen Islands herring even before colonization of the Magdalen Islands by Acadians from St. Pierre and Miquelon. By 1797, 35 American vessels were drying mainly herring on the Magdalens shores (McFarland 1911).

### Development of the Modern Herring Fisheries

By the 20th century these "distant-water" fisheries for herring had disappeared. The only remaining major fishery was a highly specialized one that developed both in Canada and the US along the shores of New Brunswick and Maine. This was a "fixed gear" fishery using inshore weirs that exploited juvenile, largely 2-year-old herring that were canned as "sardines". This industry developed rapidly in the late 19th Century and, in Canada, expanded into what was then the largest "industrial" fishery in eastern Canada. This fishery played an important part in the recent history of management (Doucet 1960) and is still one of the most important fisheries in Canada.

The lack of major markets for adult herring on the North American continent, and the distance of other mass markets meant that, apart from bait fisheries, there were only small, local and sporadic home markets for adult herring. These were supplied by correspondingly small local fisheries.

The crisis in Atlantic herring fisheries that developed in the late 1960's, one that affected both coasts of Canada as well as the east Atlantic, was the result of the development of greatly increased market opportunities.

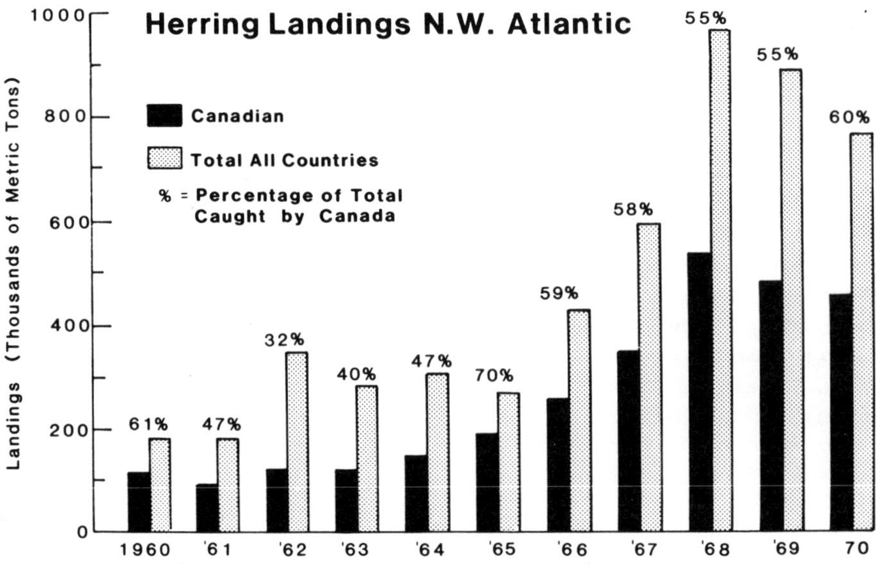


FIG. 1. Canadian and "total" herring catches in the 1960's. Source: Iles (1971a)

The late 1950's saw the beginning of a world wide spectacular expansion in the use of clupeid fish for fish meal that was stimulated by the development of the huge Peruvian anchovetta fishery (Murphy 1977).

This caused a transformation of the Canadian herring fishery because the new meal industry was of potentially almost unlimited size, and it was conducted in small processing plants capable of handling large amounts of raw material, composed of both juvenile and adult fish caught throughout the year. The chief method of capture was the purse seine, whose mesh size retained all but the smallest juveniles (Gough 1971a, b). The fishery supported a primary, large-scale fish meal industry that, in turn, provided feed for a secondary poultry industry. Colonel Sanders "Kentucky Fried Chicken" depended on the efforts of the Canadian herring fisheries to a significant degree for its expansion into eastern North America.

The opening up of the economically attractive meal market resulted in greatly increased levels of exploitation in herring fisheries on both sides of the Atlantic. In the east Atlantic, this involved the North Sea area and also the industry based on the Norwegian (Atlanto-Scandian) herring stocks (Parrish and Saville 1965). Catches increased dramatically and, after a period of about 10 years, decreased even more precipitously until the traditional fisheries of enormous economic and historical importance had almost disappeared.

In Canada, large purse seine fisheries developed, first off the west coast on Atlantic herring's sister species, the Pacific herring, and then off the east coast (Iles and Tibbo 1970). The increased exploitation involved all stages of the life history, and a particularly important development was the degree to which purse seines exploited juveniles in the Bay of Fundy area. In 1966, for example, about 60,000 tons of small herring were removed for use for meal, and in the vicinity of the traditional weir fishery that supplied the canned "sardine" market (Anon. 1972a). This would be expected both to intercept fish and interfere with the traditional fishery and increase dramatically the total numerical exploitation rates.

The opening up of the meal market to Canadian purse seiners was not the only factor leading to increased exploitation. The emergence of the USSR as a formidable "distant water" fleet in the 1950's and 1960's represented an even greater threat to the resource and impinged, to begin with, directly on the newly discovered Georges Bank herring resource. This attracted other members of the Eastern Bloc as well as other European countries (Iles and Tibbo 1970) with large domestic markets for herring that were largely unfilled because of the decline of herring resources in the northeast Atlantic (Burd 1978; Korringa 1978). Neither Canada nor the USA played any significant role in the development of these "offshore" herring fisheries.

However, the traditional "inshore" herring areas off eastern Canada were subjected to significantly increased effort due to the diversion of seiners from the Pacific herring fisheries. A major fishery for herring in the Gulf of St. Lawrence and in the fjords of southwest Newfoundland developed. Thus there was a major expansion of both the international and Canadian fisheries for herring in the Northwest Atlantic during the 1960's (Fig. 1).

## The Formative Years for Herring Management

### Pressure to Expand the Canadian Atlantic Herring Fishery

Management of the seacoast and inland fisheries became the sole responsibility of the federal government under the Constitution Act 1867, and the federal government began research activities in support of its mandate initially via the Biological Board of Canada formed in 1904, which later became the Fisheries Research Board (FRB) (Johnstone 1977).

The 1937 Fisheries Research Board Act (as amended in 1947, 1952 and 1953) gave the Board "...conduct and control of investigations of practical and economic problems connected with marine and freshwater fisheries...", a mandate that specifically included "applied" research. However, by the 1960's, at the time major changes were occurring in the nature of the herring fisheries, the Fisheries Research Board of Canada was emphasizing basic rather than applied research (Johnstone 1977).

The St. John's Biological Station, under the dynamic leadership of Wilfred Templeman, maintained its traditional emphasis on fishery-related research, as was appropriate to a province so heavily dependent on fishing. At St. Andrews, New Brunswick, the groundfish group, particularly, with Frank McCracken at its head, continued to deal with the "Maritime" areas of the Gulf of St. Lawrence, the Scotian Shelf and the Gulf of Maine. However, the Fisheries Research Board as a whole did not respond to the developing situation of increasing exploitation, that affected groundfish fisheries as well as herring fisheries, by expanding its effort on research in the threatened industries.

This left a vacuum that was filled by the Federal Department of Fisheries and Forestry, to whose Minister the Board was responsible, and whose Inspection and Protection Branch assumed the right to plan a major expansion of the herring fisheries. Unfortunately, this expansion was undertaken with little input and even less impact from the relevant scientific branch of the Canadian government. It was left to individuals to make the plea for resource protection (Anon. 1967).

The expansion of the fishery was stimulated largely by the efforts of an "Industrial Development Branch" (IDB) in collaboration with the "Federal Provincial Advisory Fisheries Committee" (FPAFC) which had been established to coordinate federal and provincial fisheries initiatives. In 1965, the IDB directed its attention to determining the degree to which herring catches could be increased to take advantage of the developing meal industry. At a federal provincial conference in Fredericton, New Brunswick, in 1966, the prospects for development of the Canadian Atlantic herring fishery were discussed (Anon. 1967). Based on IDB and provincial estimates it was concluded that Canadian catches in Canadian waters alone, i.e. mainly the Gulf of St. Lawrence, the Bay of Fundy and other inshore areas off Nova Scotia and Newfoundland, could reach at least 1 million tons by 1975; this was considered a conservative estimate which could be maintained indefinitely. What little biological input was sought was virtually ignored, and little thought was given to the possibility that expansion might have already approached the natural limits for the resource, or to the fact that the foreign "offshore" fisheries were undergoing an uncontrolled expansion. At that time, there was no scientific consensus that the decline in any herring fishery was the direct result of increased fishing effort and any scientific initiative was lost. The opinion of informed individuals, and even those in industry, carried little weight in opposition to the claims of resource availability in support of major economic development opportunities. The conference resulted in federal-provincial endorsement of the expansion of the purse seine and meal industry on an enormous scale (Anon. 1967).

### Herring Stock Collapse and its Consequences

The optimism for the future expressed at the 1966 Fredericton Meeting was soon shown to be unjustified. The Canadian west-coast herring meal fishery, which had been put forward as a model, produced only 70% of its expected yield in 1966, the very year of the conference. Within 2 years the fishery had collapsed and was closed completely by the federal fisheries minister (Stocker 1993).

This was but one of the many instances of the collapse of herring fisheries on both sides of the Atlantic and in the Pacific (Table 1) that occurred over the period and resulted, almost immediately, in diversion of effort to the east coast in the form of purse-seine boats with experienced skippers and crews to augment those already operating in the area (Iles and Tibbo 1970).

However, it did stimulate influential individuals in the newly defunct west coast fishery, anxious to recoup losses by investment in Canada's east coast fishery, to ask the then Minister, Mr. Jack Davis, for a more scientifically based review of the resource situation. As a result, in the spring of 1968, a Ministry Task Force was set up to determine the validity of the estimates of the resource potential on which the planned east coast expansion was based. An interim report was produced in June 1968 that called for proposals for an expanded research program for 1968-69 (Ketchen 1969).

TABLE 1. The temporal pattern of overexploitation of Atlantic herring fisheries. Annual landings expressed as a percentage of maximum annual landings (in year *N*). Sources: ICES, ICNAF, NAFO, Canadian and US Fisheries Statistics.

| | Pacific herring | | East Atlantic | | | | | West Atlantic | | |
|---|---|---|---|---|---|---|---|---|---|---|
| Year | Canada | USA | Iceland Spring | Iceland Summer | Norwegian spring | North Sea | East Anglian | Gulf of St. Lawrence | Fortune Bay, Nfld. | Georges Bank |
| N−9 | 70 | 30 | 7 | 15 | 68 | 53 | 45 | 7 | | |
| N−8 | 70 | 29 | 8 | 19 | 50 | 54 | 45 | 6 | | 0 |
| N−7 | 70 | 33 | 8 | 23 | 57 | 56 | 45 | 13 | | 19 |
| N−6 | 70 | 35 | 15 | 27 | 56 | 65 | 45 | 12 | | 41 |
| N−5 | 70 | 25 | 30 | 29 | 43 | 54 | 45 | 12 | 17 | 26 |
| N−4 | 70 | 14 | 31 | 41 | 43 | 55 | 64 | 15 | 58 | 35 |
| N−3 | 69 | 23 | 55 | 58 | 50 | 52 | 59 | 17 | 78 | 12 |
| N−2 | 70 | 51 | 43 | 88 | 66 | 62 | 85 | 38 | 88 | 38 |
| N−1 | 84 | 81 | 77 | 91 | 79 | 83 | 75 | 73 | 89 | 59 |
| N | 100 | 100 | 100 | 100 | 100 | 100 | 100 | 100 | 100 | 100 |
| N+1 | 99 | 32 | 38 | 79 | 86 | 73 | 60 | 88 | 82 | 83 |
| N+2 | 91 | 42 | 37 | 73 | 36 | 68 | 48 | 73 | 46 | 66 |
| N+3 | 69 | 25 | 25 | 42 | 4 | 70 | 44 | 34 | 19 | 71 |
| N+4 | 50 | 18 | 9 | 31 | 3 | 46 | 33 | 27 | 10 | 47 |
| N+5 | 5 | 11 | 6 | 15 | 1 | 45 | 28 | 16 | 6 | 54 |
| N+6 | 1 | 16 | 2 | 14 | 1 | 41 | 24 | 16 | 6 | 40 |
| N+7 | 1 | 16 | 1 | 8 | — | 40 | 35 | 17 | 8 | 40 |
| N+8 | 4 | 16 | — | — | — | 40 | 19 | 18 | 8 | 39 |
| N+9 | 17 | 16 | — | — | — | 23 | 16 | 20 | 6 | 12 |
| N+10 | 23 | 16 | | — | — | 26 | 20 | 18 | 6 | 1 |
| a) | 1963 | 1970 | 1962 | 1964 | 1966 | 1965 | 1954 | 1969 | 1971 | 1968 |
| b) | 264 | 158 | 274 | 87 | 1955 | 1425 | 280 | 384 | 15 | 374 |

a) Year of maximum catch.
b) Amount of maximum catch (thousands of tons)

This report concluded that the IDB estimate of potential yield of over one million tons from Canadian waters by 1975 was unrealistic. It also drew attention to the growing foreign offshore fishery off the east coast of North America and itemized the important research problems that needed to be addressed in the Gulf of St. Lawrence, along the Newfoundland coast and in the southwest Nova Scotia area where a large adult fishery had developed during the 1960's, and in New Brunswick coastal waters where the juvenile fishery supported the very important "sardine" industry.

The Ketchen Report (Ketchen 1969) led to a meeting in Halifax in early October 1969 attended by representatives of the herring industry. At this meeting the first public presentation of the quantitative basis for scientific management of the herring resource was made. The need for immediate management action was accepted in general terms based on arguments that the herring resource could not sustain heavy fishing pressure indefinitely.

There was no formal report of that meeting, but two important aspects can be mentioned. The argument that proved most convincing to members of both the primary and secondary branches of industry was based on the demonstration that fishing reduced the effective "capital" of a stock and thus threatened future resource prospects (Fig. 2). Also, this meeting that led to the creation of the post of Atlantic Herring Coordinator with the mandate to coordinate all aspects of the departmental operations relevant to the resolution of the critical herring resource/management situation (D. Idler, Ocean Science Centre, Memorial Univ., Nfld., pers. comm.). This post reported to the Director of Research (Atlantic). In 1972, following the successful resolution of the management situation at the international level at the International Commission for the Northwest Atlantic Fisheries (ICNAF) (the body that had the mandate for fisheries in international waters in the northwest Atlantic), the post was abolished and herring research was incorporated into the general departmental organization.

In 1970, Canada unilaterally extended its territorial sea to 12 miles (Anon. 1971) and, at the same time, the Gulf of St. Lawrence and Bay of Fundy (and other

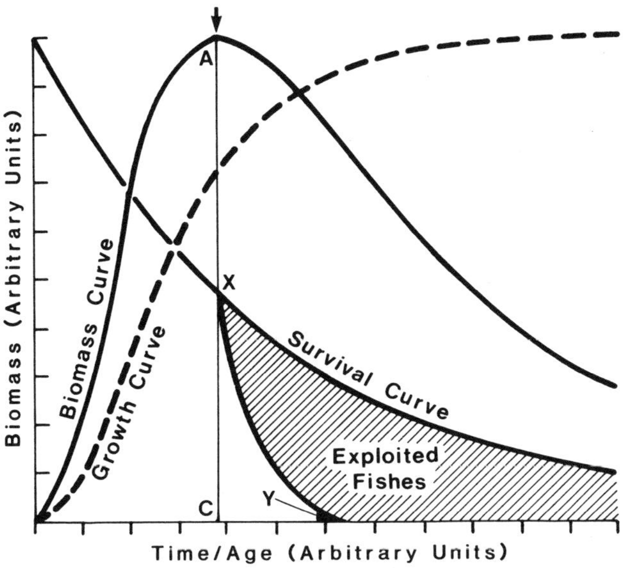


FIG. 2. The "dynamic" fisheries management model. Growth and survival curves generate a biomass curve that indicates the point (A) at which exploitation (X) can yield the maximum yield; exploitation generates a new survival curve (XY); the adult population is reduced in biomass giving lower sustained catches and, if carried too far, affecting recruitment. Source: Iles (1980).

areas on the west coast) were made inaccessible to non-Canadian fisheries by the establishment of closing lines (Parsons 1993).

In April 1970, a freeze on herring fishing licenses was announced by the Federal Minister of Fisheries and, by that time, herring research was being intensified at the St. John's Station in Newfoundland and the St. Andrews Station in New Brunswick. The St. John's program concentrated on the questions of stock identity and interactions in the Gulf of St. Lawrence and southwest Newfoundland, the site of a fishery for overwintering migrants from the Gulf of St. Lawrence (Hodder 1971) while the St. Andrews program was concerned with the Bay of Fundy-Scotian shelf stocks and their interactions with fisheries in international waters (Anon. 1972b).

By the end of 1970, the research program at the St. John's station had established that "Canadian" herring in the Gulf of St. Lawrence and off the Newfoundland coast was not vulnerable to heavy exploitation in waters outside the "closing lines" (Anon. 1971) and the territorial sea (Hodder 1971; Hodder and Parsons 1971a,b; Hodder and Winters 1970). This meant that plans for the international debate in the International Commission for the Northwest Atlantic Fisheries (ICNAF), the body that had the mandate for fisheries management in international waters in the northwest Atlantic, could concentrate on the stocks in the Bay of Fundy, the Scotian Shelf and the Gulf of Maine and Georges Bank. In these areas, the problems of demarcating stock boundaries and determining the degree of stock intermixing were difficult and complicated by the existence of the "coastal" fisheries for juvenile herring in both Canadian and US waters.

Like the Gulf of St. Lawrence, the Bay of Fundy had been reserved to Canadian fishing interests by the promulgation of the "closing lines" in April 1970 (Anon. 1971). However, it was not possible to conclude that its important weir-based industry for "sardines" would be unaffected by the new foreign fisheries, as had been demonstrated for the "internal" herring fisheries in the Gulf of St. Lawrence.

Indeed, the foreign offshore herring fisheries, particularly on Georges Bank, caused great concern to both Canada and the United States. Canada claimed sole management responsibility for stocks both within the Gulf of St. Lawrence and within the Bay of Fundy, but the major issue was the possible effect of the foreign offshore fisheries, particularly on Georges Bank. Both Canada and the US as "coastal states" were concerned for the offshore resource in which they had little direct interest at that time, but which was thought to possibly impact severely on their own inshore herring fisheries (Anon, 1972a, b).

The issue raised strong feelings in both countries. Because their respective "sardine" fisheries were contiguous along the Maine-New Brunswick coast and historically acted economically as a single entity, there was every reason for the two countries to combine their efforts to secure more effective management of the international fisheries.

In April 1971 in Washington D.C., Canada and the United States agreed on proposals to be submitted to the forthcoming June Meeting of ICNAF. These proposals addressed the question of the protection of the juvenile fisheries and catch restrictions for the international fisheries for adults on the Scotian Shelf, the Gulf of Maine, and on Georges Bank (Anon. 1972b). The Georges Bank fishery was thought to have caused a decline in the inshore juvenile fisheries in the Gulf of Maine particularly. One important result of this agreement was the elimination, during 1971 itself, of the Canadian purse-seine fishery on juveniles in the Bay of Fundy that caught more than 60,000 tons in 1966 (Anon. 1972a). Agreement on these proposals was not possible at the June 1971 meeting, partly for procedural reasons, but the seriousness of the resource situation was accepted and acted on. A Special Meeting on Herring was convened for January 1972, by which time it was expected that the procedural difficulties would be resolved (Iles 1980) and it was hoped that more specific and relevant scientific advice would be available.

During 1971 also, attention within Canada focused on the need for more direct resource management controls in "Canadian" waters. A meeting was convened in October 1971 in Dartmouth, N. S., to discuss the forthcoming Special Meeting of ICNAF, planned for January 1972, and the need for international action to protect the Canadian herring industries.

This meeting was the most representative of its kind yet to be held in Canada, and involved individuals from all branches of the department, all sections of industry and from all of the Atlantic Provinces (Pace 1971). There was broad agreement on all the important items at issue. The possible dangerous state of the stocks was recognized, as was the need for immediate and vigorous management action, and a revised estimate of the total Canadian herring resource potential was accepted; at a maximum of 400,000 tons per annum, this was little more than a quarter of the value reached at the Fredericton meeting in 1966 (Iles 1971a, b).

At this meeting, a proposal was made by industry that a representative body be set up to provide a continuing forum for advice to the government from those with a legitimate interest in the herring industry. This led eventually to the formation of the Atlantic Herring Management Committee (Pace 1971).

The success of this October 1971 meeting allowed both the integration of departmental preparations for the forthcoming ICNAF Special Meeting in January 1972 and the fullest cooperation with both industry and the provinces at the international and national levels, as well as bilaterally, with the United States. This degree of cooperation was unique and productive, and contributed significantly, and even definitively, to the success of the 1972 Special Meeting held at the FAO headquarters in Rome (Anon. 1972b).

The First Special Meeting on Herring, held between January and February at FAO headquarters in Rome, resulted in ICNAF agreement on TACs and national allocations of adults for the three recognized herring stocks on Georges Bank, in the Gulf of Maine and off Nova Scotia, and included ancillary agreements on control of exploitation of the juveniles as a single major management package.

## The Atlantic Herring Management Committee

A reduction in foreign exploitation of the northwest Atlantic herring stocks followed rapidly after 1972 as a result of ICNAF's recognition that Canada and the USA as "coastal states" enjoyed a special interest in the social and economic well being of the contiguous coastal communities. By 1976 the allocation of ICNAF Total Allowable Catches (TAC's) in international waters off the Canadian coast was made almost entirely to Canada (Anon. 1976). Such success as ICNAF achieved was not in terms of resource protection for the Georges Bank stock became economically extinct in 1977 (Table 1). Rather it provided the opportunity for the development and implementation of Canada's strategy to extend jurisdiction to 200 miles.

The herring crisis of the late 1960's had developed with remarkable speed and was dealt with equally quickly. In something less than 3 years the problems outlined in the Ketchen Report in 1969 were addressed and answered. The part played by Canadian herring biologists, particularly in the St. John's and St. Andrews Biological Stations was crucial; the period was one of close international cooperation.

Within a few days of the 1972 Special ICNAF Meeting on Herring the establishment of an Atlantic Herring Management Committee (AHMC) was announced. This Committee was to concern itself with the domestic and international problems involved in the management of Canadian east coast herring fisheries (Anon. 1972c). The terms of reference covered scientific, economic and social aspects of management in a way that recognized their interdependence, which was a natural result of the way herring management on the east coast had evolved over the previous 2 years as a cooperative effort involving the federal government, industry and provincial interests. It was the first organization of its kind in Canada and remained the only one for several years.

The AHMC was made up of senior departmental officers including the Regional Directors of the Newfoundland and Quebec Regions and the Chairman of the Fisheries Research Board Directors Committee of the Atlantic area, and was under the chairmanship of the Deputy Regional Director for the Maritimes Region. There was a large body of Advisors which included the Deputy Ministers or equivalent of the two provinces, commercial fishermen from all regions and representing all gear types, buyers and processors group, and a Technical Advisors Group that included scientists and operational personnel. The advisory group totalled 30 at the time of formation (Anon. 1972d).

Between May 23, 1972, when the first meeting was held in Moncton, New Brunswick, and 1980 when a major restructuring occurred, 19 meetings were held, as well as several special meetings. All herring stocks were dealt with, including those in the Gulf of St. Lawrence and in Newfoundland waters that had been kept out of international discussions. Quotas, closure of spawning grounds, license limitations and freezes, general licensing policy and other possible management methods were considered and discussed.

As a result, the Canadian herring industry became familiar with all aspects and levels of management. In particular, the principles underlying management were accepted by all members and were never questioned. Debate and disagreement continued, of course, but centred on the application of these principles, the level of quotas (which soon became referred to as TACs) for example, and the way in which individual interests were allocated shares of the TACs.

In the initial years following the introduction of TACs and national allocations by ICNAF, Canada took steps to ensure compliance with its share of the international quota, but in 1975 a sharp drop in the world meal commodity market threatened the Canadian purse-seine fishery with economic collapse. The actions taken to deal with this situation had far reaching consequences and will be described separately later as a special case, "The Bay of Fundy Project" (Kearney 1984).

## The Modern Era of Herring Management

Modern herring management in the northwest Atlantic began in the late 1960's in the context of international cooperation, instigated at the bilateral level with our coastal state neighbor, the United States, and under the auspices of the ICNAF convention.

Multinational agreement was first reached in January 1972 (Anon. 1972b) and thereafter, development was rapid. Following the ICNAF initiative, Canadian diplomacy was directed towards the smooth transition to the extension of management jurisdiction by Canada, and with the prior acceptance by the important "foreign" fishing interests. The Canadian extension of jurisdiction took place on January 1, 1977, and meanwhile, the AHMC operated as the agency that could both prepare for the initiatives taking place at ICNAF, and for interaction with the organization that was being set up for the expected Canadian assumption of management responsibilities.

Following 1977, scientific questions relating to stock assessment and the formulation of advice on management measures became the responsibility of the newly formed Canadian Atlantic Fisheries Scientific Advisory Committee (CAFSAC) (Anon. 1977), and within CAFSAC to the Pelagic Sub-Committee. The work of the scientific committee of CAFSAC was reviewed by a Steering Committee which passed on the revised recommendations for discussion and action by management.

Over the same period of transition, decisions were being made to reorganize the department by setting up a new regional structure that could interact more closely with the various provincial and local interests.

The situation was complicated by a Ministerial initiative begun in late 1975 and operating from 1976 on, that addressed severe economic problems that had arisen, particularly in the Scotian Shelf, Bay of Fundy area, because of a decline in landed value of the dominant product, fish meal. This led to the "Bay of Fundy Project," which resulted in the introduction of "boat quotas," the original Individual Transferrable Quotas (ITQ), the forbidding of the use of herring for meal production, and the development of a much larger gillnet fishery that emphasized the conflict between gear segments of the industry.

These developments became the subject of a meeting of all interested groups held in May 1981, at which the future direction of herring management (and of other pelagic fishes, excluding the tunas which were dealt with internationally) was debated.

This Yarmouth Seminar on Herring and Mackerel (Anon. 1981) resulted in major changes. The AHMC was replaced by four separate bodies that corresponded in their mandates to the four federal departmental regions for eastern Canada that had been set up, each with its Director General, reporting to an Assistant Deputy Minister for Atlantic Fisheries, and responsible for the promulgation of management measures and their implementation.

The departmental reorganization included the setting up of a new region, the Gulf (of St. Lawrence) Region, with its headquarters in Moncton, New Brunswick. This allowed the development of research and management programs with easier and more direct access to the area than had been possible before, resulting in better interaction with the local fisheries and the communities dependent on them (Chadwick 1988).

The resulting geographical division of responsibilities is illustrated in Fig. 3A which shows the boundary lines for the regions. Figure 3B shows how the overall Canadian east coast fishing zone, established under the extension of jurisdiction, is divided into herring areas. These areas are the basis for the allocation of catch limits and the application of other management measures resulting from the final management decisions. Table 2 shows the relationships between these herring fishing areas, the original ICNAF sub-divisions, the departmental regional organization and the provincial interests.

The Total Allowable Catches (TACs) and any other recommendations arrived at by the Pelagic Sub-Committee of CAFSAC were first considered by the CAFSAC Steering Committee which formulated the final scientific advice. At this stage, the advice was made public, as were the key scientific documents, the CAFSAC Research Documents on which it was based. The intention was to ensure that management decisions follow informed discussion on the part of the various individual interests, and this discussion takes place in appropriate bodies. For

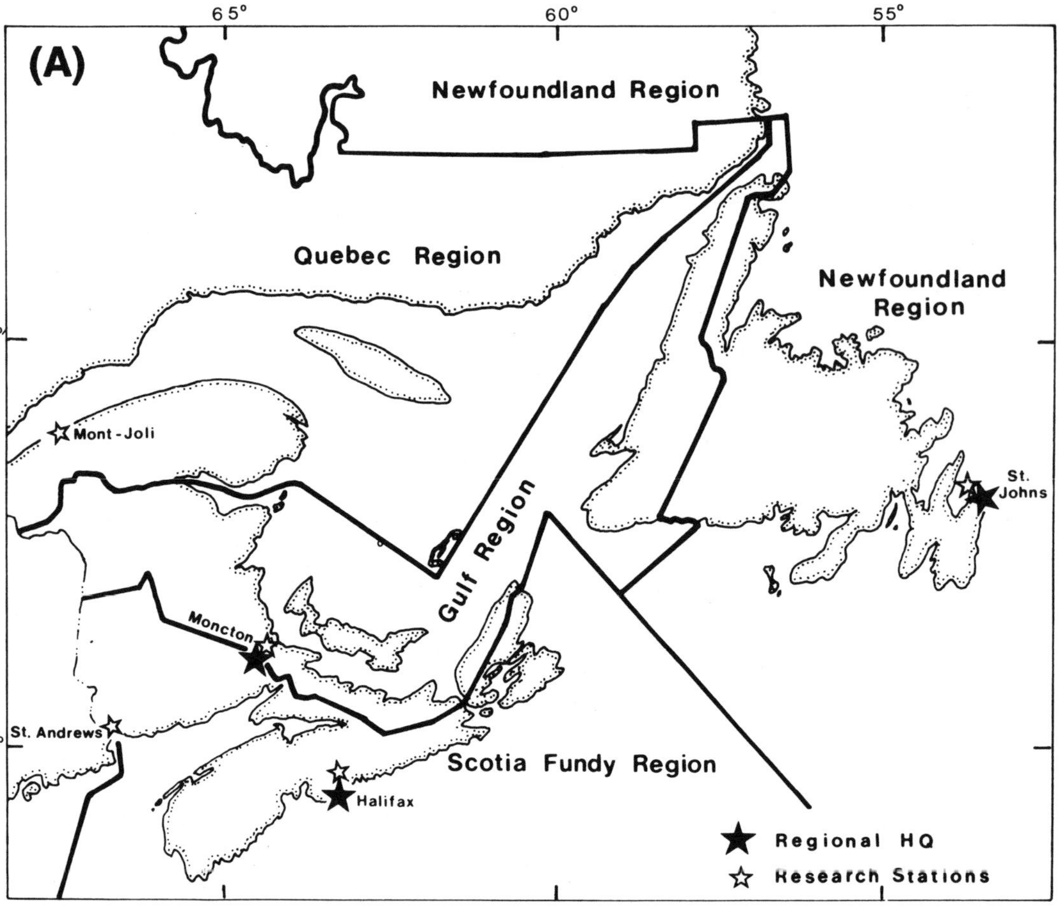


FIG. 3 (A). Canadian east coast management responsibilities (from 1984 to 1992; in 1993 the west coast of Newfoundland became part of the Newfoundland Region again).

TABLE 2. Management and assessment responsibilities for Atlantic herring.

| Herring area | | Name | ICNAF | Assessment and management region | Province |
|---|---|---|---|---|---|
| 3 & 4 | (1) | White-Notre Dame Bays | 3K–3L | Newfoundland | Nfld. |
| 5 & 6 | (1) | Bonavista-Trinity bays | 3L | Newfoundland | Nfld. |
| 7 & 8 | (1) | Conception Bay-Southern Shore | 3L | Newfoundland | Nfld. |
| 9 & 10 | (1) | St. Mary's-Placentia bays | 3L–3P | Newfoundland | Nfld. |
| 11 | (1) | Fortune Bay | 3P | Newfoundland | Nfld. |
| 13–14 | (2) | Newfoundland west coast | 4R | Quebec | Nfld. |
| 15 | | Gulf of St. Lawrence North | 4S | Quebec | Que. |
| 16A–16G | (3) | Gulf of St. Lawrence South | 4T | Gulf | Que, NS, PEI, NB |
| 17 | | Sydney Bight North | 4Vn | Scotia-Fundy | NS, NB |
| 18 | (4) | Sydney Bight South | 4Vs | Scotia-Fundy | NS, NB |
| 19 | (4) | Chedabucto Bay | 4W | Scotia-Fundy | NS |
| 20-22 | (4) | Scotian Shelf-Bay of Fundy | 4W, 4X | Scotia-Fundy | NS, NB |

(1) Areas 3–11 are dealt with as a stock complex of five different stocks, each with its own TAC.
(2) Areas 13 and 14 are covered by a single TAC.
(3) Areas 16A and 16G are covered by one TAC for spring spawning and one for fall spawning stocks.
(4) Areas 18-22 are covered by a single TAC.

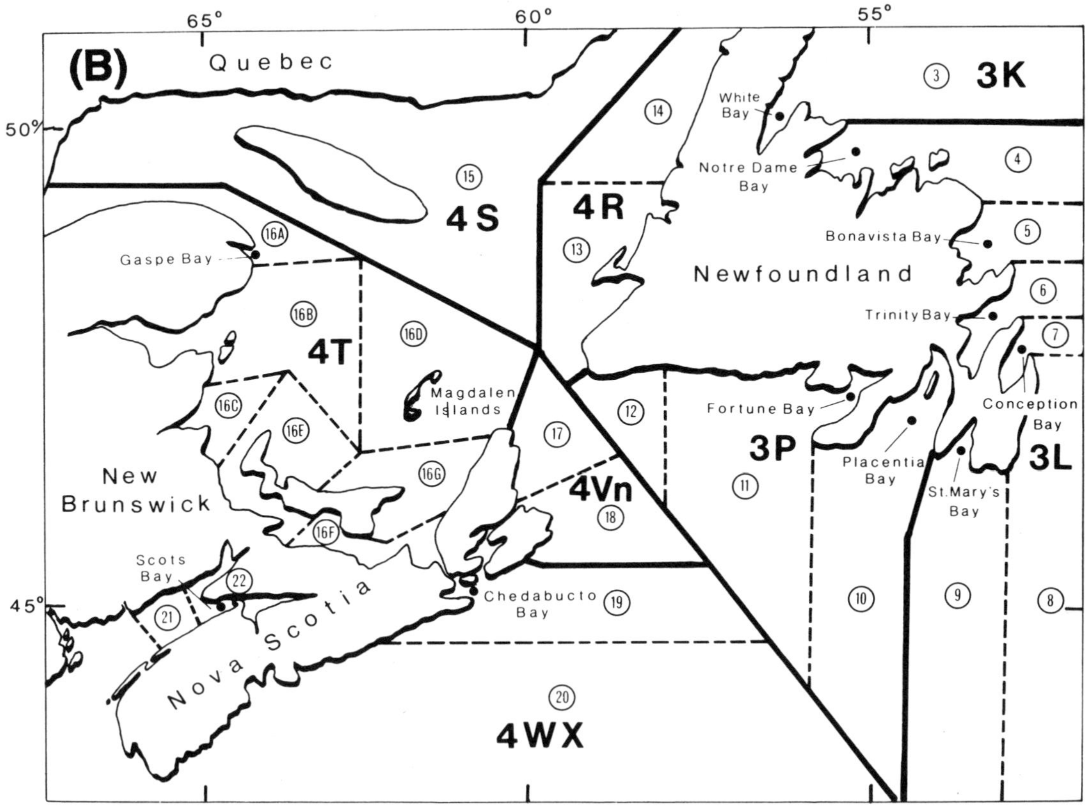


FIG. 3 (B). Herring management areas for the 1989 management plan, with NAFO (=ICNAF) Subdivisions.

the Scotia-Fundy Region, the inheritance of the infrastructure of the original AHMC made the transference relatively easy. The Gulf of St. Lawrence, as a geographical unit, and because of the complex interaction between the herring stock structure and the administrative structure, presents a more difficult situation. Each region has its own Pelagic Advisory Committee and discussion between these is carried out on an ad hoc basis for particular issues where regional interests coincide or conflict (E.M.P. Chadwick, Fisheries and Oceans, Moncton, N.B., pers. comm.).

The system has arisen over a long period as a result of interaction and discussion between the federal authority and the numerous competing interest groups. It has been said that the essence of management is the quantitative resolution of conflicts of interest. The herring management system that now exists represents the currently acceptable solution to a continually changing situation.

The Scotia-Fundy Region produces an annual "Herring Fishing Plan" and Table 3 is taken from the 1989–90 plan; it lists the way in which the overall "Quota" is subdivided by gear type, area fishery and season. For the most important gear type, the purse seine, there are quota allocations for individual boats, the latest development of the system first introduced in 1976 as part of the Bay of Fundy Project. The system has become very complex.

## A Review of the Fisheries

This brief review of the recent history of the various herring fisheries deals with stock complexes in three geographical areas: the Scotian Shelf-Bay of Fundy-Gulf of Maine-Georges Bank, the Gulf of St. Lawrence including western Newfoundland waters, and the east and southeast coast of Newfoundland.

### The Scotian Shelf — Gulf of Maine

Fisheries in the Scotian Shelf-Gulf of Maine area are of special historic interest as they were the first to be the subject of international agreement on nationally allocated shares of global TACs (Table 4) that initiated the

TABLE 3. 1989–90 herring fishing plan for Scotia-Fundy Region. Source: Fisheries and Oceans Canada pamphlet.

| Gear type | Fishery | Area | Season | Quota (tons) |
|---|---|---|---|---|
| Purse seine | Fall | 20 & 21 | Oct. 15–Dec. 31 | 9 000 |
| | Winter | 20 & 21 | Jan. 1–Feb. 28 | 6 000 |
| | Chedabucto | 19 | Nov. 1–Mar. 1 | 28 470 |
| | Summer | 20, 21, 22 | Apr. 1–Oct. 14 | 96 280 |
| Total purse seine vessel quotas | | | | 139 750 |
| | Bait | 19, 20, 21 | | 2 600 |
| Total purse seine vessel quotas | | | | 142 350 |
| Midwater trawl | Winter | 20 & 21 | Jan. 1–Mar. 31 | 850 |
| Gillnets, traps and weirs | | 17, 18, 19, 20, 21, 22 | | 8 000 |
| Total allowable catch | | | | 151 200 |

ICNAF development from 1972 on. The area currently supports the largest Canadian herring fisheries, both for the adults off southwest Nova Scotia and the northern Scotian Shelf, and for juveniles supporting the "sardine" industry along the coast of New Brunswick and Maine.

The northeast corner of Georges Bank was the spawning area of the largest herring stock in the western Atlantic (Anon. 1972b, e). It was fished largely by a "foreign" offshore fleet. The offshore fishery became virtually extinct by 1977 (Fogarty et al. 1990), but there are now indications of stock recovery. The Georges Bank stock is one of the most important "transboundary" stocks for, although the spawning groups lie on the United States side of the boundary set up by the World Court decision of 1985, the larval and juvenile stages, particularly, are found in Canadian waters.

The northwestern Maine-New Brunswick coastal waters supported the historically vital "sardine" industry for canned young herring, based mainly on the fixed-gear weir fishery (MacKenzie and Tibbo 1960), that was believed to exert more or less constant fishing effort (Anon. 1972a). Since 1977, and in United States waters, purse seines and "stop seines" working in conjunction with spotter planes were allowed access to the resource and, presumably because of increased effort and the subsequent effect on inshore availability, the traditional Maine sardine industry declined catastrophically. The New Brunswick industry is now dominant and, in 1990, produced 1.2 million or so cases of sardines with a final sales value of nearly 100 million dollars. Unregulated nearshore purse-seine activity in Maine-Massachusetts waters from 1977 on was accompanied by declining catches in the US adult fisheries, but the appearance of a large year-class that spawned in 1983 supported a resurgence of the fishery (Fogarty et al. 1990).

By the early 1980's, the Scotian Shelf and western Bay of Fundy were treated as containing a single major stock complex (Iles et al. 1984), so that a single TAC could be set for areas 19–22 (Fig. 3B).

In these areas, management involves the resolution of two kinds of problems that interact in a complex manner. Catches from particular areas and times of year and from different gear components have to be classified as "stock" or "non-stock," this referring to the spawning populations in the management area. Allocations of the TAC for the "stock" must then take into account the actual or potential conflict of interests between gear components, such as purse seine and gillnet fisheries, and the level of the fishery in relation to the concept of a limited total level of removals (TAC). Thus, the New Brunswick "fixed gear" weir fishery is believed to exploit fish originating in the Gulf of Maine and is "non-stock"; the Nova Scotia weir fishery is "stock" but it is "fixed gear" whose catch cannot be equitably controlled on an individual weir basis. As a whole, and as a limited entry fishery, it can be assumed to exert a fixed level of exploitation, and no catch limit is imposed for the total weir fishery. In the assessment procedure, an "allowance" is made which is corrected for, if necessary, during the next year's assessment. For other "stock" fishery components, fixed amounts are allocated from the global TAC. Stephenson et al. (1990) gives a detailed analysis of the procedures involved.

The dominant gear in the "stock" fisheries is the purse seine, and there are two distinct fisheries, a summer one off southwest Nova Scotia, exploiting feeding and spawning fish, and a winter fishery in the northern

TABLE 4. Stock and management information for 4WX and Subarea 5 herring (Scotian Shelf, Bay of Fundy and Georges Bank).

| Category | 1973 | 1974 | 1975 | 1976 | 1977 | 1978 | 1979 | 1980 | 1981 | 1982 | 1983 | 1984 | 1985 | 1986 | 1987 | 1988 |
|---|---|---|---|---|---|---|---|---|---|---|---|---|---|---|---|---|
| a) TAC | 90 | 90 | 90 | 89 | 109 | 110 | 99 | 65 | 100 | 80 | 82 | 80 | 125 | 98 | 127 | 151 |
| Catch Stock | 123 | 150 | 144 | 115 | 117 | 96 | 59 | 80 | 88 | 85 | 84 | 78 | 112 | 73 | 101 | 125 |
| Adjusted[d] | — | — | — | — | — | 114 | 78 | 107 | 137 | 106 | 117 | 136 | — | — | — | — |
| Catch 4Wa (Chedabucto Bay) | 8.4 | 27.0 | 27.0 | 37.2 | 23.3 | 17.3 | 14.0 | 9.0 | 18.6 | 12.3 | 8.2 | 6.3 | 8.8 | 8.4 | 8.8 | |
| Catch N.B. weirs (West Bay of Fundy) | 19.9 | 20.6 | 30.8 | 29.2 | 23.9 | 38.8 | 37.8 | 13.5 | 19.1 | 26.0 | 11.4 | 8.7 | 27.9 | 27.9 | 27.3 | 33.4 |
| Biomass 4WX (2 yr and older) | | | 458 | 374 | 279 | 243 | 376 | 403 | 373 | 381 | 427 | 505 | 590 | | | |
| b) Recruits (age 1) | 1.91 | 1.41 | 0.21 | 0.59 | 3.18 | 1.07 | 0.36 | 1.18 | 1.31 | 1.83 | 4.91 | 8.53 | 4.19 | 3.69 | 5.54 | |
| c) 4V TAC | — | — | — | 11.0 | 11.0 | 11.0 | 3.0 | 4.5 | 3.0 | 3.0 | 3.0 | 3.0 | 3.0 | 3.5 | | |
| Catch | 17.0 | 20.0 | 7.0 | 12.8 | 7.9 | 3.3 | 2.8 | 3.9 | 3.6 | 3.5 | 4.0 | 3.9 | 4.2 | 4.4 | | |
| Catch Georges Bank (5Z) | 202 | 150 | 146 | 44 | 2 | 2 | 1 | 1 | 1 | 1 | 0 | 0 | 0 | 1 | 0 | |
| Catch Gulf of Maine (5Y) Adults | 16.4 | 19.1 | 15.2 | 30.2 | 32.4 | 29.9 | 40.2 | 45.9 | 48.3 | 24.0 | 2.5 | 0.4 | | | | |
| Juveniles | 15.9 | 18.1 | 21.5 | 20.1 | 17.9 | 18.6 | 23.6 | 30.2 | 15.3 | 7.7 | 20.1 | — | | | | |

a) Catches, TAC and biomass in 1000 ton units.
b) Recruits in $10^9$ units.
c) The breakdown of 4V into management units has varied considerably over the last 15 yr.
d) Adjusted for estimated under-reporting.

Scotian Shelf areas that exploits the overwintering migrants.

As with most herring fisheries, recruitment variability has been a major factor in determining those changes in abundance that have justified modifying the level of recommended TACs. Until now, this variability appears to have been environmentally induced and there are no obvious signs of decline in recruitment levels because of overfishing leading to a too severe reduction of the parent stock. There were three above average year-classes produced in the 1960's in 1963, 1966 and 1969, but recruitment levels in the 1970's were generally poorer, although not to the same degree as in the Gulf of St. Lawrence and Newfoundland. A single large year-class of 1976 relieved a deteriorating resource situation and good recruitment in the early 1980's restored the stock to high historical levels (Stephenson and Power 1989). There are signs that the recent recruitment is much lower and caution on levels of exploitation is now advocated (Anon. 1991).

The northwestern section of the Scotian Shelf (areas 17–19) (Fig. 3B) is an area of stock mixing. Area 4V (17 and 18) was known to support a fishery of overwintering concentrations of migrating Gulf herring, but tagging experiments showed that the herring fished in Chedabucto Bay were winter migrants from the southwestern Nova Scotia where there was a summer fishery, followed by spawning of the fish in the fall (Stobo et al. 1975). Since 1978, the fishery in the Sydney Bight area (17 and 18) has been maintained at about 4000 tons as a winter fishery reserved for the Gulf of St. Lawrence purse-seine fleet, with a small inshore fishery of a few hundred tons a year in the area.

More recently, the Sydney Bight fishery moved northwards into area 17, and this allowed the setting up of a buffer zone (18) closed to all offshore fishing; this is intended to minimize the exploitation of 4WX fish outside its main management area (20–22).

Catch rates in the Sydney Bight area in the 1987–88 season for purse seines were the highest since 1971, indicating stock abundance is increasing, but the precise stock relationships for these fish are still not known. It has not been possible, therefore, to carry out a formal assessment (Stephenson and Power 1990).

## The Gulf of St. Lawrence

The Gulf of St. Lawrence presents particularly complex management problems. All five Atlantic Provinces have economic interests in the Gulf of St. Lawrence (Table 2). The biological situation is complicated by the admixture of both spring and fall spawners (Fig. 4) that require separate management. Finally, recruitment variability for the area as a whole, and between the spawning types, is high.

The biggest single event for the Gulf fisheries was the appearance and then fishing out of the extremely large year-classes of 1958 fall spawners and the 1959 spring spawners (Hodder 1971). These year-classes provided the basis for the original expansion of the purse-seine fishery in the 1960's and the older fish of these year-classes migrated to overwinter in the fjords of southern Newfoundland (areas 12 and the southern part of 11 and 13) (Hodder 1971). A large winter fishery developed in this area but this declined rapidly from a peak of about 143,000 tons in 1969 to only a few thousand tons in 1974, after which the overwinter fishery disappeared (Table 5C). The spring fishery on the returning migrants to the southern Gulf also failed and, in addition, by 1979 the major historical fishery for spring spawners in the Magdalen Islands was reduced drastically (Powles et al. 1979). The present situation is one of three more or less distinct areas with relatively highly localized fisheries. These are in 4R (13 and 14), 4S (15) and 4T (16 A–G).

### *The southern Gulf of St. Lawrence (4T)*

During the 1960's and 1970's, the large biomass resulting from the 1958 and 1959 year-classes was greatly reduced by fishing and natural mortality, and concurrent recruitment was too low to replace it (Table 5A). During the whole of the 1970's, there was but a single, moderate year-class, that of 1974.

More recently, recruitment has increased significantly to give rapidly increasing biomass estimates. Landings increased significantly in the 1980's and reached almost 80,000 tons in 1988.

In the 1950's, evidence suggested that spring spawning herring dominated the Gulf area but, by the early 1970's, fall spawners made up about 80% of the total catch. This proportion declined in the late 1970's and early 1980's but the most recent recruitment surge included a much higher proportion of fall spawning fish (Table 5B). The analysis of the factors that determine the balance between the two types is an important research issue.

Since the mid-1970's, there has been an equally significant change in the dominant gear type. Inshore effort by gillnets set on or near spawning grounds began to increase in 1977, and at an accelerating pace. This type of "local" fishing is appropriate to the complex stock situation (Fig. 4), and is facilitated by the fact that spawning close to shore and exploitation by set-nets present far less technical difficulties compared to the offshore spawning in tidally energetic areas off SW Nova Scotia.

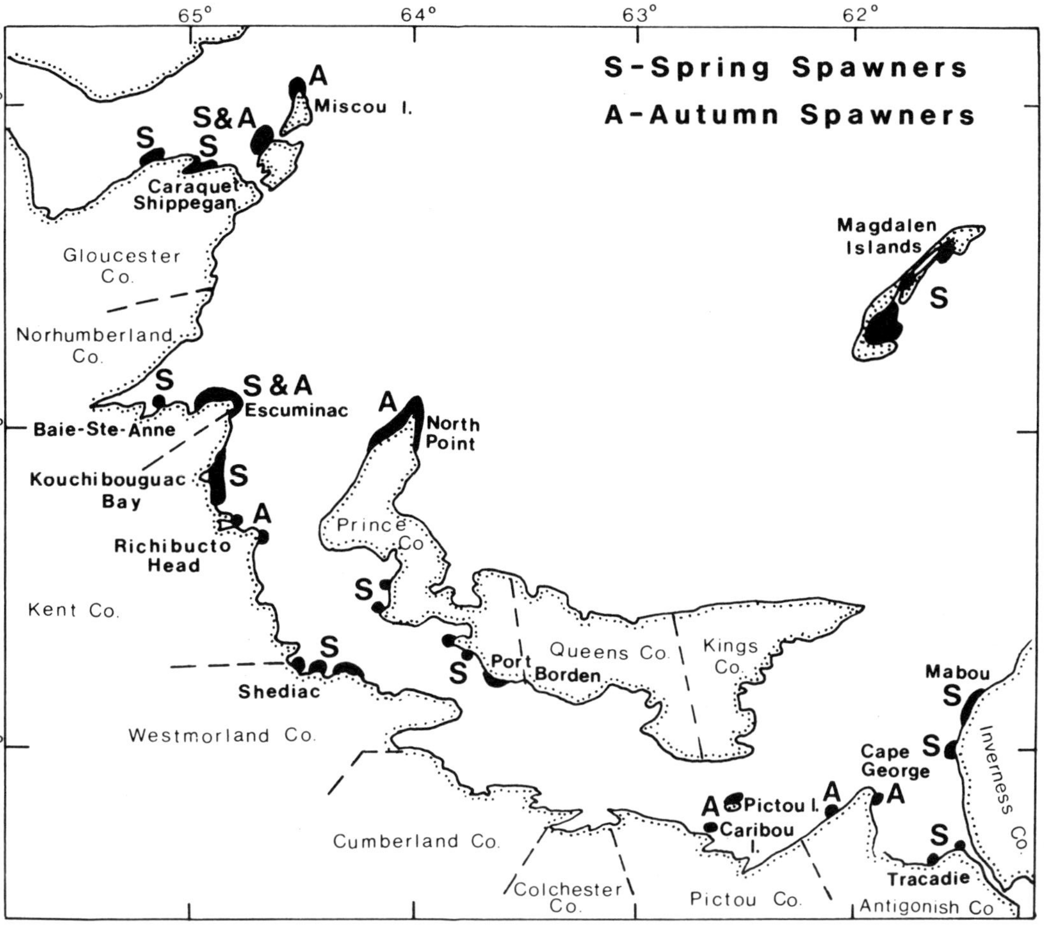


FIG. 4. Spring and fall herring spawning areas in the southern Gulf of St. Lawrence.

The recent history of assessment problems presented by this area is given special consideration.

*The West Coast of Newfoundland (4R)*

In the 1940's, and mostly in the Bay of Islands area, catches of local stocks ranged from 32,000 tons to as high as 80,000 tons but, by the 1950's and 1960's, catches had declined to only about 6000 tons (Table 5C).

The migratory components of the 1958 and 1959 year-classes of the southern Gulf represented a new feature in the fisheries of the area. They were exploited by purse seiners, following the migrating fish after the summer fishery (Hodder 1971) in the southern part of 4R in outer St. Georges Bay, as well as in the fjords further to the east. As these migrants declined in abundance in the 1970's, the fishery moved into the Bay itself and concentrated once more on local stocks.

By 1975, St. Georges Bay could be included in a west Newfoundland complex as a convenient assessment unit, and this complex became the responsibility of the Gulf Region after the 1981 reorganization. Its boundaries are from Cape Ray to Cape Norman (beginning with the 1975–76 catches), and a line from Anguille to Anticosti separates northern Gulf from southern Gulf stocks (Fig. 3B). The existence of fall spawners in the area was not recognized until the 1970's.

Both fixed gear (gillnets) and mobile gear (purse seines) are used in 4R in a seasonal fishery lasting from April-December which exploits spawning and overwintering fish (McQuinn 1989). Catches were lowest in 1969 at only 2900 tons, and reached a recent high in 1973 at 27,000 tons, the variability being largely a result of variability in purse-seine activity (Table 5C).

TACs were first applied in 1977 and an allocation procedure was instituted in 1980, with regional and seasonal quotas for both gear sectors. There was also a gear allocation under the Purse Seine Restructuring Plan by which the purse-seine component was awarded

TABLE 5A. Stock and management information for 4T herring: catches and TACs (southern Gulf of St. Lawrence).

| Category | 1971 | 1972 | 1973 | 1974 | 1975 | 1976 | 1977 | 1978 | 1979 | 1980 | 1981 | 1982 | 1983 | 1984 | 1985 | 1986 | 1987 | 1988 |
|---|---|---|---|---|---|---|---|---|---|---|---|---|---|---|---|---|---|---|
| [a] Catch | | | | | | | | | | | | | | | | | | |
| Spring | 27.4 | 9.1 | 8.9 | 17.0 | 25.7 | 20.9 | 24.8 | 16.5 | 21.3 | 19.7 | 6.6 | 6.8 | 8.6 | 5.3 | 7.1 | 7.8 | 11.7 | 12.7 |
| Fall | 107.4 | 44.5 | 31.5 | 18.2 | 18.9 | 15.6 | 20.0 | 36.6 | 26.4 | 20.7 | 14.6 | 17.2 | 17.3 | 18.7 | 27.3 | 51.2 | 65.9 | 58.7 |
| Total | 134.8 | 53.6 | 40.4 | 35.2 | 44.6 | 39.5 | 44.8 | 53.1 | 47.7 | 40.4 | 21.2 | 24.0 | 25.9 | 24.0 | 34.4 | 59.0 | 77.6 | 71.4 |
| [a] TAC total | | 150.0 | 150.0 | 150.0 | 55.0 | 55.0 | 60.0 | 55.0 | 55.0 | 55.0 | 16.0 | 15.0 | 20.0 | 19.0 | 32.2 | 43.4 | 72.8 | 78.9 |
| % fall spawners | 80 | 83 | 78 | 58 | 47 | 45 | 49 | 55 | 61 | 51 | 59 | 57 | 62 | 63 | 68 | 71 | 74 | 70 |
| % seine caught | 82 | 67 | 51 | 65 | 73 | 75 | 82 | 75 | 72 | 67 | 17 | 13 | 13 | 12 | 11 | 19 | 18 | 25 |

[a] 1000 ton units

TABLE 5B. 4T herring biomass and recruitment for spring and fall spawners.

| Category | 1974 | 1975 | 1976 | 1977 | 1978 | 1979 | 1980 | 1981 | 1982 | 1983 | 1984 | 1985 | 1986 |
|---|---|---|---|---|---|---|---|---|---|---|---|---|---|
| [a] Biomass, spring spawners | 66 | 48 | 47 | 31 | 47 | 28 | 16 | 12 | 13 | 34 | 42 | 56 | 55 |
| Biomass, fall spawners | 146 | 115 | 98 | 67 | 34 | 22 | 20 | 25 | 37 | 77 | 166 | 213 | 227 |
| Biomass, total | 212 | 163 | 145 | 98 | 81 | 50 | 36 | 37 | 50 | 111 | 208 | 269 | 282 |
| [b] Recruits, spring spawners | 131 | 47 | 330 | 50 | 52 | 116 | 109 | 240 | 176 | 149 | 97 | 118 | 118 |
| Recruits, fall spawners | 43 | 89 | 136 | 169 | 97 | 350 | 341 | 458 | 767 | 493 | 534 | 397 | 394 |
| Recruits, total | 174 | 136 | 466 | 219 | 149 | 466 | 450 | 698 | 943 | 642 | 631 | 515 | 512 |

[a] 1000 ton units.
[b] $10^6$ units.

TABLE 5C. Stock and management information for 4R herring (west coast of Newfoundland).

| Category | | 1973 | 1974 | 1975 | 1976 | 1977 | 1978 | 1979 | 1980 | 1981 | 1982 | 1983 | 1984 | 1985 | 1986 | 1987 | 1988 | 1989 |
|---|---|---|---|---|---|---|---|---|---|---|---|---|---|---|---|---|---|---|
| [a] TAC | | — | — | — | — | 12.0 | 12.0 | 12.5 | 18.0 | 16.0 | 10.0 | 10.0 | 10.0 | 10.0 | 17.0 | 30.6 | 30.0 | 30.0 |
| Catch | | 27.0[c] | 7.5[c] | 4.8 | 10.1 | 12.7 | 15.1 | 18.3 | 19.1 | 13.6 | 14.8 | 11.4 | 10.5 | 14.4 | 21.4 | 16.6 | 18.3 | 26.0 |
| % spring spawners | | 57 | 64 | 68 | 80 | 77 | 76 | 74 | 82 | 81 | 78 | 69 | 61 | 69 | 77 | 75 | 81 | |
| Biomass | Spring | 171 | 152 | 128 | 113 | 98 | 99 | 84 | 67 | 52 | 36 | 38 | 86 | 85 | 114 | 119 | 93 | |
| | Fall | 79 | 66 | 60 | 48 | 41 | 37 | 28 | 24 | 21 | 21 | 48 | 46 | 44 | 39 | 39 | 31 | |
| | Total | 250 | 218 | 188 | 161 | 134 | 136 | 112 | 91 | 73 | 57 | 86 | 132 | 129 | 153 | 158 | 124 | |
| [b] Recruits | Spring | 15 | 32 | 22 | 139 | 27 | 12 | 19 | 16 | 98 | 415 | 65 | 339 | 141 | 26 | 67 | 67 | |
| at 2 yr | Fall | 17 | 14 | 34 | 9 | 4 | 10 | 27 | 35 | 248 | 43 | 34 | 29 | 47 | 5 | 26 | 26 | |
| | Total | 42 | 46 | 56 | 148 | 31 | 22 | 46 | 51 | 346 | 458 | 99 | 368 | 188 | 31 | 93 | 93 | |

[a] TACs, catches and biomass in 1000 ton units.
[b] Recruits in millions of fish.
[c] Includes small catches of overwinter migrants from the southern Gulf of St. Lawrence.

55% of the TAC (see p. 134). Since 1985, the purse seine fleet has dominated catches because of its superiority in supplying the dominant roe market and, in 1988, this gear accounted for 90% of the total catch. There has been a corresponding change in the degree of exploitation in terms of the set TACs. These were exceeded consistently in the period 1983–86, but in 1987 and 1988, only about 60% of the TAC was taken.

Most of the catches in the early 1970's were taken in the St. Georges Bay area (Fig. 3B) (4Rd) but, more recently, the Bay of Islands and Bonne Bay fisheries were more prominent because of the small size of the St. Georges Bay fish. In the late 1980's, the spring "roe" fishery in St. Georges Bay increased and the season was extended into June and July (McQuinn 1989). TACs for 1989 and 1990 were set at 26,000 and 20,500 tons, respectively, with about 70% of these accounted for by spring spawners. Recent catches were lower than these levels (McQuinn 1989).

Catch rate data for 4R are now available from gillnet logbooks from indexed fishermen, landings from purchase slips in conjunction with estimates of total nets set and from questionnaires, and these data involve a separation of spring and autumn spawners (McQuinn 1989). Historical data go back to 1977 but the information from the dominant gear, the purse-seine, is not reliable. It has been possible to calibrate the catch-rate data using the "ADAPT" procedure (Anon. 1989a), and the 1988 F level was estimated at 0.159, a reasonably low one for herring. The $F_{0.1}$ level has varied greatly because of the variability in the recruitment patterns to the fishery as the market has demanded different sized fish.

*Northern Gulf of St. Lawrence (4S)*

The catches in this area are small, of the order of a thousand tons only. Two more or less distinct sub-areas are recognized, a western area with suspected affinities with the southern Gulf, (4T) and an eastern area thought to be related to western Newfoundland (4R); this was strongly inferred by the results of a recent tagging experiment (Anon. 1989b).

The fisheries are traditionally gillnet and highly localized but, recently, a single purse-seine has become active in the area, giving the potential for the development of larger catches for a more accessible market, as recently occurred in the Sept Iles area. (I.H. McQuinn, Fisheries and Oceans, Mont Joli, Quebec, pers. comm.).

Both spring and fall spawners are found in the area and there is some difference in the timing of the fisheries between the two areas in the spring, the western fishery taking place a little earlier than the eastern, in May, rather than June. Both fall fisheries extend from July to October.

Good year-classes have occurred in the western area in the spring of 1974 and 1981, and in the fall of 1973, 1978 and 1980; in the eastern area the spring year-classes have been above average in 1974 and 1980, and the fall in 1973, 1977 and 1979.

TACs were first set in 1983 to 1,000 tons for both areas combined and in 1988 and 1989 the two sub-areas had separate TACs at increased levels of 1,500 and 2,000 tons respectively, the result of the improved recruitment in the 1980's that was widespread throughout the whole of eastern Canada as well as in the Gulf of Maine. Uncertainty about stock definition and intermixing has led to cautious advice on possible levels of harvest (Anon. 1989b).

### The East and Southeast Coasts of Newfoundland (3K, 3L & 3P)

The area covered by this stock complex is a large one, extending from White Bay (Area 3, NAFO subdivision 3K) to Fortune Bay (Area 11, NAFO subdivision 3P). A research program initiated in the mid-1960's at the St. John's Biological Station has firmly established five main assessment sub-units. Results obtained from a successful tagging experiment (Wheeler and Winters 1984) demonstrated a high degree of stock integrity that justifies recognition of these units. Herring catches between Fortune Bay and the West Coast of Newfoundland are negligible; there is little evidence of admixture with Gulf of St. Lawrence herring. Catches north of White Bay, on the east coast are very small and indicate the northern limits for herring.

TACs for each stock unit are suballocated among inshore fixed gear, (mostly gillnets and traps) and mobile gears (bar seines and purse seines). The Management Plan in the late 1980's provides for an annual "allowance" for the fixed gear component with the "residual" of the TAC allocated to the mobile gear sector, if necessary after a reserve was set aside for contingencies (J.A. Moores, Fisheries and Oceans, St. John's, Nfld., pers. comm.); this arrangement corresponds to the "local" and "migratory" components of the stocks that are involved. The mobile allocation is divided into portions for a spring and a fall fishery.

Catches from the fisheries in this area are very variable (Table 6). Over the last twenty years they have depended almost entirely on only two year-classes, one spawned in the spring of 1968 and the other in the spring of 1982. A relatively large year-class was spawned in the spring of 1987 that appeared to be similar in size to that of 1982. Its contribution to the fisheries in 1991 was greater in the two northern areas which produced an estimated 17,000 tons in that year (Wheeler et al. 1992). The 1968 year-class initiated a fishery of some 5,000 tons in 1973 which increased in size to 26,000 tons by 1979, but subsequent stock depletion, in the absence of new recruitment, led to closure of the fisheries in 1982. The appearance of the 1982 year-class allowed the reopening of a gillnet fishery in 1984 and 1985; by 1986 purse seines and ring nets became involved (Wheeler et al. 1989).

Fall spawning herring are also found in the area but in much smaller numbers. From 1974 to 1988 they made up only about 5% of the total commercial catch and, except for the years of the fisheries closure, have been caught at remarkably constant levels of about 1,000 tons.

The future prospects and nature of these fisheries are dependent on the sporadic appearance of relatively large spring spawned year-classes. It is likely that these represent natural "boom and bust" fisheries for which a special management approach might be justified.

This extreme biological variability also presents difficulties for traditional assessment procedures, and particularly for those based on the standard assessment methods involving Sequential Population Analysis (SPA). For these herring stocks, it is virtually impossible to obtain reliable estimates of current population size using SPAs, much less of future prospects even if catch reporting procedures are ideal. Effort levels are likely to fluctuate widely and unpredictably with such variable recruitment. This means that assessment is heavily dependent on direct acoustic estimates to provide independent estimates of current population size and recruitment levels.

A research gillnet program has been in operation for ten years for both spring and fall fisheries along the northeast coast, and for the spring fishery on the southeast coast for the last 8 years. By 1989, these had indicated declines in spring-spawner catch rates for recent years in the mid-northern and the southern areas, but increases in the Conception Bay — Southern Shore areas (7 and 8) (Wheeler et al. 1989). Continuation and extension of these programs are expected to detect and quantitatively assess the crucial advent of new year-classes into the fishable biomass.

In more recent years the area from White Bay to Conception Bay has been surveyed by acoustic surveys to generate biomass estimates and, in conjunction with on the spot sampling, information on the size structure of the population that indicates future recruitment prospects (Wheeler et al. 1989).

## Two Case Histories

The complexity of the stock situation in west Atlantic herring and its migratory tendencies were important factors to be resolved in the context of individual national interests and the available management procedures in the ICNAF management era. It is remarkable that, although properly supported and sustained research programs on herring at that time were a recent phenomena in the west Atlantic, that the first major breakthrough in international fisheries management in January 1972 at ICNAF involved herring stocks with very short documented histories. This constraint on the quick and simple resolutions of management problems continued to be an issue within the purely Canadian management regime.

The "internalization" of the management process began in 1971 with the establishment of the closing lines, but became a final reality with the 1977 extension of jurisdiction.

It is impossible in a work of this scope to deal in great detail with the way the Canadian management system evolved over the last 20 years, but selected individual issues and developments can be identified as both important in themselves, and as illustrating key aspects of management in what was, and still is, a very volatile situation.

The first, involving the Gulf of St. Lawrence Biomass, is an example of difficulties caused by the lack of historical data and the importance, then, of good judgement based on sound background knowledge.

The second illustrates the impact of a major deterioration in the economic basis of an important fishery in the early 1970's and, for the industry as a whole, the unorthodox way that it was resolved by the Canadian Government, through its Minister of Fisheries, as the Bay of Fundy Project, and through a major change in the basis of the management system. This new situation became institutionalized as the result of a major policy review by a process of Fleet Restructuring.

In dealing with these case histories, much of the complexity of the management situation on the east coast of Canada will be touched upon.

### The Gulf of St. Lawrence Biomass Controversy

The Gulf of St. Lawrence herring fisheries for the most of the 20th century were small, localized and lacking reliable markets. Until the 1980's, there was no in situ research on the stocks or the fisheries. Any work that had been done was based on research centres outside the area and was not sustained for any lengthy period. The ad hoc research carried out in the late 1960's and early 1970's had documented the development of the purse-seine fishery on the 1958–59 year-classes (Hodder 1971; Iles and Tibbo 1970), and demonstrated that management could be considered a purely

TABLE 6. Stock and management information for 3K, L, P herring (east and southeast coast of Newfoundland).

| Category | | 1974 | 1975 | 1976 | 1977 | 1978 | 1979 | 1980 | 1981 | 1982 | 1983 | 1984 | 1985 | 1986 | 1987 | 1988 | 1989 |
|---|---|---|---|---|---|---|---|---|---|---|---|---|---|---|---|---|---|
| White Bay | [a] TAC | — | — | — | 10.5 | 7.9 | 11.5 | 5.3 | 5.3 | 1.2 | 0.0 | 1.5 | 2.0 | 5.5 | 32.5 | 34.7 | 14.0 |
| Notre Dame | [a] Catch | 4.0 | 5.6 | 12.5 | 11.6 | 13.4 | 15.7 | 6.5 | 4.7 | 2.0 | 0.4 | 1.5 | 1.8 | 2.7 | 14.0 | 7.3 | |
| % spring | | 96 | 98 | 99 | 99 | 98 | 98 | 96 | 98 | 96 | 89 | 86 | 87 | 92 | 96 | 96 | |
| Bonavista | TAC | — | — | — | 9.4 | 7.8 | 8.4 | 4.4 | 4.8 | 0.7 | 0.0 | 0.4 | 0.8 | 3.8 | 13.7 | 16.2 | 6.9 |
| Trinity | Catch | 2.3 | 5.9 | 9.9 | 12.0 | 8.1 | 9.8 | 5.4 | 4.0 | 0.5 | 0.1 | 0.2 | 0.6 | 1.7 | 6.9 | 10.1 | |
| % spring | | 86 | 98 | 96 | 96 | 97 | 96 | 94 | 94 | 93 | 89 | 77 | 78 | 90 | 94 | 95 | |
| Conception | TAC | | | | 2.1 | 1.8 | 0.9 | 0.4 | 0.5 | 0.2 | 0.0 | 0.1 | 0.2 | 0.6 | 3.5 | 0.6 | 1.5 |
| Southern Shore | Catch | 2.7 | 3.5 | 2.5 | 2.2 | 2.0 | 0.9 | 0.5 | 0.2 | 0.1 | 0.0 | 0.0 | 0.1 | 0.2 | 1.3 | 0.5 | |
| % spring | | 99 | 95 | 94 | 90 | 92 | 82 | 94 | 91 | 82 | 43 | 53 | 63 | 69 | 97 | 92 | |
| St. Mary's | TAC | | | | | | 3.4 | 2.5 | 1.2 | 0.0 | 0.0 | 0.0 | 0.6 | 2.1 | 2.5 | 8.9 | 1.5 |
| Placentia | Catch | 6.5 | 6.7 | 4.1 | 3.3 | 3.5 | 3.6 | 2.5 | 0.6 | 0.0 | 0.0 | 0.1 | 0.2 | 0.1 | 0.3 | 1.2 | |
| % spring | | 94 | 95 | 95 | 91 | 86 | 87 | 81 | 78 | 86 | 58 | 47 | 51 | 52 | 70 | 83 | |
| Fortune | TAC | | | | | | 1.0 | 1.0 | 0.2 | 0.0 | 0.0 | 0.0 | 0.3 | 0.7 | 2.4 | 4.7 | 1.5 |
| | Catch | 6.5 | 6.7 | 4.1 | 3.3 | 3.5 | 3.6 | 2.5 | 0.6 | 0.0 | 0.0 | 0.0 | 0.1 | 0.1 | 0.2 | 0.1 | |
| % spring | | 97 | 94 | 92 | 98 | 97 | 98 | 99 | 71 | 77 | 77 | 61 | 63 | 74 | 84 | 87 | |

[a] All TACs and catch in 1000 ton units.

Canadian responsibility, but relatively little was known of the biological interrelationships of the stocks and of their long-term fishery potential.

In particular, the spectacular expansion of the 1960's was at the expense of exceptionally large year-classes and was accompanied by a large reduction in total biomass; the year-classes were virtually fished out and the obvious question was whether "normal" recruitment could be expected to rebuild the stocks. By the late 1970's it was thought that this might in fact occur, for it was believed that exploitation had not affected long-term recruitment prospects biologically and, on this assumption, it was time for larger year-classes to appear.

In 1979 CAFSAC concluded, indeed, that the available biomass of the southern Gulf stocks (4T) had increased to a level to justify a 1979 catch of 106,000 tons, at the then accepted $F_{0.1}$ level of 0.38 for fully recruited fish. CAFSAC further suggested that while the maintenance of such a high TAC would demand improved recruitment the current TAC level of 55,000 tons ought to be sustained under current recruitment levels (Anon. 1979).

The high catches of the 1960's and early 1970's had established general expectations of high productivity. This 1979 forecast was based on expected high catches in the 1978 season, an increase in the catch rate for purse seiners in the fall fishery, and apparent trends of increasing abundance in recent years. Evidence of a decrease in the stocks was also available but was discounted; instead it was claimed that, by holding fishing mortalities well below the $F_{0.1}$ level over the last few years, the management objective of arresting the decline in stock biomass of the early 1970's, in fact had been achieved.

By 1980, this interpretation was being challenged on the grounds that fishing mortalities had been far higher than had been assumed, increases in the TAC towards "historical" levels were not justified, and the status quo of a TAC of 55,000 tons should be recommended. Certainly by the fishing season of 1979 the older and larger fish were scarce and the catch was dominated by very young fish in both the spring and fall spawning components, which was not promising (Anon. 1982a).

The individual indices of abundance available at that time were not consistent; of the six that were examined, two showed an increase in biomass and four a decrease. The resulting, somewhat cautious, advice was that the TAC for 1981 be maintained at the 1980 level. Even so, guarded optimism was expressed for it was suggested that the biomass levels would not be unduly affected even at a higher alternative TAC of 90,000 tons (Anon. 1982a). It was thought that there was sufficient time to resolve the scientific uncertainty before the stock was seriously overexploited.

By 1981, the situation was seen to be far worse than had been suspected. At the expected fishing rate of $F_{0.1}$=0.33 the biomass was estimated to be so low that the yield, projected for 1982 would be only 16,000 tons. The obvious conclusion was that annual total mortality rates had been high over recent years, the result of much higher fishing pressures than had been assumed, and that few larger and older fish were available. The stock was made up almost entirely of young fish from 2 to 4 years old, so small that they were hardly bigger than the accepted minimum size limit then in effect (Anon. 1982b,c).

By March 1981, the evidence for high levels of total mortality was convincing enough to be universally accepted but there was still disagreement as to its cause. One theory was that there had been an enormous increase in the Gulf cod stocks in the 1970's and that this had led to a greatly increased predation rate on herring (Anon. 1982c). However, a detailed analysis of cod feeding habits (Waiwood 1981; Waiwood and Majkowski 1984) showed that herring were not an important, much less dominant, target species for cod. This issue was irrelevant to the question of the current state of the herring resource. It merely suggested why the previous assessments had been so misleading as to the health of the resource. Even so, the debate on the "cod-herring" interaction continued into 1982 (Anon. 1983).

Subsequent events substantiated that there had been a very large and unsuspected deterioration in the resource situation. The degree to which this was misleading is apparent from comparisons of biomass levels as estimated and projected in 1979 and the values as finally determined in 1986, by which time information was fully available (Table 7).

The difference between the current estimate in 1979 and the situation as was finally confirmed is extreme, more than ten-fold, although the 1979 estimates did include an assumed high recruitment level for year-classes from about 1974 onwards.

With the wisdom of hindsight it is possible to be more objective in giving reasons for what must be conceded as a failure of judgment, one that in the 3 years in which it affected the assessment procedure saw a near elimination of the resource. The more important question to be addressed, however, is not whether it could have been avoided at the time but whether such an event could reoccur. Fortunately, for a number of reasons, this appears to be unlikely.

Firstly, it must be appreciated that none of the herring fisheries that became prominent in the west Atlantic in 1960's had any historical documentation. The largest, the foreign fishery on Georges Bank, lasted not much longer than a decade from discovery to extinction, which did not give much of an opportunity for realistic

TABLE 7. Biomass (thousands of tons) estimates for southern Gulf of St. Lawrence herring as estimated in 1979 and as finally confirmed in 1986.

| Biomass | 1969 | 1974 | 1975 | 1976 | 1977 | 1978 | 1979 | 1980 | 1981 |
|---|---|---|---|---|---|---|---|---|---|
| 1979 – estimated | 1591 | 295 | 303 | 255 | 235 | 231 | 377 | 311 | 267 |
| 1986 – final | | 104 | 92 | 68 | 49 | 55 | 35 | 23 | 19 |
| Final/estimated | | 0.35 | 0.30 | 0.27 | 0.21 | 0.24 | 0.09 | 0.07 | 0.07 |

assessment (Fogarty et al. 1990). Even the Magdalen Islands fishery in the Gulf, that had existed for more than 200 years, was not documented until its years of decline and near extinction in the 1970's (Powles et al. 1979). The new purse-seine fishery of the Gulf of St. Lawrence began only in the 1960's and could not be monitored and studied to the degree necessary for a rigorous assessment. Research resources were scarce and none was based in the area itself, so that first hand knowledge of the fishery was difficult to obtain.

In addition, there is an assumption, implicit in population dynamics generally and fisheries assessment in particular, that there is such a thing as an "average," "stable" or "equilibrium" population. This assumes that the historically documented variability is characteristic and can be expected to re-occur. For example, larger than average year-classes tend to appear periodically and with longer intervening periods for more northerly and long-lived stocks. In many ways the major Gulf of St. Lawrence stocks resembled such a stock, the so-called "Atlanto-Scandian" (Norwegian) herring which produced very large year-classes at about 10-year intervals (Jakobsson 1985). It would be a reasonable assumption that recruitment recovery in the Gulf of St. Lawrence would occur at similar intervals, leading to a general expectation that improvement would occur after about a decade or so.

Moreover, the herring stocks of the Gulf over the last thirty years or so have experienced radical and quite unpredictable natural changes with marked population effects (Tibbo and Graham 1963). The epizootic disease of the late 1950's was known to have decimated all herring stocks in the area, making them uncharacteristically susceptible to predation by cod, for example. This was followed by the enormous year-classes of 1958 and 1959 that gave rise to a herring population biomass of some two million tons. One might expect that future recruitment of young fish would match this demonstrated ability of the Gulf to support a large biomass of older fish.

Finally, for the 1960's and 1970's, these fisheries could be studied only from the St. John's, Newfoundland and St. Andrews, New Brunswick Biological Stations, so that on-the-spot monitoring of the new inshore fisheries that developed and expanded very rapidly, presented logistic difficulties that could not be dealt with adequately.

Even so, errors of judgment did occur. The errors of judgment were of two kinds, a failure to detect the extreme decline in biomass and a too optimistic view of future recruitment prospects, that led to "predictions" of the size of incoming year-classes that were far too high. The first misjudgment, perhaps, could and should have been avoided. There was little positive evidence for the appearance of large fish that would have demonstrated recovery. The lack of information on the prior occurrence of new year-classes as young fish, which would have signalled recovery, should have been given weight.

The improvement in recruitment for the Gulf of St. Lawrence stocks began at the end of the 1970's despite a declining biomass, indicating a major environmental effect. It continued well into the 1980's as part of a trend that appears to have been general for the whole of the northwest Atlantic herring population. This more recent information shows that the improvement began in the late 1970's (the year-class is 2 years previous to the recruit estimate) and that the resurgence continued into the 1980's; it also shows that the resurgence was initially underestimated.

## Improving the Management System

### The Bay of Fundy Project and Fleet Restructuring

The next case history involves the herring purse-seine fisheries off the southwest coast of Nova Scotia exploiting the stock that was amongst the three herring stocks that initiated the management breakthrough at ICNAF in 1972 (Anon. 1972b). By 1975, a sharp drop in the meal commodity market, the economic mainstay of the fishery, threatened its survival and caused great concern amongst the fishing communities dependent on it. The departmental response was an unusual one. The then new Federal Minister of Fisheries, Romeo LeBlanc, invited the fishermen themselves to be involved in arriving at proposals for a solution, but within the management constraints of that time.

This was a revolutionary approach in that, for the first time, fishermen were admitted to the management forum as equal partners with the processing side of industry. The overriding constraint in the fishermen's approach was that any plan must apply to all fishermen equally: this was the only way to encourage the cooperation at the primary level on which success depended. Fleet reduction was not a viable option and was never mentioned or discussed during the many meetings that took place over the winter and spring of 1975–76. It did mean that those fishermen with the most secure access to the market and/or those who were the most efficient at catching herring accepted a smaller share of the resource, in the form of boat quotas, than they could expect in a competitive situation. The solution proposed involved four measures:

(a) The directed fishery for the meal industry was to be closed. This was already provided for by Section 29 of the Fisheries Act, but this section had been in abeyance for some years.
(b) Individual boat quotas were to be established to ensure that all those already in the fishery had a chance of survival in 1976; in addition, daily total fleet quotas would be established to match catch rates to the available market.
(c) The quality of landed herring in the plants would be improved and the season extended. This required the establishment of weekly boat quotas and emphasis on improved methods of handling.
(d) An interim high priced market would be provided by selling herring "over-the-side" to the Polish state company Rybex, and at relatively attractive prices.

It was concluded that these objectives could be achieved only if the fishermen were formally organized, and an Atlantic Herring Fisheries Market Club was formed which was later established as a co-operative, with offices in Yarmouth, the centre of the fishery (Doucet 1977).

A small federal management unit was established at the St. Andrews Biological Station, responsible for day-to-day monitoring of fishing activities, and co-ordination with the departmental field staff and with Ottawa. There was also a coordinator in the Halifax headquarters who ensured contact with the Atlantic Herring Management Committee.

The complete turnaround to a food industry was achieved in 1977. There had been a restructuring of the processing industry and a continued improvement in the quality of landed herring; the price to the fisherman doubled again in 1977, as it did over the next couple of years. Prospects for the development of a share of the Japanese roe market were also being pursued.

The initial success of the Bay of Fundy Project eased the economic problem considerably but subsequently allowed the expression of inter-group rivalry that appeared on provincial lines. In 1978, this led to the formation of a break-away organization and the re-establishment of local arrangements with local processors (Kearney 1981, 1984). In 1980 the day-to-day control of the Management Plan passed from the field to the Halifax headquarters. By June of that year there was evidence of systematic under-reporting of catches involving collusion between certain plants and certain fishermen. The cornerstone of the original project, the self-disciplining of all fishermen in an atmosphere of mutual trust, was lost.

This coincided with a significant weakening of the food market, quality problems in meeting market requirements and a reduction in the level of "over-the-side" sales to foreign processing ships. To all this must be added a worsening resource situation caused by poor recruitment and the buildup of extra effort. The success of the project had attracted several hundred gillnet boats to the main summer fishery. This exacerbated the resource situation and created major problems of gear conflict. In the Gulf of St. Lawrence there had been a major increase in gillnet activity; also, the resource situation had deteriorated and similar conflict between gear components had arisen.

In the meantime the full impact of the 1977 extension of jurisdiction had been felt, CAFSAC had become fully operational and the federal departmental reorganization had resulted in the concept of four regions, Scotia-Fundy, the Gulf of St. Lawrence, Quebec and Newfoundland. It was thus an appropriate time for a full review of the management process with the object of establishing guidelines and policies for the future. The important issues were, by then, recognized as the excessive fishing effort in relation to the size of the resource, the potentially major conflict between gear components, particularly between the gillnet and purse-seine fleets, and the rapidly changing market situations involving both the decline of the European food herring market and the development of the Japanese herring roe market.

To examine these and other problems, the Department of Fisheries and Oceans, Canada, undertook a number of studies and held an industry — government Seminar at Yarmouth, Nova Scotia, from February 17–19, 1981. The studies provided the most up-to-date information available on the state of the stocks, their current and historical levels of exploitation, the size and structure of the fleet and processing sector as well as the current and expected future state of the market (Anon. 1981).

As a result of this meeting a "Management Philosophy" for the east coast herring fisheries was established with the following key elements. State ownership of the resource was reaffirmed and was thus controlled by state licensing; the principle of priority for contiguous areas was recognized; allocation by gear type (technology) was a most important factor; herring management could not be considered in isolation from other fish stocks.

When applied to the immediate situation this translated into a planned reduction in the size of the purse-seine fleet in the Atlantic area as a whole, with a re-allocation of the resource amongst the gear components, which became known as the Restructuring Plan. As a result the "Scotia Fundy" and "Gulf" components of the purse-seine fleet were separated and could not fish in each other's areas. The Gulf fleet was allocated 20% of the southern Gulf TAC, establishing the gillnet as the dominant segment there, reflecting the perceived need to encourage "local" interest in fixed gear, near-shore fisheries. In addition, the Gulf based purse seiners were awarded 55% of the West Coast of Newfoundland TAC and given sole access to the Sydney Bight area (McFarlane 1989).

The purse-seine fleet in the 4WX area then consisted of 49 vessels that had enjoyed annually determined boat quotas since 1976. They were assigned percentages of the TAC which were transferable by sale and purchase between boats. Such transactions had been features of the Bay of Fundy Project on a temporary basis since its inception, by mutual agreement amongst any self-proclaimed group who could demonstrate mutual interests. This was formalized and legalized in the Restructuring Plan, which was formally put into place in the fall of 1983 (McFarlane 1989).

A guaranteed minimum of 80%, of the 4WX TAC was allocated to the purse-seine gear segment and individual boat allocations (now referred to as Individual Transferable Quotas or ITQ's) were made at two levels, a higher level of 2.7% for the larger boats allowed to fish in both the summer fishery in the Bay of Fundy proper and in the winter fishery in the Chedabucto Bay area, and a level of 1.6% for the others. A boat could sell all of its allocation to others in its own group, with the proviso that the maximum individual boat quota in either group could not exceed 4% of the TAC. The total (minimum) fleet allocation of 80% of the TAC and the corresponding ITQ scheme were set for a period of 10 years with the proviso that at the end of that period, the situation would be reviewed and modified if necessary. The expectation was that by then the fleet would be reduced to 25 vessels each with a 4% share of the TAC.

A recent 5-year review revealed that only eleven boats had reached the 4% level and of them only four had done so by buying the quotas of other boats, and this at a lower price than had been anticipated (McFarlane 1989). However, there had been considerable investment in the boats to improve both their power and efficiency and to reduce costs while improving crew amenities.

These results were unexpected and are attributable to two developments. The mid- and late-1980's saw a significant improvement in recruitment to the resource and much higher TACs. Also, from an almost negligible level in 1983 on the east coast, a Japanese market was developed for female roe that by 1987 yielded 3,500 tons of product extracted from a catch of 70,000 tons of herring. This market improvement more than compensated for a decline in the European food market and in the still premium over-the-side sales. The net result was a much healthier primary industry in the mid- and late-1980's than in the early 1980's.

The concentration on a spawning fishery, necessarily over only a relatively small part of the potential fishing season, tended to be at the expense of a food market and was wasteful in that unused carcasses were dumped, causing environmental problems. The reliance on a single luxury market that attracted competition from Canada's west coast and the eastern Atlantic also made the fishery more vulnerable to a sudden market shock.

These unexpected results demonstrate the key importance of the market and that the future of any Canadian herring fishery is determined by market demands and the ability to anticipate changes and adjust to them. Two real possibilities for radical change exist; the resurgence of the Georges Bank stock, to provide a large adult resource increment, and a more rational use of the "unwanted" herring biomass, the side product of the roe industry, as raw material for the aquaculture industry. A third important consideration is that the 1980's may have been as overproductive in resource terms, as the 1970's were underproductive.

### Recent Management Initiatives

The 1981 Yarmouth Meeting represents a focal point in the recent history of east coast herring management (Anon. 1981). It resulted in the replacement of the original Atlantic Herring Management Committee by the "Small Pelagics Advisory Committees" in each of the four management regions. This meant that each region had the responsibility of setting up advisory bodies to assist in the preparation of Management Plans for the areas and stocks under their mandate, that implemented decisions on the level of resource removals resulting from CAFSAC recommendations, while addressing their own "local" socio-economic problems.

The Gulf of St. Lawrence as a whole presented real difficulties and can serve as an example of how the current management system has evolved. The initial situation was not promising. A serious resource problem existed there, the stock situation was very complex, the prospect of severe gear conflict was very likely and the organizational problems far more difficult. The existence of both fall and spring stocks, often in the same area, and the lack of first hand experience of the management problems on the part of newly set up regional organizations all played a part, particularly as the Gulf of St. Lawrence herring fisheries involved all four regions and, indeed all five Atlantic Provinces (Table 2).

A number of initiatives during 1985 sponsored by the Gulf Small Pelagics Advisory Committee led to a new management approach. The sub-allocation of the TAC became based on geographical areas in such a way that each of the provinces could effectively have its own allocation, but without conceding the federal mandate to regulate and manage. In 1985 a TAC was set that was higher than that recommended by CAFSAC to absorb pressures for a higher catch to take into account the current social and economic situation.

The system for the Gulf of St. Lawrence as a whole was formalized on April 22, 1985 by a Ministerial announcement that gave details of "The 1985 Herring Management Plan for the Gulf of St. Lawrence and the Sydney Bight (NAFO Divisions 4RST and 4Vn)" (Anon. 1985). This also established the gear allocations for the separate areas in accordance with the terms of the Fleet Reduction Plan, but was predicated on a coordinated approach by all regions and all provincial interests.

As a result, meetings involving the department, fishermen and processors in each province were announced to produce detailed plans tailored to the needs of local and provincial interests.

A flexible overall management system was quickly established. The Minister announced similar plans for the management of the spring fisheries of the east and southeast coast of Newfoundland, these having been closed since 1982 because of the resource situation (Wheeler et al. 1989).

The principles established at this time were:

a) conservation as the essential pre-requisite for management;
b) allocation of the TAC among all gear components;
c) guarantees under the 1983 Fleet Reduction Plan to be established for a ten year period;
d) licensed fishing only to be allowed;
e) transfer of allocations between gear types to be forbidden, and over-runs in one area not to be at the expense of another.

The management plan specified seasonal opening and closing dates, closed areas, bait permits, gear restrictions and limits, license conditions for the various gear sectors, and individual boat allocations. That such a comprehensive system was set up over a period of poor resource availability is notable and perhaps indicates that progress would have been more difficult for a large active and successful fishery. It has been claimed that the Gulf Plan represents as sophisticated a system as exists in any herring fishery anywhere.

The degree to which "local" characteristics are taken into account is an important, even vital, aspect of management plans. Conditions and complexities vary enormously from place to place. As the procedures in the regions became more established and functional, an important change was made in 1987 to increase the flexibility and timeliness in dealing with local questions. Ministerial responsibility for TAC setting was retained but within this restraint, and within recognized policy guidelines, the responsibility and authority for sub-allocation was passed to the Regional Director Generals (G. Peacock, Fisheries and Oceans, Halifax, N. S., pers. comm.).

Annual management plans are now an established feature of Canada's Atlantic coast herring fisheries. The way in which they are arrived at and promulgated varies considerably from region to region.

## Discussion and Conclusions

Despite a short historical basis and the lack of a continuous research program until the late 1960's, important management initiatives were introduced in the Canadian Atlantic herring fisheries in the early 1970's. The management regime subsequently developed into a sophisticated and complex system.

It is a dynamic system, for the interaction of the biological, economic, social and political factors that management must address (Iles 1980), means that the concept of fisheries "equilibrium", that became associated with the original "maximum sustained yield" concept, must be replaced by that of economic viability, one that is maintained by flexibility. This flexibility is essential to deal with the largely unpredictable problems and difficulties that are endemic to a variable-resource based industry which is so dependent on exports.

With the deterioration of the traditional groundfish industry in eastern Canada and its effect on shore-based employment, the scientific aspects of management, the nature of the scientific advice and the way it was incorporated into the management system came increasingly under question in the early 1990's. Herring management, from its beginning, involved a far greater interaction at the economic and socio-political interface than the groundfish industry. In practice, this has led to the realization that the single most important factor has always been the availability of a suitable market for the catch.

At present, the market for female roe is the single most important economic factor for the adult fisheries for herring on the east coast, and the degree to which TAC's are taken in the current period of stock increase is dependent very largely on access to that market. There is a single buyer, Japan, but many potential suppliers, both in Europe and on both coasts of Canada. In this buyers market, the east coast fisheries supply the low priced end, and the market situation is relatively vulnerable, so that there is a need for contingency planning both to anticipate the effects of a failure of this market, and to ensure the proper use of the "waste products" of the roe industry, which should be more closely incorporated with the growing aquaculture industry.

At the same time, the importance of the New Brunswick weir fishery must be appreciated, and a proper balance achieved between it and the use of the inshore areas in which it occurs for aquaculture. Biologically, this fishery represents an issue of great importance in Canada/US relations.

All this demands a far broader interpretation of the term management than is usually employed, and a greater degree of interaction between the scientific and operational aspects of management within the department than has normally been the case. This itself is a prerequisite for the involvement of both primary and secondary industry on an interactive basis along the lines that occurred in the late 1970's. The identification of common goals, and the agreement on minimum standards of information exchange between the managers and the managed, form the only possible basis for the continuing health of the herring industries of the area.

Over the last 20 years or so, enough information has been accumulated to address the important scientific questions relevant to management problems, if properly analyzed. These data are also of the greatest general scientific interest. The extent to which the East Coast management system has contributed to the advance of marine science and ecology is not yet realized to anything like the degree that it could be.

**Acknowledgements**

None of the developments discussed here would have happened without the decisive contribution of the Canadian east coast herring industry, through its various components and organizations. Individuals are too numerous to be mentioned.

It is a pleasure to recognize the help of the St. Andrews services, especially Bill McMullon, Frank Cunningham and Brenda Best, and to thank Rob Stephenson for his helpful comments at the draft stage. This chapter would not have been completed without the patient and competent help of Nancy Carson.

**References**

ANON. 1967. Proceedings Canadian Atlantic herring fishery conference. Can. Fish. Rep. No. 8, Dec. 1966, 268 p.

1971. Major areas closed off — exclusive fisheries zones proclaimed by Canada. Fish. Can. 23: 10–12.

1972a. The juvenile fisheries of the Bay of Fundy and the conservation of herring resources — memorandum by Canada. Int. Comm. Northw. Atl. Fish. Comm. Doc. 72/3, Ser. No. 2733.

1972b. Proceedings of the Special Meeting on herring. Int. Comm. Northw. Atl. Fish. Annu. Proc. 22, 1972: 175–215.

1972c. Atlantic Coast Herring Committee Planned New Release, Feb. 8, 1972. Environment Canada.

1972d. Minutes of the first meeting of the Canadian Atlantic Herring Management Committee and Advisors.

1972e. Report of the standing committee on research and statistics. Appendix II. Report of the herring working group. Int. Comm. Northw. Atl. Fish. Redbook Part I 1972: 43–66.

1976. Proceedings of the Seventh and Eighth Special Meeting, Sept. 1976 and Jan. 1976 and the 26th Annual Meeting, June 1976. Int. Comm. Northw. Atl. Fish. p. 304.

1977. Formation of Canadian Atlantic Fisheries Scientific Advisory Committee. Memorandum by the CAFSAC Steering Committee. Can. Atl. Fish. Sci. Advis. Comm. Advis. Doc. 77/1: 1–8.

1979. Pelagic Subcommittee Chairman's Report for 1979. Can. Atl. Fish. Sci. Advis. Comm. Ann. Rep. 1979, Vol. II, p. 31–34.

1981. The management and utilization of herring and mackerel in the 1980's. A discussion paper. Dept. Fish. Oceans, Ottawa, July 1981, 47 p.

1982a. Advice on pelagic species for 1980. CAFSAC Advis. Doc. 80/6. Can. Atl. Fish. Sci. Advis. Comm. Ann. Rep. (including Advisory Documents) Vol. 3 for 1980, p. 49–66.

1982b. Advice on pelagic stocks. CAFSAC Advis. Doc. 80/9. Can. Atl. Fish. Sci. Advis. Comm. Ann. Rep. (including Advisory Documents) Vol. 3 for 1980, p. 83–92.

1982c. Advice on pelagic stocks. CAFSAC Advis. Doc. 87/7. Can. Atl. Fish. Sci. Advis. Comm. Ann. Rep. (including Advisory Documents) Vol. 4 for 1981, p. 79–95.

1983. Advice on the management of capelin and herring stocks in the Gulf of St. Lawrence in 1982. CAFSAC Advis. Doc. 82/11. Can. Atl. Fish. Sci. Advis. Comm. Ann. Rep. (including Advisory Documents) Vol. 5, 1982, p. 67–75.

1985. 1985 Gulf herring management plan announced. News Release, Fisheries and Oceans, Monday, April 22, 1985.

1987. Report of the herring assessment working group for the area south of 62oN. ICES C.M. 1987/Assess: 9. 1989/Assess: 15.

1989a. Summary of Subcommittee activities in 1988. Can. Atl. Fish. Sci. Advis. Comm. Ann. Rep. (including Advisory Documents) Vol. II, 1988, p. 12–21.

1989b. Advice on the management of herring stocks in the Gulf of St. Lawrence. Can. Atl. Fish. Sci. Advis. Comm. Advis. Doc. 88/18. Can. Atl. Fish. Sci. Advis. Comm. Ann. Rep. (including Advisory Documents) Volume II for 1988.

1991. Herring Plan Announced. Telegraph Journal, Saint John, N. B., April 25, 1991.

BURD, A.C. 1978. Long-term changes in North Sea herring stocks. Rapp. P.-v. Réun. int. Explor. Mer 172: 137–153.

CADDY, J.F., AND T.D. ILES. 1973. Underwater observations on herring spawning grounds on Georges Bank. Int. Comm. Northw. Atl. Fish. Res. Bull. 10: 131–139.

CHADWICK, E.M.P. 1988. Herring fishermen and biologists: their role in stock assessment. Can. Ind. Rep. Fish. Aquat. Sci. 183: viii + 131 p.

DOUCET, F.J., T.D. ILES, AND P M. JANGAARD. 1977. Bay of Fundy herring fishery review of 1976 season and prospects for 1977. Memorandum and Report to Minister of Fisheries and Environment, 6 p.

DOUCET, W.F. 1960. Economic study of the herring fishery of Charlotte County, New Brunswick, 1956–1957. J. Fish. Res. Board Can. 17: 815–870.

FOGARTY, M.J., E.P. ALMEIDA, J. CHENOWETH, AND J.S. IDOINE. 1990. Population dynamics of Atlantic herring in the Gulf of Maine. NMFS, Northeast Fisheries Centre, Woods Hole Lab, SAW-9 Work. Pap. 7.

GOUGH, J. 1971a. Frankenstein's seine ... (or) the rape of the east coast herring fishery. Can. Fish. Ocean Sci. 57: 15–18.

1971b. Frankenstein's seine. Can. Fish. Ocean Sci. 57: 49–52.

GULLAND, J.A. 1974. The management of marine fisheries. University of Washington Press, Seattle, WA. 198 p.

HAEGELE, C.W., AND J.F. SCHWEIGERT. 1985. Distribution and characteristics of herring spawning grounds and description of spawning behaviour. Can. J. Fish. Aquat. Sci. 42(Suppl. 1): 39–55.

HODDER, V.M. 1971. Status of the southwest Newfoundland herring stocks, 1965-70. Int. Comm. Northw. Atl. Fish. Redbook Part III: 105–114.

HODDER, V.M., AND L.S. PARSONS. 1971a. Preliminary comparison of Scotian Shelf and southwest Newfoundland herring taken during the winter of 1971. Int. Comm. Northw. Atl. Fish. Res. Doc. 77/120: 8 p.

1971b. Some biological features of southwestern Newfoundland and northern Scotian Shelf herring stocks. Int. Comm. Northw. Atl. Fish. Res. Bull. 8: 67–73.

HODDER, V.M., AND G.H. WINTERS. 1970. Preliminary results of herring tagging in southwestern Newfoundland, March 1970. Int. Comm. Northw. Atl. Fish. Res. Doc. 70/89: 2 p.

ILES, T.D. 1964. The duration of maturation stages in herring. J. Cons. Perm. Int. Explor. Mer 29: 166–188.

1971a. A review of management strategy for Canadian Atlantic herring fisheries. Presentation to FP AFC, Charlottetown, P.E.I., Oct. 1971.

1971b. The management of Canadian herring resources. Meeting at Halifax, N.S., Oct. 7, 1971, between the Department and industry representatives. Memorandum to A/Director General, Research and Development, 4 p.

1973. The interaction of environment and parent stock size in determining recruitment in the Pacific sardine, as revealed by analysis of density-dependent 0-group growth. Rapp. P.-v. Réun. Cons. Int. Explor. Mer 164: 228–240.

1980. The natural history of fisheries management. Proc. N.S. Inst. Sci. 30: 3–19.

1984. Allocation of resources to gonad and soma in Atlantic herring *Clupea harengus* L., p. 331–347. In G. W. Potter and R. J. Wootton (ed.) Fish reproduction: strategies and tactics. Academic Press, London.

ILES, T.D., M.J. POWER, P.M. MACE, G.N. WHITE, AND F.G. PEACOCK. 1984. Assessment of the 1983 4WX herring fishery. Can. Atl. Fish. Sci. Advis. Comm. Res. Doc. 84/72: 1–42.

ILES, T.D., AND M. SINCLAIR. 1982. Atlantic herring: stock discreteness and abundance. Science 215: 627–633.

ILES, T.D., AND S.N. TIBBO. 1970. Recent events in Canadian Atlantic herring fisheries. Int. Comm. Northw. Atl. Fish. Redb. Part III 1970: 134–147.

ILES, T.D., AND R.J. WOOD. 1965. The fat/water relationship in North Sea herring (*Clupea harengus*), and its possible significance. J. Mar. Biol. Assoc. U.K. 45: 353–366.

INNIS, H.A. 1954. The cod fisheries: the history of an international economy (Revised Edition). University of Toronto Press, Toronto, Ont. 522 p.

JAKOBSSON, J. 1985. Monitoring and management of the northeast Atlantic herring stocks. Can. J. Fish. Aquat. Sci. 42(Suppl. 1): 207–221.

JOHNSTONE, K. 1977. The aquatic explorers: a history of the Fisheries Research Board of Canada. University of Toronto Press, Toronto, Ont. 342 p.

KEARNEY, J.F. 1981. A report on the breakdown of the marketing and management system of the Bay of Fundy herring purse seine fishery in 1980. Dalhousie Ocean Studies Programme. Can. Atl. Fish. Sci. Advis. Comm. Working Paper 81/79: 1–17.

——— 1984. The transformation of the Bay of Fundy herring fisheries 1976–1978: an experiment in fishermen-government co-management, p. 165–203. *In* C. Lamson and A.J. Hanson [ed.] Atlantic fisheries and coastal communities. Fisheries decision-making case studies. Dalhousie Ocean Studies Program.

KETCHEN, K.S. 1969. Atlantic herring investigations. Memorandum to Eastern Advisory Committee of the Fisheries Research Board, p. 13.

KORRINGA, P. 1978. European fisheries — an object lesson in economic and ecological mismanagement. Interdiscip. Sci. Res. 3: 335–345.

MCFARLAND, R. 1911. A history of the New England fisheries. University of Pennsylvania, 457 p.

MCFARLANE, D.A. 1989. An economic assessment of the herring purse seine fleet restructuring plan of 1983. The ten-year management plan at mid-term. Econ. Comm. Anal. Rep. No. 31, Dept. Fish. Oceans, 49 p.

MACKENZIE, R.A. 1964. Observations on herring spawning off southwest Nova Scotia. J. Fish. Res. Board Can. 21: 203–204.

MACKENZIE, R.A., AND S.N. TIBBO. 1960. Herring fishery in southern New Brunswick. J. Fish. Res. Board. Can. 17(2): 133–168.

MCQUINN, I.H. 1989. An assessment of the west coast of Newfoundland herring resource in 1988. Can. Atl. Fish. Sci. Advis. Comm. Res. Doc. 89/83.

MURPHY, G.E. 1977. Clupeoids, p. 283–308. *In* J.A. Gulland (ed.) Fish population dynamics. John Wiley and Sons, New York, NY.

PACE, L. 1971. Herring advisory council to be named within 2 months. Chronicle Herald, Halifax, Wednesday, Dec. 1, 1971.

PARRISH, B.B., AND A. SAVILLE. 1965. The biology of the northeast Atlantic herring populations. Oceanogr. Mar. Biol. Ann. Rev. 3: 323–373.

PARSONS, L.S. 1993. Management of marine fisheries in Canada. Can. Bull. Fish. Aquat. Sci. 225. 784 p.

PERLEY, M.H. 1852. Report on the sea and river fisheries of New Brunswick. Second edition. Fredericton, N.B., 294 p.

POWLES, H., J. BOULVA, AND P. SPENARD. 1979. Herring catch per unit effort in the Magdalens trap fishery and implications for Gulf herring stock assessment. Can. Atl. Fish. Sci. Advis. Comm. Work. Pap. 79/9.

RICKER, W.E. 1954. Stock and Recruitment. J. Fish. Res. Board Can. 11(5): 559–623.

SAMUEL, A.M. 1918. The herring: its effect on the history of Britain. John Murray, London, 199 p.

SINCLAIR, M., AND M.J. TREMBLAY. 1984. Timing of spawning of Atlantic herring (*Clupea harengus harengus*) populations and the match-mismatch theory. Can. J. Fish. Aquat. Sci. 41: 1055–1065.

SMITH, P.J., AND A. JAMIESON. 1986. Stock discreteness in herring: a conceptual revolution. Fish. Res. 4: 223–234.

STEPHENSON, R.L., AND D.J. GORDON. 1991. Comparison of meristic and morphometric characters among spawning aggregations of northwest Atlantic herring, *Clupea harengus* L. Proc. Int. Herring Symp., October 1990, Rep. 91–01: 279–296.

STEPHENSON, R.L., AND I. KORNFIELD. 1990. Reappearance of spawning Atlantic herring (*Clupea harengus harengus*) on Georges Bank: population resurgence not recolonization. Can. J. Fish. Aquat. Sci. 47: 1060–1064.

STEPHENSON, R.L., AND M.J. POWER. 1989. Assessment of the 1988 4WX herring fishery. Can. Atl. Fish. Sci. Advis. Comm. Res. Doc. 89/59: 1–39.

——— 1990. A summary of the 1989 4Vn herring fishery. Can. Atl. Fish. Sci. Advis. Comm. Res. Doc. 90/51: 1–12.

STEPHENSON, R.L., M.J. POWER, W.H. DOUGHERTY, D.J. GORDON AND J.B. SOCHASKY. 1990. Assessment of the 1989 4WX herring fishery. Can. Atl. Fish. Sci. Advis. Comm. Res. Doc. 90/50: 1–48.

STOBO, W.T., J.S. SCOTT, AND J.J. HUNT. 1975. Movements of herring tagged in the Bay of Fundy. Int. Comm. Northw. Atl. Fish. Res. Doc. 75/38: 24 p.

STOCKER, M. 1993. The management of the Pacific herring fishery, p. 267–293. *In* L.S. Parsons and W.H. Lear [ed.] Perspectives on Canadian marine fisheries management. Can. Bull. Fish. Aquat. Sci. 226.

TIBBO, S.N., AND T.R. GRAHAM. 1963. Biological changes in herring stocks following an epizootic. J. Fish. Res. Board Can. 20: 435–449.

TIBBO, S.N., D.J. SCARRATT, AND P.W.G. MCMULLON. 1963. An investigation of herring (*Clupea harengus L.*) spawning using free-diving techniques. J. Fish. Res. Board Can. 20: 1067–1079.

WAIWOOD, K. 1981. The predatory impact of cod in the southern Gulf of St. Lawrence ecosystem — a preliminary account. Int. Counc. Explor. Sea Comm. Meet. 1981/G:43: 16 p.

WAIWOOD, K., AND J. MAJKOWSKI. 1984. Food consumption and diet composition of cod, *Gadus morhua*, inhabiting the southwestern Gulf of St. Lawrence. Environ. Biol. Fish. 11: 63–78.

WHEELER, J.P., AND G.H. WINTERS. 1984. Homing of Atlantic herring (*Clupea harengus harengus*) in Newfoundland waters as indicated by tagging data. Can. J. Fish. Aquat. Sci. 41: 108–117.

WHEELER, J.P., G.H. WINTERS, AND R. CHAULK. 1989. Newfoundland east and southeast coast herring — 1988 assessment. Can. Atl. Fish. Sci. Advis. Comm. Res. Doc. 89/40: 86 p.

WHEELER, J.P., G.H. WINTERS AND R. CHAULK. 1992. Newfoundland East and Southeast Coast Herring — 1991 Assessment. Can. Atl. Fish. Sci. Adv. Comm. Res. Doc. 92/49: 37 p.

WHITEHEAD, P.J.P. 1985. King herring: his place amongst the clupeoids. Can. J. Fish. Aquat. Sci. 42(Suppl. 1): 3–20.Table 1. The temporal pattern of overexploitation of Atlantic herring fisheries. Annual landings expressed as a percentage of maximum annual landings (in year N). Sources: ICES, ICNAF, NAFO, Canadian and US Fisheries Statistics.

# The Management of Canadian Atlantic Salmon Fisheries

W. H. Lear

*Department of Fisheries and Oceans*
*Biological Sciences Directorate, Science Sector*
*200 Kent Street, Ottawa, Ontario, Canada K1A 0E6*

Lear, W. H. 1993. The management of Canadian Atlantic salmon fisheries, p. 151–176. *In* L. S. Parsons and W. H. Lear [eds.] Perspectives on Canadian marine fisheries management. Can. Bull. Fish. Aquat. Sci. 226.

The Atlantic salmon (*Salmo salar*) has made a significant contribution to humanity in the form of food, sporting challenge, as well as providing financial rewards to commercial fishermen and those involved with recreational fisheries. It is also an aesthetic symbol.

The management of Atlantic salmon is complex because of its anadromous life cycle. Atlantic salmon is vulnerable to the degradation of its freshwater habitat and overexploitation in the ocean and to illegal fishing at sea and in freshwater. During the last two decades, Canadian Atlantic salmon management has focused on conserving and enhancing this valuable fish. Management initiatives during the last decade, both nationally and internationally, give room for guarded optimism about future increased abundance of this valuable resource.

---

Le saumon atlantique (*Salmo salar*) a été généreux pour l'homme; depuis longtemps, il est une source d'aliment, de plaisir pour les pêcheurs sportifs et de revenus pour les pêcheurs commerciaux et les intervenants dans la pêche récréative. Il possède en outre un attrait esthétique.

La gestion du saumon atlantique est compliquée par le fait que c'est une espèce anadrome. Il est vulnérable à la dégradation de son habitat en eau douce, à la surexploitation en mer et au braconnage en mer et en eau douce. Au cours des deux dernières décennies, la gestion de cette espèce par le Canada a été axée sur la conservation et la mise en valeur. Les initiatives de gestion mises en application au cours de la dernière décennie au niveau national et international donnent lieu à un optimisme réservé quant à une augmentation future de son abondance.

## Introduction

The Atlantic salmon (*Salmo salar*) has long been highly esteemed as a food fish and as a sporting challenge. It was prized by the Romans when they settled Gaul, utilized by the early tribes in western Europe, and it was esteemed by royalty as a source of sport and as a gourmet food. Because of its unique appeal which satisfied so many various needs, it was dubbed "the king of fish." The management of Atlantic salmon, or at least its right to free passage upstream, dates from 1215 when King John signed the Magna Carta (Netboy 1968).

In Canada, Atlantic salmon has been utilized as a source of food and as a marketable commodity since colonial times. The industrialization of Atlantic salmon rivers for lumbering, sawmills, pulp and paper, and water obstruction for public, industrial uses, and hydroelectric power, as well as the overexploitation of salmon have led to a decline in salmon stocks in many rivers and even the disappearance of the species from some systems. The conflicting demands on salmon and on its habitat present a complex challenge to its successful management. The purpose of this chapter is to present a brief history of the Atlantic salmon fisheries, highlighting major features in the development of the management system for Canadian Atlantic salmon as it exists today.

## Biology

### Distribution and Stocks

In the northwest Atlantic, the Atlantic salmon ranges from eastern Hudson Bay and Ungava Bay to the northeastern United States. In Canada, the Atlantic salmon is distributed from the Nastapoka River in eastern Hudson Bay (Morin 1991) and in many rivers from Ungava Bay in northern Québec, Labrador, Newfoundland, eastern Québec, New Brunswick, Nova Scotia to a few streams in Prince Edward Island (Smith 1988). The distribution and habitat range of the Atlantic salmon have shrunk since colonial times when populations of Atlantic salmon were found in most of the accessible watercourses of Eastern North America from Ungava Bay to Long Island Sound (Dunfield 1985). In Canada, salmon stocks have disappeared from Lake Ontario, and in the St. Lawrence River are now found only east of Québec City (Fig. 1) (Muir and Bellefontaine 1988). Currently there are more than 350 rivers in Canada which contain significant populations of Atlantic salmon, and at least 200 other lesser rivers which support small populations. The annual production potential of Canadian waters currently producing Atlantic salmon is roughly 1.5 million adult large salmon and 1.5 million grilse (Anon. 1986). The production from these waters in 1980 was in excess of 1.0 million adult salmon. Cultured stocks accounted for less than 5% of Canada's Atlantic salmon production in 1980 (Ritter et al. 1980).

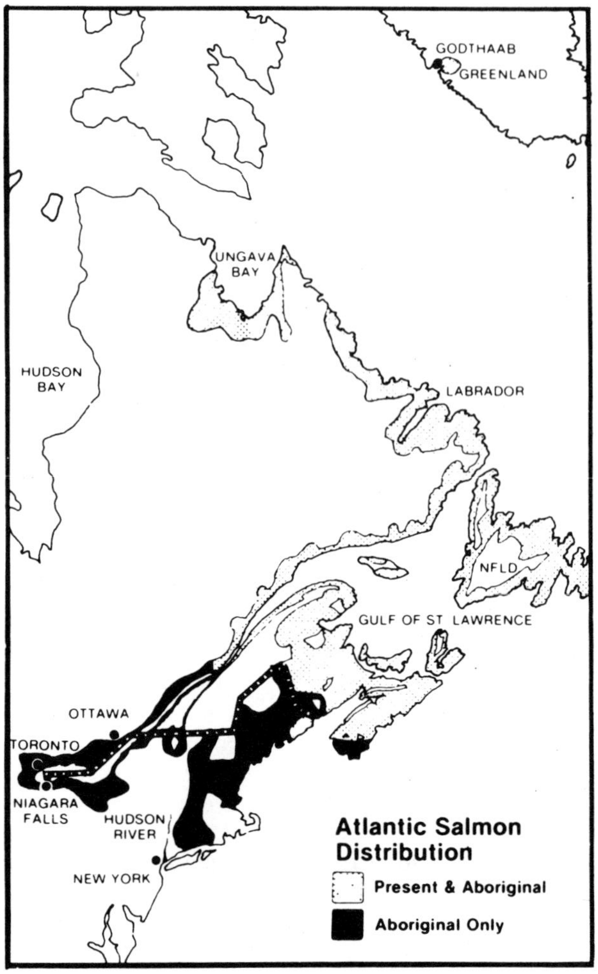


FIG. 1. Present and aboriginal distribution of the anadromous Atlantic salmon in North America (from Smith 1988).

### Life History

The Atlantic salmon is an anadromous fish. It spawns in freshwater streams, with the post-spawning surviving adults returning to sea. The young remain in fresh water as parr from two to eight years, the number of river years increasing in a cline from south to north (Lear and Misra 1978). The young salmon then descend to the sea as smolts to spend one or more years feeding and growing before returning to fresh water to spawn. Unlike Pacific salmon, which die after spawning, the Atlantic salmon may survive and spawn again in succeeding years. The proportion of repeat spawners in a population is generally less than 10% although the proportion of previously spawned salmon in commercial catches may be occasionally as high as 20% (Lear and May 1972). Salmon which return to spawn after only one year at sea are called grilse (or one sea-winter salmon) while those which spend two or more years at sea are referred to as larger or multi-sea-year salmon (Allan and Ritter 1977).

Spawning usually occurs in October and November. In northern Labrador spawning commonly occurs in early October while in more southerly regions most spawning occurs from late October to early November. The eggs incubate over the winter months in the stream beds and hatch usually in April in the Maritimes (Smith 1988) or May in Newfoundland and Labrador. Newly-hatched salmon (alevins) are about 15 mm in length and carry a large yolk sac, which nourishes them for the first few weeks while they are still buried in the gravel. After emergence from the gravel in late May – early June, the yolk sac is absorbed, the free-swimming fish (parr) now begin active feeding and establish territories of several square metres of stream. The parr remain in the river for 2 to 8 years before reaching the smolt stage and migrating downstream to sea. In the sea, the growth rate of the smolts increases due to physiological changes and a greater abundance of food. Smolts are generally 125–150 mm in length but in northern areas may be older (4 to 8 years) and larger with a minimum length of 180 mm (Power 1969). Growth at sea is very rapid since the salmon are voracious feeders and prey upon a wide variety of fish and invertebrates such as herring, capelin, alewives, launce, shrimp, squid, amphipods and euphausiids.

After leaving the coastal area, the postsmolt Atlantic salmon may move to offshore feeding areas. Postsmolt Atlantic salmon have been caught by surface gill nets in the Labrador Sea during early autumn (Reddin and Short 1991). The high catch rates compared with catch rates from Greenland suggest that there may be a large population of postsmolts in the Labrador Sea during this period of the year.

Some salmon may undergo long migrations while at sea, travelling from Maine, Bay of Fundy, the Gulf of St. Lawrence and Newfoundland and Labrador to the west coast and occasionally the east coast of Greenland where they intermingle with salmon of European origin (Fig. 2). This longer migration to West Greenland involves salmon which stay at sea for two or more years. Those that stay at sea for only one year (grilse) may make shorter migrations to the Labrador Sea or the Grand Banks. Stocks of the inner Bay of Fundy, for example, make shorter migrations and are not harvested in Newfoundland or Greenland (D. Meerburg, pers. comm.).

Records of incidental catches of Atlantic salmon by commercial trawlers (Lear 1976a) and by research vessels (Templeman 1968; May 1973) on the Grand Bank during February to June confirmed the presence of salmon in this area. These reports formed the basis of the hypothesis that the waters along the southwestern edges of the Grand Bank and St. Pierre Bank may represent an overwintering or staging area for salmon during their migration from the Labrador Sea to Canadian east coast rivers (Lear 1976a). Exploratory fishing with surface gill nets on the Grand Bank and eastward of the Bank over oceanic depths during May 1979 and 1980 revealed the presence of Atlantic salmon at most stations where surface temperatures ranged from 3.8° to 7.5° C (Reddin 1985). Analysis of sea-surface temperature charts indicated that favourable conditions (4° to 8° C water) persist for salmon during January to April in this area. Reddin (1985) suggested that the eastern and southern Grand Bank region may represent not only a route by which maturing salmon migrate from the Labrador Sea to their home rivers in eastern Canada and Northeastern United States but also a major feeding and overwintering area.

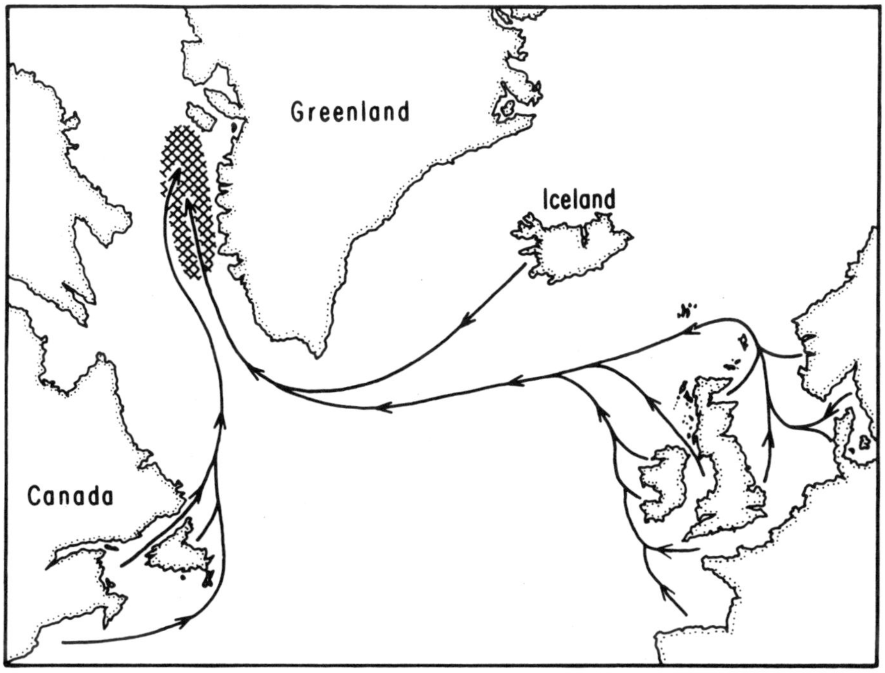


FIG. 2. Migration of Atlantic salmon from countries of origin to the feeding area off West Greenland. Directional lines do *not* signify known migration routes (adapted from Parrish 1973).

After spending one to three years at sea the mature fish return to the stream or river where they were spawned. On returning to spawn, grilse range from 1.3 to 3.0 kg in size, salmon of 2 sea years in age are generally 2.5 to 7.5 kg while salmon 3 sea years in age range from 5.5 to 12.5 kg as evidenced from commercial catch data (Lear and May 1972). Salmon are strong swimmers and because adults are large relative to parr they are not prone to predation by American mergansers, kingfishers, eels and other fish which prey on young stages (Scott and Crossman 1973). At sea, adult and young salmon have been found in the stomachs of tuna, sharks, swordfish, cod and pollack (Smith 1988).

### Historical Development of the Atlantic Salmon Fishery

Atlantic salmon has served as a food source for the indigenous people of eastern Canada for several thousand years prior to the arrival of Europeans in North America (Dunfield 1985). The early Viking explorers are reported to have utilized the Atlantic salmon as food in the tenth and eleventh centuries (Taylor 1985). These early fisheries by natives, Vikings, and the earliest colonial settlers were mainly for domestic consumption. By about 1650 several small fishing stations had been established on the lower St. Lawrence River to catch salmon for commercial profit (Dunfield 1985). Salmon fishing as a commercial venture began in the Bonavista Bay region of Newfoundland around the year 1700 (Taylor 1985). Salmon fishing continued in Newfoundland and later in Labrador as an adjunct to the cod fishery and served to augment the earnings from cod fishing and sealing especially along the northern coasts of Newfoundland.

The first major English commercial fishing venture for Atlantic salmon in New Brunswick began in 1762. Commercial fishing for salmon began on the Miramichi River in the mid-1760's and in the late 1760's commercial fisheries were established on the Restigouche River (Dunfield 1985).

By the mid-eighteenth century exports of salmon from Newfoundland had increased to as much as 790 tons (equivalent round weight). By 1793 the landings had reached 916 tons (Taylor 1985). Landings during the nineteenth century in Newfoundland and Labrador were between 195 and 1 350 tons (Taylor 1985). Newfoundland salmon catches in the early twentieth century were similar to those in the late nineteenth century (i.e. about 500 to 1 100 tons). Catches began increasing rapidly in the early 1920's, culminating in catches in excess of 2 200 and 3 800 tons in 1929 and 1930, respectively. Catches remained at about 1 800 tons until the 1950's (Taylor 1985).

Total Canadian landings of Atlantic salmon variably increased from about 2 500 tons in 1910 to about 6 100 tons in 1930 (May and Lear 1971). After 1931, landings declined in a highly variable pattern to a low point of about 1 200 tons in 1955 (Fig. 3).

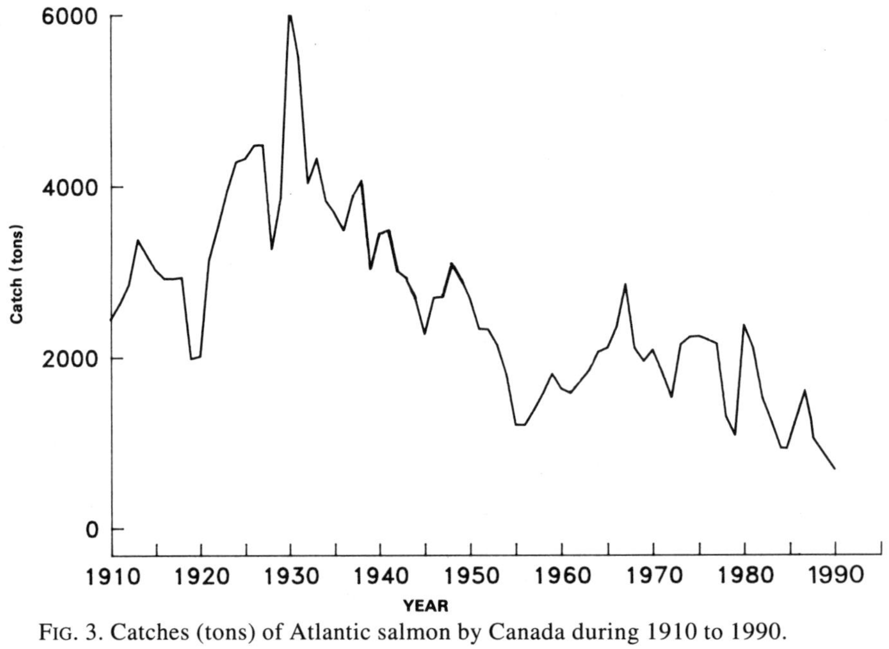


FIG. 3. Catches (tons) of Atlantic salmon by Canada during 1910 to 1990.

Prior to the beginning of the twentieth century, salmon for the export trade was usually preserved in a saturated solution of common salt (pickle), usually in large barrels called tierces (Dunfield 1985; Taylor 1985). The export of fresh salmon on ice or in "refrigerators" began around the mid-1870's in Newfoundland (Hatton and Harvey, 1883). This marked the beginning of the end of pickled salmon dominance in the market although pickled salmon continued to be exported into the early decades of the twentieth century.

Salmon fishing in Newfoundland was mainly carried out in rivers and river-mouth areas by weirs and gill nets until about 1750. In mainland Canada, the salmon fisheries have generally been in estuary areas although there were salmon fisheries conducted in areas outside estuaries such as the Gaspé, Atlantic coast of Nova Scotia and in Northumberland Strait. In Newfoundland and Labrador, competition for fishing berths forced gill nets farther away from the river mouths extending seaward towards headland locations. Fishermen discovered that salmon could be taken in areas outside of estuaries on the east and northeast coasts of Newfoundland before the regular salmon fishery began in the rivers. Taylor (1985) suggested that this "headland" fishery began in Newfoundland around 1845.

In this early fishery, large salmon from mainland Canadian, Newfoundland and Labrador rivers were intercepted on their returning migration along eastern and southern Newfoundland. The fisheries for salmon on the Canadian mainland were mostly in the rivers and estuaries especially in major systems such as the Saint John, Miramichi, Richibucto, Restigouche and the St. Lawrence. There were also local fisheries on salmon in the rivers draining into Lake Ontario since salmon were abundant there in the eighteenth and early nineteenth centuries (Dunfield 1985). The decline of salmon in the rivers around Lake Ontario stemmed from habitat destruction in addition to blatant overfishing. The dams erected on the rivers stopped the passage of fish to the spawning grounds and permitted them to be unmercifully slaughtered as they congregated at the base of the dams. The sawdust from lumber mills, lime from tanneries and other pollutants and habitat destruction also are partly responsible for the demise of salmon in rivers flowing into Lake Ontario. Pollution and habitat destruction coupled with generally unrestricted fishing, also led to the disappearance of salmon from Lake Champlain (Dunfield 1985).

In the Maritime provinces a drift-net fishery developed in the Bay of Fundy near the Saint John River and in the Gulf of St. Lawrence particularly in the Miramichi estuary around the turn of the century (Huntsman 1931). From 1907 to 1911 the boats engaged in the Miramichi drift-net fishery increased in numbers from 4 to 17, and by 1924 they numbered 50. Catches kept increasing for a number of years and the percentage of the total salmon catch of eastern New Brunswick taken in the drift-net fishery increased from 35% in 1922 to 65% in 1930. At Port-aux-Basques, Newfoundland, a drift-net fishery developed for migrating salmon enroute to Maritime, Québec and western Newfoundland rivers in 1931 and by 1936 it involved 34 boats (Belding and Préfontaine 1938).

## Modern Commercial Fishery (1950's to the Present)

### Maritimes

In the Maritime provinces, the commercial fishery had three main methods of operation: by drift nets in offshore waters, by trapnets along the shore and by trap and setnets in estuarial waters and in some instances in freshwater (Wooding 1956). A fleet of 130 drift-net boats operated in the Gulf of St. Lawrence outside Miramichi Bay (Kerswill 1971). These boats fished 40–48 km offshore for 5 nights per week during a 10 to 11-week summer period. From 1953 to 1963, annual catches ranged from 6 800 to 31 000 fish and averaged 17 000 per year (Kerswill 1971). A tagging experiment showed that only part of the drift-net catch was of Miramichi River origin (Belding and Préfontaine 1939). Many of the fish caught in the Miramichi estuary, if permitted to escape, were destined to return to rivers running into the Bay of Chaleur.

Fixed trapnets were scattered along the shores in tidal waters of the estuary from near tidehead to outer Miramichi Bay. These extended at various distances from the shore and were made in such a way as to lead the fish into a box-like net where they became trapped. In estuarial waters salmon were caught by set gill nets attached to the shore and extended outwards in a stationary position. These nets caught salmon by meshing them in a manner similar to drift nets.

The commercial fishery during the 1950's and 1960's in New Brunswick was restricted by regulation to capture of salmon weighing 5 lb (2.3 kg) and over even though the general federal regulation for Atlantic marine areas prohibited only the taking of salmon under 3 lb (1.4 kg) in weight (Kerswill 1971). In addition to the Miramichi drift-net fishery there was a substantial drift-net fishery off the Saint John River in southwestern New Brunswick. Catches in the Maritimes decreased sharply during the 1950's although they increased again during the mid 1960's. The heavy spraying of forests with DDT in northern New Brunswick has been considered responsible for a substantial portion of the severe depression in salmon fishing success, particularly in angling in the province, between 1959 and 1962 (Elson and Kerswill 1964).

In 1972, a temporary 5-year ban or moratorium was imposed upon commercial salmon fishing in New Brunswick, the Gaspé region of Quebec and the Port-aux-Basques drift-net fishery. This ban was extended and stayed in effect until 1980. When the original ban on commercial salmon fishing in New Brunswick was lifted in 1981, it was replaced by a partial ban which allowed some commercial netting in areas such as the north shore of the St. Lawrence River — where the stocks were considered to be strong (Gruenefeld 1988). The Port-aux-Basques offshore drift-net fishery was permanently closed. The commercial fishery was reopened in 1982 in the Bay of Chaleur and the entire Gaspé Peninsula in 1983. A widespread ban was reimposed in 1985 because stocks were still depressed. This is still in effect.

### Québec

The commercial catches of Atlantic salmon in Québec declined from about 360 tons in the early 1950's to about 170–190 tons in the late 1960's (May and Lear 1971). These catches were a significant decrease from the 939 tons caught in 1925 and the decade of the 1930's when catches averaged about 570 tons. In 1972 a complete ban was imposed on commercial salmon fishing in Gaspé, Québec along with New Brunswick and the drift-net fishery off Port-aux-Basques, Newfoundland. Significant by-catches of salmon in non-salmon licensed gear began shortly after the beginning of the commercial ban in the lower St. Lawrence/Gaspé region in 1972. The by-catch was apparently as high as the pre-ban catch by the licensed commercial salmon fishery (Vezina 1988). Where there was no ban on commercial fishing for salmon e.g. North Shore of the St. Lawrence, the by-catches remained at generally low levels. Commercial catches have decreased to less than 100 tons since 1977. Only 62 tons were caught in 1984 with a small increase in catch to 103 tons in 1987 and a decrease to 80 tons in 1989. The number of commercial salmon licences dropped from 501 in 1968 to 227 in 1987. The commercial fishery has been closed in Gaspé and the upper North Shore since 1985.

### Newfoundland-Labrador

Atlantic salmon landings from Newfoundland and Labrador averaged 1 769 tons during 1910–54 with high peaks of harvest of 3 129 tons, 2 993 tons, and 2 630 tons in 1930, 1931 and 1938, respectively. During 1955 to 1972, landings increased from 727 tons in 1956 to 1 814 tons in 1967 but decreased to 1 406 tons in 1972 (Lear 1976b). Commercial landings were 1 822 tons in 1980 and 745 tons in 1989 (O'Connell et al. 1991) with an average of 1 091 tons for 1980–1989.

The fishery was conducted mainly by surface gill nets set out from shore. This gear accounted for about 97% of the Newfoundland-Labrador Atlantic salmon landings prior to the early 1970's. The remaining 3% was taken by salmon traps and miscellaneous gears, mainly cod traps. During the early to mid-1970's significant increases were observed in the by-catch of salmon (primarily grilse and one-sea winter salmon) in herring and mackerel nets and cod traps. This by-catch amounted to 29% of salmon landings in Newfoundland during 1976 but only 3% of landings in Labrador (Reddin 1978). Since 1978 the herring and mackerel fishing seasons have been closed during the major portion of the salmon run. The minimum mesh size in the leaders to the cod traps was increased to 177 mm. The minimum mesh size for the walls of the trap is 89 mm. Cod traps are prohibited in some areas and in-season closures have been imposed to reduce by-catch. The regulated minimum mesh size for salmon nets and salmon traps is 127 mm for Newfoundland and Labrador except for a small portion of the south coast and the west coast where a 114 mm mesh size is applied (Vezina 1988).

A drift-net fishery for salmon once operated along a major migration route off Port-aux-Basques but this was closed in 1972. There is no drift-net fishery operating at present around the coast of Newfoundland-Labrador. The numbers of licensed commercial salmon fishermen and amount of licensed fishing gear decreased from 6 981 fishermen and 2 304 000 m of nets in 1975 to 3 130 fishermen and 1 145 600 m of nets in 1991 (Anon. 1992a). Approximately 1 500 licences were retired in 1984 and 1985 through buy-back programs.

### Greenland

The first recorded commercial exploitation of salmon at West Greenland was the landing of 14 tons by native Greenlanders in 1959 (Lear 1976b), using set gillnets attached to the shore. This shore-net fishery reached a peak of 1539 tons in 1964. From 1964 to 1982 it fluctuated around 1100 tons including catches by drift nets as well as shore nets since 1969. In addition during the late 1960's an international drift-net fishery took place along the coastline and up to 30–40 miles offshore by vessels based in Europe. The total catch peaked in 1971 near 2 700 tons (Fig. 4). Although some coastal fixed-net fishing has remained, the present fishery by Greenlanders is mainly an inshore and coastal drift-net fishery. The Greenland fishery for salmon is mainly a

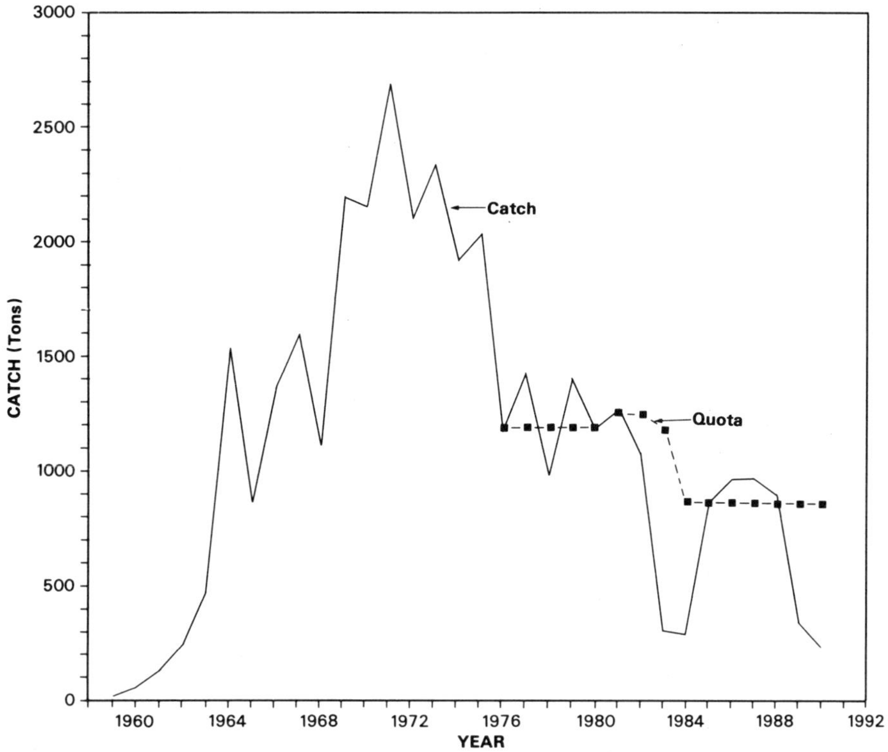


FIG. 4. Total catches (tons) of Atlantic salmon at West Greenland (1959–1990) and recent quotas (1976–1990).

small boat fishery and 80% of the landings are taken by boats smaller than 9m (Anon. 1992b). Fishing generally takes place along the west coast from 60°N to 70°N during August to November.

As a result of negotiations in the International Commission for the Northwest Atlantic Fisheries (ICNAF), regulation of the West Greenland salmon fishery started in 1972. The non-Greenlandic vessels' fishery was phased out during the period 1972–75 and since 1972 the catch by Greenland vessels has been regulated by a total allowable catch (TAC), originally set at 1 191 tons (Horsted 1988). Through negotiations in the North Atlantic Salmon Conservation Organization (NASCO), the TAC of 1 191 tons was reduced to 870 tons in 1984 (Vezina 1988). No agreement was reached in 1985, while a 850 ton quota based on an August 1 opening date for 1986 and 1987 was negotiated in 1986. In 1988, negotiations resulted in an agreement under which total catches at West Greenland were, for 1988, 1989 and 1990 not to exceed 2 520 tons. In any given year the annual catch was not to exceed the annual average (i.e. 840 tons) by more than 10%. These quotas were based on an August 1 opening. The salmon landings at West Greenland for 1988, 1989 and 1990 at 893 tons, 337 tons and 227 tons respectively (total 1 457 tons) fell far short of the 2 520 ton TAC.

## Recreational Fishery

The annual value of the recreational fishing industry in Canada is approximately $4.7 billion. The money spent by anglers to fish supports an industry which generates an estimated 37 000 jobs across Canada. Five million Canadians of all ages take part in recreational fishing every year. Another one million visitors come to Canada to fish. These six million anglers catch and consume over 45 000 tons of fish which constitutes 8% of Canada's total finfish catch, including that taken by commercial fishermen. The one million foreign anglers spend about $300 million in Canada. This accounts for 9% of Canada's foreign exchange revenues from tourism (Anon. 1988).

The recreational fishery for Atlantic salmon in Canada generates a significant economic benefit to the Atlantic Region. In 1985, an estimated 55 400 Atlantic salmon anglers spent $39 million on recreational fishing. They also invested $72 million on major items and property in conjunction with these activities, of which they reported close to $45 million was attributable to the recreational fishery. This meant that a total of $84 million was spent and invested by Atlantic salmon anglers on sportfishing in Canada in 1985 (Tuomi 1987).

Recent statistics for Atlantic salmon indicate that $42 million (84%) of the total economic value of the Atlantic salmon catch was generated in the recreational fishery, which caught only 10–15% of the harvest (Anon. 1988).

Based upon the 1985 Survey of Sportfishing in Canada, it was estimated that 55 400 persons were "active" anglers in 1985, i.e. they reported that they actually fished for, or caught, Atlantic salmon. Atlantic salmon anglers represent a small proportion (3%) of all resident (resident in the Province) anglers on the Atlantic coast, ranging from a low of 1% in Québec, to 11% in New Brunswick and 19% in Newfoundland. Non-resident (not residing in the Province) anglers fished for Atlantic salmon at significantly higher proportions than residents. Between one third and one half of all non-resident anglers fished for Atlantic salmon (Tuomi 1987). Almost three-quarters of the non-resident Atlantic salmon anglers in New Brunswick, Nova Scotia and Québec were from the United States.

The recreational fisheries for Atlantic salmon are entirely in freshwater. Their welfare from a proprietary and a revenue viewpoint is primarily a provincial concern. In the freshwater environment, there is an overlap in jurisdiction, with neither provincial or federal governments having complete authority. Thus neither level of government has complete jurisdictional responsibility or accountability. The federal government has exclusive legislative authority to make and enforce all laws in Canada for fisheries conservation and regulation. As a result of a court ruling arising from the *Queen vs. Robertson*, 1882 and as confirmed by the Privy Council in 1898, it was accepted that all provinces have legislative authority over fisheries property rights in their respective freshwaters (Tuomi 1980). As a result of this court ruling, the provinces could enact laws to govern who could fish, under what terms and how fisheries can be owned as private or common property. The provinces responded to this proprietary right in diverse ways. In the case of Québec, the administration of federal conservation laws and regulations for the province's freshwaters was delegated totally to that Province. The province of Newfoundland chose the opposite option. The operational field management of freshwater fisheries is performed by the federal Department of Fisheries and Oceans. The province of Newfoundland limits itself to setting and collecting the freshwater angling licence fees.

### Native Fisheries

Prior to European contact, the native populations of eastern Canada basically followed a summer-fishing and winter-hunting style of subsistence living (Dunfield 1985). Salmon was highly prized as a food resource by the natives. Fishing appears to have been an important activity in the subsistence economy of the earliest Indian occupations in the Debert, Annapolis and Medway river basin about 10 000 years ago. Archaeological evidence suggests that the Laurentian culture of Nova Scotia relied heavily on salmon and gaspereau resources for their food supply (Dunfield 1985). Taylor (1985) noted that the Beothucks of Newfoundland resided on the seacoast in the summer and utilized the salmon of the rivers and estuaries, notably those of the Exploits River, as a source of food. So significant was the salmon to the Beothucks of Newfoundland that dried or smoked carcasses were traditionally buried with the dead, for use while passing through the valley of darkness to the next life (Dunfield 1985). The Dorset Eskimos also spent the early part of each fishing season hunting seals on the coast and then retired to the rivers in the early summer where they would engage in salmon fishing. Indian bands from the St. Lawrence region frequently travelled to Chaleur Bay during the salmon runs to compete with local tribes who depended heavily on the salmon resource. So significant was the salmon to the Restigouche natives that they utilized its image as a tribal symbol, painting it on canoes, weaving it into their clothing and tattooing in into their persons (Dunfield 1985).

Many treaties were signed by the Indian Nations of Canada regarding their rights to fish or hunt on or occupy their traditional lands. Two treaties relevant to the aboriginal rights were signed in 1752 and 1779.

The Treaty of 1752 between the Governor of Nova Scotia and the Micmac tribes agreed that this Tribe of Indians (Micmacs) shall not be hindered from, but have free liberty of hunting and fishing as usual. They were also to have free liberty to bring for sale to Halifax or any other settlement within the province, skins, feathers, fowl, fish or any other thing that they shall have to sell, where they shall have liberty to dispose thereof to the best advantage (Hazell 1988).

The Treaty of 1779 entered into by the Micmac Indians of Nova Scotia (and New Brunswick) from Cape Tormentine to Chaleur Bay, including the Restigouche Tribe, provided that the said Indians and their constituents shall remain free in the Districts before mentioned, quiet and free from any molestation by any of His Majesty's Troops or his good subjects in their hunting and fishing (Hazell 1988).

In a 1985 decision of the Supreme Court of Canada, *Simon* v. *The Queen*, Chief Justice Dickson held that the treaty of 1752 was validly created, never terminated and cannot be limited by provincial laws. This decision confirmed earlier cases that determined that fishing and hunting provided in treaties cannot be limited or abrogated by provincial legislation such as fish and game laws. Federal legislation such as the Fisheries Act, on the other hand, does apply so as to limit the aboriginal or treaty fishing rights of Indians. Under the Indian Act, 1870, Indian Bands are empowered to enact bylaws to regulate certain local affairs on reserves including the conservation and utilization of wildlife on reserves. For a fuller treatment of the Indian Act in specific relation to the Indian Atlantic Salmon Fishery on the Restigouche River, (see Hazell 1988).

At present native food fisheries for Atlantic salmon are officially conducted from occupied reserve lands in Quebec, New Brunswick and Nova Scotia (Marshall 1988). Micmacs who settled on non-reserve lands near Conne River, Newfoundland acquired a license to fish salmon in saltwater adjacent to the river mouth in 1986. The Naskaupi and Montagnais Indian Bands of Quebec and Inuits of Labrador are also permitted to fish for food in the freshwater of Labrador.

The "Blueprint for the Future of Atlantic Salmon" (DFO 1980) priorized the allocation of Atlantic salmon to user groups who were often competing for a greater share of a diminishing resource. The first priority of allocation was for conservation of the resource and to meet optimum spawning requirements, the second allocation was to native food fisheries, followed then by commercial and recreational fisheries.

The landmark Supreme Court of Canada decision in the 1990 *Sparrow* case created a specific obligation on the part of DFO to develop management plans reflecting the needs of native people to fish for food, ceremonial and social purposes (DFO 1991b).

In the response to the *Sparrow* decision, the Minister of Fisheries and Oceans in June 1992 announced a new Aboriginal Fisheries Strategy, a 7 year, $140 million initiative to negotiate agreements with aboriginal groups, fund improvements to fish habitat and hatcheries, and test projects for the commercial sale of fish by native people. Although the initial emphasis was placed on arrangements for the native fisheries in British Columbia, projects are also underway in the Atlantic Provinces and Quebec (DFO 1992a).

## Canadian Management

### Early Attempts to Regulate the Fishery (Pre-1950)

The period from colonial times until the late nineteenth century witnessed the wholesale destruction of Atlantic salmon in most of the larger rivers in eastern Canada (Dunfield 1985). This destruction and the subsequent catch declines became a source of concern for commercial fishermen, anglers, fishery overseers and other concerned observers. In the 1850's the only region that appeared to sustain a salmon fishery in relatively undiminished proportion was Labrador. It was predicted that even stocks in that area would soon be depleted because of rapacious fishing activity. A study was commissioned in 1846 in New Brunswick to report upon the flexibility and implications of the Grand Trunk railroad scheme with respect to the fisheries. Moses Perley, later to be appointed as a fishery commissioner to enforce the Reciprocity Treaty (Dunfield 1985), compiled and presented an exhaustive report to the New Brunswick Legislature in March, 1849. Perley's principal recommendations with respect to the salmon fishery (Perley 1852) were that:

> "A 'close time' should be established, during which salmon should not be taken in any way. The taking of 'grilse' or small salmon under a certain weight should be restricted, and their sale prohibited. Spearing should be disallowed, and the regulations for this fishing for salmon should be as uniform from county to county as possible."

In addition, Perley recommended that all salmon nets be assessed, and the fee applied to the payment of overseers. He suggested that all fisheries operating from Crown land should be leased (Dunfield 1985). As a result of these recommendations, new regulations were enacted in 1851. These were intended for the sea and river fisheries in general, with some clauses specifically respecting salmon. In these regulations the salmon laws for all counties in New Brunswick were made uniform. A season of April 1 to August 31 was imposed and a fine of 10 shillings was levied on each fish captured outside this season. In addition a weekend closure was applied, from sunset on Saturday to sunrise on Monday. Spearing was forbidden during the closed season. Fishing leases were to be issued for waters adjacent to ungranted land. Mill dams were required to have fishways, and pollution of streams was forbidden. Angling was permitted to September 16 (Dunfield 1985).

Generally, although the laws were well devised and appropriate for the resource, the civil authorities responsible for their application were generally unwilling or unable to enforce them. Thus the regulations proved to be ineffectual in curbing wholesale exploitation of the resource and the deterioration of salmon habitat.

The situation in New Brunswick reflected the plight of salmon throughout eastern Canada and the Dominion of Newfoundland. Grenfell (1909) stated that illegal fishing was widespread in Labrador. The cod traps were said to be taking large numbers of salmon. The salmon trap leaders were set out from points exactly as cod trap leaders were, and being of four inches mesh instead of six inches as were the cod trap leaders, caught much smaller fish. Practically all of the Labrador rivers of any value were illegally netted (Grenfell 1909).

These early attempts to regulate the catching of Atlantic salmon in eastern Canada were consolidated in the Federal Fisheries Act of 1932. This Act, complete with amendments up to and including 1950, and the Special Fishery Regulations for each of the Atlantic provinces were summarized by the Atlantic Salmon Association in 1950 (Anon. 1950). They were also reviewed by the Sub-Committee on Regulations of the Co-ordinating Committee on Atlantic Salmon (Needler 1950). Comparisons of the regulations were made by subject area over the various provincial jurisdictions. The review demonstrated the lack of uniformity of the regulations over the entire Canadian Atlantic coast. This reflected the differences in the stocks, habits of the fish, fishing conditions and the effectiveness of various fishing methods. The regulations up to the 1950's were based on the requirements of the anglers or commercial fishermen as much as on the habits of the fish or the needs for conservation.

In New Brunswick, Nova Scotia, and Prince Edward Island, salmon could be taken only with gill nets, drift nets, trap nets, pound nets, weirs or by angling. The regulations for Québec were similar except that use of drift nets to take salmon was not permitted. In Newfoundland the regulations did not prohibit any gear type from catching salmon.

Nets could be used only under licence in Québec, New Brunswick, Prince Edward Island and Nova Scotia. In Newfoundland licences were required only in those areas under the control of local committees.

The open season for the netting of Atlantic salmon was highly variable throughout Atlantic Canada. In Québec netting was permitted from May 15 to July 31 excluding the north shore of the Gulf of St. Lawrence east of, but not including, Natashquan where the seasons extended to August 15. In New Brunswick the netting season generally extended from May 15 to August 15 with several exceptions. One major exception was in Miramichi River and Bay where netting was permitted from May 15 to August 31 but was prohibited from the first to the third Monday in July. In Prince Edward Island netting was permitted from July 1 to October 10. In Nova Scotia netting was permitted from April 16 to August 15 except for certain areas at the head of the Bay of Fundy whose open season extended to August 31.

In Newfoundland there was no general restriction on the length of the season. In certain areas the season was limited subject to the control of local committees. These were highly variable and were related to ensuring equal access or to preventing interference with other types of fishing. For example at Portugal Cove net moorings were not to be set before April 25 while at Outer Cove salmon nets were to be removed two days before cod-traps were to be set. Pouch Cove had the shortest open season, namely May 1 to June 12. Newfoundland thus showed the widest degree of local variation — all the way from the shortest open season to the whole year, the latter applying in most of the Newfoundland and Labrador area.

In every province there was a period each week when fishing was not permitted. This was intended to assure some chance each week for salmon to pass through areas where netting was occurring. These closed periods varied from 24 hours to about 50 hours and occurred on the weekend. The mesh size regulations set a 5-inch minimum mesh size for all nets in all Provinces with the exception of a 4½ inch mesh in St. Georges Bay in western Newfoundland and the area from Cape Pine to and including Terenceville in Fortune Bay in southern Newfoundland. This latter area included St. Mary's Bay, Placentia Bay and the eastern half of Fortune Bay.

Other regulations dealt with areas closed to netting, control by local committees, holding of berths, spacing of nets, lengths of nets, and angling regulations.

## Modern Era (1950–1980)

### *Licensing*

Commercial fishing for Atlantic salmon became a "limited entry" fishery in 1968. At that time there were about 9 000 commercial salmon fishermen (Fig. 5). This number increased to about 9 500 in 1970 but was reduced to about 6 200 in 1972–1973 partly as a result of a voluntary licence-buy-back policy instituted to decrease the number of fishermen harvesting salmon. The number of salmon fishermen increased to about 8 200 in 1975 mainly as a result of increases in licences to Newfoundland fishermen who held about 85% of the total Canadian Atlantic salmon licences about this time. The total number of licences decreased to 3 483 in 1990, with 3 175 of these in Newfoundland (Fig. 5). These reductions were accomplished by several policy changes. In 1975 a new Salmon Licensing Policy was implemented for the commercial fisheries of Newfoundland. The main features were: (1) a freeze on new entrants; (2) introduction of a policy of attrition; and (3) introduction of a strict transfer policy. In 1976 the licensing policy was modified to eliminate from the fishery persons employed full-time in jobs other than the fishery (Vezina 1988). In 1980 in eastern New Brunswick, the number of herring and mackerel gill net licences was frozen at the 1980 level. The aim of this regulation was to prevent increases in the salmon by-catch in herring and mackerel gill nets. In 1983 a program was implemented to standardize the amount of fishing gear per licensed fishermen. Full-time fishermen were limited to 200 fathoms (366 m) and part-time fishermen were limited to 50–100 fathoms (91–183 m) (Vezina 1988; Marshall 1988).

The recreational salmon fisheries have contributed significantly and on an increasing scale to the salmon industry since the 1960's. In 1968 there were an estimated 53 000 resident and non-resident anglers licensed in Atlantic Canada (Muir and Bellefontaine 1988). In 1965 there were 27 575 licences issued for Atlantic salmon angling in Newfoundland and New Brunswick (Fig. 6). Prior to 1976 in Québec and 1983 in Nova Scotia a special licence was not required for salmon. The number of recreational licences for salmon increased to about 71 000 by 1983 and then decreased to 54 000 in 1984. The number subsequently increased to about 72 600 in 1988 (Fig. 6).

*Area Closures*

In 1972 a complete ban was imposed on commercial fishing in New Brunswick, the Gaspé region of Québec, and the Port-aux-Basques drift-net fishery in Newfoundland. The ban was coupled with a buy-back program and annual compensation plan which amounted to over $13 million from 1972 to 1980 (Vezina 1988). The objective of the ban was to replenish stocks to provide adequate spawning escapements. There were some improvements in stocks but from a conservation viewpoint the ban was a failure because the attention was focused on only one segment of the fishery rather than fully addressing the problems of interception, and the potential increase in exploitation in home rivers by native and sports fishermen and from increases in by-catches and poaching. There was a failure to apply constraints on the annual harvest on an Atlantic-wide basis, lack of support to increase enforcement in home rivers and a lack of support to follow up and fully assess the impact of the ban in rebuilding stocks. In 1981 the commercial fishery was reopened in New Brunswick under restricted seasons and quotas with fewer fishermen due to buy back of licences. In Miramichi Bay there was an area closure to groundfish gill netting 12 miles seaward. In Newfoundland, fishing outside of two nautical miles was prohibited.

*Changes in Open Seasons*

In addition to reductions in licenses to commercial fishermen, area closures and the drift-net ban, there were reductions to the fishing season particularly in Newfoundland aimed at reducing the interception of migrating salmon enroute to the rivers of the Maritime

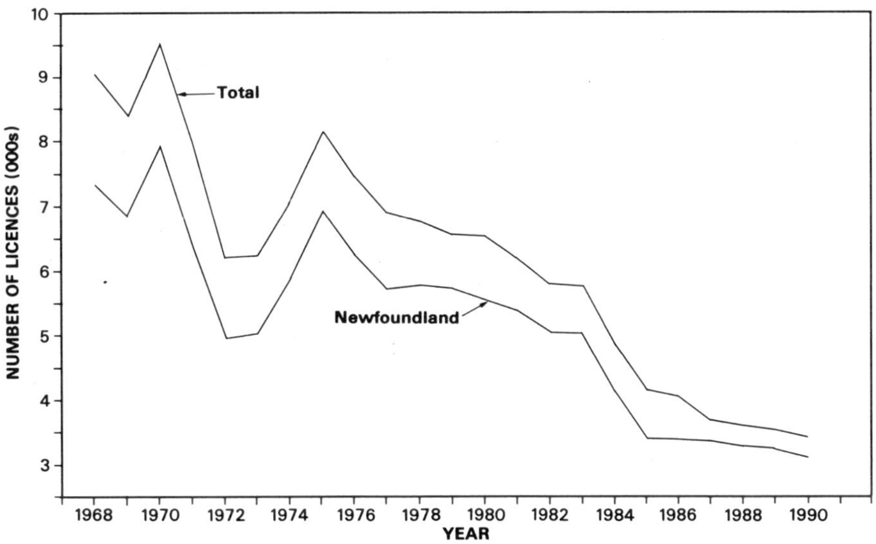


FIG. 5. Numbers of licensed commercial Atlantic salmon fishermen in Newfoundland and in Atlantic Canada in total (1968–1990).

provinces and Québec. In 1978 the fishing season for the area from Cape St. Gregory to Cape Ray, Newfoundland was changed from May 15–December 31 to June 1–July 10 and for Cape Ray to Pass Island it was changed from May 15–December 31 to May 20–July 10. There were changes in herring and mackerel fishing seasons to reduce the salmon by-catch in Newfoundland. The closed period for herring was June 15–July 31 and for mackerel it was July 1–July 31.

In 1981 the commercial salmon fishery in Newfoundland was further delayed by three days from May 15 to May 18 for all areas except Area J (Pass Island to Cape Ray) which remained May 20–July 10 (Marshall 1988). In 1984 there was a substantial delay of 2½ weeks in opening of the Newfoundland commercial fishery to June 5. In New Brunswick the commercial season was reduced to 2 weeks and in Nova Scotia to 3 weeks.

All Maritime (New Brunswick, Nova Scotia and Prince Edward Island) commercial salmon fisheries were closed in 1985. Each fishermen was offered $4 000 to compensate for lost revenue and an up-front payment on future buy-back. All commercial salmon licenses held by part-time fishermen in Newfoundland were cancelled and a buy-back program was offered. In 1986 the Newfoundland commercial fishery was closed on October 15 as agreed through the North Atlantic Salmon Conservation Organization (NASCO). The intent of this earlier closure (prior to 1986 it closed on December 31) was to reduce the interceptions of small quantities of Atlantic salmon of United States origin taken in the fall fishery off northeast Newfoundland. In the 1991 Atlantic Salmon Management Plan the commercial fishery season for the province of Newfoundland and Labrador commenced on June 5 and closed on October 15. Restrictions remained in place to prohibit commercial salmon fishing in some inner bays and estuaries of Newfoundland to allow for additional escapement to occur. The commercial salmon fisheries in the Maritime provinces remained closed (DFO 1991a). Quotas were introduced into the management of the commercial salmon fishery in insular Newfoundland in 1990 while the fishery in Labrador was still kept on an "allowance" (unregulated estimated catch).

*Local Regulations*

In addition to the general regulations various local regulations were instituted during the past several decades to reduce the exploitation rate of Atlantic salmon. In Newfoundland and Labrador, in any bay, inlet, river or sound less than six nautical miles in width, the gill net or leader of a trap net had to be removed or raised so that it was incapable of catching salmon for a 24-hour period from Saturday to Sunday. In New Brunswick the closed period in all areas and for Nova Scotia in bays or inlets less than 6 miles in width was 48 hours from Saturday to Monday morning. It was illegal to possess a salmon less than 30 cm in length.

Monofilament gill nets and trap nets were not permitted to be fished for salmon. The minimum mesh size in most areas was 127 mm with the exception of St. Georges Bay, Placentia Bay and Fortune Bay contiguous to the Burin Peninsula which had a minimum

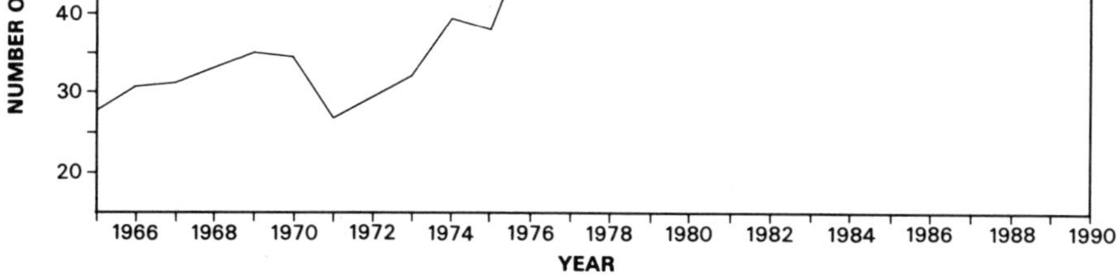

FIG. 6. Numbers of recreational Atlantic salmon licences in Atlantic Canada. Prior to 1976 in Québec and 1983 in Nova Scotia, a special licence was not required for salmon.

mesh size of 114 mm. Gill nets and trap nets (including leader) were not to exceed 183 m in length.

**1980's — Decade of Crisis: Opportunity for Change**

Serious declines took place in major Maritime and Québec salmon stocks during the late 1960's and early 1970's in spite of enhancement and enforcement activities to rehabilitate and protect them (Muir and Bellefontaine 1988). As a result steps were taken to stop these declines by instituting shortened seasons for commercial fishing in tidal waters and compensating fishermen for not fishing. These measures did not halt the decline since there was no lessening of the general exploitation of these stocks in distant waters in Greenland or Newfoundland until the mid 1970's (Muir and Bellefontaine 1988). In addition there was no control on angling or native food fisheries in the period before 1970; consequently these fisheries expanded and absorbed some of the gains achieved by closures and other regulations aimed at the commercial fishery.

*Preparing for Change*

The decline in salmon stocks precipitated a great deal of interest and concern and as a result, the International Atlantic Salmon Foundation and the Atlantic Salmon Research Trust jointly sponsored the International Atlantic Salmon Symposium at St. Andrews, New Brunswick in September, 1972, to provide a forum for discussion of the most recent knowledge on the biology and management of Atlantic Salmon. It was attended by more than 400 students, scientists, conservationists and fishermen from 12 different countries (Smith and Carter 1973). In 1971–1972, two separate task forces began compiling all the information available related to stock abundance and potential and harvest of Atlantic salmon of mainland origin in the Newfoundland commercial salmon fishery (Muir and Bellefontaine 1988). A working group was established at the April 1975 meeting of the Program Working Party on Atlantic Anadromous Fishes with the prime objective to study Newfoundland's commercial salmon fishery. The intent was to develop a series of options, based solely on existing biological knowledge and fishing patterns, which could be used to minimize Newfoundland's catch of Canadian salmon of mainland origin. The Working Group met several times between October 1975 and February 1978 and produced a report which outlined seven options involving possible regulatory approaches toward solving the problem of the harvesting of mainland salmon in the Newfoundland fishery (Pippy 1982).

In announcing a continuation into 1978 of the April, 1972 ban on commercial fishing of Atlantic salmon Fisheries Minister Roméo LeBlanc stated in December, 1977, that the recovery of salmon stocks to date had not yet reached a level to justify a continuation of commercial salmon fishing. He further added that, during the coming year the Fisheries and Marine Service would undertake a thorough review of the Atlantic salmon fishery, taking into account the biological realities and economic and social requirements within the Atlantic provinces. From this study a comprehensive plan was to be developed covering management and the further enhancement potential for Atlantic salmon stocks (DFE 1977). The Atlantic Salmon Review got underway in January 1978 with approval of the terms of reference and establishment of a Task Force of nine members. The Task Force was divided into subcommittees in order to cope with the great mass of data and varying opinions that had to be collected, assessed and reported on. Each member led a subcommittee which was given responsibility for developing input to selected portions of the salmon knowledge sphere (DFO 1979a). The subcommittees were arranged according to the following subject matter:

1) Administration of Fisheries Responsibility (2 subcommittees)
2) Fisheries and Marine Service Costs
3) Socioeconomic considerations
4) Biological conservation
5) International Factors
6) Resource Development
7) Controlling the Harvest
8) Harvesting the resource

An Atlantic Salmon Seminar was held in November 1978, (DFE 1978) following which the Minister of Fisheries and Environment, Roméo LeBlanc, requested four pilot projects be investigated for feasibility:

1) Co-ordination of the Federal and Provincial warden system,
2) Pilot transplant of coho and other Pacific salmon species,
3) Pilot mariculture operations,
4) Pilot river mouth harvesting operation.

A draft report of "The Atlantic Salmon Review" was completed and released by the Department of Fisheries and Oceans on May 1, 1979 (DFO 1979d).

Based upon the report of the Atlantic Salmon Review, Fisheries and Oceans Minister James A. McGrath issued a major statement in November, 1979, in which he announced the government's plans for the future development of Atlantic salmon. A number of Atlantic salmon management zones were established in the four Atlantic provinces. The rationale of the zones was to facilitate the tailoring of management measures to fit the biological, social and economic profile of each area.

In order to get advice and information from the grassroots level of the Atlantic community, the minister established an Atlantic Salmon Advisory Board to draw together the various components into a big picture utilizing the help and contributions of local advisory boards in each of the management zones (DFO 1979b). Mr. Walter Carter, former minister of Fisheries for Newfoundland and Dr. Wilfred Carter, Vice-chairman and Executive Director of the International Atlantic Salmon Foundation were named chairman and vice-chairman respectively, of the Board set up to oversee programs to conserve and develop Atlantic salmon resources on Canada's Atlantic coast (DFO 1979c).

The Atlantic Salmon Review was followed in 1980 by a "Blueprint for the Future of Atlantic Salmon" (DFO 1980). This document outlined various strategies to be followed for the implementation or resolution of the Atlantic Salmon Review Reports recommendations and the minister's request for the four projects announced after the November 1978 Atlantic Salmon Seminar (DFE 1978).

The Blueprint contained the following policy statements: (a) The best way to manage Atlantic salmon is through discrete stock management; (b) The interception of migrating salmon was to be reduced or eliminated; (c) The Atlantic salmon was to be managed on a "best use" basis, in consultation with the appropriate provinces and all user groups, in order to both conserve the stocks and to optimize the economic and social benefits to the people of the Atlantic region and the rest of Canada; (d) The priority for allocation of stocks would be (1) resource conservation, (2) native food fisheries, and (3) commercial and recreational fisheries.

A 1983 Study was initiated to conduct an in-depth examination of the status of Atlantic salmon stocks and longterm policy options available to effectively manage the resource. The scope of the study was to be primarily directed towards an economic and biological assessment of possible alternatives to the traditional allocation processes on the Atlantic Coast (DFO 1983a).

Muir and Bellefontaine (1988) summarized the basic common principles emerging from all these studies:

1) Conservation as the first allocation priority
2) Discrete stock harvesting near the home rivers
3) Allocate stocks to user groups that will generate optimum socio-economic benefits
4) Limit commercial salmon fishing to those most dependent on it
5) Minimize or eliminate incidental catches of salmon
6) Improve communication and consultation with user-groups and vested interests
7) Develop educational systems to promote conservation of Atlantic salmon
8) Improve catch and effort statistics
9) Strive to achieve compliance with regulations and improve effectiveness of enforcement activities
10) Strive toward development of biological and economic information to enable management by zone and where possible on a river-by-river basis.

The low returns of Atlantic salmon to home rivers continued in spite of previous management initiatives to arrest the decline. In 1983 the Canadian Atlantic Fisheries Scientific Advisory Committee (CAFSAC) advised the Minister that 1983 salmon catches should be kept as low as practically possible to ensure an adequate spawning stock for many of the major salmon stocks (DFO 1983b). The 1983 salmon fishing plan indicated that the burden of conservation had to be shared by all users of the resource. Measures in the 1983 plan included maintenance of the previous year's low commercial quotas in the three main New Brunswick rivers, the introduction of a tagging program in Nova Scotia, the ongoing standardization of numbers of nets for Newfoundland's commercial salmon fishermen, and the setting of optimal fishing effort levels for Newfoundland's salmon management zones to take into account interception levels on a year-by-year basis. As a new initiative to reduce interceptions, the Newfoundland salmon fishery outside two miles offshore was prohibited from Rose Blanche to Burgeo.

On April 6, 1984, the 1984 Atlantic Salmon Management Plan was announced which was intended to result in long-term benefits for the Atlantic Region's commercial and sport salmon fishermen. The objective of the Department of Fisheries and Oceans Atlantic salmon policy was to gradually reduce, as much as possible, interceptions of spawning salmon returning to seriously depleted river systems. To achieve this goal, the 1984 plan called for a three-week delay in the commercial salmon fishery season in Newfoundland and Labrador to permit large salmon to escape to home rivers to spawn (DFO 1984a). Commercial fishermen not licensed to fish salmon would no longer be able to retain any salmon caught in fishing gear set for other species.

On April 13, 1984, the Minister announced a voluntary commercial salmon fishing licence buy-back program for Newfoundland and Labrador (DFO 1984b).

The 1984 management plan was finalized in May 1984 with consensus that it should remain in effect for 5 years in order to cover a full life cycle (Muir and Bellefontaine 1988). Modifications to the management plan were announced on May 31, 1984. These changes included a complete closure of the key intercepting area off the south coast of Newfoundland from Burgeo west to Cape Ray (Area J2) through a mandatory buy-back policy which provided for a minimum payment of $3 500 for full-time fishermen and a minimum payment of $750 for part-time fishermen (DFO 1984c).

The voluntary commercial salmon fishery licence buy-back program announced by the Minister on April 13, 1984 was to remain in effect for the rest of Newfoundland and Labrador. The minimum payment to full-time fishermen under this program was increased to $3 000 from $2 000. Another change to the management plan stipulated that part-time commercial salmon fishing licences were not transferable under any circumstances.

One of the most significant conservation measures introduced into the management of recreational salmon fisheries in the 1980's was the mandatory release of all fish over 63cm (25 inches). This measure was designed to increase the number of multi-sea-winter-fish reaching the spawning grounds, since fish under the size limit were probably grilse, which contribute far less to egg deposition than larger salmon. Another significant conservation measure introduced into all Canadian Atlantic salmon fisheries during the 5-year-plan was the introduction of tags. These plastic devices, first used in New Brunswick recreational fishing in 1981, are attached to each salmon killed and are used to minimize the illegal harvest of fish (Bielak 1988). Nova Scotia introduced tagging to commercial and recreational fisheries in 1982 (Vezina 1988). In 1984 tagging was extended to Prince Edward Island and Quebec. A full scale commercial tagging program was extended to Newfoundland and Labrador in 1986. All salmon exported from Newfoundland and Labrador to other eastern provinces had to be tagged before leaving the province. Tagging was extended to the Newfoundland sportfishery in 1988 (Vezina 1988).

The 5-year plan of 1984 was modified and refined between 1985 and 1987. The new elements of the plan introduced during 1985 and 1986 were as follows:

1) Closure of all commercial salmon fisheries in Nova Scotia, New Brunswick and Prince Edward Island in 1985, with compensation for lost revenues.
2) All 668 part-time commercial salmon fishermen's licences in Newfoundland were cancelled and a buy-back program was offered.
3) Angling seasons were extended by two weeks (excluding Québec) where stocks permitted.
4) Anglers could not continue to hook and release after they had retained their daily bag limit.
5) In 1986 a voluntary commercial salmon licence buy-back program was offered in the Maritime Provinces
6) A 15-fish seasonal bag limit was announced for the Newfoundland and Labrador recreational fishery in 1986.
7) A full-scale tagging program was introduced in Newfoundland and Labrador in 1986.
8) The Newfoundland/Labrador fall fishery closure date was changed to October 15, consistent with Canada's obligation under NASCO (Muir and Bellefontaine 1988).

The 1984 Atlantic Salmon Management Plan appears to have increased the returns of large salmon to rivers in the Gulf of St. Lawrence (Chadwick 1989; Randall 1990).

For Newfoundland, where major changes to the commercial fishery occurred, the restrictions imposed on the commercial fisheries did not result in increased escapement of grilse in either Newfoundland (except for two areas) or Labrador. The focus of the plan was a reduction in the interception and exploitation of large salmon, primarily virgin multi-sea-winter salmon in the commercial fisheries of Newfoundland and Labrador. A spinoff of the plan with respect to Newfoundland populations other than St. Georges Bay should have been increased grilse and other large salmon runs to insular Newfoundland rivers. This did not occur as expected.

The lack of changes in escapement and catches suggested that the overall abundance of grilse did not change between pre-plan and plan years and that reductions in licensed effort had little effect on reducing commercial catches. No increases were observed for escapements of large salmon (except for one area). For the Newfoundland populations of Atlantic salmon, overall observations pertaining to river escapements did not measure up to plan expectations (O'Connell et al. 1992).

The removal of licensed gear did not necessarily mean a decrease in exploitation. It is possible that from 1984 to 1988 a greater proportion of licensed effort was directed at the small salmon component compared to pre-plan years in an attempt to compensate for the loss of harvest of large salmon. Other reasons could include: (1) many of the fishermen removed from the system may have been the least successful; and

(2) individual catch rates for fishermen remaining in the system may have increased. The end result was that overall river escapements did not increase during the plan as expected (O'Connell et al. 1992).

On May 5, 1989, Fisheries and Oceans Minister Tom Siddon announced a second 5-year Atlantic Salmon Management Plan beginning in 1989 (DFO 1989a). Conservation of salmon stocks remained the priority in the new plan which recognized a continuing role for the commercial fishery in Newfoundland and provided management flexibility for the Atlantic provinces when the resource improved.

To address the interception of migrating salmon by the Newfoundland fishery, the concept of an *allowance* (unregulated estimated catch) was introduced in the Newfoundland and Labrador commercial fisheries. The allowances were set following consultations with the various user groups as follows:

| ZONE | ALLOWANCE (Tons) |
|---|---|
| 1 | 80 |
| 2 | 350 |
| 3 | 270 |
| 4 | 170 |
| 5 | 55 |
| 6 | 45 |
| 7 | 25 |
| 8 | 25 |
| 9 | 10 |
| 10 | 35 |
| 11 | 50 |
| 12 | Closed |
| 13 | 75 |
| 14 | 110 |
| **TOTAL** | **1 300** |

After conservation, the social and cultural importance of native fisheries continued to be the second guiding principle in the 1989–1993 salmon conservation strategy (DFO 1989a).

## Quotas

A major change was introduced in the management of the commercial salmon fishery in 1990 (CAFSAC 1991a). Quotas were introduced for the first time in the Newfoundland Region in Salmon Fishing Areas (SFA) 2–11 (Fig. 7) to replace the "allowances" introduced in 1989 while in SFA 1 (northern Labrador) the catch "allowance" was kept in effect. Also, in some SFAs, caution notices were moved farther out from the mouths of the rivers. The implementation of quotas decreased commercial catches in almost all Newfoundland SFAs, the total for SFAs 1–11 being 40% lower than the average for the previous 5 years. The lower catches were not entirely related to the management because the quotas were not reached in five of the SFAs. The commercial catch for Gulf Newfoundland (SFAs 13–14) were also lower by about 40% when compared to the previous 5 years. In spite of the implementation of quotas in the commercial fishery, recreational catches in most SFAs were either less than or comparable to the 1984–89 mean. Counts of grilse at fishways and counting fences were also low in nearly all cases compared to the 1984–89 mean. This suggested that, had the 1990 management measures not been in place in Newfoundland, escapements to Newfoundland rivers might have been even lower. It has been shown that a lower marine survival could have contributed to the low abundance in 1990 (CAFSAC 1991a). This was demonstrated by studies conducted on the Conne River and Northeast Trepassey River, in southern Newfoundland (O'Connell et al. 1991).

## International Management of Canadian Salmon

### ICES/ICNAF Joint Working Party on North Atlantic Salmon

Catches of Atlantic salmon at West Greenland by native Greenlanders increased very rapidly from 14 tons in 1959 to 1539 tons in 1964 (Horsted 1988). This large catch of Atlantic salmon generated an intense international debate since these salmon originated in most of the major salmon producing countries of the North Atlantic (Anon. 1987). These salmon were predominantly 1 sea-winter fish which were destined to return to home rivers as 2 sea-winter salmon or larger and hence constituted a significant component of the spawning stock of these rivers.

The Greenland fishery represented a loss of spawning fish to these home rivers. Although the proportional continental origin varied between years and areas, there was an approximately 50:50 percent contribution from Europe and from North America (Anon. 1987). The estimated impact of the West Greenland fishery on home water stocks and yields produces varying results since the parameter values used to calculate the losses to home waters are known to fluctuate between years, resulting in a range of estimates. Recent estimates show that for each ton of European-origin salmon in the reported catch at West Greenland, there are losses of 1.22 to 1.69 tons to European stocks. The corresponding losses to North American stocks are estimated to be 1.45 to 2.02 tons for each ton of North

American-origin salmon in the reported catch at West Greenland (Anon. 1987).

In response to concern about the impact of the Greenland fishery, the International Commission for the Northwest Atlantic Fisheries (ICNAF) and the International Council for the Exploration of the Sea (ICES) initiated a scientific appraisal of the developing West Greenland Fishery, starting in 1965. The ICES/ICNAF Joint Working Party on North Atlantic Salmon, with representative membership from all North Atlantic countries interested in the Atlantic salmon resource, was established (Horsted 1988).

At its Annual Meeting in 1965, ICNAF had recognized the need for more intensive research and scientific information and advice on many biological aspects of the Greenland fishery, especially those concerning the effects of this new fishery development on future Atlantic salmon stocks and the potential losses in yield to home water fisheries and stocks (Parrish 1973). At the Annual Meeting of ICES in 1965, it was recommended that an international scientific working party be established to evaluate the problem under the joint auspices of ICES and ICNAF.

The main thrust of the Working Party was to promote the collection of the relevant scientific data, to undertake the necessary biological research, and to assess the effects of the West Greenland fishery on home water salmon stocks and catches (Anon. 1965).

The first meeting of the Working Party took place in Madrid immediately following the Annual Meeting of ICNAF in 1966 (Parrish 1973). Subsequent meetings of the Working Party were held each year thereafter to

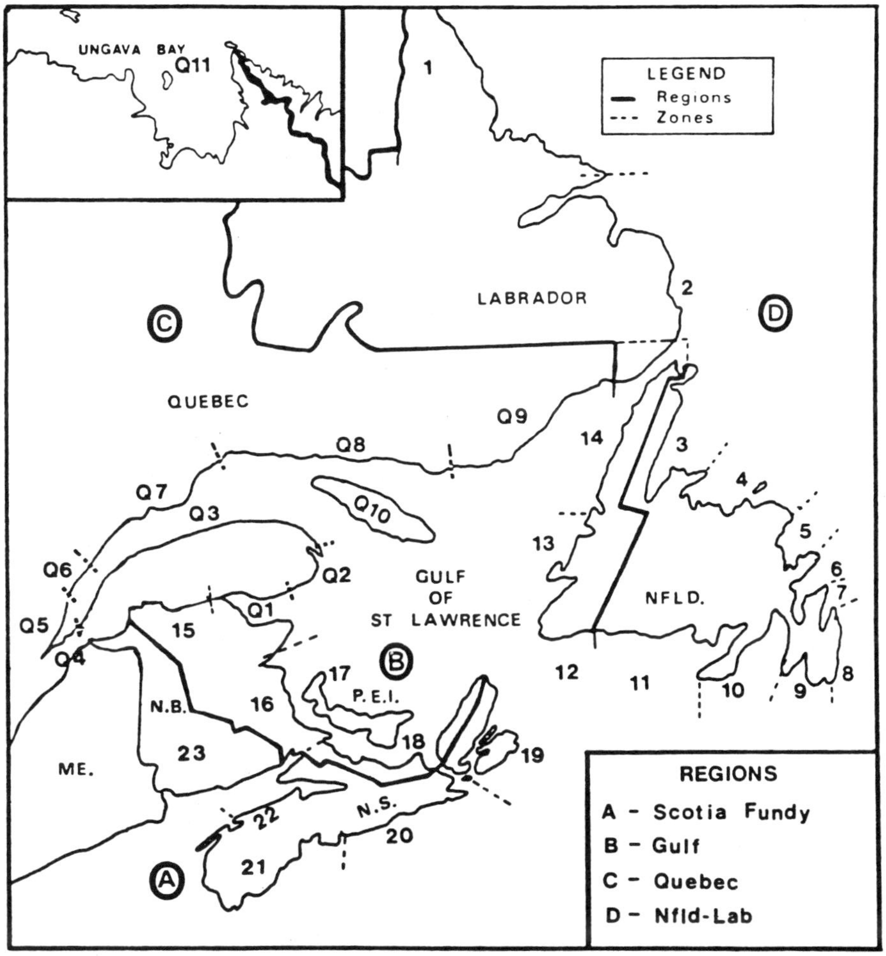


FIG. 7. Map of Atlantic Canada showing provincial boundaries, DFO Regions, Atlantic salmon fishing (zones) areas (SFAs) numbered 1 to 23, and Quebec fishing areas numbered Q1 to Q11.

1974. The group reported annually to ICNAF and ICES. The working party organized the ICES/ICNAF Salmon Tagging Experiment at West Greenland in 1972. The publication of the results of that experiment plus other related research activities essentially summarized most of the research conducted at West Greenland on Atlantic Salmon up to 1972 (Parrish and Horsted 1980).

When ICNAF was succeeded by the Northwest Atlantic Fisheries Organization (NAFO) in 1979, the former Joint Working Party was continued as the ICES Working Group on North Atlantic Salmon which first met in 1979 and has continued to meet annually since then with the exception of 1983 (D. Reddin, pers. comm.).

**North Atlantic Salmon Conservation Organization**

Developments arising from the Third United Nations Conference on the Law of the Sea in the 1970's emphasized the need for international co-operation in the conservation and management of the North Atlantic salmon.

By 1975 the Law of the Sea Conference had completed a draft of a Convention recognizing the establishment of a 200 nautical mile exclusive economic zone. Among the fisheries provisions was an article, later adopted as Article 66 of the 1982 United Nations Convention on the Law of the Sea, which established a special regime for anadromous stocks of fish (Eiriksson 1988).

On October 1, 1983 the Convention for the Conservation of Salmon in the North Atlantic Ocean entered into force. Under this convention a new international organization, the North Atlantic Salmon Conservation Organization (NASCO), was established. The objective of the organization was to contribute through consultation and co-operation to the conservation, restoration, enhancement, and rational management of Atlantic salmon (*Salmo salar*) stocks subject to the Convention, taking into account the best scientific evidence available to it (Eiriksson 1988). The present Parties to the Convention are Canada, Denmark in respect of the Faroe Islands and Greenland, the European Economic Community, Finland, Iceland, Norway, Sweden, Russia and the United States of America. The members of NASCO are representative of all the North Atlantic nations having an interest in Atlantic salmon.

The Convention applies to all salmon stocks which migrate beyond fisheries zones of coastal states of the Atlantic Ocean north of 36° N. latitude through their migratory range. The Convention bans salmon fishing beyond 12 nautical miles in all areas of the North Atlantic except in two areas: (1) within the fisheries jurisdiction of the Faroe Islands; and (2) up to 40 nautical miles off the coast of West Greenland.

The Convention does not provide for the establishment of a new scientific research organization but rather calls for co-operation with ICES.

NASCO consists of:

(a) A Council, of which all parties are members
(b) Three regional Commissions
  1) The North American Commission
  2) The West Greenland Commission
  3) The North-East Atlantic Commission

The North American Commission has a mandate to provide a forum for consultation and co-operation between its members, Canada and the United States, on matters related to minimizing catches of salmon in the fisheries zones of one member where those salmon originate in the rivers of the other, and in cases where activities of one member affect salmon originating in the rivers of the other.

Discussions on regulatory measures in the North American Commission have been concerned with the interception of United States-origin salmon in the Newfoundland fisheries. In 1986, the Commission adopted a proposal to close the Newfoundland and Labrador commercial salmon fisheries from 15 October to 31 December 1986. This initiative was implemented as a permanent measure in the 1986 Atlantic Salmon Management Plan (DFO 1986).

The West Greenland Commission adopted a proposal for the establishment of a total allowable catch for the West Greenland fishery of 870 tons for the 1984 season. This represented a 27% reduction from the ICNAF quota of 1 191 tons which had previously applied to these fisheries. In 1985, no regulatory measure was agreed in the Commission and the Greenland authorities unilaterally imposed a quota of 852 tons.

In 1986, the Commission adopted a proposal setting the total allowable catch at 850 tons for both the 1986 and 1987 fishing seasons (Eiriksson 1988). During 1988, 1989 and 1990 the total catches for the three years were not to exceed 2 520 tons while the catch in any given year was not to exceed the annual average (i.e. 840 tons) by more than 10%. The actual total salmon landings in 1988, 1989 and 1990 of 1 457 tons fell far short of the 2 520 tons TAC.

**The Future Role of Science in the International Management of Atlantic Salmon**

The first meetings of NASCO in 1984 began a new era and provided a forum for international cooperation in conserving salmon stocks and managing salmon fisheries. A series of detailed requests by NASCO to the International Council for the Exploration of the Sea (ICES) has stimulated national research and monitoring programs and has since then focused the work of the ICES North Atlantic Salmon Working Group (Doubleday 1988).

At its first annual meeting, at Edinburgh, the NASCO Council (Anon. 1984) identified six programs of research:

1. Monitoring of Fisheries
2. Data Base
3. Location of Origin
4. Index Rivers
5. Natural Mortality and Age at Maturity in the Sea
6. Geographical Distribution

Doubleday (1988) suggested that there was a need for a systematic framework for the provision of scientific advice on salmon management. This framework should be developed in consultation by NASCO and ICES. The framework should, as a minimum, address the need for conservation measures, the conservation effects of alternative management measures, and the impact of proposed management measures on affected fisheries.

Three key areas were identified by Doubleday (1988) for salmon research:

1. the relationship between spawning escapement and fresh water production;
2. identification of the location of origin of salmon caught at sea;
3. natural mortality in the marine phase, particularly in the post-smolt period.

Doubleday (1988) concluded that the first and most fundamentally important of the three was too large a problem for any one laboratory to solve but would need to be addressed by a network of compatible experiments in several North American countries. These data are required to determine the target rates for spawning escapements. At present the scientific basis for judging the adequacy of spawning is weak. Based on studies by Elson (1957) in New Brunswick in the 1950's a target spawning deposition of 2.4 eggs per square metre of rearing habitat has been adopted for some Canadian rivers (Doubleday 1988). There is high annual variation around this value even within the same river system (Elson 1975).

CAFSAC (1991b) suggested that the operational translation of conservation for Atlantic salmon be based on the potential productivity of rivers using the target egg deposition rate of 2.4 eggs/m$^2$ of fluvial rearing habitat and in addition for insular Newfoundland 368 eggs/hectare of lacustrine habitat. Because the capacity of fish habitat to produce smolts from a given level of egg deposition is variable, CAFSAC suggested there was a need to quantify conservation for each river. In the fall of 1991, CAFSAC (1991c) quantified conservation for 34 rivers in the Gulf, Scotia-Fundy and Newfoundland regions and, where possible gave first indication of potential surpluses to conservation for 1992.

The identification of location of origin of salmon caught at sea will progress in the next few years as it has already done as more sophisticated techniques such as image analysis of salmon scales are applied to the problem.

A more accurate determination of the level and pattern of natural mortality of salmon at sea will require a series of geographically separate studies to provide a sound basis for general conclusions (Doubleday 1988).

### Recent Developments in the Management of Canadian Atlantic Salmon Stocks

Following the introduction of quotas for the commercial fishery in Newfoundland in the 1990 Atlantic Salmon Management Plan (DFO 1990a), commercial catches decreased in almost all Newfoundland fisheries zones. In addition to the quota introductions, the Department of Fisheries and Oceans also increased its enforcement and surveillance activities to curb poaching and other forms of illegal activity related to salmon. In the Newfoundland region undercover surveillance teams especially trained by RCMP were deployed into areas suspected as prime targets. Officers in the field were equipped with hand-held portable radios and linked to a province-wide network enabling quick, combined response to suspected poaching activity. Helicopter surveillance added an element of surprise to the department's enforcement efforts, as did night patrols and all-terrain vehicles. The result was that as of July 27, 1990 the number of detected salmon fishing violations in the Newfoundland Region increased to 105 from 61 the previous year (DFO 1990b).

In the commercial salmon fishery, the recent restrictions upon the commercial sector, such as reduced seasons and quotas, had not been effective in rebuilding the salmon resource. Despite the prior licence retirements and other conservation measures, the commercial catch in Newfoundland and Labrador had declined over four consecutive years, 1988 to 1991. The 1991 catch was 433 tons, 39% below the lowest catch throughout the 1980's (798 tons in 1984).

In March 1992, the new Minister of Fisheries and Oceans, John Crosbie, and the Newfoundland Fisheries

Minister Walter Carter, announced a jointly-funded offer of cash payments to Newfoundland commercial salmon fishermen who voluntarily retired licences. Proposed payments to commercial licencees were in the range of $8 000 to $50 000 depending on the level of their salmon landings in the best year out of the past three. Payments were to be made after they agreed to the terms of the offer and turned in their salmon fishing gear. Estimated cost of the retirement program was about $40 million to be shared 70–30 by the federal government and the province. Minister Crosbie stated: "If the Atlantic salmon is to recover to the abundance of a generation ago, there is no alternative but to sharply curtail the commercial salmon fishery" (DFO 1992b). Coincident with the retirement offer, a minimum 5-year closure of the commercial salmon fishery for the Island portion of the Province was implemented. The fishing moratorium did not extend to Labrador, with its high dependence on salmon and limited prospects for fisheries diversification.

Commercial salmon fishing had been discontinued in Prince Edward Island, Nova Scotia and New Brunswick nearly 10 years previously. Newfoundland and Labrador were at that time the only Atlantic areas where commercial fishing was permitted. About 3 000 licences were still active in the fishery (DFO 1992b). As of March 1993, 2 739 (91%) of the commercial salmon licence holders had applied for the voluntary licence retirement program (Ken Jones, pers. comm.).

### Salmon Enhancement

In addition to the management measures aimed at increasing the spawning escapement by limiting catches in the commercial and recreational fisheries, the most promising alternative to increasing the full potential production of the Atlantic salmon resource is to increase the habitat that can be utilized by Atlantic salmon. The demands of the various user groups have increased well beyond the present production capability (Ritter et al. 1980). The Atlantic salmon resource in Canada relies on an ever-shrinking freshwater habitat.

Inventories of Canadian Atlantic rivers and salmon stocks indicate numerous opportunities both to expand the range of the species and increase salmon production within accessible streams. Approximately 90% of the potential for enhancement lies in water inaccessible to salmon because of natural or man-made barriers. The remaining 10% exists within waters accessible to salmon but requiring recolonization (Ritter et al. 1980).

Potential production from the above measures is approximately 562 000 adult salmon. This production is based on estimated stream habitat potential and assumed smolt production capabilities ranging from 0.2 smolts/100 square metres for rivers in Ungava Bay (Power 1969) to 3.5 smolts per 100 square metres for the more productive rivers in Newfoundland (Ritter et al. 1980). Full production from all rivers with enhancement potential would boost Atlantic salmon production in Canada by about 40% above the potential of its present salmon producing waters. Québec's increase would be 49%, Newfoundland and Labrador's 42% and the other three Maritime provinces' 22% (Ritter et al. 1980).

The main strategies for enhancing production of Atlantic salmon can be summarized as follows:

1) Colonization of areas upstream of natural or man-made obstructions to fish passage and recolonization of areas inaccessible to migrating salmon and which now lack stocks because of pollution and/or overexploitation.
2) Supplemental stocking by use of cultured salmon.
3) Sea ranching.

The various culture strategies which can be employed to support initiatives in salmon enhancement have been described by Ritter and Carey (1980) and are as follows:

1) Hatcheries
2) Deep substrate incubation
3) Semi-natural rearing
4) Post rearing and delayed releases of smolts
5) Kelt reconditioning.

*Colonization*

The main focus of enhancement would be to extend the range of the Atlantic salmon into inaccessible areas through provision of fish passage facilities and stocking the new area with young salmon. The fish passage to the area could be ensured by removing the obstruction, constructing a fishway, or trapping the fish at the base of the obstruction and transporting them over or around the obstruction.

Taylor and Bauld (1973) described a long-term developmental proposal by which the Exploits River, the largest on the Island of Newfoundland, which had historically been accessible to salmon for less than 10% of its watershed area would be enhanced to full production. At 267 kilometres, the Exploits River has a drainage area of 11 272 square kilometres, the island's largest watershed. Because of impassable waterfalls on the Exploits at Grand Falls and on Great Rattling Brook, fully 90% of this watershed had been closed to natural colonization by sea-run Atlantic salmon. The project involved the construction of two fishways, construction of incubation facilities (spawning channels and deep substrate incubators) in two phases. The initial stocking phase occurred during the 1957–1964 period with annual transfers followed by a second stocking phase during 1968–1979 with swim-up fry (O'Connell et al. 1991). Because of the enhancement program on the Exploits river system the return of salmon to the Exploits increased from 1000 in 1959 to 20 000 in 1984 (DFO 1989b). Although there have been annual variations in numbers, the trend over the three decades has been steadily upward. There was a decline in the counts at the fishways during 1986 to 1990 but they increased again to about 13 000 in 1992 possibly as a rcsult of the moratorium on the commercial salmon fishery (J. Pratt, pers. comm.).

From its modest beginnings in 1957, the Exploits River project has grown to become one of the largest and most successful salmon enhancement projects in Atlantic Canada. The project has a positive benefit: cost ratio, based on developmental costs versus actual and projected returns over a 25-year period. When restored to full production, the Exploits Watershed is expected to produce 100 000 adult Atlantic salmon annually (DFO 1989b).

A similar colonization of the Upper LaHave River in Nova Scotia involved the construction of a fishway in 1969 and stocking with hatchery-reared smolts and parr from 1971–1979. The annual total production to all fisheries and spawning escapement has been about 4 000 grilse and multi-sea-winter salmon.

Cultured salmon are used to rebuild diminishing-stock levels or provide stock for colonization for areas made inaccessible by industrial use. The Mactaquac Hatchery on the Saint John River in New Brunswick was built to compensate for mortalities induced by delayed smolt migration at the dams and multiple exposure to turbines that could entirely annihilate the salmon above Mactaquac where a major dam was constructed in the mid-1960's (Ritter and Carey 1980). The salmon-rearing habitat above Mactaquac represented over 60% of the total in the system and if eliminated, would have meant the loss of at least 70 tons of salmon from the commercial fishery and 3 500 fish from the recreational fishery.

The Federal Government and the New Brunswick Electric Power Commission agreed that a hatchery should be built to rear sufficient salmon smolts to replace the loss in natural production above Mactaquac and also a trapping facility in the dam from which migrating fish could be transported to the hatchery and then released upstream. The management scheme after 1967 called for the culture and annual release of 500 000 smolts. This target was reduced to 250 000 smolts as emphasis was shifted to quality rather than quantity. The smolt survival has increased tenfold over that recorded for the first three years of stocking, probably due to increased size, improved fin condition, and general health at release as rearing densities were reduced.

An alternate source of seed stock for colonization programs is the use of deep substrate incubators to produce unfed fry (Porter and Meerburg 1977). The egg-to-adult survival of fish derived from substrate incubation and unfed fry stocking appears to be lower than for fish produced by hatchery programs (Ritter and Carey 1980) which would limit their use where numbers of suitable broodstock are scarce. In some cases smolts could be stocked into natural or man-made lakes and provided with supplementary feeding.

There are other techniques which could be used to enhance survival and growth such as protecting and feeding hatchery-propagated Atlantic salmon smolts in seapens. This could potentially increase marine survival rates (Saunders 1977). Atlantic salmon kelts reconditioned in captivity could be another alternative source of eggs for culture operations supporting salmon enhancement projects and commercial aquaculture operations. Good growth, followed by maturation of both males and females and acceptable fecundities have been recorded for kelts retained in fresh, brackish and salt water (Ritter and Carey 1980). Pen-rearing of Atlantic salmon in the sea is a further stock enhancement strategy that has recently been introduced successfully into Atlantic Canada. The production from Atlantic salmon farms increased from 6 tons in 1979 to 8 182 tons in 1990 valued at C$70 million (Aiken 1993). The potential of Atlantic salmon farming from sea pens has been demonstrated very successfully in all salmon producing countries of the North Atlantic as attested by an annual production in Norway in 1990 of about 158 000 tons (Anon. 1992b) (Fig. 8). Despite past declines in the habitat of Atlantic salmon and the depleted state of some of the stocks, there is room for guarded optimism in light of recent decreased fishing activity and management initiatives. In this context Atlantic salmon enhancement will provide a worthwhile and desirable contribution to rebuilding the stocks to their potential production levels.

North Atlantic production of farmed Atlantic salmon has recently increased to about 220 000 tons in 1991 (Anon. 1992b) of which about 9 000 tons were produced in eastern Canada. This production has flooded the world market for salmon and prices have dropped accordingly. The combination of scarce resources of wild salmon and the growing availability of sea-farmed salmon at low prices may combine to make any revival of a commercial fishery for Atlantic salmon non-viable (May 1992; Bielak 1988).

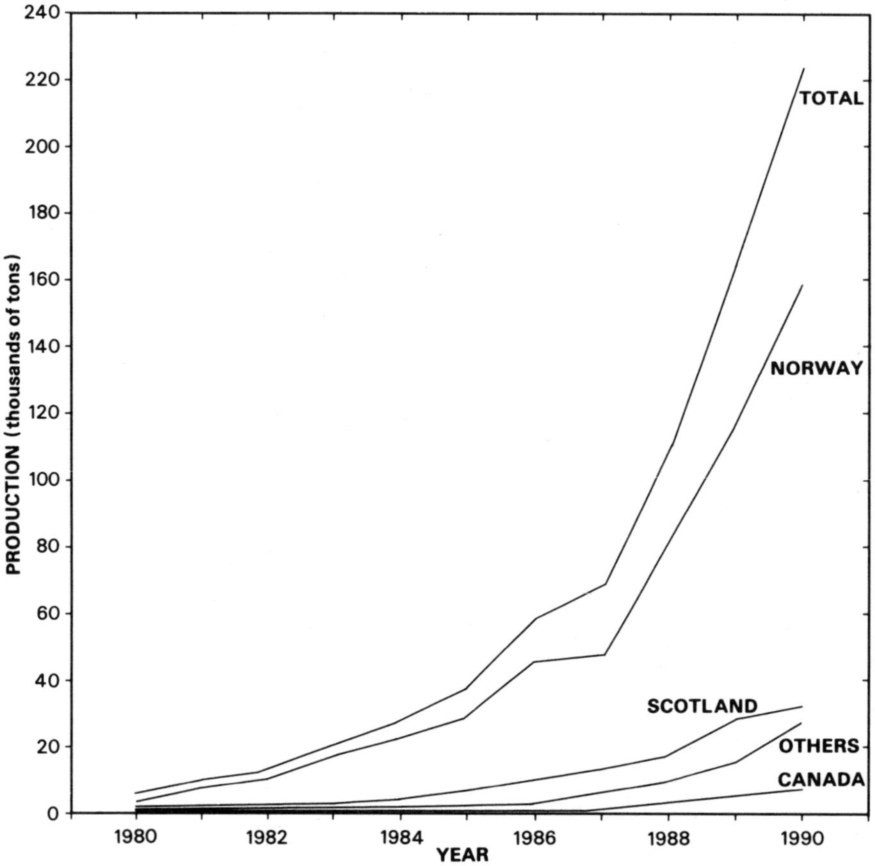


FIG. 8. Production of farmed Atlantic salmon in North Atlantic countries, 1980–1990.

On October 9, 1992, a 5-year, $21.4 million Canada-Newfoundland Cooperation Agreement to enhance and conserve Atlantic salmon and related species was announced by Minister of Fisheries, John C. Crosbie and Newfoundland Fisheries Minister, Walter Carter (Cooperation Agreement 1992).

The cooperation agreement will be financed 70/30 by the federal government and the province. Enhancement measures under the Agreement include opening and stocking new areas of fish habitat, restoring polluted or damaged habitat, or building fishways to assist passage past natural or man-made obstructions. The public were encouraged to participate actively in the program by helping to assess the success of various enhancement measures.

The window of opportunity to restore stocks to historic levels opened March 9, 1992 when Minister Crosbie announced a 5-year closure of the commercial salmon fishery off Newfoundland, following extensive federal-provincial consultations with all user groups. The moratorium announcement was accompanied by a $40 million commercial salmon licence retirement program in Newfoundland and Labrador. This set the stage to ensure that Atlantic salmon is sustained in the future as a renewable resource that will support an expanding sportfishery throughout the Newfoundland region.

## Current Status of Stocks

The status of Atlantic salmon stocks in the Atlantic area in 1990, as derived from fisheries catches, counts at enumeration facilities and from mark-recapture experiments indicated that there was a general decrease compared to the average of the previous 5–6 years (CAFSAC 1991). When 1990 recreational catches were compared to the average of the previous 5–6 years, only in Québec were they consistently higher in 1990.

The low commercial catches in the fishing zones of the Newfoundland Region, which were under quota, indicated that the decreases were not entirely related to management measures because the quotas were not reached in five of the salmon fishing areas (SFAs). There were indications from two studies in Newfoundland rivers that the decreased catches in 1990 could be related to lower marine survival. The commercial catches for Western Newfoundland were also lower by about 40% when compared to the previous 5 years.

Counting facilities and spawner counts in all Scotia-Fundy SFAs showed lower returns of grilse and larger salmon compared with 1989 or with the average for 1984–89 except in the Liscomb River where returns were significantly higher than the previous 5 years.

The commercial salmon landings at West Greenland in 1990 were also very low (229 tons), which suggested that the numbers of large salmon destined to return to Canada in 1991 would also be low. This suggestion was borne out by the low catches (173 tons) of large salmon in the commercial fishery in the Newfoundland Region in 1991. This was only 55% of the 1990 catch and was the lowest in the 18-year period from 1974 to 1991. Catches of grilse in 1991 were also the lowest in 18 years at 179 tons.

In the Newfoundland Region, commercial catches coupled with recreational catches and counts at fishways and counting fences suggested that overall there was a decrease in abundance of small salmon in 1990. Commercial catches and counts suggested the same for large salmon.

In addition to lower marine survival, it is also possible that adverse environmental conditions (low water levels and high water temperatures) in rivers in 1987 could have contributed to low returns in 1990 and 1991 by affecting juvenile survival rates.

## Discussion and Conclusions

The Atlantic salmon resource in Canada has been subjected to intense pressures since colonial times. The earlier pressures were in the form of industrial practices which produced deleterious substances harmful to salmon populations or physically barred salmon from their freshwater habitat.

Early fisheries also took their toll on stocks in smaller and generally more vulnerable areas especially in the proximity of urban areas. Later development such as the building of dams for hydroelectric projects further excluded salmon from their original habitat and reduced spawning populations and hence adult runs. In addition, heavy forest spraying with DDT in the 1950's contributed to the severe depression of the stocks in the late 1950's.

Exploitation by commercial fisheries in home waters and in international waters at the feeding area off West Greenland added incrementally more pressure on the resources. All these forces acting in concert were more than the stocks could withstand, resulting in declines. These declines led to some management initiatives such as the 5-year ban on commercial fishing in the three maritime provinces and the Gaspé region of Québec in 1972. There was some increase in catches in the late 1970's and early 1980's but the serious declines beginning in 1988 have accelerated to the lowest-ever levels at the beginning of the 1990's, resulting in a closure of the Newfoundland commercial fishery.

Despite all the management initiatives, the attempts to compensate for loss of habitat by hatchery production of smolts, recolonization, incubation facilities and stream remedial work, the legacy in most areas is still a declining resource or in a few areas at best, a stable resource. Even where stable, however, the resource is still well below the potential.

The most recent management initiatives, coupled with accelerated surveillance and enforcement activities in inland waters, may help to arrest the decline. The attainment of a greater production of Atlantic salmon will depend, to a large extent, upon the level of effort and resources directed toward salmon enhancement and the degree to which Canada can preserve and rehabilitate the freshwater habitat of the Atlantic salmon. A concerted effort is required to rebuild this valuable resource so that this natural legacy can be enjoyed to its fullest by Atlantic salmon enthusiasts and users.


## Acknowledgements

I would like to thank D.G. Reddin, D.J. Meerburg and L.S. Parsons for their constructive review of the manuscript. Art Read prepared the illustrations and Ronnie Sanford and Colleen Meloche typed various drafts of the manuscript.


## References


AIKEN, D.E. 1993. Aquaculture in Canada, p. 441–446. *In* L. S. Parsons and W. H. Lear [ed.] Perspectives on Canadian marine fisheries management. Can. Bull. Fish. Aquat. Sci. 226.

ALLAN, I.R.H., AND J.A. RITTER. 1977. Salmonid terminology. J. Cons. int. Explor. Mer. 37: 293–299.

ANON. 1950. Atlantic Salmon Legislation. Atlantic Salmon Association. Doc. No. 13: 77 p.

1965. Report of the 53rd meeting of the International Council for the Exploration of the Sea, Rome, 4–13 October, 1965. Procès-verbal de la réunion. 123 p.

1984. Report of the First Annual Meeting of the council of the North Atlantic Salmon Conservation Organization.

1986. Strategies for the long-term management of Atlantic Salmon. Report of the Special Federal/Provincial Atlantic Salmon Working Group. December, 1986, 35 p.

1987. Report of the Working Group on North Atlantic Salmon. ICES C.M. 1987/Assess: 12 p.

1988. National Recreational Fisheries Conference Proceedings 1986. Can. Spec. Publ. Fish. Aquat. Sci. 97: 174 p.

1992a. Report of the Study Group on the North American Salmon Fisheries. St. John's, Newfoundland, Canada. 17–21 February, 1992. ICES C.M. 1992/M:3. 33 p.

1992b. Report of the Working Group on North Atlantic Salmon. Dublin, Ireland, 5–12 March, 1992. ICES C.M. 1992/Assess. 15: 155 p.

Belding, D.L., and G. Préfontaine. 1938. Studies on the Atlantic Salmon. — II. Report on the Salmon of the 1937 Port-aux-Basques (Newfoundland) Drift-net Fishery. Contrib. Institute of Zoology, Univ. of Montreal, No. 3, 58 p.

Belding, D.L., and G. Préfontaine. 1939. Studies on the Atlantic Salmon — III. Report on the Salmon of the 1937 Miramichi (New Brunswick) Drift-net Fishery. Contrib. Institute of Zoology, Univ. of Montreal No. 4. 63 p.

Bielak, A.T. 1988. Canadian Atlantic Salmon Recovery Efforts. Paper presented to Trout Unlimited Conference. Moscow, September, 1988. 11 p.

CAFSAC 1991a. Status of Atlantic Salmon Stocks. CAFSAC Adv. Doc. 91/4: 17 p.

1991b. Definition of Conservation for Atlantic Salmon. CAFSAC Adv. Doc. 91/15: 4 p.

1991c. Quantification of Conservation for Atlantic Salmon. CAFSAC Adv. Doc. 91/16: 15 p.

Cooperation Agreement. 1992. Ministers announce Canada-Newfoundland Cooperation Agreement. News Release, Canada-Newfoundland Cooperation Agreement, October 9, 1992. 3 p.

Chadwick, E.M.P. 1989. An analysis of 1SW to MSW salmon returns in the Gulf of St. Lawrence Rivers in relation to the 1984–88 salmon management plan. CAFSAC Res. Doc. 89/75: 9 p.

DFE. 1977. Atlantic Salmon Fishing Ban to Continue in 1978. News Release. Dec. 28, 1977. 2 p.

1978. Major Atlantic Salmon Review Prompts November Seminar. News Release, Sept. 25, 1978. 2 p.

DFO. 1979a. Atlantic Salmon Review. Task Force For the Review. Review Summary. 27 p.

1979b. A Major Statement on Atlantic Salmon by The Honourable J.A.McGrath, Minister of Fisheries and Oceans. News Release, Thursday, November 29, 1979. I-HQ-79-009E. 6 p.

1979c. Carters Head Salmon Board. News Release, Dec. 14, 1979. NR-HQ-79-059E. 3 p.

1979d. Atlantic Salmon Report Released. News Release. May 1, 1979. NR-SJ-79-25. 3 p.

1980. Blueprint for the Future of Atlantic Salmon. A Discussion Paper prepared by Department of Fisheries and Oceans. 13 p.

1983a. A Report to the Assistant Deputy Minister, Atlantic Fisheries Service from the Atlantic Salmon Task Group. Definition of the problems and exploration of options. 51 pages + 5 Appendices.

1983b. 1983 Salmon Fishing Plan for Atlantic Region Announced. News Release. May 5, 1983. NR-HQ-083-025E. 3 p.

1984a. Long-Term Benefits for Newfoundland in 1984 Salmon Plan. News Release. April 6, 1984. NR-HQ-84-021E. 5 p.

1984b. Salmon Licence Buy-Back Program for Newfoundland Announced. News Release April 13, 1984. NR-HQ-84-026E. 2 p.

1984c. Modifications to 1984 Atlantic Salmon Plan for Newfoundland and Labrador. News Release, May 31, 1984. NR-HQ-84-042E. 3 p.

1986. The 1986 Atlantic Salmon Management Plan — Recreational and Commercial Fisheries. News Release, April 4, 1986. NR-HQ-86-18E. 3 p.

1989a. 1989 Atlantic Salmon Management plan announced. News Release, May 5, 1989. NR-HQ-89-015E. 3 p.

1989b. Salmon on the Exploits: An Enhancement Success Story.Brochure, Department of Fisheries and Oceans, Newfoundland Region.

1990a. Minister Announces 1990 Atlantic Salmon Management Plan. News Release, June 4, 1990. NR-HQ-90-15E. 4 p.

1990b. Detected Salmon Fishing Violations in the Newfoundland Region Increase by over 70% in 1990. News Release, July 27, 1990. NR-N-90-67. 2 p.

1991a. Atlantic Salmon Management Plan, 1991. Fisheries and Oceans, Ottawa. 31p.

1991b. Minister Crosbie Addresses New Brunswick Inland Fisheries Concerns. News Release, July 19, 1991. NR-HQ-91-042E. 2 p.

1992a. Notes for Remarks by the Honourable John C. Crosbie, Minister of Fisheries and Oceans, to a Media Conference to Announce the Aboriginal Fisheries Strategy, Ottawa-Vancouver, June 29, 1992. 7 p.

1992b. Canada and Newfoundland offer $40 million to Retire Commercial Salmon Fishery Licenses. News Release, March 6, 1992. NR-N-92-18E. 2 p.

DOUBLEDAY, W.G. 1988. Science needed for future management of Atlantic salmon, p. 47-53. *In* R.H. Stroud [ed.] Present and Future Atlantic Salmon Management, Measuring Progress Toward International Cooperation. Mar. Rec. Fish. 12. Atlantic Salmon Federation and National Coalition for Marine Conservation, Inc. 210 p.

DUNFIELD, R.W. 1985. The Atlantic salmon in the history of North America. Can. Spec. Publ. Fish Aquat. Sci. 80: 181 p.

EIRIKSSON, G. 1988. NASCO: Current international arrangements, p. 183–188. *In* R. H. Stroud [ed.] Present and Future Atlantic Salmon Management, Measuring Progress Toward International Cooperation. Mar. Rec. Fish. 12. Atlantic Salmon Federation and National Coalition for Marine Conservation, Inc. 210 p.

ELSON, P.F. 1957. Number of salmon needed to maintain stocks. Can. Fish. Cult. 21: 19–23.

1975. Atlantic salmon rivers, smolt production and optimal spawning: an overview of natural production, p. 96–119. *In* J. R. Bohne and L. Sochasky [ed.] New England Atlantic Salmon Restoration Conference. Boston, Massachusets, January 14–16, 1975. Int. Atl. Salmon Found. Spec. Publ. Ser. No. 6: 210 p.

ELSON, P.F., AND C.J. KERSWILL. 1964. Forest spraying and Atlantic salmon. Atl. Salmon J. (October) No. 3: 20–29.

GRENFELL, W.T. 1909. Labrador, the country and the people. Macmillan, New York, NY. 497 p.

GRUENEFELD, G. 1988. The Atlantic Salmon. Atl. Salmon Fed. Spec. Publ. Ser. No. 15: 30 p.

HATTON, J., AND M. HARVEY. 1883. Newfoundland: its history, its present condition, and its prospects for the future. Doyle and Whittle, Boston, MA. 431 p.

HAZELL, S.D. 1988. The Indian Atlantic salmon fishery on the Restigouche River: illegal fishery or aboriginal right? p.535–556. *In* D. Mills and D. Piggens [ed.] Atlantic Salmon: Planning for the Future. Proceedings of the Third International Atlantic Salmon Symposium — held in Biarritz, France, 21–23 October, 1986. Croom Helm. 587 p.

HORSTED, S.A. 1988. Past Atlantic Salmon Research: how far we have come, p. 9–18. *In* R. H. Stroud [ed.] Present and Future Atlantic Salmon Management, Measuring Progress Toward International Cooperation. Mar. Rec. Fish. 12. Atlantic Salmon Federation and National Coalition for Marine Conservation, Inc. 210 p.

HUNTSMAN, A.G. 1931. The Maritime Salmon of Canada. Bull. Biol. Board Can. 21: 99 p.

KERSWILL, C.J. 1971. Relative rates of utilization by commercial and sport fisheries of Atlantic salmon (*Salmo salar*) from the Miramichi River, New Brunswick. J. Fish Res. Board Can. 28: 351–363.

LEAR, W.H. 1976a. Migrating Atlantic Salmon (*Salmo salar*) caught by otter trawl on the Newfoundland Continental Shelf. J. Fish. Res. Board. Can. 33: 1202–1205.

1976b. Anadromous fish (Atlantic salmon, pink salmon, trout), p. 34–37. *In* A.T. Pinhorn [ed.] Living marine resources of Newfoundland and Labrador: Status and Potential. Bull. Fish. Res. Board Can. 194: 64 p.

LEAR, W.H., AND A.W. MAY. 1972. Size and age composition of the Newfoundland and Labrador commercial salmon catch. Fish Res. Board Can. Tech. Rep. 353: 70 p.

LEAR, W.H., AND R.K. MISRA. 1978. Clinal variation in scale characters of Atlantic salmon (*Salmo salar*) based on discriminant function analysis. J. Fish. Res. Board Can. 35: 43–47.

MARSHALL, T.L. 1988. Harvest and recent management of Atlantic salmon in Canada, p. 117–142. *In* D. Mills and D. Piggens [ed.] Atlantic Salmon: Planning for the Future. Proceedings of the Third International Atlantic Salmon Symposium — held in Biarritz, France, 21–23 October, 1986. Croom Helm. 587 p.

MAY, A.W. 1973. Distributions and migrations of salmon in the Northwest Atlantic. Spec. Publ. Ser., Int. Atl. Salm. Found. 4: 373–382.

1992. A review of management and allocation of the Atlantic salmon resource in Atlantic Canada. Paper presented to the Fourth International Atlantic Salmon Symposium, St. Andrews, N.B. Canada. June 14–17, 1992.

MAY, A.W., AND W.H. LEAR. 1971. Digest of Canadian Atlantic Salmon Catch Statistics. Fish. Res. Board Can. Tech. Rep. 270: 106 p.

MORIN, R. 1991. Atlantic salmon (*Salmo salar*) in the lower Nastapoka River, Quebec: distribution and origins of salmon in eastern Hudson Bay. Can. J. Zool. 69: 1674–1681.

MUIR, B.S., AND N.A. BELLEFONTAINE. 1988. Canadian Atlantic Salmon Management Program, p. 123–129. *In* R.H. Stroud [ed.] Present and Future Atlantic Salmon Management, Measuring Progress Toward International Cooperation. Mar. Rec. Fish. 12. Atlantic Salmon Federation and National Coalition for Marine Conservation Inc. 210 p.

NEEDLER, A.W.H. 1950. Atlantic Salmon Fishing Regulations. Review prepared for Sub-Committee on Regulation of the Co-ordinating Committee on Atlantic Salmon. 8 p.

NETBOY, A. 1968. The Atlantic Salmon. A Vanishing Species? Faber and Faber, London. 457 p.

O'CONNELL, M.F., J.B. DEMPSON, E.G.M. ASH, AND N.M. COCHRANE. 1991. Status of Atlantic salmon (*Salmo salar* L.) Stocks of the Newfoundland Region, 1990. CAFSAC Res. Doc. 91/16: 52 p.

O'CONNELL, M.F., J.B. DEMPSON, AND D.G. REDDIN. 1992. Evaluation of the impacts of major management changes in the Atlantic Salmon (*Salmo salar* L.) fisheries of Newfoundland and Labrador, Canada, 1984–1988. ICES J. Mar. Sci., 49: 69–87.

PARRISH, B.B. 1973. A Review of the Work of the ICES/ICNAF Joint Working Party on North Atlantic Salmon. p. 383–396. *In* M.W. Smith and W.M. Carter [ed.] International Atlantic Salmon Symposium. Proceedings of the International Symposium on the Atlantic Salmon: Management, Biology and Survival of the Species. Int. Atl. Salmon Found. Spec. Publ. Ser. 4 (1): 504 p.

PARRISH, B.B., AND SV.AA. HORSTED [ed.]. 1980. ICES/ICNAF Joint Investigations on North Atlantic Salmon. Rapp. p.-v. Réun. Cons. int. Explor. Mer. 176. 146 p.

PERLEY, M.H. 1852. The Sea and River Fisheries of New Brunswick. Report of Her Majesty's Emigration Officer at Saint John, New Brunswick. Fredericton. 34 p.

PIPPY, J. [Chairman] 1982. Report of the Working Group on the Interception of Mainland Salmon in Newfoundland. Can. Manuscr. Rep. Fish Aquat. Sci. 1654: x + 196 p.

PORTER, T.R., AND D.J. MEERBURG. 1977. Upwelling incubation boxes for Atlantic salmon, *Salmo salar*. ICES C.M. 1977/M:22, 13 p.

POWER, G. 1969. The salmon of Ungava Bay. Arctic Institute of North America, No. 22: 7–72.

RANDALL, R.G. 1990. Effect of the 1984–1988 management plan on harvest and spawning levels of Atlantic salmon in the Restigouche and Miramichi rivers, New Brunswick. CAFSAC Res. Doc. 90/45: 26 p.

REDDIN, D.G. 1978. Report on the salmon by-catch in miscellaneous marine fishing gear, 1976, Newfoundland and Labrador. Fish. Mar. Serv. Manuscr. Rep. 1461: 23 p.

1985. Atlantic salmon (*Salmo salar*) on and East of the Grand Bank. J. Northw. Atl. Fish. Sci. 6: 157–164.

REDDIN, D.G., AND P.B. SHORT. 1991. Postsmolt Atlantic salmon (*Salmo salar*) in the Labrador Sea. Can. J. Fish. Aquat. Sci. 48: 2–6.

RITTER, J.A., AND T.G. CAREY. 1980. Salmon Ranching in the Atlantic Maritime Provinces of Canada, p. 109–130. *In* J.G. Thorpe [ed.] Salmon Ranching. Academic Press. London. 441 p.

RITTER, J.A., T.R. PORTER, Y. COTÉ, AND B. TETREAULT. 1980. Issues and promises for Atlantic salmon management in Canada, p. 108–127. *In* A.E.J. Went [ed.] Atlantic Salmon: Its Future. Fishing News Books. 249p.

SAUNDERS, R.L. 1977. Sea Ranching: A Promising way to enhance populations of Atlantic Salmon for angling and commercial fisheries. Int. Atl. Salmon Found. Spec. Publ. Ser. 7: 17–24.

SCOTT, W.B., AND E.J. CROSSMAN. 1973. Freshwater fishes of Canada. Bull. Fish. Res. Board. Can. 184: 966 p.

SMITH, K.E.H. 1988. The Atlantic Salmon. Underwater World. Fisheries and Oceans. 8 p.

SMITH, M.W., AND W.E. CARTER [ed.]. 1973. International Atlantic Salmon Symposium. Proceedings of the International Symposium on the Atlantic Salmon: Management, Biology and Survival of the Species. Int. Atl. Salmon Found. Spec. Publ. Ser. 4(1): 504 p.

TAYLOR, V.R. 1985 The early Atlantic salmon fishery in Newfoundland and Labrador. Can. Spec. Publ. Fish. Aquat. Sci. 76: 71 p.

TAYLOR, V.R., AND B.R. BAULD. 1973. A program for increased Atlantic Salmon (*Salmo salar*) production on a major Newfoundland river, p. 339–347. *In*: M.W. Smith and W.M. Carter [ed.]. International Atlantic Symposium, Huntsman Marine Laboratory, St. Andrews, New Brunswick, Canada, September 20–22, 1972. Int. Atl. Salmon Found. Spec. Publ. Ser. 4(1): 504 p.

TEMPLEMAN, W. 1968. Distribution and characteristics of Atlantic salmon over oceanic depths and on the bank and shelf slope area off Newfoundland, March–May, 1966. ICNAF Res. Bull. 5: 62–85.

TUOMI, A.L.W. 1980. Canadian Atlantic Salmon: An Economic Evaluation of the New Brunswick Salmon Sport Fishery. Atl. Salmon Fed. Spec. Publ. Ser. No. 8: 35 p.

1987. Canada's Atlantic Salmon Recreational Fisheries and their Future: An Economic overview. Atl. Salmon Fed. Spec. Publ. Ser. No. 14: 29 p.

VEZINA, B.P. 1988. Historical Overview of Recent Atlantic Salmon Fisheries. Resource Allocation Branch, Atlantic Fisheries, Department of Fisheries and Oceans. 57 p.

WOODING, F.H. 1956. Canada's Atlantic Salmon. Queen's Printer, Ottawa, Ont. 23 p.

# A Retrospective Analysis of the Gulf of St. Lawrence Snow Crab (*Chionoecetes opilio*) Fishery 1965–1990

Gerard M. Hare

*Hare Fisheries & Environmental Consultants Inc.*
*Box 393, Newcastle, New Brunswick, Canada E1V 3M5*

and David L. Dunn

*Department of Fisheries and Oceans, Gulf Region*
*Box 5030, Moncton, New Brunswick, Canada, EIC 9B6*

Hare, G. M., and D. L. Dunn. 1993. A retrospective analysis of the Gulf of St. Lawrence snow crab (*Chionoecetes opilio*) fishery 1965–1990, p. 177–192. *In* L. S. Parsons and W. H. Lear [eds.] Perspectives on Canadian marine fisheries management. Can. Bull. Fish. Aquat. Sci. 226.

The snow crab (*Chionoecetes opilio*) fishery in the Gulf of St. Lawrence began in 1965 with exploratory surveys, eventually developed into an industry generating $150 million annually by 1988, and then appeared on the threshold of total collapse. Management of the fishery evolved through consultation between government and industry where concerns of economic viability and employment fuelled by increasing prices masked insufficient knowledge of snow crab biology. The fishery developed, expanded and declined in the absence of a clear appreciation for the impact of fishing pressures and the absence of effective monitoring and control over the fishing fleet and processing plants. The crisis of the late 1980's brought demands for action from all sectors of the industry for new management approaches to counter a recurring soft-shelled (white) crab problem. A retrospective analysis demonstrated that by this time consensus emerged in the scientific community that the fishery was fully exploited and dependent upon annual recruitment.

---

La pêche du crabe des neiges (*Chionoecetes opilio*) dans le golfe du Saint-Laurent a débuté en 1965 sous la forme de pêches exploratoires. Elle s'est ensuite développée pour devenir une industrie dont la production annuelle a atteint 150 millions de dollars en 1988 avant d'être menacée d'effrondrement total. La gestion de cette pêche s'est faite dans le cadre de consultations gouvernement-industrie et les préoccupations relatives à la rentabilité et à l'emploi, alimentées par des prix à la hausse, ont masqué le trop peu de connaissances que l'on avait de la biologie de ce crabe. La pêche s'est développée, a pris de l'expansion et a décliné en l'absence d'une bonne connaissance des effets de la pression de pêche ainsi que d'une surveillance et d'un contrôle efficaces de la flottille de pêche et des usines de transformation. La crise de la fin des années 1980 a amené tous les secteurs de l'industrie à exiger la prise de mesures de gestion afin de lutter contre la capture, de façon répétée, de crabes à carapace molle (crabes blans). Une analyse rétrospective a permis de montrer que, à cette époque, la plupart des scientifiques étaient d'avis que cette ressource faisait l'objet d'une exploitation maximale et que la pêche était dépendante du recrutement annuel.

## Introduction

The snow crab (*Chionoecetes opilio*) fishery in the Gulf of St. Lawrence had its inception in 1965 when the Government of Canada and the Province of Nova Scotia initiated exploratory surveys to determine the potential for a crab fishery. An industry developed which generated $150 million annually by 1988 and then appeared on the threshold of total collapse in 1989.

Snow crab fisheries in eastern Canada extend along the east coast of Newfoundland and Labrador, through the estuary and Gulf of St. Lawrence, and around Cape Breton Island (Fig. 1). The various snow crab fishing grounds are partitioned into management units based upon the location of fishing activity. In the southwestern Gulf of St. Lawrence, these management units represent the principal midshore fishery (Area 12), and the inshore fisheries of Cape Breton (Areas 18 and 19) and Prince Edward Island (Areas 25 and 26) (Fig. 2).

Development of the fishery was characterized by political strife and social unrest. Fishery management strategies identified harvesting approaches to meet the needs of industry while also attempting to conserve crab stocks. The industry was influenced by the collapse of Alaskan crab fisheries and the resulting potential demand for crab meat from the Gulf of St. Lawrence. The value of landings increased tenfold as the fishery developed. Catches were initially projected to stabilize at approximately 5,000 tons, but in fact exceeded 30,000 tons in 1982. Subsequently, catches declined and the stocks were thought to have collapsed, probably as a result of harvest levels greater than the stock could stand on a continuing basis. Throughout the 1980's, the fishery was increasingly managed to achieve socio-economic goals. Management goals in the midshore fishery generated harvesting strategies to ensure unemployment benefits to more than 8,000 plant workers. In areas where snow crab grounds approached coastal waters, inshore fishermen were provided exclusive access to productive grounds.

The purpose of this review is to describe the development and decline of the fishery from the late 1960's through 1989. It also focuses on the mitigating factors which influenced the fisheries management process, and the deficiencies in the basic understanding of snow crab biology. Although the primary focus is on the southwestern Gulf of St. Lawrence midshore fishery, a review of the related inshore crab fishing areas in the Gulf is also provided with particular emphasis on the controversial Prince Edward Island fishery.

## Development of the Gulf Snow Crab Fishery

### Evolution

Snow crab landings from the Gulf of St. Lawrence were first reported from the Gaspé area between 1960 and 1966. These landings were mainly by-catches from

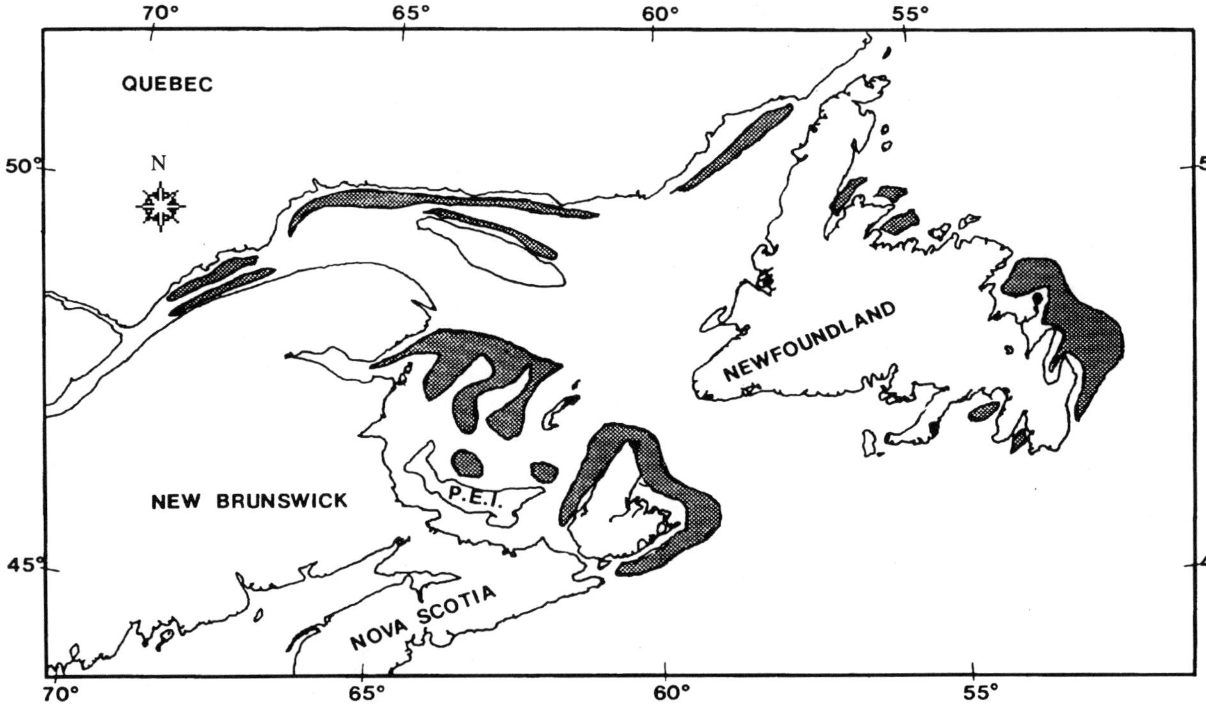


FIG. 1. Snow crab fisheries in eastern Canadian waters.

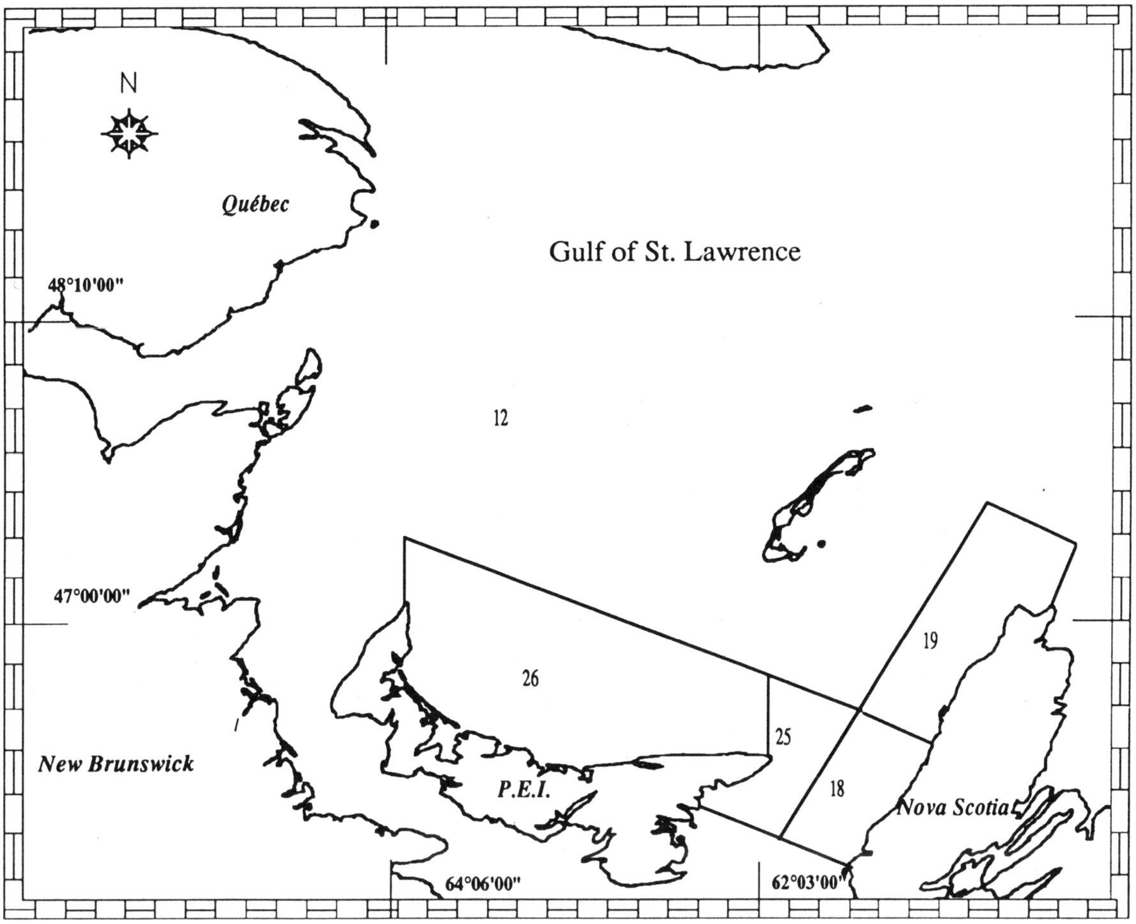


FIG. 2. Snow crab management areas in the southwestern Gulf of St. Lawrence.

groundfish draggers operating south of Anticosti Island. Fishing began to increase in 1965 as the result of a comprehensive development initiative (Deveau and McAllister 1967).

Reacting to reports of snow crab catches in Danish seines, and in light of successful developments in Japan and Alaska, a series of federal and provincial development programs were implemented over the 1965–68 period in the Maritime Provinces and Quebec. These studies focused on all aspects of fishing and processing to determine the distribution of snow crab, minimum acceptable crab size, appropriate fishing gear, vessel holding and off-loading methods, processing procedures, specialized equipment, and market demand.

In 1965 a Danish seine fishery was started for snow crab off Cheticamp, N.S., which landed 15,000 lb in the first year. The program expanded into New Brunswick and Prince Edward Island in 1966, and Quebec in 1967. By 1968 about 60 boats participated in the fishery (15 for the entire season) landing 5,000 tons to nine processing plants. Fishing effort was concentrated in two areas — off the Gaspé and the north coast of Cape Breton. Combined landings by 1969 were 7,600 tons with low prices to fishermen (i.e. 15–20 cents/kg) being a serious impediment to further development.

The results of the preliminary work on snow crab were outlined in a seminar held in March, 1969 to provide direction on the development initiative. That seminar concluded the program was a success and a fishery could be sustained. In the Chairperson's closing remarks, Dr. A.W.H. Needler expressed "reserved optimism" about expansion cautioning that this would be associated with going farther afield and declining catch per unit effort (Department of Fisheries and Forestry 1969). More important for the immediate purposes of development, however, were conclusions suggesting the crab fishery on its own could not justify new vessel construction. It did not therefore seem an attractive opportunity to many fishermen.

In Prince Edward Island, low groundfish catch rates and prices had prohibited the needed replacement of middle distance groundfish boats (45–90 feet length overall) and this fleet was in the process of disappearing from the fishery. A snow crab fishery did not offer a practical solution to this problem. In Nova Scotia, the Danish seine fleet was suitably equipped to access crab fishing grounds, but crab did not offer a competitive alternative to their seining operations. However, in New Brunswick and Quebec many boats were both relatively new and maintained relatively low capital outlays, and the new crab fishery appeared to suitably complement existing groundfish operations.

## Fishery Profile

The southern Gulf of St. Lawrence (Fig. 2) snow crab landings increased considerably from the beginning of what has become known as the "midshore" fishery in the 1960's until an abrupt decline in 1987 (Fig. 3). In 1970, some 50 boats reported catches of approximately 5,600 tons of snow crab for an approximate value of $1.2 million. By 1979, 118 boats landed about 15,700 tons with a value of almost $11 million. In 1989, the fleet had increased to 130 boats, but catches declined to 7,900 tons — a sharp decrease from 1982 when 30,500 tons were landed. However, in 1988, the landed value of these catches was approximately $62.1 million, offsetting the economic impact of a drop in landings to 12,500 tons. Unfortunately, by 1989 markets also began to weaken and prices fell about 40% to provide a landed value for the year of only $23 million (DFO, Gulf Overview, July, 1989). In 1990 catches declined to about 7,000 tons but increased to 12,695 tons in 1991 and to 15,030 tons in 1992, reflecting a modest recovery of the stock.

The inshore snow crab fishery was first introduced in 1965 off the coast of Cape Breton (Area 19 — Fig. 2) where landings gradually increased to 516 tons by 1977 (Fig. 4). In 1978, the amount of snow crab landed by Cape Breton fishermen was 1,941 tons for a total value of $6.8 million. The snow crab grounds off the southern coast of the Inverness County Management Zone (Area 18) were explored in 1979 resulting in landings of 213 tons. In 1988, 2,006 tons of crab were landed from both areas for a value of $6.85 million. About two thirds of the landings were from Area 19 (Fig. 4).

The snow crab fishery off the Prince Edward Island coast started in 1967 but, by 1969, the fishery was discontinued due to lack of interest by fishermen because of poor market conditions and low prices. In 1985, an exploratory inshore fishery was introduced with

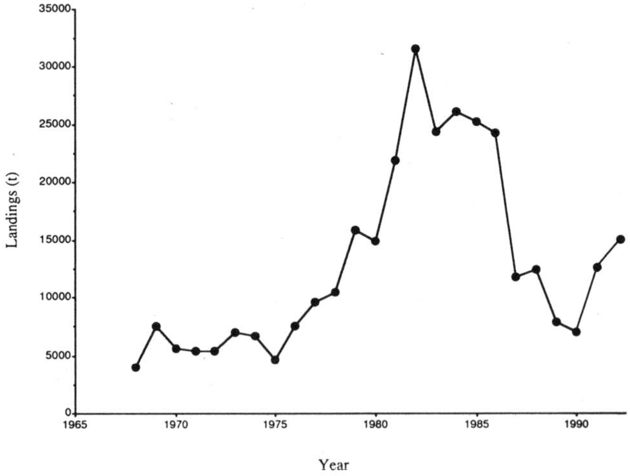


FIG. 3. Historical landings of snow crab in the southwestern Gulf of St. Lawrence fishery from 1968 to 1990.

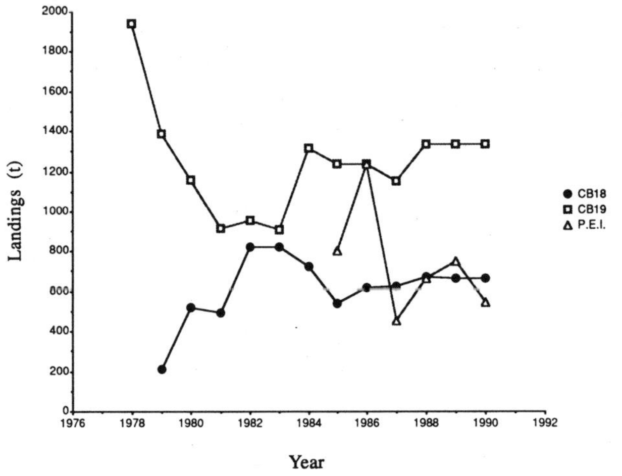


FIG. 4. Historical landings of snow crab in inshore fisheries areas of southwestern Gulf of St. Lawrence.

16 permits available. This resulted in landings of 891 tons for a value of $2.3 million. In 1986, another 14 permits were added and inshore snow crab landings increased to 1,200 tons for a value of $2.1 million. By 1988, however, landings of snow crab had decreased to 750 tons (Fig. 4).

In 1989, 130 Gulf of St. Lawrence fishermen held midshore snow crab licences. Eighty-one of these licences were held by New Brunswick fishermen while two were held by Nova Scotia fishermen, and 47 were held by Quebec fishermen. Their vessels ranged from 50 to 99 feet, with the large majority being in the 60 to 64 foot category. In 1989, the number of Gulf inshore snow crab licences and exploratory permits amounted to 260: 86 in Cape Breton, 30 in Prince Edward Island, 18 in western Newfoundland and 126 in Quebec.

The inshore fisheries along the north shore of the Gulf reported catches in excess of 1,000 tons in 1969, but fishing effort was not sustained in the 1970's (probably due to an absence of market and low prices). By the late 1970's catches were again reported leading to the eventual identification of five inshore management zones along the Strait of Belle Isle, the Lower North Shore, and the estuary of the St. Lawrence. In 1984, catches reached 6,700 tons with 175 participants; however by 1988, landings decreased to 3,200 tons, raising concern about the future of the inshore crab fishery in the Northern Gulf.

In the late 1980's the fish processing sector was severely impacted by the effects of a reduced supply of snow crab. A 75% reduction in the catches of the New Brunswick-based crab fleet between 1982 and 1989 had a major impact on employment, as had other factors such as changes in product mix and mechanization. Employment in the 31 federally registered crab processing plants in New Brunswick declined from a peak of 7,000 employees in the early 1980's to 1,800 in 1989. In Quebec, there were nine crab processing establishments in 1989 employing 800 people, down from 1,200 employees the previous year; five of these plants were located in the Gaspé and four on the Magdalen Islands. Nova Scotia had one plant which employed about 30 people.

## Biological Characteristics of the Snow Crab

Our understanding of snow crab biology has increased significantly since the mid-1980's coincident with demands for scientific advice on the management of other stocks in Atlantic Canada.

### Distribution and Habitat

Snow crab are found in deep cold water in the Northwest Atlantic and North Pacific Oceans, and in the Sea of Japan. In eastern North America, snow crab are distributed from West Greenland to Maine in waters ranging from 20 to 700 m depth. They are common in the Gulf of St. Lawrence, around Cape Breton Island, and the bays of Newfoundland from Fortune Bay to White Bay, and northward to the Hamilton Bank off Labrador (DFO 1981). Large fisheries for this species occur in the Bering Sea (Otto 1982) and in the Sea of Japan (Sinoda 1982). In the Pacific, where the genus "*Chionoecetes*" is represented by more than one species, they occur from the Sea of Japan to Alaska, British Columbia, and the states of Washington and Oregon. Snow crab are absent from the northeast Atlantic.

Snow crab live most commonly on muddy or sand-mud bottoms at temperatures ranging from −1.5 to 4.5°C. In the Gulf of St. Lawrence, they are usually found at depths of 70 to 140 m, while off Cape Breton the depth varies from 45 to 245 m, and from 170 to 380 m off Newfoundland (DFO 1981).

Snow crab in the Gulf of St. Lawrence do not undertake extensive movements. Tagging experiments (Watson 1970; Watson and Wells 1972) did not indicate any clear migration patterns by male snow crab. Most of them (80–90%) were caught within 25 km from the point of release.

### Life History

Two types of mating patterns have been observed for snow crab: (1) hard shelled terminal moult males with terminal moult females that have previously spawned and display old grasping marks (so-called multiparous females); and (2) males with soft shelled females which are spawning for the first time (so-called primiparous females). The first type of mating occurs from April to May (Taylor et al. 1985; Conan and Comeau 1986; Hooper 1986; Moriyasu et al. 1987) as soon as females hatch their previous batch of eggs. The second type of mating occurs in early February [viewed under simulated natural conditions in the laboratory (Moriyasu et al. 1987)]. The presence of thick pack ice in winter prohibits observations of this type of mating in nature. Soon after mating, sperm is deposited into the openings of the female's spermatheca located on her underside. Depending on the female's size, she then lays 20,000 to 150,000 eggs over a period of a few days. A female can lay and fertilize more than one batch of eggs without further mating since she can store spermatophores for long periods. The storage of spermatophores is considered as a safeguard mechanism against unsuccessful mating (Conan and Comeau 1986) rather than an adaptation to a single mating in their life span (Watson 1970; Watson and Wells 1972; Kon 1980). The female deposits eggs on pleopods (found on her underside) where they are carried until the following spawning season. During this period, the eggs change colour from bright orange to dark purple or black as they develop (DFO 1981). In the southern Gulf of St. Lawrence, hatching starts in May and is generally completed by the end of June (Moriyasu et al. 1987).

Newly hatched crabs (prezoae) are about 3 mm in length. They immediately rise to the surface where they are carried by the currents for several months before they settle back on the bottom. During this period, they go through three different larval stages before adopting the regular shape of crabs (DFO 1981).

Snow crab, like other crustaceans, must moult to grow. Moulting (ecdysis or shedding) is the process by which an animal gets rid of an old shell in order to grow in size. Between moults, the crab builds more organic tissues and prepares a new shell under the old one. The shell splits at the back and the crab moults by backing out of the old shell, a process which can take up to 10 hours for large crabs. The emerging wrinkled soft crab absorbs water and swells to its new size in a few hours, and the soft shell, muscle and tissue gradually harden. Based upon aquarium observations, snow crab grow about 15 to 25% in width at each moult depending upon their sex, size and maturity (Moriyasu et al. 1987).

A suspected peculiarity of the growth pattern of male crabs in the family Majidae, to which the genus *Chionoecetes* belongs, is the existence of a terminal moult in many species. Conan and Comeau (1986) suggested the existence of a terminal moult for both male and female snow crab. The terminal moult is generally assumed to coincide with the onset of morphometric maturity such as the shape of the claws in males and the abdomen in females. The moult to morphometric maturity can occur over a wide range of carapace sizes in males, eg., 45 to 140 mm in carapace width (Conan et al. 1988). The factors influencing morphometric maturity are unknown. Although terminal moult in male snow crab remains unproven (Dawe et al. 1991), the hypothesis has been accepted by consensus (Jamieson and McKone 1988).

In the southwestern Gulf of St. Lawrence, the gonads of most male snow crab are mature at a shell width of 40 mm (Comeau 1987) and they can mate successfully with soft shell females (Moriyasu and Conan 1988). Most of the small clawed males can participate in mating with primiparous females in early spring but are unable to mate with multiparous females because these males moult before the mating season of multiparous females and their carapace is still soft. Multiparous females mate only with hard shelled morphometrically mature males (Conan et al. 1988).

### Scientific Basis for Fisheries Management

Snow crab in Atlantic Canada have been managed under a strategy directed toward harvesting of large males. Harvesting of females and crab less than 95 mm carapace size has been prohibited on the assumption that snow crab stocks were protected from recruitment overfishing because females were not harvested, and males were thought to mature and mate before they reached the minimum legal size. This strategy was questioned in the late 1980's when Conan et al. (1988) observed males above the minimum legal size to be morphometrically immature (meaning that in their view some males may be harvested before they can reproduce) and some mature males below the minimum size reached a terminal moult (implying that they might never become harvestable size but could reproduce). Based upon these observations, the Canadian Atlantic Fisheries Advisory Committee (CAFSAC) expressed concern that the current management strategy may not protect entirely the reproductive capacity of the stocks (CAFSAC 1989). CAFSAC also noted that most snow crab stocks are fully exploited and largely dependent upon new recruitment each year, and that catches or stock sizes cannot be predicted because growth is not well understood and recruitment cannot be forecast.

While definitive evidence of terminal molt remains outstanding (Dawe et al. 1991), acceptance of the concept (Jamieson and McKone 1988) suggested that the reproductive potential of the southwestern Gulf of St. Lawrence stock may not have been protected by application of the minimum legal size of 95 mm CW. Most morphometrically mature males less than 95 mm CW were thought to be unable to mate with multiparous females, the most important contributors to stock reproduction, whereas most morphometrically mature males over 95 mm CW were heavily exploited as soon as they moulted to legal size, and therefore did not have the opportunity to mate before being caught (Conan et al. 1989). In the early years, the fishery had been harvesting large male crab accumulated on the fishing grounds over many years. Consequently, the size-frequency distribution of the landings had remained relatively constant whereas the catch per unit of effort and total catch had abruptly declined once all fishing grounds were covered by the fleet in the late 1980's.

Prior to 1988, CAFSAC provided advice to fisheries management on an annual quota level for southwestern Gulf snow crab despite acknowledged deficiencies in the methodology (calculations referred to as Leslie Analysis) used to estimate levels of initial biomass and exploitation. In 1988, CAFSAC emphasized the uncertainties inherent in the calculations because a major assumption of the Leslie Analysis was being invalidated by the impact of the searching activity of the fleet on catch rates (CAFSAC 1988). In the absence of a more reliable methodology for assessing snow crab stocks in 1988 and 1989, a series of Leslie Analyses were completed separately for component parts of the total fishing area to minimize the effect of searching by the fleet on catch rates (Mallet et al. 1988; Mallet et al. 1989). These analyses showed that the initial biomasses from five subareas compared favourably, thus implying similar exploitation rates.

Post-fishery bottom (Nephrops) trawl surveys of snow crab in the southern Gulf, developed in 1987 and 1988, provide a basis for estimating annual stock biomass and incoming recruitment. These surveys also identify the distribution and biomass of pre-recruits which provide an annual index of recruitment. This procedure has been applied in management of the fishery in the early 1990's.

## Fishery Management

### Management Authority

The "Constitution Act" (1982) of Canada delineates federal and provincial jurisdictional responsibilities, as initially outlined in the British North America Act (1867). Responsibility for the conservation, protection, and management of fishery resources and fish habitat throughout all of Canada and its Territorial waters has been assigned to the federal government. The "Government Organization Act" through amendment (1979) assigns this federal responsibility to the Minister of Fisheries and Oceans. The provinces have constitutional responsibility for matters relating to processing and trade within the Province, while the federal government is assigned responsibility for international or inter-provincial trade, including such matters as fish product safety. Both federal and provincial governments emphasize economic development within both policy and programming frameworks. To avoid the potential for conflict, the "Government Organization Act" also empowers the federal Minister to enter into agreement with provinces on matters relating to administration of fisheries.

In the context of the Gulf snow crab fishery, where independent federal and provincial restrictions have applied to fishing and processing respectively, social and economic objectives established for the industry often appear to cloud jurisdictional lines. This was further complicated by the assignment of the federal administrative authority for fisheries to the Province of Quebec under the 1922 Canada-Quebec Agreement and the subsequent (1984) resumption of management of the marine fisheries adjacent to Quebec by the federal government.

In essence, the federal government establishes and enforces fishery management strategies as well as guidelines pertaining to the processing of crab for the purpose of product quality and consumer safety. The provinces may regulate the number of processing facilities, the export of production out of the Province, and related factors such as the type of product pack produced for commercial sale.

### Regulatory Framework

A need for effective communication to enhance the progress of the development initiative was recognized by proponents of the various programs early in the development period. An Atlantic Crab Association was formed in 1966 to promote the development of the crab industry in an orderly fashion and to provide the best quality product available. Initial participants included the Federal Department of Fisheries, the Provincial fisheries departments in New Brunswick, Nova Scotia and Prince Edward Island, the processors involved in the development studies, and a marketing organization. The Association had expanded by 1973 to include 13 industry members. This Association was instrumental in providing the necessary co-ordination for effective and rapid development during the early part of this new fishery.

The Atlantic Crab Association served the industry well over the 1968–73 period by providing a forum for information exchange on development of this new fishery. Fishermen had met the challenge issued at a 1969 Crab seminar that expansion was contingent upon exploiting new grounds. The anticipated declining catch per unit effort was more than offset by increased prices and additional traps per boat. This effort was encouraged by low prices and oscillating catch rates, and a price stabilization program in 1970. At about the same time, a survey indicated demand in the industry for a prohibition of otter trawling for crab, a minimum trap mesh size and crab carapace size of 4" (102 mm), and a prohibition against processing dead crab. As well, biologists identified the potential impact of "ghost fishing" by lost traps.

The number of boats participating in the fishery increased significantly to about 120 in 1974 but there was a decline in catch per trap from 1,846 kg annually in 1971 to 556 kg in 1974. The contributing factors to this increased fishing effort were declining groundfish catches and prices, and a snow crab price increase from \$0.18/lb in 1968–71 to \$0.44/lb in 1973 or \$45,000 in average gross revenue per vessel (DFO 1980).

Although stock assessments were not available before 1977, by 1974 all reference indicators pointed to overexploitation. The following selected New Brunswick data were presented in 1975 to demonstrate this point:

| | 1971 | 1972 | 1973 | 1974 |
|---|---|---|---|---|
| Boats | 38 | 42 | 61 | 85 |
| Traps | 2,550 | 3,790 | 6,440 | 10,245 |
| Catch (lb)/Trap | 4,071 | 2,673 | 2,166 | 1,225 |
| Traps/Boat | 67 | 90 | 106 | 121 |
| Price (cents/lb) | 18 | 31 | 44 | 40 |

(Source: Minutes, Government-Industry Meeting, July 5, 1975).

While there was a recognition of the need for control as the fishery developed and the catch per trap decreased, the fishery proceeded with little regulation. In 1973, a government-industry meeting in Caraquet, N.B. reviewed the evidence of both biological and economic research. No consensus on management approach was achieved thereby leaving the federal government free to unilaterally control effort in the crab fishery.

Regulations were proposed setting a maximum permissible number of crab traps per fisherman (80), definitions for crab traps and mesh size, prohibition on soft shelled crab retention, minimum carapace size, fishing seasons, prohibitions on retention of female crab and a limited entry licensing policy. The regulations were initially intended to be introduced in June, 1974, but due to widely differing public opinion, they were not promulgated until August 13, 1975 (Canada Gazette 1975). Their enactment provided the management authority now found in the Atlantic Fisheries Regulations.

On May 24, 1974, the Gulf Snow Crab Management Advisory Committee was established to advise the Federal Minister on all licensing matters in order to match the fleet size to the crab stocks. At that inaugural meeting, the Committee learned that the entire snow crab research program in Atlantic Canada consisted of one biologist conducting research on Newfoundland's east coast, and that scientists believed the Gulf stock of snow crab could sustain an average yield of around 5,000 tons. In examining proposals for regulatory control, the following considerations were tabled by government for review:

- *Carapace Size Limits* — difficult to enforce and to determine due to characteristics of mating and moulting; mesh size can facilitate escape of smaller crabs.
- *Female Prohibition* — would provide more effective control and benefit than a size limit.
- *Soft-shell* — difficult to control due to problem of definition.
- *Seasons* — related to soft-shell and therefore dependent upon information from fishermen.
- *Ghost Fishing* — traps should be constructed from a biodegradable material.
- *Quotas* — insufficient information to establish appropriate quota levels.

The committee concluded that a limited entry licensing policy should be established, but it could not reach a consensus on criteria for eligibility. The Province of Prince Edward Island supported the concept of limited entry based on past participation, but requested an exemption from the policy to enable future access by Island fishermen since none were participating at that time in the snow crab fishery. This caveat would have significant implications in subsequent years.

In a meeting on June 30, 1974, the Gulf Snow Crab Advisory Committee focused on the question of regulation. It concluded that restrictions on the capture of soft-shell crab should be maintained through seasonal closures, that eligibility should be premised on past participation in the fishery, that incidental catches of snow crab in otter trawl or Danish seine need not be prohibited, that corrective measures were not required to prevent ghost fishing, that a 125 trap limit per boat be established, and that a 5 1/2" minimum trap mesh size be identified to enable retention of only larger crabs.

These proposals from the Gulf Snow Crab Advisory Committee fostered a management regime for the fishery supported by regulation in 1975. Participation in the fishery was curtailed through a limited entry policy which restricted access to those who could demonstrate participation in the fishery during any one year in the period 1970–74. This process resulted in 89 licenses in New Brunswick and 25 in Quebec.

**Temporary Assistance Program**

Promulgation of regulations in 1975 proved insufficient to respond to the crisis of the 1975 fishing season. With traditional grounds yielding up to 60% soft shelled crab at the beginning of the season, fishermen threatened to boycott processors as prices fell to $0.40/lb and catches also declined.

In an exploratory meeting with the Fisheries Prices Support Board (Chatham, N.B. in July 1975), a number of basic problems within the industry were identified by Gaspé and New Brunswick crab processors. First, even though the number of boats in the fishery decreased by 25% from the previous year, the crab stocks could not support the remaining 90 boat fleet. Second, in an attempt to increase efficiency, fishermen had been increasing the number of traps in the fishery. Third, increased fishing capacity was directly related to the demand for supply brought about by government aid which increased the number of processing plants in operation. Fourth, because of the absence of regulatory control, the industry was unable to stabilize catch and effort.

The most immediate problem, however, was related to soft-shell crab. There was no standard definition for "soft crab", resulting in a lack of discipline among buyers toward voluntary compliance with industry-established guidelines. A solution through regulation of processing soft shell crab seemed to be the consensus.

The extent of the crisis became clear by year's end. Overall landings had fallen by 50% in Quebec and 25% in New Brunswick. Landings from the fall fishery subsequently proved of high quality amounting to as much as 40% of the total compared to the usual 2% of annual landings. As a result of the July meeting, the federal Temporary Assistance Program was again applied to the crab fishery (the previous time being in 1970) in three forms: a deficiency payment of $350,000 to processors on 1974 carry-over production; a deficiency payment of $381,000 to fishermen to mitigate the impact of soft shell crab controls; and inventory financing of $206,000 to processors because of slow markets.

The Gulf Snow Crab Advisory Committee, at its June 6, 1976 meeting, reconsidered its approach to the soft-shell crab problem. A federal crab biologist indicated that intensive fishing pressure had resulted in smaller size crabs in catches and, as these are more prone to moult, the occurrence of soft-shell crab also increased. The experience of the 1975 spring fishery was not repeated in 1976 and, with a 25% increase in prices and a 50% increase in catch, the industry's resolve to control the soft shell crab problem quickly dissipated.

**Evolution of Management Measures**

The issues of fishing effort and quality continued to dominate discussions on the midshore fishery. By mid-1977, it became evident that many midshore crab fishermen were operating well in excess of the 125 trap limit and that specific control measures were necessary to identify traps and buoys at sea. As well, complaints were also surfacing about theft from traps, and about misrepresentations from fishermen gaining access to the crab fishery. Emphasis on licence limitation was again expressed in 1977 as catches exceeded 9,000 tons and many crab boats tied up early after receiving record prices of $0.62/kg. With the Advisory Committee having met four times throughout 1977 to deal with these issues, special information meetings were held in December to focus on the outstanding issue of trap control. Government officials pointed out that trap limits were not perceived as a conservation objective but as a method of distributing benefits to participants consistent with limited entry licensing policy. Government demonstrated concern about a repeat of the 1975 experience which resulted in $2 million in subsidies. Virtually all fishermen requested tougher enforcement, but they argued the trap limit should first be increased. This discussion continued throughout 1978 and into the fall of 1979 when the Department of Fisheries and Oceans proposed a new management policy as follows: (1) application of a 3 3/4" minimum carapace width; (2) mandatory discarding of small, soft shell, and female crab; (3) application of a 4 1/2" minimum trap mesh size; (4) marking and tagging procedures for buoys and traps; and (5) maintaining the 125 trap limit.

Over this period, provincial governments were attempting to maximize benefits from the crab fishery. In response to a perception that trap limits in New Brunswick had been exceeded, the Province of Quebec (under administrative authority assigned in the 1922 Canada-Quebec Fisheries Agreement) unilaterally issued 22 additional licences in 1978–79, increasing the number of Quebec midshore participants from 25 to 47. This decision so incensed federal regulators that it contributed directly to the decision to rescind the Canada-Quebec Agreement in 1984. In the Province of New Brunswick, the processing industry had suffered a major setback with collapse of the herring fisheries in the mid-1970's. To counter this problem, the Province of New Brunswick issued snow crab processing licences to approximately 10 plants, thus bringing the total number of processors to about 30. It was intended that crab provide the economic backbone of the regional economy of northern New Brunswick.

Concern about the impact of the abundance of soft shell crab catches continued throughout 1980 resulting in a 20% decline in prices. After the associated quality and market problems nearly shut down the fishery in mid-season, a second major Gulf Snow Crab seminar was held in December 1980. Conclusions of the seminar were ambiguous but consistent with the lack of consensus on measures respecting any of the four themes: the resource, harvesting, quality, and marketing (DFO 1980).

If markets and prices were a concern in the 1970's, they were quickly forgotten as the collapse in the Pacific crab fishery generated substantial demand for Gulf snow crab with associated price increases for Gulf fishermen (beginning at $0.90/kg in 1981 and rising to a peak in 1988 at $4.95/kg). Corresponding record high catches surpassed any previous forecast to reach 31,585 tons in 1982 (with a possible further 3,000 tons misreported by fishermen and processors to defer loan payments) (DFO 1982, 1983, 1984). These catch levels could not be sustained, however, and landings eventually declined to less than 8,000 tons in 1989.

Throughout the 1980's, fisheries management continued to focus on the measures necessary for a stable and sustained fishery. Measures implemented by DFO from 1979 proved impossible to maintain because of overriding social and economic pressures driving non-compliance throughout the entire industry. In 1980, the trap limit increased to 150 per boat. In 1982, a minimum mesh size of 5 1/8" was established, and the incidence of soft shell crab required a summer closure; in 1983, the fishery closed for the fall resulting in the creation of what has effectively become a spring fishery.

The Advisory Committee then established a "Quality Sub-Committee" to develop a management plan for the 1984 midshore fishery. As a result, a procedure for closure based on meat yield from catches falling below 20% (as first proposed in 1977) was established. A 24,000 ton quota for the midshore fishery was established aiming at a 50–60% exploitation rate. Following complaints from processing plant workers that they might not be able to work the 10 week minimum period required for unemployment insurance benefits, the quota was subsequently increased to 26,000 tons. Landings were reduced through implementation of weekly vessel quotas and the fishery was closed on July 2, 1984.

As the midshore crab fishing fleet gained a sound financial footing, fisheries management decisions became increasingly linked to onshore social and economic factors particularly those related to processing plant employment. During its peak year (1982), crab processing employed in excess of 8,000 workers providing unprecedented income throughout the Gaspé, Magdalen Islands, and northeastern New Brunswick. In an attempt to maintain high employment levels as landings declined, the Province of New Brunswick enacted legislation requiring processors to produce at least 60% crab meat for canning or freezing. Control proved difficult and processing workers, infuriated with plant managers' attempts to circumvent these guidelines, demonstrated and held Provincial authorities hostage to emphasize the need for enforcement. Processors were hard-pressed to meet the provincial guidelines and two of them closed after the 1984 season. Other plants were forced to reduce costs by lowering prices to fishermen and developing less labour intensive processing lines.

In an attempt to reduce the seasonal glut of landings and extend plant employment, DFO introduced weekly vessel trip limits in 1984. Both enforcement and adminstration proved difficult as fishermen and processors openly colluded to conceal accurate catch reports. Enterprise allocations were then identified as a practical alternative to the glut problem. However, this was rejected outright by Quebec fishermen concerned that efforts to modernize their fishing fleet could be impeded before they increased catches vis-a-vis New Brunswick (what they apparently did not realize was that New Brunswick fishermen were modernizing at an even faster rate).

The 1985 fishery was equally chaotic. A 26,000 ton quota was applied although weekly and trip limits introduced the previous year were not continued. Fishermen attempted to gain maximum prices for their catches by selling to plants outside northeastern New Brunswick. At the same time, plants replaced workers with processing machines. Frustrated plant workers initiated demonstrations to protest job losses, but tempers quickly erupted and on April 9, 1985 a riot ensued during which trucks loaded with crab for outside plants were damaged and processing equipment was thrown over the wharf.

As the 1985 season continued, industry collusion to thwart management measures appeared evident. Misreporting of catches was commonplace in purchase slips supplied by processors to DFO. Although never proven, it was rumoured to have been common practice for plant owners to maintain dual record books — one for DFO and one for business requirements!

Measures to protect the moulting crab population were premised on fishery closure when meat yields from plant samples fell below the established 20% level. Fishery patrols and research staff, however, identified that fishermen sorted and discarded soft-shell crab at sea to offset the sampling procedure.

The "total allowable catch" (TAC) was reached on June 18 but, prior to closing the fishery, the Minister extended the fishing season four days to enable plant workers to reach the 10-week qualifying period for unemployment insurance benefits.

Following the chaotic 1985 fishing season, the Federal Minister met with his Provincial counterparts to assess solutions to these problems. They concluded that the issues should be examined in detail at a seminar scheduled for December 1985. Participants in the seminar were unable to reach agreement on solutions to any of the problems of the fishery including establishment of fishing seasons, fishing zones, and quotas, as well as the issuance of additional licences. However, a number of important management principles were agreed upon by the participants including:

- — the need for increased stock assessment and research with more input from fishermen;
- — closure of the fishery should meat yields fall below 20%;
- — designing management plans to meet regional employment objectives;
- — use of TACs as a management tool;
- — opening dates to take into account ice conditions and white crab abundance; and
- — strengthening of control measures (DFO 1986).

These principles provided the basis for the fishery management strategy until the end of the 1989 season. The major thrust was strict control of traps based on tagging and buoy identification, a closure formula based on meat yield, a quota based on recruitment levels, and an opening date based on ice conditions.

An "ice committee" was formed to deal with the contentious issue of delaying openings (normally scheduled in early April) when ice was prevalent in portions of the fishing zone. The resulting postponement of the 1987 fishing season, beyond its scheduled opening of April 12, brought strong pressure to open. In the absence of consensus, DFO took a decision to open the fishery on April 25. Many traps were lost in the remaining ice flows and, and in the race to the fishing grounds, one vessel sank with loss of life. The subsequent enquiry was critical of the approaches of both fisheries management and industry.

The fishery could not sustain the 26,000 ton annual reference level TAC and, by 1987, only 12,300 tons were landed. In its subsequent assessment, CAFSAC concluded that it was unable to advise on a quota level as the fishery was now entirely dependent on recruitment which could not be predicted (CAFSAC 1988). In 1988, CAFSAC advised that it could not determine recruitment into the commercial fishery. Although the quota was removed from the management plan, landings in 1988 were near the 1987 level. This impact, however, was largely offset as prices soared to $4.95/kg. In 1989, the fishery yielded less than 8,000 tons, prices fell to $2.46/kg, and the fishery closed in May due to high incidence of white crab catches. The impact of this continuing catch failure, combined with falling prices, gave rise to another comprehensive review of the fishery. In a January 1990 meeting of the Gulf Snowcrab Advisory Committee, it was concluded that the following action should be initiated to protect future recruitment:

- stock conservation and enhancement objectives must supercede other management objectives;
- the moulting stock must be protected through a comprehensive sampling procedure;
- consideration should be given in future years to closure of a portion of the zone to protect areas of high moult;
- quota levels should be established below the most conservative estimates of annual recruitment in the fishery to increase future catches;
- boat quotas should be considered as a means to provide more rational exploitation patterns; and
- trap limits should be strictly enforced and new trap identification methods assessed.

## The Inshore Snow Crab Fisheries

At its 1976 meeting, the Gulf Snow Crab Advisory Committee initiated discussions on the controversial establishment of exclusive inshore crab fishing zones. Inshore fishermen were developing fisheries off Cape Breton and the Lower North Shore (with some 31 and 20 boats respectively) by combining crab fishing with complementary fishing seasons in lobster or groundfish. The Advisory Committee developed criteria to assess whether access to inshore areas should be granted to the specialized midshore fleet or to seasonal inshore fishing operations. In the background, however, was an undercurrent of conflict as the midshore fleet realized their operations could be restricted to the central portion of the Gulf.

### Northern Cape Breton Fishery (Area 19)

In the early 1970's, inshore fishermen were establishing a July through September fishery in the area off Cheticamp, N.S. using 30 traps each compared with 125 per boat for the midshore fleet. Conflicts ensued when midshore boats moved onto these grounds in the early fall and remained in the area until the arrival of the annual southern Gulf cod migration, thus causing a gear conflict with Danish seiners.

The problem festered until 1977 when the midshore crab fleet agreed to lift its gear before the cod arrived on the fishing grounds. Establishment of an exclusive inshore zone was rejected by the Advisory Committee at its August 1977 meeting, but a voluntary closure was agreed upon. The federal government then declared the Area exclusive when midshore boats began setting traps in the area in 1978.

The subsequent development of a fisheries management strategy has seen the inshore fishery off Nova Scotia flourish with 27 licences issued in 1979, a TAC (1,408 tons) and boat quotas. Licence limitation policies were then geared to the combination of price, operating costs, and catch rates. By 1984, the trap limit decreased to 20, boat quotas were set at 50,000 lb, and the number of participants expanded to 61 with the management zone broadened to facilitate administration between DFO Regions.

### Southwestern Cape Breton Fishery (Area 18)

This area was first delineated in 1979 with 14 inshore boats being provided permits to fish 30 traps. Participation then was increased to 23 boats in 1980. Three licences were offered to Prince Edward Island but these were not accepted apparently in light of low prices and catch rates.

In an attempt to develop a stable fishery, a preemptive 835 ton quota was established in 1981 to provide boat quotas consistent with those in the Northern Cape Breton zone. Subsequently, the TAC was established in the 600 ton range, but the occurrence of white crab consistently made reaching this quota difficult. The option of a fall fishery to counter the moult problem implied difficult weather conditions for inshore boats. As well, markets were limited because most processing plants closed earlier in the season when the bulk of catches throughout the Gulf had been processed.

Conflict in this zone came to a head in 1984 when the midshore fleet was prohibited from the area in an attempt to enhance stability of inshore crab operations. Additional licences were issued which fuelled arguments for creation of other inshore fisheries elsewhere — particularly in Prince Edward Island.

**Lower North Shore and Western Newfoundland — Labrador Fishery (Areas 13–17)**

Although an inshore fishery had been ongoing since 1975, a management zone was not created until 1980. The Lower North Shore was subsequently subdivided into five zones by 1983 with a major exploratory effort initiated against pre-emptive quotas in the 2,000 ton range. The exploratory effort had advanced to the Strait of Belle Isle by 1985 where a zone was created with Newfoundland and Quebec boats operating under 65,000 lb boat quotas. Catches throughout these zones declined rapidly after 1985.

Constraining factors to development of these zones proved to be distance from markets, and insufficient catches and prices to sustain industry viability.

**The Prince Edward Island Demand for Access**

Undoubtedly, the single most contentious public issue confronting management of the crab fishery has been that of access to the fishery, and nowhere has this been more evident than in Prince Edward Island. As early as June 1974, the Prince Edward Island government served notice that it accepted the principle of "limited entry" licensing policy only if the policy recognized the unique circumstances which forced Island fishermen out of the fishery in the 1960's and offered provision for subsequent re-entry.

In December 1980, the Prince Edward Island Department of Fisheries reiterated this position in a paper presented at the Atlantic Snow Crab Seminar. The Department argued that processing plants were underutilized and Island fishermen must be provided the opportunity to diversify incomes through access to fish stocks in waters contiguous to the Province. They called for the establishment of an inshore zone extending 20 miles from shore and offered supporting rationale based on similar applications elsewhere in the Gulf, the history of P.E.I. in the fishery, increasing midshore crab catches, and a right of access based on a principle of contiguity (PEI Government presentation to the 1980 Snow Crab Seminar).

The debate continued for four years. Island arguments intensified as midshore landings and the number of midshore licences in Quebec and New Brunswick increased. These were countered by midshore fishermen citing potential for overexploitation and the need for additional management controls as opposed to increased effort from new entrants. The situation came to a head in 1984 when the midshore TAC was increased in response to plant employment concerns while an exclusive inshore zone was created in Nova Scotia (Area 19).

The federal Minister announced the issuance of 16 exploratory inshore crab licences for P.E.I. on November 20, 1984. These licences were to be limited to an area north of the Island (Area 26) and a second area to the east (Area 25) where research cruises and log reports identified possible occurrences of snow crab and little fishing effort. Midshore fishermen disputed this action and, in May 1985, successfully sought a court injunction to access the inshore zone.

In 1986, an additional 14 inshore exploratory permits were issued, and a compromise reached with the midshore fleet limiting both the number of boats and the fishing season. In 1987, the initial 16 permits issued to P.E.I. were identified as permanent licences in reaction to continuous demands for access to the crab fishery, but then catches declined to 446 tons in 1987 (only about 1/3 the 1986 level). In response, a decision was taken to again restrict the midshore fleet from inshore zones in 1989.

This time, little opposition came from the midshore sector. However, with the decline in catch rates, the 1980 objective of developing a seasonal snow crab fishery for lobster fishermen appeared unattainable for the present. In light of these constraints, those additional permits issued in 1986 were not provided permanent status but maintained as exploratory permits.

## Concluding Observations

The snow crab fishery in the Gulf of St. Lawrence developed over the past 25 years. Nominal landings in the southwestern Gulf midshore fishery have exhibited a "boom" and "bust" cycle. Stock status and fishing activities were monitored through biological and economic analysis of commercial catch and landing statistics, and through research activities to estimate abundance and biological parameters. Initial regulations prohibited use of trawl gear to minimize habitat destruction and damage to crabs. Females were excluded from the catch due to their small size. Only large males over 4 inches (102 mm) carapace width (CW) were considered exploitable. Subsequently, the size limit was reduced to 95 mm CW when the industry became capable of processing smaller sized crabs.

In the mid-1970's, restrictions on fishing effort were introduced by limiting access to the fishery as well as limiting the number of traps. As catches increased, inshore zones were established and a limited number of local fishermen were granted exclusive access. In these inshore zones, it was envisioned that fishermen would not receive all of their income from snow crab but would also be involved in other fisheries. In a few zones, a pre-emptive TAC was established and each boat was given an individual quota. Total allowable catches were initially based on the estimated average annual biomass (weight) increase, then on 50–60% of the exploitable proportion of the resource, and finally on historical trends of harvest.

Increasing exploitation led to an increase in the relative proportion of soft-shelled (white) crabs due to fishing in the moulting season. In response to this problem, regulations were introduced to limit the fishing season by a summer closure. A minimum trap mesh size regulation was introduced to overcome problems in catching undersized crabs. A myriad of other regulatory measures have been introduced during the course of the fishery to compensate for biological concerns and yet, at the same time, accommodate industry concerns for economic viability and employment.

The implications of the biological features of snow crab relevant to management of the southwestern Gulf stock were first discussed at a 1987 workshop (Jamieson and McKone 1988). There was consensus that the *timing of fishing* had not necessarily been set on the basis of biology except to avoid a preponderance of white or newly moulted crab in the catch. Moulting, mating, egg hatching and spawning all occurred in spring coinciding with the fishing period in the southwestern Gulf. Secondly, there was consensus that the *minimum legal size* of 95 mm CW was not based on biological criteria. The appropriateness of the existing regulation was uncertain for maximizing yield per recruit and for avoiding recruitment overfishing. However, there was no evidence of recruitment overfishing because female snow crab were not exploited and most appeared to be carrying eggs. Thirdly, there was consensus that the *white crab problem* was the result of harvesting large terminal males over time resulting in a high proportion of white crab in the catch, and the disappearance of large males may have actively induced recruitment of sublegal males to the fishery.

The bottom line is that the fishery developed, expanded and declined in the absence of a clear appreciation for the impact of fishing pressures, and the absence of effective monitoring and control over the fishing fleet and processing plants. The crisis of 1989 brought demands for action from all sectors of the industry.

Snow crab management has since addressed these issues through new programs aimed at enhancing research, initiating a more rational approach to harvesting, ensuring accurate catch reports, and diversifying processing plants. Research is focusing on the basic biology of snow crab and on proven methodologies for estimating annual harvesting levels of snow crab and invertebrates in general. Individual boat quotas and new monitoring approaches have been introduced.

Stock recovery is now occurring only with the full co-operation of all groups interested in this fishery. Sustaining the industry has implied lowering expectations from utilization of the resource, accurate reporting of catches, and increased penalties for violators of the management plan. In the broader regional context, greater emphasis has also been placed on fostering economic diversification.

## References


CAFSAC. 1988. Advice on the management of snow crab. Canadian Atlantic Fisheries Scientific Advisory Committee Advisory Document 88/1.

1989. Advice on the management of snow crab. Canadian Atlantic Fisheries Scientific Advisory Committee Advisory Document 89/2.

CANADA GAZETTE. 1975. Part II, (1975, August 6). Atlantic Crab Fisheries Regulations, Vol. 109, No. 16, p.c. 1975–1985.

COMEAU, M. 1987. Maturité gonadique et mue prépubérale chez le crabe des neiges (*Chionoecetes opilio*) (O. Fabricius). M.Sc. thesis, Université de Moncton, Moncton, N.B. 145 p.

CONAN, G.Y., AND M. COMEAU. 1986. Functional maturity and terminal molt of male snow crab (*Chionoecetes opilio*). Can. J. Fish. Aquat. Sci. 43: 1710–1719.

CONAN, G.Y., M. MORIYASU, M. COMEAU, P. MALLET, R. CORMIER, H. CHIASSON, AND Y. CHIASSON. 1988. Growth and maturation of snow crab (*Chionoecetes opilio*), p. 45–66. *In* G.S. Jamieson and W.D McKone [ed.] Proceeds of the International Workshop on snow crab biology, December 8–10, 1987, Montréal, Québec. Can. Manuscr. Rep. Fish. Aquat. Sci. 2005.

CONAN, G.Y., R.W. ELNER, AND M. MORIYASU. 1989. Review of literature on life histories in the genus *Chionoecetes* in light of the recent findings on growth and maturity of *C. opilio* in eastern Canada, p. 163–179. *In* Proc. Int. Symp. King & Tanner Crabs, Nov. 1989. Anchorage, Alaska, USA.

DEVEAU, L.E., AND H. MCALLISTER. 1967. A report on the development of the queen crab fishery in the Maritime provinces of Canada. Industrial Development Service. Department of Fisheries of Canada, Halifax, N.S.

DAWE, E.G., D.M. TAYLOR, J.M. HOENIG, W.G. WARREN, G.P. ENNIS, R.G. HOOPER, W.E. DONALDSON, A.J. PAUL, AND J.M. PAUL. 1991. A critical look at the idea of critical molt in male snow crab (*Chionoecetes opilio*). Can. J. Fish. Aquat. Sci. 48: 2266–2275.

DEPARTMENT OF FISHERIES AND FORESTRY. 1969. Proceedings, Meeting on Atlantic Crab Fishery Development. Can. Fish. Rep. No. 13: 47 p.

DEPARTMENT OF FISHERIES AND OCEANS. 1980. Atlantic snow crab seminar, Rapporteur's Report. Quebec, Quebec. 26 p.

1981. Underwater World: The Atlantic snow crab. UW81/006E. Ottawa. 8 p.

1982. An economic assessment of the New Brunswick crab fleet. Moncton, N.B.

1983. An economic assessment of the New Brunswick crab fleet. Moncton, N.B.

1984. An economic assessment of the New Brunswick crab fleet. Moncton, N.B.

1986. Gulf of St. Lawrence snow crab seminar Final Report, Moncton, N.B. December 3–5, 1985. Moncton. DFO/GR/PUB-86-001. 47 p.

HOOPER, R.G. 1986. A spring breeding migration of the snow crab (*Chionoecetes opilio*), (O. Fabricius) into shallow water in Newfoundland. Crustaceana 50: 257–264.

JAMIESON, G.S., AND W.D. MCKONE [ed.]. 1988. Proceedings of the International Workshop on Snow Crab Biology, December 8–10, 1987, Montreal, Quebec, Can. Manuscr. Rep. Fish. Aquat. Sci. 2005: 163 p.

KON, T. 1980. Studies of the life history of the Zuwai crab (*Chionoecetes opilio*) (O. Fabricius). Spec. Publ. Sado. Mar. Biol. Stn. Niigata Univ. Ser. 2: 64 p. (In Japanese.)

MALLET, P., Y. CHIASSON, AND M. MORIYASU. 1988. A review of catch, fishing effort and biological trends for the 1987 southwestern Gulf of St. Lawrence snow crab, (*Chionoecetes opilio*), fishery. Canadian Atlantic Fisheries Scientific Advisory Committee Research Document 88/32.

MALLET, P., E. WADE, Y. CHIASSON, P. DEGRACE, G.Y. CONAN, AND M. MORIYASU. 1989. La pêche du crabe des neiges (*Chionoecetes opilio*) dans le sud-ouest du golfe Saint-Laurent en 1988: L'analyse des prises, de l'effort de pêche et l'estimation de la biomasse. Canadian Atlantic Fisheries Scientific Advisory Committee Research Document 89/25. Ministry of Aquaculture, Fisheries and Processing, (1985). Quebec Snow Crab Fisheries. Quebec, Quebec.

MORIYASU, M., AND G.Y. CONAN. 1988. Aquarium observations on mating behavior of snow crab, (*Chionoecetes opilio*). Int. Counc. Explor. Sea C.M. 1988/K:9.

MORIYASU, M., G.Y. CONAN, P. MALLET, Y. CHIASSON, AND H. LACROIX. 1987. Growth at molt, molting season and mating of snow crab (*Chionoecetes opilio*) in relation to functional and morphometric maturity. Int. Counc. Explor. Sea C.M. 1987/K:21.

OTTO, R.S. 1982. An overview of the eastern Bering Sea tanner crab fisheries, p. 83–115. *In* Proceedings of the International Symposium on the *Genus Chionoecetes*. Lowell Wakefield Fisheries Symposis Series, University of Alaska, Alaska Sea Grant Rep. 82–10.

SINODA, M. 1982. Fisheries for the genus *Chionoecetes* in southwest Japan Sea, p. 21–39. *In* Proceedings of the International Symposium on Genus *Chionoecetes*. Lowell Wakefield Fisheries Symposia Series, University of Alaska, Alaska Sea Grant Rep. 82–10.

TAYLOR, D.M., R.G. HOOPER, AND G.P. ENNIS. 1985. Biological aspects of the spring breeding migration of snow crabs, (*Chionoecetes opilio*), in Bonne Bay, Newfoundland (Canada). Fish. Bull. 83(4): 707–711.

WATSON, J. 1970. Tag recaptures and movements of adult male snow crab (*Chionoecetes opilio*) (O. Fabricius) in the Gaspe region of the Gulf of St. Lawrence. Fish. Res. Board Can. Tech. Rep. 204: 16 p.

WATSON, J., AND P.G. WELLS. 1972. Recaptures and movements of the tagged snow crabs (*Chionoecetes opilio*) in 1970 from the Gulf of St. Lawrence. Fish. Res. Board Can. Tech. Rep. 349: 12 p.

## Appendix I — Chronology of Events

1966 — Beginning of a commercial fishery;

1969 — Conference on Atlantic Queen Crab development (Fredericton, N.B.);

1972 — Issuance of a specific crab licence;

1973 — Announcement of "limited entry" policy, advisory committee;

1974 — Establishment of Gulf Snow Crab Advisory Committee;
— Development of draft Regulations;

1975 — Introduction of "The Atlantic Crab Fishery Regulations";
— Closed season January 1-January 31;
— Limited entry (participation clause);
— Restriction on female crabs;
— Trap limit of 125 for the midshore;
— Daily log records;
— Inshore licences in Quebec;

1976 — Authority to vary closed time and quotas;
— Minimum size of snow crab at 3 3/4 inches (95 mm);
— Restriction of soft-shelled crab;

1977 — Start of provision of annual CAFSAC advice;

1978 — Establishment of an exclusive inshore zone in northwest Cape Breton — Zone 1; limit of 30 traps;
— Establishment of inshore zone in Quebec;
— Requirement for tagging of gear;
— Moratorium on issuance of new licences;

1979 — Establishment of an inshore fishery in southwest Cape Breton (limit of 30 traps);

1980 — Legal recognition of inshore zone in Quebec;
— Atlantic Queen Crab Seminar (Quebec City);
— Trap limit increased to 150 for the midshore;

1982 — Establishment of Gulf Snow Crab Sub-Committee on Quality;
— Introduction of minimum mesh size of 5⅛ inches (130 mm) for crab traps;
— Closure of Cape Breton Area 7 in summer months for white crab;
— Midshore fishery closed early August for white crab and reopened early September;

1983 — Closure of midshore fishery in mid-July for white crab. No re-opening;
— Sub-Committee on Quality given mandate to develop management plan for 1984;
— Subdivision of Quebec inshore zone into 5 zones (A, B, C, D, E);
— Beginning of exploratory fishery on Lower North Shore (August 1–December 31 season);
— Re-assumption of fisheries jurisdiction in Quebec by the Federal government;

1984 — TAC of 24,000 mt established for the midshore fishery;
— Establishment of 20% meat yield for closure and reopening of the midshore fishery;
— Equivalency of Japanese conical trap of rectangular trap established at 2:1;
— TAC increased to 26,000 mt for the midshore and weekly vessel quotas established on June 1;
— Midshore fishery closed July 2 — TAC reached;
— Establishment of new boundaries for Cape Breton Area 1;
— Establishment of exclusive boundaries for inshore fishery in Cape Breton — Area 7;
— Exploratory permits issued for the Lower North Shore of Quebec (Zone D, E);
— Creation of advisory committee for the Estuary and Middle North Shore of Quebec;

1985 — Midshore TAC set at 26,000 mt;
— Midshore fishery closed on June 18 — TAC reached;
— Midshore fishery extended to June 22;
— Granting of 16 inshore exploratory permits for two exclusive inshore zones in Prince Edward Island;
— Granting of 6 inshore exploratory permits in Western Newfoundland;
— Granting of 4 exploratory permits for Quebec Zone C;
— Limit of 15.2 m (49'11") placed on inshore vessel in Quebec;
— Closure of Quebec Zone A for white crab August 1–September 15; final closure November 1;
— Creation of advisory committee for the Lower North Shore of Quebec (Zone C, D, E);
— Granting of two midshore licences to two Cape Breton fishermen;

1986 — Midshore reference TAC of 26,000 mt;
— Opening date of April 12 (traps set 4 days prior);

— 10 week season terminating on June 20;
— Extension of the fishery to June 27 and subsequently to June 29 (traps removed by July 1);
— Granting of 14 additional exploratory permits for the P.E.I. inshore fishery (total 30);
— Granting of access to 34 midshore vessels to the P.E.I. zone from May 28 to end of season;

1987 — Granting of two midshore licences to N.B. fishermen following licence appeal;
— 10 week season — opening April 12 (set traps 4 days previously) — closing June 20 (traps removed June 21);
— Reference TAC of 26,000 mt — possibility of extension of season to July 4 if not reached;
— 20% meat yield;
— Sub-committee for opening date;
— 16 inshore permits become permanent for P.E.I.;
— P.E.I. inshore zone to remain, 34 midshore vessels permitted access from May 28;
— Exploratory fishery south of Anticosti Island by 3 vessels September 1–October 31;
— Season postponed due to ice conditions (April 25–July 4);
— Sinking of N.B. vessel with loss of life;
— Trap losses;

1988 — Granting two midshore licences to Nova Scotia fishermen following licence appeal;
— Midshore TAC dropped with CAFSAC advising this is a recruitment fishery;
— Ten week season April 10–June 18;
— Sub-committee to review opening date;
— P.E.I. inshore fall fishery cancelled due to moulting crab;
— Continuation of the exploratory fishery of Anticosti Island, June 19–July 3;
— Studies initiated in the use of biodegradable material for crab traps;

1989 — Ten week season scheduled April 9–June 17;
— Removal of privileges for midshore boats to access the P.E.I. zone;
— Midshore and P.E.I. fisheries opening delayed to April 12 due to ice conditions;
— Midshore and P.E.I. fisheries close May 29 due to high incidence of soft-shelled crab.

**Source:** Resource Allocation Division files, Gulf Region, Fisheries and Oceans Canada

# Straddling Stocks — International Law and the Northwest Atlantic Problem

B. Applebaum

*Department of Fisheries and Oceans, International Fisheries 200 Kent Street, Ottawa, Ontario, Canada K1A 0E6*

Applebaum, B. 1993. Straddling stocks — International law and the Northwest Atlantic problem, p. 193–210. *In* L. S. Parsons and W. H. Lear [eds.] Perspectives on Canadian marine fisheries management. Can. Bull. Fish. Aquat. Sci. 226.

In a number of areas in the world, important fish stocks "straddle" the 200 mile limits of coastal states. These stocks are vulnerable to being overfished because any action taken by a coastal state within its 200 mile zone to provide for conservation of these stocks can be — and in some areas, is being — undermined by fisheries conducted by other states outside the 200 mile limit. International law is unclear regarding the rights and obligations of states in these circumstances.

In the Northwest Atlantic area off the east coast of Canada problems began developing in the late 1970's, shortly after Canada's 1977 extension of jurisdiction to 200 miles, and increased severely in the 1980's.

The Northwest Atlantic Fisheries Organization has not been able to resolve these problems. One possible way to find a solution is to focus on the implications of current international law. "Sub-laws" deduced from current international law oblige states fishing on the high seas to constrain their fisheries for straddling stocks within limits consistent with the special interest the relevant coastal states has in these stocks. Obtaining international recognition of these "sub-laws" remains a problem.

---

Dans un certain nombre de régions du monde, d'importants stocks de poisson « chevauchent » la zone de pêche exclusive de 200 milles établie par les États côtiers. Ces stocks sont vulnérables à la surexploitation étant donné que toute mesure prise par un État côtier en vue de conserver ses ressources halieutiques peut être minée, et dans certains cas l'est déjà, par des opérations de pêche étrangère. La loi internationale n'est pas claire quant aux droits et aux obligations des États dans ces circonstances.

En 1977, le gouvernement canadien a créé une zone de pêche exclusive de 200 milles dans l'Atlantique nord-ouest. Les problèmes qui ont commencé à faire surface peu après, soit vers la fin des années 1970, n'ont fait que s'aggraver au cours des années 1980.

L'Organisation des pêches de l'Atlantique nord-ouest (OPANO) n'a pu résoudre ces problèmes. En nous concentrant sur les conséquences de la loi internationale en vigeur, nous pourrons peut-être trouver une solution. Des « sous-lois » inférées de cette loi obligent les États dont les flottilles oeuvrent en haute mer à restreindre l'exploitation de stocks transfrontaliers à un niveau compatible avec l'importance de ces stocks pour les États côtiers concernés. Par contre, l'homologation au niveau international de ces « sous-lois » demeure un problème.

## International Fisheries Management in the Northwest Atlantic: A Case History and Legal Study[1]

### Introduction

The fishing grounds off Canada's Atlantic coast extend beyond the 200 mile limit (Fig. 1). These areas outside 200 miles, known in Canada as the Nose and Tail of the Grand Banks, are legally classified as "high seas", i.e. they are outside the area of coastal state "exclusive economic zone" jurisdiction and are thus international waters. These areas outside 200 miles include rich fishing grounds for a number of stocks which "straddle" the Canadian 200 mile limit and are fished both inside and outside this limit. The area outside the 200 mile limit is under the management authority of the Northwest Atlantic Fisheries Organization (NAFO) and is referred to in the NAFO Convention as the NAFO Regulatory Area (NRA).

NAFO was established in 1978 in the context of the international consensus that had been reached by that time in the draft United Nations Convention on the Law of the Sea (UNCLOS) and the world-wide establishment of 200 mile limits pursuant to that consensus. More specifically, it was established to, *inter alia* manage a number of fish stocks, identified as 10, which were not under the exclusive management control, within 200 mile limits, of the coastal states of the Northwest Atlantic. Of these 10 stocks, seven fall within the definition of "straddling stocks". The three additional stocks managed by NAFO are discrete stocks, entirely separate from the areas within 200 mile limits and are outside the focus of this study.

The relevant UNCLOS articles, as they were in 1978 and in their final form, left legal rights and obligations on the high seas regarding these stocks — and regarding other stocks not limited in their migratory range to areas within the exclusive control of coastal states — in an ambiguous condition. The result was that these stocks were vulnerable to the international pressures that had led, over previous decades, to overfishing and depletion of the stocks that were now under the exclusive control of coastal states within their 200 mile zones. This study provides a case history for the Northwest Atlantic, and an analysis of the relevant international legal framework.

### The NAFO Case History

The formation of NAFO was an early effort in the period following world-wide establishment of 200 mile zones to bring those pressures under control through a new international organization dedicated to conservation. NAFO comprises all the coastal states in the Northwest Atlantic, except the U.S.A., and all other states fishing in the Northwest Atlantic in 1978 both inside and outside the 200 mile limits.

The NAFO Convention (summarized in Appendix 1) was also seen, by its Parties, as a substantive implementation of the relevant UNCLOS articles, designed to reduce the legal ambiguities of the "outside 200 miles" situation and, through practical management measures, to avoid the conservation problems that could otherwise result from these ambiguities. As such, NAFO was seen by its creators as a natural transformation of its predecessor organization, ICNAF (International Commission for the Northwest Atlantic Fisheries) which had established a firm and detailed foundation in management principles and measures that NAFO inherited.

The optimism that initiated NAFO appeared to be justified in the immediately ensuing years. Two key principles, carried over from NAFO's predecessor, ICNAF, were adopted at the outset as the "pillars" of the conservation structure. These principles were "conservative" conservation in the setting of total allowable catch limits (TACs) for the stocks based on the best available scientific advice, and maintenance of the traditional proportional shares for the member countries. The "conservative" conservation principle was implemented in the form of "$F_{0.1}$" or its equivalent, a management approach that set TACs significantly below the "Maximum Sustainable Yield" (MSY) level, at 2/3 MSY in some cases. The purpose of this management approach was to provide a significant degree of stability in annual catches, larger fish, more economic catch rates, and also to minimize the dangers inherent in the possibility of errors in scientific assessments of the stocks.

This initial period was marred by a number of problems. In particular, one of the countries that had traditionally fished in the Northwest Atlantic, Spain, did not join NAFO for several years, and operated a major fishery in the area outside the NAFO conservation framework, primarily targeting one straddling stock, 3N0 cod, and fishing it at levels above those set aside for Spain by NAFO. After joining NAFO, Spain continued to fish this stock and began fishing others, above the levels set for Spain by the NAFO conservation framework. Perhaps not by coincidence 3NO cod was, during the initial NAFO period described, the only straddling stock that proved to be a significant disappointment in failing to develop to the levels the

[1]An earlier version of this paper was presented at the 23rd Annual Conference of the Law of the Sea institute, 12–15 June 1989, at Noordwijk aan Zee, The Netherlands.

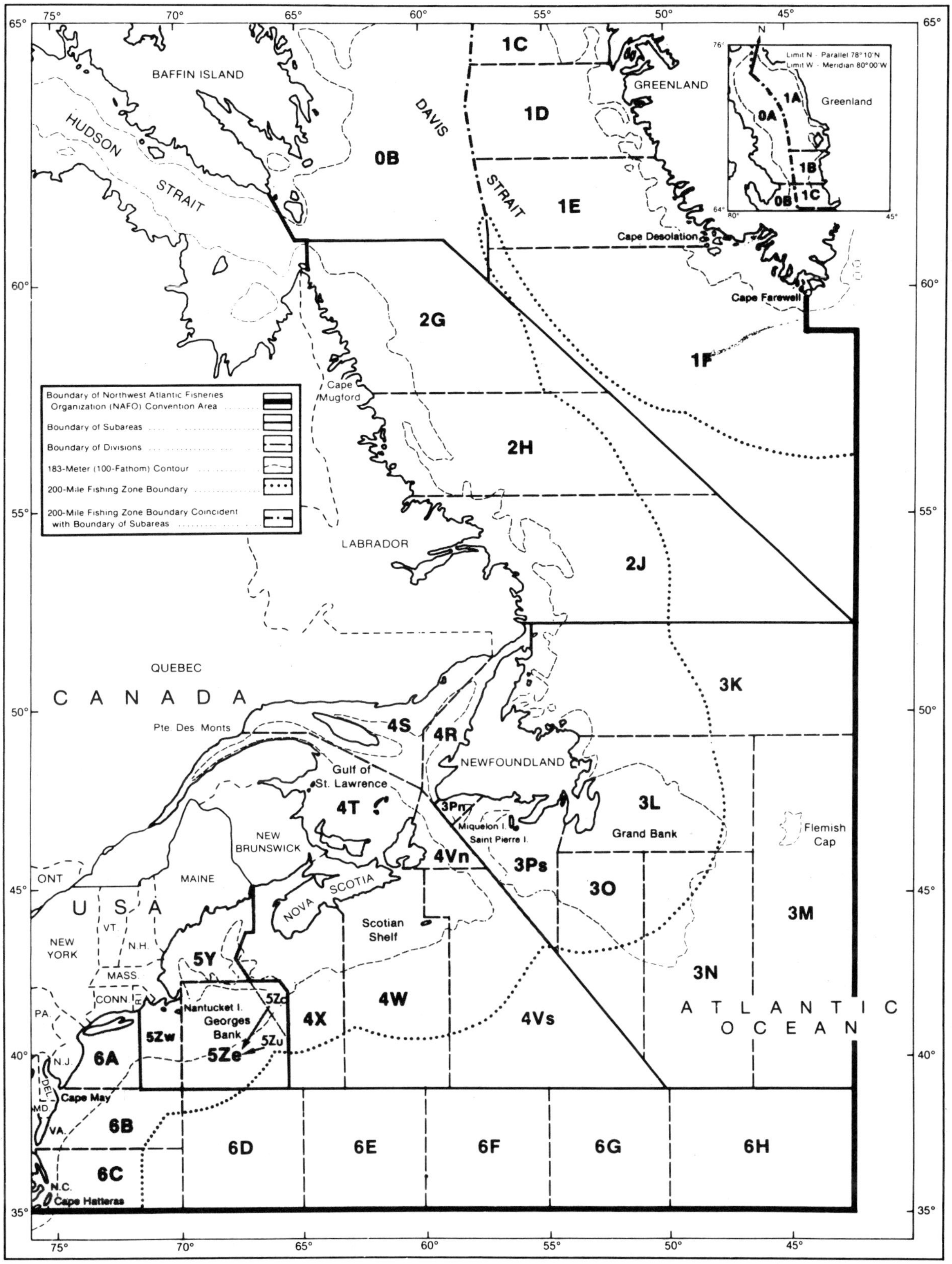


FIG. 1. NAFO's Convention Area and 200-mile fishing zone boundaries.

scientists identified as possible under proper management.

The other disturbing development during the initial period was the appearance of "new flags" — vessels flying the flags of countries which were not members of NAFO, had no traditional fishery in the Northwest Atlantic, and operated outside the NAFO conservation framework. Catches by these vessels were by definition, and often in practice, in excess of the Total Allowable Catch (TAC) limits set by NAFO for the stocks involved.

These two problems — Spain and the "new flags" — were not, however, seen during the period up to 1985, as being harbingers of serious problems for NAFO in the future. Most NAFO members considered the future in a positive light, as being one where the problems would be brought under control within the well-established NAFO management framework.

NAFO members had good reason to be positive, and the good reason was the European Community (EC). The EC, a founding member of NAFO, had been a strong supporter of the NAFO conservation framework from the outset, supporting conservative conservation in the form of $F_{0.1}$ management or its equivalent, and supporting maintenance of traditional proportional shares. With Spain about to join the EC, it seemed clear that the Spanish problem would be resolved. As regards the "new flags", many were "joint venture" operations of one kind or another connected to Spain and to Portugal which was also about to join the EC. It seemed likely that these operations would, through the EC, also be brought under control. As a means of enforcement the EC would, through its control of St.Pierre and Miquelon and the Azores, be able to close these ports to non-NAFO member vessels, as Canada had closed its ports to these vessels, cutting off their provisioning and offloading privileges. With the loss of these privileges it was likely that their operations would no longer be economically viable.

However, as of 1985, matters developed differently. With Spain and Portugal scheduled to join the EC on January 1, 1986, the EC came to the 1985 NAFO annual meeting with a radical new position for the 1986 fishing season and beyond. The EC's position was that the previous management had been too conservative, as a result of which possible catches had been lost; accordingly, the EC demanded that TACs should go up substantially thus providing higher quotas for all. When the EC's position was not accepted, the EC made it clear that it would no longer be bound by the NAFO conservation framework, and would fish above the quotas assigned to it in accordance with that framework. The EC had the legal option, under the NAFO Convention, to take this action, as the Convention provides a procedure under which a Party may lodge formal written objections to management measures adopted by the NAFO Fisheries Commission with the result that it will not be legally bound, under the Convention, by such measures. In 1986 and in the ensuing years up to 1990 the EC lodged objections to most of the management measures adopted annually by the NAFO Fisheries Commission.

One of these measures to which the EC lodged objections, the moratorium on fishing for cod in NAFO Division 3L, has been in place as of 1986 to assist in the conservation of a major stock managed, not by NAFO, but by Canada. This is the 2J3KL cod stock, the biomass of which has been identified by the NAFO Scientific Council as being, on a yearly average basis, 95–97% within the Canadian 200 mile limit. The 3–5% of the stock located outside Canada's 200 mile limit has become a target for EC fisheries, which have thus undermined the conservation regime established by Canada for this stock.

EC catches in the NRA in 1988, excluding the three discrete stocks and excluding 2J3KL cod, totalled about 48,400 metric tons (t), as compared with the EC's total relevant NAFO quotas of 15,500 t. Fishing effort by vessels flying the flags of non-NAFO members has also been increasing in recent years, totalling about 34,000 t in the NRA in 1988 excluding the discrete stocks and excluding 2J3KL cod. As compared with the total of the TACs established by NAFO for the relevant stocks, 125,000 t, the combined overfishing outcomes, 66,900 t, are significant. TAC's of the stocks normally targetted by these fleets, cod and flatfish, have been significantly reduced (Table 1) and catches have declined disastrously.

EC reported catches of 2J3KL cod expanded to very high levels of 62,000 t in 1986 and 35,000 t in 1987. The EC catches fell to about 25,000 t in 1988 and rose again to 37,000 t in 1989. The high EC catches for 1986, 1987 and 1989 reported by the EC and shown in the NAFO figures may be incorrect. There is conjecture that these figures were reported to the EC by the Spanish and Portuguese in order to justify high quotas set unilaterally by the EC for its members. The high quotas have the effect of allowing unrestricted fisheries by these members' fleets. TACs of 266,000 t in 1988 and 235,000 t in 1989 were set by Canada and fully allocated to Canadian fishermen inside the Canadian 200 mile limit.

The resulting situation has cast doubt on the future viability of NAFO to provide for conservation in the area which is supposed to be under its control. This, combined with developments in other parts of the world, has also cast doubt on the viability of similar fisheries organizations, in place or being considered. Similar

TABLE 1. Total allowable catches (metric tons) 1986 to 1990, of stocks managed by the Northwest Atlantic Fisheries Organization which may be fished on either side of Canada's 200 mile limit (Areas refer to NAFO Statistical Areas see Fig. 1).

| | Year | | | | |
|---|---|---|---|---|---|
| STOCK | 1986 | 1987 | 1988 | 1989 | 1990 |
| 3NO Cod | 33,000 | 33,000 | 40,000 | 25,000 | 18,600 |
| 3LN Redfish | 25,000 | 25,000 | 25,000 | 25,000 | 25,000 |
| 3LNO A. Plaice | 55,000 | 48,000 | 40,000 | 30,300 | 24,900 |
| 3LNO Yellowtail | 15,000 | 15,000 | 15,000 | 5,000 | 5,000 |
| 3NO Witch | 5,000 | 5,000 | 5,000 | 5,000 | 5,000 |
| 3NO Capelin | 0 | 10,000 | 15,000 | 28,000 | 30,000 |
| Finfish Totals | 133,000 | 136,000 | 140,000 | 118,300 | 108,500 |
| 3+4 Squid | 150,000 | 150,000 | 150,000 | 150,000 | 150,000 |

fisheries organizations in place in other parts of the world include ICCAT (International Commission for the Conservation of Atlantic Tunas), NASCO (North Atlantic Salmon Conservation Organization), IWC (International Whaling Commission), ICSEAF (International Commission for the Southeast Atlantic Fisheries), and IPFC (Indo-Pacific Fishery Commission). Consideration has been given to a NAFO-type organization for the "doughnut hole" fisheries in the North Pacific area.

The issues involved are broad. They encompass international relations, fisheries management approaches, and international law. This study will focus on international law, including the relevant texts of the Law of the Sea Treaty and corollary principles to be drawn from those texts, and from other sources of international law, to guide future developments. The following sections provide a legal analysis which covers these points.

## The Law

### 1. UNCLOS Treaty Texts

Central Position of Coastal State Duties, Rights and Legal and Other Interests

The starting point for this analysis of the relevant legal questions and principles is the submission that, in the case of straddling stocks, the duties, rights and interests of all states are centred on the predominant duties, rights and interests of the coastal states whose 200 mile limits are straddled by these stocks. The word "interests" is used in this study to denote two different kinds of interests involved — legal interests similar to property rights, and other interests, of the nature of wants or needs.

The following analysis supports the proposition that the UNCLOS text gives the coastal state, as regards the stocks which straddle its 200 mile limit, substantive duties, rights, and interests which have preferential status over those of other states i.e. "distant water fishing" states (DWF states) fishing outside 200 miles. These substantive duties rights and interests therefore have effect outside 200 miles.

Though it should be self-evident, it is essential at the outset to emphasize that the portion of a straddling stock that is inside a coastal state's 200 mile limit is legally a resource of the 200 mile zone, fully subject to the duties, rights and interests of the coastal state to which the zone belongs. This fact sheds important light on the meaning of the UNCLOS texts that are referred to below.

*Article 61* gives the coastal state two duties. First, it must "ensure, through proper conservation and management measures, that the maintenance of the living resources in the exclusive economic zone is not endangered by over-exploitation." Second, "as appropriate" the coastal state and competent organizations "shall cooperate to this end." The text does not limit these duties to the area inside the coastal state's 200 mile zone. The relevance of this fact becomes obvious in the straddling stock context, bearing in mind the need, in the case of such stocks, for proper conservation and management measures both inside and *outside* the zone.

One way for a coastal state to carry out its duty to establish and implement the necessary measures outside its 200 mile zone is by obtaining these measures through an international organization like NAFO, thus ensuring that the straddling stock resource in its zone is not endangered by over-exploitation. But what if this organization cannot "deliver the goods" so to speak because DWF states, either inside or outside the organization refuse to cooperate? The coastal state continues to have the duty, but cooperation with the relevant international organization does not, any longer, appear to be "appropriate". Can the UNCLOS convention have created a duty without rights through which it can satisfy that duty? Whether or not this is so, it can certainly be argued that the duty of the coastal state creates a duty on the part of DWF states to enable the coastal state to honour its duty. This point will be discussed later in this section.

*Article 62* requires the coastal state to promote the objective of optimum utilization of the living resources in its 200 mile zone. The question arises, how does a coastal state do that as regards a straddling stock, when the fisheries of DWF states outside its 200 mile zone undermine the productivity, and therefore the optimum utilization of that stock, to the point where catches and catch rates decline, and fisheries become less and less viable?

While Article 62 creates a duty to "promote" rather than "ensure", as in Article 61, it is clear that Article 62 reinforces the duties, rights and interests of the coastal state.

*Article 63* states that for straddling stocks the coastal state and the DWF states fishing outside the 200 mile limit "shall seek to agree" on conservation measures in the area outside 200 miles. There are two points to note in this article.

The first point is that the coastal state is accorded a preferential position over the DWF states outside 200 miles in two respects:

a) the obligation between the coastal state and the DWF states to "seek to agree" applies as regards a DWF state only if it is actually fishing outside the 200 mile limit, but it applies for the coastal state whether or not the coastal state is fishing the particular stock inside *or* outside 200 miles, i.e. the obligation arises solely by virtue of the coastal state's legal status as such, with duties, rights and interests as regards this stock both inside *and* outside the zone;
b) the requirement to seek to agree is limited to the area outside 200 miles, i.e. there is no obligation for the coastal state to seek to agree on measures for the stock as a whole or to coordinate the measures it takes *inside* its zone with those to be applied by an international organization or by a DFW state outside its zone — the measures inside its zone are in its sole discretion. This legal position is one of the sources of the "consistency" principle, i.e. the principle that measures taken outside the zone by an international organization or by a DFW state must be consistent with the measures taken, entirely unilaterally, by the coastal state *inside* the zone. This principle is incorporated in the NAFO Convention, Article XI, paragraph 3, reproduced as an attachment to Appendix 1.

The second point to note in Article 63 is the non-limiting effect of the obligation of the coastal state to participate in a process of seeking to agree on measures outside its limits. This is an obligation which logically follows, as a step in the process, from its duty under Article 61 to *ensure* the maintenance of the resources within its zone; however the coastal state's obligation to "seek to agree" *in no way limits* the *rights* of the coastal state that flow from its duty to *ensure* the maintenance of the resources within its zone, or the obligations of other states that flow from the coastal state's rights, duties and interests. In other words, having sought to agree, and failed, the coastal state has the duty and the right to take other steps.

*Article 116* states clearly that the right to fish on the high seas is subject to "the rights and duties as well as the interests of coastal states provided for" in the Convention. While the text makes specific reference to Article 63 paragraph 2, and Articles 64 to 67, the preceding "*inter alia*" makes clear that rights, duties and interests provided in other provisions are also applicable (Miles and Burke 1988). It should be noted that while Article 116 states that the right to fish on the high seas is subject to the rights and duties, as well as the interests, of coastal states, there is no similar "reciprocal" provision in Article 61 or 62 limiting the rights and obligations of coastal states in 200 mile zones. The preferential position of the coastal state in the management of the straddling stock as a whole should, accordingly, be obvious.

Article 116 makes clear that the right to fish on the high seas is subordinate to the preferential right of the coastal state. There should accordingly, be no doubt that the coastal state's duty under Article 61 to ensure that the maintenance of the living resources in its zone is not endangered by over-exploitation, together with its rights and interests, translates legally into a right outside its zone.

Burke (1983) has made this point in the following manner:

> "Article 116 subjects the rights to fish to Article 63(2) and other articles establishing coastal State rights, duties and interests. It is plain that the coastal State has extensive rights over the EEZ portion of the stock and is

legally competent to decide upon any necessary conservation measures as well as upon the extent and conditions of foreign access to the stock in the zone. **If fishing on the high seas is subject to the coastal State's right to establish conservation measures in the EEZ, this would appear to mean that others must recognize these measures as applicable wherever the stock in question is found on the high seas.** If the fishing State is not thus obliged to recognize and to observe coastal State measures, the prospect is that they would be made ineffective for the stock as a whole. **Whether or not failure to observe those measures on the high seas would have such an effect would depend on the extent, timing and methods of such fishing. If significant high seas harvesting occurs the probability is that coastal State measures would fail. Such an outcome seems to be inconsistent with Articles 116, 63(2) and other EEZ fisheries provisions**". (emphasis added).

*Articles 117, 118 and 119* establish the general obligations, together with specific details, concerning conservation outside 200 miles. It is implicit, of course, that the portion of a straddling stock that is outside 200 miles has the legal status of a resource of the high seas, and that states fishing this resource on the high seas have substantive duties and rights as regards this portion of the straddling stock.

However, as the preceding references and comments make clear, these rights and duties are of a secondary nature when compared to, and to the extent they come in conflict with, the rights and duties of coastal states as regards straddling stocks. Outside the 200 mile limit states do not have the same freedom of action as regards straddling stocks as they do for discrete, i.e. entirely high seas stocks. For straddling stocks, the sense of the UNCLOS articles is that the measures DWF states take outside the zone must conform, in some appropriate manner, to the measures established for these stocks by the coastal state concerned.

### Article 119 — The Non-Discrimination Provision — Legal Analysis and Implications

Current problems in the Northwest Atlantic involving one member of NAFO, the European Community, and the non-members whose vessels fish in the area, provide a basis for examining one possible application of the non-discrimination provision at the end of Article 119. One of the primary norms that have applied to the management of fisheries in the area has been the maintenance of proportional shares in accordance with a long-established percentage share formula applied to each stock. The percentage share formula was developed in ICNAF, and was carried into the initial NAFO quota table. Subsequent NAFO quota tables have reflected quota shares very close to the shares in the initial quota table, reflecting the firm establishment of the proportional sharing principle codified in the Resolution passed in 1989, a copy of which is attached as Appendix 2. The catches currently taken by the European Community and the non-NAFO members are clearly inconsistent with the sharing formulas, and have the effect of discriminating against the NAFO members that comply with these formulas. Certain implications flow from the Article 119 non-discrimination provision.

The European Community, by adopting unilateral quotas higher than the quotas assigned to it by NAFO, has thereby changed the actual proportional shares for all the NAFO members. This action discriminates severely against the other NAFO members because these measures increase the proportion of the total catch for the fishermen of the European Community, decreasing the proportional shares of the other members. In addition, by depleting the stock through overfishing, the European Community further discriminates against the other members who pay the price, in the long-term, for the benefits obtained in the short term by the overfishing member.

As regards the non-members or "new entrants" conducting fisheries without any appearance of even unilateral restrictions, a similar analysis applies. Further, given the firm establishment of the proportional sharing principle and the shares themselves, and the fact that the total needs of the countries that traditionally fish in the area exceed the TAC's that have been established (as well as the TAC's that would be established on the basis of rebuilt stocks) there would appear to be no "room", factually or legally, for any new entrant unless the NAFO members agree to reduce their proportional shares to create new shares for the new entrant. Current catches by non-members of NAFO discriminate against NAFO members by reducing their proportional shares of the total catch, and, through the resulting overfishing, by reducing their actual catches in the long term. Further, discrimination exists because the NAFO members have, over many years, restricted their catches in the interest of stock conservation, while the new entrants have paid none of the costs for the stocks they fish.

The contrary argument is, of course, that measures taken against new entrants amount to discrimination, contrary to Article 119, because the new entrants have the right to fish on the high seas. However a strong counter-argument can be made that exclusion of new entrants is not discrimination in the Northwest Atlantic and in other similar circumstances. It should be noted, first of all, that all the NAFO members are themselves exercising their right to fish on the high seas, as the NRA is high seas. As TAC's are distributed by NAFO in accordance with the customary proportional shares, the fact that the share for a specific NAFO member may be 10%, or 5% or 2% can not be considered as creating a situation of discrimination against that member. The corollary is that the fact that the customary proportional share for a non- member is zero also does not create a discriminatory situation. In fact, the question may be asked of any non-member claiming the right to start a new fishery in the NRA — what is the proportional share claimed? There can, of course, be no rational answer to this question.

Reference should be made, at this point, back to Article 116, paragraph (c), which states that the right to fish on the high seas is subject to "the provisions of this section." The analysis given above of the non-discrimination provision of this section of the LOS treaty supports the argument that:

a) states do not have the automatic right to enter into any high seas fishery that they choose, because in some cases, such entry discriminates against the fishermen of other states;
b) states that have a traditional presence in a high seas fishery may, in certain circumstances, have the right to take steps to exclude new entrants, and such steps do not constitute discrimination.

### 2. The 1958 Convention on Fishing and Conservation of the Living Resources of the High Seas

This Convention has not been ratified by any NAFO members. However the following majority of EC member states are Parties: Belgium, Denmark, France, the Netherlands, Portugal, Spain, the United Kingdom. The USA and Mexico are also Parties (vessels flying the flags of these two non-members of NAFO fish in the NRA). The possibility therefore arises that if Canada were to become party to the 1958 Convention, new legal relationships would come into force that might be used to implement Canada's coastal state rights duties and interests.

The relevance of most of the provisions of this Convention, in terms of providing binding rules applicable to the straddling stock problem described in the paper, is open to argument. This is because the 1958 Convention is written in terms of coastal state rights and interests in areas of the high seas adjacent to its territorial sea. As the subject of this paper concerns areas of the high seas adjacent to 200 mile limits, and therefore, by definition, not immediately adjacent to territorial seas, there would appear to be some room for argument as to the meaning of "adjacency". An argument can certainly be made that UNCLOS was not intended to reduce coastal states rights with regard to stocks that fell within these rights, throughout the migratory range of these stocks, under the 1958 Convention.

The key provisions of the 1958 Convention would appear to be Articles 6 and 7 which provide *inter alia*, that a coastal state has a special interest in the maintenance of productivity in the area adjacent to its territorial sea, other states must not apply conservation measures in the adjacent area opposed to those of the coastal state, and the coastal state may adopt unilateral measures in the adjacent area, subject to the requirements specified in the convention.

The basis for the argument that the 1958 Convention may continue to be applicable, rather than being superceded by UNCLOS, lies in Article 311 of UNCLOS. Article 311, paragraph 1 states:

> "This Convention shall prevail, as between the States Parties, over the Geneva Conventions on the Law of the Sea of 29 April 1958."

The word, "prevail" would appear to denote that in any circumstance where there is a conflict between provisions of the 1958 Convention and UNCLOS, the latter overrides. The word "prevail" does not denote replacement, or lapse, of the 1958 Conventions.

This argument is reinforced by Article 311, paragraph 2, which states:

> "This Convention shall not alter the rights and obligations of States Parties which arise from other agreements compatible with this Convention and which do not affect the enjoyment by other States Parties of their rights or the performance of their obligations under this Convention."

It may be argued that the Convention on Fishing and Conservation of the Living Resources of the High Seas is basically compatible with UNCLOS, and that therefore the provisions of the 1958 Convention specified above, which do not conflict with the provision of UNCLOS, remain applicable.

The underlying rationale for this argument is that it was the purpose of UNCLOS to expand the duties, rights and interests of coastal states within a specified area, 200 miles, not to detract from what the coastal state already had under the 1958 Convention (except insofar as there is a clear conflict, in which case

UNCLOS prevails). Certainly in the case of straddling stocks it would be an odd conclusion to state that one result of UNCLOS was to undermine the very limited rights that coastal states had, prior to UNCLOS, throughout the full extent of the migratory range of straddling stocks adjacent to their coasts.

One provision of the 1958 Convention is indisputably relevant in the sense of providing a binding obligation. Article 8 provides that any state with a special interest in the conservation of the living resources in an area of the high seas not adjacent to its coast may take certain action, including initiation of the compulsory dispute settlement provisions of the Convention. This clearly provides an avenue for a coastal state, even if it has no fishery in the area concerned, to compel the adoption of appropriate conservation measures in the area beyond 200 miles for a stock which straddles its 200 mile limit.

### 3. The NAFO Convention as a Source of Norms of International Law

The possibility that the NAFO Convention contains norms of international law in the fisheries field should not come as a surprise. Born of UNCLOS, the NAFO Convention represents a serious effort by key members of the international community to deduce, from the broad norms of the LOS treaty, subsidiary norms essential to implement the UNCLOS texts.

Two provisions of the NAFO Convention, both contained in Article XI (reproduced as an attachment to Appendix 1) of the Convention, are particularly relevant in this context.

Paragraph 3 of Article XI is the consistency provision, and requires the NAFO Fisheries Commission, in establishing management measures for straddling stocks, to seek to ensure that such measures are consistent with those established by the coastal state inside the zone for these stocks. It has been submitted, in a previous section of this study, that the consistency requirement is implicit in the UNCLOS provisions. The 15 parties (all the current Parties plus the USA, which did not become a Party) that negotiated the NAFO Convention may have thought so too, and decided to make this explicit in the NAFO Convention.

Paragraph 4 of Article XI recognizes the special interest of Canada in the stocks of the NAFO Regulatory Area (both the straddling stocks and the discrete stocks). The provision specifies the reasons for this recognition — the dependence of its coastal communities, and its extensive surveillance and inspection activities. The special legal position of Canada as the relevant coastal state, as described in this study, was obscured to avoid prejudicing the legal positions of other countries that took the position that coastal states, as a general rule, do not have any "special interest" in areas outside their 200 mile limits.

The consistency provision has been applied in NAFO in two ways.

First, the Fisheries Commission has followed the lead of the coastal state in adopting the $F_{0.1}$ formula, or its equivalent, for the setting of TAC's. As a result there has been a single TAC established for each straddling stock as a whole, with member quotas also applied to the stock as a whole, so there has been no need to consider the connection between two separate sets of TAC's.

Second, as regards the 2J3KL cod stock which is under Canada's sole management authority, the Commission has recognized that the TAC is fully subscribed by Canada and accordingly that the appropriate measure to adopt, for the purpose of consistency between the area outside and the area inside 200 miles, is a moratorium on fishing this stock outside 200 miles.

Arguments can, of course, be made that other approaches are possible for the NAFO-managed stocks and the Canadian managed 2J3KL cod stock, within the concept of consistency. There is no doubt that this is true, as a hypothetical concept. This does not detract from the validity of the current approaches in NAFO which are considered, by the majority of its members, as valid applications of the consistency principle.

However, as regards the establishment of management measures for straddling stocks managed by NAFO, it appears that even if a different consistency system applied, and separate management measures were implemented outside the 200 mile limit, the result would be the same in terms of the TAC's for the stocks. The management strategy adopted by the coastal state, e.g. $F_{0.1}$, would provide the overriding rule and the Commission, to ensure consistency, could not adopt a management strategy that would provide a higher level of exploitation. If the system produced, at the outset, two separate $F_{0.1}$ TAC's, for each stock, one applying to the portion of the stock inside 200 miles and the other applying to the portion of the stock outside, the net result would be a single $F_{0.1}$ TAC.

### 4. Bilateral Treaties for the Northwest Atlantic Fisheries as a Source of Norms of International Law

In the periods immediately preceding and immediately following Canada's 1977 extension of jurisdiction to 200 miles, bilateral treaties were negotiated between Canada and most of the countries that had been fishing in the Northwest Atlantic. All the countries involved were then members of ICNAF, and subsequently became members of NAFO. In the period immediately preceding the 1977 extension of jurisdiction the treaties with Norway, Poland, the USSR, Spain and Portugal were negotiated and came into force. In the period immediately following the 1977 extension, the treaties with Cuba, the German Democratic Republic, Bulgaria, Romania, Japan and the Faroe Islands were negotiated and came into force. The last in the series was the treaty negotiated with the European Community and brought into force on December 30, 1981. All these treaties (except the treaty with Spain which was terminated by Canada), some with amendments, have remained in force at the time of preparation of this study.

In the period immediately preceding the 1977 extension of jurisdiction the primary purpose of the treaties was to pave the way for the extension by establishing clear understandings on rights and obligations for the period following extension. While the United Nations Conference on the Law of the Sea had, by that time, advanced to the stage that it was clear 200 mile zones would, at some point, become incorporated into international law, it was not clear that this would be the case by January 1, 1977, the time Canada was planning to extend jurisdiction. Similarly, the Conference had, by that time, developed the key provisions of UNCLOS regarding the rights and obligations pertaining to the 200 mile zones, but these were not yet legally binding under international law. Accordingly, Canada and the five countries that were taking the largest non-Canadian catches in the Northwest Atlantic — Norway, Poland, the USSR, Spain and Portugal — considered it useful to establish bilateral treaties that would recognize Canada's right to extend jurisdiction to 200 miles whenever this took place, and would also contain the provisions on 200 mile zone rights and obligations that had already been developed in the United Nations Conference. In doing so it was also considered appropriate to include ancillary provisions governing, *inter alia*, fisheries in the area beyond and adjacent to the Canadian 200 mile limit.

In the period following Canada's extension of jurisdiction Canada received requests from all the other countries that had been fishing on the Northwest Atlantic (except Iceland), and from the European Community, for equivalent bilateral treaties. Canada acceded to these requests.

The provisions of these treaties relevant to straddling stocks are reproduced in Appendix 3. A review will quickly indicate that the norms that emerge are consistent with and support three of the norms described in this paper as implicit, and explicit, in the other relevant sources of international law, i.e.

a) the general obligation to ensure the conservation of the living resources in areas outside the jurisdiction of coastal states;
b) recognition of the special interest of the coastal state in the portion of a straddling stock outside its 200 mile limit; and
c) recognition of the requirement that measures adopted for the area adjacent to a coastal state's 200 mile limit are to be consistent with the measures adopted by the coastal state for the same stock within its zone.

## Options for Action by Coastal States

The broad spectrum of options for action by coastal states based on their preferential rights outside 200 miles regarding straddling stocks may be reviewed for the purposes of this paper under four categories:

— incorporation in coastal state law of provisions related to preferential rights over straddling stocks, and their enforcement;
— action in conformity with International Law relating to retorsion;
— dispute settlement;
— action in conformity with International Law relating to reprisals.

### *Incorporation in Coastal State Law of Provisions Related to Preferential Rights Over Straddling Stocks, and Their Enforcement*

The starting point for this approach is provided by Miles and Burke:

> "If efforts to agree on a conservation regime are unsuccessful, although all parties have negotiated in good faith to secure such a regime, what further procedural steps may be taken? Does the treaty permit further actions by the coastal state to seek recognition of its right to exercise sovereign rights over the straddling stocks? Is the situation beyond effective action under the treaty? The answer to this is clearly no. The treaty provides that the coastal state has superior rights over straddling stocks. In the absence of agreed measures, under the treaty the coastal state might prescribe measures for observation by all who fish the straddling stocks, including on the high seas; demand that these states observe these measures; and if refused, seek a remedy through the compulsory dispute mechanism."

Although Miles' suggestion that "the coastal state might prescribe measures for observation by all who fish the straddling stocks, including on the high seas" leads, in his view, to dispute settlement, there are other possibilities. One possibility open to the coastal state is to incorporate in its domestic legislation, in some manner consistent with its obligations under international law, provisions relating to its preferential rights over straddling stocks outside its 200 mile limit. Such incorporation could provide for regulations applicable outside its 200 mile limit and for enforcement of these regulations outside 200 miles in, for example, cases where there are bilateral or multilateral agreements providing for such enforcement. Provision for non-flag state enforcement on the high seas has already been made in at least one international fisheries management convention, the International Convention for the High Seas Fisheries of the North Pacific Ocean in force since 1953 (Article X).

*Action in Conformity with International Law Related to Retortion*

Kelsen (1959) defines retorsion in the following manner:

> "the conduct by which a state violates some interest of another state may not be a delict, that is to say, the state whose interest is violated may not be authorized to execute a sanction by taking an enforcement action against the state which has violated its interest; but it may react by a similar violation of an interest of the latter state. Such a reaction is called a retorsion. It is no sanction, for it is not an enforcement action — the employment of physical force in case of resistance not being permitted".

Elements of retorsion are included in the following list of recourses provided by Miles and Burke (1988):

> "Diplomatic action (protests), domestic remedies (embargoes on fishery or other trade, refusal of access to ports for logistic support, denials of economic assistance, suspension of particular benefits), international sanctions (remedies available under international agreements, including trade agreements) are all possible instrumentalities. Whether any of these is available and, if so, feasible to employ is another question — the point is that a state whose interests are harmed by refusal of a high seas fishing state to take necessary conservation measures is not necessarily helpless."

In the Northwest Atlantic situation, Canada has employed the following:

- diplomatic action (protests) to the European Community and to the non-NAFO members fishing in the NAFO Regulatory Area
- refusal of access to ports to the same parties, except in cases of force majeure.

*Dispute Settlement*

Quoting from Miles and Burke (1988) again:

> "If the 1982 Convention on the Law of the Sea were in force...between the states concerned, disputes about the application of articles concerning high seas fishing would be subject to compulsory dispute proceedings in Part XV"

In the interim there are, of course, other possible approaches to dispute settlement. These are diverse and require no exhaustive commentary in this paper. Reference has already been made, previously in this paper, to the compulsory dispute settlement provision of the 1958 Convention on Fishing and Conservation of the Living Resources of the High Seas, and the fact that, at the present time, Canada is not party to this Convention.

In any event, governments are generally very reluctant to engage compulsory dispute settlement processes for a number of reasons, including the heavy costs involved in personnel and money, and the time required to obtain results.

*Action in Conformity with International Law Relating to Reprisals*

Kelsen's (1959) definition of the law relating to reprisals is as follows:

> "Reprisals are acts which, although normally illegal, are exceptionally permitted as reaction of one state against a violation of its rights by another state. Typical examples are...nonfulfillment of treaty obligations in relation to that state."

The starting point for consideration of this option is the submission that, under certain circumstances, a state which permits its nationals to fish straddling stocks outside the 200 mile limit of a coastal state has violated the rights of that coastal state, and its own obligations under international law to that state, and has thus committed an international delict.

The range of possible reprisal actions by the coastal state is, of course, very broad in theory. The spectrum includes imposition of trade sanctions which would, *prima facie* be contrary to the General Agreement on Tariffs and Trade (GATT), and physical interference with the operations of vessels on the high seas for example, by arrests or by inflicting damage to their fishing gear. However, the range of possible reprisal actions is very limited in terms of practicability in the prevailing geopolitical situation in which states try to minimize disruptions to their normal relations.

### International Legal Norms Applicable to Straddling Stocks

The following is an attempt to articulate specific norms that may be deduced from the comments above and from texts referenced, with commentaries summarizing points made earlier in this paper.

1. *The right to fish on the high seas is subject to specific obligations and limitations, including the obligation to respect the duties, rights and interests of coastal states in stocks which extend, beyond their 200 mile limits, into the high seas.*

These obligations include the following specifics:

a) to take the necessary conservation measures (Article 117 UNCLOS);
b) to cooperate with other states in taking such measures (Article 117 UNCLOS);
c) to ensure that the measures adopted are non-discriminatory against the fishermen of any state (Article 119 UNCLOS).

2. *States have the duty to take, or cooperate with other states in taking, such measures for their respective nationals as may be necessary to ensure consistency between:*

a) *the measures in place for their nationals regulating their fisheries on the high seas on stocks that occur both on the high seas and within an area under the fisheries jurisdiction of a coastal state, and*
b) *any measures taken by the coastal state for the management and conservation of the same stocks with respect to fishing activities conducted within the area under its fisheries jurisdiction.*

3. *Where fisheries take place in an area adjacent to the 200 mile limit of a coastal state, the state whose nationals conduct such fisheries:*

a) *shall not adopt for the regulation of such fisheries, measures which are opposed to those which have been adopted by the coastal state for a stock or stocks which extend beyond the coastal state's 200 mile limit, and which undermine the objectives of such coastal state measures;*
b) *shall prevent fisheries by its nationals which undermine the objectives of such coastal state measures.*

While this provision would appear to be the logical obverse of the previous provision, it is a necessary re-statement of the primary norm, if only because the "consistency" concept in the previous text is open to distortive interpretation.

4. *States have an obligation to prevent harm to coastal states resulting from fishing on the high seas.*

It is a recognized principle of international law that no state has the right to use its territory, or permit it to be used, in a way that causes environmental harm to another state (*sic utere tuo alienum non laedas*). This obligation is expressed in the pollution context in Article 194(2) of UNCLOS.

In recent years, it has become clear that the obligation to prevent harm extends to activities under a state's control in any location (e.g., liability for satellites falling from outer space).

In light of the above, it is submitted that a fishing state has an obligation to prevent fishing of straddling stocks to the extent that such fishing can be shown to damage a coastal state's ability to manage and conserve fisheries in its 200-mile zone in accordance with the LOS Convention.

5. *A state which permits its nationals to fish, in an area adjacent to the 200 mile limit of a coastal state, at levels in excess of those established for that state by the international organization established to manage fisheries in the relevant area, is engaged in a form of discrimination against other states fishing in accordance with the levels established for those states by the international organization concerned.*

UNCLOS Article 119(3) provides that "(s)tates concerned shall ensure that conservation measures (for high seas fisheries) and their implementation do not discriminate in form or in fact against the fishermen of any State". If any state takes straddling stocks in excess of agreed levels, while the catches of other interested states remain at the customary levels in accordance with conservation arrangements, then the "excess-fishing" state has discriminated in fact against the other states involved in the fishery. This is so because, if the other states are to adhere to their duty to conserve the fishery, they are unable to increase their catch as the "excess-fishing" state has done.

6. *Where the nationals of a state fish, outside another state's 200 mile limit, a stock which occurs both inside and outside that limit, in the event of failure between the coastal state and the other state to reach agreement on management measures outside 200 miles for that stock, the coastal state has the right to take unilateral measures to conserve the stock.*

UNCLOS Article 118 calls for the establishment of subregional or regional fisheries organizations for the purpose of taking measures necessary for the conservation of high seas fisheries. UNCLOS Article 63(2) also refers to such organizations, as a mechanism to "seek ... to agree upon the measures necessary for the conservation of (straddling stocks)". These provisions leave in doubt the means by which conservation of stocks is to be ensured if direct negotiations between the coastal state and fishing states should fail, or if the (sub)regional organization is unable to agree on conservation measures.

The 1958 Geneva Convention gave coastal states the right, in certain circumstances, to adopt unilateral conservation measures (Article 7) in the area adjacent to its territorial sea. It could be argued that the circumstances dealt with in the Geneva Convention have been addressed by the extension of coastal state fisheries jurisdiction to 200 miles. This argument seems weak, however, since the nature of the problem, i.e. the fishing of stocks adjacent to the outer limit of coastal state jurisdiction, remains the same. In addition, the fundamental principle of fisheries conservation should not be limited by arbitrary geographical lines.

The provisions of UNCLOS support the view that the duties, rights and interests of the coastal state with respect to straddling stocks are superior to those of other states. The Convention clearly defines the sovereign rights and duties of the coastal state in its 200 mile zone, with respect to fisheries (UNCLOS, Part V). The actions of other states beyond 200 miles should not be allowed to derogate, in terms of their effects, from the sovereign rights of the coastal state within its 200-mile zone.

In view of the above, it may be permissible for a coastal state, in the absence of agreed measures, to take unilateral measures for the conservation of straddling stocks. The nature of such unilateral measures might vary according to the circumstances.

An argument can be made that such measures could not include enforcement against foreign flag vessels outside the coastal state's 200 mile limit, such enforcement being reserved exclusively to the flag state under international law. The question may well be asked, however, whether such enforcement might be justified under customary international law, in circumstances equivalent to situations where a state faces a serious threat from action taken outside its boundaries where such action has the effect, in some way, of crossing these boundaries. Further analyses of this point would be useful.

7. *States have the duty to prevent their nationals from fishing in areas of the high seas for which international conservation regulations have been promulgated by an international organization established to manage the fisheries of the area concerned, and the available catches have been fully allocated to the members of the organization.*

There is clearly an international legal norm favouring, for states that have traditionally fished a particular area, priority over new entrants to the fishery. For example, under UNCLOS Article 62(3), "the need to minimize economic dislocation in States whose nationals have habitually fished in the (200 mile) zone" is a relevant factor to be taken into account by a coastal state in giving access to the fisheries in its 200 mile zone. This provision reflects the customary practice of "grandfathering" traditional fisheries when the coastal state extends its fisheries jurisdiction.

Traditional rights are also given recognition in other conventions. For example, Article XI(4) of the NAFO Convention obliges the Fisheries Commission to take into account the interests of Commission members whose vessels have traditionally fished in the Regulatory Area in allocating catches.

These examples, and the discrimination analysis provided previously in this study support the view that a fully subscribed high seas fishery, subject to internationally agreed conservation measures, is closed to new entrants unless the agreed measures are adjusted to provide an allocation to the new entrants.

## Conclusion

The 1982 United Nations Convention on the Law of the Sea, in establishing an outer limit of 200 miles for coastal state jurisdiction over fisheries, has provided the key to conservation of most of the world's fish stocks. However it has left major stocks, in a number of areas of the world, exposed to international fishing pressures similar to those which affected, prior to the late 1970's, those stocks now under clear coastal state control. The straddling stocks in a number of areas of the world are a significant sub-category of these exposed stocks because of their substantial volume and value and because of the sovereign rights established by UNCLOS as regards these stocks inside the 200 mile limits of coastal states. The availability of these stocks outside 200 mile limits undermines, to a significant degree, these sovereign rights.

The international community, in developing the LOS Convention, knew that as regards fisheries outside 200 miles they were leaving a jurisdictional gap and an environmental protection gap, which would have to be dealt with in some way at some point in the future.

One way that looked promising at the outset was through international organizations that would establish sound conservation structures and make them work. This study has shown how one such organization, NAFO, has run into serious problems in this respect. The limitations of such organizations, dependent on international consensus among its members, and on the willingness of other countries to stay out of the fisheries concerned, undermine the ability of these organizations to do their job.

Current developments in a number of areas of the world suggest the possibility that the time has come for a reassessment of the international legal principles in play regarding fisheries conservation and environmental protection, in order to determine if these principles can be developed so as to strengthen international institutions and make them more effective.

The United Nations is currently addressing these issues through the 1993–94 UN Conference on Straddling Fish Stocks and Highly Migratory Fish Stocks.

## References

BURKE, W.T. 1983. Convention on the Law of the Sea: provisions on conditions of access to fisheries subject to national jurisdiction, Expert Consultation on the Conditions of Access to the Fish Resorces of the Exclusive Economic Zone. FAO Fisheries Report No. 293: p. 39.

KELSEN, H. 1959. Principles of International Law. Rinehardt and Company Inc., New York, NY. p. 25.

MILES, E.L., AND W.T. BURKE. 1988. Pressures on the United Nations Convention on the Law of the Sea of the 1982 Arising from New Fisheries Conflicts: The Problem of Straddling Stocks. Paper presented to the Joint Soviet Maritime Law Association/Law of the Sea Institute Symposium on the Law of the Sea, Moscow, 28 November–2 December, 1988.

## APPENDIX 1

### The NAFO Convention — Relevant Elements

The NAFO membership comprises 12 Parties as follows: Bulgaria, Canada, Cuba, Denmark in respect of the Faroe Islands and Greenland, the EEC, the German Democratic Republic, Iceland, Japan, Norway, Poland, Romania, the USSR.

NAFO is composed of 3 bodies — a General Council, a Fisheries Commission, and a Scientific Council. The Fisheries Commission is the body that manages the NAFO Regulatory Area (NRA), and is composed of all the NAFO members except Iceland.

NAFO holds a regular annual meeting in September of each year, to establish the management measures for the NRA for the following year. This meeting is preceded by a regular annual meeting in June of each year of the Scientific Council, which provides scientific advice to the Fisheries Commission in accordance with the terms of reference previously specified by the Commission or on its own initiative under certain circumstances. Additional meetings of any of the bodies can be and have been held at other times of the year as required.

The key guidelines for the adoption of management measures are contained in Article XI of the Convention (copy attached). They comprise optimum utilization, management measures for straddling stocks outside 200 miles are to be consistent with management measures inside 200 miles, respect for traditional fisheries, and special consideration for Canada in allocations.

The system for implementing management measures is as follows: the Fisheries Commission, based on the advice of the Scientific Council, adopts, by majority vote, TAC's and other relevant measures for the NRA; it then adopts, by majority vote, the respective quotas for each stock for each of the NAFO members. Following the NAFO meeting the Executive Secretary mails the relevant texts to the members, who then have 60 days within which to lodge written objections to any or all of the measures adopted. When an objection is lodged, the member that lodges it is not legally bound by the relevant measures, and further objection periods are opened up for other NAFO members, within which they can lodge objection to the same measure.

**Article XI**

1. The Fisheries Commission, hereinafter referred to as "the Commission", shall be responsible for the management and conservation of the fishery resources of the Regulatory Area in accordance with the provisions of this Article.

2. The commission may adopt proposals for jointaction by the Contracting Parties designed to achieve the optimum utilization of the fishery resources of the Regulatory Area. In considering such proposals, the Commission shall take into account any relevant information or advice provided to it by the Scientific Council.

3. In the exercise of its functions under paragraph 2, the Commission shall seek to ensure consistency between:

a) any proposal that applies to a stock or group of stocks occurring both within the Regulatory Area and within an area under the fisheries jurisdiction of a coastal State, or any proposal that would have an effect through species interrelationships on a stock or group of stocks occurring in whole or in part within an area under the fisheries jurisdiction of a coastal State; and
b) any measure or decisions taken by the coastal State for the management and conservation of that stock or group of stocks with respect to fishing activities conducted within the area under its fisheries jurisdiction.

The appropriate coastal State and the Commission shall accordingly promote the coordination of such proposals, measures and decisions. Each coastal State shall keep the Commission informed of its measures and decisions for the purpose of this Article.

4. Proposals adopted by the Commission for the allocation of catches in the Regulatory Area shall take into account the interests of Commission members whose vessels have traditionally fished within that Area, and, in the allocation of catches from the Grand Banks and Flemish Cap, Commission members shall give special consideration to the Contracting Party whose coastal communities are primarily dependent on fishing for stocks related to these fishg banks and which has undertaken extensive efforts to ensure the conservation of such stocks through international action, in particular, by providing surveillance and inspection of international fisheries on these banks under an international scheme of joint enforcement.

5. The Commission may also adopt proposals for international measures of control and enforcement within the Regulatory Area for the purpose of ensuring within that Area the application of this Convention and the measures in force thereunder.

6. Each proposal adopted by the Commision shall be transmitted by the Executive Secretary to all Contracting Parties, specifying the date of transmittal for the purposes of paragraph 1 of Article XII.

7. Subject to the provisions of Article XII, each proposal adopted by the Commission under this Article shall become a measure binding on all Contracting Parties to enter into force on a date determined by the Commission.

8. The Commission may refer to the Scientific Council any question pertaining to the scientific basis for the management and conservation of fishery resources within the Regulatory Area and shall specify terms of reference for the consideration of that question.

9. The Commission may invite the attention of any or all Commission members to any matters which relate to the objectives and purposes of this Convention within the Regulatory Area.

APPENDIX 2

Northwest Atlantic Fisheries Organization

Serial No. N1682

NAFO/GC Doc. 89/3

ELEVENTH ANNUAL MEETING - SEPTEMBER 1989

Resolution of the General Council
of the Northwest Atlantic Fisheries Organization

The General Council,

Recalling the obligations inscribed in the Law of the Sea Convention of 1982 as regards international cooperation to provide for the conservation and optimum utilization of the living resources of the sea;

Recalling that the NAFO Convention provides that the object of the Organization shall be to contribute through consultation and cooperation to the optimum utilization, rational management and conservation of the fishery resources of the NAFO Convention Area;

Noting that implementation of the NAFO Convention from the outset has been based on the principles of conformity with scientific advice and relative stability in the proportionate shares of the total catches of NAFO-managed stocks, subject to changes through decisions taken by NAFO;

Noting the commitment of all NAFO Contracting Parties to the principles of conservation and rational management of living marine resources;

Calls for compliance with the NAFO management framework in place since 1979, and compliance with NAFO decisions in order to provide for conservation and maintain the traditional spirit of cooperation and mutual understanding in the Organization.

# APPENDIX 3

## Agreement between the Government of Canada and the Government of Spain on Mutual Fisheries Relations

**Madrid, June 10, 1976**

**In force June 10, 1976**

### Article IV

The Government of Canada and the Government of Spain undertake to cooperate directly or through appropriate international organizations to ensure proper management and conservation of the living resources of the high seas beyond the limits of national fisheries jurisdiction, including areas of the high seas beyond and immediately adjacent to the areas under their respective fisheries jurisdiction, taking into account their interest in such resources.

## Agreement between the Government of Canada and the Government of the Union of Soviet Socialist Republics on Mutual Fisheries Relations

**Moscow, May 1, 1984**

**In force May 1, 1984**

The Government of Canada and the Government of the Union of Soviet Socialist Republics undertake to cooperate directly or through appropriate international organizations to ensure proper management and conservation of the living resources of the high seas beyond the limits of national fisheries jurisdiction including the high seas beyond and immediately adjacent to the areas under their respective fisheries jurisdiction, taking into account their interests in such resources.

### Article VII

Government of Canada and the Government of the Union of Soviet Socialist Republics undertake to cooperate pursuant to the convention on Future Multilateral Co-operation in the Northwest Atlantic Fisheries, in particular, Article XI, paragraph 4, thereof duty without rights through which it can satisfy that duty. Whether or not this is so, it can certainly be argued that the duty of the coastal state creates a duty on the part of other states to make it possible for the coastal state to honour its duty. This is a point which will be picked up later in this section.

# An Overview of Canada's Arctic Marine Fisheries and their Management with Emphasis on the Northwest Territories

R. McV. Clarke

*Department of Fisheries and Oceans,*
*Central and Arctic Region, Freshwater Institute,*
*501 University Crescent, Winnipeg, Manitoba, Canada R3T 2N6*

Clarke, R. McV. 1993. An overview of Canada's Arctic marine fisheries and their management with emphasis on the Northwest Territories, p. 211–241. *In* L. S. Parsons and W. H. Lear [eds.] Perspectives on Canadian marine fisheries management. Can. Bull. Fish. Aquat. Sci. 226.

Canada's Arctic marine fisheries are conducted primarily by aboriginal peoples and are of great importance to them. The subsistence fisheries provide for the continuation of their traditional cultures and supply a considerable portion of their food, the replacement value of which is about $15 million. The commercial and recreational fisheries provide one of the few sources of employment and income for northern communities, about $6.3 million to the Canadian economy, and about $1 million in benefits to Canadian consumers. The main species harvested in the fisheries are scallops (*Chlamys islandicus*), shrimp (*Pandalus* spp.), Atlantic salmon (*Salmo salar*), Arctic charr (*Salvelinus alpinus*), whitefishes (*Coregonus* spp.), Greenland halibut (*Reinhardtius hippoglossoides*), beluga (*Delphinapterus leucas*), narwhal (*Monodon monoceros*), walrus (*Odobenus rosmarus*) and seals (Phocidae). The fisheries are managed cooperatively by resource managers and resource users. Formal cooperative management boards that have specific decision-making responsibilities exist in some areas. Further development of fisheries is a goal of native groups and governments; however, biological productivity is low, economic potential is limited and infrastructure is lacking. Sound management and conservation of Arctic marine fisheries requires better understanding and information on the biology of fishery resources and on Arctic marine ecosystems, expanded cooperative management, increased fishery planning, and integrated resource management and planning so that non-renewable resource developments in the Arctic and throughout the world do not adversely affect the fisheries.

---

Les peuples autochtones sont les principaux exploitants des ressources marines des eaux canadiennes de l'Arctique, ressources qui jouent un rôle déterminant dans leur vie quotidienne. Cette pêche de subsistance assure leur mode de vie traditionnelle et fournit une grande partie de leurs aliments dont la valeur de remplacement se situe à environ 15 millions $. Les pêches commerciales et récréatives s'inscrivent parmi les quelques sources de revenus et d'emplois pour les collectivités du nord, contribuent environ 6,3 millions $ à l'économie canadienne et rapportent environ 1 million $ en avantages pour les consommateurs canadiens. Les principales espèces exploitées sont le pétoncle d'Islande (*Chlamys islandicus*), la crevette (*Pandalus* spp.), le saumon de l'Atlantique (*Salmo salar*), l'omble chevalier (*Salvelinus alpinus*), le corégone (*Coregonus* spp.), le flétan du Groenland (*Reinhardtius hippoglossoides*), le béluga (*Delphinapterus leucas*), le narval (*Monodon monoceros*), le morse (*Odobenus rosmarus*) et les phoques (Phocidae).

Les gestionnaires et les exploitants de la ressource gèrent la pêche selon un régime de coopération. Il existe dans certains secteurs des conseils officiels de gestion coopérative chargés de responsabilités précises concernant la prise de décisions. Certains groupes et gouvernements autochtones désirent développer ces pêches; par contre, la productivité biologique est faible, le potentiel économique est limité et l'infrastructure est absente. La saine gestion et la conservation des pêches maritimes dans l'Arctique requièrent une compréhension plus profonde que l'on obtiendra grâce à de plus amples données sur la biologie des ressources halieutiques et les écosystèmes marins de l'Arctique, une gestion coopérative élargie, une planification accrue des activités de pêche ainsi que l'intégration de la gestion de la ressource et de la planification de la pêche afin que l'exploitation des ressources non renouvelables de l'Arctique et du globe n'ait pas un impact négatif sur les pêches.

## Introduction

Canada's Arctic marine fisheries and their management differ in many respects from the fisheries of the Atlantic and Pacific coasts. The subsistence fisheries for fish and marine mammals are culturally and nutritionally important to the lives of northerners, especially the Cree, Dene and Inuit. Commercial and recreational fisheries are small but provide a major source of income and employment in northern communities. Northern residents hope that fishery development will provide increased economic returns and food and cultural experiences for a growing population. Fishery development, however, is hampered by low biological productivity, a lack of much necessary biological information and limited economic potential. Because fisheries are small and scattered and management costs are high, fishery management is not intensive. Much reliance is placed on self-compliance by the resource users for the majority of the fisheries. Fishery management is evolving into formal cooperative management arrangements in which resource users share responsibility for fishery management with government.

Knowledge of Arctic fishery resources and their exploitation has been summarized by McCart and Den Beste (1979), Davis et al. (1980), Davies et al. (1986), Stewart et al. (1986) and Crawford (1989), while various aspects of their management have been described by Clarke et al. (1989), Strong (1989), Yaremchuk et al. (1989), Yaremchuk and Wong (1989a, 1989b) and Cosens et al. (1990). In this paper, the physical, biological and human settings and the fisheries for anadromous and marine fish and marine mammals are described, and the benefits from them are outlined. Their management is discussed, especially aspects related to conservation of fishery resources, regulation and compliance, fishery development and cooperative management. Finally, impacts of other human activities on Arctic fishery resources and their management are discussed. Whenever possible harvest and associated information is provided for 1987/88.

For the purpose of this paper Canada's Arctic seas are considered to be Canada's marine waters north of 60°N plus the waters of Hudson, James and Ungava bays. Excluded from the overview are the offshore (beyond 12 miles) fisheries in Baffin Bay and Davis Strait which are in the purview of the Northwest Atlantic Fishery Organization. Some information is provided on the offshore shrimp fishery conducted by Baffin Island and northern Quebec interests.

## The Setting of the Fisheries

### The Physical Setting

Canada's Arctic seas are diverse. They are largely surrounded by land and are a comparatively closed system. They include open waters in the Arctic Ocean, Beaufort Sea, Baffin Bay and Davis Strait, a large almost enclosed area in Hudson Bay and a complex of interconnecting channels through the Arctic Archipelago (Fig. 1). Their total area, estimated to be about 3.0 million km$^2$ (Table 1), is two to three times that of Canada's Atlantic and Pacific seas combined. The Arctic and Hudson Bay coastline comprises 71% of Canada's total coastline (Table 1).

The main input of waters to the Arctic Ocean is from the Atlantic Ocean between Greenland and Eurasia (Fram Strait), mainly at depths below 200 m, and from the Pacific Ocean through Bering Strait (Table 1). Rivers provide only about 3% of the water input into the Arctic Ocean and of this the Mackenzie River, the main Canadian source of water, provides about 11% (Anderson et al. 1983). The water exits through the Arctic Archipelago, primarily through Lancaster Sound, Jones Sound and Smith Sound, and east of Greenland mainly at depths above 100 m (Table 1). The main circulatory pattern of the flow in the more open water

TABLE 1. Some characteristics of Canada's Arctic seas and their drainage basins.

| Characteristics | Size | Reference |
|---|---|---|
| Marine area ($km^2$) | | |
| Total | 3 000 000 | L. Johnson, pers. comm. |
| Hudson Bay | 822 324 | Statistics Canada (1981) |
| Coastline length (km) | | Statistics Canada (1981) |
| Total | 172 979 | |
| Mainland | 35 646 | |
| Islands | 137 333 | |
| Arctic Ocean water inputs ($10^6\ m^3 \cdot s^{-1}$) | | Anderson et al. (1983) |
| Atlantic: Fram Strait | 2.43 | |
| Pacific: Bering Strait | 1.35 | |
| Rivers | 0.11 | |
| Precipitation | 0.03 | |
| Arctic Ocean water outputs ($10^6\ m^3 \cdot s^{-1}$) | | Anderson et al. (1983) |
| Canadian Arctic Archipelago | 2.0 | |
| East Greenland: Fram Strait | 1.8 | |
| Sea ice | 0.10 | |
| Evaporation | 0.02 | |
| Hudson Bay water inputs ($10^6\ m^3 \cdot s^{-1}$) | | Barber (1968) |
| Atlantic: Hudson Strait | 0.5 | |
| Arctic: Fury and Hecla Strait | 0.05 | |
| Rivers and precipitation | 0.05 | |
| Hudson Bay water outputs ($10^6\ m^3 \cdot s^{-1}$) | | Barber (1968) |
| Hudson Strait | 0.6 | |
| Drainage basin area ($km^2$) | | Statistics Canada (1987) |
| Arctic | 3 576 900 | |
| Hudson Bay | 3 945 700 | |

bodies is anti-clockwise (Department of Fisheries and Oceans 1982), with the major exception being the clockwise circulation of the Beaufort Sea gyre (Giovando and Herlinveaux 1981).

Hudson Bay receives most of its water from the Atlantic through Hudson Strait, and the remainder from the Arctic Ocean through Fury and Hecla Strait, rivers and precipitation (Table 1). The water circulates anti-clockwise and exits to the Labrador Sea through the southern part of Hudson Strait (see Milko 1986).

Canada's Arctic seas receive fresh water draining from 74% of the land area of Canada (Statistics Canada 1987) plus 148 000 $km^2$ of the USA (Department of Energy, Mines and Resources, and Information Canada 1974; Fig. 1; Table 1).

The major physical difference between Canada's Arctic seas and the Atlantic and Pacific seas is their low temperatures and the presence of ice. Extending from about 52°N to 85°N, the Arctic seas receive a low annual energy input which is highly pulsed. The majority of the energy input occurs during the short summer and is absorbed in melting ice. The result is ice covering the majority of the waters from six to twelve months each year (Department of Fisheries and Oceans 1982). Most of the area is ice-covered in winter but tides and currents keep the pack-ice, with 7 to 10 tenths ice cover, in the Arctic Ocean, Baffin Bay, Davis Strait and Hudson Bay in motion. The timing and extent of breakup varies considerably from year to year. By early September in some years virtually no ice occurs south of the permanent polar pack, whereas in other years 6 or more tenths ice covers most of the area except Hudson Bay, eastern Baffin Bay, Davis Strait and some coastal areas. Normally in September the permanent polar pack is 110–200 km from the Beaufort Sea coast but lies close to the northwest coasts of the Queen Elizabeth Islands. New ice usually begins to form in early September.

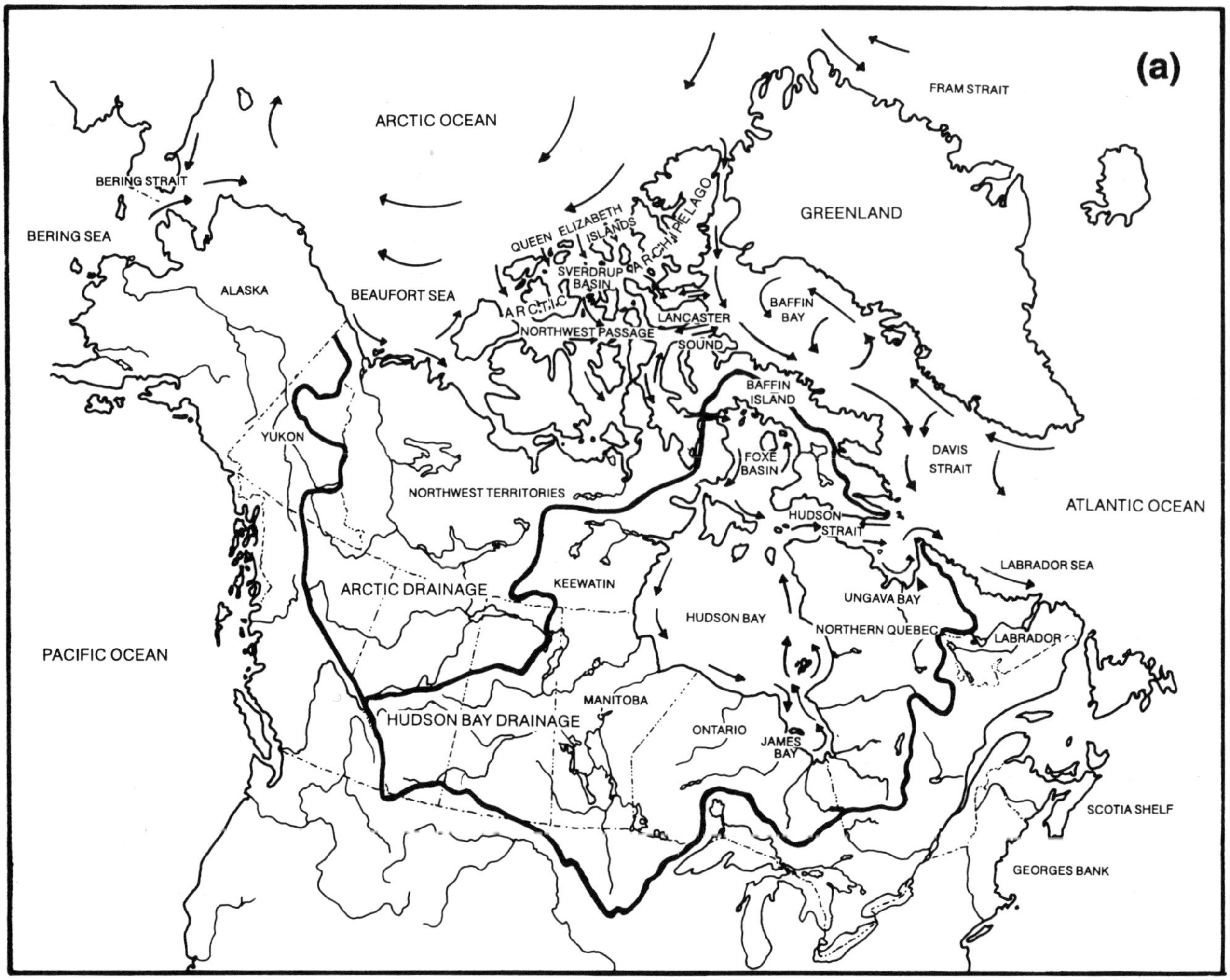


FIG. 1. Features of the Canadian Arctic. (a) major ocean currents, drainage basins and features; (b) other places mentioned in the text.

One important feature of the ice cover is the occurrence of polynyas and shore leads. These areas of open water or brash ice recur in the same locality in most years (see Stirling and Cleator 1981).

## The Biological Setting

Basic knowledge exists on the distribution and abundance of anadromous fish and marine mammals in the Arctic. However, there is little knowledge of the basic biology of marine fish and invertebrates, little understanding of the functioning of Arctic marine ecosystems, and little understanding of the long-term impacts of industrial activities and shipping. Knowledge, in general, is better for coastal areas than for offshore areas and is better for the Beaufort Sea and Lancaster Sound than for offshore Hudson Bay, the channels of the central Arctic Archipelago, Sverdrup Basin and the Arctic Ocean.

The inter-relationships amongst water, land and ice result in diverse habitats for marine and anadromous fishery resources. The habitats include open seas, channels, estuaries, shallow bays, deep fjords, lakes, rivers, fast ice, pack ice, ice edges, polynyas and shore leads. Despite the habitat diversity, biological diversity is low, compared with temperate and tropical seas, with relatively few species of plankton (Menzies 1975), bivalve molluscs (Lubinsky 1980) and marine fish (Hunter et al. 1984). The marine fauna is considered to be of recent origin (Dunbar 1968), as evinced by the low level of endemism (Menzies 1975).

The most important biological feature is the low productivity resulting from the low, pulsed energy input and

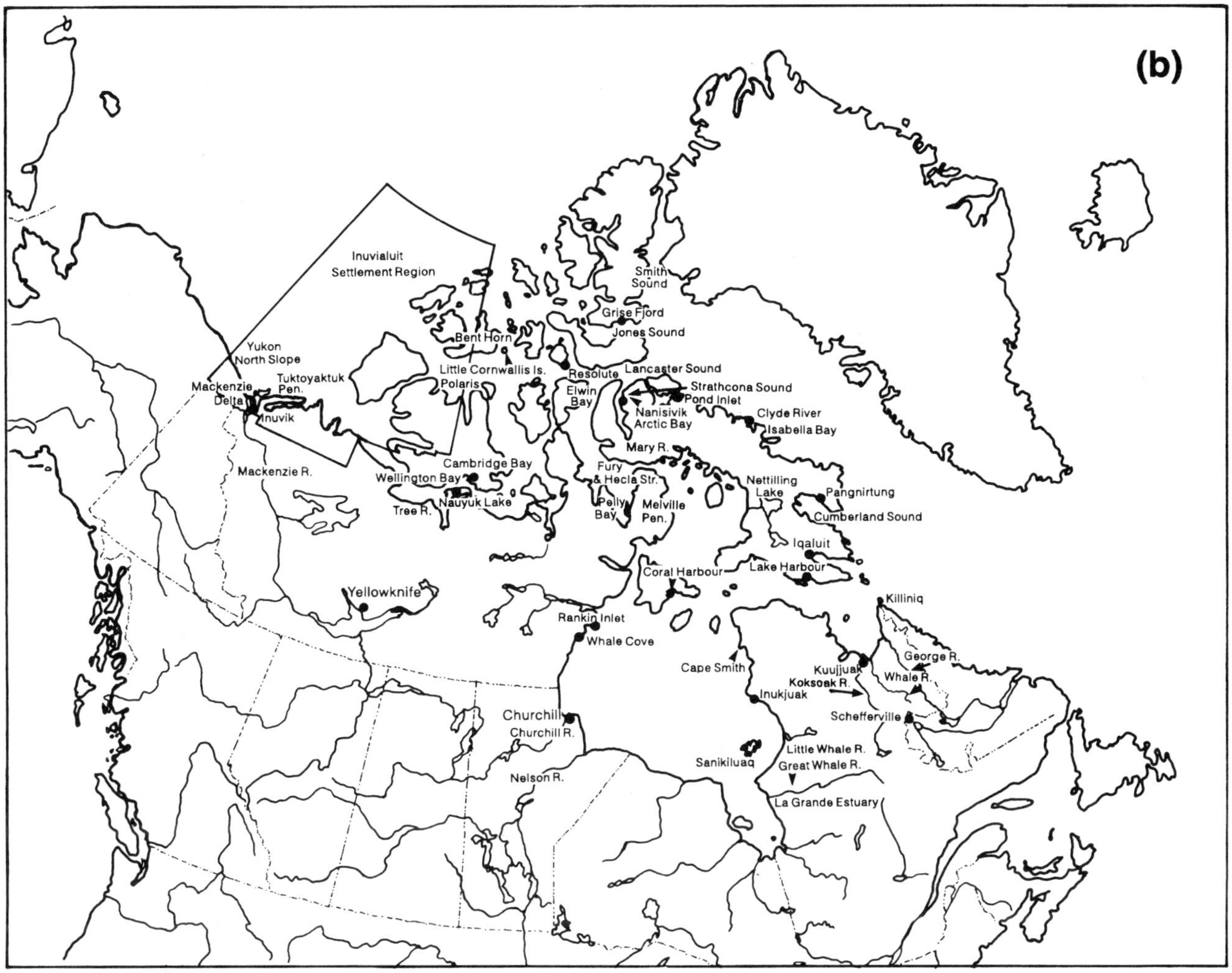


FIG. 1. (*Concluded.*)

the presence of ice. Three main sources of primary productivity occur in Arctic marine waters: phytoplankton, epontic algae and benthic plants (Davis et al. 1980). Productivity from phytoplankton in the water column mainly occurs during the open water season from May to October, with a peak just after the ice melts; it ranges from 9 to 40 g $C \cdot m^{-2} \cdot yr^{-1}$ (Davis et al. 1980). Productivity from epontic algae contributes about another 5 g $C \cdot m^{-2} \cdot yr^{-1}$ to total primary production (Davis et al. 1980; Welch and Bergmann 1989). Epontic algae bloom from about April to June with algal development being controlled by the depth of snow cover. Primary productivity from benthic plants is limited temporally by ice cover and the long period of low light in winter, and spatially by ice scour which results in an intertidal zone barren to a depth of 2-3 m. In some areas kelp (*Laminaria* spp.) is a major contributor to primary production. In contrast, primary production on the Scotia Shelf and Georges Bank is 102-450 g $C \cdot m^{-2} \cdot yr^{-1}$ (Mills and Fournier 1979).

Food webs in Arctic seas (Fig. 2) are relatively simple (Davis et al. 1980). Energy passes from phytoplankton and epontic algae to zooplankton and epontic and littoral amphipods. These are eaten by fish, especially Arctic cod (*Boreogadus saida*) and Arctic charr (*Salvelinus alpinus*), and some marine mammals and sea birds. Arctic cod, a key link in the food web (Bradstreet and Cross 1982), are eaten by Arctic charr, seals, beluga (*Delphinapterus leucas*), narwhal (*Monodon monoceros*) and sea birds (Fig. 2). Benthic macroinvertebrates are a major source of food for walrus (*Odobenus rosmarus*) and bearded seal (*Erignathus barbatus*). Large pelagic and benthic marine fish are absent except in Baffin Bay and Davis Strait. Limited data exist for secondary productivity but it is expected to be low, whereas the biomass is high.

Different parts of the food web (Fig. 2) are more important at different seasons or at different places. Production from the epontic community is available earlier in the year than production from the phytoplankton community (Bradstreet and Cross 1982). In ice-free offshore waters, the phytoplankton-supported portion of the food web dominates. The benthic component of the food web is most important in nearshore areas beyond the intertidal and ice scour zone.

Like the Arctic seas, fresh waters supporting anadromous fish generally have low productivity, simple food webs and high biomass (Rigler 1974; de March et al. 1978; Welch et al. 1989). However, primary production in the shallow and warmer lakes of the Mackenzie Delta and Tuktoyaktuk Peninsula, which support important stocks of anadromous coregonids (Bond and Erickson 1985), is higher because of the major contribution to overall production by aquatic macrophytes (Fee et al. 1988; Bodaly et al. 1989). On the mainland, especially further south, energy input is greater and longer, lake ecosystems are more diverse and more fish species occur.

The Arctic seas are not uniform in their biological importance. Riverine discharges, ocean currents, mixing of water masses, upwellings and ice patterns result in areas of increased production or concentrations of fish and marine mammals. Estuaries, such as those of the Mackenzie River and of rivers flowing into Hudson Bay, support major concentrations of beluga and anadromous coregonids (*Coregonus* spp.). Areas where water masses meet or upwelling occurs have increased productivity; shrimps (*Pandalus montagui* and *P. borealis*) concentrate in eastern Hudson Strait in such an area. Polynyas, leads and ice edges are areas where marine mammals and sea birds concentrate (Davis et al. 1980; Stirling and Cleator 1981; Bradstreet and Cross 1982).

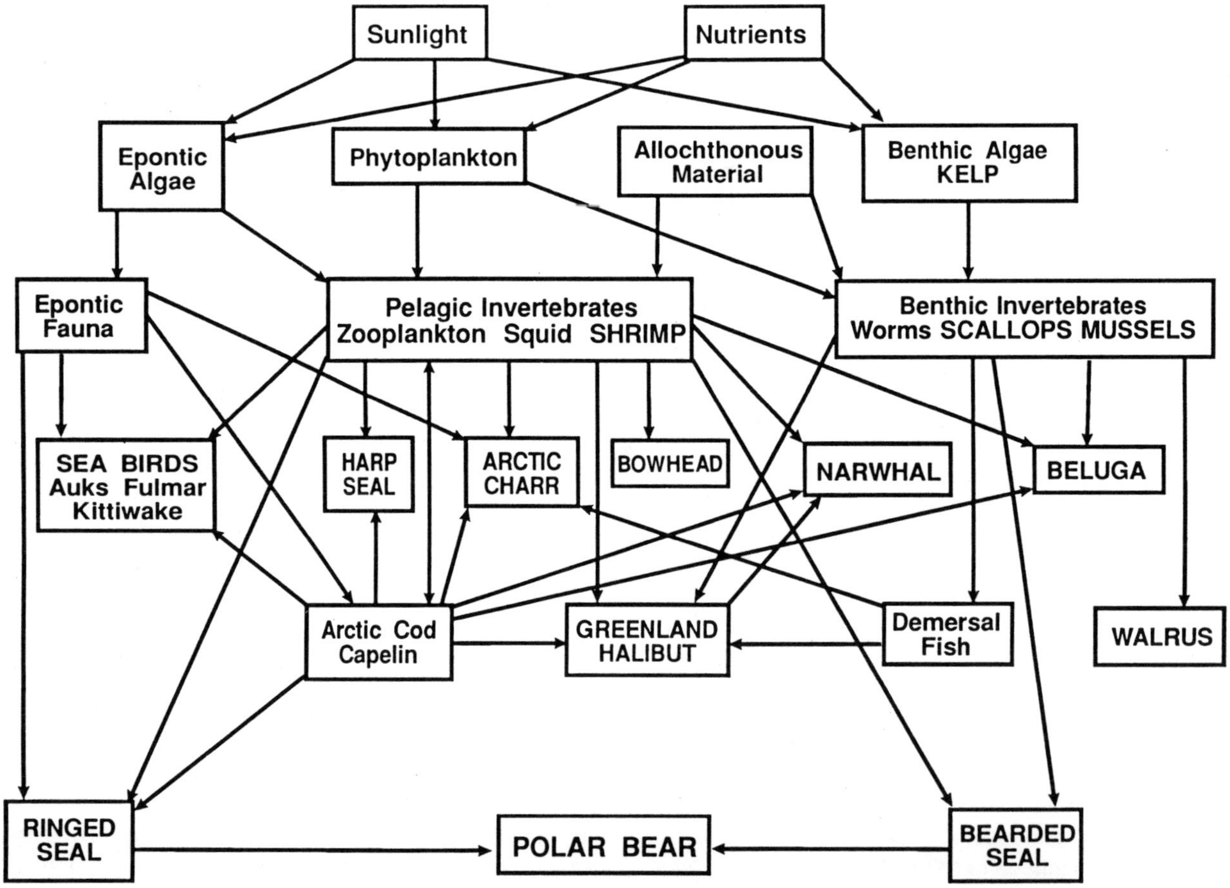


FIG. 2. A composite Arctic marine food web showing main feeding relationships and emphasizing harvested fishery resources (in capitals).

In general, the waters of the Beaufort Sea, Lancaster Sound, Baffin Bay, Davis Strait and eastern Hudson Strait contain the greatest concentrations of fishery resources. Other areas, such as offshore Hudson Bay, the channels of the central Arctic Archipelago, Sverdrup Basin and the Arctic Ocean, are not known to contain any large concentrations of fishery resources.

The fishery resources of the Arctic are primarily the terminal predators of the food web. These fish and marine mammals share common features such as low productivity, great biomass, low reproductive rates with reproduction often not occurring annually, and possibly a stable population structure in the absence of external perturbations (Johnson 1983a). The population structure of Arctic charr, for example, exhibits uniformity in adult size, highly variable age-at-length, individuals attaining great age, a low proportion of potential recruits to the adult population, and a constant unimodal or bimodal length frequency curve (Johnson 1983a). These features promote stability in population structure, a consequent lack of early warning of population decline through age and length changes, and a slow recovery from any perturbation.

Another feature shared by many of the terminal predators is that they avoid the Arctic seas in winter. Anadromous fish migrate to estuaries, rivers and lakes. Most marine mammals migrate out of the ice covered areas to more favourable water and ice conditions in southern Davis Strait or the Bering Sea. However, ringed seal (*Phoca hispida*), bearded seal and some walrus stocks are essentially non-migratory. Ice conditions may affect the non-migratory stocks; for instance a 50% drop in abundance of ringed and bearded seals in the eastern Beaufort Sea between 1974 and 1975 was attributed to abnormally heavy ice conditions (Stirling et al. 1982).

The stocks of beluga, narwhal, bowhead (*Balaena mysticetus*) and walrus that occur in Canada's Arctic are of international importance in that they constitute major proportions of the world's total populations of these species. Also several marine mammal and fish stocks are shared with either Alaska (U.S.A.) or Greenland (Denmark).

### The Human Setting

The Arctic coastal area is occupied primarily by aboriginal peoples. Canada's Arctic was first occupied about 30 000 years ago when Siberian hunters reached the northern Yukon (McGhee 1978). As the glaciers retreated about 9 000 years ago the Canadian Arctic was inhabited by Indian peoples (McGhee 1978). Today, Cree inhabit the southern coast of Hudson Bay and Gwich'in (Loucheux Dene) inhabit the Mackenzie Delta.

Part of the coast is populated by Inuit. The first wave of Eskimo people migrated throughout the Canadian Arctic about 4 000 years ago (McGhee 1978). A second wave, the Thule, spread into the area about the year 1 000 and are the ancestors of the modern Inuit.

The distribution of Indian peoples and Inuit throughout the area varied with the warming and cooling trends of the climate. Both peoples led a nomadic life until the 1950's when they gradually moved into communities from their camps. The federal government also established new communities by moving people from far away; for instance in 1953 Inuit from Inukjuak and Pond Inlet were moved to Grise Fjord and Resolute (Outcrop Ltd. 1984). Today, about 40 000 people, 78% aboriginal, occupy 60 communities along the Arctic coast and in the Mackenzie Delta (Statistics Canada 1988a, b, c, d). Their population is increasing rapidly; in the Northwest Territories (NWT) the natural population increase is 24 per 1 000, the highest in Canada where the national average is 7.7 per 1 000 (Statistics Canada 1987). Unemployment is high; for example the 1984 unemployment rate in the Keewatin was 42% mainly among Inuit (Irwin 1989). Many Inuit are employed only part-time and would like full-time jobs.

The harvesting of fish and marine mammals always has been a dominant human activity in the Canadian Arctic. Subsistence fisheries occur in all areas where people live or travel and provide a major and essential source of food and a significant contribution to the cultural life of the residents. Up to 75% of the population participates in the subsistence harvest, but only 51% of the potential native labour force of all the Northwest Territories (NWT) is engaged in traditional pursuits (Stabler 1989). Commercial and recreational fisheries and the commercial use of marine mammal products provide one of the few sources of employment and income to the economies of the communities.

Protein self-sufficiency and further development of commercial and recreational fisheries are major goals for regional, territorial and provincial governments and various native organizations. Cooperative management of fishery resources by resource users and managers is a major and expanding feature of fishery management in the Arctic.

## The Fisheries

### The Anadromous Fishery

Important anadromous fishery resources occur along much of Canada's Arctic coast (Table 2). Arctic charr is the most widespread species, occurring throughout the area except in the southern part of Hudson Bay (Johnson 1980). At least 176, and possibly 300 or more, assumed

TABLE 2. Anadromous fishery resources of the Canadian Arctic. Key: S-subsistence resource; C-commercial resource; R-recreational resource.

| Species | Region: Beaufort Sea | Central Arctic | Baffin | Northern Hudson Bay and Foxe Basin | Northern Quebec | Southern Hudson Bay | Reference |
|---|---|---|---|---|---|---|---|
| Atlantic salmon, *Salmo salar* | | | | | S C R | | Crawford (1989); Power (1981) |
| Arctic charr, *Salvelinus alpinus* | S C R | S C R | S C R | S C R | S C R | | Crawford (1989) |
| Brook trout, S. fontinalis | | | | | | S R | Roy (1989) |
| Chum salmon, *Oncorhynchus keta* | S | | | | | | V. Gillman, pers.comm. |
| Lake cisco, *Coregonus artedii* | | | | | | S | Crawford (1989); Roy (1989) |
| Arctic cisco, *C. autumnalis* | S C[a] | | | | | | Crawford (1989); V. Gillman, pers. comm. |
| Least cisco, *C. sardinella* | S | | | | | | Crawford (1989) |
| Lake whitefish, *C. clupeaformis* | S | | | | | S | Crawford (1989); Roy (1989) |
| Broad whitefish, *C. nasus* | S C | | | | | | Crawford (1989) |
| Inconnu, *Stenodus leucichthys* | S | | | | | | Crawford (1989) |
| Other species | S | | | | | S | Kemp et al (1989); Lawrence et al. (1984) |

[a]Commercial fishery occurs in Colville River, Alaska.

discrete stocks of anadromous Arctic charr exist. Many are harvested in subsistence fisheries, 28 in recreational fisheries and up to 142 in commercial fisheries (Clarke et al. 1989; R. Moshenko, Department of Fisheries and Oceans (DFO), Winnipeg R3T 2N6, pers. comm.).

Anadromous coregonids are important in the Mackenzie Delta region (Bond 1982; Lawrence et al. 1984; Bond and Erickson 1985;). They are harvested in subsistence fisheries (Table 2). In addition, broad whitefish (*Coregonus nasus*) is harvested in a developmental commercial fishery (Table 2). The Arctic cisco (*C. autumnalis*) stock spends part of its lifecycle in Alaskan waters where it supports a commercial fishery. Chum salmon (*Oncorhynchus keta*) also is harvested in the Mackenzie Delta (Table 2).

Atlantic salmon (*Salmo salar*) occurs in rivers of southern Ungava Bay and occasional fish have been taken in eastern Hudson Bay between Inukjuak and Cape Smith (Power 1981). The Ungava Bay stocks support subsistence, commercial and recreational fisheries (Table 2). Anadromous lake whitefish (*C. clupeaformis*), lake cisco (*C. artedii*) and brook trout (*Salvelinus fontinalis*) stocks occur along the coast of southern Hudson Bay (Morin et al. 1980; Bodaly et al. 1989; Kemp et al. 1989; Roy 1989) and support subsistence fisheries (Table 2).

Other freshwater and anadromous fish species occur in coastal and estuarine waters of the Mackenzie Delta region and Hudson Bay (Morin et al. 1980; Lawrence et al. 1984; Kemp et al. 1989) and may be taken in subsistence fisheries.

Subsistence fisheries have occurred since the area was occupied by humans. Arctic charr bones have been found at camp sites 4 000 years old (McGhee 1978). An estimated 12 000 or more people from 60 Arctic communities now participate in the subsistence harvest (R. Allen, DFO, Inuvik X0E 0T0, pers. comm; M. Breton, DFO, Quebec City G1K 7Y7, pers. comm; V. Gillman, DFO, Burlington L7R 4A6, pers. comm.; D. Moshenko, DFO, Winnipeg R3T 2N6, pers. comm.). The subsistence harvest occurs throughout the year with peaks in the late summer and autumn when the fish are migrating upstream, in the spring during downstream migrations, and during the summer in estuarine and coastal areas. Although information on the amount of the subsistence harvest exists (McCart and Den Beste 1979; Corkum and McCart 1981; Gamble 1988; Crawford 1989; Berkes 1990), it is not comprehensive. Subsistence harvests of anadromous Arctic charr and coregonids occur in the NWT and northern Quebec, chum salmon in the Mackenzie Delta and Atlantic salmon in northern Quebec (Table 3). Probably the subsistence harvest is the major portion of the total catch of anadromous fish.

Throughout the Arctic, about 500 people participate in commercial fisheries for various anadromous species. A typical investment in the fishery is $6,000 to $8,000 for a 6 m canvas and wood canoe, a 25–35 hp outboard motor and 45 m of gillnet. This investment also is used for harvesting fish and marine mammals for subsistence purposes.

The oldest commercial anadromous fishery in the Arctic is that for Atlantic salmon in Ungava Bay (Power 1976). The Hudson Bay Company began the fishery in 1867 on the Koksoak River and it continued on the Koksoak, Whale and George rivers until the Second World War. The average annual catch from 1894 to 1915 was 21 600 kg (Power 1976). The fishery was reestablished at Kuujjuak in 1961 (Power 1976). The Koksoak River is the most productive river and has an annual commercial and subsistence quota of 18 000 kg for Atlantic salmon (Roy 1989). In 1987/88 the harvest was 10 900 kg (J. Chouinard, DFO, Quebec City G1K 7Y7, pers. comm.; Table 3). A winter fishery is being developed.

Commercial fisheries for anadromous coregonids have occurred in the Mackenzie Delta region since the mid 1950's but not with great success (Davies et al. 1986). In 1987/88 6 613 kg of broad whitefish were harvested commercially (Table 3).

The commercial fishery for anadromous Arctic charr (Davies et al. 1986; Crawford 1989) began in 1932 when 1 000 kg of Arctic charr were caught along the Keewatin coast. Major fisheries were started at Iqaluit in 1947, Cambridge Bay in 1960, Rankin Inlet in 1961, Pelly Bay in 1969 and Nettilling Lake in 1974. The total commercial harvest in the NWT has varied from 3 765 kg in 1958 to 182 685 kg in 1977 (Yaremchuk et al. 1989). In 1987/88, 77 stocks of anadromous Arctic charr were fished commercially in the late summer and autumn. The harvest was 142 000 kg or more (Tables 3 and 4). This fishery is a major source of income to 428 commercial fishermen, fish plant workers and their communities. The fisheries at Cambridge Bay and Rankin Inlet are the most important, providing 76% and 14% respectively of the harvest sold through the Freshwater Fish Marketing Corporation. In 1988, 18 fishermen harvested 48 000 kg of Arctic charr from the six stocks fished in the Cambridge Bay area, and 123 fishermen harvested 45 200 kg from the twelve stocks in the Keewatin (Carder and Stewart 1989). The catch from the commercial fisheries was processed at two registered and 24 unregistered processing plants. In northern Quebec, quotas of 14 000 kg were established for Arctic charr in each large river; within 10 years, yields and mean size of fish had declined so much that the fisheries stopped (Roy 1989).

Arctic charr also is harvested by anglers in many areas. Twenty-three lodges offer Arctic charr fishing. The Tree River Arctic charr fishery is world renowned (Moshenko et al. 1984). Information on the 1987/88 recreational harvest of Arctic charr is sparse, but 17 304 angling licences were sold for the NWT (L. Anderson, DFO, Winnipeg R3T 2N6, pers. comm.). In 1985, Arctic charr fishing accounted for 7% of the 172 692 angler-days (including 51 957 by non-residents of the NWT) recorded for the NWT by 16 907 licensed anglers and 3 922 children who spent $15,574,537 in the NWT (Department of Fisheries and Oceans 1988). Arctic charr was the most preferred species of 17.7% of the anglers, and 16 460 Arctic charr were caught and retained.

Arctic charr, brook trout and Atlantic salmon stocks are harvested recreationally in northern Quebec. One sport fishing camp exists on the Koksoak River (Robitaille et al. 1986). However, similar information to that for the NWT is not available because data from the Canada sport fishery survey are aggregated for all of Quebec.

The whitefishes of the Mackenzie Delta are, perhaps, the only major Arctic anadromous fish resource which could sustain greater harvests. Some Arctic charr stocks may be under-exploited, but some others, close to communities, have been over-exploited.

## The Marine Fishery

The marine fish and invertebrate fishery is of less importance at present than the anadromous and marine

TABLE 3. 1987/88 subsistence and commercial harvests of anadromous fishery resources.

| Species | Region | Harvest | | Reference |
|---|---|---|---|---|
| | | Subsistence | Commercial | |
| Atlantic salmon | Northern Quebec | 10 227 fish | 10 900 kg | J. Chouinard, pers. comm. |
| Arctic charr | NWT | 350 000 kg[a] | 142 000 kg | Clarke et al. (1989); L. Anderson, pers. comm. |
| | Northern Quebec | 93 977 fish | | J. Chouinard, pers. comm. |
| Chum salmon | Mackenzie Delta | 800 kg | | V. Gillman, pers. comm. |
| Anadromous coregonids | Mackenzie Delta | 394 467 kg | | V. Gillman, pers. comm. |
| | Northern Quebec | 84 649 fish | | J. Chouinard, pers.comm |
| Broad whitefish | Mackenzie Delta | 6 613 kg | | V. Gillman, pers. comm. |

[a]1984/85 harvest.

mammal fisheries in the NWT. However, some species such as Arctic cod and capelin (*Mallotus villosus*) are major links in Arctic marine food chains (Fig. 2 and see Mikhail and Welch 1989). Information on the distribution of marine fish in the Canadian Arctic ( Morin et al. 1980; Hunter et al. 1984; Lawrence et al. 1984) shows a general absence of medium and large-sized fishes, except in Baffin Bay and Davis Strait where species such as Greenland halibut (*Reinhardtius hippoglossoides*) and Greenland shark (*Somniosus microcephalus*) occur. Smaller fishes, such as Arctic cod, sculpins (Cottidae) and eelpouts (*Lycodes* spp.), occur in most areas, as do marine invertebrates and plants. Concentrations of shrimp occur at the eastern end of Hudson Strait and off southeastern Baffin Island. Little is known about the stock identity or stock size of harvested marine fish, invertebrates and plants.

Available marine fish, invertebrates and plants are harvested for subsistence use by residents of coastal communities (Table 5). Fewer residents are involved in the subsistence marine fishery than in the subsistence anadromous and marine mammal fisheries. Little is known about the size of the subsistence harvest. However, in 1987/88, 1 880 kg of Pacific herring were caught in the Beaufort Sea region by about 10 residents (V. Gillman, pers. comm.) and about 30 000 kg of Atlantic cod and 10 000 kg of sculpins were harvested in northern Quebec (J. Chouinard, DFO, pers. comm.).

TABLE 4. Information for 1987/88 on the commercial Arctic charr fishery in the Northwest Territories (from L. Anderson, pers. comm.).

| | |
|---|---|
| Total harvest | 142 000 kg |
| Exported to Freshwater Fish Marketing Corporation | 59 228 kg |
| Landed value | $510 723 |
| Market value | $542 528 |
| Contribution of Cambridge Bay fisheries | 76% |
| Contribution of Rankin Inlet fisheries | 14% |
| Number of fishermen and fish plant workers | 428 |

Commercial offshore fisheries exist for shrimp, Atlantic cod, redfish (*Sebastes* spp.), Greenland halibut and roundnose grenadier (*Coryphaenoides rupestris*) in Davis Strait, primarily by southern Canadian and foreign fishermen. These fisheries are considered part of the Atlantic fisheries although they occur in NWT waters. There is increasing participation of Baffin Island and northern Quebec Inuit in these fisheries.

Residents of the Arctic communities have little history of commercial marine fishing. The main exception is the Killiniq fishery for Atlantic cod which began in 1950 (Crawford 1989). The fishery was active during the 1960's with catches up to 27 000 kg. However, the settlement at Killiniq was gradually abandoned, and the fishery closed in 1975. The fishery was resumed in 1983 by Makivik Corporation, representing the Inuit of northern Quebec and Labrador, but catches were only 439 kg in 1984 and 704 kg in 1985.

Recently, commercial marine fisheries have begun in Cumberland Sound and Hudson Strait for Greenland halibut, Iceland scallop (*Chlamys islandicus*) and shrimp. In 1988, 40 residents from the Baffin region participated in these fisheries. The catch of Greenland halibut from Pangnirtung was 10 600 kg valued at about $21,000 (D. Topolniski, DFO, Winnipeg R3T 2N6, pers. comm.), and eight Greenland sharks were caught as well. This fishery is conducted through the ice from January to April using snowmobiles and long-lines (R. Allen, pers. comm.). The following year the fishery had expanded to about 60 fishermen who caught 180 000 kg. The catch of Iceland scallops from Pangnirtung was about 1 720 kg of meats with a value of $17,200 (D. Topolniski, pers. comm.). This is an open water fishery using two scallop-draggers (R. Allen, pers. comm.). The scallop catch from northern Quebec was 10 000 kg of product (G. Chantingny, DFO, Quebec City

TABLE 5. Subsistence marine fishery resources of the Canadian Arctic (from Crawford 1989).

| Species | Region: Beaufort Sea | Central Arctic | Baffin | Northern Quebec | Hudson Bay and Foxe Basin |
|---|---|---|---|---|---|
| Kelp, *Laminaria* spp. | | x | | | |
| Invertebrates[a] | | x | x | x | x |
| Pacific herring, *Clupea harengus* | x | | | | |
| Capelin, *Mallotus villosus* | | | | | x |
| Polar cod, *Arctogadus glacialis* | x | x | | | |
| Arctic cod, *Boreogadus saida* | x | x | x | | |
| Saffron cod, *Eleginus gracilis* | x | | | | |
| Atlantic cod, *Gadus morhua* | | | x | x | |
| Greenland cod, *G. ogac* | | x | x | x | x |
| Sculpins, Cottidae | | x | x | x | x |
| Fourhorn sculpin, *Myxcocephalus quadricornis* | x | | | | |
| Lumpfish, *Cyclopterus lumpus* | | | | | x |
| Flatfish, Pleuronectidae | x | | | | |
| Greenland halibut, *Reinhardtius hippoglossoides* | | | x | | |

[a]Invertebrates include green sea urchin (*Strongylocentrotus droebachiensis*), clams, scallops, crabs, shrimps and brittle stars where available.

G1K 7Y7, pers. comm.) with a value of about $100,000. The shrimp catch from Hudson Strait and the waters off south-east Baffin Island was 455 000 kg, valued at about $3 million (J. Chouinard, pers. comm.).

Attempts have been made to establish a Pacific herring fishery along the Beaufort Sea coast (Crawford 1989). Catches of 8 000 kg were made in 1963 but the fishery was abandoned because of high marketing costs. The possibility of a herring roe fishery was examined in the 1980's and in 1983 4 580 kg of Pacific herring were caught and yielded 398 kg of roe. However, no herring stock of sufficient size has been found to support a fishery.

## The Marine Mammal Fishery

Beluga, narwhal, bowhead, walrus, ringed seal, harp seal (*Phoca groenlandica*), harbour seal (*P. vitulina*), bearded seal and hooded seal (*Cystophora cristata*) occur in the eastern Arctic and Hudson Bay. Beluga, bowhead, ringed seal and bearded seal occur in the western Arctic. Present knowledge indicates that there are 25 or more distinct stocks of these species in the Arctic (R. Moshenko, pers. comm.). All stocks of these species, except for endangered stocks of the bowhead, are harvested in subsistence fisheries in the Canadian Arctic. Several of the stocks also are harvested outside the Canadian Arctic. Shared stocks of bowhead and beluga are harvested in Alaskan waters, beluga, narwhal and probably walrus stocks are harvested in Greenland waters, and harp seal and hooded seal stocks are harvested in Atlantic Canada and Greenland.

The conduct of the marine mammal fishery and its products are an extremely important source of food and an important part of the cultural and economic life of most Inuit and some Cree and Dene. Marine mammal bones have been found at campsites as old as about 3 500 years (McGhee 1978), suggesting that seals may have been taken throughout human occupation of the coast. Hunting of bowhead probably did not begin until the Thule inhabited the Canadian Arctic about 1 000 years ago (McGhee 1978).

Undoubtedly commercial whaling has been the major human influence on whales. In the Eastern Arctic, European whalers began to harvest whales about the year 1600, with major peaks of harvest from 1729 to 1738 and from 1825 to 1834 (Mitchell and Reeves 1981). Bowhead was the main quarry although opportunistic harvests of narwhal, walrus, beluga and seals occurred (Table 6).

About 29 000 bowhead were taken from the Davis Strait and Hudson Bay areas between 1719 and 1915. At least 22 bowhead were killed from 1919 to 1976 in the Eastern Arctic and Hudson Bay, primarily by Inuit; meat and muktuk (the edible epidermis and dermis) were consumed locally, and some oil and baleen were sold commercially (Mitchell and Reeves 1982).

TABLE 6. Period and amount of commercial harvest of bowhead, beluga, narwhal and walrus.

| Species | Region | Period of harvest | Confirmed harvest | Reference |
|---|---|---|---|---|
| Bowhead | Davis Strait, Hudson Bay | 1719–1915 | 29 000 | Mitchell and Reeves (1981) |
| | Western Arctic | 1848–1915 | 18 650 | Bockstoce and Botkin(1983) |
| Beluga | Canadian Arctic | 1719–1969 | Unknown | Mitchell and Reeves (1981) |
| | Eastern Arctic, Hudson Bay | 1868–1911 | 20 431 | Mitchell and Reeves (1981) |
| | Eastern Arctic, Hudson Bay | 1923–1939 | 4 486 | Mitchell and Reeves (1981) |
| | Eastern Hudson Bay | 1752–1905 | Unknown | Reeves and Mitchell (1987) |
| | Great Whale R., Little Whale R. | 1854–1868 | 8 294 | Reeves and Mitchell (1987) |
| | Ungava Bay | 1881–1890 | 778 | Findley et al. (1982) |
| | Cumberland Sound | 1868–1909 | 1 438 | Mitchell and Reeves (1981) |
| | | 1922–1939 | 4 551 | Mitchell and Reeves (1981) |
| | | 1948–1960 | 2 225[a] | J.T. Strong, pers. comm. |
| | | 1964 | 150[a] | J.T. Strong, pers. comm. |
| | Churchill | 1950–1960 | 450 $yr^{-1}$ | Smith and Taylor (1977) |
| | Whale Cove | 1961–1969 | 200 $yr^{-1}$ | Smith and Taylor (1977) |
| Narwhal | Canadian Arctic | 1719 — ? | Unknown | Mitchell and Reeves (1981) |
| | Canadian and Greenland waters | 1914–1924 | 10 970 | Mitchell and Reeves (1981) |
| Walrus | Canadian Arctic | 1719–1952 | Unknown | Davis et al. (1980) |
| | Northern Hudson Bay | 1900–1915 | 2 750 | Davis et al. (1980) |
| | Davis Strait, Baffin Bay | 1907–1910 | 3 200 | Davis et al. (1980) |
| | Davis Strait | 1949–1952 | 2 000 | Mansfield (1973) |

[a]Totals may include a few beluga taken for subsistence purposes.

In the western Arctic, commercial whaling for bowhead began in 1848 in the Bering Sea and was extended into the Beaufort Sea from 1888 to 1914 (Bockstoce and Botkin 1983). Some 18 650 bowhead were taken from the western Arctic between 1848 and 1915. Incidental catches of beluga occurred but, unlike the eastern Arctic, no specific commercial whaling existed for beluga.

With the decline in abundance of the bowhead in the 19th and 20th centuries, emphasis switched to narwhal and beluga. No acceptable estimate has been made of total commercial narwhal harvest, but significant numbers of narwhal were taken commercially (Table 6). In the eastern Arctic, beluga were taken commercially from Baffin Bay, Davis Strait, Hudson Bay and Ungava Bay between 1868 and 1911, mainly from Cumberland Sound and Elwin Bay, and periodically after 1911 (Table 6). Beluga were also taken in commercial harvests in Hudson Bay (Table 6). Harvesting occurred in eastern Hudson Bay from 1752 until about 1905 (Reeves and Mitchell 1987). In western Hudson Bay, commercial harvests occurred at Churchill between 1950 and 1960, and at Whale Cove between 1961 and 1969 (Table 6). The last Canadian commercial whaling for beluga occurred in 1969.

The total size of the commercial walrus harvest is unknown. Considerable numbers were taken at various times (Table 6). The last commercial harvest of walrus occurred in Davis Strait between 1949 and 1952 (Mansfield 1973).

Today, there are 6 000 active subsistence harvesters of marine mammals in the Canadian Arctic. Information on the annual harvest is incomplete, especially for seals, and the amount of harvest varies annually (Table 7). In 1987/88 reported harvests were 869 beluga, 181 narwhal, 398 walrus and 50 000 seals. Since 1979, harvesting of bowhead in Canada has been prohibited without a licence; the first licence was issued in 1991 and a bowhead was harvested in the Beaufort Sea.

By-products of the marine mammal subsistence harvest are a major source of income for the communities. Local hunters in northern Canada harvested about 61 000 seals each year before 1983 and about 67% of the harvested skins were sold commercially for at least $520,000 (Stewart et al. 1986). With the collapse of the

TABLE 7. The subsistence harvest of Arctic marine mammals.

| Species | Region | Date | Reported annual harvest | Reference |
|---|---|---|---|---|
| Beluga | NWT | 1948–87 | 145–1 094 | Strong(1989) |
| | Northern Quebec | 1975-78 | 197–682 | Findley et al. (1982) |
| | NWT and northern Quebec | 1987/88 | 869 | J. Chouinard, pers. comm.; R. Moshenko, pers. comm. |
| Narwhal | NWT | 1948-87 | 40–406 | Strong (1989) |
| | NWT and northern Quebec | 1987/88 | 181 | J. Chouinard, pers. comm.; R. Moshenko, pers. comm. |
| Walrus | NWT | 1948-87 | 7–912 | Strong (1989) |
| | NWT and northern Quebec | 1987/88 | 398 | J. Chouinard, pers. comm.; R. Moshenko, pers.comm. |
| Seals | NWT and northern Quebec | 1987/88 | 50 000 | J. Chouinard, pers. comm.;R. Moshenko, pers.comm. |
| Ringed seal | NWT | 1984/85 | 37 000 | Clarke et al. (1989) |
| Harp seal | NWT | 1984/85 | 2 000 | Clarke et al. (1989) |
| Other seals[a] | NWT | 1984/85 | 2 000 | Clarke et al. (1989) |

[a]Includes bearded seal, harbour seal and hooded seal.

market for seal skins, both the average price per skin and the number of skins sold have dropped dramatically; in 1984/85 5 418 skins were sold for $54,351. Other marine mammal products, such as narwhal and walrus ivory, baleen and whale bone, are used as raw materials for art and sculpture. The wholesale value of Inuit sculpture and crafts based on marine mammal products probably was about $1 million in 1988/89 with the retail value being up to about $3 million (S. Kirwan, DFO, Winnipeg R3T 2N6, pers. comm.).

In addition to the commercial and subsistence harvests of marine mammals, sport hunting of beluga occurred at Churchill in 1965 and 1966 with 16 taken (Strong 1989). A live-capture program also has occurred at Churchill with 61 beluga being captured from 1967 to 1989 for aquaria in North America, West Germany and Japan (R. Moshenko, pers. comm.).

Whale-watching and associated tourism is increasing. Current centres are Churchill for beluga and Pond Inlet and Arctic Bay for narwhal.

As a result of historical commercial harvest of marine mammals, the bowhead stocks and the beluga stocks of eastern Hudson Bay, Ungava Bay and southeast Baffin are reduced in numbers (Findley et al. 1982; Mansfield 1985; Richard 1991). No evidence exists that harvesting has affected the numbers or range of other beluga and narwhal stocks (Strong 1988; Richard et al. 1990) or other seal species, nor is there evidence that these stocks and species are being harvested at levels which cannot be sustained. The walrus is reduced in range and probably in numbers (see Richard and Campbell 1988), and the harbour seal is reduced in both range and numbers. Only the ringed seal, and possibly the harp seal, appear to have potential for supporting an increased harvest.

## Benefits from the Fisheries

The Arctic marine fisheries provide cultural, social and economic benefits to individuals and communities, to regional economies and to Canada's economy as a whole. Cultural and social benefits of the fisheries are described in qualitative terms, while economic benefits are described by calculated economic impact and economic efficiency indicators (Anderson 1977). The benefits calculated below are first-order approximations.

### *Cultural and social benefits*

The subsistence fisheries make an essential and major contribution to continuation of traditional cultures, provision of traditional foods and other materials, and local self-sufficiency. Marine mammal meat and muktuk have a special cultural and nutritional significance that cannot be replaced by southern foods. In addition, commercial and recreational fisheries contribute to the maintenance of traditional lifestyles, especially where people participate not to earn a livelihood but to supplement their cash incomes (Yonge 1988, 1989).

### *Gross benefits*

The distribution of gross benefits (economic impact benefits), is an indicator of the fisheries' contribution to national income (value added) and employment. Economic impacts arise from the expenditures of harvesters, anglers and government. Value is added to the economy by spending of wages, salaries, profits and rents created at all stages of the fisheries and the subsequent income and employment in supplier services.

The most important economic impact is the participation of 12 000 or more Arctic residents from the 60 communities in subsistence fisheries. Subsistence fisheries continue to contribute substantially to the economic base and local self-sufficiency of many Arctic communities even though the per capita subsistence harvest probably has decreased since the early 1970's. More southern food is now available, less dogfood is required with the wide use of snowmobiles (Crawford 1989), and the human population and the importance of the wage economy have increased.

The value added to the export-based commercial fisheries can be estimated from a model which shows that $1 expenditure in the harvesting sector is worth $1.29 to the Canadian economy (Shaffer and Associates 1989). The resultant value is about $4.7 million (Table 8) of which $3.1 million (calculated at 66.7%) accrues as wage income, providing the equivalent of 152 person-years (at $20,650 per person) of employment.

Similarly, the value added to recreational fisheries in the NWT can be estimated from a model which shows that $1 expenditure in the lodge industry is worth $1.44 to the Canadian economy (D. Topolniski, pers. comm.). Thus, the value added to the Arctic charr recreational fisheries in the NWT is about $1.6 million (Table 8), of which approximately $1 million accrues as wage income, providing the equivalent of 51 person-years of employment.

### *Net benefits*

The difference between the gross value of output and the cost of production (economy efficiency), is a measure of the net benefits of the fishery to Canadian consumers, producers and governments on behalf of society at large.

The price of protein equivalent items has been used to estimate the gross value of all NWT subsistence anadromous fish and marine mammals, but not marine fish, invertebrates and plants (D. Topolniski, pers. comm.). The estimated value was $9.4 million in 1982/83. Using this approach, the gross value of the Arctic subsistence fisheries was about $15 million in 1987, of which about $6 million was derived from anadromous fishes and $8.5 million from marine mammals (Table 8). However, a large degree of uncertainty exists with this estimate since it may differ considerably from values estimated using local exchange values or alternative market prices.

Net economic value has been calculated for selected subsistence fisheries (Usher 1971), but no Arctic-wide estimate exists. Uncertainty surrounds gross values, and complete and current information on effort, harvest, harvesting costs and the costs of management is lacking. As a result, the net economic value of subsistence production to consumers, producers and society at large is unknown.

The local and inter-settlement trade of commercial fishery products at sales prices less than those of comparable imported products provides net consumer benefits. The approximate value of these consumer benefits, conservatively estimated at $1 per kg, from anadromous fish was at least $83,000 in 1987/88 (Table 8). Arctic charr, Atlantic salmon, scallops and shrimp are specialty products. Their export trade provides about $600,000 in net consumer benefits to other Canadian consumers (Table 8). Other commercial fishery products, such as Greenland halibut, are more readily available and net consumer benefits are less.

The export trade of marine mammal ivory products, seal skins and walrus hides, with an annual export value of about $1 million (S. Kirwan, pers.comm.), provides an unquantified net consumer benefit.

The Arctic charr recreational fishery provides net consumer benefits to anglers because access to fishing opportunities is provided at a cost far below its value to anglers. In 1987/88, angler benefits were at least $300,000 (Table 8) based on 12 000 angler days valued at $25 per day.

The fisheries contribute other unquantified recreational and non-consumptive benefits. Net economic values to consumers arise from their participation in whale-watching activities. Canadian consumers derive benefits from the knowledge that Arctic marine mammal populations are conserved and managed even though they may never observe them nor consume products of the fishery. Canadians and others derive educational and scientific benefits from beluga and other species held in aquaria.

Limited knowledge exists of the financial and economic performance of the commercial and recreational fishing sectors. Harvesting, processing and lodge operations probably provide, at best, normal economic returns for the investments of labour and capital (D. Topolniski, pers.comm.) because of the open-access management of the fishery, and the logistical and economic constraints on production for inter-settlement trade and export from the NWT and northern Quebec. The ongoing requirement for capital and operating subsidies to supplement market revenues of commercial fisheries indicates marginal economic performance. The financial and economic performance of businesses engaged in the trade of marine mammal products is not documented well.

### *Summary*

In summary, the Arctic fisheries provide a wide range of consumptive and non-consumptive benefits to Canadian producers, consumers and Canadian society at large (Table 9). The cultural and social benefits are of most importance. Arctic fisheries provide for continuation of aboriginal cultures by supplying traditional food and other essential materials. The fisheries are also one of the few sources of income in the

TABLE 8. Summary of calculated economic benefits from Arctic fisheries for 1987/88.

| Fishery | Harvest (kg) | Landed value ($)[a] | Calculated economic benefits: Gross benefits($)[b] | Calculated economic benefits: Net benefits($)[c] |
|---|---|---|---|---|
| *Subsistence* | | | | |
| Anadromous | | 6,000,000 | | |
| Marine | | ? ? | | |
| Marine Mammal | | 8,500,000 | | |
| Total | | 14,500,000+ | | |
| *Commercial* | | | | |
| Anadromous | | | | |
| Arctic charr | 142 000+ | 510,723 | 658,833 | |
| Export | 59 228 | — | — | 59,228 |
| Local trade | 83 000+ | — | — | 83,000+ |
| Broad Whitefish | 6 613 | ? | ? | 6,613 |
| Atlantic salmon | 10 900 | ? | ? | 10,900 |
| Marine | | | | |
| Scallop meats | 11 720 | 117,200 | 151,188 | 70,000 |
| Shrimp[d] | 455 000 | 3,000,000 | 3,870,000 | 455,000 |
| Greenland halibut[e] | 10 600 | 20,988 | 27,075 | ? |
| Total | 779 061+ | 3,648,911+ | 4,707,096+ | 684,741+ |
| *Recreational* | | | | |
| NWT Arctic charr | | | 1,569,913 | 302,211 |
| Northern Quebec | | | ? | ? |

[a]Calculated value for subsistence fishery is gross value of production from price of protein equivalent items.
[b]Calculation: commercial fishery — landed value × $1.29 (a "value added" multiplicative factor); recreational fishery — $15,574,537 (anglers' expenditures) × 7% (angler days for Arctic charr) × $1.44 (a "value added" multiplicative factor).
[c]Calculation: commercial fishery — harvest × $1 $kg^{-1}$; recreational fishery — $172,692 (anglers' expenditures) × 7% (% angler days for Arctic charr) × $25 $d^{-1}$ ( a value for each angler day).
[d]Benefits to the Canadian economy from shrimp are overestimates since foreign vessels are involved and a portion of the benefits accrue to other nations.
[e]Greenland halibut is not a specialty product and net consumer benefits are less than $1 $kg^{-1}$.

TABLE 9. Summary of producer and consumer benefits from Canada's Arctic marine fisheries.

| Item | Benefits | | |
|---|---|---|---|
| | Cultural and social | Gross | Net |
| Subsistence fisheries | Cultural continuation<br>Provides country foods<br>Helps local self-sufficiency | 12 000 participants | $15 million benefits from price of protein equivalent items<br>Net economic value unknown<br>Unknown net consumer benefits of marine mammal ivory, skins and hides for export |
| Commercial fisheries | Cultural continuation<br>Provides cash income<br>Subsidizes subsistence harvest | $4.7 million value added from export | $700,000 net consumer benefits from local trade and export trade of specialty products<br>At best normal economic returns on investments harvest of labour and capital |
| Recreational fisheries | Cultural continuation<br>Provides cash income<br>Subsidizes subsistence harvest | $1.6 million value added | $300,000 net consumer benefits from Arctic charr angling<br>Unquantified benefits from whale watching<br>At best normal economic returns on investments of labour and capital |
| Other | | | Unquantified benefits from conservation of Arctic fish and marine mammals<br>Unquantified educational and scientific benefits from beluga and other species in aquaria |

communities, and provide employment-like activity for 50–75% of the population.

The gross economic benefits contribute substantially to regional development and employment. The economic impacts of subsistence fisheries arise from the participation of 12 000 or more residents in an activity that contributes to the economic base and local self-sufficiency. The commercial and recreational fisheries are estimated to be worth about $6.3 million, of which $4.1 million accrues as wage income, the equivalent of 203 person-years of employment.

The net economic benefits, although not well documented, are of lesser importance. The net economic benefits of the subsistence fishery have not been quantified, but the protein equivalency has been valued at $15 million. Canadian consumers obtain net consumer benefits of approximately $1 million annually from the commercial and recreational fisheries and an unquantified benefit from other recreational and non-consumptive activities and the trade of marine mammal products. At best, normal economic returns are earned from investments of labour and capital in the Arctic fisheries.

## Fishery Management

Management of Arctic marine fisheries has evolved and is evolving in response to the nature of the fisheries and the fishery resources and the requirements of Arctic people. Until the 1950's, with the exception of commercial whaling, the fisheries were small-scale and conducted virtually only for subsistence purposes. Since then, commercial and recreational fisheries have increased considerably. Now, fishery management is changing in response to the need for cooperative management and the desire for fishery development.

Responsibility for management of these fisheries is divided among several agencies. The Department of Fisheries and Oceans (DFO) is responsible for the overall conservation and management of all fishery resources and with the following exceptions: in northern Quebec, Ontario and Manitoba, the provincial governments manage anadromous fishes in fresh waters; in the Northwest Territories and along the Yukon North Slope, the territorial governments administer the recreational fishery.

## Fish and Marine Mammal Conservation

The essence of conservation of the Arctic fish and marine mammal resources is to manage their harvesting and their habitats so that the value of the social and economic benefits received by Canadians is maintained or enhanced through the continued existence of exploitable stocks.

One major management issue is to ensure that the harvest does not exceed the ability of a stock to support the level of harvest. This requires the development and use of appropriate models to predict the safe harvesting level for each stock, collection of data for estimating each parameter of the model, and compliance of harvesters with safe harvesting levels. At present, the development and application of appropriate models, the collection of better data on stock identity, stock size and vital parameters such as recruitment, reproduction and natural mortality, the compilation of accurate and more complete harvest statistics and the collection of long-term data series are required (Clarke et al. 1989; Cosens et al. 1990). Despite these shortcomings, current management of Arctic fish and marine mammal stocks is as scientifically sound as existing information and theories permit (Clarke et al. 1989).

In general, stock identities are not well known. It has been assumed for management purposes that Arctic charr and Atlantic salmon of different river systems form discrete stocks, an assumption which has been supported by tagging studies and by the collapse of fisheries on specific river systems. However, in some areas a small proportion of Arctic charr is known to use more than one river system (Kristofferson et al. 1984; Dempson and Kristofferson 1987). The stock structure of anadromous coregonids in the Mackenzie Delta is unknown and is complicated by the occurrence of several life history patterns for each species except Arctic cisco (Reist and Bond 1988; Bodaly et al. 1989). No information exists on the identity of marine fish and invertebrate stocks (Crawford 1989). At present, Greenland halibut and shrimp are managed as though they are part of a larger Davis Strait stock, whereas other species, such as scallops, are considered to constitute local stocks for management purposes.

Marine mammal stock identity is better understood through studies of distribution, morphology and catch history. Sufficient information exists to suggest that six beluga stocks (Clarke et al. 1989), two narwhal stocks (Strong 1988) and two to three bowhead stocks (Mitchell and Reeves 1982; Cosens et al. 1990) exist in the Canadian Arctic. Uncertainty exists about the interrelationship of beluga occurring in Hudson Bay, Ungava Bay, Hudson Strait and off southeast Baffin Island, and of bowhead in Hudson Bay and the eastern Arctic (Clarke et al. 1989). The stock identity of seals (Stewart et al. 1986) and walrus (Richard and Campbell 1988) is unknown, but several stocks are assumed to exist as a practical management measure (Clarke et al. 1989).

Several approaches have been used to set harvest levels for fish and marine mammal stocks. Simple yield indices, such as basing commercial quotas for Arctic charr in the Northwest Territories on the size of the river system (Clarke et al. 1989), have been used. In northern Quebec, a modification of Ryder's morphoedaphic index has been applied to rivers by using a river harvest/lake harvest ratio (Roy 1989).

Frequently, exploratory (test) fisheries are conducted on anadromous or marine stocks to set harvest levels before a commercial fishery is established (Kristofferson and McGowan 1981). A provisional, conservative quota for the exploratory fishery is based on comparative information from other stocks or water bodies. The exploratory fishery may be conducted for two or three years. A condition of the licence is reporting catch and biological information. The information from the exploratory fishery is used to determine whether a commercial fishery is feasible and to establish an initial commercial quota and other necessary management conditions for the fishery. Monitoring of the yield and biological information from the catch may lead to modification of the quota or to other management changes.

Various models exist that can be used to set harvest levels and manage Arctic fishery stocks. Different models require different data. For instance, a Schaefer model requires only catch and effort information, a Beverton-Holt model requires growth and mortality data, a self-regenerating model requires stock and recruitment information, and multispecies models in addition require estimates of species interactions (Larkin 1983). The availability of data dictates to a large degree which models can be used. Those that are used have been derived from basic fishery management models or from models for non-Arctic species or stocks. These require proving or adapting for Arctic species. As and example, an initial model has been developed for narwhal by Kingsley (1989).

Most models require information on stock size, but knowledge of the size of Arctic fish and marine mammal stocks is limited. Information exists for some stocks of Arctic charr and Atlantic salmon. However, no information exists for other anadromous species nor for most marine species, although some incomplete information exists for shrimp in Hudson Strait, scallops in Hudson Strait and Ungava Bay, and Atlantic cod near Killiniq. The information for marine mammals is better with population estimates for five of the six beluga stocks (Davis et al. 1980; Clarke et al. 1989; P. Richard, DFO, Winnipeg R3T 2N6, pers. comm.), two narwhal stocks (Strong 1988) and one bowhead stock (Cosens et al. 1990), and some information for the other two bowhead stocks (Mitchell and Revees 1982; Cosens et al. 1990), walrus (Richard and Campbell 1988) and seals (see Stewart et al. 1986).

Vital parameters, such as age, length and weight relationships and maturity are required for some models.

Values for some of these are known for anadromous fish stocks for which stock assessment or harvest sampling have occurred. Similar information is beginning to be gathered for harvested marine fish and invertebrate stocks. Vital parameters, such as age at maturity and reproductive rate, are beginning to be known for some Arctic marine mammal stocks (see Clarke et al. 1989; Cosens et al. 1990).

Information also is incomplete for harvest levels and effort. Generally, information on commercial harvests exists (Yaremchuk et al. 1989), but it is not always available for each stock. For instance, some Arctic charr harvests are reported for a geographical area rather than separately for each stock. Harvests of beluga, narwhal and walrus have been recorded (Strong 1989) but these are not necessarily complete. Subsistence harvest levels of seals, anadromous fish and marine fish and invertebrates are poorly known; some estimates are available for total subsistence harvests in some areas (e.g. Gamble 1988) but these usually are on a community basis rather than a stock basis. Similarly, the approximate level of recreational fishing and harvest is known (Department of Fisheries and Oceans 1988) but the information is rarely known for each stock.

The result of the lack of models and data has been the adoption of a relatively simple approach to establishing safe harvest levels and quotas. Quotas have been established only for commercial harvests of anadromous species, exploratory marine fisheries, one recreational fishery (Tree River Arctic charr), subsistence salmon fisheries, and some beluga, narwhal and walrus stocks, but not for other subsistence or recreational fisheries or other marine mammal stocks.

Development of a model for Arctic charr life history and productivity is progressing, but its features preclude the application of mathematical models except to a limited extent (Clarke et al. 1989). Normally, commercial quotas for anadromous Arctic charr have been established from the size of the river system and, if available, the history of fishing including subsistence harvests (Clarke et al. 1989). Experimental work at Nauyuk Lake and review of exploitation histories show that the total annual harvest should be less than 10% of stock size, and harvest rates now are set conservatively at less than 5% for stocks of known size (Ayles 1987). In addition an overall quota was tried for Wellington Bay, part of the Cambridge Bay fishery, where several stocks occur; however, this proved to be impracticable, and quotas now have been established for each river system entering Wellington Bay (Kristofferson et al. 1984).

Quotas for marine fish and invertebrates harvested by Arctic communities are not yet based on mathematical models. Instead they are based on allocations from the total allowable catches established by the Atlantic Groundfish Advisory Committee or the Northern Shrimp Advisory Committee, knowledge of the species and areas to be fished, and the potential yield of the equipment to be used.

Quotas for beluga, narwhal and walrus also are not based on any mathematical model. In the NWT, quotas were established only if the stock was thought to be declining, and then they were set at or near historical subsistence harvest levels (Yaremchuk and Wong 1989a). Also, the quotas were set by community rather than by stock. The exception is the quotas recently established for the south-east Baffin beluga stock which were set well below recent harvest levels. In contrast, an agreement exists in northern Quebec between DFO, the Cree and Inuit on the guaranteed harvesting levels for beluga; for 1986 the level was 199 beluga which, because of conservation concerns, was considerably less than the level of harvesting guaranteed as a result of the James Bay Northern Quebec Agreement (Department of Fisheries and Oceans 1986).

Even if biologically safe harvesting levels are established, an additional biological problem is uncertainty of the way in which Arctic fishery resources respond to fishing pressure. Johnson (1983a, 1989) suggests that Arctic charr stocks may maintain a stable population configuration in response to fishing pressure until the configuration switches suddenly to a second stable configuration. Similarly, marine mammals may not show significant changes in population parameters until the stock is threatened. Consequently, monitoring stock size rather than mean length may be more appropriate (Ayles 1987).

**Regulation and Compliance**

To date the basic approach to Arctic fishery management by government is minimal intervention by government with a low level of investment. For instance, DFO's responsibilities for the fisheries conducted by the 60 Arctic coastal communities are discharged by a staff of 14 located in Inuvik, Iqaluit, Rankin Inlet and Yellowknife and none in northern Quebec. This staff is supported by additional DFO biological and other fishery management personnel located in Yellowknife, Manitoba and Quebec, and by provincial and territorial personnel and the Royal Canadian Mounted Police. As a result, effective fishery management relies on the cooperation and compliance of the fishermen, especially for subsistence harvesting.

A key feature of fishery management in the Arctic is consultation between management agencies and resources users. All plans, management proposals and research studies are discussed and consensus usually is reached on changes necessary to improve the management of a fishery.

Subsistence fisheries, the largest component of Arctic marine fisheries, are essentially unregulated and unmanaged. Harvesting rights are assigned or are being assigned in land claim settlements. Native subsistence fishermen receive the first priority in the allocation of fish and invertebrate stocks; they do not require a licence whereas non-native subsistence fishermen do (Yaremchuk and Wong 1989b). There is no restriction on the amount of harvest or on the time and location of fishing, and virtually no restriction on the method of fishing (Yaremchuk and Wong 1989b). Except for catches of beluga, walrus and some anadromous species in northern Quebec, there is no requirement for subsistence fishermen to report their catch; the size of the harvest is, therefore, largely unknown. One result is regulation that limits the use of the catch; thus fish from subsistence fisheries in the NWT cannot be sold, given to non-natives or wasted (Yaremchuk and Wong 1989b).

A similar approach exists for management of the marine mammal fisheries, but quotas do exist and management effort is much greater. Some harvests of beluga and narwhal are monitored regularly. Native marine mammal harvesters receive the first priority in the allocation of marine mammals (Yaremchuk and Wong 1989a) and harvest rights are assigned or are being assigned in land claim settlements. Natives do not require a licence to harvest marine mammals except for bowhead. Non-natives may harvest seals for food without a licence and may harvest beluga and walrus with a licence. Some quotas exist for beluga, narwhal and walrus but not for other species (Yaremchuk and Wong 1989a). Northern Quebec communities have a self-imposed quota for beluga. Quotas exist for beluga harvested by the communities of Iqaluit, Lake Harbour and Pangnirtung. Narwhal quotas, administered through a tag system, exist for communities that normally harvest narwhal. An individual can harvest only four walrus and, in addition, community walrus quotas exist for Arctic Bay, Clyde River, Coral Harbour and Sanikiluaq. As in the case of fish, regulations restrict the disposal of the harvest (Yaremchuk and Wong 1989a). The sale of meat, muktuk and blubber is restricted or prohibited, but intersettlement trade or barter is permissible. Permits are required to transport narwhal products and walrus hides and tusks from the NWT.

The greatest management effort is for the established commercial fisheries. About 60% of the potential anadromous Arctic charr commercial fisheries are fished in a given year (Clarke et al. 1989). These fisheries are managed through quotas and appropriate time and gear restrictions. Access to the fisheries is open, but fishermen must have a licence for which they need to be a Canadian citizen or permanent resident. Licences are issued to either individual fishermen or groups from the community closest to the fishery resource. Licences are not transferrable and are issued for one year. The major fisheries are monitored closely, but smaller fisheries are monitored rarely. The majority of the catch of anadromous species exported from the NWT has to be sold to the Freshwater Fish Marketing Corporation, except the catch from the Inuvialuit Settlement Region. Developing commercial fisheries are managed through exploratory fishing licences, which establish catch limits and other fishing and reporting requirements.

Management effort on recreational fisheries in the NWT is limited. Access is open but participants require licences. The catch is managed on the basis of limits for each species on the daily catch and possession by anglers. Some water bodies have more restrictive limits if conservation concerns exist. For example, at Tree River, which is managed as a trophy recreational fishery (Moshenko et al. 1984), an overall quota exists for Arctic charr caught by angling in addition to the daily catch and possession limits. Limited patrols are undertaken of the recreational fisheries. Information on catches by angling is obtained by creel census and periodic surveys of anglers (Department of Fisheries and Oceans 1988).

Issues related to the allocation of fishery resources among competing fishery sectors or the allocation of licences to individuals have not been a major factor in managing Arctic marine fisheries.

This management system has worked reasonably well, but it is becoming increasingly less effective as fishing pressure increases, as more complex fisheries develop on single stocks, and as increasing emphasis occurs on marine fisheries.

**Fishery Development**

The development of commercial and recreational fisheries is a major goal of governments, communities and peoples in the Canadian Arctic. Fishery development is viewed as a major approach to maintain or expand traditional activities, stimulate regional growth and community development and reduce unemployment (D. Topolniski, pers. comm.). The federal, Quebec and territorial governments have assigned a high priority to renewable resources development; this is reflected in economic development programs such as the Canada-Quebec Auxiliary Agreement on Fisheries Development, the Special Agricultural and Rural Development Agreement and the Economic Development Agreement in the Arctic (Indian and Northern Affairs Canada 1988).

Fishery development is administered differently in the NWT than in northern Quebec. In the NWT, fishery development is aided by the Government of the NWT (GNWT), the Department of Indian Affairs and Northern Development (DIAND) and the Department of Industry, Science and Technology (DIST) through various development incentives and strategies. The GNWT has developed a renewable resources development strategy. In 1987/88 about $500,000 was provided for resultant fishery development projects through the Canada-Northwest Territories Economic Development Agreement, the NWT Special Agricultural and Rural Development Agreement and the Native Economic Development Program (Topolniski et al. 1987). The Economic Development Agreement is administered by DIAND. The latter two funding programs now are combined into the Canadian Aboriginal Economic Development Strategy which is administered by DIST. Funds are available for capital investment, operation and training. In addition, GNWT's freight assistance program helps subsidize transportation costs of shipping products to market (Topolniski et al. 1987). The Department of Fisheries and Oceans provides advice on development proposals, manages the harvesting aspects of projects and inspects processing plants and products as necessary.

In northern Quebec, funds for fishery development are administered by the Department of Fisheries and Oceans jointly with the Ministère de l'Agriculture, des Pêcheries et de l'Alimentation du Québec. In 1987/88 about $500,000 was provided through the Canada-Quebec Auxiliary Agreement on Fisheries Development.

The demand for fishery development is high. At least 19 communities throughout the Arctic have expressed a desire to establish commercial marine fisheries. The greatest interest is in shrimp, scallop, and Greenland halibut fisheries off southeast Baffin Island and in Ungava Bay and Hudson Strait. Interest also exists in developing fisheries for Atlantic cod, Greenland cod and Greenland shark off southeast Baffin Island, and for mussels, sea urchins and sea cucumbers around the Belcher Islands. At least 17 communities would like to establish new commercial or subsistence fisheries for Arctic charr and anadromous coregonids, and to establish or expand recreational Arctic charr fisheries.

Much interest exists in expanding the harvesting and non-consumptive use of marine mammals. Interest exists to reestablish the subsistence bowhead hunt on all stocks, to harvest beluga at locations where no current harvest occurs, to market beluga muktuk, to reestablish commercial sealing with leather products being produced from skins, to introduce sport hunting of narwhal and walrus, to expand whale watching, and to establish a bowhead sanctuary at Isabella Bay which would attract tourists.

Although a great demand exists for fishery development, there are significant constraints. First, the biological potential is limited by the extremely low productivity of Arctic ecosystems and by the current utilization of fishery resources. Most accessible stocks of fish and marine mammals are harvested in existing subsistence, recreational or commercial fisheries. Coregonids of the Mackenzie Delta are perhaps the only major anadromous fish resource with potential to support greater harvests. The extent of marine fish and invertebrate resources is largely unknown, but commercial potential, although limited, probably exists only in the Hudson Strait and southeast Baffin Island areas. Among marine mammals, only the ringed seal has potential for supporting a greater harvest.

Second, knowledge of population sizes, safe harvesting levels and food web relationships of new fishery stocks is virtually non-existent. This could result in delays in fishery development or the introduction of quotas which might be either too low, resulting in reduced income and economic viability, or too high, resulting in non-sustainable fisheries. In addition, harvesting of one species may affect the harvesting of other species; for instance, overharvesting of shrimp in Hudson Strait could result in depletion of beluga stocks which may feed on the shrimp in winter.

Third, development could result in allocational conflicts. Fishery development requires unutilized or underutilized stocks or reallocation from existing subsistence, commercial or recreational use. The subsistence fisheries are an integral part of the culture of the Inuit and Indian peoples. If reallocation of fishery resources to new fisheries is to occur, the need of the subsistence sector of the renewable resource economy has to be balanced against the aspirations of the developing commercial or recreational sector.

Fourth, the economic potential for developing new commercial fisheries is limited. Shrimp fisheries are especially capital intensive (Crawford 1989). Small fisheries frequently are either economically marginal or unviable, requiring subsidization, and unrestricted access has resulted in greater labour and capital investment than is necessary for economic efficiency (Cauvin 1979; Thompson 1981). The most easily accessible resources already are exploited at or close to the biological limits (Crawford 1989). Local communities offer very small opportunities for marketing fishery products. Markets in southern Canada and elsewhere do exist for products from Arctic fisheries, but the high cost of transportation makes such products uncompetitive in the market-place unless they fill a specialty niche.

Fifth, the infrastructure for developing new commercial fisheries in the Arctic hardly exists. This is especially true for marine fisheries which are receiving the greatest attention for development at present. The Inuit and Indian people have no history of exploiting marine fish offshore, and they have limited entrepreneurial and business experience in developing viable marine fisheries. Vessels, docks and processing plants for marine fisheries and associated infrastructures are largely absent. In the NWT two inshore vessels exist at Pangnirtung, one at Iqaluit and one at Clyde River. Twenty-five or so 14 m vessels exist in northern Quebec, and other vessels from Atlantic Canada are employed by Makivik Corporation for fishing waters adjacent to northern Quebec. No small craft harbours for marine fishing vessels exist in the NWT or northern Quebec. Some problems incurred in developing an infrastructure for processing Arctic charr and marine mammal products in the Keewatin are reviewed by Iredale (1984).

Despite these constraints, opportunities do exist. At present the Pangnirtung Greenland halibut fishery and the Qiqitallik-Makivik shrimp fishery appear to be potential successes provided that the stocks persist, markets expand, and the products can compete in cost with those of other suppliers. The Pangnirtung Greenland halibut fishery harvested 180 000 kg in the late winter of 1989. Sixty to 70 fishermen and processors participated in the fishery, during which welfare payments in the community dropped 20–30% and the crime rate dropped 60–70% (P. Sutherland, DFO, Winnipeg R3T 2N6, pers. comm.). The shrimp fishery is much larger and makes a correspondingly greater economic impact on northern Quebec and south-east Baffin.

Expansion of inter-settlement trade in products such as Arctic charr, marine mammal meat and muktuk is a possibility, especially if expansion of the wage economy occurs and leads to a decline in subsistence fishing (Topolniski et al. 1987). An expanded export trade could occur for specialty products, such as Arctic charr, Atlantic salmon, scallops, shrimp and Greenland shark (the skins of which sell for $60 to $80), and for products, such as Greenland halibut, that could be available at a time when they are unavailable from other sources. However, the economic viability of marine fisheries would depend on providing an infrastructure at virtually no cost to the participants.

Economic returns from existing fisheries might be increased. For instance, the introduction of weirs for the commercial harvest of anadromous Arctic charr would allow operational flexibility and better aircraft utilization in transporting a top quality product to processing facilities (Kristofferson et al. 1986). Introduction of the pulse fishery concept, in which a large quota is harvested only once in several years, could improve the economics of some fisheries, such as those for Arctic charr, without creating stock conservation problems (McCart and Den Beste 1979). Johnson (1989) also suggests that heavy exploitation of Arctic charr stocks for 2–3 years followed by a 10-year respite might be an appropriate management strategy. However, pulse fishing only lowers harvesting and transportation costs, and its biological implications are poorly known (Reid and Momot 1985; Bodaly et al. 1989).

Perhaps the greatest potential, socially, economically and biologically, is in developing or expanding recreational fisheries (Crawford 1989) and whale watching. The development of recreational fisheries for marine mammals, such as seals and walrus, is another possibility but this may create political problems and management costs may exceed benefits. The local cost of developing such opportunities would be relatively low and the economic benefits to the communities would be relatively high. The development, if possible, of some recreational fisheries for especially large (trophy) fish could provide even greater economic returns than non-trophy fisheries. The major drawback is the cost of access by anglers and tourists from outside the Arctic, but people are willing to pay the cost and more people are travelling to the Arctic.

At present, no comprehensive approach exists to fishery development in the NWT. As a result, proposals are pursued on an *ad hoc* basis and funds for fishery development frequently are provided before the biological potential is known. Fishery development planning should be integrated more with fishery management planning. Fishery management plans should be developed for each stock and should identify the safe harvest level and existing harvest levels. Then an analysis can be conducted to determine whether development or expansion of fisheries would increase the benefits from the stock. Part of the analysis could include comparison of various combinations of subsistence, commercial and recreational fisheries and tourism to determine the mix that would provide the greatest benefits to the culture and lifestyle of the communities, regional development, employment and economic efficiency while ensuring the conservation of the stock. The resultant fishery management plan also could address other requirements for management of the stock including research, habitat protection and compliance.

### Cooperative Management

Cooperative management in the Arctic is a necessity because of the settlement of land claims and the limited resources available for fishery management. It always has been an integral part of Arctic fishery management. Cooperative management began with consultations between fishery managers and fishermen, community hunter and trapper organizations or regional associations on quotas, measures for conserving fishery resources, the requirement for new information and proposed research. Whenever possible the views of the resource users have been incorporated by fishery managers into regulatory and management approaches and research proposals.

Now, the settlement of aboriginal land claims is helping to formalize cooperative management and to give specific fishery management responsibilities to aboriginal peoples. Three cooperative land claims have been signed in the area: the James Bay Northern Quebec Agreement with the Grand Council of the Crees of Quebec and the Northern Quebec Inuit Association in 1975; the Northern-eastern Quebec Agreement with the Naskapis of Schefferville in 1978 (Editeur Officiel du Québec 1976); and the Inuvialuit Final Agreement with the Inuvialuit of the Western Arctic in 1984 (Indian and Northern Affairs Canada 1984). An agreement in principle has been reached for the Tungavik Federation of Nunavut claim which covers the remaining coastal area in the NWT. The final agreement for the Gwich'in claim, the part of the Dene/Metis claim (Task Force to Review Comprehensive Claims Policy 1985) that extends into the Mackenzie Delta, was ratified by the Gwich'in in 1991 and awaits ratification by government. From a fishery management perspective, the features of a successful land claim settlement include responding to claimant aspirations, guaranteeing conservation of fishery resources, establishing an effective fishery management system, and reflecting the interests and aspirations of other Canadians (G. Yaremchuk, DFO, Ottawa K1A 0E6, pers. comm.).

The fishery management responsibilities in land claim settlements are an extension of existing cooperative management. Cooperative management recognizes that the conservation, development and sustained utilization of Arctic fishery resources is achieved best by increasing the understanding and participation of user groups and government in all aspects of fishery management. Management is exercised through cooperative arrangements between governments and resource users which can range from advisory roles to sharing the responsibility for decisions about the use of fishery resources. Legislative accountability for the conservation of the fishery resources remains with government. The process implemented within an area will reflect land claim settlements and local and regional needs and aspirations.

Currently, three formal cooperative management committees exist for Arctic marine fisheries. In northern Quebec, the Hunting, Fishing and Trapping Coordinating Committee was established under the James Bay Northern Quebec Agreement to review, manage, supervise and regulate the hunting, fishing and trapping regime, and advise the various ministers on wildlife issues. In the western Arctic, the Fisheries Joint Management Committee was established under the Inuvialuit Final Agreement to manage Inuvialuit fishing rights under the agreement and advise the Minister of Fisheries and Oceans on the management of fish and marine mammal resources within the claim area. The Nunavut Wildlife Management Advisory Board has been established for the remaining NWT Arctic coastal and marine areas in advance of the settlement of the Tungavik Federation of Nunavut land claim.

The formal cooperative management structures have advisory and decision-making responsibilities. The more recently established structures have more decision-making responsibilities. The two main features of cooperative management are that government relinquishes considerable control to resource users, and the resource users have to take responsibility for their decisions. The cooperative management structure should be involved in all aspects of fishery management such as establishing safe harvest levels, allocating resources, licensing, developing regulations, monitoring harvests, carrying out stock assessments and identifying research needs (Yaremchuk and Wong 1989a, b).

One example is the Fisheries Joint Management Committee which has a mixture of advisory and decision-making responsibilities (Indian and Northern Affairs Canada 1984). The Committee has the responsibility to allocate subsistence fishery resources among the Inuvialuit communities and to regulate fishing on certain Inuvialuit lands. It advises the Minister of Fisheries and Oceans on overall subsistence quotas, commercial and recreational fishing and fishing regulations. In addition, the Committee is responsible for determining current harvest levels and reviewing information on fishing in the Inuvialuit Settlement Region. In contrast, the Hunting, Fishing and Trapping Coordinating Committee in northern Quebec is a consultative body to the responsible governments and may submit recommendations to the responsible federal or provincial minister. Whatever the responsibilities assigned to a cooperative management structure, the Minister of Fisheries and Oceans retains ultimate authority for the conservation of the fishery resources.

Cooperative management in the Arctic has considerable advantages. A key aspect is that the cooperative management board or committee has responsibility for making allocational decisions among subsistence, recreational and commercial fisheries and among different communities. Resource users are provided an opportunity for involvement, directly or through their representatives, in all aspects of fishery management. They, through cooperative management boards, have clearly defined responsibilities including responsibility for certain categories of decision. They have a clear process for stating their knowledge and their concerns. For the fishery management agencies, cooperative management involves relinquishing responsibility for making allocational decisions, and provides a source of local knowledge and advice which should decrease the need for compliance and enforcement activities as consensus occurs. Cooperative management also provides a focus for planning and coordinating fishery development. For all participants, cooperative management provides an opportunity for common planning of fishery management and development, developing a single information base that integrates local and scientific knowledge, communicating effectively, and educating and training resource users to play a greater role in fishery management.

To be most effective, the management structures should include representatives of all affected interests (Yaremchuk and Wong 1989a, b). In the widest sense this would involve representatives of the fishermen, possible third party interests, the management agency and, as necessary, representatives of other agencies involved in economic and social development, public health and protection of the environment. Also the process should be open, with all affected people able to learn about decisions.

However, disadvantages do exist. Cooperative management, especially formal cooperative management, is expensive. For instance, in 1988/89 the Fisheries Joint Management Committee received $665,000 for its operations, monitoring and stock assessments (Fisheries Joint Management Committee 1989), and the Department of Fisheries and Oceans had significant expenses for interacting with the Committee. Even informal cooperative management is expensive since considerable travel to meetings is necessary. Cooperative management is time-consuming, requiring commitment of people's time and effort for preparing, exchanging and explaining information, and reaching decisions especially if these are made through consensus. Finally, because resource users and managers are inexperienced with cooperative management at present, much learning is involved and effective interpersonal relationships have to be established. Despite these disadvantages, the advantages far exceed the disadvantages.

## Other Human Activities

The conservation of Arctic fishery resources is dependent upon effective management of other human activities as well as effective management of harvesting. The Arctic marine environment is not pristine. It has undergone physical and chemical changes resulting from industrial development in its drainage basins and in other parts of the world. Such changes are known to have the capability of affecting fishery resources and their habitats.

Hydrocarbon exploration and development has been the greatest industrial activity in the Arctic (Table 10). Most activity has occurred in the Beaufort Sea/ Mackenzie Delta region and the Sverdrup Basin. Hydrocarbon exploration began in the Beaufort Sea/ Mackenzie Delta region in the 1950's, and drilling began offshore in 1972 (Brouwer et al. 1988). Some 70 wells have been drilled offshore (Table 10). In 1986, 50 400 $m^3$ of oil was shipped from the Beaufort Sea (Canada Oil and Gas Lands Administration 1988). In the Sverdrup Basin about 39 wells have been drilled offshore (Table 10). Oil has been produced from Bent Horn in the Sverdrup Basin since 1985, and 136 400 $m^3$ have been shipped to market. Considerable interest exists in expanding production of oil and gas from both regions. Also, two wells were drilled in southern Davis Strait and three in Hudson Bay, and drilling has been proposed for Lancaster Sound and Baffin Bay.

Hydrocarbon exploration and production involves many activities that could affect fishery resources. Habitat in the southern Beaufort Sea has been altered by dredging to form artificial islands and to improve harbours; from 1959 to 1982 approximately 46.5 million $m^3$ of seabottom materials were dredged (Taylor et al. 1985). Other disturbances have occurred. Between 1980 and 1986 in the southern Beaufort Sea, at least 22 861 km of seismic lines were shot and 7 000 vessel trips occurred (Brouwer et al. 1988). Contaminants have been discharged, including 14 160 tonnes of drilling mud plus formation water in the Sverdrup Basin (Sackmann and Smiley 1985). The drilling muds contained barite, clay, caustic soda, salts and heavy metals such as zinc, lead, copper, chromium, nickel, cadmium and mercury (Sackmann and Smiley 1985). From 1972 to 1985 about 1.6 million litres of petroleum products entered Arctic waters from 180 reported spills. The specific effects of these activities are unknown but they could alter habitat, could cause toxic effects on fishery resources or could result in the accumulation of metals by fishery resources.

Two lead-zinc mines are located on the Arctic coast, the Nanisivik mine on Strathcona Sound and the Polaris mine on Little Cornwallis Island. They produce lead and zinc concentrates which are stockpiled during the winter and shipped through Lancaster Sound during the open-water season. Both mines discharge their tailings into lakes which drain directly into the sea. Large iron ore deposits exist near Mary River on Baffin Island and on the Melville Peninsula; mines could be developed at some future time at these sites. Heavy

TABLE 10. Hydrocarbon exploration, production and reserves in the offshore Canadian Arctic.

| Region | Exploration | Production | Discovered oil and gas |
|---|---|---|---|
| Beaufort Sea[a] | 70 wells | 1986: 50 400 $m^3$ oil | 253.2 million $m^3$ oil<br>308.5 billion $m^3$ gas |
| Sverdrup Basin[a] | 39 wells | 1985–88:136 400 $m^3$ oil | 65.7 million $m^3$ oil<br>416.4 billion $m^3$ gas |
| Lancaster Sound | Proposed | | |
| Baffin Bay | Proposed | | |
| Southern Davis Strait | 2 wells | | |
| Hudson Bay | 3 wells | | |

[a]From Canada Oil and Gas Lands Administration (1988).

metal contamination is the main concern about mining in the Arctic, and has been recorded in the sediments, seaweed and benthic invertebrates (Fallis 1982) and ringed seals (Wagemann 1989) in the vicinity of the Nanisivik mine.

Shipping in the Arctic is associated with hydrocarbon and mineral exploration and production, resupply of 44 communities, grain export from Churchill, tourist travel and research. Canadian Coast Guard ice-breakers frequently support these voyages or make their own trips. In 1989, about 40 vessels planned to operate in the Canadian Arctic, visiting 56 communities and localities. Shipping is restricted, for the most part, to the open-water season or to the late winter period of ice cover. Hydrocarbon production could result in increased tanker traffic including possible year-round shipping of oil through the Northwest Passage or past Alaska. Concerns relate to the effects of shipping, especially during the season of ice-cover, on the behaviour and distribution of marine mammals through ice breaking and noise production (Mansfield 1983) as well as spills of oil or other cargoes. Beluga and narwhal along the Lancaster Sound ice edge have reacted to ships at distances of 45 to 60 km (Cosens and Dueck 1988). Ice breaking also may impede travel and harvesting activities of Inuit. Also, dredging has occurred at Churchill and in the Beaufort Sea in support of shipping.

Other activities in the area may affect fishery resources and their habitats. The discharge of sewage from communities and ocean dumping may result in the addition of nutrients or contaminants.

Activities within the Arctic and Hudson Bay drainage basins may affect the Arctic marine environment. Major hydroelectric power development has occurred on the Churchill and Nelson rivers which flow into western Hudson Bay, on rivers in northern Quebec that flow into eastern Hudson Bay, in the Mackenzie Basin and elsewhere (Wright 1980; Rosenberg et al. 1987; Bodaly et al. 1989). Proposals exist to divert water from James Bay to southern Canada (Milko 1986), from the headwaters of the Mackenzie River and from rivers in Siberia. The effects from hydroelectric power development and diversions on fresh waters generally are known, although much uncertainty exists in predicting environmental effects especially at higher trophic levels (Hecky et al. 1984; Rosenberg et al. 1987). The planned hydroelectric power developments in the Hudson Bay and James Bay drainage will increase winter runoff rate by 52% and decrease the summer runoff rate by 6% (Prinsenberg 1980). Impacts on adjacent estuarine and marine ecosystems are less well known. Filling the LG-2 reservoir affected the La Grande estuary resulting in delayed breakup, an influx of marine species and disruption of fisheries (Berkes 1982). Such developments also may alter temperature, salinity and nutrient patterns, and change primary production (Bugden et al. 1982) and may affect survival of larval fish. Development of reservoirs also mobilizes mercury (Bodaly et al. 1984; Hecky et al. 1984) which may accumulate in Arctic marine biota.

Agricultural, urban, forestry, mining, hydrocarbon and pulp and paper developments in the Arctic and Hudson Bay drainage basins (Wright 1980; Bodaly et al. 1989; Brousseau and Goodchild 1989; Roy 1989) may result in contaminants reaching Arctic seas. Such contaminants include organochlorines, pesticides, hydrocarbons and heavy metals.

More insidious than even the riverine inputs is the introduction of contaminants into the Arctic marine environment through long range atmospheric and oceanic transport. Artificial radionuclides, originating from fallout and from liquid discharges from a nuclear fuel reprocessing plant in England (Livingston et al. 1984), have been detected in the Arctic Ocean. Organic contaminants are widespread (for example see Wagemann and Muir 1984). For instance, PCBs, chlordane-related compounds, DDT, hexachlorocyclohexane, toxaphene and chlorobenzenes have been detected in Arctic cod and ringed seals in the Lancaster Sound region (Muir et al. 1988). The organochlorine compounds probably are deposited in the Arctic after long-range atmospheric transport.

Each activity or input poses a potential threat to the fishery resources and their habitats or to the suitability of fishery products for human consumption. Perhaps the greatest threat is not the effect of one such activity, or even one major spill, but is the cumulative effects of continued minor impacts from different sources (Johnson 1983b). This could affect the functioning of Arctic marine ecosystems and even result in the elimination of important stocks (Johnson 1983b). The most vulnerable species from this perspective are considered to be the bowhead and walrus (Johnson 1983b).

Such stresses on Arctic fishery resources are occurring when further stress is being added through global warming. Evidence, but not proof, exists that increasing atmospheric carbon dioxide will cause global warming of several degrees by the end of the next century (Ripley 1987). Several models predict global warming, and most predict greater warming in polar regions than in tropical regions. Ripley (1987) based a model on a doubling of atmospheric carbon dioxide by 2085, an average scenario; the model predicted Arctic temperatures rising by up to 4°C, precipitation increasing by up to 200 mm $yr^{-1}$, primarily in summer, and run-off increasing in the Arctic drainage basin by 95% and in the Hudson Bay drainage by 20%. The temperature and precipitation increases would vary in different parts of the Arctic. Other models have predicted greater temperature increases in the Arctic (Manabe and Wetherald 1986). Some specific changes expected from the warming would be longer ice-free periods, higher water temperatures, retreat of the Arctic Ocean ice pack and increased sediment discharge from rivers (Ripley 1987). Also, the sea level may rise.

Effects of climate change on fisheries are well established. Climate changes affect fish recruitment, the relationship between larval fish productivity and food availability, and species distribution (Cushing 1982; Frank et al. 1988). In the Arctic, primary productivity may increase because of the longer ice-free periods and higher water temperatures. Arctic fishery resources may be affected directly. The southern distribution limits of marine mammals, such as beluga, narwhal and walrus, and possibly of Arctic charr may move further north. Atlantic fishery resources, such as Atlantic cod and other groundfish species may extend further north into Davis Strait and Baffin Bay (Frank et al. 1988). Coastal habitats could be changed significantly. Harvesting practices may be affected since winter ice would be of less use as a platform for travel and harvesting. The location of coastal communities could be affected by a rise in sea level. Concentrations of fishery resources could be further away from communities. Such changes could have profound effects on subsistence harvesting and other existing fisheries but may provide opportunities to establish new fisheries.

## Required Management Approaches

Sustainable development now is the accepted approach for managing and balancing various human developments so that they occur in an environmentally acceptable manner (World Commission on Environment and Development 1987). However, the interpretation of what constitutes sustainable development varies (Ekins 1989; Simon 1989) from immediate economic growth in an environmentally sound manner to consideration of whether development should occur at all.

Sound management of the Arctic and its fishery resources requires considering whether industrial development should occur, when it should occur and under what conditions it should occur. The Arctic Marine Conservation Strategy discussion paper advocates integrated resource planning and management as a major cornerstone for ensuring "the future health and well-being of Arctic marine ecosystems ... and ... the sustained utilization of marine resources ..." (Department of Fisheries and Oceans 1987; Snider 1987). Integrated resource planning and management requires participation of and cooperation among resource users, developers and resource managers. It requires identification of objectives of different groups, the building of consensus and resolution of conflicts. Most of all, successful integrated resource planning and management requires people making hard decisions so that any development is sustainable and long-term environmental security is not sacrificed for shorter-term economic gain.

The major approach to integrated resource planning instituted in the Northwest Territories and Yukon is Northern Land Use Planning. The intent is that land use plans will be developed for all land, freshwater and marine areas, and that the plans will anticipate, resolve and minimize competing demands, identify best uses and ensure integrated resource management. The conservation and sustained utilization of fishery and other renewable resources are accepted as the priorities, but other activities could occur if they are consistent with this priority.

The Arctic Marine Conservation Strategy discussion paper also identifies other approaches necessary for the conservation of the Arctic marine environment

(Department of Fisheries and Oceans 1987; Snider 1987). Further development of scientific knowledge is required; for instance to obtain the understanding and information necessary for managing fish stocks or predicting the effects of industrial activities on these stocks. The quality of the Arctic marine environment must be protected from waste disposal at sea and on-land, from alteration of rivers' flow regimes and from long-range atmospheric and oceanic transportation of contaminants. A system of protected marine areas must be established for some especially important areas, such as Isabella Bay where it is proposed to establish a sanctuary for the bowhead; it has been recommended that 12% of lands and waters should be given protected area status (World Commission on Environment and Development 1987). International cooperation is a necessity because pollution of Canada's Arctic frequently originates outside Canada, and because stocks of fish and marine mammals are shared with Alaska (U.S.A.) and Greenland (Denmark). Finally, communication must occur among all groups to promote understanding of the Arctic marine environment and to develop a common understanding of different groups' perceptions and needs.

## Conclusions

Although Arctic marine fisheries are small compared to Atlantic and Pacific fisheries, they play an integral and major role in the lives of northern Canadians. Harvesting fish and marine mammals is part of the traditional culture of the native peoples and provides a considerable portion of their food. The commercial and recreational fisheries provide one of the few sources of income and/or occupation for 50–75% of the population in northern communities. The replacement value of food from the subsistence harvest is about $15 million, other quantified consumer benefits are valued at $1 million, and the value added to the Canadian economy is about $6.3 million.

The importance of Arctic marine fishery resources is increasing because protein self-sufficiency and fishery development are priorities of native groups and federal, provincial and territorial governments. Careful planning of fishery development must occur on a stock-by-stock basis so that fishery resources will provide the greatest continuing and long-term benefits to the cultural life of the community, regional development, employment and economic efficiency while the fish and marine mammal stocks are conserved. Although the biological and economic potential is limited and infrastructure is largely absent, recreational fisheries, whale watching, inter-settlement trade and export of specialty products offer some opportunities for increasing the benefits derived from fishery resources.

Better information is required to ensure conservation of fishery resources and to aid fishery development. Requirements include more understanding and information on the structure and functioning of Arctic marine ecosystems, better understanding and information on stock identity, stock size and vital parameters, development of better models for determining safe harvesting levels, and more accurate estimates of subsistence, commercial and recreational harvests from each stock.

Cooperative management is an integral part of fishery management in the Arctic. It is becoming more formalized through land claim settlements which are establishing management boards that have specified decision-making and advisory responsibilities. Although it is expensive, cooperative management enables all participants to develop common understanding, a common information base and a common plan for fishery management and development, and to educate and train resource users to play a greater role in fishery management. Also, cooperation is required with Greenland (Denmark) and Alaska (U.S.A.) on management of and research on shared stocks of fish and marine mammals.

Even successful cooperative management, better information and sound fishery management planning cannot ensure the conservation and sustainable utilization of Arctic fishery resources. Integrated resource planning and management is required to ensure that activities, such as hydrocarbon development, mining and shipping, in Arctic seas or along the Arctic coast occur without adverse effects on fishery resources. A greater challenge is to manage developments, especially waste discharges and hydroelectric power developments, in the Arctic and Hudson Bay drainage basins so that Arctic marine fishery resources and their habitats are not affected. However, the greatest challenge is to achieve international cooperation to stop the long-range transport of contaminants to the Arctic and to reverse or halt the trend to global warming, either on both of which could result in the elimination of today's Arctic marine fisheries.

## Acknowledgements

I gratefully thank the many people who have provided me with suggestions, information, help and comments in the preparation of this paper: R. Allen, M. Breton, G. Chantigny, J. Chouinard, R. Crawford, L. de March, B. Dempson, V. Gillman, L. Johnson, S. Kirwan, A. Mansfield, D. Moshenko, R. Moshenko, E. Snider, J. Stein, R. Stewart, P. Sutherland and H. Welch.

Especially I thank D. Topolniski for providing advice and information, his patience in explaining the economics of benefits to me and his help in writing that part of the paper.

## References

ANDERSON, L.G. 1977. The economics of fisheries management. John Hopkins University Press, Baltimore, MD. 214 p.

ANDERSON, L.G., D.W. DYRSSEN, E.P. JONES, AND M.G. LOWINGS. 1983. Inputs and outputs of salt, fresh water, alkalinity, and silica in the Arctic Ocean. Deep-Sea Res. 30: 87–94.

AYLES, G.B. 1987. Arctic fisheries research in Canada: an informal perspective. Northern Perspectives 15(4): 13–15.

BARBER, F.G. 1968. The water and ice of Hudson Bay, Part I, The water, p. 287–318. *In* C.S. Beals [ed.] Science, history and Hudson Bay, Vol. 1. Department of Energy, Mines and Resources, Ottawa, Ont. 501 p.

BERKES, F. 1982. Preliminary impacts of the James Bay hydroelectric project, Quebec, on estuarine fish and fisheries. Arctic 35: 524–530.

——— 1990. Native subsistence fisheries: a synthesis of harvest studies in Canada. Arctic 43: 35–42.

BOCKSTOCE, J.R., AND D.B. BOTKIN. 1983. The historical status and reduction of the western arctic bowhead whale (*Balaena mysticetus*) population by the pelagic whaling industry, 1848–1914. Rep. Int. Whaling Comm. Spec. Issue 5: 107–141.

BODALY, R.A., R.E. HECKY, AND R.J.P. FUDGE. 1984. Increases in fish mercury levels in lakes flooded by the Churchill River diversion, northern Manitoba. Can. J. Fish. Aquat. Sci. 41: 682–691.

BODALY, R.A., J.D. REIST, D.M. ROSENBERG, P.J. MCCART, AND R.E. HECKY. 1989. Fish and fisheries of the Mackenzie and Churchill river basins, northern Canada, p. 128–144. *In* D.P. Dodge [ed.] Proceedings of the International Large River Symposium. Can. Spec. Publ. Fish. Aquat. Sci. 106.

BOND, W.A. 1982. A study of the fishery resources of Tuktoyaktuk Harbour, southern Beaufort Sea coast, with special reference to life histories of anadromous coregonids. Can. Tech. Rep. Fish. Aquat. Sci. 1119: vii + 90 p.

BOND, W.A., AND R.N. ERICKSON. 1985. Life history studies of anadromous coregonid fishes in two freshwater lake systems on the Tuktoyaktuk Peninsula, Northwest Territories. Can. Tech. Rep. Fish. Aquat. Sci. 1336: vii + 61 p.

BRADSTREET, M.S., AND W.E. CROSS. 1982. Trophic relationships at high Arctic ice edges. Arctic 35: 1–12.

BROUSSEAU, C.S., AND G.A. GOODCHILD. 1989. Fisheries and yields in the Moose River basin, Ontario, p. 145–158. *In* D.P. Dodge [ed.] Proceedings of the International Large River Symposium. Can. Spec. Publ. Fish. Aquat. Sci. 106.

BROUWER P., J.W. MCDONALD, W.J. RICHARDSON, AND R.A. DAVIS. 1988. Arctic industrial activities compilation: volume 3. Canadian Beaufort Sea: seismic and sounding surveys, vessel movements, helicopter traffic and site-specific activities 1980 to 1986. Can. Data Rep. Hydrogr. Ocean Sci. 32(3): xvi + 170 p.

BUGDEN, G.L., B.Y. HARGRAVE, M.M. SINCLAIR, C.L. TANG, J.-C. THERRIAULT, AND P.A. YEATS. 1982. Freshwater runoff effects in the marine environment: the Gulf of St. Lawrence example. Can. Tech. Rep. Fish. Aquat. Sci. 1078: ix + 89 p.

CANADA OIL AND GAS LANDS ADMINISTRATION. 1988. Annual Report 1988. Canada Oil and Gas Lands Administration, Ottawa, Ont. 46 p.

CARDER, G.W., AND D.B. STEWART. 1989. Data from the commercial fishery for Arctic charr, *Salvelinus alpinus* (L.), in the central and Keewatin regions of the Northwest Territories, 1987–88. Can. Data Rep. Fish. Aquat. Sci. 766: vi + 47 p.

CAUVIN, D.M. 1979. Regulating access in Canada's inland fisheries. J. Fish. Res. Board Can. 36: 827–836.

CLARKE, R.MCV., L. JOHNSON, G.D. KOSHINSKY, A.W. MANSFIELD, R.W. MOSHENKO, AND T.A. SHORTT. 1989. Report of the Arctic Fisheries Scientific Advisory Committee for 1986/87 and 1987/88. Can. Manuscr. Rep. Fish. Aquat. Sci. 2015: iv + 68 p.

CORKUM, L.D., AND P.J. MCCART. 1981. A review of the fisheries of the Mackenzie Delta and nearshore Beaufort Sea. Can. Manuscr. Rep. Fish. Aquat. Sci. 1613: v + 55 p.

COSENS, S.E., J.F. CRAIG, AND T.A. SHORTT. 1990. Report of the Arctic Fisheries Scientific Advisory Committee for 1988/89. Can. Manuscr. Rep. Fish. Aquat. Sci. 2063: iv + 40 p.

COSENS, S.E., AND L.E. DUECK. 1988. Responses of migrating narwhal and beluga to icebreaker traffic at the Admiralty Inlet ice-edge, N.W.T. in 1986, p. 39–54. *In* W.M. Sackinger and M.O. Jeffries [ed.] Port and ocean engineering under Arctic conditions, Vol. II, Symposium on Noise and Marine Mammals. Univ. Alaska, Fairbanks, AK. 111 p.

CRAWFORD, R. 1989. Exploitation of Arctic fishes. Can. Manuscr. Rep. Fish. Aquat. Sci. 2002: v + 43 p.

CUSHING, D.H. 1982. Climate and fisheries. Academic Press, London, England. xiv + 373 p.

DAVIES, S., K. KROEKER, AND D. MACDONNELL. 1986. Commercial fisheries of the Northwest Territories. An historical perspective. Department of Economic Development and Tourism, Government of the Northwest Territories, Yellowknife, NWT. vii + 54 p.

DAVIS, R.A., K.J. FINLEY, AND W.J. RICHARDSON. 1980. The present status and future management of Arctic marine mammals in Canada. Science Advisory Board of the Northwest Territories, Yellowknife, NWT. Rep. 3: 93 p.

DE MARCH, L., B. DE MARCH, AND W. EDDY. 1978. Limnological, fisheries and stream zoobenthic studies at Stanwell-Fletcher Lake. Preliminary Report 1977. Canada Department of Indian and Northern Affairs, Ottawa, Ont. ESCOM Rep. AI-04: 84 p.

DEMPSON, J.B., AND A.H. KRISTOFFERSON. 1987. Spatial and temporal aspects of the ocean migration of anadromous Arctic charr. Am. Fish. Soc. Symp. 1: 340–357.

DEPARTMENT OF ENERGY, MINES AND RESOURCES, AND INFORMATION CANADA. 1974. The national atlas of Canada. Ottawa, Ont. 267 p.

DEPARTMENT OF FISHERIES AND OCEANS. 1982. Sailing directions: Arctic Canada. Vol. 1. The Department, Ottawa, Ont. 284 p.

1986. Northern Quebec beluga management plan, 1986. The Department, Quebec City, Que. 11 p.

1987. Canadian Arctic Marine Conservation Strategy discussion paper. The Department, Ottawa, Ont. 21 p.

1988. Sport fishing in the Northwest Territories, 1985. The Department, Ottawa, Ont. 12 p.

DUNBAR, M.J. 1968. Ecological development in polar regions: a study in evolution. Prentice-Hall Inc., Englewood Cliffs, NJ. 119 p.

EDITEUR OFFICIAL DU QUÉBEC. 1976. The James Bay and Northern Quebec Agreement. Editeur Officiel du Québec, Québec. vii + 455 p.

EKINS, P. 1989. Beyond growth: the real priorities of sustainable development. Environ. Conserv. 16: 5–6, 12.

FALLIS, B. 1982. Trace metals in sediments and biota from Strathcona Sound, NWT; Nanisivik marine monitoring program, 1974–1979. Can. Tech. Rep. Fish. Aquat. Sci. 1082: v + 34 p.

FEE, E.J., R.E. HECKY, S.J. GUILDFORD, C. ANEMA, D. MATHEW, AND K. HALLARD. 1988. Phytoplankton primary production and limnological data for lakes and channels in the Mackenzie Delta and lakes on the Tuktoyaktuk Peninsula, N.W.T. Can. Tech. Rep. Fish. Aquat. Sci. 1614: v + 62 p.

FINDLEY, K.G., G.W. MILLER, M. ALLARD, R.A. DAVIES, AND C.R. EVANS. 1982. The belugas (*Delphinapterus leucas*) of northern Quebec: distribution, abundance, stock identity, catch history and management. Can. Tech. Rep. Fish. Aquat. Sci. 1123: v + 57 p.

FISHERIES JOINT MANAGEMENT COMMITTEE. 1989. Fisheries Joint Management Committee annual report 1987/88 1988/89 combined. Fisheries Joint Management Committee, Inuvik, NWT. 24 p.

FRANK, K.T., R.I. PERRY, K.F. DRINKWATER, AND W.H. LEAR. 1988. Changes in the fisheries of Atlantic Canada associated with global increases in atmospheric carbon dioxide: a preliminary report. Can. Tech. Rep. Fish. Aquat. Sci. 1652: v + 52 p.

GAMBLE, R.L. 1988. Native harvest of wildlife in the Keewatin Region, Northwest Territories for the period October 1985 to March 1986 and a summary for the entire period of the harvest study from October 1981 to March 1986. Can. Data. Rep. Fish. Aquat. Sci. 688: v + 85 p.

GIOVANDO, L.F., AND R.H. HERLINVEAUX. 1981. A discussion of factors influencing dispersion of pollutants in the Beaufort Sea. Pac. Mar. Sci. Rep. 81–4: 198 p.

HECKY, R.E., R.W. NEWBURY, R.A. BODALY, K. PATALAS, AND D.M. ROSENBERG. 1984. Environmental impact prediction and assessment: the Southern Indian Lake experience. Can. J. Fish. Aquat. Sci. 41: 720–732.

HUNTER, J.G., S.T. LEACH, D.E. MCALLISTER, AND M.B. STEIGENWALD. 1984. A distributional atlas of records of marine fishes of Arctic Canada in the National Museums of Canada and Arctic Biological Station. Syllogeus 52: 35 p.

INDIAN AND NORTHERN AFFAIRS CANADA. 1984. The western Arctic claim. The Inuvialuit final Agreement. Ottawa, Ont. 115 p.

1988. A northern political and economic framework. Ottawa, Ont. 14 p.

IREDALE, D.G. 1984. Commercial processing of charr in Canada's eastern Arctic, p. 365–369. *In* L. Johnson and B.L. Burns [ed.] Biology of the Arctic charr, Proceedings of the International Symposium on Arctic Charr, Winnipeg, Manitoba, May 1981. University of Manitoba Press, Winnipeg, Man.

IRWIN, C. 1989. Lords of the Arctic: wards of the state. Northern Perspectives 17: 2–12.

JOHNSON, L. 1980. The arctic charr, *Salvelinus alpinus*, p. 15–98. *In* E.K. Balon [ed.]. Charrs: salmonind fishes of the genus *Salvelinus*. Dr. W. Junk Publishers, The Hague, The Netherlands. Perspect. Vertebr. Sci. 1: viii + 928 p.

———— 1983a. Homeostatic characteristics of single species fish stocks in Arctic lakes. Can. J. Fish. Aquat. Sci. 40: 987–1024.

———— 1983b. Assessment of the effects of oil on Arctic marine fish and marine mammals. Can. Tech. Rep. Fish. Aquat. Sci. 1200: viii + 15 p.

———— 1989. The anadromous Arctic charr, *Salvelinus alpinus*, of Nauyuk Lake, N.W.T., Canada. Physiol. Ecol. Japan Spec. Vol. 1: 201–227.

KEMP, A., L. BERNATCHEZ, AND J.J. DODSON. 1989. A revision of coregonid fish distribution and abundance in eastern James-Hudson Bay. Environ. Biol. Fishes 26: 247–255.

KINGSLEY, M. 1989. Population dynamics of the narwhal *Monodon monoceros*: an initial assessment (Odontoceti: Monodontidae). J. Zool. Lond. 219: 201–208.

KRISTOFFERSON, A.H., AND D.K. MCGOWAN. 1981. Data on Arctic charr, *Salvelinus alpinus* (Linneaus), collected from test fisheries in the Baffin Region, Northwest Territories, 1975–79. Can. Data Rep. Fish. Aquat. Sci. 255: vi + 43 p.

KRISTOFFERSON, A.H., D.K. MCGOWAN, AND G.W. CARDER. 1984. Management of the commercial fishery for anadromous Arctic charr in the Cambridge Bay area, Northwest Territories, Canada, p. 447–461. *In* L. Johnson and B.L. Burns [ed.] Biology of the Arctic Charr, Proceedings of the International Symposium on Arctic Charr, Winnipeg, Manitoba, May 1981. University of Manitoba Press, Winnipeg, Man.

KRISTOFFERSON, A.H., D.K. MCGOWAN, AND W.J. WARD. 1986. Fish weirs for the commercial harvest of searun Arctic charr in the Northwest Territories. Can. Ind. Rep. Fish. Aquat. Sci. 174: iv + 31 p.

LARKIN, P.A. 1983. How much is enough? An essay on the structure of fisheries management agencies, p. 229–245. *In* B.J. Rothschild [ed.] Global fisheries perspectives for the 1980s. Springer-Verlag New York Inc., New York, NY.

LAWRENCE, M.J., G. LACHO, AND S. DAVIES. 1984. A survey of the coastal fishes of the southeastern Beaufort Sea. Can. Tech. Rep. Fish. Aquat. Sci. 1220: x + 178 p.

LIVINGSTON, H.D., S.L. KUPFERMAN, V.T. BOWEN, AND R.M. MOORE. 1984. Vertical profiles of artificial radionuclide concentrations in the Central Arctic Ocean. Geochim. et Cosmochim. Acta 48: 2195–2203.

LUBINSKY, I. 1980. Marine bivalve molluscs of the Canadian central and eastern Arctic: faunal composition and zoogeography. Can. Bull. Fish. Aquat. Sci. 207: 111 p.

MANABE, S., AND R.T. WETHERALD. 1986. Reduction in summer soil wetness induced by an increase in atmospheric carbon dioxide. Science 232: 626–628.

MANSFIELD, A.W. 1973. The Atlantic walrus *Odobenus rosmarus* in Canada and Greenland. IUCN Publ. New Series, Suppl. Paper 39: 69–79.

———— 1983. The effects of vessel traffic in the Arctic on marine mammals and recommendations for future research. Can. Tech. Rep. Fish. Aquat. Sci. 1186: x + 97 p.

———— 1985. Status of the bowhead whale, *Balaena mysticetus*, in Canada. Can. Field-Nat. 99: 421–424.

MCCART, P.J., AND J. DEN BESTE. 1979. Aquatic resources in the Northwest Territories. Science Advisory Board of the Northwest Territories, Yellowknife, NWT. vii + 54 p.

MCGHEE, R. 1978. Canadian Arctic prehistory. Van Nostrand Reinhold Ltd., Toronto, Ont. 128 p.

MENZIES, R.J. 1975. Origin and evolution of the Arctic marine system, p. 15–25. *In* Procedings of the Circumpolar Conference on Northern Ecology, Ottawa, Ont.

MIKHAIL, M.Y., AND H.E. WELCH. 1989. Biology of Greenland cod, *Gadus ogac*, at Saqvaqjuac, northwest coast of Hudson Bay. Environ. Biol. Fishes 26: 49–62.

MILKO, R. 1986. Potential ecological effects of the proposed GRAND canal diversion project on Hudson and James Bays. Arctic 39: 316–326.

MILLS, E.L., AND R.D. FOURNIER. 1979. Fish production and the marine ecosystems of the Scotian Shelf, eastern Canada. Mar. Biol. 54: 101–108.

MITCHELL, E., AND R.R. REEVES. 1981. Catch history and cumulative catch estimates of initial population size of cetaceans in the eastern Canadian Arctic. Rep. Int. Whaling Comm. 31: 645–682.

———— 1982. Factors affecting abundance of bowhead whales *Balaena mysticetus* in the eastern Arctic of North America, 1915–1980. Biol. Conserv. 22: 59–78.

MORIN, R., J. DODSON, AND G. POWER. 1980. Estuarine fish communities of the eastern James-Hudson Bay coast. Environ. Biol. Fishes 5: 135–141.

MOSHENKO, R.W., R.F. PEET, L.W. DAHLKE, AND D.H. DOWLER. 1984. The Arctic charr sport fishery at Tree River, Northwest Territories, Canada, 1964–78, p. 349–364. *In* L. Johnson and B.L. Burns [ed.] Biology of the Arctic charr, Proceedings of the International Symposium on Arctic Charr, Winnipeg, Manitoba, May 1981. University of Manitoba Press, Winnipeg, Man.

MUIR, D.C.G., R.J. NORSTROM, AND M. SIMON. 1988. Organochlorine contaminants in Arctic marine food chains: accumulation of specific polychlorinated biphenyls and chlordane-related compounds. Environ. Sci. Tech. 22: 1071–1079.

OUTCROP LTD. 1984. NWT data book 84/85. Outcrop Ltd., Yellowknife, NWT. 240 p.

POWER, G. 1976. History of the Hudson's Bay Company salmon fisheries in the Ungava Bay region. Polar Record 18: 151–161.

1981. Stock characteristics and catches of Atlantic salmon (*Salmo salar*) in Quebec, and Newfoundland and Labrador in relation to environmental variables. Can. J. Fish. Aquat. Sci. 38: 1601–1611.

PRINSENBERG, S.J. 1980. Man-made changes in freshwater input rates of Hudson and James bays. Can. J. Fish. Aquat. Sci. 37: 1101–1110.

REEVES, R.R., AND E. MITCHELL. 1987. History of white whales (*Delphinapterus leucas*) exploitation in eastern Hudson Bay and James Bay. Can. Spec. Publ. Fish. Aquat. Sci. 95: vi + 45 p.

REID, D.M., AND W.T. MOMOT. 1985. Evaluation of pulse fishing for walleye, *Stizostedion vitreum vitreum*, in Henderson Lake, Ontario. J. Fish. Biol. 27 (Suppl. A): 235–251.

REIST, J.D., AND W.A. BOND. 1988. Life history characteristics of migratory coregonids of the lower Mackenzie River, Northwest Territories, Canada. Finn. Fish. Res. 9: 133–144.

RICHARD, P.R. 1991. Status of the belugas, *Delphinapterus leucas*, of southeast Baffin Island, Northwest Territories. Can. Field-Nat. 105: 206–214.

RICHARD, P.R., AND R.R. CAMPBELL. 1988. Status of the Atlantic Walrus, *Odobenus rosmarus rosmarus*, in Canada. Can. Field-Nat. 102: 337–350.

RICHARD, P.R., J.R. ORR, AND D.G. BARBER. 1990. The distribution and abundance of beluga, *Delphinapterus leucas*, in eastern Canadian waters: a review and update, p.23–38. *In* T.G. Smith, D.J. St. Aubin and J.R. Geraci [ed.] Advances in research on the beluga whale, *Delphinapterus leucas*. Can. Bull. Fish. Aquat. Sci. 224.

RIGLER, F.H. 1974. Char Lake project. Canadian Committee for the International Biological Program. Final Rep. PF-2.

RIPLEY, E.A. 1987. Climatic change and the hydrological regime, p. 137–178. *In* M.C. Healey and R.R. Wallace [ed.] Canadian aquatic resources. Can. Bull. Fish. Aquat. Sci. 215.

ROBITAILLE, J.A., Y. COTÉ, G. SHOONER, AND G. HAYEUR. 1986. Growth and maturation patterns of Atlantic salmon, *Salmo salar*, in the Koksoak River, Ungava, Quebec, p. 62–69. *In* D.J. Meerburg [ed.] Salmonid age at maturity. Can. Spec. Publ. Fish Aquat. Sci. 89.

ROSENBERG, D.M., R.A. BODALY, R.E. HECKY, AND R.W. NEWBURY. 1987. The environmental assessment of hydroelectric impoundments and diversions in Canada, p. 71–104. *In* M.C.Healy and R.R. Wallace [ed.] Canadian aquatic resources. Can. Bull. Fish. Aquat. Sci. 215.

ROY, D. 1989. Physical and biological factors affecting the distribution and abundance of fishes in rivers flowing into James Bay and Hudson Bay, p. 159–171. *In* D.P. Dodge [ed.] Proceedings of the International Large River Symposium. Can. Spec. Publ. Fish. Aquat. Sci. 106.

SACKMANN, T., AND B.D. SMILEY. 1985. Arctic industrial activities compilations Vol. 2. Sverdrup Basin: hydrocarbon exploration 1974 to 1984. Can. Data Rep. Hydrogr. Ocean Sci. 32(2): xiii + 181 p.

SHAFFER, M., AND ASSOCIATES LTD. 1989. Economic impact analysis of the salmonid enhancement program. Department of Fisheries and Oceans Economic and Commercial Analysis Rep. 44: 113 p.

SIMON, D. 1989. Sustainable development: theoretical construct or attainable goal? Environ. Conserv. 16: 41–48.

SMITH, T.G. , AND D. TAYLOR. 1977. Notes on marine mammal, fox and polar bear harvests in the Northwest Territories, 1940 to 1972. Can. Fish. Mar. Serv. Tech. Rep. 694: vi + 37 p.

SNIDER, E. 1987. The Arctic Marine Conservation Strategy. Northern Perspectives 15(4): 11–12.

STABLER, J.C. 1989. Jobs, leisure and traditional pursuits: activities of native males in the Northwest Territories. Polar Record 25: 295–302.

STATISTICS CANADA. 1981. Canada year book 1980–81. Statistics Canada, Ottawa, Ont. 1004 p.

1987. Canada year book 1988. Statistics Canada, Ottawa.

1988a. Census Canada 1986. Populations and dwelling characteristics — census divisions and subdivisions. Quebec: Part 2, Vol. 2. Profiles. Statistics Canada, Ottawa, Ont. 94–110.

1988b. Census Canada 1986. Populations and dwelling characteristics — census divisions and subdivisions. Ontario: Part 2. Profiles. Statistics Canada, Ottawa, Ont. 94–112.

1988c. Census Canada 1986. Populations and dwelling characteristics — census divisions and subdivisions. Manitoba: Part 2. Profiles. Statistics Canada, Ottawa, Ont. 94–114.

1988d. Census Canada 1986. Populations and dwelling characteristics — census divisions and subdivisions. Northwest Territories: Part 2. Profiles. Statistics Canada, Ottawa, Ont. 94–124.

STEWART, R.E.A., P. RICHARD, M.C.S. KINGSLEY, AND J.J. HOUSTON. 1986. Seals and sealing in Canada's northern and Arctic regions. Can. Tech. Rep. Fish. Aquat. Sci. 1463: iv + 31 p.

STIRLING, I., AND H. CLEATOR [ED.]. 1981. Polynyas in the Canadian Arctic. Can. Wildl. Serv. Occas. Pap. 45: 73 p.

STIRLING, I., M. KINGSLEY, AND W. CALVERT. 1982. The distribution and abundance of seals in the eastern Beaufort Sea, 1974–79. Can. Wildl. Serv. Occas Pap. 47: 25 p.

STRONG J.T. 1988. Status of the narwhal, *Monodon monoceros*, in Canada. Can. Field–Nat. 102: 391–398.

1989. Reported harvests of narwhal, beluga and walrus in the Northwest Territories, 1948–1987. Can. Data Rep. Fish. Aquat. Sci. 734: iv + 14 p.

TASK FORCE TO REVIEW COMPREHENSIVE CLAIMS POLICY. 1985. Living treaties: lasting agreements. Report of the Task Force to Review Comprehensive Claims Policy. Department of Indian Affairs and Northern Development, Ottawa, Ont. 132 p.

TAYLOR, D.A., M.G. REED, B.D. SMILEY, AND G.S. FLOYD. 1985. Arctic industrial activities compilation. Volume 1. Beaufort Sea: marine dredging activities 1985 to 1982. Can. Data Rep. Hydrogr. Ocean Sci. 32(1): xiii + 192 p.

THOMPSON, P.C. 1981. The economic performance of the commercial skiff fishery in western Canada. Can. Tech. Rep. Fish. Aquat. Sci. 1037: v + 21 p.

TOPOLNISKI, D., S. KERWAN, AND A. KRISTOFFERSON. 1987. Economic opportunities in the commercial fisheries in the Northwest Territories. Northern Perspectives 15 (4): 8–12.

USHER, P.J. 1971. The Bankslanders: economy and ecology of a frontier trapping community. Volume 2. Economy and ecology. Information Canada, Ottawa, Ont. xv + 169 p.

WAGEMANN, R. 1989. Comparison of heavy metals in two groups of ringed seals (*Phoca hispida*) from the Canadian Arctic. Can. J. Fish. Aquat. Sci. 46 1558–1563.

WAGEMANN, R., AND D.C.G. MUIR. 1984. Concentrations of heavy metals and organochlorines in marine mammals of northern waters: overview and evaluation. Can. Tech. Rep. Fish. Aquat. Sci. 1279: v + 97 p.

WELCH, H.E., AND M.A. BERGMANN. 1989. Seasonal development of ice algae and its prediction from environmental factors near Resolute, N.W.T., Canada. Can. J. Fish. Aquat. Sci. 46: 1793–1804.

WELCH, H.E., J.A. LEGAULT, AND H.J. KLING. 1989. Phytoplankton, nutrients and primary production in fertilized and natural lakes at Saqvaqjuac, N.W.T. Can. J. Fish. Aquat. Sci. 46: 90–107.

WORLD COMMISSION ON ENVIRONMENT AND DEVELOPMENT. 1987. Our common future. Oxford University Press, Oxford, England. xv + 400 p.

WRIGHT, D.G. 1980. A twenty-five year scenario of anticipated resource developments that may have an impact on fish and fish habitat in the prairie provinces and the Northwest Territories. Can. Fish. Mar. Serv. Manuscr. Rep. 1546: iv + 31 p.

YAREMCHUK, G.C.B., M.M. ROBERGE, D.K. MCGOWAN, G.W. CARDER, B. WONG, AND C.J. READ. 1989. Commercial harvests of major fish species from the Northwest Territories, 1945 to 1987. Can. Data Rep. Fish. Aquat. Sci. 751: iv + 129 p.

YAREMCHUK, G.C.B., AND B. WONG. 1989a. Issues in the management of marine mammals in the Northwest Territories, and Yukon North Slope. Can. Manuscr. Rep. Fish. Aquat. Sci. 2009: v + 10 p.

1989b. Issues in the management of native domestic fishing in the Northwest Territories. Can. Manuscr. Rep. Fish. Aquat. Sci. 2010: iv + 10 p.

YONGE, L. 1988. An economic analysis of the 1987 Whale Cove commercial charr fishery. Whale Cove Hunters and Trappers Association, Whale Cove, NWT. 33 p.

1989. An economic analysis of the Eskimo Point and Maguse River commercial charr fisheries, summer 1988. Department of Economic Development and Tourism, Government of the Northwest Territories, Yellowknife, NWT. 33 p.

# The Management of Pacific Salmon Fisheries in British Columbia

M. C. Healey

*University of British Columbia*
*Westwater Research Centre,*
*Vancouver, British Columbia, Canada V6T 1W5*

Healey, M. C. 1993. The management of Pacific salmon fisheries in British Columbia, p. 243–266. *In* L.S. Parsons and W. H. Lear [eds.] Perspectives on Canadian marine fisheries management. Can. Bull. Fish. Aquat. Sci. 226.

The salmon fisheries are the most valuable and the most complex on Canada's Pacific coast. The fisheries are based on five species and hundreds of different populations of Pacific salmon. The species differ widely in their ecology and productivity but all make extensive use of both freshwater and marine habitats. Salmon are harvested by a variety of user groups whose interests range from commercial to recreational to cultural. Because the species are highly migratory, international cooperation has been necessary to prevent unrestrained overharvesting. Management of the fisheries has involved approaches ranging from monopolistic licencing to unrestricted access at different times. Current management is based on ensuring an escapement of salmon to natural spawning beds that will maximize subsequent recruitment and on augmenting natural production through artificial propagation and improvements to spawning and nursery habitat. Priorities for allocation of harvestable surpluses are first to satisfy native Indian domestic needs and second to provide for commercial and recreational harvest. Restructuring of fishery management over the past two decades has emphasized economic rationalization and organizational accountability. As user groups have become more knowledgable and sophisticated, management of salmon has become more democratic. These trends are likely to continue into the future.

---

La pêche du saumon vient en tête de file des activités de pêche en termes de valeur et de complexité sur la côte ouest du Canada. Elle est axée sur cinq espèces et des centaines de différentes populations de saumons du Pacifique. Bien que ces espèces soient nettement différentes au niveau de l'écologie et de la productivité, elles fréquentent toutes des eaux douce et marine. Elles sont exploitées par divers groupes d'utilisateurs à vocation commerciale, récréative et culturelle. Étant donné qu'elles effectuent de grandes migrations, une coopération internationale a été nécessaire afin d'éviter une surexploitation outrancière. Les mesures de gestion de la pêche sont allées de la délivrance de permis de pêche exclusif à l'accès illimité au cours de certaines periodes de pêche. À l'heure actuelle, la gestion est basée sur une échappée qui assurera la maximisation du recrutement ultérieur et sur l'augmentation de la production naturelle par la ponte artificielle et l'aménagement des frayères et des zones d'alevinage. En ce qui concerne l'allocation des surplus exploitables, les besoins alimentaires des autochtones ont priorité sur la pêche commerciale et récréative. La restructuration de la gestion de la pêche au cours des deux dernières décennies a mis l'accent sur la rationalisation économique et la responsabilité organisationnelle. La gestion de la ressource est devenue plus démocratique en réaction à des utilisateurs plus avertis et plus sophistiqués. Il est probable que ces tendances se poursuivront à l'avenir.

## Introduction

The Pacific salmon fisheries are the richest fisheries on the Pacific coast of Canada. Their management is also among the most complex of any fisheries in the world. This complexity derives from the large number of species and stocks within a species that contribute to the fishery, from the diversity of user groups with a vested interest in salmon, and from the fish's considerable historic, cultural. and economic value. As I shall show, salmon management has had a tempestuous history and the future outlook is no less stormy. Yet salmon management also provides outstanding examples of resource management and bioengineering triumphs.

Pacific salmon are endemic to the North Pacific Ocean. They are all members of the genus *Oncorhynchus*. Five species, pink (*0. gorbuscha*), sockeye (*O. nerka*), chum (*O. keta*), coho (*O. kisutch*), and chinook (*O. tshawytscha*) are found in Pacific North America. This number may be increased considerably if a recent revision of the genera *Oncorhynchus* and *Salmo*, becomes accepted (Smith and Stearley 1989). All five species support active and valuable commercial fisheries in both the United States and Canada. From 1980–1985 British Columbian fishermen landed, annually, an average of 69 000 metric tons of Pacific salmon valued at $157 million (DFO 1986a–1988). In addition, coho and chinook and, to a lesser extent, pink and sockeye support valuable recreational fisheries. Between 1980 and 1985 sportfishermen harvested about 900 000 salmon[1] a year valued at more than $75 million (Michelle James, Dept. of Fisheries and Oceans, pers. comm. 1989).

During the last decade, pen rearing of salmon has become established on the British Columbian coast. In 1986 salmon farmers marketed 391 metric tons of product valued at $2.65 million and by 1989 this had increased to 12 385 metric tons (Hall and Franks 1988; Ron Genetz, Dept. of Fisheries and Oceans, pers. comm. 1990). The pen culture industry projects that its production will equal the wild harvest within 10 years (B.C. Salmon Farmers Association, January 1989). Coho and chinook salmon are, at present, the principal species used in pen culture. There is, however, considerable interest in expanding the production of exotic species such as Atlantic salmon (*Salmo salar*) and large future increases in production are likely to be based on sockeye salmon rather than chinook and coho (Ron Ginetz, Dept. of Fisheries and Oceans, pers. comm., 1990).

Salmon were a staple food of native Indians in British Columbia long before European settlements were established. The natives employed sophisticated technology including weirs, traps, and nets in the capture of salmon and had systems of management based on ownership of fishing rights, centralized authority over fishing activities, protection and restoration of salmon habitat, and reverence for salmon (Morrell 1989; Stewart 1977). Salmon played a prominent role in the culture and traditions of many tribes.

The historic importance of salmon as a food stuff and a cultural symbol, and their continuing importance as commercial and recreational species, derive from the fact that they are abundant, easy to catch, aesthetically attractive, and command a high price in a variety of marketed forms (fresh, frozen, smoked, canned). Because of their prominence in Pacific coast fisheries, salmon have engaged the attention of research and management more than any other group of species. The fact that several species are involved, that all have complex life cycles involving both fresh water and marine habitats, and that the species and separate spawning populations are often intermingled at the time of harvest, impart a complexity to the management of salmon not present in other fisheries. In addition, the preservation of salmon spawning and nursery habitat in fresh and coastal marine waters is often in conflict with other potential uses of these waters. Yet the cultural value of salmon is so great that any decision to sacrifice salmon habitat to some other use evokes considerable emotional debate. Conserving and managing the salmonid resource is, therefore, a highly politicized activity.

In this paper I shall describe British Columbia's salmon fisheries and their management from both an historic and a current perspective. My emphasis will be on modern salmon management as it is practised in British Columbia. but modern management has deep historic roots and these must be examined if current practice is to be properly understood.

## Characteristics of Pacific Salmon Species

The five species of Pacific salmon are all characterized by spending their egg and juvenile stages in freshwater but migrating to sea to grow and mature (i.e., they are anadromous) and by dying after spawning (i.e., they are semelparous). Within the constraints of this general pattern the different species show considerable variation in life cycle characteristics. To comprehend fully the complex task of managing modern salmon fisheries one must understand the extent to which the species differ in both their freshwater and their marine ecology.

[1]Based on average Strait of Georgia landings of 767 thousand from sports fish creel census and assuming that Strait of Georgia landings represent about 85% of total Provincial landings.

Fresh water is where breeding and early development occur. Management in fresh water is directed toward protecting the spawning salmon as well as toward maintaining, rehabilitating, or enhancing the quality of spawning and nursery areas. The species differ markedly in the distribution of their spawning areas and in the nature of their freshwater nurseries. Since a high proportion of British Columbia's rivers and streams provides spawning and nursery habitat for one or more salmon species, conflicts over salmon conservation and alternative uses of land and water are commonplace.

The ocean is where the species grow to marketable size and where most are harvested. Management during this stage is directed toward protecting known coastal nursery habitats and toward regulating fisheries to prevent overfishing and to achieve an equitable sharing of the harvest among competing user groups. Successful regulation of salmon fisheries depends on knowledge of their coastal and oceanic distribution, their migration routes and timing, their abundance. their productivity, and their size and age when mature. All of these characteristics differ among species and stocks within a species. Unfortunately, characteristics such as coastal distribution and migration routes and timing are not sufficiently discrete that individual stocks can be managed independently. Consequently, many species and stocks must be harvested jointly and differences in their abundance and productivity make it impossible to maximize the yield from all (Paulik et al. 1967).

Canada is also concerned that salmon production from its rivers should be mainly harvested by Canadian fishermen. Understanding the salmon's ocean migrations is the key to assessing the extent of interception of Canadian salmon by fishermen of other nations.

Because the ecological characteristics of each species are so important in determining and constraining the kinds of management measures that are possible, I shall begin with a brief description of the natural history of each species. Detailed descriptions of each species may be found in Groot and Margolis (1991).

### Pink Salmon

The pink salmon is numerically the most abundant species in British Columbian waters. On the Pacific coast the pink salmon has a fixed 2-year life cycle so that even and odd year brood lines do not exchange genetic material. Pink salmon spawn in both large and medium sized rivers all along the coast, and there are some 700 separate spawning populations in British Columbia. Pink salmon usually spawn within a few km of the river mouth although some stocks spawn far upstream in the larger rivers. Fry emerge in the early spring (March–May) and migrate to sea immediately after emerging from the nest. The young salmon remain in coastal waters throughout their first summer but in the early autumn move offshore to feeding areas in the Gulf of Alaska (Healey 1982a, 1986; Hart 1973). Pink salmon from British Columbia probably do not move much further west than 150 W during their ocean migrations (Fig. 1). Maturing adults return to the spawning rivers in August and September in southern B.C. but somewhat earlier, in late July and August in northern B.C.

Landings of this species by B.C. fishermen ranged from 2.7 to 23.9 million fish between 1980 and 1985. Because even and odd year brood lines are reproductively isolated, abundance in even and odd years can differ considerably. This is particularly obvious in the Fraser River, which has an extremely large odd year run but virtually no even year run. The overwhelming predominance of odd year runs is typical of pink salmon populations from rivers south of Johnstone Strait. Elsewhere in the province the difference between even and odd year runs is less pronounced, but is still evident. Even and odd year lines also differ in average size, individuals of the even year line averaging about 1.7 kg and those of the odd year line about 2.2 kg (Healey 1986).

Pink salmon are generally capable of withstanding a moderate exploitation rate of about 60%. Exploitation rates estimated for various stock groupings on the B.C. coast ranged from 14 to 84% but most were in the range 50–70% (Walters et al. 1982; Beacham 1984a). Current management of pink salmon, and all other salmon species, is not designed to control exploitation rate, however, but to permit an "optimal" escapement (the escapement that, it is assumed, will maximize long-term yield).

Exploitation rates, thus, vary depending on the strength of the annual "run" of salmon. Nevertheless, they are indicative of the intensity of exploitation that each species can support.

### Sockeye Salmon

Sockeye salmon is the second most abundant species in British Columbian waters. The sockeye matures at ages ranging from 3 to 6 years in British Columbia. The relative abundance of different ages in the spawning runs varies considerably among populations (Foerster 1968; Healey 1986) but the average age increases toward the north.

Sockeye spawn in medium to large sized rivers, and in lakes, throughout the Province. There are about 300 separate spawning populations of sockeye in British Columbia and some spawning grounds are more than 1000 km from the river mouth (Healey 1982a). The lake or the spawning river serves as a nursery for the sockeye fry for several months to a year or more before the fish migrate to sea. In British Columbia, lakes are the most common nurseries for fry, and the most common age of seaward migration is 1 or 2 years. On entry to the ocean, usually between April and June, sockeye move northwestward along the coast. In the autumn they move away from shore into the open waters of the Gulf of Alaska where they remain for 2–3 years before maturing. The sockeye's ocean migrations are the most extensive of any British Columbian salmon, extending at least to 175°W (French et al. 1976; Hartt and Dell 1986) (Fig. 1).

Maturing sockeye return to the spawning rivers from July through September and sockeye runs typically peak about 2 weeks earlier than pink salmon runs. There is, however, considerable variation in run timing. For example, spawning runs to the southern part of the coast generally peak during August but the run to Barkley Sound on the west coast of Vancouver Island peaks in late June.

From 1980 to 1985 sockeye landings by B.C. fishermen ranged from 3.1 to 12.2 million fish. Because of its larger size (2.5–4.0 kg), its bright red flesh, and its outstanding flavour and qualities as a canned product, the sockeye is much more valuable than the pink salmon. The Fraser River is the most important producer of sockeye in B.C., followed by the Skeena River.

Sockeye, in general, do not show strong cyclic variation in abundance as pink salmon do. In the Fraser River, however, a 4-year cycle in abundance is common. The extreme example of this is the Adams River population, in which the 1902 cycle year[2] is an order of magnitude more abundant than the other cycle years. Sockeye runs are typically harvested at rates between 50% and 75% (Walters et al. 1982).

**Chum Salmon**

Chum salmon is the third most abundant species in British Columbian waters but landings of this species rank fourth. In British Columbian waters chum salmon mature at ages ranging from 2 to 5 but most are mature at ages 3 and 4. Chum salmon spawn in virtually every river and stream in coastal British Columbia and there are about 880 different spawning populations in the province (Healey 1982a). The spawning grounds of chum salmon are seldom located far up stream. In the Yukon River, however, chum salmon migrate more than 2000 km upstream and long upstream migrations are typical of Asian stocks.

Like pink salmon, chum migrate to sea immediately after emergence from the gravel spawning nest in the spring. The timing of their migration and their pattern of early ocean residence are similar to pink salmon (Healey 1982b). Chum salmon range further afield in the Gulf of Alaska than pink salmon but not as far as sockeye (Fig. 1). Mature chum salmon typically return to their spawning river later than pink salmon, so that the peaks in sockeye, pink, and chum runs are usually somewhat separated in time. There is still considerable overlap, however, so that in many fishing areas it is not possible to regulate the harvest of these species independently.

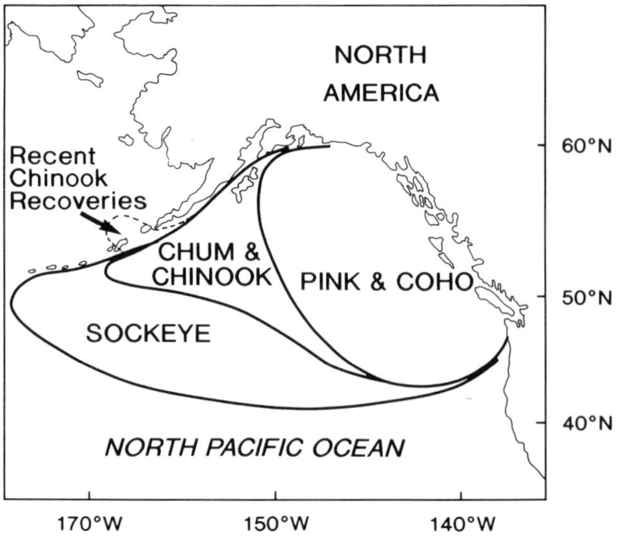


FIG. 1. The known ocean distribution of British Columbia stocks of Pacific salmon species. Modified from Hart (1973).

Annual landings of chum salmon by British Columbian fishermen for the period 1980 to 1985 ranged from 1.0 to 5.5 million fish, each weighing between 5 and 6 kg. Chum, together with sockeye and pink salmon, are termed the "net" species because they are typically captured by purse seine and gillnet rather than by hook and line. In recent years, however, substantial numbers of all three species have been captured in the commercial hook and line fisheries. Chum cannot withstand exploitation rates as high as the other species. Walters et al (1982) estimated that exploitation rates for chum ranged from 10 to 68% among stock groupings with most less than 40%. Beacham (1984b) estimated that, for all B.C. chum stocks combined, exploitation rate varied from 40 to 60% between 1951 and 1981.

**Coho Salmon**

Coho salmon is the fourth most abundant species in British Columbian waters but ranks third in landings. Most coho salmon in British Columbia are 3 or 4 years

[2]Here I follow Ricker's (1987) convention for naming the cycle years of Fraser sockeye.

old when mature. Like chum salmon, coho spawn in virtually every stream and river in British Columbia and there are more than 970 recorded spawning populations in the province (Healey 1982a). Some spawning areas are as much as 1000 km upstream. Unlike chum, coho spend 1 or 2 years in freshwater as juveniles before migrating to sea. Seaward migration occurs in May or June and coho smolts quickly disperse in coastal waters.

Although the recorded dispersion of British Columbian coho salmon in the Gulf of Alaska is similar to that of pink salmon (Fig. 1), most coho remain in continental shelf waters throughout their ocean life. Coho appear not to undertake so directed a northward migration as sockeye, pink, or chum salmon. Their dispersal away from their natal river is mainly to the north, however, so that most recaptures of fish tagged as smolts are from north of the spawning river. The coho spawning run often overlaps with the chum spawning run and they often spawn in similar areas of the same rivers. Coho can have a very protracted spawning period with some populations spawning in September and others in the same watershed not until December or January.

Greatest landings of coho are by the commercial hook and line, or troll, fishery and this species is also the mainstay of the recreational salmon fishery. Total commercial landings between 1980 and 1985 ranged from 2.8 to 4.1 million fish and an additional 0.5 million were taken in the recreational fishery. Coho typically weigh 3 to 5 kg at maturity and those that remain in sheltered inside waters, such as the Strait of Georgia, throughout their ocean life are somewhat smaller than those that migrate to the open coast (Healey 1980, 1982b, 1986; Godfrey 1965).

Coho are capable of withstanding high rates of exploitation and various stock groupings in British Columbia are exploited at rates ranging from 51 to 95% (Walters et al. 1982).

### Chinook Salmon

Chinook salmon is the least abundant of the five species in British Columbian waters. Chinook also has the most varied age structure of any of the species in B.C. Age at maturity ranges from 2 to 6 years but the most common ages are 3 to 5. Chinook spawn in medium to large sized rivers and there are about 300 separate spawning populations in British Columbia. Chinook will spawn from near the river mouth to thousands of kilometres upstream in the headwaters of the largest river systems (Healey 1982a, 1986).

There are two behavioural forms of chinook in British Columbia. The most abundant form is termed ocean-type. This type enters the spawning rivers from July through October and typically spawns in the lower to middle reaches of the river system. Fry of ocean-type chinook migrate to sea during their first year of life, either immediately upon emergence from the spawning nest in the spring or after about 2 months of river residence, in June. In the sea, ocean-type chinook are like coho in that they disperse gradually northward from their natal stream and remain concentrated in continental shelf waters throughout their ocean life.

The second, less abundant form, is termed stream-type. Stream-type chinook enter spawning rivers in the spring and early summer and most migrate to the headwaters of the larger river systems to spawn. Although they enter the spawning rivers much earlier than ocean-type, stream-type spawn in September and October about the same time as ocean-type. Young of stream-type chinook remain in freshwater for at least 1 year before migrating to sea. In the sea they migrate quickly offshore and complete their ocean life mainly in the open waters of the Gulf of Alaska (Healey 1983) (Fig. 1).

Chinook are captured mainly in the commercial troll and recreational fisheries. Because of their protracted ocean life (up to 5 years in the ocean) and their concentration in coastal waters, many are captured while still immature. Size at capture ranges from about 2 to 20 kg. Commercial landings by British Columbian fishermen ranged from 0.87 to 1.23 million fish between 1980 and 1985 and recreational landings contributed about another 300 thousand. Exploitation rates for chinook are relatively high, ranging from 46 to 89% for various stock groupings but with most rates greater than 70% (Walters et al. 1982),

## Historical Fisheries

Before European settlement and the development of colonial fisheries for salmon, native Indian fisheries took a substantial harvest for domestic food needs and inter-tribal trade. Natives employed a variety of techniques for capturing salmon including weirs and traps in spawning rivers, gillnets. and hook and line (Stewart 1977). Early European settlers also made extensive use of the abundant fishes for domestic food supplies. The first commercial fishery by Europeans in British Columbia was begun at Fort Langley in 1829 when salmon were dried and salted for export (Lyons 1969). Extensive harvesting for commercial purposes did not begin, however, until canning technology was perfected. The first salmon cannery was built at Annieville, near the Surrey-Delta boundary, in 1870. By the early 1880's canneries were dotted all along the coast (Lyons 1969).

## Early Commercial Fisheries

Early commercial fisheries were concentrated in the vicinity of the canneries, and canneries were located where there was a concentration of fish to ensure an abundant supply of raw product. Fishing was generally conducted from small boats and skiffs and the gear used included gillnets, beach seines, and traps.

When British Columbia entered Confederation in 1871, fisheries became the responsibility of the Dominion government. No special provisions were made at that time for the management of British Columbia's developing fisheries. With the explosive increase in canneries in the 1870's, however, it became evident that some regulation must occur. Consequently, in 1876 the Fisheries Act was extended to include B.C. Initial fishery regulations included the allocation of exclusive canning licences for certain British Columbian waters, effectively providing a local fishing monopoly to the cannery. The receipt of exclusive cannery rights often carried some conservation, administrative, or social responsibilities with it. For example, the cannery might be required to construct and operate a hatchery to replenish stocks of fish, to hire local residents in the plant and Indians as fishermen, to provide records of cannery and fishing operations, or to provide services to any local community. The policy of allocating exclusive fishing rights prevailed until 1920 when all waters were opened to anyone who wished to gamble on the fishery,

In addition to cannery licensing, regulations to control net mesh size and depth and weekly fishing closures to permit some salmon to escape to spawn were also in place by 1890. Limits on the number of licences allocated for various types of fishing gear were also employed in the late 1800's and early 1900's to control fishing effort.

## Historic Catches

Records of salmon harvest by species extend back to about 1904 (Dept. of Marine and Fisheries 1927–28; INPFC 1969, 1980–1985). Early records consist only of the number of 48 lb cases of salmon canned each season. These can be converted into estimates of the number of salmon processed by multiplying by the number of each species required to produce a case. The early records provide a minimum estimate of landings because they do not include salmon caught and discarded because they could not be processed or those preserved by means other than canning. Most chum, coho, and chinook, for example, were salted. smoked, or sold fresh rather than canned. The records indicate that total salmon landings in the province have ranged between about 12 million and 35 million fish since about 1915 with no apparent long-term trend (Fig. 2). In fact, some recent catches have been the highest on record. Individual species have waxed and waned in abundance over the period of record with chum generally declining in abundance and chinook and coho showing recent reductions in catch. Landings are a very imperfect indicator of abundance, however, and it is likely that total abundance was much higher in the early part of the century than it is now (see e.g., Ricker 1987, 1989). The high landings early in the development of the fisheries testify to the ease with which salmon are captured and to the rapid expansion of processing capacity after canning was perfected. The number of canneries operating on the coast reached a peak in the early 1920's and it seems likely that most of the major runs were close to being fully exploited by this time.

## Effects of Habitat Degradation

By the 1920's, also, degradation of spawning habitat and blockage of spawning migrations by logging operations and other human activities were limiting the productivity of various stocks. The Quesnel stock of sockeye on the Fraser River system was decimated by a dam built for gold mining purposes that blocked the run in 1898. A logging dam built across the Adams River just below Adams lake blocked the migration of spawners to the productive upper Adams River. All Fraser River salmon stocks spawning above the Fraser Canyon were seriously reduced by the Hells Gate slide of 1913, a result of railway construction (Lyons 1969; Ricker 1987, 1989). Skeena River sockeye and pink salmon were significantly reduced in numbers by a rock slide on the Babine river in 1951 (Godfrey et al. 1954, 1956). Fishways installed at Hells Gate and elsewhere, removal of the Babine River rock slide, protection of, and improvements to, natural spawning and rearing areas, construction of artificial spawning channels, and the natural resilience of the salmon populations have helped to bring some of these stocks back to historic levels of production. Others. such as the upper Adams River stock, have not been reestablished, although recent efforts at restocking this river have met with some success (Williams 1989).

# Modern Fisheries

At present virtually all commercial harvesting of salmon in British Columbia is conducted in salt water or in the estuaries of some of the largest rivers. Only three gear types are permitted in salt water and estuarine fisheries: the purse seine, the gill net, and troll gear. There are currently about 550 seiners, 1550 gillnetters, and 1350 trollers in the salmon fleet. In addition,

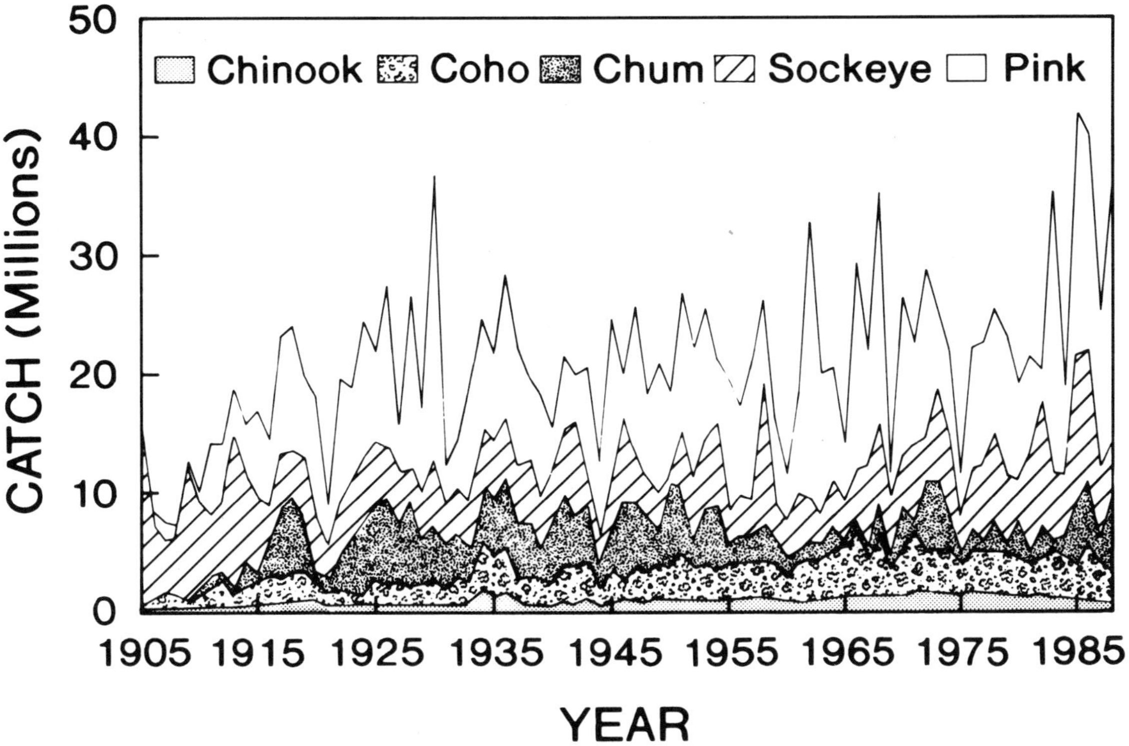


FIG. 2. Catch of salmon by British Columbia fishermen 1905–1988. Data from INPFC 1979, 1980–85; DFO 1986–88; Dept of Marine and Fisheries 1927–28.

about 1000 vessels are licenced to fish with either gillnet or troll (Wendy Grider, Dept. of Fisheries and Oceans, pers. comm. 1989). A limited number of commercial fisheries occur in freshwater. Some of these are in rivers that originate in Canada but traverse the Canada-U.S. boundary before entering the ocean, such as the Yukon, the Taku, and the Stikine. Others are designed to harvest surpluses from salmonid enhancement projects, such as the "jack" sockeye fishery at the Babine River counting fence on the Skeena River system.

Although the majority of salmon are captured in the commercial fisheries, the harvest by sport fisheries has increased substantially over the past two decades. Coho and chinook salmon are the principal species captured in the sport fisheries but pink and sockeye are captured as well. The majority of sport fishing takes place in salt water, but there are a few rivers in which limited sport fishing for chinook and coho is permitted. More than 350 000 salt water sport fishing licenses were purchased from the Federal Department of Fisheries and Oceans in 1986 and sales have continued at a similar level since that time (Fisheries and Oceans, Licencing Division, unpubl. data). Most salt water sport fishing licences are purchased for the purpose of salmon fishing.

In addition to licensed commercial and sport fisheries, there are licensed native Indian domestic fisheries. These are intended to provide for food and the continuance of native cultural traditions. Indian domestic fisheries are conducted in both fresh water and salt water and employ either traditional or modern fishing gear. In recent years these fisheries have become contentious. Disputes are common over the numbers of fish that will satisfy traditional needs and whether historic fishing for trading or other "commercial" purposes ought to be a part of the domestic fisheries.

The proportion of total landings of each species taken by different gear types has varied over the years. For the period 1980–1985 seiners took the majority of sockeye, pink, and chum (48–58% of catch in numbers) followed by gillnetters (25–38%). Trollers took between 4 and 15% of these species and native domestic and sport fishermen very small amounts, except for the native Indian catch of sockeye which was about 5% of the total (Fig. 3). Commercial trollers took by far the majority of coho and chinook (66 and 59%, respectively). Sport fishermen took 15%, seiners 11%,

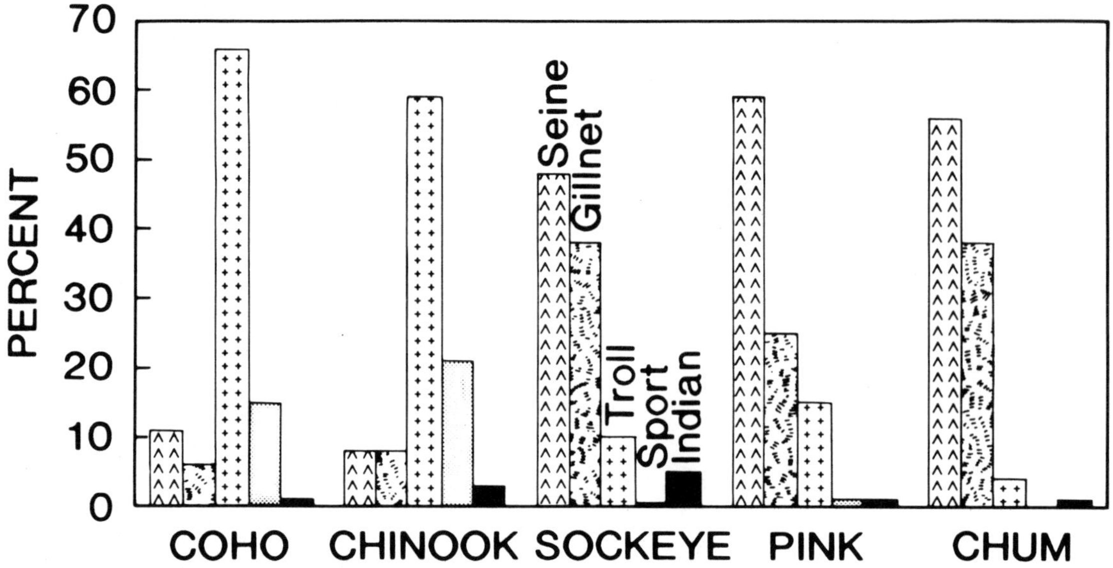


FIG. 3. The percentage distribution of catch among gear types for each species of Pacific salmon. Percentages based on catch data for 1981–1985.

gillnetters 7% and native Indians 1% of coho. Sport fishermen took 21% of chinook followed by gillnetters and seiners at 8% and native Indians at 7% (Fig. 3). Thus, commercial fisheries take 80% or more of the combined catch of chinook and coho and 90% or more of the combined catch of sockeye, pink, and chum. The remainder is divided among recreational and native Indian domestic fisheries.

### Modern Salmon Management, 1950 to present

Fisheries for salmon have probably always been managed in the broadest sense of the term. Historically, native Indian fisheries were regulated under a variety of patterns of ownership rights to preferred fishing locations, centralized authority over the start and end of fishing activities, and numerous rules governing the treatment of fish. As I have described above, a variety of management tools were employed in the early European salmon fisheries. The era of scientific management of Pacific salmon in British Columbia, however, dates from about 1950.

The transition to the scientific era was not abrupt. Some important things happened in the late 1940's and early 1950's, however, that make this time an appropriate landmark. Intermittent discussions between Canada and the U.S. over sharing the catch of Fraser River sockeye ultimately led to ratification of the Sockeye Salmon Fisheries Convention in 1937. It was not until 1946, however, that the International Pacific Salmon Fisheries Commission (IPSFC) actually began to manage the sockeye run. The management program implemented by the IPSFC was based on extensive research and a systematic evaluation of run strength so that the agreed upon quantitative sharing of the run between Canadian and U.S. fishermen could be accomplished. In 1949 and 1950 the Canadian Department of Fisheries was reorganized to implement a broad fisheries development program the main objectives of which were:

> " to increase scientific research and facilitate practical application of research findings; to extend conservation and development activities through planned biological-engineering projects; to institute economic studies; to expand inspection services and encourage greater consumption in Canada of fish products; to study means of extending capital assistance for industrial expansion, and to develop a greater understanding of the importance of the resource and the industry in the social and economic life of the country." (Can. Dept. of Fish. 20th Ann. Rept.).

Finally, in 1952 Canada, Japan, and the U.S. signed the International Convention for High Seas Fisheries of the North Pacific Ocean (INPFC) to address, among other issues, the potential interception of North American salmon by the Japanese mothership fishery. All of these developments required an underpinning of science and the systematic collection and interpretation of fisheries information.

A system of 27 statistical areas along the B.C. coast was created in 1927 to facilitate management of salmon by permitting opening and closing of small areas to fishing. These areas have been adjusted somewhat over the years but the concept of area and subarea control of fishing effort remains central to the management process (Fig. 4). It was not until 1950, however, that catch statistics began to be routinely collected for each of these areas through implementation of a fish sales slip system. At the same time, stream side enumeration and recording of salmon escapement to the many streams and rivers of the province were standardized. Research was initiated under the auspices of the INPFC into the oceanic distribution of major stocks of Pacific salmon. The stage was set to provide the data that would permit systematic application of the developing theory of stock and recruitment to the management of Pacific salmon populations. The organization was in place to apply the rapidly accumulating knowledge about salmon biology and ecology to conserving and enhancing the productivity of salmon populations.

The goals of salmon management at this time were to ensure that escapement of adult fish to the spawning grounds was sufficient to maintain a high fishery productivity and to harvest any surplus in an economic and equitable manner. The priorities for the allocation of any returning fish were first to provide for escapement, second to satisfy native Indian domestic fishery needs, and third to provide for commercial and recreational fisheries (Sprout and Kadowaki 1987). The sequence of allocation proceeded in the reverse direction, however, with the commercial and recreational fisheries taking fish first, the native Indian domestic fisheries second, and any remaining uncaught fish providing the spawning escapement. To ensure sufficient escapement either an accurate forecast of the number of returning salmon or a measure of the number present in a fishing area at any particular time was needed. Research has been conducted on both needs for more than 40 years. A model that will forecast salmon runs accurately has proved elusive (Healey 1982a; Henderson et al. 1989). Consequently, most management decisions are based on successive estimates, made while the fishery is in progress, of the number of salmon in the various fishing areas and the number that have escaped to the spawning rivers. In the early development of the fisheries, abundance in the fishing areas was assessed from commercial catch rates. The fishing power of the fleet soon increased to the point that commercial catch statistics could not be gathered quickly enough to prevent overharvesting. Under these circumstances, test fisheries were implemented in areas where commercial fisheries were most intense. Interpretation of test fisheries and in-season management decision making have become quite sophisticated over the years. An example of this process will be given later.

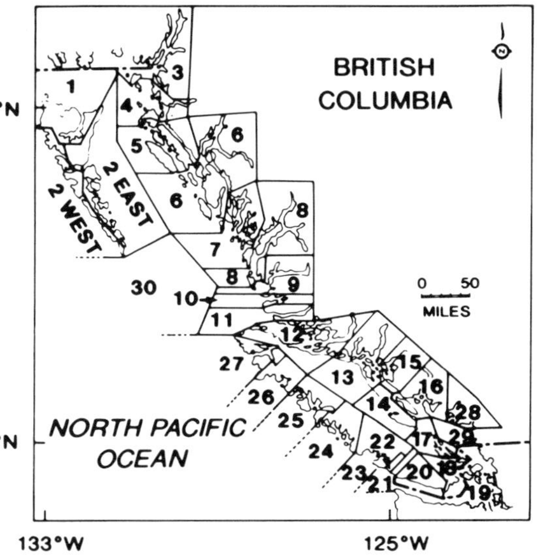


FIG. 4. The British Columbia coast showing the fishery management statistical areas, 1–30.

The piece of information greatly desired by salmon managers is the spawning escapement for each stock that will generate maximum adult recruitment. Even for well studied stocks it has not proven possible to determine this figure precisely (Healey 1982a; Beacham 1984a, b). The strong relationship between spawning stock and subsequent recruitment that would have identified the escapement for maximum yield, and that was anticipated in the early literature on stock and recruitment (Ricker 1954), has not materialized. Recruitment has proven to be an imperfect function of spawning stock as illustrated by almost 50 years of observation on Skeena River sockeye salmon (Fig. 5). For most stocks, not even this kind of a relationship can be determined. The surrogate measure of optimal escapement that is used is either the escapement that produced the largest historical run size or an estimate based on the ratio of area of spawning gravel to area occupied by one spawning pair. Such estimates provide an anchor for management decisions but they can not be easily defended scientifically either as a basis for maximizing yield from the fish stocks or as a basis for conservation.

Techniques of resource husbandry, in-season management, and habitat protection have evolved considerably in response to the growth in knowledge about

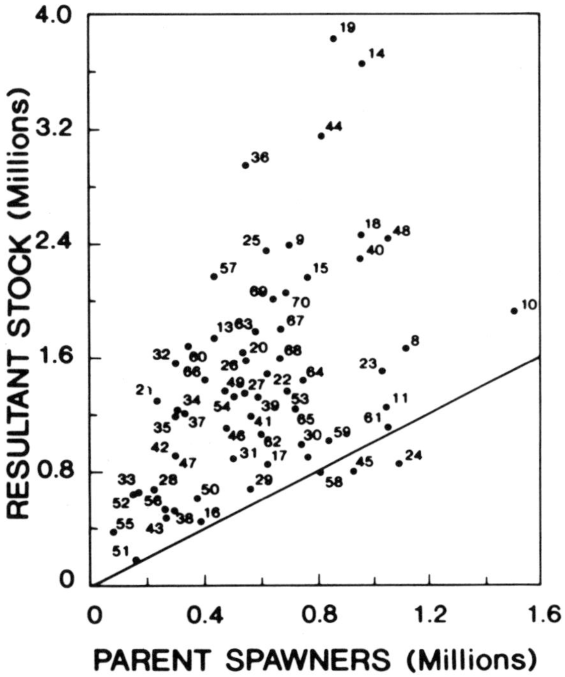


FIG. 5. A plot of recruitment from a given spawning stock for Skeena River sockeye salmon, 1908–1970. Data after 1970 are affected by artificial enhancement of sockeye from Babine Lake. Diagonal line indicates a recruitment equal to the spawning stock. Recruitment above this line indicates a surplus available for harvest. Adapted from Walters (1986).

salmon and their habits. Nevertheless, these techniques seem always to lag behind the capacity of the salmon fleet to discover and exploit the resource. The number and complexity of regulations has grown and the times and locations in which fishing is permitted have shrunk, yet even the strong stocks are often overharvested (Walters et al. 1982).

## The Davis Plan

Concern about overcapacity in the fleet and its contribution to management problems dates from early in the century when the Stanford Evans Commission of 1917 asserted that too much capital and labour were employed in the fishery (Sinclair 1978). Relatively little was done specifically to counter this situation, however, until the 1960's. In 1960 the Fisheries and Marine service commissioned a study of the fishery by economist Sol Sinclair (Sinclair 1960). Sinclair's recommendations were mainly concerned with economic rationalization of the fishery to permit more effective management and to improve the economic well being of the fishermen. Ultimately, most of Sinclair's recommendations were incorporated into what became known as the "Davis Plan" after then Minister of Fisheries, Jack Davis.

The Davis Plan, announced in 1968, was designed to improve economic conditions in the fishery by limiting participation in the salmon fishery, reducing existing capacity in the fleet, and improving the quality and value of the landed product. A secondary objective was to minimize hardship and dislocation among fishermen while accomplishing the economic goals. It consisted of four main phases:

1. freeze fleet size at the current level;
2. effect a gradual reduction in fleet size to a desirable level
3. upgrade and maintain vessels in the fleet at seaworthy standards
4. introduce economically optimal gear and area regulations.

Phases 1–3 were implemented from 1969–1973. Phase 4 was never fully implemented.

The first phase was accomplished, with a view toward the second phase, by introducing a two tiered vessel licensing system. The kind of licence that a vessel could obtain was governed by the amount of that vessel's landings during 1967 or 1968. To be eligible for a class A licence the vessel must have landed more than 10 000 lbs (4545 kg) of pink or chum salmon, or an equivalent value in other species, in either year. Those which had captured less than this amount were eligible for a class B licence. The cutoff point for class A licences was, therefore, landings of about $1250 (Fraser 1979). The intent of the two licence system was to distinguish serious fishermen from those who merely used fishing to supplement income from some other source and gradually to phase out this latter group. The requirement that a vessel have a minimal catch of salmon to be eligible for a class A licence was subsequently relaxed to permit a vessel that had catches of any species greater in value than $1250 to obtain a salmon licence.

All salmon vessel licence fees were set at $10 in 1969. The fee for a class A licence was increased to $100 or $200 in 1970 depending on the size and capacity of the vessel and further increased to $200 and $400 in 1971. Class A licence fees are currently $400 to $800 depending on vessel size. Class B licence fees remained at $10 until the last of these licences was phased out in 1984.

Class A licences could be held in perpetuity provided the fee was paid each year and some salmon (one was sufficient) were landed at least once every 2 years. Class A licences could be freely transferred among individuals and, for the first few years following their introduction, could also be transferred to any new vessel provided the old vessel was retired from the fleet. Class B licences could also be freely transferred among individuals but could not be transferred to a new vessel unless the old vessel was lost due to fire or sinking. Class B licences were slated to be phased out over a 10 year period. This period was subsequently extended to 15 years.

Other features of phase 1 of the Davis plan were intended to encourage continued or increased participation by native Indians in the fishery. The Department of Indian and Northern Affairs (DINA) provided funds to purchase vessels from the salmon fleet and create a "tonnage bank" to be administered under the Indian Fisherman's Assistance Program. Indians wanting to enter the fishery were allocated the required tonnage from the bank. In 1969 and 1970, when an Indian applied for a restricted B licence rather than the ordinary A licence, DINA would attempt to arrange for the higher class A licence fee to be paid so that no Indian would be denied access to an A licence due to financial hardship. In 1971 an Indian licence, AI, was introduced. Any Indian could convert his class B licence into this category and pay a nominal $10 licence fee rather than the $200 charged other fishermen. Vessels with AI licences were not eligible to participate in the buy back program implemented in phase 2 of the Davis plan. Class AI licences were freely transferable to other native Indians and could be transferred to a non-native provided the back licence fees applicable to an ordinary A licence were paid. In 1972 all class B licences held by native Indians were converted to AI licences. After 1979 it was no longer permissable to convert class AI licences into ordinary class A licences for sale to non-natives.

Phase 2 of the Davis plan, the reduction in fleet size, was initiated in 1970. It included a program to buy salmon fishing vessels out of the fleet funded by money from the salmon vessel licence fees, substantial increases in class A licence fees as noted earlier, and the continued phase out of class B licences. The program to buy salmon fishing vessels out of the fleet remained in effect only until 1973. It was terminated due to the rapidly escalating cost of purchasing a licenced vessel.

Under phase 2, the boat for boat replacement permitted under phase 1 was replaced by a ton for ton replacement. This change was instituted because fishermen began replacing their old vessels with new vessels that were substantially larger and more powerful and had correspondingly greater fishing power. After 1970, a fisherman wishing to replace his existing vessel with a larger vessel was required to purchase a second licenced vessel so that the combined tonnage of old vessels to be replaced was equal to or greater than the tonnage of the new vessel.

Finally, to prevent a concentration of fishing rights in the hands of processing companies, the Department of Fisheries and the fishing companies agreed that company owned fleets would be scaled down by the same proportion as the owner operated fleet.

Phase 3, the upgrading of vessel safety standards, occurred naturally as fishermen had greater disposable income. New vessels were larger, faster. better built, and better equipped. Old vessels were upgraded to make them safer and more comfortable. A set of vessel standards relating mainly to storage and preservation of fish was introduced in 1973. The effect of these was to improve the quality of product delivered to the processors and, thus, its landed value.

Phase 4 of the Davis Plan, which was intended to move the fishery toward more economically rational methods of harvesting, was never implemented. A committee, The West Coast Salmon Fleet Development Committee, consisting of fishermen, processors, Federal Government, and university representatives, was struck to advise the minister on gear and area regulations to effect phase 4. After more than a year of meetings the committee found itself unable to agree on a set of recommendations, and so Phase 4 was abandoned.

The Davis Plan has been subject to a number of analyses to determine whether it achieved its goals (Sinclair 1978; Fraser 1979; Pearse 1981; Rettig 1984). The consensus seems to be that, in terms of its secondary objective of minimizing disruption and hardship in the industry, the plan was successful. Although the goal of restricting access to the fishery was regarded as revolutionary by fishermen of the day (it was not revolutionary in the context either of the historic salmon fishery or of fisheries in general) it was accomplished without major dislocation of existing fishermen. No one who had even a minor dependency on the fishery was initially denied a licence. Vessels licensed as class A could always be downgraded to class B if the owner found the class A licence fees onerous. Phase out provisions, licence transfer provisions, vessel replacement provisions, and support for native Indian fishermen were all generous. No one was coerced into leaving the fishery to meet the fleet reduction goal of phase 2. Given that the Davis Plan was the first large scale attempt at economic rationalization through licence limitation and fleet reduction, implementation was remarkably successful. The British Columbia experience provided a model for many other overcapitalized fisheries.

In terms of equity the Davis plan was also modestly successful. Fishermen were intended to be the principal beneficiaries of any increased economic well being

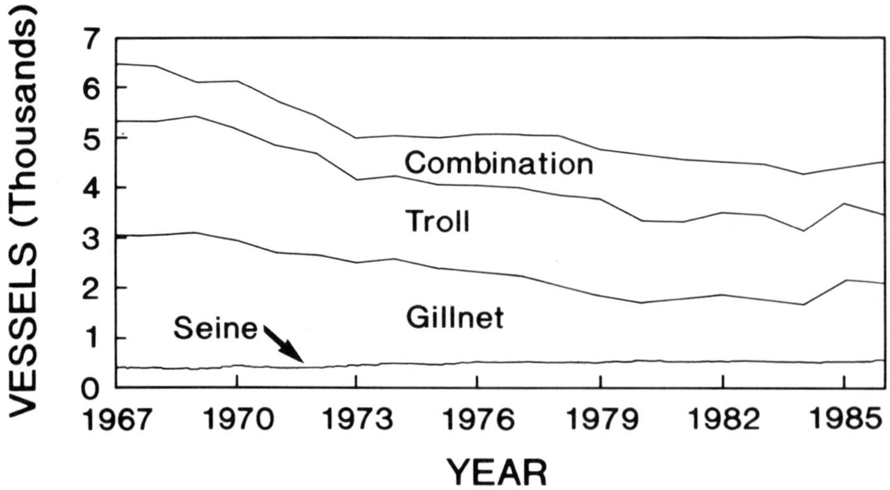


FIG. 6. The number of licensed salmon fishing vessels of different types that reported landings each year between 1967 and 1986.

resulting from the plan, and Fraser (1979) acknowledged that this was the case, Considerable effort was expended to ensure that remote communities continued to benefit from the fishery and, in particular, that native Indians had the opportunity to participate. These latter efforts were not entirely successful. However, it is arguable that both native Indians and remote communities would have lost much more if the plan had not been made sensitive to their needs.

Opinion is divided as to whether the plan succeeded in significantly reducing capital and labour in the fishery. By the late 1970's the number of vessels in the fleet was reduced by about 1100 vessels (from 6122 in 1969 to 5084 in 1977) and has since been further reduced by about another 600 (Fig. 6). Also, the number of persons employed in the fishery has declined. It is doubtful, however, that 10 years after implementation of the Davis Plan capital and labour in the fishery had been reduced at all. Following implementation of the plan the average size of vessels in the fleet gradually increased as fishermen replaced older vessels with new larger vessels (Sinclair 1978). Even the ton for ton replacement rule did not inhibit the increase in vessel size as fishermen found ingenious ways to design a larger vessel that still had the formal tonnage measure or their old vessel. There was a shift in the proportion of vessels of different types toward more seiners and fewer gillnetters (Fig. 6). Also, the number of small vessels licensed to fish with either troll or gillnet increased. Overall, the average vessel in 1978 was larger, faster, more powerful, and better equipped than when the plan was implemented (Sinclair 1978). These changes have continued to the present. The changes in fleet characteristics were all to the good in so far as vessel safety and product quality were concerned, but they worked against any reduction in overcapitalization. Fraser (1979) estimated that the capital value of the fleet in 1978 had, in fact, increased by about 50% over its value in 1968. What the Davis Plan seemed to have accomplished was to substitute technological competition for numerical competition in the scramble for fish with no reduction in the dissipation of resource rents by the fleet.

Since the fleet was reduced in size but not in fishing power, managing the fisheries to conserve salmon stocks and maintain a high yield was no less a problem in 1979 than it had been in 1969. There was still too much fishing power chasing too few fish. Nor had the increased licence fees done much to offset the public investment in managing the resource, In 1979 management costs for the salmon fishery were on the order of $20 million against licence fees of about $1 million. The people of Canada were still far from realizing any resource rent from the fishery (Fraser 1979).

### Introduction of the Optimal Yield Policy

Over this same 10 year period the general policy of the Federal Government regarding fisheries management continued to evolve. Historically the focus had been very much on resource conservation and maximizing yield from fisheries. These goals remained valid. But the first was becoming less clearly defined as society's concept of mankind's position in the natural ecosystem evolved. And the second was proving impossible to measure, if it existed at all as a unique number. Beginning with its introduction at the first Law of the Sea conference in Geneva in 1959, an alternative objective, that of optimizing rather than maximizing yield, was becoming fashionable (Roedel 1975; Healey 1984). When applied to resource management the concept of optimal yield implies that more than one (usually many) goals are to be satisfied by any management regime. Under an optimal yield policy the principal goal of management may not be to harvest the maximum sustainable number (or weight) of fishes. Rather, it is to ensure that some prescribed mix of social. economic, and biological goals is satisfied. This new concept was formally introduced as the policy for Canadian fisheries in 1976 (Environment Canada 1976).

In large measure the new policy simply articulated what was common practice among fishery managers. The Department of Fisheries' priorities with respect to salmon allocation (escapement > native Indian domestic > commercial/recreational) was regularly ignored to permit small commercial fisheries and native Indian domestic fisheries even when runs were below escapement targets. Thus, the goal of maximizing yield was often sacrificed to prevent economic or social hardship. The Davis Plan was an acknowledgement that managing only to achieve resource conservation and maximum yield is insufficient if fishermen are to prosper and the country as a whole is to derive benefits from fisheries. The 1976 policy document enshrined the principle of multiobjective management, however, it did not provide any clear guidelines as to how managers were to deal with the trade-offs and value judgements inherent in the determination of an optimal yield. Nor did the various academic analyses of the concept that appeared concurrent with the policy provide the necessary operational framework (Roedel 1975; Orbach 1977). Some authors worried that the emphasis on multiple goals and economic efficiency would distract managers from their primary goal of conserving fish stocks (Larkin 1979). Consequently, although the terminology changed (e.g., in the salmon fishery target escapements were now termed "optimal escapements") management practice remained largely unaffected.

### The Salmonid Enhancement Program

The second major innovation of the 1970's affecting salmon was the salmonid enhancement program. For a long time managers had supposed that salmon stocks were significantly depleted in comparison with their historic abundance and that great benefits could be realized if runs were rehabilitated. The conventional wisdom was that a doubling of salmon runs was possible. Recent analyses by Ricker (1987, 1989) are consistent with this view. Assisting salmon production by artificial means was not a new idea. The first hatchery for salmon was constructed in B.C. in 1882 at Bon Accord creek near Port Mann on the Fraser River. As I noted earlier, receipt of a cannery licence often was conditional on the cannery operating a hatchery or otherwise taking steps to replenish fish runs. Hatcheries were very popular early in the century until research at Cultus Lake produced strong evidence that they did not contribute enough to salmon returns to justify their operating costs (Evidence on which this decision was based is reviewed by Foerster 1968). As a consequence, all hatcheries in British Columbia were closed in 1937 (Lyons 1969).

Interest in rehabilitating and enhancing salmon runs revived after the Second World War. Initial emphasis was on fishways to assist salmon passage around obstructions in rivers. The most famous and, perhaps, the most successful of these were the fishways at Hells Gate on the Fraser River, completed in 1945. Later, interest turned to spawning channels to increase the amount of spawning habitat and to improve egg to fry survival. The largest of these projects was constructed at Babine Lake in the 1960's and has resulted in an average 70% increase in sockeye catch since 1973 (West and Mason 1987).

Although there had been great improvements in hatchery technology, and hatcheries were in common use in the United States, British Columbia was slow to reintroduce hatcheries after the disappointments of the past. In the late 1960's, however, an experimental hatchery was constructed at Big Qualicum River on Vancouver Island and this was followed by hatcheries on the Capilano River near Vancouver and the Quinsam River near Campbell River on Vancouver Island. In addition, a Province wide program of salmonid enhancement, funded jointly by the Federal and Provincial governments, was announced in 1974. This program included the construction of spawning channels and hatcheries as well as habitat improvement to increase the production of juvenile salmon in fresh water. Included among the habitat improvements was the fertilization of juvenile sockeye nursery lakes to increase smolt production. The long-term goal was to double salmon production from British Columbia's river systems. Secondary goals included the incorporation of salmonid production projects into regional economic development, the increasing of public awareness about salmon and their conservation, and the increasing of native involvement in fisheries and fish production.

Now well into its second phase, the salmonid enhancement program lists some notable achievements toward all of its goals. Twenty-nine large scale and 28 small scale enhancement projects together with lake enrichment for sockeye contribute almost 4 million fish to British Columbia's salmon catch. There are more than 700 public involvement projects throughout the province from which ordinary citizens are discovering the intricacies of salmon biology and salmon husbandry. Native Indians are involved in the program in many different capacities from egg incubation through to harvesting surplus brood stock (DFO 1989).

The program has not been without its problems, however, not the least of which is that created by the mixing of enhanced and unenhanced stocks during harvest. Enhanced stocks are typically able to support a higher exploitation rate than unenhanced stocks. Consequently, when they are harvested together, the manager must either permit an overescapement of enhanced fish or deliberately overexploit the wild, unenhanced stocks. Neither option is attractive. The surplus of enhanced fish may sometimes be harvested at the hatchery or near the spawning ground. But at this stage the fish are usually of a significantly poorer quality than if they had been harvested earlier in their migration. Overharvesting of wild stocks is contrary to the goals of the enhancement program, which was to supplement not replace natural production. The tendency, however, has been to exploit the enhanced stocks as fully as possible at sea provided any wild stocks harvested jointly were not seriously overexploited. Consequently, it is likely that some of the currently enhanced production has simply replaced wild production rather than supplemented it.

## The Pearse Commission

As I noted earlier, the Davis Plan was unsuccessful in significantly reducing the fishing capacity of the salmon fleet. Nor was the salmonid enhancement program able to increase the abundance of salmon sufficiently to alleviate the continuing pressure toward overexploitation. Other Pacific coast fisheries were in similar distressed state at the end of the 1970's. Consequently, in March, 1981. Peter Pearse was appointed as a one man commission of inquiry to investigate and make recommendations concerning the condition, management, and utilization of the fisheries of the Pacific coast of Canada (Pearse 1982).

After extensive public hearings and review of 193 written submissions on Pacific fisheries, Pearse submitted his analysis and recommendations. His most severe criticism of the management of Pacific fisheries was that it was not based on cohesive, consistent, and forward looking policies. Rather, the Department managed the fisheries in an ad hoc and reactive manner. In Pearse's view the problems were pervasive.

In his wide ranging final report Pearse (1982) made 292 recommendations on topics ranging from procedures for the review of fisheries research projects to licensing of vessels fishing combinations of gear. Many of his recommendations, particularly those relating to the assessment of stock status and upgrading of information systems for management, have been implemented. However, Pearse was also concerned with overcapacity in the fleet and a great many of his recommendations focused on ways to reduce this overcapacity. His more controversial recommendations, those regarding reductions in fleet size through area and gear licensing and competitive bidding for licenses, and those regarding the capture of resource rents through taxes on landings have not been adopted.

Pearse was sensitive to the fact that salmon productivity could only be maintained if freshwater habitats were conserved. He made 23 recommendations concerning management of fish habitat, most of them directly related to salmon. His key recommendations were that the productive capacity of fish habitat must be maintained, and that if loss of habitat occurred in one area it should be compensated by a gain in habitat in another. These key recommendations as well as some others were incorporated into The Department of Fisheries and Oceans Policy for the Management of Fish Habitat (DFO 1986a). This policy articulates a goal of increasing the productive capacity of fish habitat by: (1) maintaining the productive capacity of existing habitat through application of the principle of no net loss; (2) rehabilitating habitat where economic or social benefits can be achieved; and (3) creating or improving fish habitat where economic or social benefits can be achieved.

Pearse was also sensitive to the need for consistency and even-handedness in the application of regulations and to the need for documenting procedures of management. He criticized the Department for its failure to act professionally in this regard and made numerous recommendations for improving the technological process of scientific management. Following the Pearse report a salmon stock assessment unit was created within the regional research branch and projects relating to stock assessment were consolidated within this unit. A thorough scientific assessment of B.C. salmon stocks was undertaken by the stock assessment unit, Regional working groups with responsibility for fishery assessment and planning were reactivated and a senior resource management executive committee was created to ensure consistent application of policy and regulations. Local fishery officers now document their actions and the reasons for them in a Record of Management strategies. The documentation of past management actions together with assessments of the historic performance of salmon stocks provides an invaluable source book for regional fishery officers (A. Wood, Dept. of Fisheries and Oceans, pers. comm. 1989).

Perhaps the most significant change in the management process following the Pearse Report was the creation of the Pacific Stock Assessment Review Committee (PSARC). The purpose of this committee is to undertake formal stock assessments for Pacific fisheries after the pattern of the Canadian Atlantic Fisheries Scientific Advisory Committee on the east coast. Salmon stock assessments were incorporated into the PSARC process in 1986.

The terms of reference of PSARC are to review and evaluate biological and technical advice on the status and management of Pacific fisheries and methods used for estimation and assessment; to advise the Resource Management Executive Committee on sound methods for assessment and decision making, and on fishing statistics, sampling of catches, and other information needed for stock assessment; to identify resource-assessment related research priorities; and to ensure liaison with other regional committees. The salmon stock assessment subcommittee of PSARC addresses these terms of reference with a view to providing biological advice consistent with sound conservation principles and optimization of production for all salmon stocks (Stocker 1987).

### The Canada–U.S. Salmon Interception Agreement

Improvements in the forecasting of run sizes, in-season assessment of stock size, monitoring of fisheries, and handling of statistical data were also necessary requirements of the Canada-U.S. Pacific Salmon Treaty (DFO 1985). This bilateral agreement, many years in negotiation, dealt with the interception of salmon produced from rivers in one country by fishermen of the other country. Details of the management of specific fisheries to minimize interceptions were specified in annexes to the treaty which are amended annually. Principal goals of the Treaty are to enable each country, through better conservation and enhancement, to increase its production of salmon and to reap the benefits of that increase. The Treaty created a Pacific Salmon Commission to oversee its implementation. Under the Treaty each country is responsible for assessing the status of its own salmon stocks, for assigning total allowable catches for each stock, and for designing fishing plans to harvest the allowable catch in ways consistent with the agreed upon sharing of stocks.

Various articles of the Treaty detail specific restrictions on harvest to control interceptions. For example, quotas are set on the commercial harvest of chinook salmon in southeastern Alaska, northern and central British Columbia, the west coast of Vancouver Island, and the Strait of Georgia. These quotas are to limit interception in Canada of chinook of Columbia River and Puget Sound origin and interception in Alaska of chinook of Canadian and Columbia River origin.

The Treaty included a formula for sharing sockeye and pink salmon returns to the Fraser River that ensured that Canada would be the principal beneficiary of any increased production of these two species. This formula was a critical component of the treaty from Canada's point of view. Under the International Pacific Salmon Fisheries Commission, which formerly regulated the catch of Fraser River sockeye and pink salmon, the harvestable surplus of these two species within the Fraser River Convention area was shared equally between Canadian and U.S. fishermen. This way of sharing the Fraser River sockeye and pink salmon had caused Canada to postpone its enhancement of these salmon, even though they were considered to have great enhancement potential.

The Treaty also detailed measures to limit other interception fisheries and to effect conservation of overfished stocks such as: sockeye, chum, and coho in transboundary rivers of British Columbia and Alaska; chum salmon in Portland Canal and the Fraser River; sockeye and pink salmon from central and northern British Columbia that are intercepted in southeastern Alaska; and chum and coho salmon from Washington State that are intercepted in British Columbia.

Although many contentious issues are not fully resolved by the treaty, it represents an important step toward a more equitable allocation of salmon to fisheries in their country of origin and provides a mechanism to avoid wasteful "tit-for-tat" overfishing of shared stocks. The terms of the Treaty do, however, limit the options open to Canadian fishery managers in regulating their fisheries. Management plans for fisheries that intercept U.S. bound salmon (which includes all of British Columbia's major south coast fisheries) must be consistent with the agreement and, in most instances, must be approved by the Pacific Salmon Commission. However, Canada now has virtually complete control over the assessment, enhancement, and management of Fraser River sockeye and pink salmon, British Columbia's two most valuable salmon runs. This was not the case under the old International Pacific Salmon Fisheries Commission.

### Management of the Skeena River Sockeye Fishery

The evolution of policy, legislation, fishing technology, and international agreements since 1950 has been extensive. The fishery management process is now much more structured and analytic than it was in the past. I can best illustrate the Progress by an example, the management of Skeena River sockeye. I have taken details of the process for the Skeena from Sprout and Kadowaki (1987) but have generalized them a bit to relate the process to other fisheries.

The Skeena River sockeye are harvested commercially in eight fishery statistical areas in northern British Columbia and southeastern Alaska as well as in northern sports and native Indian domestic fisheries. The annual harvest ranges from about 200 000 to 2 million fish. The most intensive fishery is in British Columbia statistical area 4 with smaller fisheries in statistical areas 3, 5, and 1 (Fig. 4). In southeastern Alaska, Skeena River sockeye are harvested mainly in fisheries at Cape Fox and Noyes Island and in Sumner and Clarence Straits.

Most sockeye are captured in gillnet and seine fisheries. Relatively few are caught in commercial troll or sport fisheries. The native Indian fishery harvests about 100 000 fish annually either in marine waters or in the River at points distributed from the estuary upstream to the spawning grounds.

The first step in the annual management process is to review management objectives for the salmon stock. Although general objectives for salmon fisheries are formulated at the international (Pacific Salmon Commission), national (Environment Canada 1976), and regional levels, with few exceptions these are not prioritized. For the Skeena River sockeye. as for other stocks, the primary objectives are to ensure optimal escapement and maintain stock diversity. Once these primary objectives, particularly that of ensuring optimal escapement, are met, the first priority in allocating any harvestable surplus is to native Indian domestic fisheries. Beyond these general goals. there are no specific priorities and any allocation decision will be contentious and subject to criticism.

Once management objectives have been confirmed, the next step is to forecast the expected size of the salmon run for the coming year. Three methods are used for Skeena River sockeye: (1) the historic average rate of return per spawner multiplied by the number of spawners in the parent generation and apportioned among return years according to the historic maturity schedule; (2) the size of the smolt run multiplied by the average smolt-to-adult survival rate apportioned according to the historic maturity schedule; and (3) a prediction of age 4 returns from age 3 returns the previous year and of age 5 returns from age 4 returns the previous year based on regression relationships. For other species and stocks the forecast may be based on a model as simple as the average of recent returns, or cycle year returns for stocks with discrete generations, or it may be based on a complex model incorporating estimates of juvenile production, ocean conditions, and other factors thought to influence ocean survival (See Smith et al. 1987 for examples). Where more than one forecasting model is used, the estimates commonly differ considerably in magnitude and some guesswork is required in deciding on a particular forecast.

In the past, forecasting and the fishing plans based on the forecast were all made by regional fisheries management staff. Now, however, some forecasts are reviewed in PSARC and passed to the Senior Management Committee with recommendations for management. Once approved these forecasts are returned to regional staff for the development of preliminary fishing plans. The forecast return and preliminary fishing plan are distributed to various interest groups for comment.

The third step in the management process is consultation with regional user groups. On the Skeena this is the Skeena River Salmon Management Committee. The Committee is chaired by a representative from the Department of Fisheries and Oceans and includes representatives from the commercial, sport, and native Indian fisheries. The terms of reference for the Committee are not defined and consensus among user groups on major questions such as catch allocation is rare. Consequently, it is often difficult for the managers to incorporate advice from the committee into regional management plans. Analogous consultations take place in relation to all other major fisheries along the coast (e.g., Hilborn and Luedke 1987).

After pre-season fishing plans are finalized and published no substantial actions occur until the fishery begins. Management during the fishery is critical to accomplishing the goals established at the start of the annual planning and management cycle because managers generally have little confidence in pre-season run predictions. Once the fishery begins, pre-season planning takes second place to in-season decision making. For the Skeena sockeye, in-season management has four major components: (1) data collection; (2) run assessment; (3) decision making; and (4) implementation (Fig. 7).

Data on the strength of the run during the fishery each year are provided by an escapement test fishery, the commercial fishery itself, and the native Indian food fishery in marine waters. In addition, the commercial and test fishery catches are sampled for size, age, and sex composition.

The escapement test fishery is located in the river estuary just upstream from the most intense commercial fishery and provides an estimate of the number of sockeye escaping through the commercial fishery. This test fishery has been calibrated against counts at a weir on the Babine River, through which most of the Skeena River sockeye pass on their way to spawning grounds, and it is known to provide an accurate estimate of escapement in most years.

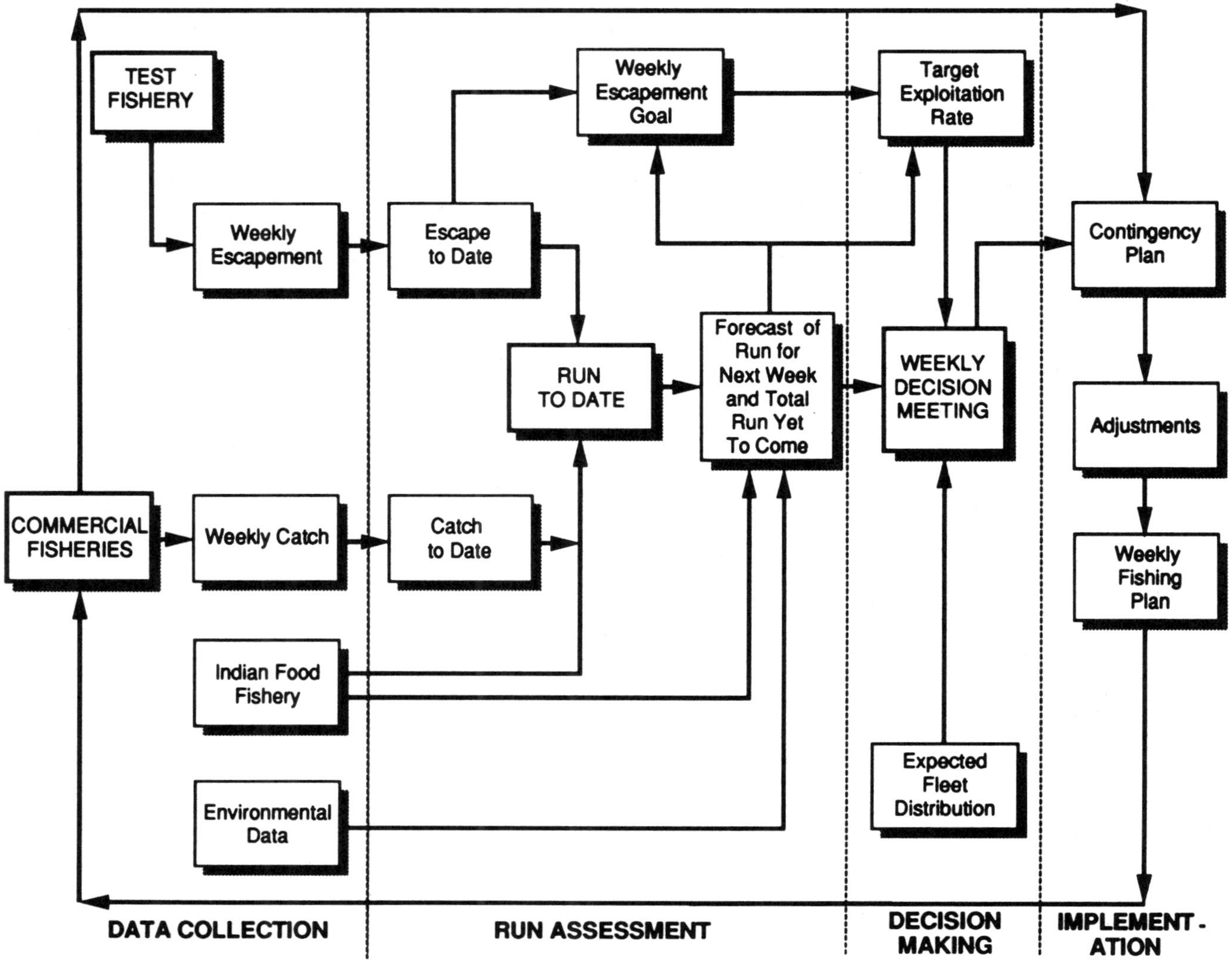


FIG. 7. A flow chart illustrating the in-season management decision process for Skeena River sockeye salmon management. Data collected from the various on-going fisheries and environmental data are used to estimate the size of the run to date, the number of sockeye that have escaped through the fisheries into the River, and the number of sockeye yet to come. From these estimates a forecast of the run size and an escapement goal for the coming week are determined. The forecast run and the target escapement define the target exploitation rate for the coming week. The exploitation rate together with the expected number and type of vessels in the fishery lead to a weekly fishing plan. Data, calculations, assessments, and fishing plans are updated weekly. Adapted from Sprout and Kadowaki (1987).

Commercial catch and effort data provide important information on run strength and timing. The catch from area 4, closest to the river mouth, is given greatest weight in daily decision making whereas catches from more distant fishing areas in British Columbia and southeastern Alaska are of secondary importance. The native Indian food fishery can be useful but it can also be misleading because it is conducted in preferred fishing areas and during preferred tides when other fisheries are closed. The age and sex composition of the catch provide further clues about the progress of the run because age 5 sockeye tend to peak in numbers before age 4 and males before females. Environmental data assist with interpretation of catch data because it is known that catchability is greatly affected by tide, wind, and rainfall (Sprout and Kadowaki 1987).

Analysis of commercial and test fishery catches leads to an assessment of the size of the run to date and what the escapement into the River has been, a forecast of the total run still to come, and a forecast of the run for the coming week. On the basis of this analysis the managers set an escapement target for the coming week which, together with the forecasted run for the week, establishes what the harvestable surplus should be. A weekly fishing plan is then decided on that will result in harvest of the allowable catch assuming there is no drastic change in the fleet size or catchability of the fish.

The decision making process is an iterative one (Fig. 7). Expectations of run size, harvestable surplus, and escapement targets are all updated weekly as the run progresses. If the run is particularly strong then total exploitation rate may be held below that for maximum yield of the stronger stocks to permit greater escapement to the weaker stocks. Since the run size is initially very uncertain, but this uncertainty declines as the run progresses, decision making may lean more heavily on the pre-season forecast early in the run, but ignore this forecast completely in the later weeks of the run.

The process of managing the sockeye fishery is complicated by the fact that other species are migrating to the river at the same time, and the catch of these must be regulated as well (Fig. 8).

Extensive overlap of species run timing is typical of major river systems such as the Skeena and Fraser (Fig. 8). Of particular concern in the Skeena is the fact that chinook and coho salmon and steelhead trout return to the river at the same time as the sockeye and pink salmon. Chinook and coho are able to withstand a terminal exploitation rate considerably less than sockeye or pink because they have already been heavily exploited by Canadian and U.S. troll fisheries. Steelhead are also able to sustain only a moderate exploitation rate

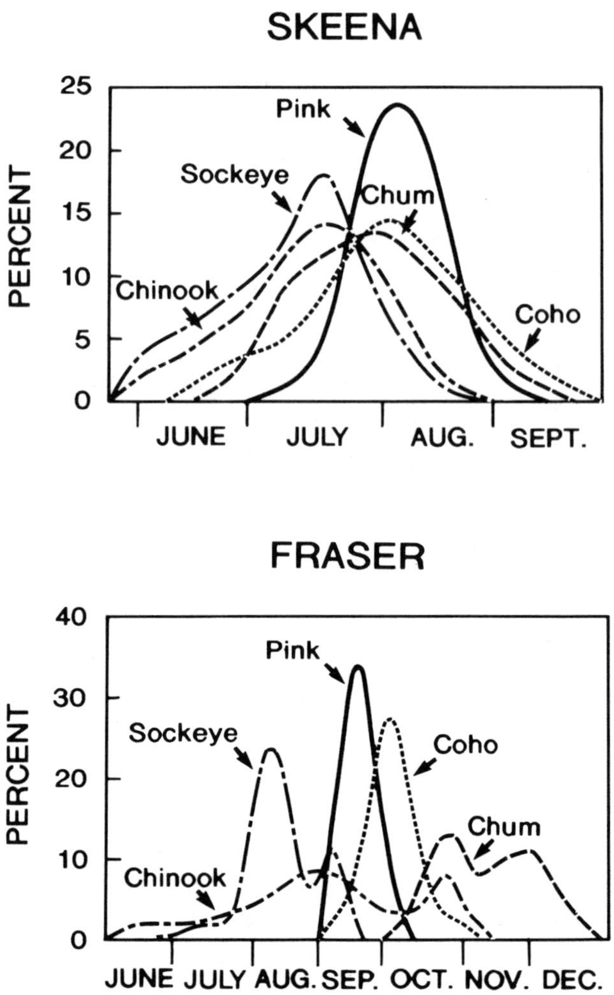


FIG. 8. The seasonal pattern arrival of each salmon species at the Skeena and Fraser River estuaries. Curves show the proportion of the annual run arriving each week.

and are most valuable when harvested by recreational fishermen. Although gear restrictions can help to direct the catch toward one or another species to some extent, the manager must still perform an almost impossible juggling act to keep track of harvest rates and optimize harvest of all these species at once (Healey 1982a).

## Management of the West Coast Troll Fishery

The pattern I have described above is fairly typical for purse seine and gillnet fisheries in British Columbia. The troll fishery, however, has been managed on quite a different basis. Argue et al. (1987) reviewed the history of troll fishery regulations and Shardlow et al. (1988) described the application of recent regulations. This summary is based on their accounts.

Prior to the mid 1960's there were few restrictions on troll fishing except for a general minimum size limit of 1.4 kg (about 48 cm total length) for commercially caught salmon, and closed seasons for coho and chinook. The chinook closed season, which was only 2 months long, was regularly waived. From the mid 1960's on, however, successively more and more regulations have been applied to the troll fishery to meet conservation goals, allocation goals, and obligations under the U.S.-Canada salmon treaty. These regulations have affected times and areas of fishing, species quotas and legal size limits, and types and amounts of gear that could be employed.

The first major area closures to troll fishing occurred in 1967 when fishing Area 19 and Area 20 east of Sheringham Point were closed to commercial fishing as a result of lobbying by sports fishermen. In 1968 Burrard Inlet and Howe Sound were similarly closed. Recently, area closures have been used extensively to target the fishery on particular species, to effect conservation of weak stocks or undersized fish, to effect equitable allocation of allowable catches, or to prevent quota overruns under the Pacific Salmon Treaty.

Season lengths have also been restricted for similar reasons, and open seasons for all species are now usually limited to the July to September period. Size limits on chinook salmon have been progressively increased to the current limit of 66 cm fork length (62 cm in the Strait of Georgia). In 1981 all trolling vessels were restricted to six fishing lines, or gurdies, whereas formerly they had been permitted to fish eight. Barbless hooks are required to be used in some areas and at some times to reduce the hooking mortality of species that must be released. Also, in 1981 troll fishermen were required to opt for a licence to fish either within or outside of the Strait of Georgia. Either licence permits fishing in the Johnstone-Queen Charlotte Strait area.

Under the Pacific Salmon Treaty, Canadian catch of chinook is restricted in the Strait of Georgia to 275 000 fish and on the west coast of Vancouver Island to 360 000 fish. The coho catch is also restricted on the west coast of Vancouver Island to 1.75 million fish. Within the Strait of Georgia, 50 000 of the allowable chinook catch were initially allocated to the commercial troll fishery but this was reduced to 30 000 in 1988. The remainder of the quota is allocated to the sport fishery.

The increasing restrictions on catch have necessitated an in-season management process for the troll fishery analogous to that for the net fisheries. Fisheries and Oceans' biologists develop a pre-season management plan in consultation with the industry. During the fishery, the weekly troll catch is monitored closely. If the cumulative catch departs significantly from the amount specified at the start of the fishing season, regulations are adjusted to increase or reduce the rate of capture. Because trollers range widely while fishing and do not return to port at specific times, monitoring the cumulative catch and communicating changes in regulations to the fishery is more difficult than it is with the net fisheries. For the west coast of Vancouver Island,a system of weekly conference calls has been instituted and appears to be working satisfactorily (Shardlow et al. 1988).

### Management of the Salmon Sport Fishery

For many years the salt water sport fishery for salmon was considered to be such a minor component of total harvest that it could virtually be ignored. During the 1970's and early 1980's, however, the sport catch grew explosively, particularly in the strait of Georgia. In the mid 1970's, the traditional dockside estimates of sport catch were evaluated by means of returns of tagged chinook and coho from hatchery releases. The evaluation suggested that the traditional estimates may have been only 1/2 to 1/3 the true catch (Argue et al. 1977). Not only were catch statistics for the sport fishery probably considerably in error, other statistics, such as numbers of sport fishermen, were non-existent. Consequently, a tidal waters sport fishing licence was introduced in 1981. During the first year of licensing more than 280 000 licences were sold. In 1980 a creel survey was implemented in the Strait of Georgia to provide a more accurate estimate of total sport catch. The creel census indicated that almost 900 000 salmon, mainly coho and chinook, were taken annually by sports fishermen in the Strait of Georgia. The sport fishery, thus, accounted for 50% and 68% of the landings of chinook and coho, respectively, in the Strait of Georgia and 21% and 15% of the coastwide landings of these two species. The sport fishery could no longer be ignored in the management of chinook and coho.

Historically the regulation of the sport fishery had been through bag and size limits and limitations on the type of gear that could be used (Argue et al. 1983). Recently, area closures have also been imposed and. under the terms of the U.S.-Canada Salmon Treaty, sport catch of chinook salmon in the Strait of Georgia is limited by quota. Recent poor escapements of chinook salmon to rivers tributary to the southern Strait of Georgia have resulted in even greater restrictions on sport fishermen. For the Strait of Georgia, Johnstone, and Queen Charlotte Straits daily bag limits for chinook salmon are currently 2, the annual bag limit is 15, and the minimum size limit is 62 cm. Elsewhere on the south coast size and bag limits for chinook are less restrictive, but the daily limit for any combination of salmon species is 4. There is no annual limit for species

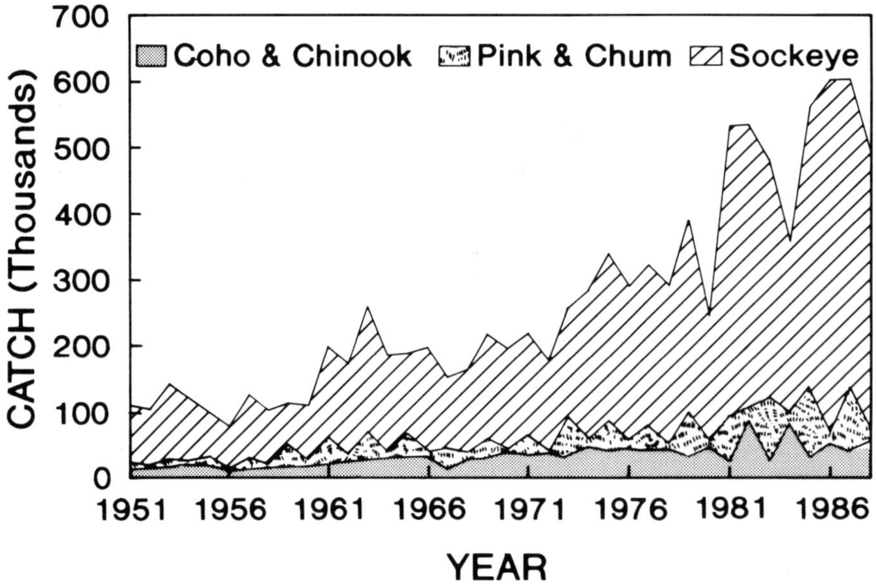


FIG. 9. Changes in the numbers of salmon harvested in native Indian domestic fisheries on the Fraser River from 1951 to 1988. Data from Macdonald (1989).

other than chinook. In addition to size and bag limits, there are a number of favorite fishing areas within the Georgia/Johnstone/Queen Charlotte Strait area that are closed during certain specified times to reduce the catch of chinook salmon.

### Management of Native Indian Domestic Fisheries

Providing for native Indian domestic fisheries continues to be a priority and a problem for fishery managers in British Columbia. Native bands argue that they have aboriginal rights to salmon and should not be subject to restrictions on their fishing activities for what they define as domestic purposes. Pearse (1982) argued that specific allocations of salmon for "domestic" purposes should be negotiated with native Indian bands and that once these allocations were agreed, the Department of Fisheries and Oceans should have no voice in how the allocation was used by the band. Fisheries and Oceans has not been able to reach agreement with all bands on either an equitable allocation or on how native Indian domestic fisheries should be administered.

Disagreement over what constitutes an equitable allocation of salmon to native bands is often difficult to separate from the perception that all catches must be documented for management and conservation purposes. Native Indian harvesting of some stocks is substantial and is increasing (Fig. 9). Although every effort is made to record the landings from native Indian fisheries, there is concern that these records are not accurate. Recent court rulings with regard to native rights to fish have, to some extent, clarified and reaffirmed the special rights of natives in the salmon fisheries and the fiduciary responsibilities of Federal agencies toward native Indians. Consequently, Fisheries and Oceans has entered into various co-management agreements with native Indians with the goal of increasing native Indian involvement in all aspects of salmon fishery management (Paisley et al. 1993). These initiatives by Fisheries and Oceans, which represent a significant departure from historic management practice in British Columbia, have raised considerable concern among other user groups and some resource managers that resource conservation will be compromised (Pearse 1992). It seems likely that allocations of salmon to native Indian fisheries and the way in which these fisheries are managed will remain contentious for some time to come.

### Prospects for the Future

The British Columbia salmon fisheries are fascinating in their biological, economic, and sociological complexity. It is always tempting, when reviewing the management of salmon fisheries, to prescribe what needs to be done to ensure a healthy and prosperous future for the industry (Larkin 1979). Experience suggests, however, that such prescription is futile. Rather than focus on prescription, therefore, I shall attempt to forecast the issues that will dominate the next two decades of salmon management in British Columbia. Whatever issues come to dominate the immediate future, I expect that salmon management in British Columbia will challenge academics and practical managers for decades to come.

The future of salmon management is likely to be just as tumultuous as its past has been. Excess fishing power in the fleet will continue to complicate management. The need for precise management created by this excessive fishing power will, however, require ever greater cooperation among resource managers and the fishing industry. This is because the Department of Fisheries and Oceans no longer has a monopoly on the information and skills needed for fisheries management. Co-management by government and industry will evolve well beyond the current, largely advisory, arrangements in both the production and harvest components of the fishery. In my view, both industry and the resource will benefit from arrangements in which industry takes on responsibility for resource management and conservation.

The inclusion of rights to some segment of the wild salmon resource as a component in the settlement of native Indian land claims is likely in the near future. When this occurs it will add another layer of complexity to the problem of managing for catch allocation and conservation. The inclusion of extensive fishing rights within land claims settlements should, however, stimulate innovative co-management arrangements between the Department of Fisheries and Oceans and native Indians. The alternative to such voluntary arrangements will be a court imposed settlement perhaps along the lines of the Boldt decision in the United States (Cohen 1989).

Competition for markets between wild harvests and salmon raised in sea pens may impact adversely on the fishing industry. The demand for salmon on world markets is believed to exceed by far the current production. However, poor prices for salmon in 1989 were blamed on an oversupply in 1988 combined with reduced Japanese consumption following Emperor Hirohito's death. The impact of reduced prices in 1989 was greater on the fledgling pen culture industry than on the traditional fisheries, however, this may not always be the case. A properly constituted pen culture industry should be able virtually to tailor salmon to market demand as well as being able to supply product all year long.

It is also conceivable that success in the pen culture industry could weaken the current policy of conserving all wild populations of salmon. Economically valuable uses of water resources on the Pacific coast have been postponed or foregone because they threatened salmon production (e.g., power production potential of the Fraser). Foregoing such alternative uses of freshwater salmon habitats may seem less an imperative if production can be maintained by pen culture.

Climate change could have a dramatic impact on the productivity of wild salmon stocks but the particular effects of climate change are still conjectural (Healey 1990). Average production of each species may be increased in some areas and decreased in others. However, it is changes in the year to year variability in production that are more likely to create crises in management. Significant changes in freshwater and oceanic habitats of salmon due to climate change seem likely to increase the variability in recruitment, at least in the short term. Whatever the future crises, however, the diversity of the resource and its resilience in the face of both habitat alteration and exploitation, and its prominent place in Pacific coast culture, auger well for its long-term survival.

**Acknowledgments**

I wish to express my appreciation to Al Wood, Dave Schutz, and Sandy Argue for critical and insightful comments on an early draft of this manuscript. Wendy Grider, Bob Wowchuck, and Michelle James provided useful information.

**References**

ARGUE, A.W., J. COURSLEY, AND G.D. HARRIS. 1977. Preliminary revision of Georgia Strait and Juan de Fuca strait tidal salmon catch statistics, 1972–1976, based on Georgia Strait head recovery program. Fish. Environ. Can., Fish. Mar. Serv. Tech. Rep. No. PAC/T-77-16: 68 p.

ARGUE, A.W.,R. HILBORN, R. PETERMAN, M. STALEY, AND C. WALTERS. 1983. strait of Georgia chinook and coho fishery. Can. Bull. Fish. Aquat. Sci. 211.

ARGUE, A.W., M.P. SHEPARD, T.F. SHARDLOW, AND A.D. ANDERSON. 1987. Review of salmon troll fisheries in southern British Columbia. Can. Tech. Rep. Fish. Aquat. Sci. 1502: 150 p.

BEACHAM, T. 1984a. Catch, escapement, and exploitation of pink salmon in British Columbia, 1951–1981. Can. Tech. Rep. Fish. Aquat. Sci. 1276: 215 p.

1984b. Catch, escapement, and exploitation of chum salmon in British Columbia, 1951–1981. Can. Tech. Rep. Fish. Aquat. Sci. 1270: 201 p.

BRITISH COLUMBIA SALMON FARMERS ASSOCIATION. 1989. 1989 B.C. farmed salmon forecast. Vancouver, B.C.

CANADA DEPARTMENT OF FISHERIES. 1950. 20th annual report, 1949–50. Edmond Clouter, King's printer, Ottawa.

COHEN, F.G. 1989. Treaty Indian tribes and Washington State: The evolution of tribal involvement in fisheries management in U.S. Pacific Northwest, p 37–48. *In* E. Pinkerton [ed.] Cooperative management of local fisheries, University of British Columbia Press, Vancouver, B.C.

DFO. 1985. Pacific salmon Treaty. Dept. Fisheries and Oceans, Canada, Ottawa, Ont.

1986a. The Department of Fisheries and Oceans policy for the management of fish habitat. Dept. of Fisheries and Oceans, Ottawa, Ont.

1986b. Annual summary of British Columbia commercial catch statistics 1985. Fisheries and Oceans, Canada. 555 W. Hastings St., Vancouver, B.C.

1987. Annual summary of British Columbia commercial catch statistics 1986. Fisheries and Oceans, Canada. 555 W. Hastings St., Vancouver, B.C.

1988. Annual summary of British Columbia commercial catch statistics 1987. Fisheries and Oceans, Canada, 555 W. Hastings St., Vancouver, B.C.

1989. Salmonid enhancement program 1987/1988 update. Fisheries and Oceans, Canada. 555 W. Hastings St., Vancouver, B.C.

DEPT. MARINE AND FISHERIES 1927–28. sixty-first annual report of the fisheries branch. Ottawa. Ont.

ENVIRONMENT CANADA. 1976. Policy for Canada's commercial fisheries. Environment canada, Fisheries and Marine Service, Ottawa, Ont.

FOERSTER, R.E. 1968. The sockeye salmon. Fish. Res. Board Can. Bull. 162: 1–422.

FRASER, G.A. 1979. Limited entry: experience of the British Columbia salmon fishery. J. Fish. Res. Board Can. 36: 754–763.

FRENCH, R., H. BILTON, M. OSAKO, AND A. HARTT. 1976. Distribution and origin of sockeye salmon (*Oncorhynchus nerka*) in offshore waters of the North Pacific Ocean. Int. N. Pac. Fish. Comm. Bull. 34: 113 p.

GODFREY, H. 1965. Coho salmon, p. 1–40. *In* Salmon of the North Pacific part 9. Coho, chinook, and masu salmon in offshore waters. Int. N. Pac. Fish. Comm. Bull 16. Vancouver, B.C.

GODFREY, H., W.R. HOURSTON, J. W. STOKES, AND F.C. WITHLER. 1954. Effects of a rock slide on Babine River salmon. Bull. Fish. Res. Board Can. 101: 100 p.

GODFREY, H., W.R. HOURSTON, AND F.C. WITHLER. 1956. Babine River salmon after removal of the rock slide. Bull. Fish. Res. Board Can. 106: 41 p.

GROOT, C., AND L. MARGOLIS. 1991. Pacific salmon life histories. University of British Columbia Press, Vancouver, B.C.

HALL, L., AND C. FRANKS. 1988. British Columbia 1986 fish aquaculture statistics. B.C. Ministry of Agriculture and Fisheries. Aquaculture Information Bull. 23: 1–8.

HART, J.L. 1973. Pacific fishes of Canada. Bull. Fish. Res. Board Can. 180: 740 p.

HARTT, A., AND M. DELL. 1986. Early oceanic migrations and growth of juvenile Pacific salmon and steelhead trout. Int. N. Pac. Fish. Comm. Bull. 46: 1–105.

HEALEY, M.C. 1980. The ecology of juvenile salmon in Georgia Strait, British Columbia, p 203–229. *In* W.J. McNeil and D.C. Himsworth [ed.]. Salmonid ecosystems of the North Pacific. OR State U. Press, Corvallis, OR.

1982a. Multispecies, multistock aspects of Pacific salmon management, p. 119–126. *In* M.C. Mercer [ed.]. Multispecies approaches to fisheries management advice. Can, Spec. Publ. Fish. Aquat. Sci. 59.

1982b. Juvenile Pacific salmon in estuaries: the life support system. p 315–341 In: V.S. Kennedy [ed.]. Estuarine Comparisons. Academic Press, New York, NY.

1983. Coastwide distribution and ocean migration patterns of stream-and ocean-type chinook salmon, *Oncorhynchus tshawytscha*. Can. Field-Nat. 97: 427–433.

1984. Multiattribute analysis and the concept of optimum yield. Can. J. Fish. Aquat. Sci. 41: 1393–1406.

1986. Optimum size and age at maturity in Pacific salmon and effects of size selective fisheries, p. 39–52. *In* D.J. Meerburg [ed.]. Salmonid age at maturity. Can. Spec. Publ. Fish. Aquat. Sci. 89.

1990. Implications of climate change for fisheries management policy. Trans. Am. Fish. Soc. 119: 366–373.

HENDERSON, M.A., D. PEACOCK, AND R. GORUK. 1987. Evaluation of the reliability of several models used to forecast the size of adult sockeye salmon (*Oncorhynchus nerka*) runs, p. 266–272. *In* H.D. Smith, L. Margolis, and C.C. Wood [ed.] Sockeye salmon (*Oncorhynchus nerka*) population biology and future management. Can. Spec. Publ. Fish. Aquat. Sci. 96.

HILBORN, R., AND W. LUEDKE. 1987. Rationalizing the irrational: a case study in user group participation in Pacific salmon management. Can. J. Fish. Aquat. Sci. 44: 1796–1805.

INPFC. 1979. Historical catch statistics for salmon of the North Pacific Ocean. Int. N. Pac. Fish. Comm. Bull. 39. Vancouver, B.C.

1980–1985. Statistical yearbooks. Int. N. Pac. Fish. Comm. Vancouver, B.C.

LARKIN, P.A. 1979. Maybe you can't get there form here: a foreshortened history of research in relation to management of Pacific salmon. J. Fish. Res. Board Can. 36: 98–106.

LYONS. C. 1969. Salmon, our heritage. Mitchell Press, Vancouver, B.C.

MACDONALD, A. L. 1989. The Indian food fishery of the Fraser River: 1988 summary. Can. Data Rep. Fish. Aquat. Sci. 725: 107 p.

MORRELL, M. 1989. The struggle to integrate traditional Indian systems and state management in the salmon fisheries of the Skeena River, British Columbia, p. 231–248. *In* E. Pinkerton [ed.] Cooperative management of local fisheries. University of British Columbia Press, Vancouver, B.C.

ORBACH, M. 1977. Report of the national workshop on the concept of optimum yield in fisheries management, Houston, Texas. Available from the office of fisheries management, National Marine Fisheries Service, Washington, D.C.

PAISLEY, R.K., M.C. HEALEY, AND T. MCDANIELS. 1993. Analysis of strategies for co-operative fisheries resource management between government and first nations. Westwater Research Centre, University of B.C., Vancouver, B.C. 51 p.

PAULIK, G.J., A.S. HOURSTON, AND P.A. LARKIN. 1967. Exploitation of multiple stocks by a common fishery. J. Fish. Res. Board Can. 24: 2527–2537.

PEARSE, P.H. 1981. Conflict and opportunity. Toward a new policy for Canada's Pacific fisheries. Preliminary report of the Commission on Pacific Fisheries Policy, Vancouver, B.C.

1982. Turning the tide. A new policy for Canada's Pacific fisheries. Final report of the commission on Pacific fisheries policy. Vancouver, B.C.

1992. Managing salmon in the Fraser. Report to the minister of fisheries and oceans on the Fraser River salmon investigation. Department of Fisheries and Oceans, 36 p.

RETTIG, R.B. 1984. License limitation in the United States and Canada: an assessment. N. Am. J. Fish. Manage. 4: 231–248.

RICKER, W.E. 1954. Stock and recruitment. J. Fish. Res. Board Can. 11: 559–623.

1987. Effects of the fishery and of obstacles to migration on the abundance of Fraser River sockeye salmon (*Oncorhynchus nerka*). Can. Tech. Rep. Fish. Aquat. Sci. 1522: 1–75.

1989. History and present state of the odd-year pink salmon runs of the Fraser River region. Can. Tech. Rep. Fish. Aquat. Sci. 1702: 1–37.

ROEDEL, P.M. 1975. Optimum yield as a concept in fisheries management. Am. Fish. Soc. Spec. Publ. 9: 89 p.

SHARDLOW, T.F., T.G. HOYT, AND A.D. ANDERSON. 1988. Review of 1986 south coast salmon troll fisheries. Can. Tech. Rep. Fish. Aquat. Sci. 1560: 99 p.

SINCLAIR, S. 1960. License limitation — British Columbia. Report to the Department of Fisheries, Ottawa, Ont.

1978. A licensing and fee system for coastal fisheries of British Columbia. Report to the Department of Fisheries and Oceans, Vancouver, B.C.

SMITH, H.D., L. MARGOLIS, AND C.C. WOOD. 1987. Sockeye salmon (*Oncorhynchus nerka*) population biology and future management. Can. Spec. Publ. Fish. Aquat. Sci. 96.

SMITH, G.R., AND R.F. STEARLEY. 1989. The classification and scientific names of rainbow and cutthroat trouts. Fisheries 14: 4–10.

SPROUT, P.E., AND R.K. KADOWAKI. 1987. Managing the Skeena River sockeye salmon (*Oncorhynchus nerka*) fishery — the process and the problems, p. 385–395. *In* H.D. Smith, L. Margolis, and C.C. Wood [ed.] Sockeye salmon (*Oncorhynchus nerka*) population biology and future management. Can. Spec. Publ. Fish. Aquat. Sci. 96.

STOCKER, M. 1987. Pacific Stock Assessment Review Committee (PSARC) annual report for 1986. Can. MS Rep. Fish. Aquat. Sci. 1951: 97 p.

STEWART, H. 1977. Indian fishing. Early methods on the northwest coast. Univ. Washington Press, Seattle, WA.

WALTERS, C.J. 1986. Adaptive management of renewable resources. MacMillan Publ. Co., Toronto, Ont.

WALTERS, C.J., R. HILBORN, M. STALEY, AND F.C. WONG. 1982. An assessment of the major fish and invertebrate stocks on the westcoast of Canada with recommendations for management. Report to the Commission on Pacific Fisheries Policy, January 1982. Environmental and Social Systems Analysis Ltd. Vancouver, B.C.

WEST, C.J., AND J.C. MASON. 1987. Evaluation of sockeye salmon (*Oncorhynchus nerka*) production from the Babine Lake development project, p. 176–190. *In* H.D. Smith, L. Margolis and C.C. Wood [ed.] Sockeye salmon (*Oncorhynchus nerka*) population biology and future management. Can. Spec. Publ. Fish. Aquat. Sci. 96.

WILLIAMS, I.V. 1987. Attempts to re-establish sockeye salmon (*Oncorhynchus nerka*) populations in the upper Adams River, British Columbia, 1949–84, p. 235–242. *In* H.D. Smith, L. Margolis, and C.C. Wood [ed.] Sockeye salmon (*Oncorhynchus nerka*) population biology and future management. Can. Spec. Publ. Fish. Aquat. Sci. 96.

# Recent Management of the British Columbia Herring Fishery

Max Stocker

*Department of Fisheries and Oceans*
*Biological Sciences Branch, Pacific Biological Station*
*Nanaimo, British Columbia, Canada V9R 5K6*

Stocker, M. 1993. Recent management of the British Columbia herring fishery, p. 267–293. *In* L. S. Parsons and W. H. Lear [eds.] Perspectives on Canadian marine fisheries management. Can. Bull. Fish. Aquat. Sci. 226.

Pacific herring (*Clupea harengus pallasi*) is a pelagic species which forms large schools. It is susceptible to large fluctuations in biomass, and prone to stock collapses. Pacific herring migrate onshore to spawn inter- and sub-tidally during the months of February to May. Currently two methods are used to model population dynamics and assess abundance: (1) Age-structured model analysis, and (2) an escapement based method. These models give a reasonable understanding of past productivity of British Columbia herring populations. The current herring fishery management objective in British Columbia, formalized in 1982–83, is based on a fixed harvest rate of 20% in conjunction with a predetermined CUTOFF level. Management involves cooperation among four groups. The Pacific Stock Assessment Review Committee (PSARC) Herring Subcommittee establishes relative stock trends to establish harvestable surplus. The Herring Management Working Group takes the information to draw up fishing plans. The Herring Industry Advisory Board reviews proposed fishing plans, and provides input for the final fishing plans. The in-season management teams (fisheries officers and management biologists) sound for herring (using echo sounders), take roe samples from test sets from chartered seine vessels, enforce fisheries regulations, and fish to quota. Successful management of this resource is critical, as overfishing, natural fluctuations in biomass, oceanic conditions, and environmental degradation can cause collapse of the stocks.

---

Espèce pélagique, le hareng du Pacifique (*Clupea harengus pallasi*) forme de grand banc dont la biomasse fluctue grandement. De plus, les stocks sont enclins à l'effondrement. De février à mai, le hareng migre vers les eaux côtières pour frayer dans la zone intertidale et infratidale. A l'heure actuelle, on utilise deux méthodes pour la modélisation de la dynamique des populations et l'évaluation de l'abondance : (1) l'analyse de modèles établis en fonction des âges et (2) la méthode basée sur l'échappée. Ces modèles nous donnent une bonne idée de la productivité passée des populations de hareng de la Colombie-Britannique. L'objectif actuel de la gestion de la pêche du hareng dans les eaux de cette province, entériné en 1982–1983, est basé sur un taux d'exploitation fixe de 20 % assorti d'une limite de capture prédéterminée. La gestion repose sur la coopération de quatre entités. Le sous-comité responsable du hareng du Comité d'examen de l'évaluation des stocks du Pacifique établit les tendances relatives des stocks afin de déterminer le surplus exploitable. Le groupe de travail sur la gestion de la pêche du hareng se sert ensuite de cette information pour élaborer les plans de pêche. Ceux-ci sont examinés par le Conseil consultatif de l'industrie du hareng qui participe aussi à l'élaboration des plans définitifs. Enfin, les équipes saisonnières de gestion (composées d'agents des pêches et de biologistes gestionnaires) effectuent des relevés par sondage acoustique, prélèvent des échantillons de rogue lors de la pêche expérimentale effectuée par des senneurs affrétés, appliquent les règlements de pêches et voient au respect des contingents. Il est critique que la gestion de cette ressource soit efficace étant donné que la surpêche, les fluctuations naturelles de la biomasse, les conditions océaniques et l'appauvrissement du milieu peuvent entraîner l'effondrement des stocks.

## 1. Introduction

The British Columbia herring fishery is one of Canada's most exciting fisheries and has the most innovative management regime. The demands on scientists to support this management regime are great and pose many challenges. The main objective of herring research is to provide a scientific basis for conservation and rational management of this renewable resource.

British Columbia herring show patterns of wide fluctuations in abundance typical of clupeoid populations. These dramatic fluctuations have over the years frustrated scientists, fishery managers, fishermen, and processors. Interannual herring abundance can fluctuate widely due to the addition of first time spawning recruits. Recruits can account for more than half the spawning stock size and their numbers are difficult to predict. The factors affecting the abundance of recruits are believed to include changes in predator abundance, zooplankton biomass, and oceanic conditions (Ware 1990).

Pacific herring migrate inshore from their summer feeding grounds on the continental shelf in the early fall and winter. They hold in sheltered bays and inlets to prepare for spawning in March and April. During this time three commercial herring fisheries take place: the food-and-bait fishery in November/December; the sac-roe fishery, and the spawn-on-kelp fishery in March/April (Chalmers 1989). The major challenge to herring fishery managers today is to control the roe herring fishery. Characteristically roe herring fishing is frenzied. The seine or gillnet fisheries are conducted on, or near, spawning grounds. Herring are highly vulnerable to capture when they are about to spawn. Generally there are many more fishing vessels than necessary for full utilization of the stocks. Some seine fishery openings may last only several hours or even minutes. These conditions make this a very difficult fishery to manage.

Tester (1935, 1945) and Taylor (1964) reviewed the early history of the fishery. Hourston (1980a) reviewed in detail the decline and recovery of Pacific herring stocks from the 1950's through the 1970's. An extensive literature review relevant to the fisheries oceanography of Pacific herring in the northeastern Pacific has recently been published by Grosse and Hay (1988).

This paper documents the recent management of Pacific herring in British Columbia with a brief treatment of the historical evolution of the current management system. The basis for management recommendations requires a knowledge of herring biology, knowledge of the industry and methods of stock assessment, awareness of the history of the fishery and previous management, and the impact of current management. These are all reviewed in the following sections of this paper.

## 2. Biology and Life History

Pacific herring (*Clupea harengus pallasi*) are a pelagic species living 10 to 20 years and forming large schools. Like other Clupeiformes, Pacific herring are susceptible to large fluctuations in biomass, and are prone to stock collapse. These biological characteristics make herring a difficult species to manage.

### a) Distribution

Pacific herring range from Korea to northern Kamchatka on the Asian coast, and from southern California to the Mackenzie River Delta on the north American coast (Hart 1973; Blaxter 1985). The abundance of the species off the western north American coast is greatest from the Bering Sea south to British Columbia. Abundance declines significantly from British Columbia southward through Washington, Oregon and California. In southern B.C. information on the summer offshore distribution of herring comes from summer fisheries, a fur seal food habit study, hydroacoustic cruises (Taylor 1971, 1984), and research surveys (Shaw et al. 1987, 1989a, 1989b). These studies indicate that schools of herring are confined mainly to the continental shelf.

## b) Life History

In British Columbia the major herring populations migrate onshore to spawn in protected inter-and sub-tidal waters during March–April (Taylor 1964). Eggs are adhesive and are laid on eelgrass (*Zostera* sp.), rockweed (*Fucus* sp.), filamentous red algae, and kelp (*Macrocystis* sp.) (Haegele and Schweigert 1985). Pacific herring spawn once a year, and they are exclusively spring spawners (Haegele et al. 1981). During the peak spawning period, spawning may occur in two or more waves separated by a 2- to 4-week period (Hay 1985; Ware and Tanasichuck 1989, 1990).

Larvae generally hatch after 6–18 days. Robinson and Ware (1988) observed the main hatch of a cohort 15 days after peak spawning in the Strait of Georgia. The hatching continued for 7 or 8 days. Herring nursery areas are close to spawning grounds. As larvae grow older they gradually disperse (Stevenson 1962; Alderdice and Hourston 1985). The dispersion mechanism is related to tidal flows and offshore advection (McGurk 1989). Larval research on B.C. herring has been conducted recently to establish whether initially high larval mortality is due to starvation (McGurk 1984; Robinson and Ware 1988). A study of a larval cohort in the Strait of Georgia, concluded that starvation probably only indirectly affected the developing larvae during a period of about 11 days (Robinson and Ware 1988).

In about mid July of their first year, larvae metamorphose to become juveniles. At completion of metamorphosis juveniles are about 35 mm long. During the summer juveniles gather in large schools near spawning grounds, and by the fall, schools gradually move to offshore wintering grounds (Hourston and Haegele 1980). Some juveniles, however, remain inshore, as indicated by the presence of juveniles in winter reduction fishery catches (Taylor 1964). Recruitment to the adult fishable stock occurs in the 3rd year of life, although some recruitment occurs during their 4th year (Taylor 1964).

## c) Reproduction

Mature herring populations include some 2-year old but mostly 3-year old herring spawning for the first time, and repeat spawners (ages 4–9). They can live to age 20, but few are found beyond the age of 9 (Hay 1985; Ware 1985). The average size of 3-year old herring is 185 mm, weighing about 90 g. There is a positive relationship between proportion of 3-year old mature and average size (Haist and Stocker 1985). Sizes of mature fish taken by the fishery range from 175–250 mm in length, and from 70–200 g in weight (Hourston and Haegele 1980). Number of eggs per female is proportional to body weight (Hay 1985; Ware 1985). A female herring produces between 20,000–40,000 eggs in 1 year. The relative fecundity for British Columbia herring is about 200 eggs/female gram (Hay 1985). Thus for an even sex ratio about $10^8$ eggs are deposited per tonne of fish.

## d) Mortality

The main source of natural mortality in herring is predation. As soon as herring eggs are deposited an intense struggle for survival begins against predators and adverse environmental conditions (Ketchen 1986). An egg predation study in Barkley Sound showed that 20% of eggs deposited were lost to predators. The predators included gulls, diving ducks, starfish, snails, and grey whales (Haegele and Schweigert 1989). Larval herring are subject to predation by filter feeding fish, zooplankton, amphipods, and jellyfish (Stevenson 1962; Westernhagen et al. 1979; Arai and Hay 1982; Purcell 1986). Harbour seals (*Phoca vitulina*), fur seals (*Callorhinus alascanus*), and Stellar and California sea lions (*Eumetopias jubatus* and *Zalophus californianus*) are major predators of herring (Taylor 1971; Bigg 1985; Olesiuk et al. 1990). Fish predators in B.C. include Pacific cod (*Gadus macrocephalus*), dogfish (*Squalus acanthias*), Pacific hake (*Merluccius productus*), lingcod (*Ophiodon elongatus*), sablefish (*Anoplopoma fimbria*), and salmonids (*Oncorhynchus tshawytscha*, *O. kisutch*) (Outram and Haegele 1972; Healey 1976; Jones and Geen 1977; Westrheim and Harling 1983; Ware and McFarlane 1986; Westrheim et al. 1989).

Taylor (1955) estimated annual total mortality rates of adult herring as 60% for ages 6–9. These calculations were from 1915 to 1917, a period when fishing in Barkley Sound was negligible. Tester (1955) calculated annual natural mortality rates for herring age 3–4, 4–5, and 5–9 in the Queen Charlotte Islands as 32%, 47%, and 60%, respectively .

## e) Behaviour

Spawning behaviour studies by Hourston et al. (1977) and Stacey and Hourston (1982) showed that spawning can be initiated by the presence of small amounts of sperm in the water. Schooling is standard behaviour in herring populations and seems to be an antipredator device (Blaxter 1985). Schooling behaviour of pelagic fishes results in increasing catchability as stock size decreases. This phenomenon makes herring stocks highly susceptible to overfishing. Adult herring are

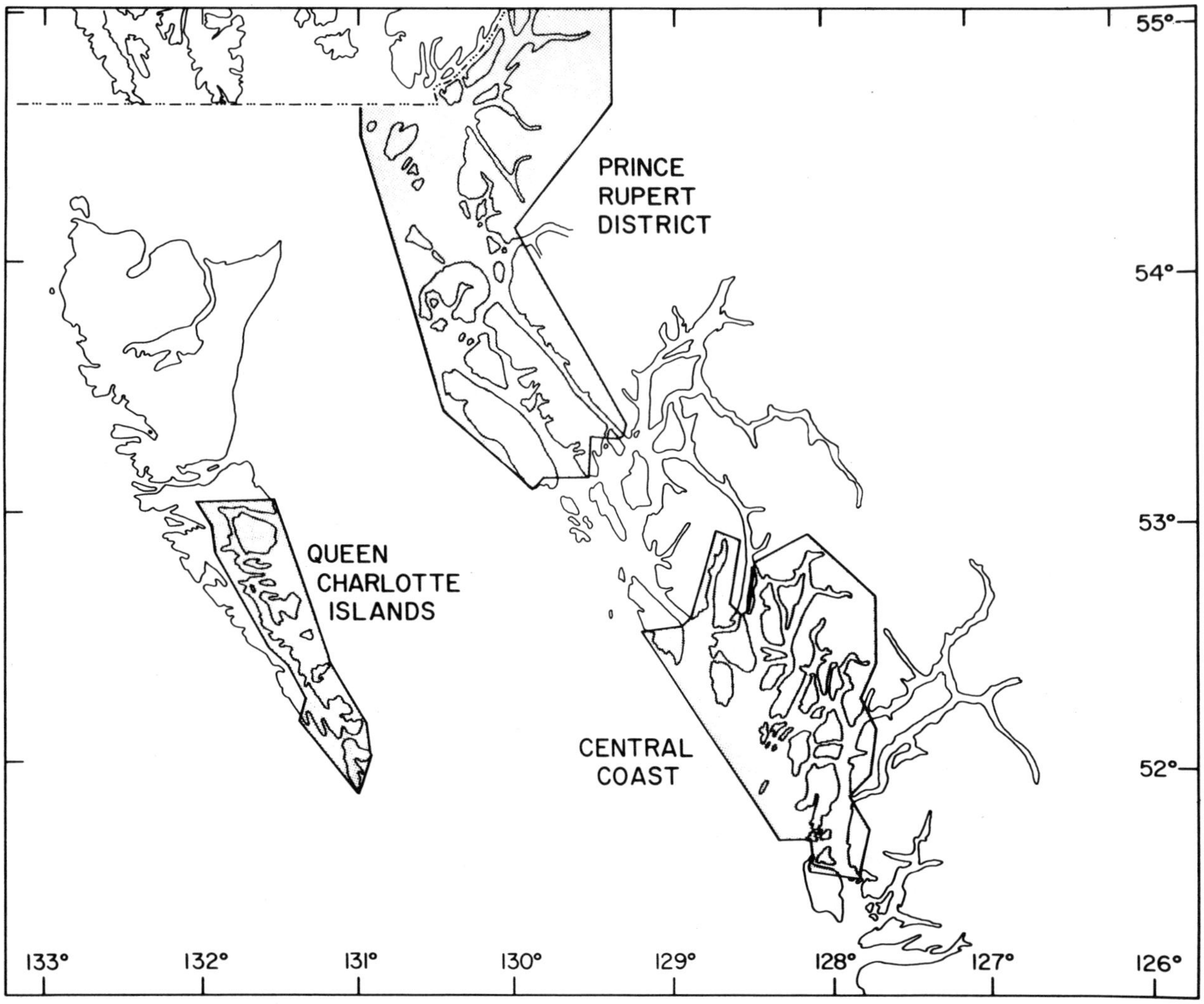


FIG. 1. Herring stock assessment regions in northern British Columbia (Source: Haist and Schweigert 1990).

pelagic planktivores, feeding mainly on euphausiids, copepods and decapod larvae.

### f) Migration

Knowledge of the movement of Pacific herring in B.C. comes largely from tagging studies (Stevenson 1955). From October through March, dense schools of maturing herring move from offshore feeding grounds on the continental shelf to the inshore spawning grounds (Taylor 1964). Adult herring aggregate near the spawning grounds several weeks to months before spawning (Haegele and Schweigert 1985; Taylor 1983). After spawning, from February through May, adult herring migrate back to offshore feeding grounds. For example, a herring tagged just prior to spawning in Lambert Channel (Strait of Georgia) travelled 350 km in 16 days to be recovered in Juan de Fuca Canyon (Haegele et al. 1982).

### g) Stock Delineation

The stock structure and genetic relationship among stocks within the distributional range are not clearly understood (Grosse and Hay 1988). Several techniques have been applied over the years to delineate stocks. The first attempts to delineate herring stocks relied on morphometric and meristic measurements (Tester 1937a,b,1948). A different approach was based on a long term, large scale tagging study (Stevenson 1955). Other stock identification studies of B.C. herring used morphometrics and meristics (Schweigert 1981; Meng and

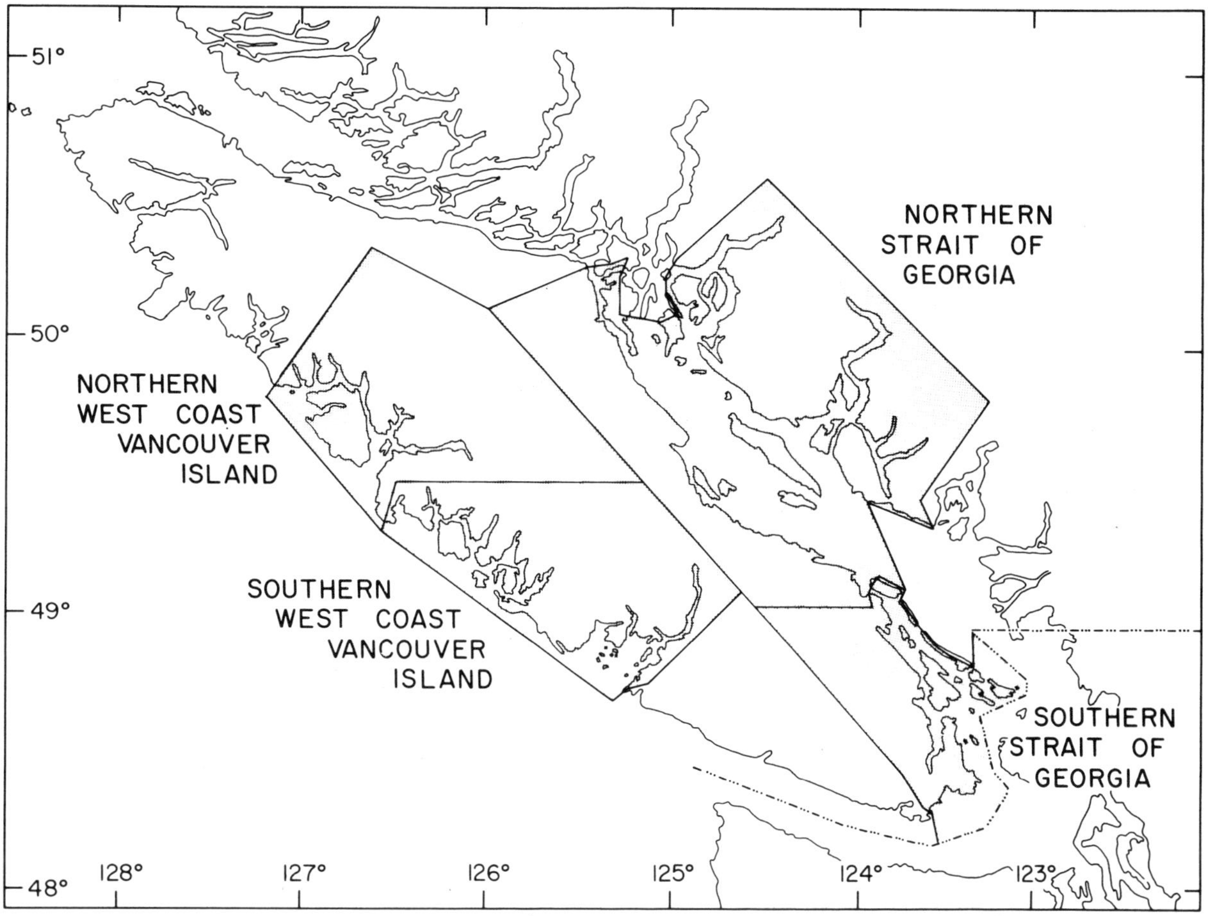


FIG. 2. Herring stock assessment regions in southern British Columbia (Source: Haist and Schweigert 1990).

Stocker 1984), as well as electrophoresis (Grant and Utter 1984), and an analysis of the time series of herring spawn data (Haegele and Schweigert 1985; Hay and Kronlund 1987). The most recent attempts to differentiate B.C. herring stocks have used truss network analysis (Schweigert 1990), mitochondrial DNA analysis (Schweigert and Withler 1990), and size and age information (Schweigert 1991).

The stock concept for B. C. herring has changed over time in response to the need of fisheries management. The two types of herring populations in B.C. are major migratory stocks, and smaller resident stocks usually found at the head of inlets (Taylor 1964). From studies of age compositions and vertebral counts of herring samples, Tester (1937a,b,1948) concluded that there were at least six more or less discrete stocks. Results of a large scale long term tagging study led to the recognition of nine major migratory stocks (Stevenson 1955; Taylor 1964). This nine stock concept was in use during much of the course of the reduction fishery (Taylor 1964). In recent years, the migratory stocks are the target of the roe herring fishery, while local stocks support bait fisheries in some coastal areas. With the shift to the roe-fishery came a more detailed breakdown of stocks into 35 coastal management units (Hourston 1982). At present seven migratory stocks are recognized for the purpose of stock assessment and fishery management (Figs. 1-2).

## 3. Harvesting History, Management History and Industry Profile

### a) Catch History

The commercial herring fishery in British Columbia dates back to 1877. The fishery expanded over time in response to various market demands including dry salted herring in the Orient, herring meal and oil, food and bait herring, and roe herring. Over the years herring fisheries have shown erratic catch levels. In 1962–63,

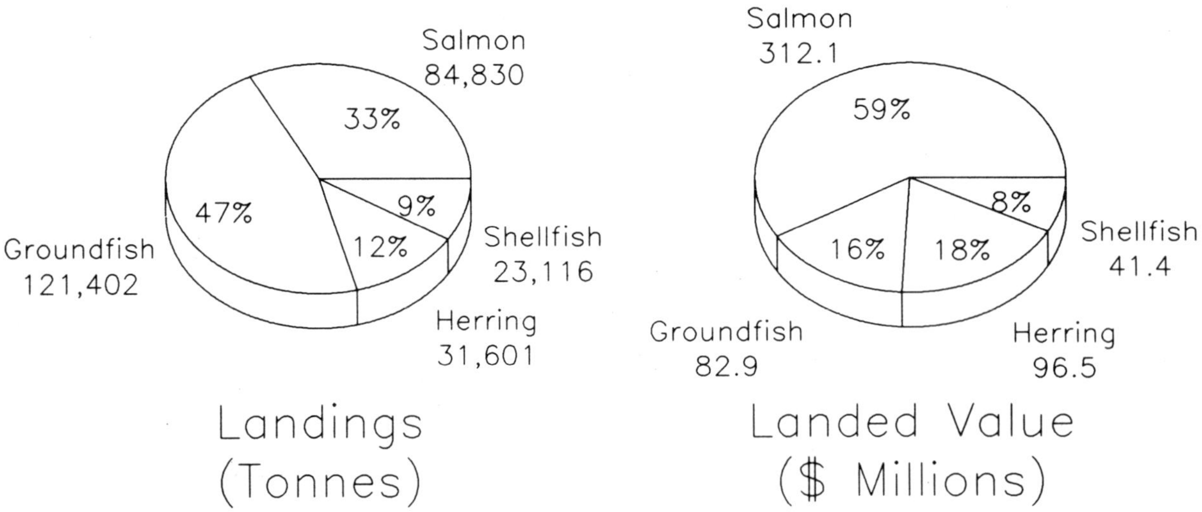


FIG. 3. Quantity (tonnes) and value ($ Millions) of British Columbian fish landings in 1988. Salmon, groundfish and shellfish categories are made up of several species.

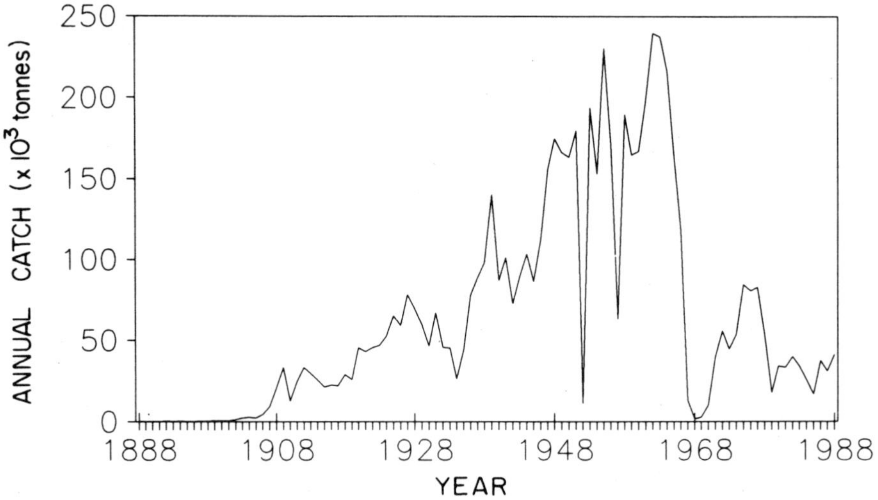


FIG. 4. Annual catch (tonnes) of British Columbian herring, 1888–1988.

during the reduction fishery period, a record catch of 240,000 t was landed. In 1988 the landed value of Pacific herring was 18% of all B.C. marine species, making it the most valuable single species caught (Fig. 3).

Reports of herring catches in British Columbia date back to 1877 (Taylor 1964). Catches were small until 1900 but increased rapidly (Fig. 4) when a market for dry salted herring opened in the Orient. This fishery operated from 1904–1935. By 1927 the catch increased to 77,000 t and fell to 27,000 t in 1935 with the decline of the dry salted herring market. A domestic reduction industry developed for herring in 1935 after the collapse of the pilchard fishery (in 1945). Catches increased sharply in response to an expanding meal and oil market. The main herring fishing took place from October to March on dense schools of maturing herring. At this time of the year herring move from their offshore summer feeding grounds to the inshore spawning grounds. Almost all the catch was by purse seine (Taylor 1964). In the 1950's catches averaged around 180,000 t per year. Low catches during the 1952–53 and 1957–58 seasons (Fig. 4) were due to industrial disputes (strikes). Reduction fishing peaked in the 1962–63 season, when a record catch of 240,000 t was landed. After 1965 the herring stocks collapsed, and the reduction fishery was closed indefinitely in 1968 (Pearse 1982; Hourston 1980a). Only the traditional minor food and bait fisheries remained open for the next 4 years (Hourston 1980a).

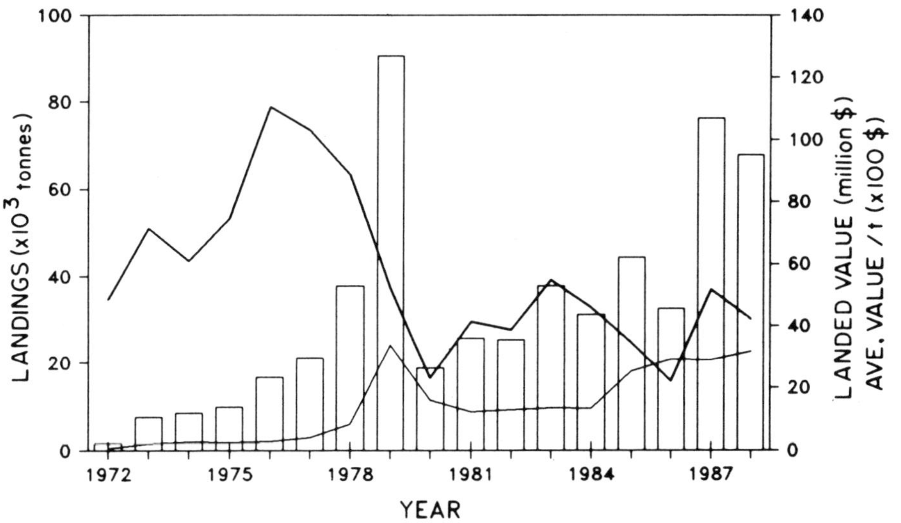


FIG. 5. British Columbia roe herring landings (thick line), landed value (including roe-on-kelp) (columns), and average value per t (thin line) (Source: Fisheries Production Statistics of British Columbia, Province of British Columbia, Ministry of Agriculture and Fisheries).

Cessation of the intense reduction fishery led to stock recovery and commencement of the roe herring fishery. The fishery began in 1972 and took an average of 48,000 t over the next 4 years (Hourston 1980a). The seine and gillnet fishery takes place in late February through April. Herring caught on or near the spawning grounds supply herring roe to the lucrative Japanese Kazunoko market. Since 1977 the roe herring fishery has been a major, but unstable fishery on the west coast. The catch varied within the range 14,500–73,600 t over the period 1977–89, averaging 34,800 t per year (Fig. 5). The fishery is sometimes short and intense with openings measured in hours and minutes. The fleet has excess catching capacity and is difficult to manage due to this tremendous fishing power.

Two other minor fisheries also exploit herring. A small food and bait fishery has persisted since the 1960's taking herring mainly in the Strait of Georgia. A commercial fishery for spawn-on-kelp emerged in British Columbia in 1975.

### b) Management History

Management restrictions were initially confined to closures and mesh regulations to protect spawning and juvenile herring (Hourston 1980b). During the reduction fishery the exploitation of herring stocks off the Pacific coast was limited by regulation for the purpose of maintaining maximum sustained productivity (Taylor 1955). Regulations were in two forms: catch quotas, and closures during the spawning season. The first fishery quotas were introduced in 1936 in response to a declining herring population on the west coast and the lower east coast of Vancouver Island. During the 1940/41 season, quotas were established for the middle east coast Vancouver Island. The following season saw quotas established for the northern and central coast populations. There never was a quota for the Queen Charlotte population during the reduction fishery. The quota was dropped in 1946/47 for the west coast of Vancouver Island population. Quotas, however, had a limited effect on catches as extensions were frequently granted (Hourston 1980a).

From the start of the roe fishery in 1972 through the 1982 season the roe herring fishery was managed on the basis of an optimum escapement policy (Trumble and Humphreys 1985). An optimum spawning escapement level was calculated for each management unit and additional herring were considered harvestable surplus (Hourston 1980b). This policy resulted in very large catches of roe herring during years of strong recruitment off the west coast of Vancouver Island from 1975–77. The policy exacerbated biomass declines during periods of low recruitment.

### c) Harvesting Sector

From 1981–88 commercial fishermen received an average of $60.5 million for their herring catches. Over the 8-year period, landings of all herring products averaged 31,600 t. In 1988 the value of herring catches was $96.5 million, down by 10% from the previous year.

The bulk of the 1988 landings (30,070 t) and value ($95.1 million) came from the roe fisheries.

Both the landings and landed values of roe herring (including roe-on-kelp) have been extremely variable (Fig. 5). From 1972–88 the average annual roe fishery catch was 40,500 t bringing an average of $45.6 million a year to commercial fishermen. During the "Bonanza Year" of 1979 the average value of roe herring was $3,400 per t, yielding a record 127 million dollars for the year (Fig. 5). In 1980, a weak market combined with industrial disputes drove the value to half its 1979 value, yielding a total of only 26 million dollars. From then until 1984 prices stabilized at lower levels, but threafter increased gradually to a maximum of $3,200 per t in 1988 (Fig. 5).

The roe herring fishery was restricted in 1974 by introducing limited entry. The original target figures were to issue 150 seine and 450 gillnet licenses. This was, however, greatly exceeded, and by 1977, the year entry was closed, 246 seines and 1,328 gillnet licenses had been issued. In 1987 there were 1,607 licenses issued for commercial herring fisheries. These included 1,327 roe herring gillnet licenses, 252 roe herring seine licenses, and 28 spawn-on-kelp licenses. Herring licenses accounted for 19% of the total number of commercial fisheries licenses issued in British Columbia in 1987. These numbers are far in excess of the capacity needed to harvest the allowable catch. Roe herring ("H") licenses were issued to persons, and were non-transferable. The roe herring license was intended to reduce the number of licenses by attrition. The net effect of all changes to the original licensing scheme is that no reduction has occurred, and in 1990 there are still 252 seine and 1,327 gillnet licenses. Most members of the industry are satisfied with the overall performance of area licensing implemented in 1981. Other aspects of the licensing program, however, have sparked a heated debate in the harvesting and processing sectors. The major issues debated are nontransferability of herring licenses, leasing of roe herring licenses, employment loss due to multiple licensing, and lease cost recovery from crew members of lease licensed vessels. These factors are the main impetus for the current licensing policy review with the goal to make the industry perform more efficiently.

### d) Processing Sector

The roe herring fishery supplies the very lucrative Kazunoko market that exists in Japan. Kazunoko is used on festive occasions, primarily as a family gift. From 1983–87 B.C. roe comprised an average of 28% of the market share of imports of salted and frozen herring roe. Other suppliers of herring roe in order of importance included Atlantic Canada, the United States, South Korea, China, and Ireland.

Processing roughly doubles the landed value of fish catches (Pearse 1982). In 1987 the total wholesale value of herring product was 173.7 million dollars. This represents an increase of 97% over the previous year. From 1983–87 the total wholesale value of herring products averaged 111.9 million dollars.

Fish processors and buyers are licensed by the Province of British Columbia. In 1987 there were 37 roe herring processing licenses, and 185 roe herring buying licenses issued to persons and/or companies by the B.C. Ministry of Agriculture and Fisheries.

The B.C. fish processing industry is the third-largest employer in B.C. after forestry and tourism. It is difficult to isolate employment generated by the herring resource alone. For this reason the data referred to describes the entire processing industry. During 1987 processing companies reported an average of 5,500 employees a month on their payroll. Because of the seasonal nature of fish harvesting, employment ranged from a low 3,400 in January to a high of 8,300 in August. In addition to the direct employment the industry provides spinoff employment in the packaging, and transportation industries, as well as in fish product wholesale companies.

## 4. Classification of Stock Assessment Procedures

This section provides a synthesis of where we are and how much we have progressed in the management of herring stocks relative to the past. This is accomplished by classifying the models and analytical procedures of past and present management systems. The major breakthroughs in the progression of herring management systems are highlighted.

Research on Pacific herring in British Columbia began in 1929 (Taylor 1985). Over the years, the major focus of herring research has been on population dynamics. Work has included the development of analytical stock assessment methods, studies on recruitment processes, studies on stock identification, and the development of various enumeration techniques. Fisheries data on B.C. herring have been collected since the early 1930's. Since then, stocks have been observed under a wide range of harvest rates. As a result, the herring fishery is well documented for biological assessment (Pearse 1982).

Holling (1978) classified nine case studies of resource management and environmental assessment using three broad measures: complexity, data, and understanding. Similarly, the progression of the herring management systems can be shown in the context of these three measures. These measures can be thought of as three

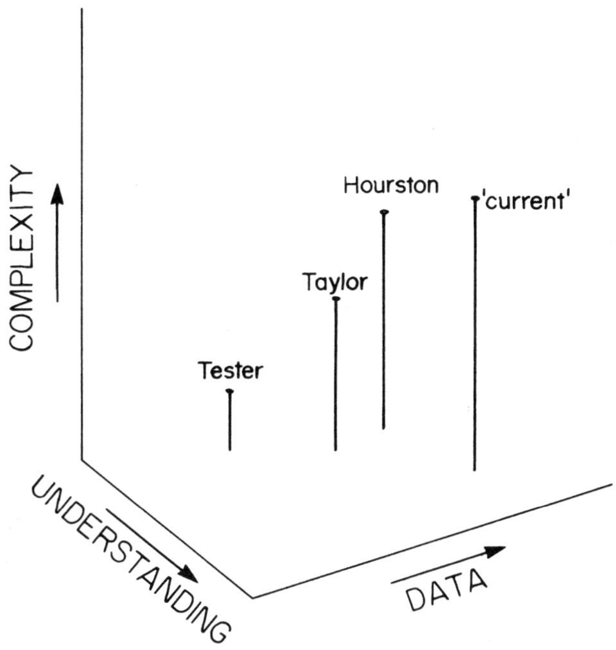


FIG. 6. Placements of four models and analytical procedures used to manage B.C. herring fisheries within the dimensions of complexity, data and understanding.

axes of a graph on which we can place current and past systems used to manage B.C. herring stocks. Whether a system is simple or complex is a matter of opinion. Components that contribute to complexity are the number of variables, number of spatial units, number of management acts, and the time resolution which in these cases was always one year (Table 1). Data are required to calculate parameters that describe relationships of the system. Understanding refers to the extent of basic understanding we have of the behavior of the system. Figure 6 shows subjective placement of four management systems within the three dimensions.

Tester (1948) examined the efficacy of catch limitations in regulating the B.C. herring fishery off the west coast of Vancouver Island. Fluctuations in herring abundance were modeled with few state variables including simple indices of spawning population size, relative abundance of year-classes, and catch and effort data. Detailed life history studies of herring had been made. Significant data were available on growth (in weight), natural mortality, reproduction, and the extent of stock mixing. An asymptotic relationship was postulated between egg deposition and recruitment. It was concluded that for the herring stocks the greatest yield from an incoming year-class was produced by an intense fishery. To determine by the most direct method whether or not quota restrictions were necessary, a large scale "west coast experiment" was launched in 1946/47. The experiment entailed abandoning catch quotas on the west coast of Vancouver Island reduction fisheries. Tester (1948) realized that a risk of overfishing was involved, and that this risk must be taken if the question was to be answered.

A more complex management system was described by Taylor (1964). More state variables were used to described the system than in Tester's model described

TABLE 1. Components of complexity for four B.C. herring management systems.

| | Number of state variables | Number of spatial units | Management acts |
|---|---|---|---|
| Tester (1948) | Few | 6 | closed areas<br>closed seasons<br>catch quotas[a]<br>mesh regulations |
| Taylor (1964) | Moderate | 9 | closed area<br>closed seasons<br>catch quotas<br>mesh regulation |
| Hourston and Schweigert (1981) | Many | 35 | closed area<br>closed seasons<br>catch guidelines<br>(i.e., optimal escapement) |
| "Current" Stocker et al. (1983); Haist et al. (1986) | Many | 7 | closed areas<br>closed seasons<br>(i.e., CUTOFF) catch quotas<br>(i.e., 20% harvest rate)<br>mesh regulations |

[a]Beginning in the 1946/47 season the quota for the west coast of Vancouver Island was removed.

above. Tag recoveries indicated that B.C. herring could be grouped into nine discrete major migratory populations (Table 1). A broad range of data was available including data on growth, natural mortality, tag recoveries, extent of spawn deposition, and estimates of year-class strengths. The sheer volume of data available was not necessarily helpful in understanding the system behavior. The result of the "west cost experiment" lead to the conclusion that the quota system was ineffective in stabilizing abundance. The understanding was that most populations would be able to withstand 60% harvest rates. Taylor (1964) developed stock-recruitment relationships to estimate "surplus" stock the fishery could take. Variations in year-class strength were mostly explained by natural causes.

Hourston assessed the status of individual stocks (i.e. 35 management units) annually on the basis of 3 different data sources: catch, biological samples, and spawn (Hourston and Schweigert 1981). With the advent of computing power, the herring data base was computerized during this phase. Data were fed into a computer model producing estimates of tonnes of fish at age. Average recruitment was assumed to forecast future stock sizes. The optimum spawning escapement was estimated for each stock. Total Allowable Catch (TAC) was calculated as forecast stock size less optimum stock size.

Currently two methods are used to annually assess the status of herring stocks: (1) an age-structured model, and (2) an escapement model (see section 5a below). Of the four systems described, the current system is the most complex. It also has the most supporting knowledge of the biological processes. One element of the complexity of the current management sytem is the prediction of spawn timing. Ware and Tanasichuck (1990) developed a stock specific procedure to forecast accurately when female herring become ripe. This procedure is useful to field managers concerned with staging seine roe and gillnet fisheries on herring that are about to spawn. The industry also benefits from this procedure considering that the highest prices are paid for top quality (ripe) roe.

## 5. Current Stock Assessment Procedure

### a) Assessment Models

Forecasting the potential catch that can be removed from herring stocks requires assessing the status and determining the factors affecting the stock dynamics. Traditionally, yield models have assumed equilibrium, implying no changes in age structure, growth, or mortality over time. Herring are, however, strongly affected by changes in environmental conditions, thus making equilibrium models of limited use. The methods used to estimate current stock conditions of Pacific herring have the aim of making catch recommendations that will not have deleterious effects on the stocks (Stocker et al. 1983).

Two analytical stock assessment methods are used: (1) an age-structured model (Fournier and Archibald 1982), and (2) an escapement-based model (Schweigert and Stocker 1988). These models provide descriptions of the dynamics of the herring populations. Both methods use catch, spawn, age composition, and weight-at-age data beginning in 1950.

*Age-structured Model.* The use of models utilizing age structure data for stock reconstruction is well documented in the fisheries literature. A modified version of the age-structured model described in Fournier and Archibald (1982) has been used for B.C. herring stock assessment since 1982 (Stocker et al. 1983). The model has undergone several revisions since 1982 making it more consistent with the life history of herring and the fisheries. The current version described in Haist and Schweigert (1990) includes auxiliary information in the form of spawn survey data available for B. C. herring. The model separates catch-at-age data by gear type. Seines are assumed to be non-selective while gillnets are selective for larger, older fish. An availability term models partial recruitment to the spawning stock.

The objective function used to estimate model parameters includes both measurement (observational) and process (deviations from modelled assumptions) errors. A maximum likelihood method is used to simultaneously estimate all model parameters. Input data used for age-structured model analysis include, for each season, proportions caught-at-age, average weight-at-age, numbers of fish aged, and the spawn index. Numbers of fish in the catch are calculated using the average fish weight from biological samples of the catch. The spawn index sums the lengths of spawns multiplied by a widths/intensity factor (product of spawn width and the weighted intensity), standardized by geographical section of coastline (Hay and Kronlund 1987). Relative importance of spawn can be adjusted in the age-structured model depending on degree of belief in the accuracy of the spawn information.

Herring on the B. C. coast have experienced strong fluctuations in recruitment success. Estimated number of recruits as 3-year olds of the seven major migratory populations are shown in Figs. 7 and 8 (Haist and Schweigert 1990). Generally, strong year-classes may dominate occasionally (like the 1977 year-class in the north coast of B.C., which recruited to the fishery in the 1979/80 season; Fig. 7), and several seasons of weak year-classes often follow a season of a strong year-class.

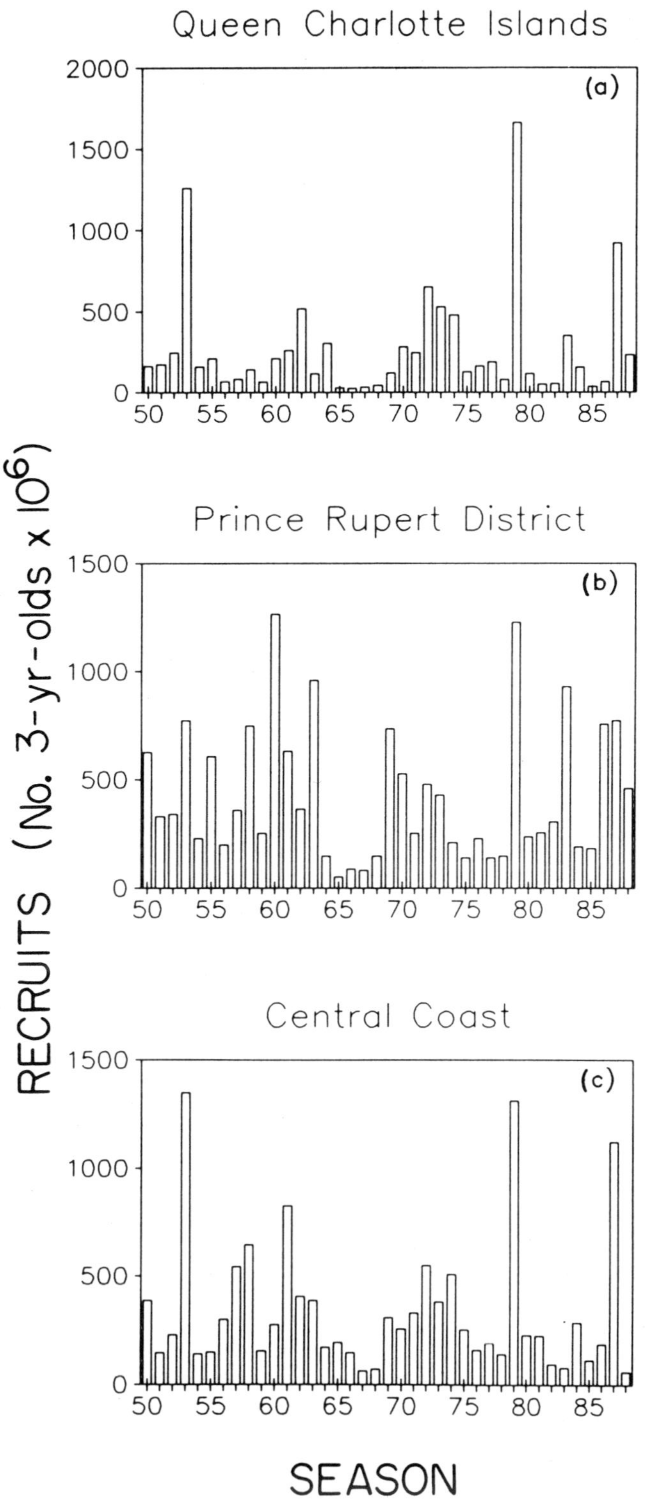


FIG. 7. Estimates of recruitment of 3-year-old herring from age-structured model analysis for the northern stock groupings, 1950/51–1988/89 seasons (i.e. 50 refers to 1950/51 season; source: Haist and Schweigert 1990).

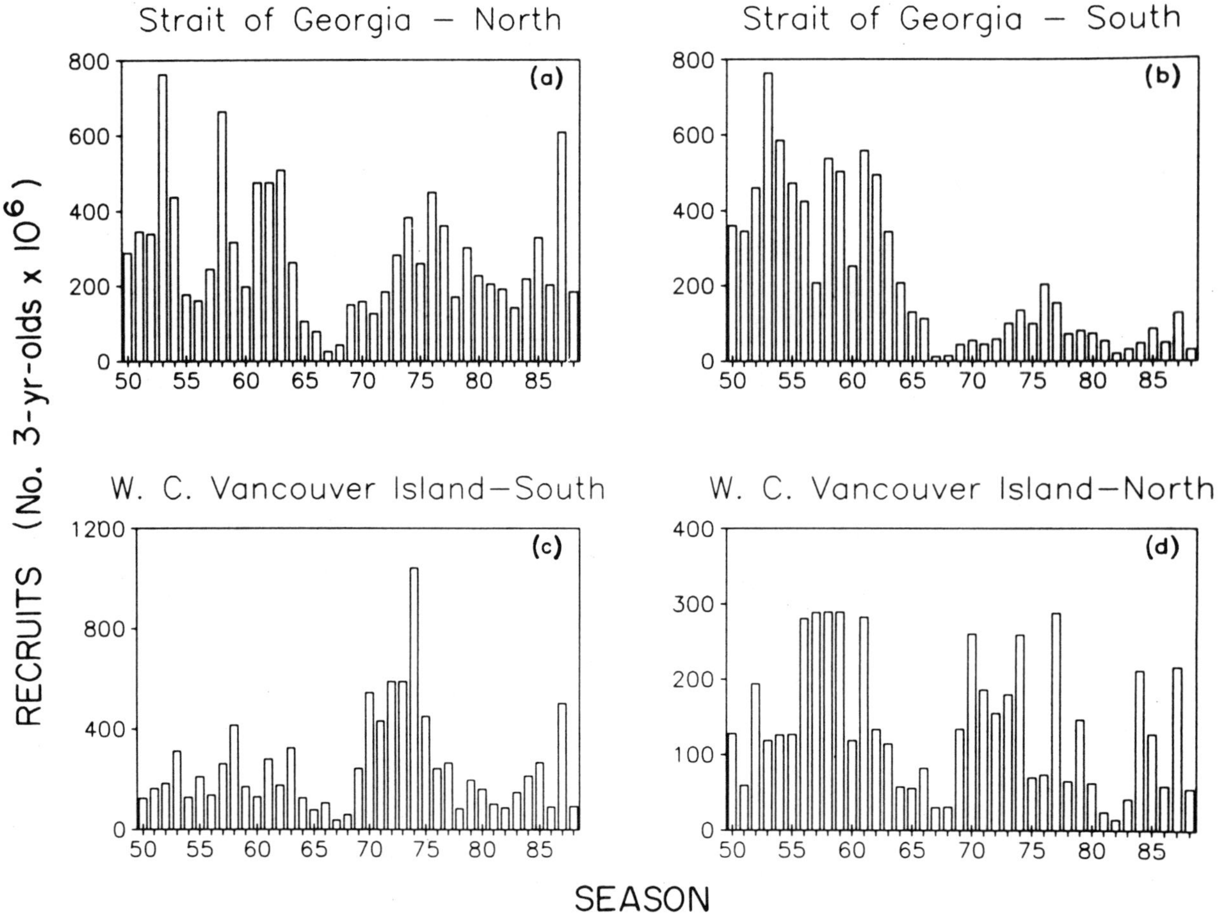


FIG. 8. Estimates of recruitment of 3-year-old herring from age-structured model analysis for the southern stock groupings, 1950/51–1988/89 seasons (i.e. 50 refers to 1950/51 season; source: Haist and Schweigert 1990).

The three northern stocks show highly variable recruitment estimates (Fig. 7). Generally, low recruitment occurred in the mid 1960's, prior to the collapse of the reduction herring fishery. In the Queen Charlotte Islands, estimated recruitment varies consistently within the range 30–920 million 3-year olds over the period 1950–88 (Fig. 7a). The exceptions were the 1953/54 and 1979/80 seasons when the estimates were 1260 and 1660 million 3-year olds, respectively. These years represent the extremely strong 1951 and 1977 year-classes. Estimates of recruitment for the Prince Rupert District stock show a 25 fold difference between the peak in the 1960/61 season and the trough in the 1965/66 season (Fig. 7b). The very strong year-class of 1977 is again evident as shown by 1230 million 3-year olds in the 1979/80 season. Similar fluctuations in recruitment are shown in the Central Coast (Fig. 7c). At least one very strong year-class has emerged in this area in each of the last four decades, including again the 1977 year-class.

Comparable recruitment patterns emerge for the four southern stocks (Fig. 8). The Strait of Georgia stocks, and the southern west coast of Vancouver Island stock show similar levels of recruitment estimates, while the northern west coast of Vancouver Island stock shows much smaller estimates. Very weak recruitment (expressed as millions of 3-year-olds) was estimated during the 1965–68 period. Stronger recruitment began to appear in the 1970/71 season in three of the four southern stocks (Fig. 8a, c, d). Low estimates of recruitment have persisted for the southern Strait of Georgia stock since the collapse of the reduction fishery (Fig. 8b).

When historical escapement (spawning biomass) and catch trends are added to show total abundance estimated from the age-structured model for the seven major migratory herring stocks a common pattern emerges (Figs. 9 and 10). High catches and low spawning escapement prevailed through the 1950's and early

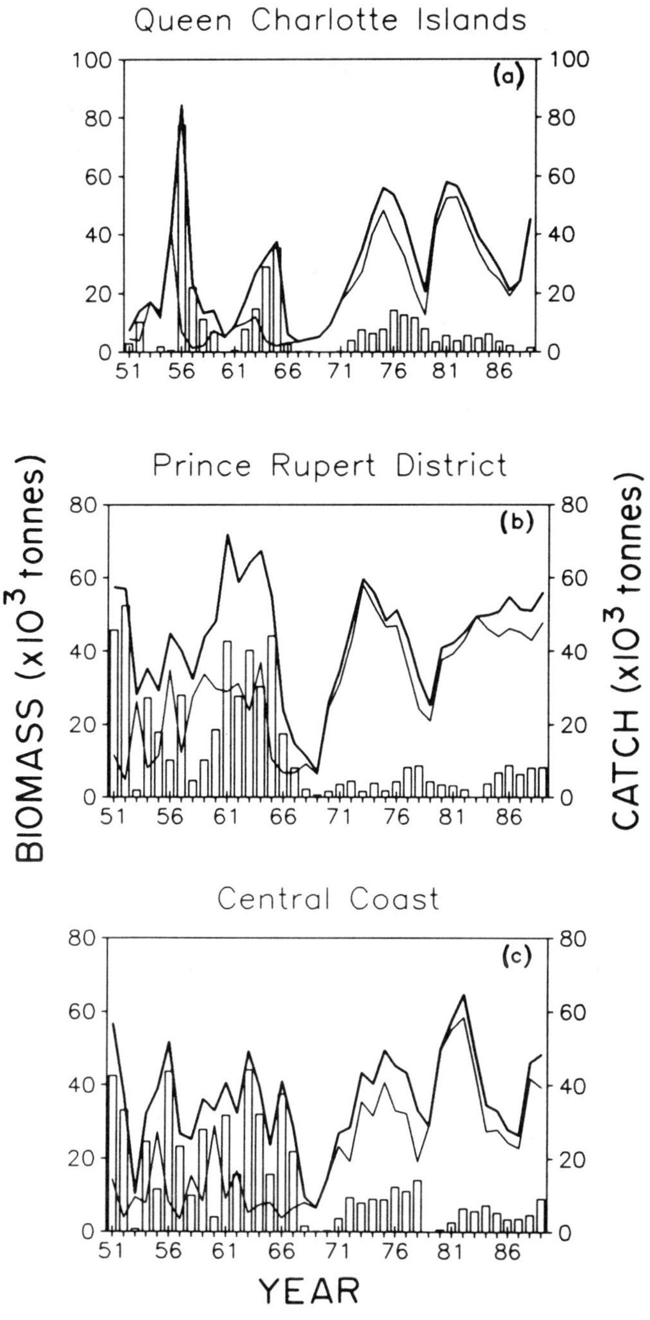


FIG. 9. Estimates of total biomass (thick lines), and escapement (spawning) biomass (thin lines) for the three northern stock groupings, 1951–89. Columns represent catch statistics.

1960's. Low production in the form of recruitment occurred during the period 1965–68, and by 1968 the fishery had collapsed. Spawning escapements increased sharply following the moratorium on fishing, and stocks generally recovered to near or above the 1950's levels by the mid 1970's. The exception to this recovery pattern is the southern Strait of Georgia stock, which remained at relatively low levels until 1989 (Fig. 10b). In contrast to the reduction fishery, the roe herring fishery shows low catches and high escapement levels.

Herring stocks show natural fluctuations in abundance due to interannual variations in recruitment described

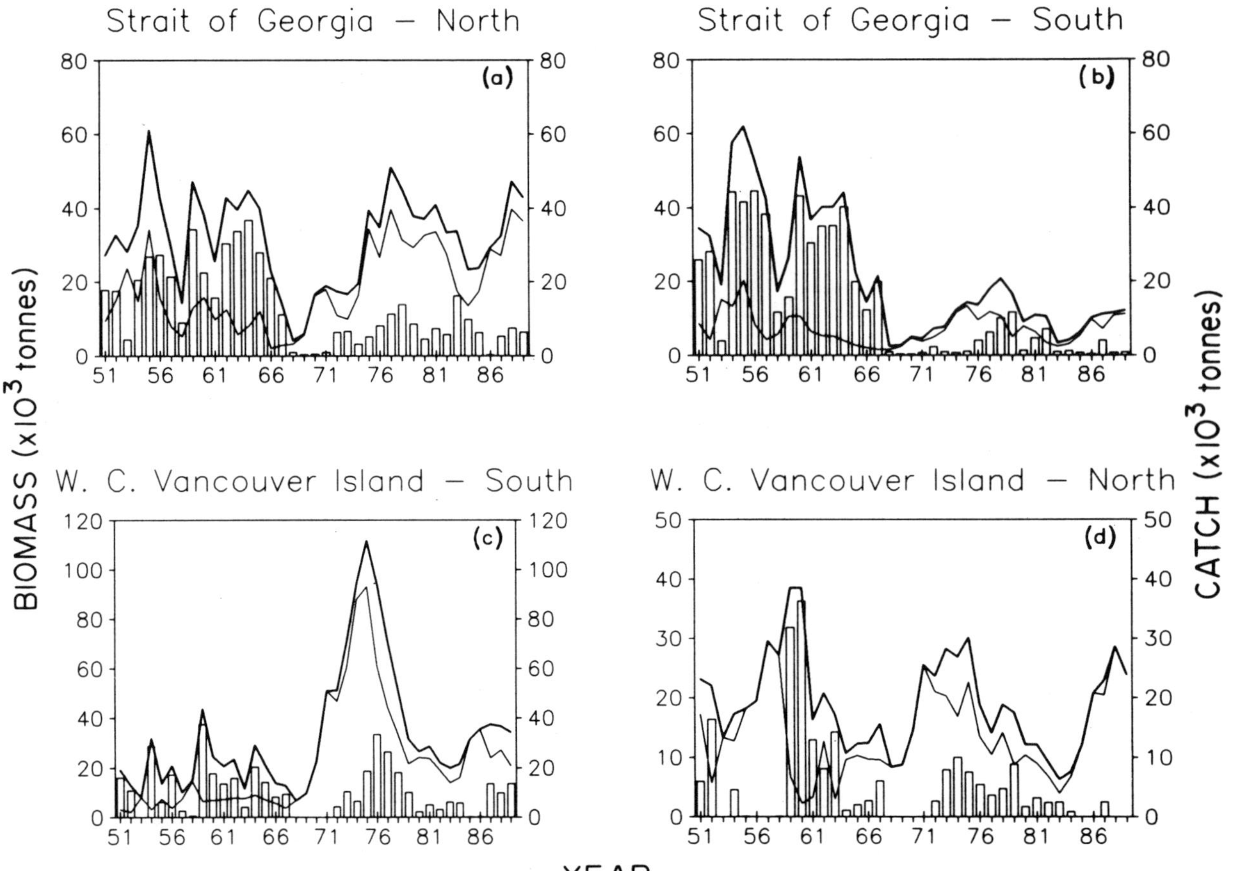


FIG. 10. Estimates of total biomass (thick lines), and escapement (spawning) biomass (thin lines) for the four southern stock groupings, 1951–89. Columns represent catch statistics.

above. The pattern of fluctuations is, however, not always in synchrony between stocks. This is indicated by results of a correlation analysis of herring recruitment time series of the seven major stocks (Ware and McFarlane 1989). The northern stocks (Fig. 9) exhibit wider fluctuations than the southern stocks (Fig. 10). This characteristic stems from the occurrence of very large but irregular year-classes (e.g., 1977) in the north that support the fishery for a number of years. Some hypotheses linking herring recruitment patterns to ocean conditions are emerging. The northern and southern stocks exist in different production domains (Ware and McFarlane 1989). These domains have different ocean environment conditions and thus result in different recruitment patterns over time. A discussion of the more recent status of the major migratory herring populations based on results of the 1989 herring stock assessment follows (Haist and Schweigert 1990).

The Queen Charlotte Islands stock has been declining from 1982–88 as the extraordinarily large 1977 year-class moved through the fishery (Fig. 9a). The estimated 1989 spawning biomass was 44,000 t. This is a substantial increase, and is due to the apparently strong 1985 year-class. The stock in the Prince Rupert District has shown substantial increases since 1979 (Fig. 9b). The estimated spawning biomass was 47,800 t for 1989. The Central Coast stock, also characterized by a very strong 1977 year-class, has declined from 1982–87 (Fig. 9c). A strong 1985 year-class recruiting in the 1987/88 season has nearly doubled the spawning biomass to 42,000 t from 1987–88. The 1989 spawning biomass was estimated at 39,000 t.

The northern Strait of Georgia stock has shown a strong oscillation since the start of the roe fishery (Fig. 10a.) The southern Strait of Georgia has been at low levels over the last few years (Fig. 10b). Spawning biomass estimates for 1989 indicated a slight decrease in the northern Strait of Georgia stock, and a slight increase in the southern Strait of Georgia stock from 1988 levels. In 1975 the southern west coast of

Vancouver Island stock was at its highest level since the crash (Fig. 10c). The stock has declined substantially since then, and the 1989 spawning biomass has been estimated at 21,000 t, about 25% of the high in 1975. A major increase in stock size is indicated for the northern west coast of Vancouver Island stock since 1983 (Fig. 10d). The estimated spawning biomass for 1989 was 23,800 t.

*Escapement Model.* Pacific herring are unique in that they lay demersal eggs in the intertidal and upper subtidal zones. This makes it possible to estimate the extent of spawning area from the surface. Hart and Tester (1934) first proposed spawn surveys to assess herring stocks in British Columbia. Fishery officers of the Department of Fisheries and Oceans have monitored the extent of spawn deposition in most areas along the B. C. coast since 1937 (Tester 1948). They estimate the total length and average width of the spawning area, and the intensity of spawn deposition. From this information spawn indices have been calculated and published since 1937 (Hay and Kronlund 1987).

Tester (1948) calculated two indices of spawn deposition for the west coast of Vancouver Island from 1931–46 spawn data. Stevenson and Outram (1953) determined that a km of spawning area adjusted to medium intensity was equivalent to 1,633 t of spawners. Outram and Taylor (1964) derived similar values based on counts of egg samples.

A new assessment procedure was developed when the roe herring fishery began. Hourston et al. (1972) used surface spawn survey estimates of length, width and intensity to estimate egg deposition. These estimates were then incorporated into a model which also considered biological sampling data to estimate spawning escapement and forecast future run sizes (Hourston 1981; Hourston and Schweigert 1981).

SCUBA diving surveys of spawn were initiated in 1975 to evaluate the accuracy of Fishery Officer surface spawn surveys, and to derive the most cost-effective diver surveys of major spawning areas (Humphreys and Haegele 1976; Schweigert et al. 1985). The diver information is comparable to the historical Fishery Officer data, permitting comparisons that have led to intercalibration formulae and estimates of Pacific herring spawning stock sizes (Schweigert and Stocker 1988; Schweigert et al. 1990).

The escapement model estimates spawning-stock biomass by calibrating fishery officer indices of spawn deposition with data collected during SCUBA surveys of spawn depositions (Schweigert and Stocker 1988). Estimates of egg depositions from diver surveys are based on a two-stage sampling design (Schweigert et al. 1985). Dive survey observations are used directly, while surface survey observations where no dives are conducted are adjusted to make them comparable to dive survey estimates. Calibration equations derived from duplicate dive and surface surveys are used for this purpose.

The escapement model, estimating total egg deposition, was first applied in 1984 (Haist et al. 1985). The total egg deposition is converted to tonnes of spawners based on an average fecundity estimate of 100 eggs per gram of spawning fish (sex ratio 50:50). Escapement plus catch provides an estimate of the total pre-fishery spawning stock biomass (Schweigert and Stocker 1988).

In summary, the record of roe catch for the total B.C. coast vs roe quotas indicates good correspondence (Fig. 11). The analytical stock assessments provide a reasonable understanding of past performance of B.C. herring populations. The methods, however, need continuous refining. With this approach we hope to avoid pitfalls that led to severe stock declines in the past.

### b) Forecasting

Forecasts of next year's stock abundance from the age-structured model is made in two steps. First, numbers at age of herring returning to spawn next year are the estimates of numbers at age escaped this year multiplied by survival. Variance of estimated numbers at age is introduced by applying bootstrap techniques (Haist and Schweigert 1989). Second, recruitment is projected for three scenarios (poor, average, good) based on estimates of numbers at age 3 for the entire time series. Poor, average, and good recruitment is based on the means of the lowest, the mid, and the highest one-third of historic numbers at age 3 (Haist and Schweigert 1989).

The escapement model forecasts run size for each stock as the product of the estimated escapement in the previous year and an average annual survival rate of 0.64, plus predicted recruitment. Recruitment is calculated for poor, average, or good levels observed during the historical time series (Haist and Schweigert 1990).

As shown above, recruitment biomass can vary greatly among years. For Pacific herring new recruits can represent a considerable (>20%) proportion of the total stock biomass (Fig. 12). Accurate forecasting of the size of year-classes recruiting to the fishery is of fundamental importance in herring fisheries management (Stocker and Noakes 1988). New recruits may significantly influence future quotas established by fisheries managers.

Current hypotheses on factors affecting year-class abundance of Pacific herring have recently been summarized by Grosse and Hay (1988). Tester (1948) first

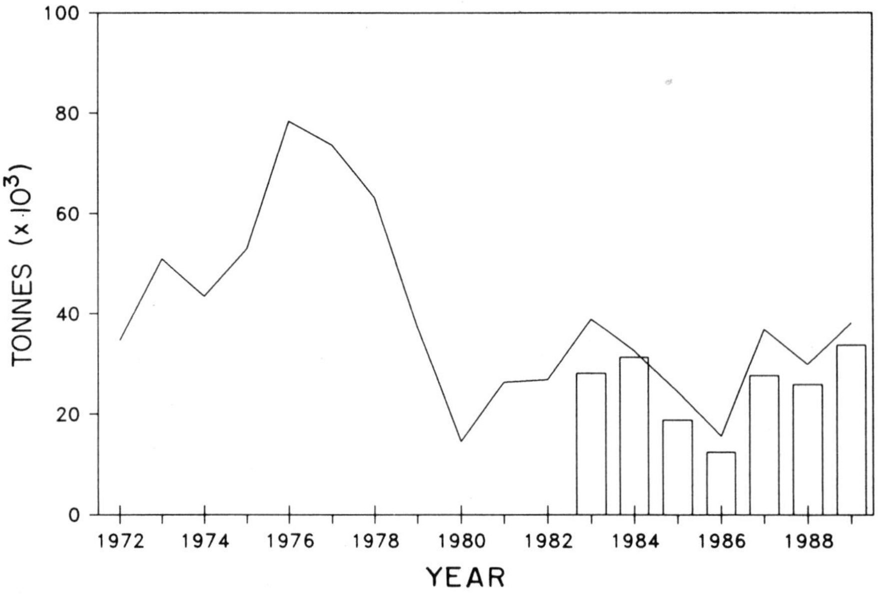


FIG. 11. British Columbia roe herring fishery landings (1972–89), and quotas (bars, 1983–89). Prior to 1983 fishery was managed by an escapement policy.

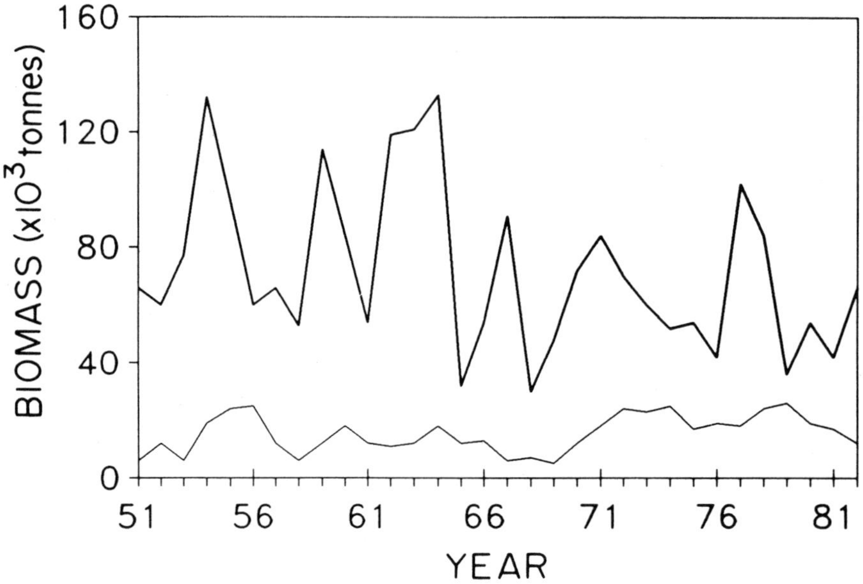


FIG. 12. Temporal trends in adult production (ages 4+) (thin line), and recruit biomass (age 3) (thick line) for herring in the Strait of Georgia, 1951–82 (Source: Haist and Stocker 1985).

theorized that catch trends might reflect some cyclic phenomenon. Factors that might cause such oscillations have received much attention. Stocker et al. (1985) correlate year-class strengths with sea surface temperature, river discharge, and sea level. Ware and McFarlane (1986) obtained a negative correlation for adult Pacific hake (*Merluccius productus*) and sea surface temperature with juvenile herring abundance on the west coast of Vancouver Island. In northern B. C. herring abundance cycles might be caused by predator-prey interactions between Pacific cod (*Gadus macrocephalus*) and herring (Walters et al. 1986). Compelling evidence is mounting that long-term variations in ocean conditions seem likely to affect cyclical changes in herring biomass in the northeastern Pacific (Mysak et al. 1982, Ware 1990). We still do not know, however, what the precise mechanisms are that cause these cycles.

To assign an overall stock forecast for a given year, we assign subjective probabilities to the two alternative stock assessment models. Based on intuition and past experience we believe that the age-structured model makes the most likely predictions of forecast run sizes, and thus in general we assign higher probabilities to this model (Haist and Stocker 1984). The assigned probabilities are used to weight the forecast run sizes obtained from each stock assessment model to provide a single "weighted run" for each of the herring stocks (Haist and Stocker 1984). The recommended catch levels are based on these run sizes (see section 6b below).

### c) Data Sources

The three primary data sources are: (1) commercial catch landing data, (2) biological samples, and (3) spawn survey data. These data are kept in the herring data base at the Pacific Biological Station, Nanaimo. Herring data have been coded geographically using a location system that is unique to herring management (Haist and Rosenfeld 1988, Hourston and Hamer 1979). Herring data are collected by "season" (seasons are from July 1–June 30). Data are available from the 1950–51 season to the 1988–89 season. This time period includes the reduction fishery period prior to 1968 and the subsequent roe fishery period beginning in 1972 (Haist and Schweigert 1990).

Catch information is obtained from the DFO sales slips. Sales slips are forms used between commercial buyers and fishermen to record the details of a fishing vessel's landings. Records are submitted by buyers to the Statistical Division of the Fisheries Branch.

Biological samples are collected from pre-fishery charter vessels, commercial fishery landings, and from DFO research cruises (Hamer 1989). Roe fisheries are sampled to obtain at least 10 samples from each seine fishery and six samples from each gillnet fishery. Each sample contains at least 120 individual fish. This sampling scheme is designed to provide a representative estimate of the biological structure of the catch (Hamer 1989; Schweigert and Sibert 1983; Humphreys and Hourston 1978). Sample information includes date, and locality of capture, and type of gear used. Information on length, weight, sex, maturity, and gonad weight is recorded for individual fish (Hamer 1989). A scale is taken from each fish for age determination. Summaries of herring sampling data by season are published every year (e.g., Hamer and Haist 1989).

TABLE 2. Forecast and observed herring stock biomass (tonnes) and relative difference (%) for B.C. herring stocks, 1986–89.

| Year | Stock[e] | Forecast stock biomass | Observed stock biomass | Relative difference (%) |
|---|---|---|---|---|
| 1986[a] | QCI | 19900 | 15910 | 25.08 |
| | PRD | 33200 | 36068 | −7.95 |
| | CC | 17200 | 23339 | −26.30 |
| | SOG | 21600 | 50112 | −57.55 |
| | WCVI | 19300 | 45466 | −12.54 |
| 1987[b] | QCI | 15310 | 17505 | −12.54 |
| | PRD | 32050 | 46358 | −30.86 |
| | CC | 23000 | 27270 | −15.66 |
| | SOGN | 42950 | 37678 | 13.99 |
| | SOGS | 10000 | 12966 | −22.88 |
| | WCVIS | 33700 | 34218 | −1.51 |
| | WCVIN | 14620 | 18217 | −19.75 |
| 1988[c] | QCI | 12125 | 16369 | −25.93 |
| | PRD | 43750 | 44328 | −1.30 |
| | CC | 23800 | 49838 | −52.25 |
| | SOGN | 36300 | 41439 | −12.40 |
| | SOGS | 10350 | 10238 | 1.09 |
| | WCVIS | 23000 | 37673 | −38.95 |
| | WCVIN | 16560 | 16604 | −.26 |
| 1989[d] | QCI | 15400 | 25539 | −39.70 |
| | PRD | 37600 | 38328 | −1.90 |
| | CC | 48450 | 51712 | −6.31 |
| | SOGN | 41150 | 51593 | −20.24 |
| | SOGS | 12600 | 8250 | 52.73 |
| | WCVIS | 31750 | 40817 | −22.21 |
| | WCVIN | 18940 | 13630 | 38.96 |
| Overall | | 654605 | 811465 | −19.33 |

Sources:
[a]Haist et al. 1986 (forecast); Haist et al. 1987 (observed).
[b]Haist et al. 1987 (forecast); Haist et al. 1988 (observed).
[c]Haist et al. 1988 (forecast); Haist and Schweigert 1989 (observed).
[d]Haist and Schweigert 1989 (forecast); Haist and Schweigert 1990 (observed).
[e]QCI, Queen Charlotte Isl.; PRD, Prince Rupert District; CC, Central Coast; SOG, Strait of Georgia (N. North; S, South); WCVI, West Coast Vancouver Island (S. South; N, North).

Detailed records of the date, location, length, width and intensity of herring spawns in British Columbia are collected by fishery officers (Chalmers 1989). Coverage and detail of these observations has changed substantially over time (Schweigert and Stocker 1988). Spawning records from 1928–50 have been added to the retrievable data base (Hay and Kronlund 1987). A detailed 50-year (1937–1986) geographical review of

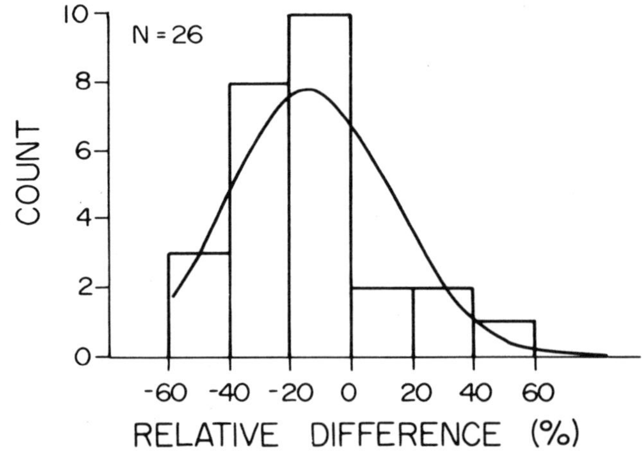


FIG. 13. Histogram of relative difference (%) between forecast and observed stock sizes for 26 area-stock combinations of B.C. herring, 1986–89. A normal curve is superimposed on the histogram.

the history of spawnings for all areas of the B.C. coast has recently been published (Hay et al. 1989).

**d) Evaluation of Procedure**

Two criteria are used to evaluate the performance of the current herring stock assessment procedure and the fishery management objective. Firstly, a retrospective analysis is used to evaluate the accuracy of the forecasts made for B.C. herring stock size. Secondly, an evaluation on how closely the 20% harvest rate objective has been met is undertaken. To evaluate the performance of past forecasts of herring stock biomass, the 1985–88 forecasts were compared to estimates of stock biomass calculated (observed) the following year (1986–89). The observed stock sizes were calculated using the same weighting factors for the two stock assessment models as were used to provide a single "weighted run" for the forecasts (see section 5b above). The measure used to compare the difference between forecast and observed stock size is the relative percent difference:

$$\text{Relative difference (\%)} = \frac{\text{Forecast} - \text{Observed}}{\text{Observed}} \times 100$$

There were considerable differences between forecast stock sizes and observed stock sizes (Table 2). For the 26 forecasts made during the period 1985–88, four deviations are greater than 50%. The overall relative difference for the 26 year-stock combinations analysed was −19.3% (Table 2). There was a tendency to under predict stock sizes according to the criteria used in this evaluation. This tendency is shown by the frequency distribution of the deviations (Fig. 13), as the peak of the normal curve superimposed on the frequency distribution is below zero. Deviations in forecasting stock size result largely from the uncertainties associated with forecasting the size of the recruitment component of the stock.

To evaluate how closely the 20% harvest rate objective was met, the 1986–89 harvest rates were estimated. For the 4-year period only 19 of 26 area-stock combinations had a 20% quota established under the current decision rules. The observed stock sizes were calculated using the same weighting factors for the two stock assessment models as were used to provide a single "weighted run" for the forecasts (see section 5b above). The harvest rate was calculated as the catch divided by the observed stock size multiplied by 100. The results indicated that of 19 observations, 15 fell between 10–30% (Fig. 14). Because there is uncertainty about the immediate state of the herring resource and uncertainty of its future behaviour, these results are encouraging.

## 5. Current Management System

There are four criteria fundamental to the current management system: (1) the stock assessment must be consistent with all available information; (2) there must be a management structure for planning and implementing the fishery; (3) there must be a framework for establishing Total Allowable Catch (TAC); and 4) the regulatory framework must be consistent with sound

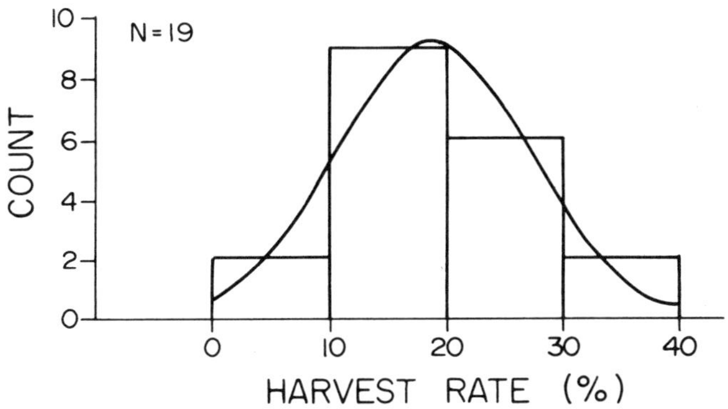


FIG. 14. Histogram of harvest rates (%) observed for 19 area-stock combinations of B.C. herring, 1986–89. A normal curve is superimposed on the histogram.

conservation principles. I begin with a discussion of criteria 2–4. A separate section (section 5) summarized the approach taken to assess stock status.

### a) Management Structure

The herring management structure of DFO (Fig. 15) consists of four interacting groups: (1) PSARC (Pacific Stock Assessment Review Committee) Herring Subcommittee, (2) Herring Management Working Group, (3) Herring Industry Advisory Board, and (4) In-season DFO Management Teams.

The annual cycle is as follows (Fig. 15). Beginning with the fishery, data are collected, stocks are assessessed and forecasts of potential catches are made. The PSARC Herring Subcommittee reviews analytical results, identifies stock trends, biological characteristics, and trends in field observations to determine stock status and harvestable surplus. The Herring Management Working Group takes this information and drafts the various fishing plans. The Herring Industry Advisory Board (HIAB) reviews the previous fishery results, proposed management plans, and provides input to the final fishing plans. In-season management of the herring roe fishery is similar among areas and involves proximal assessment with echo sounding, test sets for roe content, tagging and biological sampling for stock characteristics. When roe content reaches an acceptable level, fishing to annually adjusted District quotas occurs. In-season management controls include restrictions on fishing periods, enforcement of mesh sizes, and managing to quota for each gear type. The fishery is reviewed by Department staff and pertinent details of its success or failure are entered into a record of management strategies.

The process formalized in 1982–83 is overseen by the Regional Herring Coordinator. The key to the process is that the Coordinator must seek a consensus from the many perspectives of DFO and industry. A significant contribution to this consensus system of management is a permanent record of planning and decision making. This historical record of management forms an integral part of later management decisions and helps in the development of formal management protocols. A more detailed description of the four groups and their activities follows.

The Herring Stock Assessment Subcommittee of PSARC provides biological advice consistent with sound conservation principles, and a view to optimizing production of herring stocks (Stocker 1987). The subcommittee's chair rotates every 2 years between a member of DFO's Biological Science Branch and a member of the Fisheries Branch. The subcommittee is composed of scientific staff from the Herring Section of the Pacific Biological Station, management biologists and Fisheries Officers from Districts, the Herring Coordinator, an economist from the Program Planning and Economics Branch, and a senior program/policy advisor from Ottawa. DFO also invites representatives from industry to participate as observers. The participation of industry observers has been a very significant and positive step in informing industry about how assessments are done and all the factors that are taken into consideration in deriving stock status. The subcommittee meets in early September to review stock assessment documents, to discuss herring stock status, and make recommendations of TACs on sound biological grounds. The subcommittee report is reviewed by the PSARC Steering Committee and forwarded to the Regional Fisheries Resource Management Executive Committee in the form of a biological advisory document for approval.

The Herring Management Working Group (HMWG) is the Departmental group charged with developing

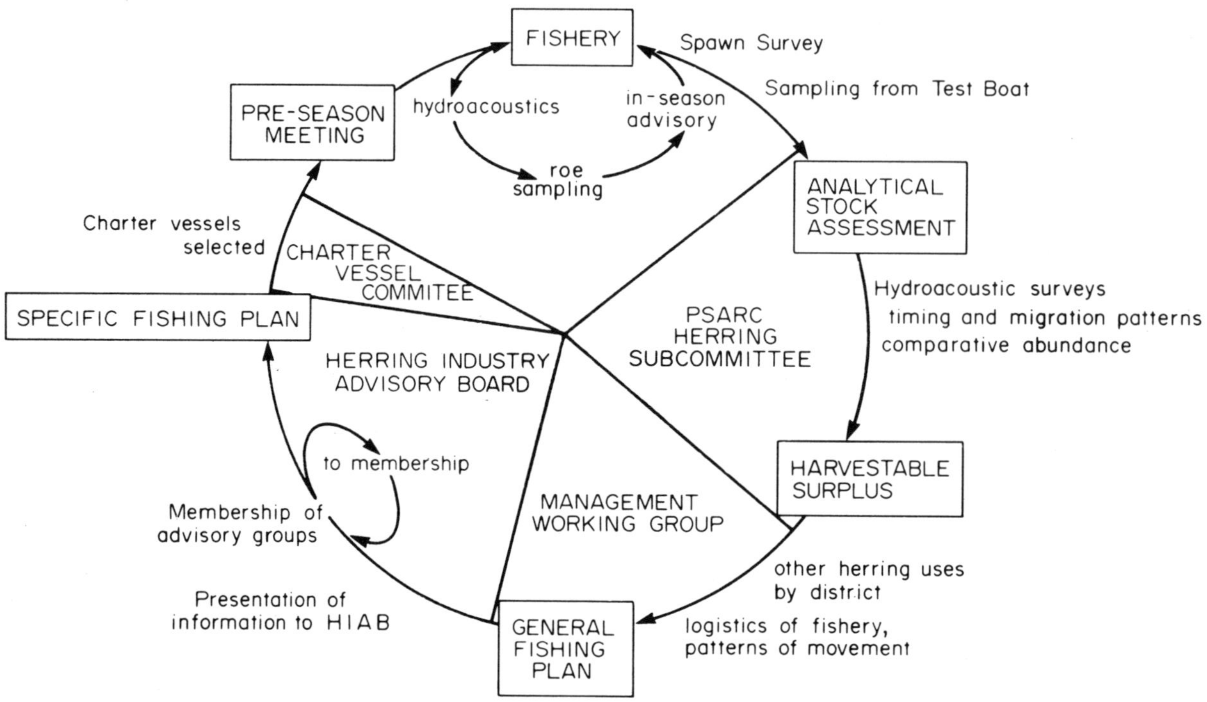


FIG. 15. Structure of herring management in British Columbia (modified from Farlinger 1986).

policy for managing herring fisheries. The group chaired by the Herring Coordinator consists of key District personnel and licensing personnel from Fisheries Branch, an economist from the Program Planning and Economics Branch, and scientists from the Pacific Biological Station. DFO also invites representatives from industry to participate as observers. The group is charged with resource allocation, licensing, policy development, in-season assessment and management, reviewing regulations, and communications with the fishing industry.

The HMWG meets in September following the PSARC review process. The purpose of the meeting is to review the approved biological advisory document, allocate total allowable catches by fishery, and develop options for fishing plans. In general herring is allocated to their most valuable use. Throughout the 1980's this policy has resulted in allocating the majority of the allowable catch to the roe herring fishery. For most of the license areas allocations are made to minor fisheries such as the spawn-on-kelp fishery, the winter food and bait fishery, and charity fisheries, with the balance of the quota being allocated to the roe herring fishery. For example, the 1989/90 season's coastwide allocation (%) of total allowable catch to herring fisheries was:

| | |
|---|---|
| Roe Herring | 82.9 |
| Spawn-on-Kelp | 5.7 |
| Food and Bait | 3.8 |
| Commercial Bait | 1.8 |
| Sports Bait | 1.3 |
| Personal Use, Charity Zoo/Aquarium, etc. | 4.5 |

The HMWG works out options for license area quotas for the roe net fisheries. These quotas are allocated by adhering to a coastwide 55%/45% seine/gillnet catch split. When working out options each coastal license area is given consideration for its suitability to each gear type. Options are reviewed by DFO Senior Management and presented to the Herring Industry Advisory Board (HIAB). The fishing plans for the food and bait, roe herring, and spawn-on-kelp fisheries are completed after consultation with HIAB. The "Roe Herring Fishery Management Plan" describes the licensing, inspection and management system employed for the upcoming fishing season.

A number of Committees with specialized tasks report to the HMWG: the Herring Charter Committee, the Spawn Committee, the Sounder/Sonar Committee.

The Herring Charter Committee selects seine charter vessels to conduct echo soundings, herring roe test fishing, tagging and research programs. These operations are conducted throughout most of the B.C. coast. The charters begin prior to and continue through the herring roe fishing season. Charter vessels operating in each license area determine herring abundance and roe maturity conditions in support of roe fishery management. Payment of the charter program is made in roe herring. In 1985 a Spawn Committee was struck. The committee's function is to ensure continuity in spawn survey data, to coordinate spawn surveys regionally, and to develop guidelines for surface surveys. The Sounder/Sonar Committee has a regional responsibility to review all proposals involving the use of hydroacoustics. This committee meets regularly to review current advances in hydroacoustics and to discuss results of projects under way. At the present time DFO's acoustic equipment is equal to that of the fishing fleet. Continual upgrading of the acoustic equipment (sounders and sonars) is required to maintain this capability.

DFO has developed a consultative process to assist in the planning and management of herring fisheries. The Herring Industry Advisory Board (HIAB) is the consultative body. There are 15 member organizations that make up the Herring Industry Advisory Board (HIAB). The organizations represent native and non-native commercial fishermen, the processing industry, and the Provincial Ministry of Agriculture and Fisheries.

DFO sets up meetings with HIAB chaired by the Herring Coordinator. DFO representatives include District staff, Management Biologists, and scientific staff from the Pacific Biological Station. Meetings in the spring are held to review the conduct of the year's roe herring fishery. The meetings also provide a forum for industry to voice any concerns they may have on any issue related to managing the herring resource. Meetings in the fall are part of the planning process for next season's fisheries. At this time the biological research being conducted in the region is also summarized to keep industry informed. Industry advisors are asked to comment on management options worked out by the HMWG, and recommend preferred options to DFO. Thoroughly prepared minutes of meetings provide the medium of accountability for HIAB's efforts. The minutes also communicate conclusions and advice to Senior Management.

HIAB member groups nominate in-season on-the-ground advisors. These advisors assist DFO staff in executing roe herring fisheries on the grounds in a manner that respects both the resource and provides quality fisheries for industry.

Pre-fishery public meetings take place in Prince Rupert and Nanaimo in February, and lead into in-season roe herring fishery operations (Fig. 15). The purpose of these meetings is to update seine and gillnet fishermen, and members of the processing industry on the upcoming roe herring fisheries. DFO staff review quotas, opening procedures, safety procedures, regulations, policies, and pertinent research data.

Successful in-season management of roe herring fisheries depends on the District field manager getting timely and accurate information on stock movement, abundance, and roe maturity (Armstrong 1986). In a license area, the required information is gathered by patrol vessels and chartered seine vessels. A DFO vessel is the main management platform from which the field manager coordinates the operations of these vessels.

Normally there are two chartered seiners in a license area to locate pre-spawning concentrations of herring, estimate biomass through echo sounding, and determine roe yields. The charter operation begins two weeks or so before anticipated fisheries for a duration of about 26 days. Test sets are taken to sample fish length frequency distributions and to monitor roe maturity. Results of these tests are communicated to the fishing fleet by the District fishery manager and to Regional headquarters on a daily basis from the coordinating management vessel. These data provide information on where and when fisheries are likely to take place.

Once the roe yield, the proportion of ripe roe in the catch, exceeds 10% an opening becomes imminent. At this time on-ground industry advisors may be called to meet with the District management team on the coordinating vessel. The purpose of the meeting is to discuss roe maturity, roe quality, availability of stocks, and proposed fishing boundaries and times. At the conclusion of the meeting, the results are communicated to the local fishing fleet. Once an opening is announced fishing to quota takes place.

The field manager also coordinates the enforcement team. The responsibilities of the enforcement team are: checking licenses, ensuring orderly openings and closures, enforcing net regulations, and assisting in the collection of hailed catch, which is used to determine when the quota has been reached.

In view of the importance of in-season management of roe herring fisheries DFO devised a recording system termed "Record of Management Strategies". These documents provide a written record of the management strategies, the reasons and considerations affecting decisions, descriptions of fisheries, and results of the management process. The records also provide a basis for post-season evaluation of the management of a fishery, and a mechanism for transferring knowledge and information to current and future DFO staff.

### b) Decision Rules for TACs

The current herring fishery management objective is based on a fixed harvest rate of 20% with a predetermined CUTOFF level. CUTOFF is defined as a level of biomass for each of the major migratory stocks below which no catch is recommended. In 1982 we adopted, for the 1983 season, a conservative fixed exploitation rate harvest policy of 20% of forecast run size (Stocker et al. 1983). It was anticipated that such low exploitation rates would reduce variability in catch, and perhaps dampen any large fluctuations due to dome shaped stock-recruitment relationships. The latter point follows from the notion that natural fluctuations may be greatly exaggerated if we fish harder following the appearance of a strong year-class. In 1985 a safety measure to protect endangered stocks was introduced in the form of a CUTOFF policy. CUTOFFs should minimize the risk of reaching very low biomass where reproductive potential is severely limited, and from which the stock requires a long time to rebuild (Haist et al. 1986).

CUTOFFS have been defined using a two-state variable dynamic model (Haist et al. 1986). The production relationship was estimated from stock-recruitment observations derived from stock reconstructions. The unfished equilibria were determined for each stock through simulations. A critical biomass is arbitrarily set at 25% of the unfished equilibrium. The CUTOFF policy, in effect since 1986, stipulates that no fishing occurs if the forecast stock biomass level is below the critical stock biomass. This approach provides a substantial unfished reserve stock to protect against accidental overfishing and drastic declines that occur naturally (ie. El Nino events). CUTOFF has been invoked in the south coast of B.C. in 1986, and in the Queen Charlotte Islands in 1988.

The CUTOFF policy provides a documented method for a preseason justification of a fishing closure (Hall 1986). The policy insures a biomass buffer against overharvesting. For the seven stock assessment regions the CUTOFF levels are (Haist and Schweigert 1990):

| | |
|---|---|
| Queen Charlotte Islands | 10,600 t |
| Prince Rupert District | 12,100 t |
| Central Coast | 10,600 t |
| Strait of Georgia — north & | |
| Strait of Georgia — south | 22,100 t |
| W. C. Vancouver Island — south | 15,100 t |
| W. C. Vancouver Island — north | 8,700 t |

These biomass levels should be high enough to allow large population growth for quick stock rebuilding. The exact population level defining the CUTOFF is a matter of acceptable probabilities, rather than guarantees. Thus the levels defined here only serve as a guide to establish quotas for future herring seasons.

For stocks which are marginally above CUTOFF a reduced catch level of run size less CUTOFF is utilized. This will provide for smaller fisheries for stocks where a 20% harvest rate would bring the escapement to levels below the CUTOFF. If stocks are forecast below CUTOFF, no fishery is recommended.

### c) Regulations

The basic statute under which regulations for fisheries management are made is the Fisheries Act. Specific regulations respecting fishing in the Province of British Columbia are set out in the British Columbia Fishery Regulations. The statute regulating fishing for herring in Canadian waters on the Pacific coast is set out in the Pacific Herring Fishery Regulations. These regulations outline fishing gear restrictions, terms and conditions of issuing licenses, and provisions for respecting the conservation and protection of herring and their spawning grounds.

The Fish Inspection Act provides for control of the quality of herring fishery product. All vessels used in the roe herring fishery must meet the constructional and operational requirements of the Fish Inspection Regulations.

Area licensing is an integral part of the herring management system. Area licensing was introduced in response to management problems associated with the large roe herring fleet, on an experimental basis in 1981. Under the program, each holder of a roe herring license was able to obtain a license for one of three predesignated geographical areas on the B.C. coast. The three original areas remained in effect until 1984. The closure of the west coast Vancouver Island area in 1985, prompted redefinition of areas as a condition of each fishing licence. The area licensing program has generally prevented large concentrations of boats at individual openings, and greatly reduced manoeuvring between coastal locations for anticipated openings. In sum, it is believed that area licensing has improved the management of the roe herring fishery.

## 7. Future Prospects

There are many challenges ahead to maintain the exceptionally valuable Pacific herring resource. Management of this resource is critical, as overfishing, biological factors, changing environmental conditions, and environmental degradation can cause collapse of the stocks. Herring fisheries management leaves much scope for improvement. The first requirement for herring fisheries management is knowledge about the species. Science must be the foundation of managing herring fisheries aimed at benefiting the entire industry. We must ensure the continuation and improvement of our data base upon which herring resource management is based. We must also ensure that scientific assessments and advice continues to be received by industry with a high degree of confidence. To this end fisheries scientists and managers must work on the fishing grounds alongside fishermen, and continue with the well established consultative process. There are a number of pressing issues that affect the current herring management system. These issues are the herring licensing policy, the Canada and United States herring trade dispute, and herring markets. The following comments address each of these issues.

The current herring licensing policy does not appear to meet the approval of the industry at large. We need an effective licensing policy for managing the herring resource. A new herring licensing policy must have the following objectives: (1) promote proper management and conservation, (2) ensure a stable and prosperous industry to optimize benefits to Canadians, (3) be equitable to individual fishermen, (4) avoid concentration, corporate or otherwise, of license holdings, (5) preserve native participation (6) ensure licenses are held by bona fide fishermen, and (7) must be simple to administer and enforce. In practical terms, such a policy would reduce fleet size, increase harvesting efficiency, and improve management through correcting difficulties caused by excessive fishing gear on the grounds.

The lifting of export restrictions and landing requirements on herring create uncertainty about future data collection. Under these changes we must make sure that we continue to provide conservation, development and sustained economic utilization of the herring resource. Reports of all herring landings, and access to landings for biological sampling must be ensured in perpetuity.

Currently herring is harvested primarily for its roe for the Japanese Kazunoko market. This single market dependency for B.C. roe products is risky, even though its quality is highly-favoured by Japanese consumers. The markets are changing as alternative products are introduced that de-emphasize the Kazunoko product. These alternative products can be supplied by competitors. The industry must seek new markets for herring products if the resource is to provide high levels of social and economic benefits to Canadians in the future.


## Acknowledgements

I thank Vivian Haist, Dan Ware, Lloyd Webb, and Doug Hay for their review of an earlier draft of the manuscript. Their many comments were useful and appreciated. Two anonymous referees have made many valuable comments to improve the manuscript. I also thank L.S. Parsons for his continuous encouragement throughout the project.


## References


ALDERDICE, D.F., AND A.S. HOURSTON. 1985. Factors influencing development and survival of Pacific herring (*Clupea harengus pallasi*) eggs and larvae to beginning of exogenous feeding. Can. J. Fish. Aquat. Sci. 42 (Suppl. 1): 56–68.

ARMSTRONG, R.W. 1986. The 1986 roe herring charter vessel monitoring and sampling program. Can. Ind. Rep. Fish. Aquat. Sci. 172: 110 p.

ARAI, M.N., AND D.E. HAY. 1982. Predation by medusae on Pacific herring (*Clupea harengus pallasi*) larvae. Can. J. Fish. Aquat. Sci. 39: 1537–1540.

BIGG, M.A. 1985. Status of the Stellar sea lion (*Eumetopias jubatus*) in British Columbia. Can. Spec. Publ. Fish. Aquat. Sci. 77: 20 p.

BLAXTER, J.H.S. 1985. The herring: a successful species? Can. J. Fish Aquat. Sci. 42 (Suppl. 1): 21–30.

CHALMERS, D.D. 1989. Review of the 1987–88 British Columbia herring fishery and spawn abundance. Can. Ind. Rep. Fish. Aquat. Sci. 198: 83 p.

FARLINGER, S. 1986. Herring management in British Columbia, p. 11–19. *In* C.W. Haegele [ed.] Proceedings of the fifth Pacific coast herring workshop, October 29–30, 1985. Can. Manuscr. Rep. Fish. Aquat. Sci. 1871.

FOURNIER, D., AND C.P. ARCHIBALD. 1982. A general theory for analyzing catch at age data. Can. J. Fish. Aquat. Sci. 39: 1195–1207.

GRANT, S.W., AND F.M. UTTER. 1984. Biochemical population genetics of Pacific herring (*Clupea pallasi*). Can. J. Fish. Aquat. Sci. 41: 856–864.

GROSSE, D.J., AND D.E. HAY. 1988. Pacific herring, *Clupea harengus pallasi*, in the northeast Pacific and Bering Sea, p. 34–54. *In* N.J. Wilimovsky, L.S. Incze, and S.J. Westrheim [eds.] Species synopsis life histories of selected fish and shellfish of the northeast Pacific and Bering Sea. Washington Sea Grant Publication, University of Washington, Seattle, WA. 88-2.

HAEGELE, C.W., R.D. HUMPHREYS, AND A.S. HOURSTON. 1981. Distribution of eggs by depth and vegetation type in Pacific herring (*Clupea harengus pallasi*) spawnings in southern British Columbia. Can. J. Fish. Aquat. Sci. 38: 381-386.

HAEGELE, C.W., AND J.F. SCHWEIGERT. 1985. Distribution and characteristics of herring spawning grounds and description of spawning behaviour. Can. J. Fish. Aquat. Sci. 42 (Suppl. 1): 39–55.

1989. Egg loss from Pacific herring spawns in Barkley Sound in 1988. Can. Manuscr. Rep. Fish. Aquat. Sci. 2037: 37 p.

HAEGELE, C.W., C.E. TURNER, L. HOP WO, AND D.C. MILLER. 1982. Tagging of herring in British Columbia during the 1980–1981 herring season. Can. Ind. Rep. Fish. Aquat. Sci. 132: 95 p.

HAIST, V., AND L. ROSENFELD. 1988. Definitions and codings of Localities, Sections, and Assessment Regions for British Columbia herring data. Can. Manuscr. Rep. Fish. Aquat. Sci. 1994: 123 p.

HAIST, V., AND J.F. SCHWEIGERT. 1989. Stock assessments for British Columbia herring in 1988 and forecasts of the potential catch in 1989. Can. Manuscr. Rep. Fish. Aquat. Sci. 2028: 48 p.

1990. Stock assessments for British Columbia herring in 1989 and forecasts of the potential catch in 1990. Can. Manuscr. Rep. Fish. Aquat. Sci.: 2049: 62 p.

HAIST, V., J.F. SCHWEIGERT, AND D. FOURNIER. 1987. Stock assessments for British Columbia herring in 1986 and forecasts of the potential catch in 1987. Can. Manuscr. Rep. Fish. Aquat. Sci. 1929: 63 p.

1988. Stock assessments for British Columbia herring in 1987 and forecasts of the potential catch in 1988. Can. Manuscr. Rep. Fish. Aquat. Sci. 1990: 63 p.

HAIST, V., J.F. SCHWEIGERT, AND M. STOCKER. 1986. Stock assessments for British Columbia herring in 1985 and forecasts of the potential catch in 1986. Can. Manuscr. Rep. Fish. Aquat. Sci. 1889: 48 p.

HAIST, V., AND M. STOCKER. 1984. Stock assessments for British Columbia herring in 1983 and forecasts of the potential catch in 1984. Can. Manuscr. Rep. Fish. Aquat. Sci. 1751: 50 p.

1985. Growth and maturation of Pacific herring (*Clupea harengus pallasi*) in the Strait of Georgia. Can. J. Fish. Aquat. Sci. 42 (Suppl. 1): 138–146.

HAIST, V., M. STOCKER, AND J.F. SCHWEIGERT. 1985. Stock assessments for British Columbia herring in 1984 and forecasts of the potential catch in 1985. Can. Tech. Rep. Fish. Aquat. Sci. 1365: 53 p.

HALL, D.L. 1986. Alternative harvest strategies for Pacific herring. M.Sc. Thesis, University of British Columbia, B.C., 83 p.

HAMER, L. 1989. Procedures for collecting and processing British Columbia herring samples. Can. Manuscr. Rep. Fish. Aquat. Sci. 2030: 27 p.

HAMER, L., AND V. HAIST. 1989. Summary of British Columbia herring sampling data for the 1987–88 season. Can. Data Rep. Fish. Aquat. Sci. 727: 121 p.

HART, J.L. 1973. Pacific fishes of Canada. Fish. Res. Board Can. Bull. 180: 740 p.

HART, J.L., AND A.L. TESTER. 1934. Quantitative studies on herring spawning. Trans. Am. Fish. Soc. 63: 307–313.

HAY, D.E. 1985. Reproductive biology of Pacific herring (*Clupea harengus pallasi*). Can. J. Fish. Aquat. Sci. 42 (Suppl. 1): 111–126.

HAY, D.E., AND A.R. KRONLUND. 1987. Factors affecting the distribution, abundance, and measurement of Pacific herring (*Clupea harengus pallasi*) spawn. Can. J. Fish. Aquat. Sci. 44: 1181–1194.

HAY, D.E., P.B. MCCARTER, R. KRONLUND, AND C. ROY. 1989. Spawning areas of British Columbia herring: a review, geographical analysis and classification. Volumes I–VI. Can Manuscr. Rep. Fish. Aquat. Sci. 2019: 1140 p.

HEALEY, M.C. 1976. Herring in the diets of Pacific salmon in Georgia Strait. Fish. Res. Board Can. Manuscr. Rep. 1382: 38 p.

HOLLING, C.S. [ED.] 1978. Adaptive environmental assessment and management. John Wiley and Sons, New York, N.Y. 377 p.

HOURSTON, A.S. 1980a. The decline and recovery of Canada's Pacific herring stocks. Rap. P.-v. Réun. Cons. int. Explor. Mer 177: 143–153.

1980b. The biological aspects of management of Canada's west coast herring resource, p. 69–90. *In* B.R. Melteff and V.G. Wespestad [eds.] Proceedings of the Alaska Herring Symposium, February 1980. Alaska Sea Grant Rep. 80-4.

1981. Forecasting British Columbia herring stocks, 1973–1981. Can. Tech. Rep. Fish. Aquat. Sci. 1047: 43 p.

1982. Homing by Canada's west coast herring to management units and divisions as indicated by tag recoveries. Can. J. Fish. Aquat. Sci. 39: 1414–1422.

HOURSTON, A.S., AND C.W. HAEGELE. 1980. Herring on Canada's Pacific coast. Can. Spec. Publ. Fish. Aquat. Sci. 48: 23 p.

HOURSTON, A.S., AND J.M. HAMER. 1979. Definitions and codings of localities, sections, management units and divisions for British Columbia herring data. Fish. Mar. Serv. Manuscr. Rep. 1533: 91 p.

HOURSTON, A.S., D.N. OUTRAM, AND F.W. NASH. 1972. Millions of eggs and miles of spawn in British Columbia herring spawnings, 1951 to 1970 (Revised 1972). Fish. Res. Board Can. Tech. Rep. 359: 154 p.

HOURSTON, A.S., H. ROSENTHAL, AND N. STACEY. 1977. Observations on spawning behaviour of Pacific herring in captivity. Meeresforschung 25: 156–162.

HOURSTON, A.S., AND J.F. SCHWEIGERT. 1981. Procedures for assessing and forecasting British Columbia herring stocks from catch, sampling, and spawn deposition data. Can. Tech. Rep. Fish. Aquat. Sci. 1000: 58 p.

HUMPHREYS, R.D., AND C.W. HAEGELE. 1976. An evaluation of herring spawn survey techniques used in British Columbia waters. Can. Fish. Mar. Serv. Tech. Rep. 613: 142 p.

HUMPHREYS, R.D., AND A.S. HOURSTON. 1978. Sampling design for assessing the biological characteristics of British Columbia herring stocks. Fish. Mar. Serv. Tech. Rep. 848: 7 p.

JONES, B.C., AND G.H. GEEN. 1977. Food and feeding of spiny dogfish (*Squalus acanthias*) in British Columbia waters. J. Fish. Res. Board Can. 34: 2067–2078.

KETCHEN, K.S. 1986. The spiny dogfish (*Squalus acanthias*) in the Northeast Pacific and a history of its utilization. Can. Spec. Publ. Fish. Aquat. Sci. 88: 78 p.

MCGURK, M.D. 1989. Advection, diffusion and mortality of Pacific herring larvae *Clupea harengus pallasi* in Bamfield Inlet, British Columbia. Mar. Ecol. Prog. Ser. 51: 1–18.

MCGURK, M.D. 1984. Effects of delayed feeding and temperature on the age of irreversible starvation and on the rates of growth and mortality of Pacific herring larvae. Mar. Biol. 84: 13–26.

MENG, H.J., AND M. STOCKER, 1984. An evaluation of morphometrics and meristics for stock separation of Pacific herring (*Clupea harengus pallasi*). Can. J. Fish. Aquat. Sci. 41: 414–422.

MYSAK, L.A., W.W. HSIEH, AND T.R. PARSONS. 1982. On the relationship between interannual baroclinic waves and fish populations in the Northeast Pacific. Biol. Oceano. 2: 63–103.

OLESIUK, P.F., M.A. BIGG, G.M. ELLIS, S.J. CROCKFORD, AND R.J. WIGEN. 1990. An assessment of the feeding habits of harbour seals (*Phoca vitulina*) in the Strait of Georgia, British Columbia, based on scat analysis. Can. Tech. Rep. Fish. Aquat. Sci. 1730: 135 p.

OUTRAM, D.N., AND C.W. HAEGELE. 1972. Food of Pacific hake (*Merluccius productus*) on an offshore bank southwest of Vancouver Island, British Columbia. J. Fish. Res. Board Can. 29: 1792–1795.

OUTRAM, D.N., AND F.H.C. TAYLOR. 1964. A quantitative estimate of the number of Pacific herring in a spawning population. J. Fish. Res. Board Can. 21: 1317–1319.

PEARSE, P.H. 1982. Turning the tide: a new policy for Canada's Pacific fisheries. The Commission on Pacific Fisheries Policy, Vancouver, Canada. 292 p.

PURCELL, J.E. 1986. Jellyfish as predators of larval herring at spawning grounds in British Columbia. p. 139–140. *In* C.W. Haegele [ed.] Proceedings of the fifth Pacific coast herring workshop, October 29–30, 1985. Can. Manuscr. Rep. Fish. Aquat. Sci. 1871.

ROBINSON, S.M.C., AND D.M. WARE. 1988. Ontogenetic development of growth rates in larval Pacific herring, *Clupea harengus pallasi*, measured with RNA-DNA ratios in the Strait of Georgia, British Columbia. Can. J. Fish. Aquat. Sci. 45: 1422–1429.

SCHWEIGERT, J.F. 1981. Pattern recognition of morphometric and meristic characters as a basis for herring stock identification. Can. Tech. Rep. Fish. Aquat. Sci. 1021: 13 p.

1990. Comparison of morphometric and meristic data against truss networks for describing Pacific herring stocks. Am. Fish. Soc. Spec. Publ. 7: 47–62.

1991. Multivariate description of Pacific herring (*Clupea harengus pallasi*) stocks from size and age information. Can. J. Fish. Aquat. Sci. 48: 2365–2376.

SCHWEIGERT, J.F., C.W. HAEGELE, and M. STOCKER. 1985. Optimizing sampling design for herring spawn surveys in the Strait of Georgia, B.C. Can. J. Fish. Aquat. Sci. 42: 1806–1814.

1990. Evaluation of sampling strategies for assessing Pacific herring spawn deposition using SCUBA surveys. N. Am. J. Fish. Manage. 10: 185–195.

SCHWEIGERT, J.F., AND J.R. SIBERT. 1983. Optimizing survey design for determining age structure of fish stocks: an example from British Columbia Pacific herring (*Clupea harengus pallasi*). Can. J. Fish. Aquat. Sci. 40: 588–597.

SCHWEIGERT, J.F., AND M. STOCKER. 1988. Escapement model for estimating Pacific herring stock size from spawn survey data and its management implications. N. Am. J. Fish. Manage. 8: 63–74.

SCHWEIGERT, J.F., AND R.E. WITHLER. 1990. Genetic differentiation of Pacific herring based on enzyme electrophoresis and mitochondrial DNA analysis. Am. Fish. Soc. Spec. Publ. 7: 459–469.

SHAW, W., R. TANASICHUCK, D.M. WARE, AND G.A. MCFARLANE. 1989a. Biological and species interaction survey of Pacific hake, sablefish, spiny dogfish and Pacific herring off the southwest coast of Vancouver Island, August 1987. Can. Manuscr. Rep. Fish. Aquat. Sci. 2027: 149 p.

1989b. Biological and species interaction survey of Pacific hake, sablefish, spiny dogfish and Pacific herring off the southwest coast of Vancouver Island. F/V CALEDONIAN, August

12–25, 1986. Can. Manuscr. Rep. Fish. Aquat. Sci. 2012: 134 p.

SHAW, W., R. TANASICHUCK, D.M. WARE, D. DAVENPORT, AND G.A. MCFARLANE. 1987. Biological and species interaction survey of Pacific hake, sablefish, spiny dogfish and Pacific herring off the southwest coast of Vancouver Island. F/V EASTWARD HO, August 10–22, 1985. Can. Data Rep. Fish. Aquat. Sci. 651: 159 p.

STACEY, N.E., AND A.S. HOURSTON. 1982. Spawning and feeding behaviour of captive Pacific herring, *Clupea harengus pallasi*. Can. J. Fish. Aquat. Sci. 39: 489–498.

STEVENSON, J.C. 1962. Distribution and survival of herring larvae (*Clupea harengus* Valenciennes) in British Columbia waters. J. Fish. Res. Board Can. 19: 735–810.

1955. The movement of herring in British Columbia waters as determined by tagging. Rapp. P.-V. Réun. Perm. Int. Explor. Mer 140: 33–34.

STEVENSON, J.C., AND D.N. OUTRAM. 1953. Results of investigation of the herring populations on the west coast and lower east coast of Vancouver Island in 1952–53. Rep. B.C. Dep. Fish. 1953: 57–84.

STOCKER, M. [EDITOR]. 1987. Pacific Stock Assessment Review Committee (PSARC) Annual Report for 1986. Can. Manuscr. Rep. Fish. Aquat. Sci. 1951: 97 p.

STOCKER, M., V. HAIST, AND D. FOURNIER. 1985. Environmental variation and recruitment of Pacific herring (*Clupea harengus pallasi*) in the Strait of Georgia. Can. J. Fish. Aquat. Sci. 42(Suppl. 1): 174–180.

1983. Stock assessments for British Columbia herring in 1982 and forecasts of the potential catch in 1983. Can. Tech. Rep. Fish. Aquat. Sci. 1158: 53 p.

STOCKER, M., AND D.J. NOAKES. 1988. Evaluating forecasting procedures for predicting Pacific herring (*Clupea harengus pallasi*) recruitment in British Columbia. Can. J. Fish. Aquat. Sci. 45: 928–935.

TAYLOR, F.H.C. 1955. The Pacific herring (*Clupea pallasii*) along the Pacific coast of Canada. Bull. Internat. North Pacific Fish. Comm. 1: 105–128.

1964. Life history and present status of British Columbia herring stocks. Bull. Fish. Res. Board Can. 143: 81 p.

1971. Offshore distribution of Pacific herring (*Clupea pallasi*) on basis of records from fur seal stomachs. J. Fish. Res. Board Can. 28: 1663–1665.

1983. The hydroacoustic assessment of herring distribution and abundance in Barkley Sound, February 18 to March 12, 1982. Can. Tech. Rep. Fish. Aquat. Sci. 1197: 106 p.

1984. Distribution and abundance of herring and other pelagic fish off the west coast of Vancouver Island in September, 1981. Can. Tech. Rep. Fish. Aquat. Sci. 1333: 43 p.

1985. Introduction to the international symposium on the biological characteristics of herring and their implication for management. Can. J. Fish. Aquat. Sci. 42(Suppl. 1): 2.

TESTER, A.L. 1935. The herring fishery of B.C. — past and present. Biol. Board Can. Bull. 47: 37 p.

1937a. Populations of herring (*Clupea pallasii*) in the coastal waters of British Columbia. J. Fish. Res. Board Can. 3: 108–144.

1937b. The length and age composition of herring (*Clupea pallasii*) in the coastal waters of British Columbia. J. Fish. Res. Board Can. 3: 145–168.

1945. Catch statistics of the B.C. herring fishery to 1943–44. Fish. Res. Board Can. Bull. 67: 47 p.

1948. The efficacy of catch limitations on regulating the British Columbia herring fishery. Trans. R. Soc. Can. 42: 135–163.

1955. Estimation of recruitment and natural mortality rate from age-composition and catch data in British Columbia herring populations. J. Fish. Res. Board Can. 12: 649–681.

TRUMBLE, R.J., AND R.D. HUMPHREYS. 1985. Management of Pacific herring (*Clupea harengus pallasi*) in the eastern Pacific Ocean. Can. J. Fish. Aquat. Sci. 42(Suppl. 1): 230–244.

WALTERS, C.J., M. STOCKER, A.V. TYLER, AND S.J. WESTRHEIM. 1986. Interactions between Pacific cod (*Gadus macrocephalus*) and herring (*Clupea harengus pallasi*) in the Hecate Strait, British Columbia. Can. J. Fish. Aquat. Sci. 43: 830–837.

WARE, D.M. 1985. Life history characteristics, reproductive value, and resilience of Pacific herring (*Clupea harengus pallasi*). Can. J. Fish. Aquat. Sci. 42(Suppl. 1): 127–137.

1990. Climate, predators and prey: behaviour of a linked oscillating system, p. 279–292. *In* T. Kawasaky et al. [eds.] Long-term variability of pelagic fish populations and their environment. Proceedings of the International Symposium. Pergamon Press.

WARE, D.M. AND G.A. MCFARLANE. 1986. Relative impact of Pacific hake, sablefish and Pacific cod on west coast of Vancouver Island herring stocks. Int. N. Pac. Fish. Comm. Bull. 47: 67–77.

1989. Fisheries production domains in the Northeast Pacific Ocean, p. 359–379. *In* R.J.

Beamish and G.A. McFarlane [eds.] Effects of ocean variability on recruitment and an evaluation of parameters used in stock assessment models. Can. Spec. Publ. Fish. Aquat. Sci. 108.

WARE, D.M., AND R.W. TANASICHUK. 1989. Biological basis of maturation and spawning waves in Pacific herring (*Clupea harengus pallasi*). Can. J. Fish. Aquat. Sci. 46: 1776–1784.

1990. A method for forecasting the timing of ripeness in Pacific herring (*Clupea harengus pallasi*) females. Can. J. Fish Aquat. Sci. 47: 2375–2379.

WESTERNHAGEN, H.V., H. ROSENTHAL, S. KERR, AND G. FURSTENBURG. 1979. Factors influencing predation of *Hyperoche medusarum* (Hyperiida: Amphipoda) on larvae of the Pacific herring *Clupea harengus pallasi*. Mar. Biol. 51: 195–201.

WESTRHEIM, S.J., AND W.R. HARLING. 1983. Principal prey species and periodicity of their incidence in stomachs of trawl-caught Pacific cod (*Gadus macrocephalus*), rock sole (*Lepidopsetta bilineata*), and petrale sole (*Eopsetta jordani*) landed in British Columbia, 1950–80. Can. Manuscr. Rep. Fish. Aquat. Sci. 1681: 38 p.

WESTRHEIM, S.J., P.B. MCCARTER, AND D.E. HAY. 1989. Stomach contents of commercially important fishes landed in British Columbia during 1982–83. I. Pacific cod (*Gadus macrocephalus*), September–November 1982. Can. Manuscr. Rep. Fish. Aquat. Sci. 2025: 69 p.

# Fish Habitat and the Impact of Human Activity with Particular Reference to Pacific Salmon

Michael Waldichuk[1]

*Department of Fisheries and Oceans*
*West Vancouver Laboratory, 4160 Marine Drive*
*West Vancouver, British Columbia, Canada V7V 1N6*

Waldichuk, M. 1993. Fish habitat and the impact of human activity with particular refence to Pacific salmon, p. 295–337. *In* L. S. Parsons and W. H. Lear [eds.] Perspectives on Canadian marine fisheries management. Can. Bull. Fish. Aquat. Sci. 226.

Fish habitat, particularly of Pacific salmon, is examined in terms of human impact on it and how it can be managed to minimize such impact. The life cycle of salmon, because they are anadromous, involves the freshwater, estuarine, coastal marine and high-seas environments, in each of which human activities can be detrimental. Some of the critical habitat variables, such as dissolved oxygen, suspended solids, temperatures and acidity, are discussed in relation to their effects on salmonids and how their alteration due to human activity can have an adverse effect on fish stocks. Among the aquatic contaminants that can affect fish stocks in British Columbia, the most serious are pulpmill effluents, especially since recent findings of dioxins and furans in fish and shellfish near the mills have led to closures of crab, prawn and shrimp fisheries in coastal waters. Logging, agriculture, mining, dams and water diversions for hydroelectric power, structural developments on streams and their estuaries and ocean dumping have also had adverse impacts on fish habitats. The egg and alevin stages of salmon are vulnerable to logging, road-building, agriculture and mining, which introduce sediments into streams that may clog spawning gravel.

Management of habitat for Pacific salmon is done largely by inference and subjective judgement based on experience. The scientific basis for developing universally-applicable models for habitat management is still inadequate, albeit site-specific and stock-specific models appear to have some utility. Among the federal and provincial legislation for protecting fish habitats, the Fisheries Act is the most effective. The federal Fisheries and Oceans policy with a no-net-loss principle of the productive capacity of fish habitats has been in effect since 1986.

---

L'auteur examine l'habitat du poisson, en particulier des saumons du Pacifique, en termes des incidences des activités humaines dont il est l'objet et des mesures de gestion visant à minimiser ces incidences. Étant donné que le saumon est une espèce anadrome, son cycle biologique se déroule en eau douce, en milieu estuarien, en eau salée, près de la côte et en haute mer, soit des environnements où les activités humaines peuvent avoir des incidences négatives. L'auteur examine certaines des variables critiques de l'habitat, dont la teneur en oxygène dissous, les matières solides en suspension, la température et l'acidité, en fonction de leurs incidences sur les salmonidés et des répercussions de leur modification, entraînée par des activités humaines, sur les stocks de poisson. Parmi les contaminants aquatiques qui peuvent affecter les stocks de poisson de la Colombie-Britannique, les plus

[1]Michael Waldichuk died on May 4, 1991, one month after submitting the final revision of this manuscript. C.D. Levings provided information to complete the publication process.

menaçants sont les effluents d'usines de pâtes, surtout depuis que la récente découverte de dioxine et de furanne chez les poissons, mollusques et crustacés recueillis près de ces usines a mené à la fermeture de la pêche côtière du crabe et de la crevette. L'exploitation forestière, l'agriculture, l'exploitation minière, la construction de barrages et de structures de divagation des eaux dans le cadre de travaux d'aménagement hydro-électrique, la construction de structures en bordure de cours d'eau et d'estuaires ainsi que les déversements de déchets en mer ont aussi eu des incidences néfastes sur les habitats du poisson. Les oeufs et les alevins du saumon sont des stades vulnérables aux conséquences de l'exploitation forestière, la construction des routes, l'agriculture et l'exploitation minière car ces activités entraînent le déversement de sédiments dans les cours d'eau où ils encrassent le gravier des frayères.

La gestion de l'habitat des saumons du Pacifique s'effectue en grande partie par déduction et jugements subjectifs basés sur l'expérience. La base scientifique nécessaire à l'élaboration de modèles de gestion de l'habitat universellement applicables est encore inadéquate, bien que des modèles particuliers à certains sites et certains stocks semblent être utiles dans une certaine mesure. La Loi sur les pêches est la plus efficace parmi les instruments législatifs provinciaux et fédéraux visant la protection de l'habitat du poisson. La politique du ministère fédéral des Pêches et des Océans, qui vise aucune perte nette de la capacité de production des habitats du poisson, est en vigueur depuis 1986.

## 1. Introduction

The Oxford Dictionary defines "Habitat" as the "Natural home of plant or animal". According to the Canadian federal Fisheries Act (Section 31(5)), the term "fish habitat" covers a great variety of environments where fish live and defines them as: "Spawning grounds and nursery, rearing, food supply and migration areas on which fish depend, directly or indirectly, in order to carry out their life processes." This means that fish habitats include not only the waters in rivers, lakes, streams, estuaries and the oceans, but also the quality of these waters. It means also that the total surroundings in which plants and other life forms interact to make fish life possible are an integral part of fish habitat. Human activities can have adverse impacts on any one of the components of fish habitat including water, bottom, shoreline, vegetation and organisms that serve as food for the fish.

In this chapter, an effort will be made to examine the various products of human activities that affect the quality of the water in which fish live at various stages in their life cycle. Developments that destroy or unfavourably modify fish habitats will be discussed. Emphasis will be placed on Pacific salmon not only because they represent a major fisheries resources in the Pacific Region of Canada, but also because their life cycle includes habitats in freshwater, estuaries and the marine environment. Some critical aspects of habitats for Pacific salmon will also be relevant for other species.

An important objective of this chapter is to review fish habitat management in Canada, as carried out by the Department of Fisheries and Oceans under the Fisheries Act. Again, this will be done particularly with reference to Pacific salmon. Section 35(1) of the Fisheries Act states: "No person shall carry on any work or undertaking that results in the harmful alteration, disruption or destruction of fish habitat." The Fisheries Act provides a powerful weapon for enforcement of the various requirements to ensure that fish habitat is protected. Fish habitat managers can take advantage of this legislation in their role of preventing harmful alteration, disruption or destruction of fish habitat. The no-net-loss policy, with respect to fish habitat, has been developed recently by the Department of Fisheries and Oceans, and if properly implemented, will help to stem the loss of vital fish habitat through human developments. A great deal of knowledge and experience is required to recognize when habitat damage will occur. Some of the approaches to such assessment will be discussed.

There is a diverse literature on the impact of human activities on fish habitats scattered in numerous reports and many scientific journals. Results of studies on the effects of habitat changes on salmonid stocks, primarily in Canada, are given in the Proceedings of a recent national workshop on that subject (Levings et al. 1989).

## 2. The Life Cycle of Pacific Salmon and Habitat Requirements in Each Life Stage

The Pacific salmon, *Oncorhynchus* spp., are anadromous and spend portions of their life cycle in freshwater, estuaries, coastal waters and the high seas. The impacts of human activities on salmon are recognized as being most serious in the freshwater environment, where salmon spawn and spend their early life. Natural and human factors can also affect salmon adversely, but in the other habitats frequented by them during their life cycle. The magnitude of the effects of human activities, except harvesting, becomes less well known as one leaves the spawning grounds. On the high seas, natural oceanographic variations may have a large effect on survival of juvenile salmon and on recruitment to the fishery, but the cause-effect relationship between oceanographic variables and salmon survival is still poorly understood. The causes and effects of habitat damage by human activity in different stages of the life cycle of Pacific salmon are shown in Table 1. The life cycle of Pacific salmon, with different stages when major natural and man-made impacts occur, is schematically illustrated in Fig. 1 (DFO 1985a). The vital life processes that can be affected in each life stage of salmon and the effects of pollutants on the life processes are shown in Fig. 2.

## 2.1. The Life Cycle of Pacific Salmon and Sea-Run Trout

The life histories of Pacific salmon and sea-run trout are well described by Hart (1973). The following is a brief summary of their life cycle, particularly with reference to the impact of human activities.

Pacific salmon in the Pacific Region of Canada are represented by five species of the genus *Oncorhynchus*: sockeye, *O. nerka*; pinks, *O. gorbuscha*; chinook, *O. tshawytscha*; coho, *O. kisutch*; and chum, *O. keta*. In addition to these salmonids, there are two species of sea-run trout: steelhead, *O. mykiss* (formerly *Salmo gairdneri*) and cutthroat, *O. clarki* (formerly *Salmo clarki*). All are anadromous, that is, they spawn in freshwater and spend most of their adolescent and adult life at sea. They differ, however, in the amount of time they spend in freshwater after their fry emerge from the gravel on the spawning grounds and before proceeding to sea. In general, pink fry spend only a few days in freshwater, chum from a few days to a few months, sockeye spend a year in lakes, chinook from 3 months to a year (sometimes 2 years) in rivers, coho spend a year or sometimes 2 years in rivers and smaller streams, steelhead spend a year or two in streams, and coastal cutthroat may spend 1–5 years in streams before moving to the estuaries and nearshore coastal waters. These salmonids differ also in the length of their life cycle, ranging from 2 years for pinks to as much as 7 years for chinook, and in the amount of time they spend at sea before returning to their spawning grounds in freshwater. All species of Pacific salmon, however, are distributed widely in the Northeast Pacific in their marine phase, covering the area from the British Columbia coast to the International Date Line (180°W longitude) and from the latitude of the Oregon/California border (42°N latitude) to the Gulf of Alaska (Hart 1973). This is true also for steelhead but not so for coastal cutthroat, which spend most of their marine life in the estuary of their native stream or along the coast within 15 km of such estuary.

## 2.2. Environments in Different Life Stages of Pacific Salmon and Human Impacts in Each

Because of their anadromous nature, Pacific salmon move through environments with widely differing physical and chemical characteristics. These characteristics range from low dissolved constituents in the freshwater environment, which can undergo wide temperature fluctuations from near 0°C to 25°C, to a 3.5% salt content in seawater with a comparatively narrow temperature range of 8–15°C. The transition occurs in the estuaries, and the stresses of the changing characteristics of the water add to the stresses of human activity in the estuarine environment. In the following, the different environments through which salmon move in fulfilling their life cycle will be discussed.

### *2.2.1. Freshwater*

The freshwater environment is where adult salmon spawn, the eggs hatch, the salmon larvae or alevins in the gravel grow as they feed on the yolk-sac, then move up through the gravel to become free-swimming fry. They remain in the stream or lake environment from a few days to two or more years, depending on the species, before proceeding to sea. It is in the freshwater environment where the largest seasonal and in-season changes can occur. The effects of large water flow fluctuations, temperature changes, and sedimentation can have the most severe impacts on the most sensitive stages of salmon, the eggs and alevins. Often the size of a run of salmon is determined by conditions in the freshwater environment during the periods of upstream adult migration, spawning, incubation, alevin feeding, fry emergence, and fry feeding.

Extremely low stream flows during adult spawning migration may result in inaccessibility of the spawning grounds for a large proportion of spawners. As a consequence, large numbers of unspawned salmon may result in inadequate seeding for the next crop. If after the adults have successfully spawned, torrential rains cause heavy runoff with silt-laden water, the eggs in the gravel may become buried with mud and fine sand. The permeability of the gravel to percolating waters becomes greatly reduced. The transport of dissolved

TABLE 1. Causes and effects on fish stocks of different types of habitat damage.

| Habitat Damage | Causes | Effects on Fish Stocks |
|---|---|---|
| **Freshwater systems:** | | |
| Low flow | Logging in the watershed. Stream diversion for irrigation, industry, drinking water supply or for hydroelectric power. | Spawning grounds may have insufficient water to replenish dissolved oxygen to eggs and alevins in gravel and to remove waste products. In an extreme situation, eggs may become desiccated or freeze. Adults may have insufficient water for upstream migration. |
| High flow | Logging in the watershed. Diversion of water into the stream from other sources. | Spawning grounds may get washed out or silted in. Salmonid fry may be displaced downstream and washed out to sea prematurely. |
| High suspended solids | Logging in the watershed. Gravel washing. Poor agricultural practices. Road building and railway construction. Industrial wastes, e.g., from pulpmills. | Spawning grounds may become silted in with fine materials, resulting in reduced permeability of gravel and mortality of incubating fish eggs. Respiration in fish may be impaired by clogging of gills. Abrasion of gills may occur as a result of sharp, angular particles. |
| High temperature | Logging in the watershed. High atmospheric temperatures. | Mortality of juvenile and adult fish due to excessively high temperatures. Sublethal stress on juvenile and adult fish to high temperatures. Vulnerability of stressed fish to diseases. |
| Low dissolved oxygen | Input of sewage, organic-rich industrial wastes and over-enrichment with nutrients | Fish eggs in gravel may fail to hatch. Mature fish and juveniles may die owing to suffocation or suffer distress due to deficient oxygen. |
| Toxic conditions | Industrial waste discharges, e.g., from pulpmills and mines. Accidental spills on toxic substances, e.g., oil and anti-sapstain chemicals. Input of pesticides. | Both mature fish and juveniles may die of acute toxicity. Fish eggs and alevins exposed to highly toxic conditions on spawning grounds would die also. Both mature and young fish may suffer from sublethal effects, if exposed to sub-acute toxic conditions, and become debilitated and vulnerable to predation, disease, parasites, etc. |
| Loss of spawning ground | Logging in the watershed. Gravel removal from the stream bed. Road construction. | Adult fish go unspawned because of lack of suitable spawning ground. Any spawning that occurs may be only marginally successful or totally unsuccessful. |
| Loss of rearing habitat | Channelization of the stream. Construction and other development adjacent to the stream. | Side channels and quiescent sections of a stream may have disappeared, resulting in loss of suitable area for rearing of salmonid fry and smolts. Elimination of large boulders and wood debris renders habitat less suitable for rearing. |
| Stream blockage | Input of logging debris. Construction of dams. Landslides during road or railway building. | Upstream passage by adult salmonids is rendered impossible by dams, log and debris jams, rock slides and other impenetrable obstacles in a stream. |
| **Estuaries:** | | |
| Loss of rearing habitat | Dredging and filling during construction. | Loss of prime estuarine rearing habitat, e.g., eelgrass beds, marshes, benthic algae, eliminates or reduces area where juvenile salmonids and other species can feed. With salmonids, feeding during estuarine rearing may be critical for survival when they reach the open sea, and subsequent growth and recruitment to the fishery. |
| Low dissolved oxygen | Input of sewage and industrial waters. Dredging of anoxic sediments. | Mortality of adult and juvenile fish when dissolved oxygen level is depressed sufficiently. Distress to mature and young fish when low dissolved oxygen is at a sub-acute level — fish may become susceptible to disease and parasites and predation by birds. |

TABLE 1. (*Continued*)

| Habitat Damage | Causes | Effects on Fish Stocks |
|---|---|---|
| Toxic condition | Spills of oil or toxic chemicals from industry, ships and other sources. Dredging of high-$H_2S$ sediments. | Mortality in acute cases of high toxicity. Debilitating effect in chronic toxicity conditions, with fish rendered susceptible to predation, disease, parasites and inability to cope with other stresses, natural and man-made. |
| High suspended solids | Upstream sources, dredging locally and input of sewage and industrial wastes. | Effects are similar to those in freshwater systems, but in estuaries, flocculation helps to settle the suspended solids into the sediments. Flushing effect of the tide also removes the suspended solids from the estuary. |
| **Other coastal waters:** | | |
| Low dissolved Oxygen | As in estuaries | As in estuaries |
| Toxic condition | As in estuaries | As in estuaries |
| High suspended solids | As in estuaries | As in estuaries |
| Alteration of coastline configuration | Jetty installation for ferry terminals and coal ports and for navigational channel protection. | May block along-shore migration of fish. May also block along-shore movement of water, which is used by fish as a directional clue in their seaward or riverward migration. |
| Turbidity | Ocean dumping | Fish may avoid migration routes and spawning and rearing habitats when water is turbid. |
| Intertidal area and nearshore water littered by debris. | Solid wastes from ships blown ashore by onshore winds. | Fugitive nets could conceivably capture adult fish. |
| Oil pollution | Oil spills by vessels offshore can be blown ashore by onshore winds. Oil spills from vessels or land-based sources up coast may be transported onto valuable habitats by coastal currents. | Feeding by adult salmon and juvenile could be adversely affected. Tainting of adult salmon affects marketability of the fish. |
| Toxicity by toxic chemicals other than oil | Spills of toxic substances from vessels offshore can be blown ashore by onshore winds. Toxic substances from vessel spills or land-based sources up coast transported onto fish habitats by coastal currents. | Acute and sublethal toxicity to adult and juvenile fish and invertebrates may occur. Potential tainting of adult salmon could affect the marketability of the fish. |

oxygen to the eggs and removal of waste products may cease or become greatly curtailed. The eggs will likely die from lack of oxygen and/or toxicity from ammonia and carbon dioxide in their waste products. A similar problem may arise owing to heavy precipitation and runoff during the period of alevin feeding on the yolk-sac in the gravel. A heavy deposition of silt and sand can cause suffocation of the alevins and/or prevent emergence of the fry from the gravel. When the salmon are in the fry stage, heavy runoff and suspended solids do not normally cause suffocation of the fry, but the swift flow of the stream can displace them seaward, and in extreme cases, wash them out to sea prematurely.

Large temperature fluctuations can also have a serious impact on the embryonic and juvenile stages of salmon in streams. For example, during periods of low stream flow in winter, sub-zero temperatures can lead to freezing and heavy mortality of salmon eggs in the gravel. It is conceivable that under unusual weather conditions alevins and fry could also suffer freezing. This is not very likely to occur, however, because freeze-ups seldom occur after salmon eggs hatch in late winter or

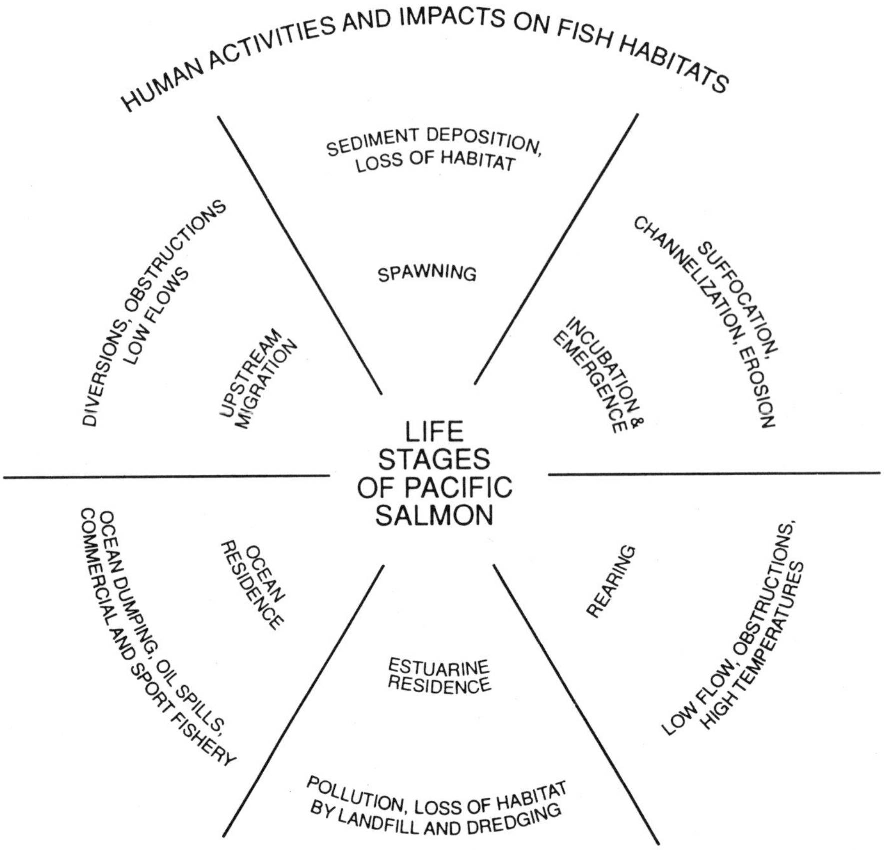


FIG. 1. Schematic diagram showing adverse impacts of human activities on the different stages in the life cycle of salmon (adapted from DFO 1985a).

early spring. Unusually high winter temperatures will usually accelerate hatching of salmon eggs. This may advance the timing of the movement of fry seaward with possibly unfavourable consequences.

An important function of salmon streams and lakes is in rearing of the fry and smolts. Suitable pools must be available for feeding of fry in streams. The shelter provided by large boulders and large wood debris is vital for their survival (Hartman and Scrivener 1990). The value of side channels and ponds for winter feeding by coho, chinook and steelhead has been recently shown to be high (Swales et al. 1986). Such rearing areas must be protected.

Sockeye lakes are vulnerable to a host of problems that can arise from human intervention, ranging from pollution to flooding for hydroelectric developments. Fortunately, sockeye lakes in British Columbia have not yet suffered severely from the kinds of degradation that could stem from a larger population and greater industrial development than exist there at present.

The foregoing conditions in the freshwater environment may occur naturally. Human activities, however, can exacerbate them and cause undesirable extremes. For example, logging in the watershed usually causes accelerated runoff following a rainstorm. Because soil is disturbed by both the logging operation and associated road building, suspended solids in runoff waters will increase. As noted in the Carnation Creek study (Hartman and Scrivener 1990), temperatures in the creek water were found to increase following logging, the time of emergence of fry was advanced, and the fry got a head start for their spring and summer feeding and growing period.

The conditions that prevail in the freshwater regime, from the time adult spawners migrate upstream to spawn to the time the progeny leave for the sea, often determine how well the salmon will prosper in the remaining stages of their life cycle. Fish that, on the one hand, have not been nourished well in the streams and lakes may not have the necessary stamina to evade predators and disease. Their rate of survival may be low. On the other hand, fish that had an adequate supply of food, and were reared in streams with ideal conditions

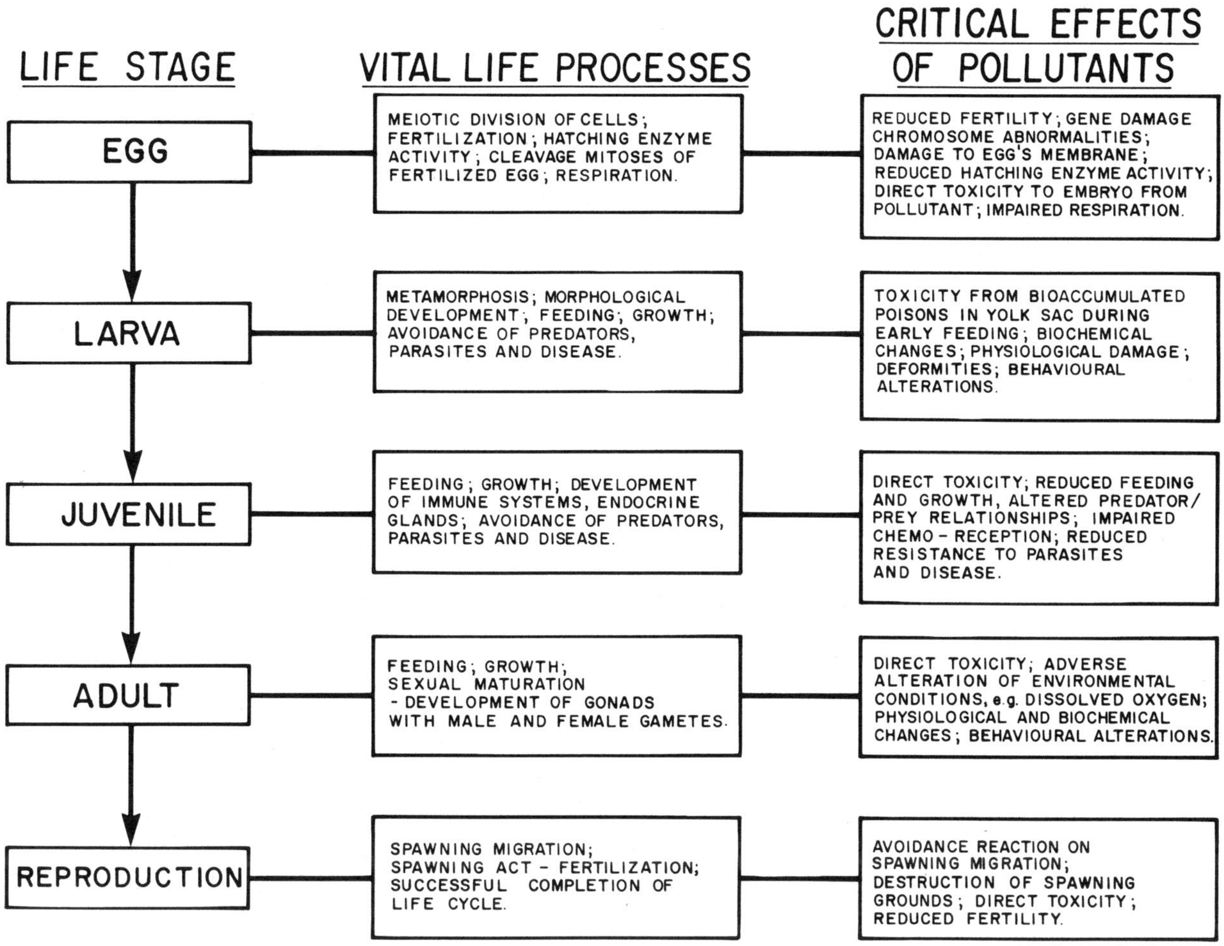


FIG. 2. Block diagram schematically representing vital life processes in different stages of the life cycle of salmon and the effects of pollutants on those processes.

for robust development and growth, will more likely cope better with any unfavourable conditions they encounter in the estuarine, coastal and high-seas environments.

### 2.3. Estuary

Next to the riverine environment, estuaries present the harshest conditions for aquatic organisms. Some ecologists say that estuaries pose harsher conditions for creatures living there than the rivers that drain into them, because they not only reflect the widely fluctuating river conditions, but also the rapid environmental changes caused by the daily tidal oscillations of the sea. Moreover, salinity now becomes a variable, and organisms can be exposed to a wide range of salinity, depending on where they are located in the estuary, the range and stage of the tide and the discharge of the river. Organisms dwelling in an estuary become adapted to these rapid and widely-ranging variations. Diversity of species may be quite low, but the number of organisms representing a given species may be large. The primary productivity of estuaries is usually high. The biomass, represented by resident populations in an estuary, can be enormous, compared to that in other non-estuarine coastal areas.

For salmonids, the value of an estuary is that it provides a nursery area where the juveniles can feed on amphipods and other invertebrates before moving out to sea. The duration of stopover in an estuary for juvenile Pacific salmon varies according to species, with very brief periods of only a few days or less for pink and some chμm juveniles, and much longer stays of up to two months or more for sockeye, coho and chinook. There are some exceptions in certain areas where pinks and chums may spend longer periods in estuaries. Eelgrass beds and freshwater marshes are preferred habitats in estuaries for salmon juveniles. The salmon not only use the attached vegetation for shelter but also feed on invertebrates that colonize the stems of the aquatic plants.

Estuaries are extremely vulnerable as sites for assorted coastal developments. On the coast of British Columbia, they often constitute the only suitable sites for construction, the remaining area usually being steep-sloped as part of the hilly and mountainous coastal terrain. Consequently, sawmill and pulpmill companies that process the timber harvested on the coastal slopes into lumber and pulp and paper products, often favour estuaries for their installations. A further attraction for such forest-product operations is the availability of cheap transportation by sea. In addition to modification of the shoreline by dredging and filling to install wharves and warehouses, there is the potential of spillage of diesel fuel oil and other toxic substances from construction equipment, and later, from vessels used for shipping products and raw materials. If a pulpmill is established, there is usually the prospect of effluents from pulp and paper production being discharged into the estuary or adjacent waters.

The Fraser River estuary, through which pass some of the biggest salmon runs in the world, is adjacent to the largest centre of population in British Columbia. Numerous industries have chosen the banks of the lower river for their operational sites. Fortunately, few of these discharge toxic wastes. But there are always the hazards of spills of petroleum hydrocarbons and other noxious substances, whenever the operations are serviced by trucks or vessels. Periodically, dredging has to be conducted in the river adjacent to these sites, and the dredged material is usually disposed of at one of the designated dumpsites, either off Point Grey or Sand Heads. Expansion of existing facilities or establishment of a new industrial or commercial operation on the shores of any one of the three estuarine channels of the Fraser River may necessitate destruction of valuable estuarine habitat by dredging or filling. The Fraser River estuary poses a special challenge in habitat management, because the resource to be protected is so valuable and the risks for habitat damage in the midst of an urban, industrial environment are so high.

The estuary of the Skeena River, also a major salmon stream in British Columbia, could also be seriously affected by development and waste disposal. Fortunately, it is well removed from any major centres of population. Prince Rupert is the nearest city, and its sphere of influence at present is just beyond the area where the Skeena River estuary could be affected. The same thing generally applies to the industry nearest to the estuary, the pulpmill on Watson Island, although this mill with its ancillary activities, such as log booming, could be a greater threat to the estuarine ecosystem than the city of Prince Rupert. A favourable feature of the oceanography of this area, from the point of view of protection of the Skeena River estuary, is that the net flow of surface water is seaward, in conformity with the normal estuarine movement. Moreover, the tidal range is comparatively large, so that tidal flushing is highly effective.

The comparatively small proportion of the British Columbia coastline that comprises estuaries, and the importance of estuaries as nursery areas for juvenile salmonids makes it extremely vital that estuaries be given the utmost protection against developmental degradation and pollution. This is where the challenges for fish habitat managers are perhaps the greatest.

### 2.4. Coastal Waters

Habitats in coastal waters are more commonplace than those in estuaries. For that reason, coastal waters cannot be regarded with the same degree of concern for their protection as estuaries. Nevertheless, there are some sections of the coast that are more important than others as habitats for salmon. They may have a larger input of nutrients and a higher productivity than other sections of the coast owing to convergences caused by the configuration of the coastline. The coastal waters along a given section of the coast, such as a point or promontory, may be in the path of the migratory route of salmon. Under such circumstances, it might be as important to protect the waters along a protruding coastline as the waters in an estuary.

Otherwise, coastal waters require a moderate degree of protection so that fish and invertebrates normally frequenting these waters are not exposed to hazardous environmental conditions, e.g., low dissolved oxygen concentrations, toxic substances and chemicals that might be bioaccumulated. A large industrial operation on the coast, indiscriminately discharging its wastes, could conceivably lead to such conditions. A coastal mining operation for sand and gravel or for minerals could have an undesirable effect in destroying coastal habitats and creating highly turbid conditions. A large oil spill that is carried ashore by wind and tidal currents could be devastating not only to coastal invertebrate populations but also to herring spawning areas. Although salmon in coastal waters would not likely be directly affected by an oil slick, the oil mixed into the water through which they swim and consumed oil-contaminated food organisms could cause tainting of the flesh.

### 2.5. High Seas

It is difficult to visualize high-seas fisheries habitats being seriously affected by human activities, except for such long-term modifications as climate change due to the greenhouse effect caused by carbon dioxide and other radiatively-active gases. There are human ecological impacts at present, however, even on the high seas. The most obvious one is the disposal of garbage, including plastics, from ships. From the point of view of fish conservation, the most serious effect stems from loss of fishing nets and the deliberate discard of damaged nets by fishermen. With the entry into force on 31 December 1988 of Annex V of the Convention on the Prevention of Pollution from Ships, 1973, as modified by the Protocol of 1978, there may be a decline in the amount of solid refuse, particularly plastics, discharged from ships. Annex V deals specifically with garbage and other solid refuse released from ships and prohibits the discharge of persistent plastics. The problem of entanglement of mammals, birds and fish in fugitive netting and monofilament line will undoubtedly continue as long as ships and fishing vessels ply the ocean. The use of drift nets on the high seas for flying squid, as was practised by Japan, Taiwan and South Korea, undoubtedly added more fugitive netting to the high seas that entangles fish and wildlife.

Other high-seas pollution problems could arise from accidental spills of oil and hazardous chemicals carried by ships. Eventually, deep-sea mining for metals from crusts and manganese nodules could pose a pollution problem, caused by the dredging operation for minerals as well as by tailings disposal from deep-sea concentrator ships.

Deep-sea habitat problems induced by human activities are expected to remain small compared to the natural year-to-year variations in ocean conditions. Years of a major El Niño event, such as the one in 1982–83 for example, will continue to exhibit far greater natural environmental changes, with dramatic effects on fisheries, than anything envisaged due to human activity.

## 3. Critical Environmental Characteristics for Survival of Fish

The aquatic environment has certain natural properties that are vital for the survival of fish. These properties can be altered by the introduction of contaminants or by certain changes in the physical nature of a body of water, e.g., restriction of flow causing reduced rate of replacement of water. Fish can usually tolerate small changes in these aquatic characteristics without serious ill effects. But large changes can lead to mortality or debilitating sub-lethal effects. In the following a few of the major environmental variables are briefly discussed in terms of their implications for fish.

### 3.1. Dissolved Oxygen

Aquatic animals require oxygen for respiration, and ultimately for survival, in the same way that land animals do. Whether they get the oxygen directly from the atmosphere, as in the case of air-breathing marine mammals, or from the water through their gills in the fashion of fish and invertebrates, animals cannot survive for long without oxygen. Normally, uncontaminated water contains dissolved in it a plentiful supply of oxygen to meet the needs of living animal organisms. The amount of oxygen that can be dissolved in a given volume of water decreases as the temperature rises. It also decreases with increasing salinity of the water, so that more oxygen can dissolve in freshwater than in seawater. In some waters, particularly in coastal inlets, heavy input of natural organic material coupled with the lack of regular replacement of the water, leads to oxygen depletion. This commonly occurs in the deep water of marine inlets with a shallow threshold sill that are highly stratified with freshwater inflow at the surface. It may also occur in lakes that are stratified by temperature and salt in bottom water and do not undergo an annual turnover with replacement of bottom water. Side channels of rivers, where the flow-through of water is not rapid and/or there is intrusion of seawater, may also experience a natural oxygen deficiency in bottom waters.

Dissolved oxygen deficiency can be caused by human activities in both the freshwater and marine environments. The input of biodegradable organic constituents often leads to depletion of dissolved oxygen, if dilution and dispersion in the receiving waters is not rapid enough. Such effects can result from discharges of sewage and organic-rich industrial wastes. A major cause of low dissolved oxygen in fjord waters of British Columbia has been effluent from the pulp and paper industry. The dissolved oxygen requirements under different levels of protection for salmonids and other species in aquatic environments at various temperatures, as developed by Davis (1976), are shown in Fig. 3.

### 3.2. Suspended Solids

Suspended solids in the aquatic environment have a variety of ecological impacts, which range from siltation of salmon spawning beds to direct effects on juvenile and adult salmonids. The benthic habitats of fish and invertebrates can be severely altered by heavy deposition of organic and inorganic sediments, with an adverse impact on benthic animals.

On the spawning grounds, the fine sand and silt from various human activities may soon fill up the

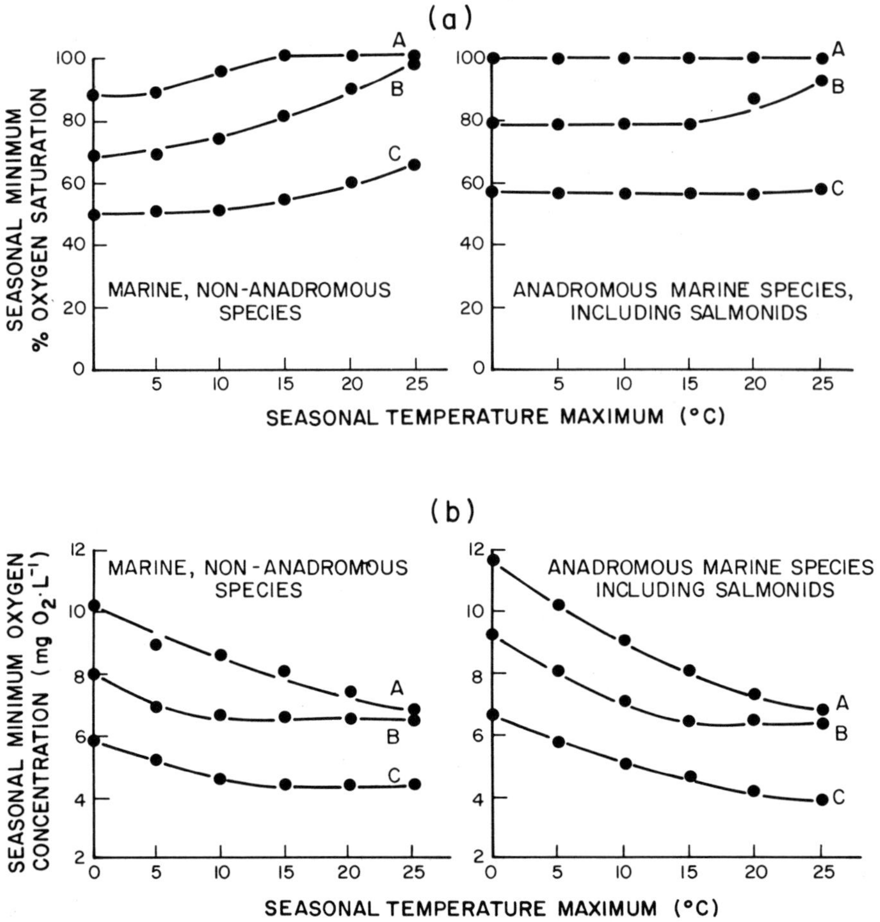


FIG. 3. Dissolved oxygen requirements for fish at various temperatures for three levels of protection. Level A: high safeguard for very important aquatic populations; Level B: possibility of moderate risk to aquatic organisms due to some degree of oxygen depression; Level C: large portion of a given fish population may be affected by low oxygen, especially on a "chronic exposure" basis (from Davis 1976).

spaces between the stones in the gravel of the redds where salmon have spawned, seriously affecting the gravel's permeability to water flow. A female salmon, before she deposits her eggs in the gravel, will fan it with her tail to remove the fines. But if the eggs have already been laid and the fines are then deposited, there will be difficulty for oxygen-rich water to reach the eggs, to replenish the oxygen supply and to remove the waste products (ammonia and carbon dioxide) from around the eggs. If the transfer of oxygen from stream water to the eggs is insufficient, the eggs will die (Alderdice et al. 1958).

Suspended solids in ambient water may affect respiration in both mature and juvenile salmon through an effect on the gills. While suspended solids are present naturally in such streams as the Fraser River, which contains so-called glacial flour derived from the glaciated headwaters, these are usually particles of only a few micrometres or less in diameter, many of which are present in colloidal suspension. Salmon have adapted over time to both the particle size and the concentration of this natural, suspended material. Addition of suspended solids from human activities, however, adds stress for fish exposed to them. It has been found in laboratory work that the adverse effect of suspended solids in water increases not only with concentration but also with particle size (Fig. 4) (Servizi and Martens 1987).

The effects of suspended solids on fish are extremely difficult to quantify. In situations where sediments have been added through human activity, the magnitude of the impact has often had to be drawn from inference.

An example of such a situation is the effect of sediments from gravel washing on salmon populations in the Coquitlam River. Although the salmon populations in this stream have declined over time, it has been difficult to ascertain whether this has been due to the sediments introduced by gravel washing or from over-fishing of the stocks. When the company that was discharging the gravel washing into the Coquitlam River was taken to court on a charge of water pollution under the Fisheries Act, the effect on salmon stocks had to be drawn mainly from inference and some circumstantial evidence.

There has been a proposal, with strong scientific support, that duration of exposure of fish to suspended sediment is as important as concentration (Newcombe and MacDonald 1991). This may be a key consideration in a situation, such as gravel washing, where the input of particulate material is episodic rather than continuous.

### 3.3. Temperature

It is quite obvious that temperature is an important property of water that affects all aquatic organisms. Fish in temperate climates have adapted to considerably lower temperatures than fish in the tropics. Salmonids prefer comparatively cool water, and in the marine environment they will migrate northward in summer to avoid the higher temperatures in the southern portion of their range.

Brett (1952) conducted the classical work on the tolerance of young Pacific salmon to different temperatures (Fig. 5). Among the 5 species tested, there was variation from species to species in their resistance to high and low temperature. When the salmon were acclimated at temperatures ranging from 5 to 24°C, they exhibited different degrees of resistance to high temperature. The most resistant species were chinook and coho. The next most resistant species was sockeye. Pink and chum salmon were least resistant. Brett (1952) found that no species of Pacific salmon could withstand temperatures exceeding 25.1°C when exposed for one week.

Tolerance of Pacific salmon to low temperatures was found to be influenced by the size of fish and the salinity of the water (Brett 1952). When taken from holding troughs with water at temperatures as low as 5°C, coho and sockeye could not tolerate exposure for four days to a temperature of 0°C. The lower lethal temperatures for the highest acclimation temperature, 23°C, were: chinook, 7.4°C; coho, 6.4°C; sockeye, 6.7°C; and chum, 7.3°C (Brett 1952). In later studies, Brett and Alderdice (1958) found that cultured young chum and sockeye salmon could survive for 3 days at a temperature of −0.5°C. Lower temperatures (−1.0 and −1.5°C) showed more severe effects on sockeye than on chum.

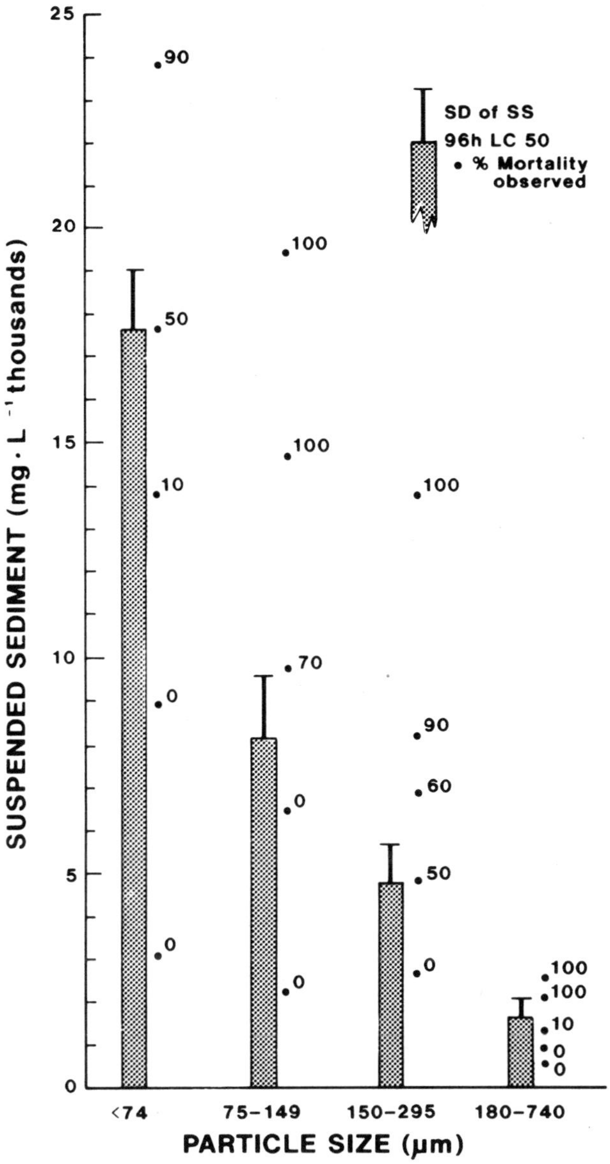


FIG. 4. Acute toxicity of suspended inorganic solids to juvenile sockeye salmon, as shown by mortality in relation to particle size (from Servizi and Martens 1987).

Laboratory bioassays generally provide results of somewhat artificial conditions that may not be fully representative of those in the field. Nevertheless, they do provide an indication of the response of fish to extreme temperatures in this case. Brett (1956) discussed some of the principles of thermal requirements of fishes, where he noted that the majority of systematic experiments performed to determine the resistance of fish to

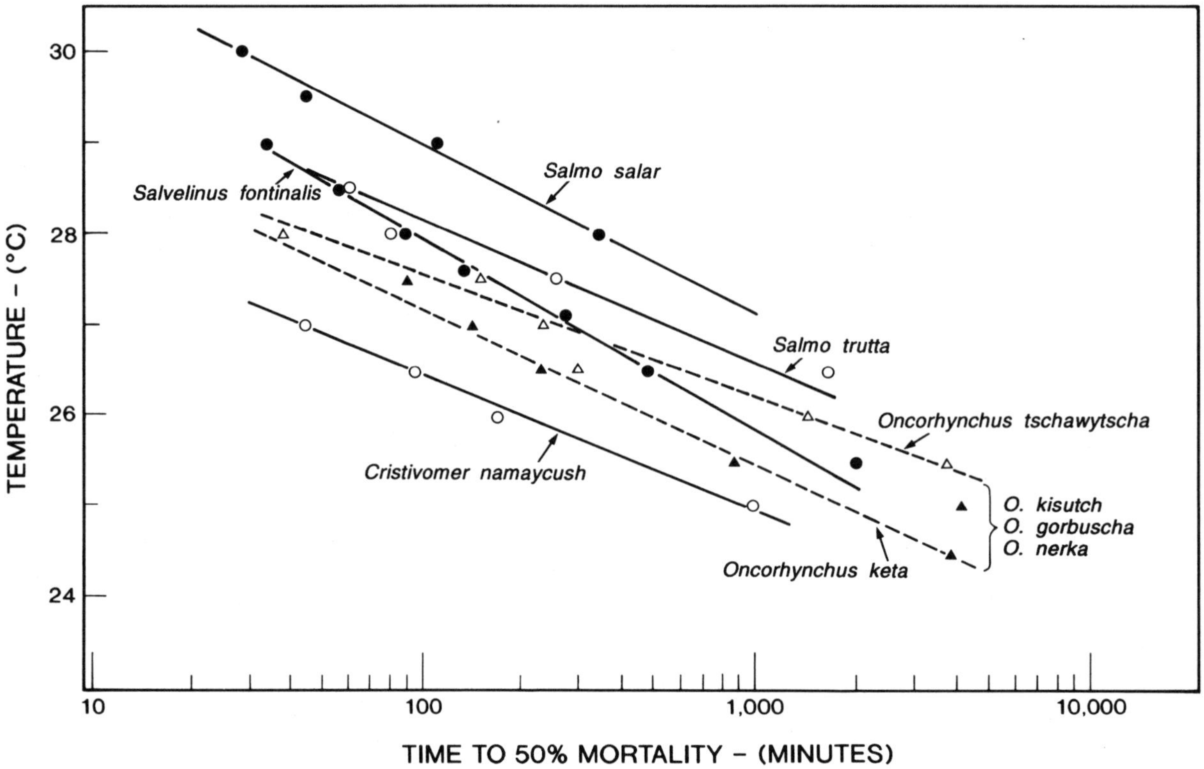


FIG. 5. Median resistance times for different salmonids acclimated to 20°C (from Brett 1952).

extremes of temperature have been conducted in freshwater. Because salmon spend a large proportion of their life cycle in estuaries and in the marine environment, it is clear that laboratory studies of the resistance of salmon to low temperature should be conducted also in seawater, which freezes at a lower temperature than freshwater.

There are important effects of temperature on salmonids at different stages of their life cycle other than those commonly used in bioassay experiments (yearling juveniles). Small changes in temperature of the water on spawning grounds can accelerate or slow down the rate of incubation of salmon eggs in the gravel. A slight elevation of temperature in streams following the emergence of salmon fry may increase the rate of growth of the fry and reduce the length of their residence in the stream. A drop in temperature can have the opposite effect. In the marine environment, salmon generally shift their distribution with the seasons — northward in summer when temperatures are higher and southward in winter when temperatures are lower. It is believed that some species of Pacific salmon may not fare too well when there is an unusual oceanographic event, such as the 1982–83 El Niño, that caused an increase in oceanic temperature.

### 3.4. Acidity

The acidity of water, as represented by pH, can have a profound effect on the aquatic life in a body of water. Some of the biological implications of changing pH in lakes and streams have been described by Park (1987). At low pH, bodies of freshwater can become virtually sterile. Fortunately, seawater is chemically buffered with weak-acid salts, and extreme pH changes do not occur.

While the problem of acidified lakes and streams has been one often considered to be confined to eastern Canada, where the granitic geology causes the lakes and streams to be sensitive to acid precipitation, there are areas in western Canada as well that are vulnerable (NRCC 1981). These western areas include parts of Manitoba, Saskatchewan, Northwest Territories and a coastal strip of British Columbia. Regions with average precipitation more acidic than normal "clean" rain (i.e., below pH 5.6) cover most of British Columbia. Virtually all of Vancouver Island and the extreme southwestern tip of mainland British Columbia have average precipitation of pH 5.0 or less. Effects of low pH on salmonids and other fish life and on their food have not yet been identified in British Columbia, but the potential exists for damage to occur. Fish are affected not only by the acidity of the water, but they may also suffer from the effects of toxic metals such as mercury, aluminum, lead and manganese released by acidic water from the surrounding soils or lake and stream beds. A leaflet issued by the Department of Fisheries and Oceans (DFO 1985b) states: "Lakes are considered to be in serious condition when their pH reaches or falls below 5.5 — the level at which many organisms on which fish feed cannot survive."

Acid precipitation is caused by the presence of acidic constituents, such as sulphur dioxide, in the atmosphere. So far, there are no large sources of such constituents that can be transported by winds for deposition in British Columbia, so that serious effects of acid precipitation are not expected to arise in the near future. Minor, sub-lethal effects could already be present, however, in parts of the province.

A more serious effect of acidification arises from acidic mine drainage. Sulphidic ores exposed to the atmosphere are oxidized to sulphates which leach into water courses as sulphuric acid. These cause problems not only in active mining areas, but also at abandoned mine sites. The acidic water is capable of dissolving metals, such as copper, zinc and lead. An example of such an abandoned mine that continues to release acidic drainage is at Mt. Washington, located north of Courtenay on Vancouver Island. It discharges its drainage into the watershed of the Tsolum River, an important salmon and trout stream. Before the copper mine went into operation, there were coho and pink salmon and steelhead and cutthroat trout in the Tsolum River system. By 1987, there was apparently only a small run of pink salmon remaining (Waldichuk 1987).

### 3.5. Salinity

The salt content of water can be both beneficial and detrimental to fish, depending on whether their habitat is in freshwater or seawater, or whether their different life stages exist in these two media. A salinity higher than that normally found in seawater may be lethal to all stages of salmonids. The egg, alevin and early fry stages of most species of Pacific salmon would not be expected to survive for long in seawater. The fry of pink and chum salmon are physiologically adapted to make an early transition from freshwater to seawater. Other species, such as coho and sockeye, normally require at least a year to reach the process of "smoltification", which prepares the young salmon physiologically to move from freshwater to seawater. As adults prepare to return from the sea to their native streams to spawn, they are physiologically prepared to move from seawater of comparatively high salinity (35‰) to freshwater of zero salinity.

Juvenile salmonids, taken from a freshwater environment, can be artificially acclimated to seawater of moderate salinity (20–25‰). This is often done when young rainbow trout or coho salmon are used for bioassays in seawater. It has to be recognized, however, that exposure of salmon in any life stage to salinity outside its normal range adds a stress that renders the salmon less capable of coping with stresses caused by other factors, such as depressed dissolved oxygen and contaminants in the water (Alderdice 1972). Salinity in the right range for salmon and for other species of fish, however, is part of the life support system in the marine environment.

## 4. Aquatic Contaminants

Contaminants are regarded as being hazardous to aquatic ecosystems owing to: (1) toxicity; (2) bioaccumulation; and (3) persistence. Certain contaminants, such as metals and organochlorines, possess all three of the foregoing attributes. Some are merely toxic but do not bioaccumulate and are not persistent. Others, such as plastic, are persistent but are not toxic and do not bioaccumulate.

The following groups of contaminants have caused pollution problems in Canada, including the Pacific Region. Their effects on various members of the aquatic biota are briefly noted. Not all have been a threat to Pacific salmon, partly because salmon are mobile after they emerge from the gravel as fry and can avoid contaminants, and partly because their comparatively short life span and anadromous nature does not allow them to remain in a particular area for long. Pacific salmon are particularly vulnerable to toxic substances. They are less vulnerable to substances that bioaccumulate, and if bioaccumulation does occur, it is usually not in the muscle, but rather, in the liver, kidney and spleen, which are not normally consumed by humans.

Although some effort has been made to group the contaminants chemically, certain contaminants, e.g., pulpmill effluents and solid wastes, cannot be classified in this way. They are examined on the basis of the damage they cause the aquatic ecosystem, related to their toxicity, persistence and other characteristics. Much of the information comes from examination of species other then Pacific salmon, but the environmental contamination reflected by these studies is relevant to environments inhabited by Pacific salmon.

### 4.1. Petroleum Hydrocarbons

Many volumes have been written on the effects of petroleum hydrocarbons on aquatic organisms (e.g., Smith 1968; Hoult 1964; Cowell 1971; Malins 1977;

Wolfe 1977; Geyer 1980; NAS 1975a, 1985; Johnson 1984; Engelhardt 1985). No attempt will be made here to summarize the enormous amount of information that has been generated by investigations on the ecological impacts of oil pollution. Only highlights relevant to the Canadian situation, particularly the Pacific Region, will be given. Because crude oil and refined petroleum products have been transported in huge quantities by sea, and tanker oil spills and their effects have been highly dramatized by the media, much of the available information on the impact of oil spills deals with the marine environment. Oil spills have occurred in freshwater systems, however, often without much publicity, and these have been devastating to local fish populations. A large volume of petroleum hydrocarbons enter the marine environment from runoff (NAS 1975a, 1985).

Canada has had comparatively few large marine oil spills. A major spill (17.2 million L) of bunker C oil from the tanker *Arrow* in Chedabucto Bay, Nova Scotia, in February 1970, received nationwide and international attention (Thomas 1973). Its persistence and effects continued to be studied for at least 5 years (e.g., Vandermeulen and Gordon 1976). On the Canadian Pacific coast there have been assorted small spills over the course of the last three decades. The one with the greatest impact was the 875 000 L spill of bunker C oil from the barge *Nestucca* off the southwest coast of Washington State on 22 December 1988 (Waldichuk 1989b; Harding and Englar 1989). It resulted in closure of molluscan and crab fisheries off the southwest coast of Vancouver Island and threatened colonies of transplanted sea otters at Bajo Reef and the Bunsby Islands.

Canada has had a great deal of concern about potential oil spills resulting from oil exploration and exploitation in the Arctic, particularly the Beaufort Sea (Milne and Childerhouse 1978). To determine the impact of an oil spill in the Arctic, a unique experiment, the Baffin Island Oil Spill (BIOS) Project, with multidisciplinary field studies, was conducted between May 1980 and August 1983 in Canada's eastern Arctic at Cape Hatt on the northern end of Baffin Island. A release of 45 000 L of a sweet, medium-gravity crude oil was made in a typical coastal Arctic environment for the purpose of conducting comprehensive scientific investigations on the problems arising from marine oil spills in the Arctic. The results have been published (Sergy and Blacksall 1987), and the impact of oil on various components of the Arctic ecosystem have been documented; for example, the effects of oil on Arctic marine bacteria are given by Bunch (1987).

Being a highly visible contaminant, oil receives a great deal of attention from the aesthetic and amenities point of view. Immediate action is always taken to clean up an oil spill. Early victims of a marine oil spill are sea birds. Following the 41-million L oil spill from the *Exxon Valdez* in Prince William Sound, Alaska, on 24 March 1989, in addition to over 30 000 dead seabirds counted, about 1000 sea otters were known to have perished from the spilled oil (Hodgson 1990). Problems for fish and shellfish from oil pollution are less clear cut. If there is a risk of contamination of harvested fish or shellfish, authorities will usually close the fisheries, as happened for molluscan shellfish and crabs on the west coast of Vancouver Island, following the *Nestucca* spill, and for herring and salmon in Prince William Sound, following the *Exxon Valdez* spill. Tainting of the seafood product is a major concern.

Shellfish beds receiving oil will usually suffer the greatest damage of all the fisheries resources. If it is a refined petroleum product, such as diesel fuel, the mortality can be very high, because of the severe toxicity of the so-called "white oils". The more viscous Bunker C oil does not penetrate as easily or deeply into the substrate, so that infauna, such as clams, may not be as seriously affected as organisms on the surface of the sediments, such as oysters. Fish, such as herring (*Clupea harengus*), that spawn in the intertidal or in the shallow sub-tidal zone may receive a serious impact from an oil spill if it coincides with the period of spawning and contaminates either the eggs or the vegetation on which the eggs are deposited. Species that have pelagic eggs which remain at or near the sea surface, such as the petrale sole (*Eopsetta jordani*) and English sole (*Parophrys vetulus*), could also be adversely affected by an oil spill forming an oil slick offshore. Oil that has sunk because of increased density due to absorption of particulate materials can contaminate bottom species like crabs. This was observed with Bunker C oil off the west coast of Vancouver Island in early 1989. There are not many known examples in the marine environment where there has been permanent or long-term damage from an oil spill, except for possibly the huge *Amoco Cadiz* crude oil spill off the coast of Britanny, France, in 1978, when coastal oyster beds were severely polluted by oil and the effect persisted for several years.

In freshwater systems, an oil spill can be particularly devastating when fish are spawning or juveniles are present in high concentrations. An example was the 11 000 L diesel fuel spill from a railway locomotive into the Brunette River, near Vancouver, British Columbia, in May 1989. Just prior to the spill, the Sapperton Fish and Game Club had introduced 35 000 coho fry into the stream in an effort to rehabilitate the run. Early reports indicated that 2 000 of these fry had been killed by the spilled diesel fuel.

One group of components of petroleum hydrocarbons that is considered a particular hazard environmentally is the polycyclic aromatic hydrocarbons (PAHs). Some of these constituents are considered to be carcinogenic as well as being quite toxic. The PAHs stem not only from crude oil but also from incineration, creosoted pilings, coke-producing ovens in steel manufacture and aluminum smelters.

Recent studies in Vancouver Harbour (Goyette and Boyd 1989) showed the presence of skin tumours and pre-cancerous liver lesions in English sole, associated with comparatively high levels of PAHs in the sediments. These findings led to regional guidelines being developed by the Regional Ocean Dumping Advisory Committee, Pacific and Yukon Region, under Part VI of the Canadian Environmental Protection Act, on allowable levels of PAHs in sediments proposed for ocean dumping.

### 4.2. Chlorinated Organic Compounds

The classical chlorinated organic compounds that contaminated the environment are the chlorinated pesticides, such as DDT, and the industrial chemicals, the polychlorinated biphenyls (PCBs), which have been used in electrical transformers and capacitors, as well as in printing inks, plasticizers and other similar applications. DDT was phased out in the early 1970's in most developed countries because of its undesirable ecological effects, particularly in reproductive failure of certain birds high in the food chain. It had an adverse effect also on freshwater fishes, causing reproductive failure in some species (Burdick et al. 1964). Large-scale forest spraying with DDT in New Brunswick had a devastating effect on Atlantic salmon in streams that were contaminated (Kerswill et al. 1960; Kerswill 1967; Kerswill and Edwards 1967). DDT, like other chlorinated hydrocarbons, tends to concentrate in tissues with high lipid concentrations and undergoes biomagnification through the food chain, so that the top predators end up with the highest concentrations. Most organochlorine pesticides have been banned or restricted for agricultural and forestry applications in Canada, as they have in other developed countries of the northern hemisphere. They have been replaced in many cases by the less persistent organophosphates and carbamates. There are still traces of organochlorine pesticides found, however, in marine mammals from residues of earlier use (Massé et al. 1986), and possibly through atmospheric transport from areas where they have not been banned. There are indications that the use of DDT and other organochlorine pesticides has shifted to the tropics and the southern hemisphere (Tanabe et al. 1989), as noted earlier by Goldberg (1975). Arctic studies (DINA 1989), on the other hand, have shown unexpectedly high levels of certain organochlorine compounds in the blubber of narwhals and beluga whales, as well as in seals and polar bears, that appear to have been transported there by tropospheric winds from Eurasian sources.

The picture with PCBs in the Canadian environment is somewhat different than that for the organochlorine pesticides. Although the use of PCBs in new installations was banned in the 1970's, existing electrical transformers and capacitors containing PCBs were allowed to operate for as long as they were needed and functioned properly. Many transformers taken out of service were stored with their PCB contents, because of lack of suitable PCB-destruction facilities. Consequently, PCBs continue to enter the environment from leaks, PCB-storage warehouse fires and accidents during transport of PCB-containing transformers and drums. Comparatively high levels of PCBs can still be found in marine mammal tissues, such as the blubber of belugas in both the Gulf of St. Lawrence (Béland 1988) and the Arctic (DINA 1989). A concern has been expressed in Japan (Tanabe et al. 1987; Kannan et al. 1989) that the highly toxic co-planar PCB congeners are transported far afield by tropospheric winds and are bioaccumulated in the blubber of whales and porpoises on the high seas.

A new problem emerged in the late 1980's with respect to toxic constituents in pulpmill effluents. The highly toxic polychlorodibenzo-p-dioxins (PCDDs) and polychlorinated dibenzofurans (PCDFs) were found in crustacean shellfish in the vicinity of coastal pulpmills of British Columbia (Waldichuk 1990). Fisheries for crabs, prawns and shrimps were closed by the Department of Fisheries and Oceans in Howe Sound in the vicinity of the pulpmills at Woodfibre and Port Mellon, and adjacent to the pulpmill in the Prince Rupert area on 30 November 1988, because concentrations of the extremely toxic dioxin cogener, 2,3,7,8-tetrachlorodibenzo-p-dioxin (2,3,7,8-TCDD) exceeded 20 ppt (parts per trillion) in these edible species. The closures were imposed on the advice of the Health Protection Branch of the Department of National Health and Welfare. A number of other coastal British Columbia areas adjacent to pulpmills, and one in the vicinity of a large sawmill, were closed for crab, prawns and shrimp harvesting on 23 November 1989, after further analyses showed excessively high concentrations of dioxins and furans in these crustaceans. Overwintering juvenile chinook salmon exposed to bleached kraft mill effluent in the upper Fraser River were found to be contaminated with 2,3,7,8-TCDD and 2,3,7,8-TCDF (tetrachlorodibenzofurans) (Rogers et al. 1989). Comparatively high levels of dioxins and furans were also found in the non-anadromous species of fish in the Fraser River near pulpmills at Prince George. Advisories were issued by the Health Protection Branch warning individuals, particularly the native Indians, against eating too many of these species in a given week.

The presence of dioxins and furans in pulpmill effluents arises from the use of wood chips treated with chlorophenols that contain dioxins and furans as impurities, the use of certain defoamers, and/or the combination of chlorine used in bleaching the pulp with phenolic organic compounds originating in the wood. It was not until the early 1980's that analytical technology had advanced sufficiently to allow measurement of dioxins and furans in the parts per trillion range. The pulpmills were all required by the B.C. Ministry of Environment to take immediate measures to eliminate production of chlorophenol-contaminated woodchips by B.C. sawmills and the use by pulpmills of defoaming agents that produce dioxins. Under new regulations of the B.C. Waste Management Act, the pulpmills were required to reduce their discharge of chlorinated organic compounds to a maximum of 2.5 kg per air-dry ton of pulp production by 31 December 1991 and to 1.5 kg per ton by 31 December 1994. They were also given a target of 31 December 1991 to provide secondary treatment for their effluent.

While the foregoing actions have been taken to protect human consumers against the health hazards of consuming dioxin- and furan-contaminated seafood, dioxins and furans are known to be toxic to fish and other aquatic species. The effects of exposure of young salmon to 2,3,7,8-TCDD in water at concentrations greater than 23 $ng \cdot g^{-1}$ (ppb) for 24–96 h were found to be irreversible, and resulted in death within 10–80 days (Miller et al. 1973). Significant effects of 2,3,7,8-TCDD on survival, growth and behaviour of rainbow trout were noted during both the exposure and depuration phases (Mehrle et al. 1988). The no-observed-effect concentration was lower than the lowest exposure concentration of 38 $pg \cdot L^{-1}$ (parts per quadrillion). The average measured bioconcentration factor at 28 days was 26 707. The no-observed-effect concentration for exposure of rainbow trout to 2,3,7,8-TCDF was 1.79 $ng \cdot L^{-1}$ (ppt), based on survival, and 0.41 $ng \cdot L^{-1}$, based on growth. The measured bioconcentration factor was 6 049 in fish exposed to 0.41 $ng \cdot L^{-1}$, and 2 455 in fish exposed to 3.93 $ng \cdot L^{-1}$ for 28 days (Mehrle et al. 1988).

The entry of unwanted organochlorine by-products and impurities into Canadian aquatic systems will probably continue. As our analytical capability improves, we may discover other toxic substances, some of which may be organochlorine compounds, being inadvertently released by industry into our aquatic systems.

### 4.3. Metals

Metals became a serious issue in Canada in 1970 when it was found that fish in some of our streams contained a relatively high concentration of mercury (Wobeser et al. 1970). This finding appeared to be particularly alarming because it followed on the heels of the tragic Minamata disease, caused by consumption of seafood contaminated by mercury in Minamata Bay, Japan (Irukayama 1967). The comparatively high mercury levels in Canadian fish were soon noted to be due mainly to mercury releases by chlor-alkali plants, using mercury as the cathode in their cells for production of chlorine and caustic. In addition, some mercury was being released by pulpmills using organic mercury compounds for slime control.

On the advice of the Food and Drug Directorate of the Department of National Health and Welfare, fish containing more than 0.5 ppm mercury, wet weight, in muscle tissue were banned from the market. Chlor-alkali plants were required by the federal government to eliminate or substantially reduce their release of mercury to the environment. Pulpmills in British Columbia had already by 1970 switched from mercurial slimicides to sodium pentachlorophenate or other similar slime-controlling agents. Other pulpmills across Canada that had not yet eliminated the use of mercurial compounds were required to do so by regulation in 1970. In general, there has been a reduction of mercury in freshwater species of fish since controls were imposed. Some species of marine fish, however, particularly the older age groups of swordfish, tuna, halibut and dogfish (Peterson et al. 1973; Forrester et al. 1972), continued to exhibit comparatively high levels of mercury. It soon became apparent that these fish were concentrating mercury from natural sources.

In 1972, the pulp and paper industry of British Columbia was advised that oysters in the vicinity of some coastal mills were exhibiting unusually high concentrations of zinc, in some cases exceeding 10 000 ppm, dry weight (Waldichuk 1974). The industry had been using zinc dithionite (referred to as zinc hydrosulphite in the trade, $ZnS_2O_4$) as a brightening agent for groundwood pulp used in newsprint production. To ameliorate the problem of zinc contamination of oysters, and possibly other shellfish, the industry voluntarily substituted sodium dithionite for zinc dithionite as a brightening agent for groundwood in 1973.

Cadmium is another metal about which human health concerns, as a result of contamination, were first raised in Japan (Shigematsu 1973). This arose out of the *Itai itai* disease, a painful bone-weakening disorder due to the lack of calcium in the bone caused by cadmium interference. The high cadmium levels were due to mining effluent that contaminated irrigation water used in rice cultivation. In Canada, a cadmium pollution problem occurred in Belledune Harbour, New Brunswick, in the early 1980's as a result of leaching of a tailings pile from a zinc-lead smelter (Uthe and Zitko 1980). Lobsters were found to have high concentrations of cadmium (~500 ppm) in their tomalley (digestive gland), which is consumed by many local people. Harvesting of lobsters in Belledune Harbour was closed by the Department of Fisheries and Oceans in 1980. A 97% reduction in the metals discharged in the effluent was achieved by certain corrective measures after 1981. Metal concentrations in the mussels and lobsters decreased in response to the reduction of metal input (Hildebrand 1984). The lobster fishery was reopened in 1985 under an initial arrangement whereby only the white meat would be marketed.

Elemental phosphorus is not normally regarded as a metal of pollution concern. But its release into Placentia Bay, Newfoundland, from a fertilizer plant in Long Harbour resulted in the infamous "red herring" episode of the spring of 1969 (Idler 1969; Jangaard 1970). It resulted in mortality of a certain proportion of the herring exposed to the phosphorus, as well as toxicity to other species such as Atlantic salmon, lobsters and beach fleas (Zitko et al. 1970). Following corrective action to prevent the release of phosphorus from the fertilizer plant into the sea, the herrings ceased to exhibit the red discolouration. Other evidence of contamination by phosphorus in Long Harbour also disappeared (Addison et al. 1972).

A metal problem that has arisen in the last three decades stems from the use of metal-containing antifouling paints on hulls of boats. To replace copper, which was considered to be highly toxic to certain marine organisms, tributyl tin (TBT) was chosen as an anti-fouling ingredient for marine paints during the 1960's. TBT was soon found to be not only an effective anti-fouling compound but also toxic to other marine organisms in the vicinity of boats treated with it. In the United Kingdom, it led to a significant reduction in the reproduction success of oysters (Thain and Waldock 1986). Shell thickening in oysters, due to TBT contamination, rendered them unmarketable for the half-shell trade. Residues of TBT were also found in Atlantic salmon held in pens with nets treated by TBT-based paint in the UK (Davies and McKie 1987). Tributyl tin is now either banned or restricted in several countries, including Canada, for use as an anti-fouling additive for paints applied to hulls of boats <25 m in length and nets of fish pens.

The foregoing examples of metal contamination usually involved bioaccumulation, which resulted in controls to protect human health. It should be noted that metals also have various effects on the aquatic organisms themselves that range from acute toxicity to assorted sublethal effects (Waldichuk 1974). The sublethal effects are the most difficult to assess (Cole 1979), but they are probably the most important in the long term, because they could lead to reproductive failure and a decline in population of a given species.

### 4.4. Pulp and Paper Mill Effluents

Because of Canada's extensive forest resources, the pulp and paper industry features strongly in its economy. It is a major water user in processing wood into pulp and paper. The effluent, which is discharged into streams, lakes and coastal waters, has varying degrees of toxicity according to the process used and the amount of treatment provided. There has been a great deal of investigation on the toxicity of pulpmill effluents on both the Atlantic and Pacific coasts of Canada (e.g., Alderdice and Brett 1957; Sprague and McLeese 1968a,b; Sprague and Drury 1969; Walden 1976). In addition to measurement of the acute effects of pulpmill effluents to provide LC50 values (the concentration at which 50% of the test species die during a specified exposure period), there have been numerous studies also on the sublethal effects of pulpmill effluents (Sprague 1971; Davis 1976; McLeay 1973, 1977; McLeay and Brown 1979). Because pulpmill effluents are very complex mixtures of dissolved organic substances derived from wood that is pulped, and chemicals used in the digestion process, as well as chemical compounds formed during the pulping and bleaching processes, it has been difficult to attribute their toxicity to specific constituents. Efforts have been made, nevertheless, to identify constituents in kraft mill effluent that are toxic to salmon (Leach and Thakore 1973; Rogers et al. 1975). Evidently, resin acids, particularly dehydroabietic acid, various sulphur-containing organic compounds and chlorinated guaiacols contribute significantly to the toxicity of pulpmill effluents. The presence of toxic dioxins and furans in pulp and paper mill effluents has already been noted in section 4.2, but these compounds have been identified only recently.

### 4.5. Hydrogen Sulphide

Often referred to as foul-smelling, rotten-egg gas, hydrogen sulphide is highly toxic and can be produced in nature under anaerobic conditions, if there is sufficient organic material and a source of sulphur. It may be a constituent of industrial wastes, such as those from pulp and paper mills, chemical manufacturing plants, tanneries and gas works. As a chemical used in precipitating metals as sulphides in treatment systems for metal-rich effluents, hydrogen sulphide may appear in minute residual concentrations in such treatment plant effluents. Hydrogen sulphide combines rapidly with oxygen, however, and does not remain freely for too long in waters with an ample supply of dissolved oxygen.

Hydrogen sulphide increasingly dissociates into sulphide and hydrogen ions as the pH of the water increases. Thus, at pH 5, about 99% is present as hydrogen sulphide. At pH 7, it is about equally divided between hydrogen sulphide and hydrosulphide ion ($HS^{-1}$); and at pH 9, about 99% of the sulphide is present in the form of $HS^{-1}$. Since it is the undissociated hydrogen sulphide which is most toxic, pH of the water has a major effect on its toxicity. Seawater has a high buffering capacity, because of the presence of weak-acid salts, and the pH is usually on the alkaline side, near 8.0, so that the toxicity of hydrogen sulphide is not nearly as great in seawater as it is in freshwater.

The concern about hydrogen sulphide in both freshwater and marine environments stems from the potential release of this gas from sludge beds. Dredging in an area where fibres and other wood solids from a pulpmill had been deposited in Alberni Harbour, for example, led to fish kills (Hourston and Herlinveaux 1957). The mortality was undoubtedly due to the sudden release of hydrogen sulphide by the dredging operation while fish were in the area. Similar problems could arise in lakes or streams if dredging occurred there. A major concern in freshwater environments has been the potential effect of hydrogen sulphide released from sludge beds on fish eggs and larvae. On the basis of available data to 1972, a level of undissociated hydrogen sulphide assumed to be safe for all aquatic organisms, including fish, was given as $0.002\ mg \cdot L^{-1}$ (NAS 1972).

### 4.6. Cyanide

Sodium cyanide is used in the mining industry to extract gold from ore. Cyanide-bearing wastes may be released also from gas works, coke ovens, metal-plating operations, gas scrubbing in steel plants and chemical industries. Cyanide is highly toxic to aquatic animals as well as to humans. Its toxicity varies widely with temperature, pH and dissolved oxygen concentration. To protect aquatic species, it has been recommended (NAS 1972) that cyanide concentrations do not exceed $0.005\ mg \cdot L^{-1}$.

In British Columbia, a gold tailing leach operation, using sodium cyanide, 70 km north of Grand Forks, was stopped by the B.C. Ministry of Environment in April 1989 when cyanide was detected in test wells. The concern was mainly about contamination of groundwater drinking supplies for area residents. But there was also a possibility that the cyanide could have contaminated Burrel Creek, which feeds the Kettle River, a major water system in the area. A controversial proposal by a mining company to test a novel technique of pumping a solution of sodium cyanide into a placer deposit and then recovering it, for gold extraction, upstream from the community of Cottonwood in the Cariboo region of British Columbia was strongly opposed by residents and went to a public hearing in 1988. The gold-bearing gravel is about 50 m beneath the surface of Lightning Creek, 50 km east of Quesnel. Again, the residents of Cottonwood were worried that their well water supply would be contaminated.

### 4.7. Other Toxic Chemicals

Many other toxic chemicals enter our aquatic environment. Over 4 million chemicals are presently registered with about 70 000 in commercial use (Somers 1982). Environment Canada is now developing a list of about 44 priority chemicals which pose toxicity, bioaccumulation and persistence problems in the environment. A scientific discussion of these types of substances, and risks involved with their presence in the environment, was conducted in a Priority Substances Scientific Forum, held at the Canada Centre for Inland Waters, Burlington, Ont., 22–23 February 1989 (Waterston 1989). Particular attention was given to the human health and environmental assessment of materials enumerated in the Priority Substances List, being developed under authority of the Ministers of Environment and of National Health and Welfare, as a requirement of the Canadian Environmental Protection Act. Toxicity of these chemicals to aquatic organisms would have to be an important consideration. But there was also potential for contaminating salmon and trout-bearing streams.

### 4.8. Solid Wastes

Solid wastes assume various forms, including wood debris, mine tailings, household garbage, plastic sheets, rope, nets and containers, and ordinary litter. They have been often regarded in the past as being aesthetically unpleasant and mainly an amenity problem. More recently, however, they have been shown to pose also ecological problems, particularly in the marine environment. Plastic bands, fugitive nets, rope and six-pack plastic yokes have been found to entangle marine mammals, birds, sea turtles and fish. The magnitude of the problem in the sea was presented in a number of papers given at a Workshop on the Fate and Impact of Marine Debris, held in Honolulu, Hawaii, in November 1984 (Shomura and Yoshida 1985).

Much of the marine debris originates from ships, fishing vessels and recreational boats. A study conducted by the U.S. National Academy of Sciences estimated that 6.36 million tons of marine litter are generated by all vessels globally every year (NAS 1975b). It is hoped that this volume is being reduced by the entry into force on 31 December 1988 of Annex V of the International Convention for the Prevention of Pollution from Ships, 1973, as modified by the Protocol of 1978 relating thereto (MARPOL 1973, 1978). This Annex is intended to prevent pollution by the dumping of garbage, which includes all kinds of victual, domestic and operational waste generated during the normal operation of a ship. An important provision of Annex V is a ban on the dumping into the sea of all plastics, such as synthetic ropes, synthetic fishing nets and plastic garbage bags. Contracting Parties to the Convention are obliged to provide facilities in ports for the reception of garbage.

The problem of solid wastes at sea has been exacerbated in the last few years with the massive driftnet fishery for flying squid conducted by some Asian countries. Lost and discarded nets may continue "ghost fishing" for a long time, taking various species of fish including Pacific salmon.

In streams, large wood debris has been found to be useful as shelter by juvenile salmon (Hartman and Scrivener 1990). However, debris torrents during heavy rainstorms have been devastating to spawning and rearing habitats, particularly in steep-gradient streams. Log and debris jams in streams have also impeded at times the migration of juvenile and adult salmon. Fortunately, the cessation of log drives in streams and the close attention of fishery officers to stream conditions has virtually eliminated log and debris jams of any significant duration. The impact of suspended mineral solids has already been discussed in section 3.2.

## 5. Impacts of Human Activities on Fish Habitats

Human activities may not only introduce contaminants into the aquatic environment but also destroy or degrade habitats. The introduction of harmful effluents from industries is often the main concern of the habitat manager. But the installation of the effluent-discharging industries in the first place could have destroyed or severely impaired fish habitat. Some of the ongoing activity after production commences, such as log-booming near a pulpmill, can also degrade fish habitat.

In this section, some of the human activities which have affected fish habitats, particularly of Pacific salmon, in British Columbia will be reviewed. The environmental and ecological effects of different types of contaminants and activities on aquatic systems are given in Table 2.

### 5.1. Logging Operations

Logging activities inevitably disturb the soil in the drainage basins of streams and lead to particulate materials, both organic and inorganic, being washed into the streams. The process of felling trees, trimming off the branches, cutting the trees into logs, dragging the logs and removing them from the forest all contribute to destruction of the protective vegetative cover on the forest floor and loosening of the soil. Road building through the forest adds a further source of soil, silt and sand that can be washed into the nearest stream by runoff from rain and snowmelt. Because logging roads are seldom paved, the gravel surface allows the erosion of silt and sand from the road surface, as well as from exposed banks on the sides of the roads, with every heavy rainfall and by runoff from snowmelt in spring. Culverts under these logging roads channel the silt-laden runoff water to the nearest natural watercourse. Improperly designed and constructed culverts can cause road washouts, which may prevent upstream migration of fish.

A further logging activity that exacerbates the erosion of soil and release of sedimentary material from a logging area in a watershed into nearby watercourses is the burning of slash. Such burning not only destroys the wood debris remaining from logging but also leads to combustion of the binding root systems and other organic materials in the soil, exposing the inorganic sand and gravel to rapid erosion by runoff. Forest areas denuded by clear-cut logging, followed by slash burning, lose their capacity to hold moisture. Heavy precipitation in the form of rain is rapidly followed by heavy runoff and severe erosion. Rapid snowmelting in the spring likewise causes increased flow of meltwater and accelerated erosion. A combination of rain and melting snow can lead to enormous erosion of soil, silt and sand from a watershed and a heavy load of suspended solids in any nearby streams. The slash from logging can enter streams and contribute to debris torrents during heavy rainstorms, which can be devastating to spawning grounds. The effect of logging on the hydrology of a watershed can persist for years, even decades. It is not until the area becomes reforested, and more importantly, the lower vegetal cover is regenerated that water added to the watershed by rain or snow can be absorbed and held for slower release.

The effects of logging on salmon streams in British Columbia have been observed for many years. The increase in suspended solids in streams receiving runoff from logged areas is one of the many effects known to occur. This increase in suspended load occurs in pulses following periods of heavy rainfall and runoff. As the rate of stream flow decreases and turbulence in the

TABLE 2. Aquatic environmental and ecological effects of different types of contaminants and activities.

| Contaminant or Activity | Bacterial/ Viral Contamination | Oxygen Depletion | Toxicity | Bioaccumulation | Habitat Degradation | Living Resource Stock Depletion |
|---|---|---|---|---|---|---|
| Logging | | | | | x | x |
| Sewage | x | x | | x | x | |
| Urban and agricultural runoff | x | x | x | x | x | |
| Pulp and paper mill wastes | x | x | x | x | x | |
| Mining wastes | | | x | x | x | x |
| Ocean dumping | x | x | x | x | x | (x)[a] |
| Coastal developments | | | | | x | x |
| Petroleum hydrocarbon Exploration and exploitation | | x | x | x | x | |
| Agriculture | x | x | x | x | x | |
| Pesticides | | | x | x | | |
| Petroleum refineries | | | x | x | x | |
| Chlor-alkali plants | | | x | x | | |
| Chemical spills and leaks | | (x)[a] | x | x | | |
| Atmospheric emissions | (x)[a] | | x | x | | x[b] |
| Food and beverage processing | | x | | | x | |
| Litter | | | | | x | x |

[a]Potential effect.
[b]Acid rain.

water subsides, the suspended solids settle to the stream bottom. The worst damage occurs if the sedimentation covers spawning grounds and reduces the permeability of the gravel used by spawning salmon for laying their eggs. Usually such deposited material becomes resuspended with the next freshet, following another heavy rainfall, and is carried further downstream. By such intermittent transport with succeeding freshets, the introduced sedimentary material is eventually carried to the stream's estuary, where it is deposited more permanently at the delta.

The impact of suspended solids, among other effects from logging, on a salmon stream has been examined in Carnation Creek on the southwest coast of Vancouver Island in a study covering conditions during the pre-logging, logging and post-logging periods (Hartman and Scrivener 1990). As the authors note in their conclusions to this study, the six-year post-logging assessment is probably of inadequate duration to fully evaluate the long-term impact of logging, including that of suspended solids, on the salmonid stocks of Carnation Creek. A useful guide for the protection of fish habitats during logging has been prepared by Toews and Brownlee (1981). A summary of the impacts of timber harvesting and how these can be mitigated is given in a Department of Fisheries and Oceans brochure (DFO 1985c).

### 5.2. Pulp and Paper Production

Pulp and paper mills are known mainly for their discharges of noxious effluents. However, these forest-product operations contribute also to habitat degradation in other ways. Usually located near streams or estuaries, pulpmills and their ancillary facilities may occupy what was originally prime shoreline habitat. The woodroom and its jack ladder for hauling logs into the mill may have been built on marshland in an estuary. A deep-sea wharf may extend through estuarine fish habitat. Log-booming for a constant supply of logs to the mill is a continuing source of wood debris, which leads not only to dissolved oxygen removal from the bottom water, but also alters in a negative way the quality of the benthic habitat for fish and invertebrates.

Pulp and paper mills use large volumes of water in the processing of their products. The wastewater must be disposed of, and it is usually discharged, after minimal treatment, into the nearest water body, whether this be a freshwater stream, lake, estuary or coastal marine waters. The effluent from pulp and paper mills always contains organic matter from the wood and bark, both in dissolved and particulate form.

The amount of organic material released from a given production of pulp depends on the process used, the amount of recovery practised and the degree of treatment given the effluent prior to discharge. The conventional sulphite mill provided little recovery in the past and has been characterized by the release of a great deal of dissolved, biodegradable organic matter in the form of "red liquor", or "spent sulphite liquor", from the digestors. More recently, most sulphite mills have introduced recovery systems to reduce the amount of dissolved organic matter released into the aquatic environment. These soluble organic compounds, which consist largely of wood sugars with a high biochemical oxygen demand (BOD), undergo rapid decomposition by bacteria and lead to accelerated dissolved oxygen depletion in the receiving waters. On the other hand, sulphate (kraft) mills have always practised recovery of the organic constituents and the inorganic salts from the "black liquor", arising from digestion of the wood chips under pressure at high temperature. In fact, the economy of the kraft process in pulp production has been based, in part, on energy utilization from combustion of the evaporated black liquor and on salt recovery from the ash. Regardless of the degree of recovery and treatment provided waste streams in the sulphite and kraft processes, there is always a certain BOD in the effluent finally discharged into the aquatic environment.

A third process of pulp production consists of mechanically grinding blocks of wood with rotating circular stone grinders to produce a pulp of short fibres, commonly referred to as groundwood. There are variations of this form of pulp production by applying high temperature and/or some chemical digestion to produce thermo-chemical mechanical pulp. This pulp is mixed with a small proportion of long-fibred kraft or sulphite pulp, in order to give it strength, to produce newsprint paper. In all forms of mechanical pulp production, there is some degree of leaching of the organic constituents of the wood, and this is enhanced by higher temperatures and chemical treatment. Although the BOD of the effluent from mechanical pulping is usually lower than that from the sulphite and kraft processes, there can be a significant impact on the dissolved oxygen concentration in receiving waters from the discharge of a large volume of effluent from mechanical pulping, especially if replacement of the receiving waters is slow.

One of the undesirable properties of pulp and papermill effluents is often a high concentration of wood particles, ranging in size from fibres to chips and fragments of bark (Waldichuk 1988a). While some of the fine particles may be carried long distances from the point of discharge, a high proportion of the suspended wood solids is deposited near the outfall in coastal waters. A description is provided by Bewers and Pearson (1972) of the behaviour of particulate material from a bleached kraft pulpmill in a freshwater treatment lagoon, where coarse materials (>10 μm dia.) had good settling characteristics. But fine particles (<1 μm dia.) remained in suspension throughout the treatment system. Once released into brackish coastal waters, however, the increased salinity promoted the combination of fine particles in the process of flocculation and improved their settling characteristics.

Fibres decompose more rapidly than wood chips because of their larger surface-area-to-volume ratio, which is more favourable for bacterial attack. Sludge beds of fibres and woody debris undergo decomposition, robbing the sediments and overlying water of dissolved oxygen. Gases, such as methane and hydrogen sulphide, may also be released from the sludge (Werner 1968). It is obvious that benthic habitats can become rapidly impaired from deposition of wood solids, not only as a result of alteration of bottom substrate but also because of lack of dissolved oxygen.

Although high concentrations of wood particles suspended in water can be damaging to fish through clogging of gills and interference with respiration, fish can generally cope quite well with low to moderate concentrations of suspended wood solids, judging from results of controlled bioassays on such material (Werner and Robinson 1978). Wood solids generally do not remain in suspension for too long. They either become water logged or become associated with mineral particles and settle in quiescent waters by gravitation (Werner 1965). The main ecological impact, therefore, occurs on the bottom. Heavy deposition can eliminate both the benthic infauna and the epifauna and render the bottom water anoxic, or nearly so, and unsuitable for bottom fish.

Most pulpmills now have mechanical clarifiers and/or settling ponds, where wood solids are removed prior to discharge of the effluent into receiving waters. This has led to considerable improvement in bottom conditions adjacent to pulpmill outfalls. In mills where solids-removal facilities are unavailable, fibre beds can form in the most unexpected places, such as experienced in Northumberland Channel, adjacent to the Harmac pulpmill, near Nanaimo, British Columbia (Waldichuk 1988a). In this case, a large area on the bottom was covered by a fibre bed and the shrimp habitat was severely impacted. Fishermen aboard shrimp trawlers reported during a 2-year period from 1979 to 1980 finding clumps of unusual material from 5 to 100 cm in diameter over a large area of Northumberland Channel. The clumps were aggregates of wood debris, wood fibres, shell and sand, and had a strong odour of hydrogen sulphide.

Dredging areas with sludge beds or heavy deposition of fibres may lead to fish kills, because of the sudden large release of highly toxic hydrogen sulphide (e.g., Hourston and Herlinveaux 1957). Aside from these acute situations, however, even a thin distribution of fibres on the bottom of shallow estuarine coastal waters, such as Alberni Harbour, may contribute to oxygen depletion in waters overlying the sediments. Specifically in Alberni Harbour, where there have been low dissolved oxygen concentrations in the bottom water even before the pulp mill was constructed in Port Alberni, the wood solids, some of which also come from log storage in the area, have no doubt contributed to the present summertime low-oxygen conditions which have essentially destroyed the benthic population of macroinvertebrates (see section 6.2.3). Degradation of the benthic environment by wood solids can also lead to a reduction in the number of species of certain organisms, as observed for polychaetes in Cousins Inlet, British Columbia, owing to effluent discharge from the now-closed pulpmill at Ocean Falls (Fig. 6) (Fournier and Levings 1982).

### 5.3. Municipal Wastewater Disposal

Municipal wastewater is made up mainly of domestic sewage and liquid wastes from restaurants and other commercial establishments, hospitals, dental clinics, laboratories, and often small industries and sometimes larger industries in cities. Consisting basically of human wastes, as well as kitchen and laundry washwaters and occasionally garburated organic substances from the kitchen, domestic sewage contains a comparatively high concentration of both particulate and dissolved organic material. Depending on the degree of treatment municipal wastewater receives prior to discharge to receiving waters, the effluent may have a comparatively high BOD.

Most municipal wastewaters in British Columbia receive at least primary treatment, which involves removal of solids by screening and settling. Some sewage is still discharged raw into coastal waters, however, with only simple break-up of the solid matter, by what is commonly called comminution. In some areas, particularly if the effluent is discharged into a river or lake, secondary treatment may be provided, where additional screening, filtration, aeration and other processes are applied to remove suspended and dissolved organic matter. Tertiary treatment, involving removal of nutrients, may be applied in cases where the effluent could cause excessive enrichment (eutrophication) and lead to undesirable algal growth. Such treatment is only provided in those unusual situations where the quality of lake and river water could be seriously impaired by over enrichment with nutrients.

Usually there are no problems of depletion of dissolved oxygen in receiving waters if the degree of treatment is higher than primary. Raw municipal wastewater and effluent from primary treatment can lead to dissolved oxygen depletion both in the water column and in the sediments, depending on how rapidly they are diluted and dispersed in the receiving waters. The effect in sediments tends to be cumulative as organic materials are deposited there through settling of both particulate organic matter and of flocs of natural silt from a river, which scavenge fine suspended solids and dissolved organic materials from the water column as they descend to the bottom. Such a situation prevailed on Sturgeon Bank of the Fraser River estuary, as a result of the surface discharge of primary-treatment effluent from the Iona Island Sewage Treatment Plant. The organic-enriched surface sediments on Sturgeon Bank were found to remove oxygen from the overlying water, particularly during the warm period of summer when bacteriological activity was highest. This led to severe stress, and occasionally mortality, to bottom fish during the summer months. It was this adverse ecological effect, more than anything else, that eventually led to installation of a deep outfall off Sturgeon Bank for the effluent from the Iona Island Sewage Treatment Plant (Birtwell et al. 1981).

### 5.4. Agricultural Runoff

A large volume of organic material from a variety of sources is contributed by agriculture to runoff that drains from so-called non-point sources into lakes, streams, estuaries and coastal waters. Most of this organic substance is in the dissolved form, but a proportion is particulate, and this fraction increases during periods of heavy rainfall or irrigation. Agricultural contributions of organic material stem from feed lots, the application of organic fertilizer (manure) on fields and leaching from manure piles. Intensive feeding of livestock on feed lots leads to a large amount of animal waste products and uneaten food being washed away during heavy rains. Good house-keeping practices can substantially reduce the organic losses from feed lots and manure piles. Also, precautions taken in farming methods can greatly minimize the losses of fertilizer and soil from fields during periods of rainfall and snow melt.

As in the case of untreated or only primary-treated domestic sewage, the waste products from domestic animals have a comparatively high BOD. If runoff from feedlots and manure piles enters a slow-flowing stream of small flow, low dissolved oxygen conditions can arise in the water of the stream or its estuary, particularly during the warm period of summer. This has been observed in the Serpentine and Nicomekl rivers, which drain

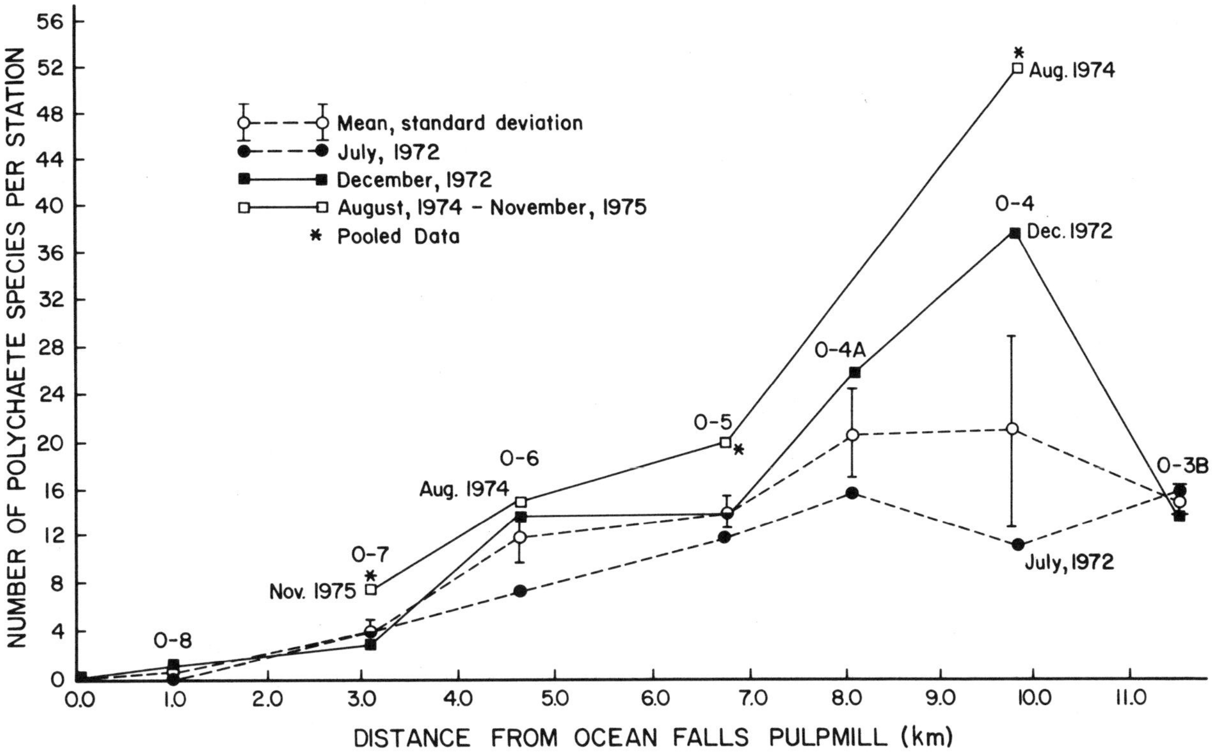


FIG. 6. Variations in number of polychaete species with distance from the former pulpmill at Ocean Falls in Cousins Inlet, British Columbia, during surveys from 1972 to 1975 (from Fournier and Levings 1982).

the farmlands of the lower Fraser Valley and discharge into Mud Bay and Boundary Bay, on the lower mainland coast of British Columbia.

Livestock grazing has been identified as one of the more serious land-use impacts on fish. It is obvious that streams must be protected in some way from the adverse effects of grazing by cattle and sheep. Strips of forest cover adjacent to a stream bordering on grazing land could prove to be effective in filtering runoff to prevent soil and manure from entering the stream.

As in logging, it is the disturbance of the soil in agriculture that causes the release of both organic and inorganic particulate materials into nearby bodies of water. To produce crops, a farmer must till the soil, plant the seeds, cultivate the growing plants and irrigate in some cases, fertilize, apply pest-control measures, and then harvest the crop. All those activities disturb the soil in one way or another and make it vulnerable to erosion.

There are certain practices in agriculture that can be followed to minimize the loss of soil and fertilizers from farmland. These include the ploughing of sloping fields along the contours rather than along the steeper grades, so that furrows do not serve as channels for runoff waters. It is wise also to leave strips of fallow land at certain intervals on sloping farm land. This would tend to act as a barrier to water flow down-slope on cultivated land. It would also act as a filter to remove suspended solids from any runoff water that passes through. Any ditches on farmland should pass through a settling basin before they reach a stream. Periodic dredging of such a settling basin would help to keep it functional. The way that different crops are planted and the way that irrigation is practised can influence the amount of soil that is eroded from farmland into neighbouring streams. Feedlots should be drained with drainage ditches equipped with weirs or sumps to trap particulate materials and prevent them from entering the streams. To be effective, of course, sumps and the sections of ditches behind weirs must be cleaned out periodically, and the dredged material returned to an appropriate place on the farmland.

In general, even if farmers are not aware of the ecological damage soil and fertilizer eroded from their farms can do, they will make an effort to prevent loss of these materials from their farmland, because in the long run, such losses can be costly. In some cases, however, it may be necessary to inform farmers that erosion of soil from their farmland is causing damage in streams and that certain corrective action must be taken.

## 5.5. Mine Waste Disposal

The extraction of metals from ores ends up with a small percentage of the ore as the desirable metal (about 1% for such base metals as copper), and a large proportion of waste rock. In open pit mining, there may be also a substantial volume of rock from the overburden that is not processed for metal extraction. The pulverized waste rock from metal extraction, known as tailings, is usually deposited in tailings ponds adjacent to the mining operation. After all the available ore is extracted from the mine and the mining operation ceases, an effort is made to cover the tailings pile with a suitable soil and to plant appropriate vegetation on it. In the more recent tailings ponds, an effort has been made to take precautionary measures to prevent leaching of metal-laden water from both the active pond and from the tailings pile that has been planted with vegetation. In older mining operations, these precautions were not taken and leaching of such metals as copper, zinc and lead has been a continuing problem (Sprague 1964; Sprague et al. 1965; Saunders and Sprague 1967). In addition, oxidation of pyritic (sulphide-containing) tailings leads to acidic mine drainage, which is another threat to fish.

In coastal mining operations, there has been some disposal of mine tailings into the sea, particularly in coastal British Columbia. One of the longer-lasting mines that discharged its tailings into the sea was the Britannia Mine, located about 50 km north of Vancouver. It discharged its tailings into Britannia Creek which then carried them into Howe Sound, over the course of three-quarters of a century from about 1900 for 1975. A molybdenum mine in Kitsault, B.C., operated originally for about 5 years from 1968 to 1972, discharging its tailings into Lime Creek, which carried them into Alice Arm on the northern British Columbia coast. A plan to expand the mine and discharge the tailings into Alice Arm during the late 1970's became a high-profile environmental issue (Ellis 1982; Littlepage et al. 1984). The mine finally went into operation in April 1981, but lasted for only 18 months, owing to declining molybdenum prices.

The one remaining mine that discharges tailings directly into the sea is the open-pit Island Copper Mine, which has been operating since 1971 at Port Hardy on the northeast coast of Vancouver Island. It discharges about 50 tons per day of tailings from an outfall at 50 m depth in Rupert Inlet. The impact of tailings disposal from this mine on the marine ecosystem is discussed in section 6.3.

A mine at Nanisivik on northern Baffin Island in the Northwest Territories discharges its tailings into a nearby lake, which drains into Strathcona Sound. While some metal-containing leachate enters Strathcona Sound, the suspended solids from the tailings are generally confined to the lake (Fallis 1982).

## 5.6. Dams and Water Diversions for Hydroelectric Power

A host of problems is created for fish habitat with hydroelectric power developments. The routes of migration of both adult salmon and juveniles may be blocked by dams. Important spawning grounds for salmon may be flooded. Salmon rearing areas may be destroyed or severely impaired. Diversions may introduce species, including parasites and diseases from another watershed.

Some moves were made to install dams for hydroelectric power and flood control in the Fraser River system during the 1950's and in the early 1970's (e.g., proposed Moran Dam). There has been generally strong opposition from fisheries managers, scientists and environmentalists to dams on salmon rivers in British Columbia. A program of research on the effects of dams on salmon was undertaken at the Pacific Biological Station during the 1950's (Brett 1957). The problems of passage of adult salmon upstream beyond dams are difficult enough, but can be overcome with fish ladders, lifts and simple transport by tank truck around dams. The problems of movement of fry and smolts downstream, however, appear to be still largely insurmountable. The juvenile salmon are usually drawn through the penstocks and turbines of power-generating stations or swept over spillways. They suffer not only physical injury, but also gas bubble disease, owing to exposure to water supersaturated with dissolved gases (Fickeisen and Schneider 1976).

## 5.7. Developments on Streams and their Estuaries

Shorelines of streams and estuaries are preferred areas for assorted industrial developments, partly because water is available for cheap transportation, and such areas usually represent low cost in construction. The presence of water, in some cases, provides also a desirable attribute for residential condominiums and marinas for recreational boats. Unfortunately, such areas are also prime habitat for salmon.

Shoreline developments invariably involve dredging and filling. Marshes may be destroyed. Intertidal mudflats may be filled in with dredged material or other rock, sand and gravel to provide support for buildings. Juvenile salmon are often found in large numbers feeding on invertebrates in the shoreline marshes of estuaries. The destruction of these rearing habitats can have a severe impact on salmon populations. Recognition of the ecological value of these habitats has led managers to require mitigation, or in important salmon-feeding habitats, compensation. In some vitally important areas for salmon, for which compensation is not possible, developers may be denied authorization to proceed with a development.

### 5.8. Ocean Dumping

Up to 1975, there was little control of ocean dumping in Canada, except possibly under the Navigable Waters Protection Act to maintain navigable channels at an adequate depth, and under the Fisheries Act, where there was deemed to be a threat to fisheries resources. This changed in 1975 when the federal government promulgated the Ocean Dumping Control Act (ODCA). The ODCA was drafted to conform with the Convention on the Prevention of Marine Pollution by Dumping of Wastes and Other Matter (commonly known as the London Dumping Convention), which was brought into force in 1975, and for which Canada is a Contracting Party. No ocean dumping was allowed in Canadian coastal waters without approval after the ODCA came into effect. Applications for ocean dumping were reviewed by a Regional Ocean Dumping Advisory Committee (RODAC), and a permit was issued under orders of the Minister of the Environment, only if a RODAC recommended it.

Substances that could be dumped at sea were controlled by regulations under the ODCA. Certain substances, such as mercury and cadmium and their compounds, organohalogen compounds, petroleum hydrocarbons and persistent plastics, were prohibited for ocean dumping, except when present in trace amounts in other materials, e.g., dredge spoils. Trace amounts of different prohibited substances were defined by upper limits given by chemical analyses or effects on organisms as determined by bioassays. Other metals and toxic organic compounds could only be dumped at sea if special precautions were taken.

The total number of ocean dumping permits issued in Canada during 1987 were 165, divided up among regions as follows: 101 in the Atlantic Region; 37 in the Pacific & Yukon Region; 25 in the Quebec Region; and 2 in the Western & Northern Region (Environment Canada 1989). Of the 165 permits issued and used in 1987, 122 (73.9%) covered the dumping of dredged material. The other 43 permits were issued for disposal of a variety of other materials, including vessels (5 permits), fish offal (27 permits), scrap metal (1 permit), fire arms and prohibited weapons (1 permit) and excavated construction rubble (2 permits).

Ocean dumpsites are selected on the basis that there will be minimal impact on fisheries resources, and an effort is made not to proliferate dumpsites. For example, estuarine areas through which salmon migrate and cod-spawning areas in coastal waters are avoided in selection of dumpsites. Information is regularly compiled on research and monitoring at dumpsites in the Pacific and Yukon Region (e.g., Sullivan 1987). One dumpsite that receives a great deal of material from Vancouver Harbour and the surrounding area is the Point Grey dumpsite, which had been used unofficially since the 1930's. Monitoring and specific investigations have been conducted there since it was officially designated as a dumpsite in 1975. Some of the results have been reviewed by Waldichuk (1989a). Although ecological changes have occurred there as a result of dumping, no alarming trends in terms of impacts on fisheries in the Strait of Georgia or in the Fraser River estuary have been apparent.

The Ocean Dumping Control Act was subsumed as Part VI in the new Canadian Environmental Protection Act (CEPA) on 30 June 1988. Existing ODCA regulations were incorporated into CEPA without substantive change in the following year. Among other associated federal legislation, the Fisheries Act plays an important role in protection of fish habitats with the implementation of Part VI of CEPA.

## 6. Case Histories in British Columbia of Effects of Human Activities on Fish Habitats

There is a wide range of experience connected with various human activities, which have environmental impacts, and with efforts to protect fish habitats in Canada. Some of this experience can be used to provide case histories on assorted developments and strategies used to prevent or mitigate the adverse effects of such developments on habitats. In the following are selected examples, taken from the Pacific Region, of activities that have created conflicts with maintenance of fisheries resources and, in some cases, have inflicted injuries on habitats.

### 6.1. Forestry — The Carnation Creek Experience

The subject of fish-forestry interaction has always been controversial to some extent. While there was general scientific agreement that logging has an unfavourable effect on the freshwater phase of salmonids, there was little quantitative information prior to 1970 to support this contention. Therefore, in the early 1970's an experiment was designed to examine a watershed and fisheries production in it for a period of 5 years before logging, 5 years during logging and 5 years following logging (Narver and Chamberlin 1976). The experiment received the cooperation and involvement of many government agencies, universities, and private enterprise, including the forest industry. It was multi-disciplinary. The project covered the 16-year period from 1971 to 1986 in three phases: (1) pre-logging, 1972–1976; (2) logging, 1977–1981; and (3) post-logging, 1982–1986. A variety of effects was observed during the course of the study which has been thoroughly documented by Hartman and Scrivener (1990).

Very briefly and simply, the effects of logging appeared to have a positive effect initially on the production of coho smolts, but there was a later decline because recruitment of coho to the adult population decreased. In more detail, again highly simplified, removal of the forest canopy over the creek by logging increased solar insolation and water temperature. The fish-egg incubation period was reduced, leaving a longer period for the fry and smolts to feed in the creek. Coho fry reached a larger size earlier than before logging, and the proportion of 1- to 2-year-old smolts leaving Carnation Creek increased (Fig. 7). Because 1+ smolts were smaller than 2+ smolts, and smaller smolts survived poorly in the ocean compared with larger ones, overall survival in the ocean declined. This would have reduced spawning escapement unless increased numbers of smolts were sufficient to offset the increased mortality. Coho escapement after logging (1982–87) was lower than before logging (1971–76), but whether this was due to increased mortality or increased exploitation was unknown. It has been noted that during the course of the Carnation Creek study, marine conditions have not been particularly favorable for salmonids, with a major El Niño event, giving increased ocean temperatures, occurring during 1982–83. It was noted also that streamside logging activities decreased the stability of the stream channel and of its large wood debris.

The effects of both logging and ocean conditions have been especially severe on chum salmon. The composition of the spawning gravel, along with the foregoing stream changes, reduced survival and numbers of chum salmon fry. Escapement of chum salmon was only 200 adults in 1989, which was 10% of the pre-logging mean (2100). Chum salmon returns during the years 1986–89 were less than 375 annually for three of those years, and no chum fishery occurred in Barkley Sound during those years. Models indicated that 74% of the negative impact can be attributed to recent ocean conditions and 26% to logging. Details on the effects of climatic variability, clear-cut logging and fishing on the numbers of coho and chum salmon returning to Carnation Creek have been given by Holtby and Scrivener (1989). Hartman and Scrivener (1990) suggested that the greatest value of long-term case history studies is that they reveal ecosystem processes that permit separating effects caused by climate variability from man-induced impacts.

### 6.2. The Pulp and Paper Industry in British Columbia

There are 11 pulpmills on the coast and 9 in the interior of British Columbia. Most of the production is

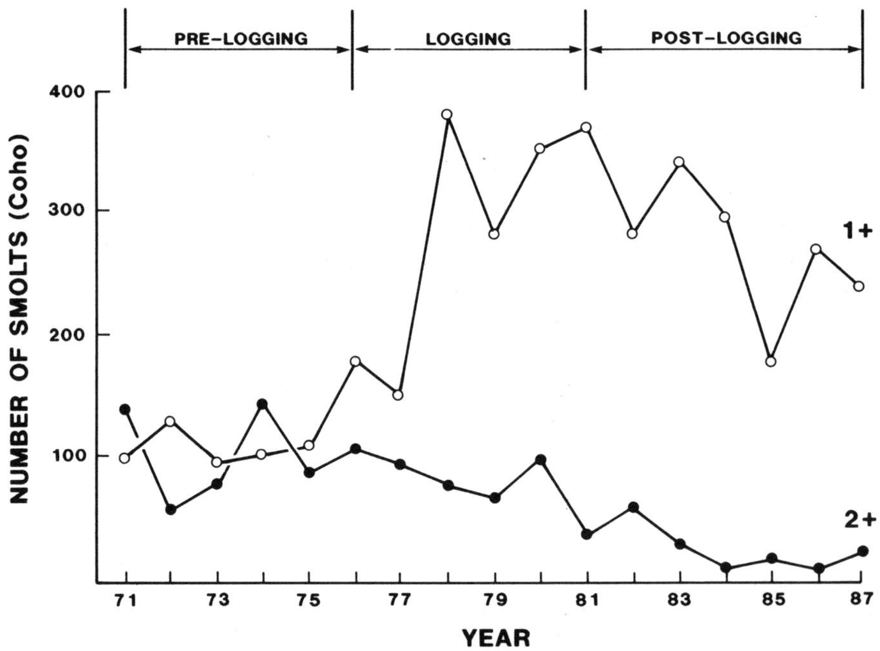


FIG. 7. Numbers of 1+ and 2+ coho salmon smolts migrating from Carnation Creek, during the period 1971 to 1987 (from Hartman and Scrivener 1990).

in kraft pulp, but there is still one mill (Port Alice) that produces sulphite pulp. Some of the mills produce groundwood (mechanical) pulp for newsprint. A few mills have gone into production of thermo-chemical mechanical pulp, which is a combination of mechanical and chemical disintegration of the wood into fibres at high temperature. Effluents from the coastal mills are discharged into estuaries or coastal seawater. Interior mills discharge their effluents into nearby rivers. All interior mills provide at least primary treatment for their effluent, which usually constitutes settling out of the wood solids in a treatment lagoon. Most of the coastal mills discharge their effluents untreated into the sea.

The case-history studies given here will be confined to coastal mills (Fig. 8). The problem of dioxins and furans has been discussed already in section 5.2 and will not be repeated. The traditional problems of pulpmill effluents relate to dissolved oxygen depletion, toxicity and benthic habitat alteration from deposition of suspended solids. Most of the problems caused by effluent discharges from coastal mills in British Columbia have been a result of their high biochemical oxygen demand (BOD). These have been primarily due to effluents from sulphite pulpmills. One area that presents a perennial low-oxygen problem, however, is Alberni Harbour, and it receives effluent from a kraft mill.

### *6.2.1. Port Alice*

The pulpmill at Port Alice, on the northwest coast of Vancouver Island, has been in operation since 1917, producing sulphite pulp. The low-oxygen conditions in Neroutsos Inlet, which received the effluent from this mill, were first identified in 1957 (Waldichuk 1958). Dissolved oxygen concentrations were near 1 $mg \cdot L^{-1}$ within 10 km of the mill in late summer. To overcome this situation the mill, in 1970, changed the chemical used for digestion of wood chips from non-recoverable calcium bisulphite to recoverable ammonium bisulphite, which produces a spent liquor that can be burned. This conversion was not immediately effective in restoring the dissolved oxygen levels in Neroutsos Inlet, because the recovery system was not installed at that time. Before the recovery system was installed in 1975, the "red liquor" from the digestors was barged out to sea.

After installation of the recovery system, intensive studies were conducted in Neroutsos Inlet by the pulpmill company environmental staff (Tollefson 1982) and by their consultants (Vigers et al. 1978), during the period of 1978–80. The improvements in dissolved oxygen concentrations in the top 2 m of Neroutsos Inlet from 1969, when there were no controls, through the period 1973–77, when the effluent was barged for deep-ocean disposal, to 1978, when the recovery system was in operation, are shown in Fig. 9 (Tollefson and Tokar 1978). Although there was a marked improvement in 1978, there was still a significant depression in dissolved oxygen in the southern end of Neroutsos Inlet, particularly near Port Alice. The situation did not improve in the 1980's. During June–October 1985, low dissolved oxygen concentrations were observed again, which were attributed to an increase in production, a changeover in the mill to a high-BOD product and a combination of warm weather and a low freshwater inflow into the inlet (Colodey and Pomeroy 1985, unpublished).

### *6.2.2. Port Edward*

The pulpmill at Port Edward on Watson Island, near Prince Rupert, was installed as a sulphite mill in 1950. Low dissolved oxygen concentrations in waters (Wainwright Basin and Porpoise Harbour) receiving effluent from the mill were first noted in 1961 (Waldichuk 1966), following reports that herring were being netted there either dead or moribund. A new kraft mill was added to the sulphite mill in 1967, and its effluent was discharged into Wainwright Basin. An outfall for the spent sulphite liquor was installed off the west coast of Ridley Island also in 1967. In 1976, the total effluent of the sulphite mill was piped into Chatham Sound. Subsequently that year, the sulphite mill was closed, and the full production of the mill on Watson Island was devoted to kraft pulp. A new kraft mill was brought on-line in 1978. The pipeline conveying sulphite mill effluent to Chatham Sound was abandoned, and a new outfall with diffuser for the enlarged kraft mill was installed in Porpoise Harbour in 1978.

The changes in the dissolved oxygen concentrations in Porpoise Harbour during the course of the foregoing modifications in the mill and the effluent disposal systems are shown in Fig. 10 (Kay 1987). The dissolved oxygen concentrations were restored to near-normal levels by 1979. In Wainwright Basin, where dissolved oxygen concentrations were near zero from top to bottom in 1961 (Waldichuk 1966), they ranged from 6.9 $mg \cdot L^{-1}$ at 15 m to 10.4 $mg \cdot L^{-1}$ at 10 m depth, and from 7.0 to 10.0 $mg \cdot L^{-1}$ at 30 m in 1985 (Dwernychuk 1986). These changes clearly show the small impact of the low-BOD kraft effluent in contrast with the large oxygen-depressing effect of high-BOD sulphite effluent.

### *6.2.3. Port Alberni*

The original kraft pulpmill at Port Alberni went into operation in 1947 with a production of 220 tons of unbleached kraft pulp per day. It was decided by the mid-1950's to expand production of this mill. Studies

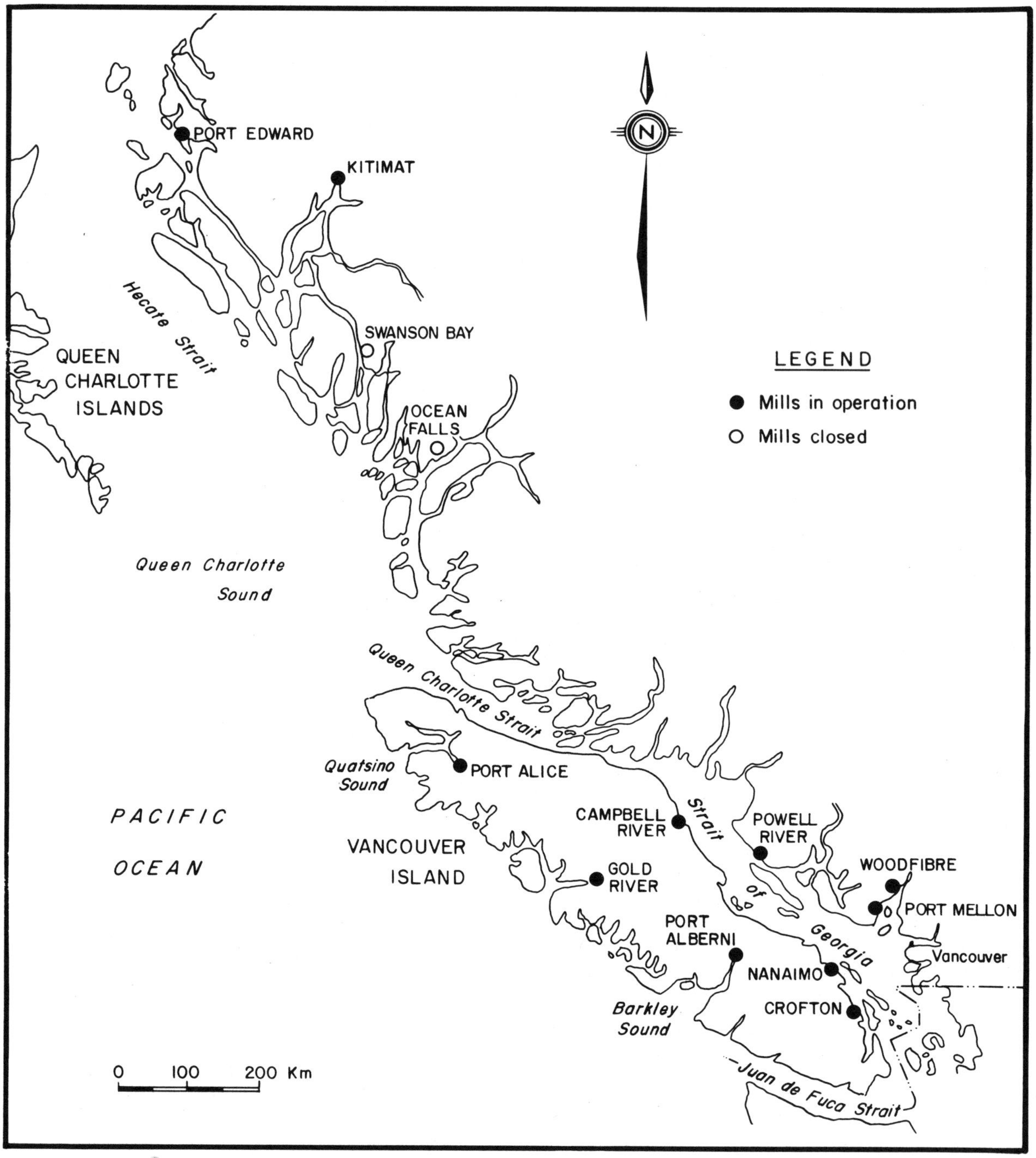


FIG. 8. Locations of existing and closed coastal pulpmills in British Columbia.

on the oceanography of Alberni Inlet (Tully 1949), on toxicity of a typical full-bleach kraft-mill effluent (Alderdice and Brett 1957) and on its BOD (Greer et al. 1956) indicated that if the dissolved oxygen requirement for salmonids in Alberni Inlet were met, the acute (lethal) toxicity of the effluent would not be a serious problem. Tully (1949) demonstrated that the dissolved oxygen supply in the surface zone of Alberni Inlet could be maintained at an adequate level for salmonids during the summer by increasing the discharge of the

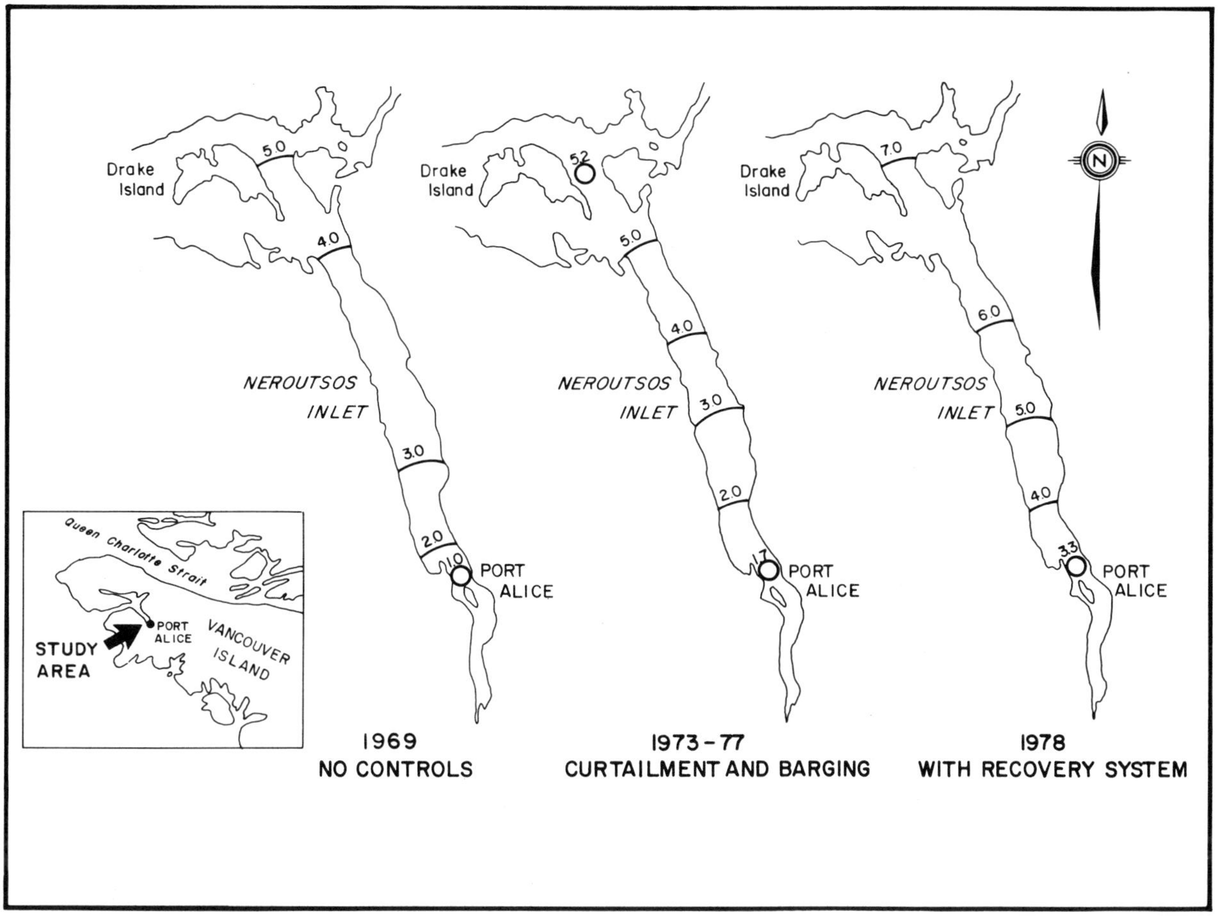


FIG. 9. Changes in mean dissolved oxygen concentrations (mg·L$^{-1}$) in the upper 2 m of Neroutsos Inlet during July, August, September, and October of different years, under various operating conditions (from Kay 1987).

Somass River. This was done with a dam on the Stamp River, a tributary of the Somass River, at its exit from Great Central Lake.

The original plan for expansion to a 500-ton-per-day full-bleach kraft mill was abandoned, because the dissolved oxygen supply in the surface zone of Alberni Inlet would be over-taxed. Instead, the expansion that took place in the late 1950's was mainly into groundwood pulp for newsprint. By the summer of 1965, dissolved oxygen levels in Alberni Harbour dropped to critical levels (see Fig. 9), in spite of maintenance of Somass River discharge at a level to supply adequate dissolved oxygen, as predicted by Tully's (1949) model. As noted earlier, the company soon undertook corrective action to remove wood solids, which were responsible for about 10% of the oxygen uptake from harbour waters. A clarifier was installed in 1966 and a biobasin in 1970. Improvement in the summertime dissolved oxygen concentrations was noted in Alberni Harbour for several years after these installations were made (Fig. 11). There was a serious decline in dissolved oxygen at a monitoring station in mid-harbour during the warm, dry summer of 1985 (see Fig. 11), and this was repeated in 1986 (Waldichuk 1988b). Dissolved oxygen concentrations appeared to have increased in August 1987, possibly owing to a cooler and wetter summer. There is a genuine threat to the salmonid stocks in the Alberni Inlet system, inasmuch as the valuable sockeye run occurs at the time of year that the dissolved oxygen in Alberni Harbour is depressed to its lowest level. This was demonstrated during the warm, dry summer of 1990, when even in the absence of a commercial fishery, escapement objectives were not met.

### 6.3. Mining — The Rupert Inlet Experience

The main environmental impact of this mine on the Rupert Inlet-Holberg Inlet ecosystem has been the

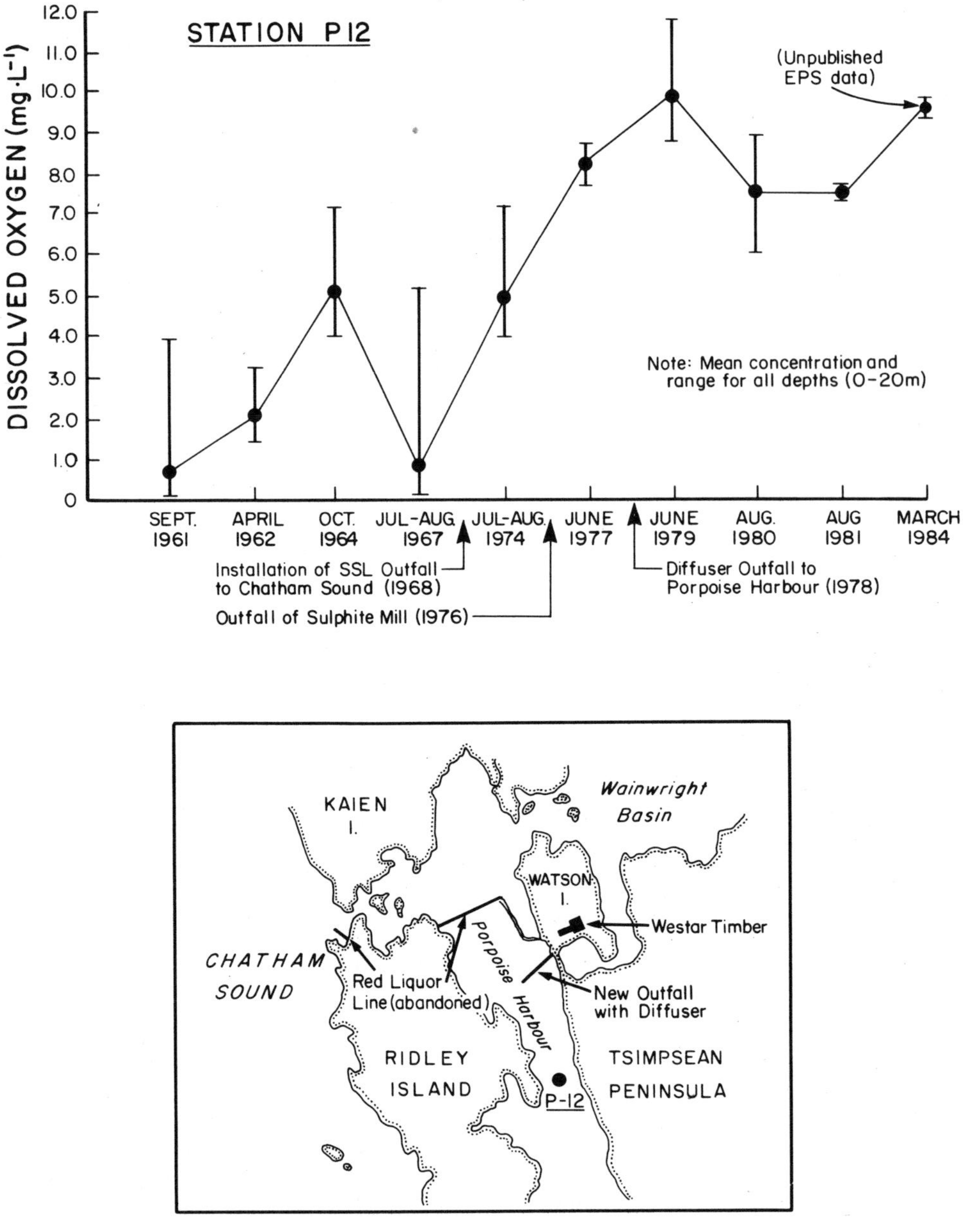


FIG. 10. Changes in dissolved oxygen concentrations ($mg \cdot L^{-1}$) in Porpoise Harbour, British Columbia, from 1961 to 1981, arising from modifications in effluent disposal and pulp production at the Port Edward pulpmill (from Kay 1987).

deposition of a large volume of tailings on the bottom, and an even larger volume of waste rock deposited in the intertidal and subtidal zones on the northeast side of Rupert Inlet. The tailings have covered the deepest parts of the Rupert Inlet-Holberg Inlet system of Quatsino Sound. Some of the suspended rock particles have also been carried by strong tidal currents through Quatsino Narrows into the more open waters of Quatsino Sound. The physical impact of deposition on the bottom of a large volume of pulverized rock has been confined mainly to Rupert and Holberg Inlets. The bottom habitat, especially of the deeper parts of this inlet system, has been essentially obliterated by the deposited tailings. Species living on the bottom or in the bottom sediments of the deepest part of the inlets would be expected to be most affected.

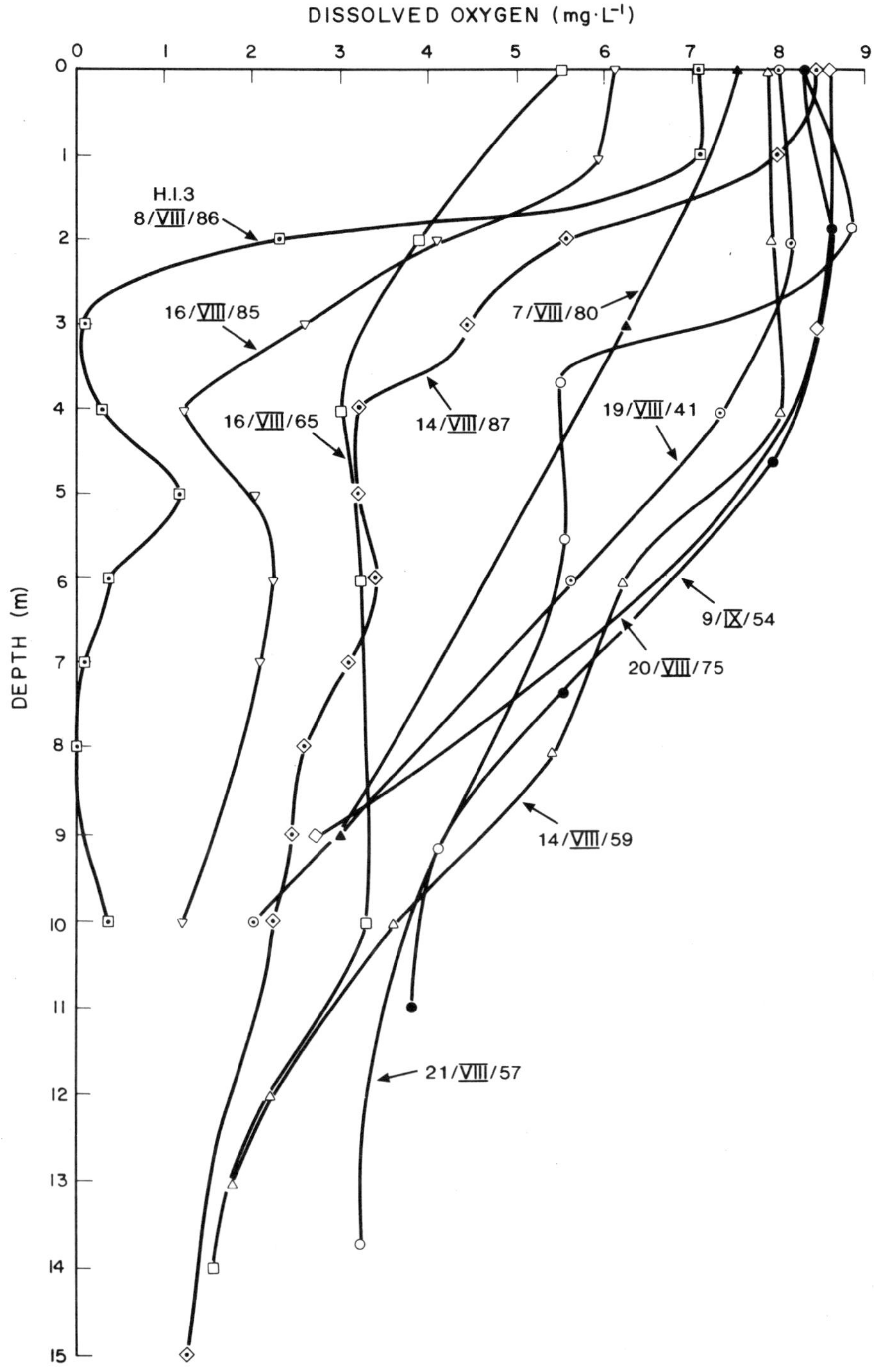


FIG. 11. Dissolved oxygen profiles at a monitoring station near Hohm Island in Alberni Harbour, British Columbia, during late summer in selected years from 1941 to 1987 (updated from Waldichuk 1988b).

The runs of Pacific salmon species to streams tributary to Rupert Inlet do not appear to have been affected. This is probably due to the fact that the tailings are largely confined to waters deeper than 50 m and the salmon frequent mainly the upper layer. Little is known about the effect the tailings have had on bottom fish in the inlets, but fish are capable of swimming out of the active tailings deposition areas. The Dungeness crab, *Cancer magister*, population in the inlets may not have suffered too greatly, because the preferred depth range of this crab is from the intertidal zone to about 30 m, which is well above the zone of suspended tailings particles or deposited tailings. The one species that may have suffered the most from the tailings discharge is the prawn, *Pandalus platyceros*, which prefers a deepwater, rocky habitat (Waldichuk and Buchanan 1980). Its population appears to have declined during the course of the mining operation, and the one fisherman harvesting prawns in Rupert Inlet moved on to other prawn-fishing areas outside of Quatsino Sound by 1980. While the inlet system will probably become recolonized with its indigenous species after the mining operation closes down, it may be necessary to emplace large boulders on the bottom to restore the bottom habitat for prawns.

The main impact of suspended solids in tailings from the Island Copper Mine is on the bottom of the inlets after they have been deposited. An effort was made by the mine to achieve rapid deposition of the tailings by pre-mixing them with seawater for flocculation before discharge. Hence, only a small portion of the solid particles in the tailings are in actual suspension for any appreciable length of time.

The principal conclusions that can be drawn from the review by Waldichuk and Buchanan (1980) on the significance of the environmental changes due to waste disposal from the mine are: (1) deep-water benthic habitats have been drastically modified by the tailings; (2) intertidal and shallow subtidal habitats on the northeast side of Rupert Inlet have been eliminated by the waste rock; (3) there has been no significant bioaccumulation of metals by marine organisms from the tailings or the waste rock; (4) the bottom habitat of the prawn population in Rupert Inlet has been undoubtedly severely impacted by the tailings, and the habitat of Dungeness crabs at the lower depth of that range has been probably also adversely affected; and (5) there is no evidence of a significant effect on salmonid runs to tributaries of the Rupert-Holberg Inlet system.

The mine waste disposal into Rupert Inlet has been allowed to continue, because there has been no indication of a human health hazard from consumption of contaminated seafood, and the impact on commercial fisheries of the area has been rather small, if any. Biological studies in the area (e.g., Ellis and Heim 1985) suggest that recolonization of the bottom of the inlet system after tailings disposal has ceased will be comparatively rapid, except for such species as prawns that prefer a rocky bottom. Although there was little opposition from local citizens and environmentalists at the time the Island Copper Mine went into operation, and subsequently, this situation could drastically change in another locale at another time, as witnessed at Kitsault, B.C., 10 years later (Ellis 1982; Littlepage et al. 1984).

### 6.4. Dams and Diversions for Hydroelectric Power Development — The Nechako River Experience

The most controversial issue concerning hydroelectric power in British Columbia in recent times has been the proposal by Alcan Smelters and Chemical Ltd. (Aluminum Company of Canada Ltd.) to expand hydroelectric power production at Kemano for their aluminum smelters at Kitimat. Phase I, with construction of the Kenney Dam at the Grand Canyon on the Nechako River to form the Nechako reservoir, and a tunnel through the Coast Range to divert the water to the powerhouse on the Kemano River, draining into Gardner Canal, was completed in 1957. Phase II, or Kemano II, as proposed, would require a dam on the Nanika River, now draining into the Bulkley River and then to the Skeena River, a diversion tunnel from Nanika Lake to Tahtsa Lake, which is part of the Nechako Reservoir, and a further diversion of Nechako River flows. A second tunnel, parallel to the original one through the Coast Range, would carry the additional water to the second powerhouse on the Kemano River.

Alcan was proceeding on the basis of a water licence granted by the B.C. government in 1949, which is valid to 1999. During the winter of 1979–80, the Nechako River discharge was reduced to 12.7 $m^3 \cdot s^{-1}$ (450 cfs), which was about one-third of that considered necessary to safeguard chinook salmon production. The Department of Fisheries and Oceans, therefore, requested a minimum of 28.3 $m^3 \cdot s^{-1}$ (1000 cfs) in the Upper Nechako River, plus any additional flow necessary to maintain acceptable water temperatures in the lower river. The company would not comply. Court action was initiated under Section 20(10) of the Fisheries Act. This resulted in an injunction being served on Alcan on 5 August 1980, ordering the release of specified volumes of water, as requested by the Department. Alcan complied.

Alcan continued to comply with the court order until 1985, when it refused to allow the injunction to be renewed. The court case, with Alcan challenging the powers of the federal government under S.20(10) of the Fisheries Act, was set to commence on 31 March 1987. Postponements followed and the case was finally settled out of court in September 1987 (Nechako Fisheries Conservation Program 1988). Alcan was given the right to use additional Nechako River water, provided it protected salmon in the river but forfeited the right, in perpetuity, to use water from the Nanika River.

Environmentalists were not satisfied with the decision and took the federal government to court[2]. Besides the impact of the diversions on sockeye and chinook salmon in the Nechako River, there were concerns about the effects of joining the Fraser River and Skeena River drainage basins, with possible transfer of parasites and diseases between the two watersheds.

Some of the biological concerns with respect to the consequences of the Kemano hydroelectric project for natural salmon production are given by Mundie and Bell-Irving (1986).

### 6.5. Agriculture — The Serpentine and Nicomekl River Experience

Agriculture contributes to a host of aquatic environmental problems in the interior of British Columbia, but few streams exemplify the effects of agriculture at their estuaries like the Serpentine and Nicomekl rivers. They both drain agricultural land of the lower Fraser Valley and flow into Mud Bay, which adjoins Boundary Bay, just north of the USA/Canada border. They both support runs of coho salmon, steelhead and cutthroat trout. Summaries are given on environmental conditions in both streams in the Greater Vancouver Liquid Waste Management Plan Stage 1 (GVRD 1987).

Both streams suffer from periodic low dissolved oxygen concentrations. High nutrient levels in runoff stimulate algal growth. When the algae die down they decompose and deplete the dissolved oxygen in the water. A number of fish kills have occurred in the Serpentine River, and these have been attributed primarily to low dissolved oxygen levels. Losses of spawning coho salmon, as a percentage of the total run (350 to 1100 coho counted annually since 1980), were reported as follows: 42% in 1980; 36% in 1983; and 35% in 1984 (Moore 1984). Concentrations of dissolved oxygen as low as 1.1 $mg \cdot L^{-1}$ were recorded.

In the Nicomekl River, 1500 to 3500 adult coho have been counted each year since 1970. Dissolved oxygen levels have been found to be depressed from July to October, because of decay of heavy algal blooms (Bourque and Hebert 1982). The spawning coho salmon can be expected to be under considerable stress as a result of the low oxygen levels.

Both the Serpentine and Nicomekl rivers contribute high concentrations of faecal coliforms from agricultural runoff, which have been largely responsible for closure of shellfish beds for commercial harvesting in Boundary Bay since about 1950.

### 6.6. Foreshore Development and Urbanization — The Fraser River Estuary Experience

The Fraser River is one of the great salmon rivers in the world. Adult salmon must pass through the estuary of this river on their way to the spawning grounds which may be a long way upstream. Juvenile salmon spend a certain amount of time, depending on the species, in the estuary feeding before they proceed out to sea. The Fraser River estuary is surrounded by urban settlement, with the City of Vancouver to the north and numerous suburban communities to the south and east. There is continuing pressure from municipalities bordering the estuary to develop foreshore areas for assorted purposes, ranging from marinas to residential condominium complexes.

Much of the wetland delta area was dyked for agricultural purposes early in the century. Marshes, eelgrass beds and tideflats that remain are vitally needed for wildlife and fish (Hoos and Packman 1974). A development that took place in the Fraser River estuary during the early 1980's, namely, an expansion of an existing coal port terminal, was preceded by a review by a Federal Environmental Assessment Panel (FEARO 1979). This Panel took into account all the ecological values in the area where the expansion was proposed. It concluded that eelgrass beds provide prime habitat for both juvenile salmonids and birds. The saltmarsh area, on the other hand, was considered to be suitable for the birds but of little value to the fish, inasmuch as it was bathed only by the highest tides. The impact of the development was determined in terms of the effect it would have on eelgrass density and distribution. The final size and configuration of the terminal enlargement was designed to have a minimum impact on eelgrass beds as habitat for fish and wildlife. Follow-up studies after the construction for the enlarged terminal was completed proved that protection of eelgrass beds as a priority was probably a wise decision (Levings 1985; Waldichuk 1985). Chinook and chum resided between the port and the Tsawwassen B.C. Ferry Terminal causeways, and chinook fed on invertebrates, especially harpacticoids, produced in eelgrass and sandflat habitats.

With the new federal habitat policy of "no net loss" (DFO 1986), it is essential to compensate with newly created habitat of equal value to that destroyed or alienated by development in an area like the Fraser River estuary.

[2]The Agreement was struck down by a court order in May 1991. The ruling was successfully appealed by Alcan and the Federal Government, but this decision was then put before the Supreme Court of Canada. In February 1993, the Supreme Court declined leave to appeal. The agreement thus stands.

## 7. Fish Habitat Management

Habitat management for fisheries is based in part on science, but to a large extent on experience. In the present state of knowledge, a habitat manager works more on the basis of inference and intuition than on habitat science in protecting fish habitats. Ideally, it would be valuable to have conclusive documentation on the impact of various human activities on fish stocks. This kind of information has been rarely obtained in Canada for a number of reasons. Research to acquire it requires careful experimental design to take into account all natural factors that might affect a fish population, as well as harvesting. Such research is of necessity long-term in nature to adequately cover at least one, but preferably several, life cycles of a particular species. It is obvious that such research is difficult to justify, when many projects, especially if they are done by private consultants under contract, must be of short-term duration, requiring completion within a fiscal year.

Even in the case of monitoring escapement of salmonids, some useful data could be acquired when there are potential effects on a stock from human activities in a stream or in its watershed. Unfortunately, as runs decline in such situations, there is a tendency among fishery officers to "write the stream off" and to abandon further escapement estimations. Thus, a data set, which could have been useful for further reference, is often terminated before its full potential value is achieved.

### 7.1. Fish Habitat Management Models

There are no universally applicable rules or formulae for protection of fish habitats. Any physical change imparted to a fish habitat will lead to an ecological change, which may be positive or negative. The physical and ecological linkages in an aquatic ecosystem are complex and not too well understood. Experience gained in managing fish habitat is usually of greater value for protecting habitats than documented information on effects of particular disturbances. A fishery officer with a long history of maintaining streams in good order for spawning salmon may be in a better position to protect fishery habitats in streams than a highly qualified biologist newly graduated from a university.

This is not to say that research on the relationship of various physical and chemical characteristics of a stream with fish populations is of little value. There is merit in pursuing the development of relationships between fish populations and various important habitat variables. Much has already been done in this regard, particularly in the Rocky Mountains area of the USA for non-anadromous trout. A referee for this chapter noted that a recent report lists 85 different models that relate fish populations to various habitat variables. Shirvell (1989) pointed out, however, that models are generally applicable only for the local situations for which they were developed. He tested six models and found that they explained 50–96% of the variation in fish numbers for the data sets from which they were derived and only 7–30% of the variation in fish numbers or biomass for data sets from which they were not derived. He concluded that "because different habitat variables limit different fish populations, habitat models may not be widely transferable". Shirvell (1989) noted that models may be "location or environment-specific".

One can examine some scenarios in streams where different degrees of habitat modification occur. In the one extreme, a salmonid stream can be eliminated by filling it in and diverting the water elsewhere. Any runs of salmon native to that stream will either vanish or colonize other nearby streams. In any event, the stocks of salmon involved will likely become extinct. At the other end of the habitat-modification spectrum, no significant change in habitat due to human intervention has occurred. The stocks of salmon should remain unchanged, all other factors being equal. In between these two extremes, a wide variety of changes could be imposed on the salmon through human activities. Some may cause marked changes in the salmon's habitat, which lead to significant effects on the salmon's survival and dramatic impacts on the population. Others cause only subtle changes that may be obscured by other stresses in the salmon's life cycle, such as poor ocean conditions for survival or excessive harvesting.

Once the potential impact of a particular activity is identified, then there are various measures that can be taken to ameliorate impact, or to stop it entirely. Certain habitat variables will produce predictable impacts on fish populations. The key is to identify those habitat variables and to monitor them carefully.

Another approach is the so-called "limiting factors" analysis for the various species. For example, pink and chum salmon, on the one hand, are typically limited in freshwater by the amount of suitable spawning habitat. On the other hand, most chinook, coho and trout populations are limited by the amount of juvenile rearing habitat. A human activity that directly impacts a population's limiting factor is much more critical than one that has an impact on some other life-history stage.

### 7.2. Water Quality Criteria and Standards

On the basis of available information on the effects of various substances on aquatic organisms, particularly fish, criteria have been developed for the maximum concentrations allowed in aquatic systems without the risk of damaging fisheries resources. Bioassay data usually provide an LC50 value for a particular substance, which is the concentration of the substance that kills 50% of the test organisms over a given period of exposure, usually 96 hours. The approach for deriving water quality criteria with both acute (lethal) and chronic (sub-lethal) bioassays is illustrated in Fig. 12. Various documents have been prepared giving the maximum permissible concentrations of substances for fish-bearing waters, the most notable one being "Water Quality Criteria 1972" (NAS 1972). The US Environmental Protection Agency periodically updates this document (e.g., EPA 1976). Guidelines for water quality used for various purposes, including fish, have been prepared under the auspices of the Canadian Council of Resource and Environment Ministers (CCREM) (Leggatt 1986).

Standards for water quality have legislative backing through regulations. For example, maximum permissible faecal coliform concentrations in oyster-growing waters are enforced by regulation. Likewise, drinking water standards are legislated.

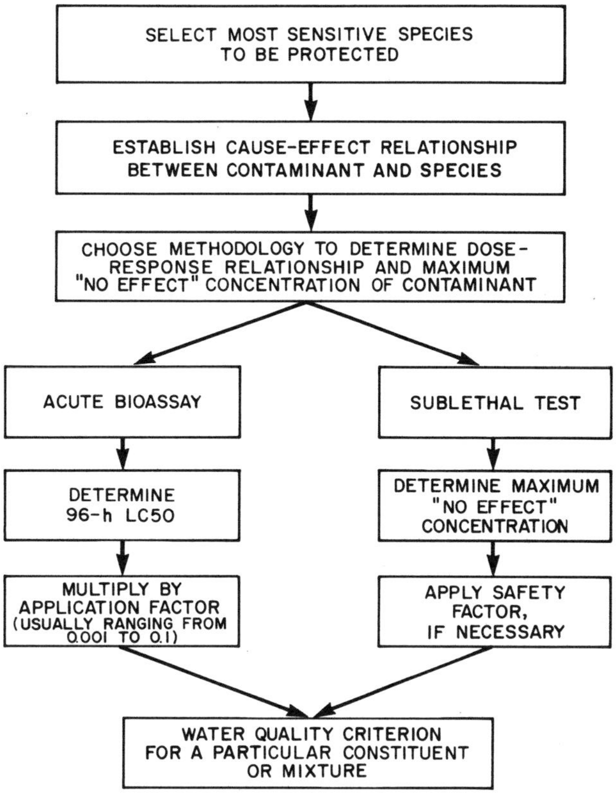


FIG. 12. Schematic diagram showing stepwise procedure for obtaining water quality criteria for aquatic contaminants.

### 7.3. Application of Laboratory Bioassay Results to Field Situations

There is always a question of validity in applying laboratory bioassay data to field situations. Conditions in the field cannot be fully simulated in the laboratory. For this reason, an application factor of at least 0.1 is applied to LC50 values derived in a laboratory from acute bioassays. This is essentially a safety factor that takes into account certain factors in the field that were not represented in the laboratory. It also allows to some extent for sublethal stress that may arise from nonlethal concentrations of a substance. It is important when applying laboratory data to the field to first appropriately validate the laboratory data in the field.

### 7.4. Monitoring Water Quality, Habitat Change and Fish Populations

Ideally, a program of water quality monitoring should be conducted if there is any input of contaminants into waters frequented by fish. This is particularly important if there is any possibility of damage to a fishery directly from the contaminant through its toxicity or indirectly through its effect on the aquatic environment. The effect of effluent from the pulpmill at Port Alberni on the dissolved oxygen concentration in Alberni Harbour, as discussed in section 6.2.3, is an example of the latter. Monitoring dioxin levels in crustacean shellfish in Howe Sound will signal when the uptake of dioxins by crustaceans has been reduced to an acceptable level and the crustaceans are safe to eat. The input of dioxins from the pulpmills has to be reduced to meet the requirement for safe human seafood consumption. It is important to select the right variables to monitor in the right media, in order to provide the essential information for water quality management. Analytical techniques should not be modified during the course of the monitoring program, in order that the data from one period are comparable to another. A system of sound quality assurance/quality control should be applied in all analyses to ensure that reliable data are being acquired.

Monitoring habitat change is somewhat more difficult than monitoring water quality. It must take into account not only the water but also the bottom. Perhaps the best technique for monitoring benthic habitat change is to note the changes in benthic communities. In the case of estuaries, where juvenile salmon feed, a decline in populations of amphipods may be a significant adverse change for salmon. Changes in biomass and in species diversity do not always provide the best type of information to quantify habitat alteration (Waldichuk 1973).

Monitoring fish populations is one activity for which fishery managers are best experienced. There are many pitfalls in monitoring fish populations that should be either avoided or properly accounted for. Usually there are good data on harvesting of a fish stock. Unless there is a fish fence on a stream, the quality of salmon escapement data is often poor. Tag and recovery programs can improve escapement data over those from strictly visual sightings. But even with careful counts of spawning adults and downstream-migrating smolts, the survival of salmon at sea may not be known. Thus, as shown in the Carnation Creek experience, it is sometimes difficult to pin down changes in fish populations that are attributable to anthropogenic effects.

### 7.5. Habitat Management Based on Monitoring Data

A good set of consistent monitoring data can be extremely useful for habitat management. A decline of a particular characteristic, suitably chosen to represent the quality of a habitat, would provide a warning that some corrective measures must be taken. For example, a reduction in the area of eelgrass beds or freshwater marshes would alert habitat managers that something is wrong ecologically. Perhaps contaminants are causing the decline, and suitable measurements should be initiated to trace the sources. A decline in fish production that cannot be attributed to overharvesting or underescapement would be a signal that there are unfavourable conditions in a stream for fish egg incubation and fry and smolt rearing. Other indices could be used as monitors of habitat quality to provide an early warning of habitat degradation.

### 7.6. Habitat Management Based on Inference

Unfortunately, there are few situations where monitoring data of the type discussed above are available to provide clear trends on habitat quality. There are still very few quantitative cause-effect relationships established that would allow managers to select specific measurements that would tell them how the quality of a habitat is changing or precisely how it might be affected by a particular development. Prediction of an effect from introduction of a particular contaminant, based on bioassays and water quality criteria, can be comparatively reliable. But the impact on rearing of juvenile salmonids of a shoreline development is much more difficult to estimate. In most such cases, managers base their decisions on experience and inference (defined as "a process of deduction based on limited knowledge"). Very often, if freshwater marshes or eelgrass beds in intertidal areas are involved, there will be strong opposition from habitat managers to developments that encroach on these ecosystems. On the basis of the no-net-loss policy, compensation would be required with creation of new eelgrass beds or freshwater marshes of equal extent and utilizability as those destroyed by dredging or filling. Boyd et al. (1989) discuss the permissible levels of inference in habitat management for salmonids, and how the use of inference differs in habitat management from that in scientific circles.

From a practical point of view, habitat managers base their decisions, in part, on information acquired from referrals to other agencies. For example, a fish habitat manager in the Department of Fisheries and Oceans in British Columbia will usually refer an application for waste disposal from a particular industry to the Waste Management Branch in the British Columbia Ministry of Environment. Habitat managers are urged to be proactive, that is, anticipate habitat damage before it occurs. While this is a highly desirable approach to habitat management, it is sometimes difficult to achieve when the essential data are unavailable. A case in point was the availability of data on dioxins and furans in crustacean shellfish near pulpmills only after sensitive analytical techniques were available for measuring these compounds in very low concentrations.

## 8. Legislation and Policy for Control of Pollution and Habitat Destruction and Degradation

In Canada, there is provincial and federal legislation to control water pollution and fish habitat damage. The provinces generally control the discharge of contaminating effluents into freshwater systems, and in some provinces such as British Columbia, into coastal waters as well. The federal government has jurisdiction over commercial fisheries, and therefore, has the mandate to control pollution and habitat damage for the commercially fished species.

### 8.1. Provincial Legislation

Every province in Canada has legislation for protecting water quality and aquatic living resources. In British Columbia, the main legislation to control contaminants entering aquatic systems is the Waste Management Act and the regulations thereunder. The Waste Management Branch of the Ministry of Environment administers this Act and its regulations. Permits are issued by the Waste Management Branch to dischargers of effluent, with specific conditions that have to be met under this legislation. British Columbia also has a Water Act that deals with freshwater resources, and a Health Act that protects consumers against consumption of contaminated seafood and bathers against infection from bathing in sewage-contaminated waters.

### 8.2. Federal Legislation

There are several federal Acts that are designed to directly or indirectly protect water quality and fish habitats. The principal federal legislation is the Fisheries

Act, which dates back to 1867 and has undergone numerous amendments since then. Sections 31 and 33 of that Act specifically deal with protection of fish habitats and control of deleterious substances deposited in waters frequented by fish. The Fisheries Act is powerful in its provisions for protection of fish habitats. The key in enforcement and prosecution is proof of damage caused by a particular human activity. Under the Fisheries Act, regulations have been developed for effluent disposal into fish-bearing waters from such industries as pulp and paper manufacture and mining.

The Canadian Environmental Protection Act, especially Part VI dealing with ocean dumping, has been discussed already in section 5.8. Other federal legislation that has elements of water pollution control and fish habitat protection includes the Coastal Fisheries Protection Act, Territorial Seas and Fishing Zones Act, Arctic Waters Pollution Prevention Act and the Canada Shipping Act. The Food and Drug Act under the Department of National Health and Welfare controls the levels of toxic substances allowed in human foods, including seafood.

### 8.3. Fish Habitat Management Policy

The new Fisheries and Oceans "Policy for the Management of Fish Habitat" (DFO 1986) is, according to the Honourable Tom Siddon, Minister of Fisheries and Oceans at the time, "an explicit recognition by the federal government that fish habitats are national assets." It helps to fulfill Canada's commitment to the United Nations' World Conservation Strategy, which among other things calls for "the maintenance of the support systems for fisheries and for the control of pollution". It fulfills also the need for sustainable development, as recommended by the Brundtland Commission (World Commission on Environment and Development 1987).

The objective of the new policy for fish habitat management is a net gain of productive capacity for fisheries resources in Canada. The goals of the policy are: (1) fish habitat conservation; (2) fish habitat restoration; and (3) fish habitat developments. The guiding principle is "no net loss of productive capacity of habitats" (DFO 1986).

The implementation strategies for fish habitat management as given by Fisheries and Oceans (DFO 1986), are as follows:

"1. ***Protection and compliance***. Protect fish habitats by administering the Fisheries Act and incorporating fish habitat protection requirements into land and water use activities and projects.

2. ***Integrated resource planning***. Participate in and encourage resource planning and management to incorporate fish habitat priorities into air, land and water use plans.

3. ***Scientific research***. Conduct scientific research to provide the information and technology necessary for the conservation, restoration and development of fish habitats.

4. ***Public consultation***. Consult the public on major or controversial fish habitat issues and on the development of new policies and legislation for fish habitat management.

5. ***Public information and education***. Promote public awareness in the conservation, restoration and development of fish habitats.

6. ***Cooperative action***. Encourage and support involvement by government agencies, public interest groups and the private sector to conserve, restore and develop fish habitats.

7. ***Habitat improvement***. Initiate projects and provide advice to other interested groups to restore and develop fish habitats, in support of the net gain objective.

8. ***Habitat monitoring***. Evaluate the effectiveness of decisions taken and techniques used to conserve, restore and develop fish habitats."

One of the difficulties in achieving the principle of "no net loss of productive capacity of habitats" is in establishing a quantitative approach to compensation for habitats lost to development. Some habitat managers and scientists will argue that certain habitats, e.g. spawning grounds, are irreplaceable and that compensation, therefore, would not be possible. But even for less vital habitat for which compensation may be valid, it is not yet possible to apply a formula that would provide a quantitative value for a particular habitat. What criteria could be used to assess the value of a habitat? For example, a particularly valuable eelgrass bed in an estuary has been used by juvenile salmonids for feeding and rearing. How much salt marsh would have to be provided in compensation for the loss of one hectare of that eelgrass bed?

## 9. Conclusions

1) Freshwater habitats for salmonids are the most vulnerable to human impact, because it is here that the adults spawn, the eggs incubate and the fry and smolts rear. The success of a year-class of a Pacific salmon species is usually determined by conditions during its freshwater phase.
2) Dissolved oxygen, suspended solids, temperature, acidity and salinity are considered to be aquatic characteristics that may be affected by human activities. Among these, dissolved oxygen and suspended solids have been found to be most

critical for survival of fish in freshwater, estuaries and coastal waters of British Columbia.

3) Among aquatic contaminants, pulpmill effluents have caused the greatest concern in freshwater and coastal marine environments of British Columbia. This concern increased with findings in the late 1980's of dioxins and furans in crabs, prawns and shrimps near coastal pulpmills and in fish species downstream of pulpmills in the Fraser River system.
4) The impact of logging on fish habitats in the watersheds of salmon streams of British Columbia is perhaps the most severe among human activities. It contributes to large fluctuations in streamflow, increased suspended sediment load, loss of canopy along stream banks, increased temperatures, and loss of large wood debris used by juvenile salmonids for shelter in the stream.
5) The 16-year (1971–1986) Carnation Creek project is perhaps the best long-term study on the effects of logging on fish populations in Canada. It was carried out in 3 phases: pre-logging, logging and post-logging. The impact of logging appeared to be greater on chum salmon than on coho, but the long-term effect will not be known for several decades. The effect of unfavourable ocean conditions (high temperatures) during the 1982–83 El Niño confounded the analysis of the results of the impact of logging on fish populations.
6) Other industries in British Columbia which have had an impact on fish habitats include pulpmills, mines, agriculture and hydroelectric power development. The best documented effects on the aquatic environment and aquatic living resources are those stemming from the pulp and paper industry.
7) Foreshore developments in such areas as the Fraser River estuary are believed to have seriously affected rearing areas for juvenile salmonids, but there is little documented evidence to prove this.
8) There are no universally applicable rules or formulae for fish habitat management. Much of the action taken to protect fish habitats is based on experience, inference and intuitive judgement. Models relating habitat variables to production of fish have been usually site-specific and stock-specific.
9) It is desirable to take an anticipatory or proactive approach to habitat management. Inadvertent production and release of toxins into the aquatic environment, however, make proactive habitat management difficult, because such releases cannot be anticipated, based on available knowledge. A good example of such releases are the dioxins and furans noted in organisms in the vicinity of pulpmills in British Columbia only after analytical technology was available to detect these highly toxic substances at parts per trillion levels.
10) There is both provincial and federal legislation that can be used for protection of fish habitat. But the oldest and most powerful is the federal Fisheries Act, in which Sections 35 and 36 are specifically designed to protect all aspects of fish habitats.
11) The Fisheries and Oceans "Policy for the Management of Fish Habitat", in which the guiding principle is no net loss of the productive capacity of habitats, along with other measures to protect fish habitats, will greatly reduce the incremental loss of valuable fish habitat and fulfills the needs of "sustainable development" as given by the Brundtland Commission (World Commission on Environment and Development 1987).

## References


ADDISON, R.F., M.E. ZINCK, R.B. ACKMAN, P. CHAMUT AND A. JAMIESON. 1972. Results of the elemental phosphorus monitoring program, Long Harbour, Nfld. July 1970–April 1971). Fish. Res. Board Can. Tech Rep. 303: 25 p.

ALDERDICE, D.F. 1972. Factor combinations. Responses of marine poikilotherms to environmental factors acting in concert, p. 1659–1722. In O. Kinne [ed.] Marine ecology, Vol. 1, Environmental Factors Part 3. Wiley (Interscience), New York, NY.

ALDERDICE, D.F., AND J.R. BRETT. 1957. Some effects of kraft mill effluent on young Pacific salmon. J. Fish. Res. Board Can. 14: 783–795.

ALDERDICE, D.F., W.P. WICKETT, AND J.R. BRETT. 1958. Some effects of temporary exposure to low dissolved oxygen levels on Pacific salmon eggs. J. Fish. Res. Board Can. 15: 229–249.

BÉLAND, P. 1988. Witness for the prosecution. Nature Can. 17: 28–36.

BEWERS, J.M., AND G.J. PEARSON. 1972. The behavior of particulate material in the treatment lagoons of a bleached kraft mill. Water Air Soil Pollut. 1: 347–358.

BIRTWELL, I.K., J.C. DAVIS, M. HOBBS, M.D. NASSICHUK, M. WADICHUK, AND J.A. SERVIZI. 1981. Brief presented to the British Columbia Pollution Control Board Inquiry pertaining to the Annacis Island Sewage Treatment Plant and municipal waste discharges into the Lower Fraser River. Can. Ind. Rep. Fish. Aquat. Sci. 126: xi + 40 p.

BOURQUE, S., AND G. HEBERT. 1982. A preliminary assessment of water quality and biota in the Serpentine and Nicomekl Rivers and Mahood

Creek 1974–1975. Environment Canada, Environmental Protection Service, Pacific Regional Program Report 82-17.

BOYD, F.C., L.H. BAIN, R.J. BEAMISH, J.C. DAVIS, C.D. LEVINGS, C.D. MCALLISTER, R. MORLEY, A.R. THOMPSON, M. WALDICHUK, AND H. WALDOCK. 1989. Report of the Working Group on Permissible Levels of Inference in Habitat Management of Salmonids, p. 194–195. *In* C.D. Levings, L.B. Holtby, and M.A. Henderson [ed.] Proceedings of the National Workshop on Effects of Habitat Alteration on Salmonid Stocks. Can. Spec. Publ. Fish. Aquat. Sci. 105: 199 p.

BRETT, J.R. 1952. Temperature tolerance in young Pacific salmon, genus *Oncorhynchus*. J. Fish. Res. Board Can. 9: 265–323.

1956. Some principles in the thermal requirements of fishes. Q. Rev. Biol. 31: 75–87.

1957. Salmon research and hydroelectric power development. Bull. Fish. Res. Board Can. 114: 26 p.

BRETT, J.R., AND D.F. ALDERDICE. 1958. The resistance of cultured young chum and sockeye salmon to temperatures below 0°C. J. Fish. Res. Board Can. 15: 805–813.

BUNCH, J.N. 1987. Effects of petroleum releases on bacterial numbers and microheterotrophic activity in the water and sediment of an Arctic marine ecosystem. Arctic 40(Suppl. 1); 172–183.

BURDICK, G.E., E.J. HARRIS, H.J. DEAN, T.M. WALKER, J. SKEA, AND D. COLBY. 1964. The accumulation of DDT in lake trout and the effect on reproduction. Trans. Am. Fish. Soc. 93: 127.

COLE, H.A. [ED.]. 1979. The Assessment of sublethal Effects of Pollutants in the Sea. The Royal Society of London, University Press, Cambridge, U.K. 235 p.

COLODEY, A.G., AND W.M. POMEROY. 1985. Low dissolved oxygen in Neroutsos Inlet, British Columbia: observed impact of sulphite pulpmill effluent July–October 1985. Environmental Protection Service, West Vancouver, B.C., Interim Report (unpublished).

COWELL, E.B. [ED.]. 1971. The Ecological Effects of Oil Pollution on Littoral Communities. Adlard & Son Ltd., Bartholomew Press, Dorking, U.K. 250 p.

DAVIES, I.M., AND J.C. MCKIE. 1987. Accumulation of total tin and tributyltin in muscle tissue of farmed Atlantic salmon. Mar. Pollot. Bull. 18: 405–407.

DAVIS, J.C. 1976. Minimal dissolved oxygen requirements of aquatic life with emphasis on Canadian species: a review. Can. J. Fish. Aquat. Sci. 32: 2295–2332.

DFO. 1985a. The life cycle of Pacific salmon. Department of Fisheries and Oceans, Ottawa, Ont. 8 p.

1985b. Acid rain and fisheries. Department of Fisheries and Oceans, Communications Directorate, Ottawa, Ont. 4 p.

1985c. Fish habitat and forestry. Department of Fisheries and Oceans, Communications Directorate, Ottawa, Ont. 6 p.

1986. Policy for the management of fish habitat. Department of Fisheries and Oceans, Ottawa, Ont. 30 p.

DINA. 1989. Contaminants in Northern ecosystems and native diets: Summary of an Evaluation Meeting held in Ottawa, February 28 to March 2, 1989. Department of Indian and Northern Affairs, Ottawa, Canada, 7 p. + Appendix I.

DWERNYCHUK, L.W. 1986. Water quality and biological studies in the marine environment near the Westar Timber Pulp Mill 1985. Prepared for Westar Timber by Hatfield Consultants Limited, West Vancouver, B .C.

ELLIS, D. 1982. Kitsault, B.C. — Technical communication in a non-technical world. Mar. Pollut. Bull. 13: 78–80.

ELLIS, D.V., AND C. HEIM. 1985. Submersible surveys of benthos near a turbidity cloud. Mar. Pollut. Bull. 16: 197–203.

ENGELHARDT, F.R. [ED.]. 1985. Petroleum Effects in the Arctic Environment. Elsevier Applied Science Publishers, Amsterdam, The Netherlands, 281 p.

ENVIRONMENT CANADA. 1989. Keeping the Ocean Clean. Ocean Dumping Control Act. 1987/88 Annual Report. Minister of Supply and Services Canada, Ottawa, Ont., 32 p.

EPA. 1976. Water Quality Criteria for Water. U.S. Environmental Protection Agency, Washington, D.C. 256 p.

FALLIS, B.W. 1982. Trace metals in sediments and biota from Strathcona Sound, NWT, Nanisivik marine monitoring programme, 1974–1979. Can. Tech. Rep. Fish. Aquat. Sci. 1082: v + 34 p.

FEARO. 1979. Roberts Bank Port Expansion. Report of the Environmental Assessment Panel. Environmental Assessment and Review Process, Ottawa, Ont.

FICKEISEN, D.H., AND M.J. SCHNEIDER [ED.]. 1976. Gas bubble disease. Technical Information Center, Office of Public Affairs, Energy Research and Development Administration, Oak Ridge, TN. 123 p.

FORRESTER, C.R., K.S. KETCHEN, AND C.C. WONG. 1972. Mercury content of spiny dogfish (*Squalus acanthias*) in the Strait of Georgia, British Columbia. J. Fish. Res. Board Can. 29: 1487–1490.

FOURNIER, J.A., AND C.D. LEVINGS. 1982. Polychaetes recorded near two pulp mills on the coast of northern British Columbia: a preliminary taxonomic and ecological account. National Museum of Natural Sciences, National Museum of Canada, Syllogeus Series No. 40: 91 p.

GEYER, R.A. [ED]. 1980. Marine Environmental Pollution. 1. Hydrocarbons, Elsevier Scientific Publishing Co., Amsterdam, The Netherlands. 591 p.

GOLDBERG, E.D. 1975. Synthetic organohalides in the sea. Proc. Roy. Soc. Lond. B. 189: 277–289.

GOYETTE, D., AND J. BOYD. 1989. Distribution and environmental impact of selected benthic contaminants in Vancouver Harbour, British Columbia, 1985 to 1987. Environment Canada, Conservation and Protection, Environmental Protection, Pacific and Yukon Region, Reg. Prog. Rep. 89-02: 99 p.

GREER, B.A., R.E. GILLESPIE, AND P.C. TRUSSELL. 1956. Biochemical oxygen demand of total effluent from full-bleached kraft mill. Tappi 39: 599–603.

GVRD. 1987. Greater Vancouver Liquid Waste Management Plan Stage 1, Greater Vancouver Receiving Water Quality Conditions. Prepared by Coastline Environmental Services Ltd. and Envirochem Services, Greater Vancouver Regional District, Burnaby, B.C. 331 p.

HARDING, L.E., AND J.R. ENGLAR. 1989. The *Nestucca* oil spill: fate and effects to May 31, 1989. Environment Canada, Conservation and Protection, Environmental Protection, Pacific Region, Reg. Prog. Rep. 89-11: 52 p. + App. 1–4.

HART, J.L. 1973. Pacific Fishes of Canada. Bull. Fish. Res. Board Can. 180: 740 p.

HARTMAN, G.F., AND J.C. SCRIVENER. 1990. Impacts of forestry practices on a coastal stream ecosystem, Carnation Creek, British Columbia. Can. Bull. Fish. Aquat. Sci. 223: 148 p.

HILDEBRAND, L.P. 1984. An assessment of environmental quality in the Baie des Chaleurs. Environment Canada, Environmental Protection Service, Atlantic Region, Dartmouth, N.S., Rep. EPS-5-AR-84-8.

HODGSON, B. 1990. Alaska's Big Spill — Can the Wilderness Heal? Nat. Geogr. 177: 5–43.

HOLTBY, L.B., AND J.C. SCRIVENER. 1989. Observed and simulated effects of climatic variability, clearcut logging, and fishing on the numbers of chum salmon (*Oncorhynchus keta*) and coho salmon (*O. kisutch*) returning to Carnation Creek, British Columbia, p. 62–81. *In* C.D. Levings, L.B. Holtby, and M.A. Henderson [ed.]. Proceedings of the National Workshop on Effects of Habitat Alteration on Salmonid Stocks. Can. Spec. Publ. Fish. Aquat. Sci. 105: 199 p.

HOOS, L.M., AND G.A. PACKMAN. 1974. The Fraser River estuary: status of environmental knowledge to 1974. Dep. of the Environment, Special Estuary Series No. 1, West Vancouver Laboratory, West Vancouver, B.C. 518 p.

HOULT. D.P. [ED.]. 1964. Oil on the sea. Plenum Press, New York, NY. 114 p.

HOURSTON, A.S. AND R.H. HERLINVEAUX. 1957. A "mass mortality" of fish in Alberni Harbour, B .C. Fish. Res. Board Can. Prog. Rept. Pac. Coast Stn. 109: 3–6.

IDLER, D.R. 1969. Coexistence of a fishery and a major industry in Placentia Bay. Chem. Can. 21: 16–21.

IRUKAYAMA, K. 1967. The pollution of Minamata Bay and Minamata disease. Adv. Water Pollut. Res. 3: 153–183.

JANGAARD, P.M. 1970. The role played by the Fisheries Research Board of Canada in the "red" herring phosphorus pollution crisis in Placentia Bay, Newfoundland. Circular of the Office of the Fisheries Research Board's Atlantic Regional Director, Research , Halifax, N.S. No. 1: 20 p.

JOHNSTON, R. 1984. Oil pollution and its management, p. 1433–1582. *In* O. Kinne [ed.]. Marine ecology. Vol. V. Ocean Management. Part 3. Pollution and Protection of the Seas. Radioactive Materials. Heavy Metals and Oil. John Wiley & Sons, Chichester, U.K. p. 1098–1618.

KANNAN, N., S. TANABE, M. ONO, AND R. TATSUKAWA. 1989. Critical evaluation of polychlorinated biphenyl toxicity in terrestrial and marine mammals: Increasing impact of non-*ortho* and mono-*ortho* coplanar polychlorinated biphenyls from land to ocean. Arch. Environ. Contam. Toxicol. 18: 850–857.

KAY, B.H. 1987. West coast marine environmental quality: summary report. Environment Canada, Conservation and Protection, Environmental Protection, Pacific and Yukon Region, West Vancouver, B.C., Reg. Prog. Rep. 86-04.

KERSWILL, C.J. 1967. Studies on effects of forest sprayings with insecticides, 1952–63, on fish and aquatic invertebrates in New Brunswick streams: Introduction and summary. J. Fish. Res. Board Can. 24: 701–708.

KERSWILL, C.J., AND H.E. EDWARDS. 1967. Fish losses after forest sprayings with insecticides in New Brunswick, 1952–63, as shown by caged specimens and other observations. J. Fish. Res. Board Can. 24: 709–729.

KERSWILL, C.J., P.F. ELSON, M.H.A. KEENLEYSIDE, AND J.B. SPRAGUE. 1960. Effects on young salmon of forest spraying with DDT. Trans. 2nd Seminar on Biol. Problems in Water Pollution, 1959, p. 71. Robert A. Taft Eng. Center, Tech. Rept. No. W60.3, Cincinnati, OH.

LEACH, J.M., AND A.N. THAKORE. 1973. Identification of the constituents of kraft pulping effluent that are toxic to juvenile coho salmon (*Oncorhynchus kisutch*). J. Fish. Res. Board Can. 30: 479–484.

LEGGATT, E.A. 1986. Canadian water quality guidelines, p. 2–7. *In* CCREM. 1986. Environment Canada, Ottawa, Ont.

LEVINGS, C.D. 1985. Juvenile salmonid use of habitats altered by a coal port in the Fraser river estuary, British Columbia. Mar. Pollut. Bull. 16: 248–254.

LEVINGS, C.D., L.B. HOLTBY, AND M.A. HENDERSON [ED.]. 1989. Proceedings of the National Workshop on effects of habitat alteration of salmonid stocks. Can. Spec. Publ. Fish. Aquat. Sci. 105: 199 p.

LITTLEPAGE, J.L., D.V. ELLIS, AND J. MCINERNEY. 1984. Marine disposal of mine tailing. Mar. Pollut. Bull. 15: 242–244.

MALINS, D.C. [ED.]. 1977. Effects of petroleum on arctic and subarctic marine environments and organisms. Vol. 1. Nature and fate of petroleum. Vol. II. Biological effects. Academic Press Inc., New York, NY. 321 p. and 500 p.

MARPOL. 1973. International Conference on Marine Pollution, 1973. International Marine Organization. Publ. Sect. London.

1978. Protocol relating to the International Convention for the Prevention of Pollution from Ships, 1973 (MARPOL PROT 1978). International Marine Organization. Publ. Sect. London.

MASSÉ, R., D. MARTINEAU, L. TREMBLAY, AND P. BÉLAND. 1986. Concentrations and chromatographic profile of DDT residues in stranded beluga whales (*Delphinapterus leucas*) from the St. Lawrence Estuary. Can. Arch. Environ. Contam. Toxicol. 15: 567–579.

MCLEAY, D.J. 1973. Effects of a 12-h and 25-day exposure to kraft pulp mill effluent on the blood and tissues of juvenile coho salmon (*Oncorhynchus kisutch*). J. Fish. Res. Board Can. 30: 395–400.

1977. Development of a blood sugar bioassay for rapidly measuring stressful levels of pulpmill effluent to salmonid fish. J. Fish. Res. Board Can. 34: 477–485.

MCLEAY, D.J., AND D.A. BROWN. 1979. Stress and chronic effects of untreated and treated bleached kraft pulpmill effluent on the biochemistry and stamina of juvenile coho salmon (*Oncorhynchus kisutch*). J. Fish. Res. Board Can. 36: 1049–1059.

MEHRLE, P.M., D.R. BUCKLER, E.E. LITTLE, L.M. SMITH, J.D. PETTY, P.H. PETERMAN, D.L. STALLING, G.M. DE GRAEVE, J.J. COYLE, AND W.J. ADAMS. 1988. Toxicity and bioconcentration of 2,3,7,8-tetrachlorodibenzodioxin and 2,3,7,8-tetrachlorodibenzofuran in rainbow trout. Environ. Toxicol. Chem. 7: 47–62.

MILLER, R.A., L.A. NORRIS, AND C.L. HAWKES. 1973. Toxicity of 2,3,7,8-tetrachlorodibenzo-p-dioxin (TCDD) in aquatic organisms. Environ. Health Perspect. 5: 177–186.

MILNE, A., AND R.J. CHILDERHOUSE. 1978. Oil, ice and climate change: The Beaufort Sea and the search for oil. Department of Fisheries and Oceans, Institute of Ocean Sciences, Patricia Bay, Sidney, B.C. 103 p.

MOORE, B. 1984. Serpentine River, dissolved oxygen survey 1981–1983. Waste Management Branch, British Columbia Ministry of Environment, Victoria, B.C.

MUNDIE, J.H., AND R. BELL-IRVING. 1986. Predictability of the consequences of the Kemano hydroelectric proposal for natural salmon production. Can. Water Resources J. 11: 14–25.

NARVER, D.W., AND T.W. CHAMBERLIN. 1976. Carnation Creek — an experiment towards integrated resource management. Fish. Mar. Serv. Pac. Biol. Sta. Circ. 104: 20 p.

NAS. 1972. Water Quality Criteria 1972. Committee on Water Quality Criteria, Environmental Studies Board, National Academy of Sciences/National Academy of Engineering, Washington, DC. 594 p.

1975a. Petroleum in the Marine Environment. National Academy of Sciences, Washington, DC. 107 p.

1975b. Assessing Potential Ocean Pollutants. National Academy of Sciences, Washington, DC. 438 p.

1985. Oil in the Sea: Inputs, Fates, and Effects. National Research Council, National Academy of Sciences, National Academy Press, Washington, DC. 601 p.

NECHAKO FISHERIES CONSERVATION PROGRAM. 1988. Annual report September 14, 1987–March 31, 1988. P.O. Box 1630, Vancouver, B.C. 13 p. + app.

NEWCOMBE, C.P., AND D.D. MACDONALD. 1991. Effects of suspended sediments on aquatic ecosystems. N. Am. J. Fish. Manage. 11: 72–82.

NRCC. 1981. Acidification in the Canadian Aquatic Environment: Scientific Criteria for Assessing the Effects of Acidic Deposition on Aquatic Ecosystems, National Reseach Council Canada, Associate Committee on Scientific Criteria for Environmental Quality, NRCC No. 18475: 369 p.

PARK, CRIS C. 1987. Acid rain: rhetoric and reality. Methuen & Co. Ltd., London, U.K. 272 p.

PETERSON, C.L., W.L. KLAWE, AND G.D. SHARP. 1973. Mercury in tunas: a review. Fish. Bull. 71: 603–613.

ROGERS, I.H., J.C. DAVIS, G.M. KRUZYNSKI, H.W. MAHOOD, J.A. SERVIZI, AND R.W. GORDON. 1975. Fish toxicants in kraft effluents. Tappi 58: 136–140.

ROGERS, I.H., C.D. LEVINGS, W.L. LOCKHART, AND R.J. NORSTROM. 1989. Observations on overwintering juvenile chinook salmon (*Oncorhynchus tshawytscha*) exposed to bleached kraft mill effluent in the upper Fraser River, British Columbia. Chemosphere 19: 1853–1868.

SAUNDERS, R.L., AND J.B. SPRAGUE. 1967. Effects of copper-zinc mining pollution on a spawning migration of Atlantic salmon. Water Res. 1: 419–432.

SERGY, G.A., AND P.J. BLACKSALL. 1987. Design and conclusions of the Baffin Island Oil Spill Project. Arctic 40(Suppl.): 1–9.

SERVIZI, J.A., AND D.W. MARTENS. 1987. Some effects of suspended Fraser River sediments on sockeye salmon (*Oncorhynchus nerka*), p. 254–264. *In* H.D. Smith, L. Margolis, and C.C. Wood [ed.]. Sockeye salmon (*Oncorhynchus nerka*) population biology and future management. Can. Spec. Publ. Fish. Aquat. Sci. 96.

SHIGEMATSU, I. 1973. Cadmium poisoning. Environmental Health Report. Japanese Society of Public Health, Tokyo, Japan. (Translation available from Information Services Division, U.S. Enviromental Protection Agency, Washington, DC.)

SHIRVELL, C.S. 1989. Habitat models and their predictive capability to infer habitat effects on stock size, p. 173–179. *In* C.D. Levings, L.B. Holtby, and M.A. Henderson [ed.] Proceedings of the National Workshop on effects of habitat alteration on salmonid stocks. Can. Spec. Publ. Aquat. Sci. 105.

SHOMURA, R.S., AND H.O. YOSHIDA [ED.]. 1985. Proceedings of the Workshop on the Fate and Impact of Marine Debris, 26–29 November 1984, Honolulu, Hawaii, U.S. Dep. Commer., NOAA Tech. Memo. NOAA-TM-NMFS-SWFC-54: 580 p.

SMITH, J.E. [ED.]. 1968. Torrey Canyon — pollution and marine life. Cambridge University Press, Cambridge, U.K. 196 p.

SOMERS, E. 1982. The risk from environmental chemicals, p. 25–41. *In* Proceedings of the Seminar on Toxic Chemicals Management, Moncton, N.B., 28 April 1982.

SPRAGUE, J.B. 1964. Avoidance of copper-zinc solutions by young salmon in the laboratory. J. Water Pollut. Contr. Fed. 36: 990–1004.

1971. Measurement of pollutant toxicity to fish — III. Sublethal effects and "safe" concentrations. Water Res. 5: 245–266.

SPRAGUE, J.B., AND D.E. DRURY. 1969. Avoidance reactions of salmonid fish to representative pollutants, p. 169–179. *In* S.H. Jenkins [ed.]. Advances in Water Pollution Research. Pergamon Press, Oxford and New York.

SPRAGUE, J.B., P.F. ELSON, AND R.L. SAUNDERS. 1965. Sublethal copper-zinc pollution in a salmon river — a field and laboratory study. Int. J. Air Water Pollut. 9: 531–543.

SPRAGUE, J.B., AND D.W. MCLEESE. 1968a. Toxicity of kraft mill effluent for larval and adult lobsters, and juvenile salmon. Water Res. 2: 753–760.

1968b. Different toxic mechanisms in kraft pulp mill effluent for two aquatic animals. Water Res. 2: 761–765.

SULLIVAN, D.L. 1987. Compilation and assessment of research, monitoring and dumping information for active dump sites on the British Columbia and Yukon coasts from 1979 to 1987. Department of Environment, Conservation and Protection, Environmental Protection, Pacific and Yukon Region, Manuscr. Rep. 87-02: 263 p.

SWALES, S., R.B. LAUZIER, AND C.D. LEVINGS. 1986. Winter habitat preferences of juvenile salmonids in two interior rivers in British Columbia. Can. J. Zool. 64: 1506–1514.

TANABE, S., N. KANNAN, M. FUKUSHIMA, T. OKAMOTO, T. WAKIMOTO, AND R. TATSUKAWA. 1989. Persistent organochlorines in Japanese coastal waters: An introspective summary from a Far East developed nation. Mar. Pollut. Bull. 20: 344–352.

TANABE, S., N. KANNAN, A. SUBRAMANIAN, S. WATANABE, AND R. TATSUKAWA. 1987. Highly toxic coplanar PCBs: occurrence, source, persistency and toxic implications to wildlife and humans. Environ. Pollut. 47: 147–163.

THAIN, J.E., AND M.J. WALDOCK. 1986. The impact of tributyl tin (TBT) antifouling paints on molluscan fisheries. Water Sci. Tech. 18: 193–202.

THOMAS, M.L.H. 1973. Effects of Bunker C on intertidal and lagoonal biota in Chedabucto Bay, Nova Scotia. J. Fish. Res. Board Can. 30: 83–90.

TOEWS, D.A.A., AND M.J. BROWNLEE. 1981. A handbook for fish habitat protection on forest lands in British Columbia. Department of Fisheries and Oceans, Vancouver, B.C. 173 p.

TOLLEFSON, R. 1982. Environmental improvement at Neroutsos Inlet, B.C. Summary Report. Western Forest Products Ltd., Vancouver, B.C.

TOLLEFSON, R., AND E. TOKAR. 1978. Environmental improvements at Neroutsos Inlet, B.C. Vol. 1:

Summary Report. Rayonier Canada Ltd., Port Alice, B.C.

TULLEY, J.P. 1949. Oceanography and prediction of pulp mill pollution in Alberni Inlet. Bull. Fish. Res. Board Can. 83: 169 p.

UTHE, J.F., AND V. ZITKO [ED.]. 1980. Cadmium pollution of Belledune Harbour, New Brunswick, Canada. Can. Tech. Rep. Fish. Aquat. Sci. 963: v + 107 p.

VANDERMEULEN, J.H., AND D.C. GORDON, JR. 1976. Reentry of 5-year-old stranded Bunker C fuel oil from a low-energy beach into the water, sediments, and biota of Chedabucto Bay, Nova Scotia. J. Fish. Res. Board Can. 33: 2002–2010.

VIGERS, G., W. CAVE, AND R. JANSSEN. 1978. Environmental Improvement at Neroutsos Inlet, B.C. Vol. 6: Acute and chronic effects of Port Alice sulphite mill effluent to fish. Rayonier Canada Ltd., Port Alice, B.C.

WALDEN, C.C. 1976. The toxicity of pulp and paper mill effluents and corresponding measurement procedures. Water Res. 10: 639–664.

WALDICHUK, M. 1958. Some oceanographic characteristics of a polluted inlet in British Columbia. J. Mar. Res. 7(Suppl.): 536–551.

1966. Effects of sulfite wastes in a partially enclosed marine system in British Columbia. J. Water Poll. Contr. Fed. 38: 1484–1505.

1973. Trends in methodology for evaluation of effects of pollutants on marine organisms and ecosystems. Crit. Rev. Environ. Contr. 3: 167–211.

1974. Some biological concerns in heavy metals pollution, p. 1–57. *In* F.J. Vernberg and W.B. Vernberg [ed.]. Pollution and physiology of marine organisms. Academic Press, New York, NY. 492 p.

1985. Roberts Bank Coal Port, British Columbia: How did the Environmental Impact Assessment fare? Mar. Pollut. Bull. 16: 179–185.

1987. Pollution by abandoned mines. Mar. Pollut. Bull. 18: 422–423.

1988a. Effects of solid wood wastes on marine benthic organisms and habitats, p. 193–208. *In* O.A. Wolfe and T.P. O'Connor [ed.]. Oceanic Processes in Marine Pollution, Vol. 5, Urban Wastes in Coastal Marine Environments. Robert E. Krieger Publishing Company, Malabar , Florida, U.S.A. 273 p.

1988b. Alberni Inlet, British Columbia: a multi-use fjord system with fisheries resources and a large waste-disposal component, p. 107–135. *In* Proceedings Seventh International Ocean Disposal Symposium, Wolfville, Nova Scotia, 21–25 September 1987, Environment Canada, Ottawa, Ont.

1989a. Point Grey Dumpsite: Effects of waste disposal and constraints under Canadian regulations, p. 187–208. *In* D.W. Hood, A. Schoener and P.K. Park [ed.]. Oceanic Processes in Marine Pollution. Vol. 4. Scientific Monitoring Strategies for Ocean Waste Disposal. Robert E. Krieger Publishing Company, Malabar, Florida, U.S.A. 286 p.

1989b. The *Nestucca* oil spill. Mar. Pollut. Bull. 20: 419–420.

1990. Dioxin pollution near pulpmills. Mar. Pollut. Bull. 21: 365–366.

WALDICHUK, M., AND R.J. BUCHANAN. 1980. Significance of environmental changes due to mine waste disposal into Rupert Inlet. Canada Department of Fisheries and Oceans and British Columbia Ministry of Environment, West Vancouver and Victoria, B.C. 56 p.

WATERSTON, J. [ED.]. 1989. A Forum Report. Canadian Environmental Protection Act Priority Substances Science Forum, Canada Centre for Inland Waters, Burlington, Ontario, 22–23 February 1989. Rawson Academy of Aquatic Science, Ottawa, Ont. 137 p.

WERNER, A.E. 1965. Suspended solids from mill effluents. Can. Pulp Pap. Ind., April, p. 109–114.

1968. Gases from sediments in polluted coastal waters. Pulp Pap. Mag. Can. 69: 127–136.

WERNER, A.E., AND J. ROBINSON. 1978. Acute effects of wood-pulp on sockeye salmon (*Oncorhynchus nerka*). Water Air Soil Pollut. 9: 69–81.

WOBESER, G., N.O. NIELSEN, R.H. DUNLOP, AND F.M. ATTON. 1970. Mercury concentrations in tissues of fish from the Saskatchewan River. J. Fish. Res. Board Can. 27: 830–834.

WOLFE, D.A. (ED.). 1977. Fate and Effects of Petroleum Hydrocarbons in Marine Ecosystems and Organisms. Pergamon Press, Oxford, U.K. 478 p.

WORLD COMMISSION ON ENVIRONMENT AND DEVELOPMENT. 1987. Our Common Future. Oxford University Press, Oxford and New York. 400 p.

ZITKO, V., D.E. AIKEN, S.N. TIBBO, K.W.T. BESCH AND J.M. ANDERSON. 1970. Toxicity of yellow phosphorous to herring (*Clupea harengus*), Atlantic salmon (*Salmo salar*), lobster (*Homarus americanus*) and beach flea (*Gammarus oceanicus*). J. Fish. Res. Board Can. 27: 21–29.

# A Review of Federal Assistance to the Canadian Fishing Industry, 1945–1990

R. W. Crowley

*School of Policy Studies, Queens University*
*350 Sparks Street, 5th Floor, Ottawa, Ontario K1P 5R5*

B. McEachern, and R. Jasperse

*Department of Fisheries and Oceans, Market Development, Fisheries Operations*
*200 Kent Street, Ottawa, Ontario K1A 0E6*

Crowley, R. W., B. McEachern, and R. Jasperse. 1993. A review of federal assistance to the Canadian fishing industry, 1945–1990, p. 339–367. *In* L. S. Parsons and W. H. Lear [eds.] Perspectives on Canadian marine fisheries management. Can. Bull. Fish. Aquat. Sci. 226.

Government programs to assist the fishing industry have attempted to address real problems. In fact, however, they have been shaped by the government philosophy of the day and have not always contributed to long- term industry viability. Since 1945, many of these programs have exacerbated excess capacity in the industry and, in fact, have sometimes led to calls for further assistance.

Assistance objectives have changed over time, but they have seldom been in complete harmony with fisheries management objectives. Stated fisheries management objectives since the 1970's have been to reduce domestic and foreign fishing effort and to improve economic efficiency in the industry. Assistance programs have often detracted from these objectives, especially by compromising industry efficiency.

---

Les programmes d'aide à l'industrie halieutique visaient la résolution de problèmes réels. Par contre, ils étaient de fait modelés par la philosophie du parti au pouvoir et n'ont pas toujours contribué à la viabilité à long terme de l'industrie. Depuis 1945, nombre de ces programmes n'ont fait qu'exacerber la surcapacité de l'industrie et ont même mené à des demandes d'aide supplémentaire.

Les objectifs des programmes d'aide ont changé avec le temps, mais ils ont rarement été en complète harmonie avec les objectifs de la gestion des pêches. Les objectifs énoncés depuis les années 1970 ont visé une baisse de l'effort de pêche canadienne et étrangère et l'amélioration du rendement économique de l'industrie. Les programmes d'aide se sont souvent éloignés de ces objectifs au détriment de l'efficacité de l'industrie.

## I. Introduction

Canada's fishing industry seems always to have been either in crisis or between crises. Indeed, crises in the fisheries have existed as long as fishermen have fished. Economic conditions in the industry fluctuate with markets, costs of production, structural factors as well as fish catches. Cycles are the industry's fortune, perhaps misfortune. They vary in duration and have many origins. For groundfish, the mainstay of the Atlantic fishery dominated by frozen fish exports to the United States, many observers have noted a regular 7-year boom and bust cycle. Because of the importance of the groundfish fishery in Atlantic Canada, the repercussions of these cycles have often extended throughout the fishing industry and sometimes beyond.

Cycles usually imply fluctuations in individual incomes, and this has provided a rationale for government intervention. Assistance programs have often been rationalized in terms of stabilizing incomes or industrial development or regional development. For this reason, some have claimed that fisheries policy *is* regional development policy in Atlantic Canada (though no one makes the same claim for the Pacific Coast). Whatever the case, fisheries managers (and governments) have too often responded to crises as they occur, rather than tackling head-on the difficult task of managing the recurrent cycles that are an integral part of the fisheries.

Fisheries are especially important to Atlantic Canada because few alternative employment opportunities exist. In Atlantic Canada, 75% of all communities have a fishing base and about 20% are fully dependent on fishing. Excluding centres with populations above 10 000, there is almost total dependence on fisheries throughout Atlantic Canada. On the Pacific Coast, economic dependence is less pronounced but cycles are also dominant. Freshwater fisheries, largely in land-locked provinces, are important for many small isolated communities and they, too, are subject to the ubiquitous cycles.

When economic conditions are favourable, the fisheries provide good livelihoods for many Canadians. Canada has some of the world's most productive fisheries and good access to the richest food markets in the world. The central issue has not been the wealth generated from the fisheries, but how it is distributed. Who is to receive benefits from the fisheries? Are the benefits to be dissipated as widely as possible or are they to accrue to a smaller group? This central question — seldom posed — is what government involvement is all about. And reflecting government's attempts to distribute benefits as widely as possible, assistance programs often have been used to "top-up" incomes in the troughs of cycles because too many people are involved in the fisheries and wish to remain for the "good times". This no-win situation has characterized the fisheries throughout the period analyzed in this paper.

This paper traces the history of major government assistance programs affecting the fishing industry since 1945. The first section discusses conditions in the industry in the period to 1976, and the programs that were put in place. The second section deals with the period from 1977 to 1984 when fishing jurisdiction was extended to 200 miles with all of the potential that that single action represented. The final section briefly traces developments since 1984.

Discussion is limited to federal government programs for several reasons. The federal government is responsible for fisheries management and undertakes most fisheries-related expenditures in Canada. Only while fish are being processed in land based plants (and before they are exported) is there explicit provincial jurisdiction. This provincial jurisdiction has nonetheless often been the basis on which direct and indirect provincial assistance has been provided, usually encouraging growth in processing (and sometimes harvesting) capacity.

Provincial assistance has generally accentuated trends already in place at the federal level. Provincial loan boards have provided individual fishermen with substantial financing for a variety of fishing and nonfishing purposes. Provincial grants have supported new or expanded processing capacity, frequently camouflaged as part of other programs (e.g., industrial development). In any case, because of the dearth of data on these programs and their complexity, it was decided to exclude provincial programs from the purview of this paper.

As with other parts of the federal government, fisheries administration has undergone many name changes since Confederation. The following designations for the current Department of Fisheries and Oceans are used intermittently in the text.

| | |
|---|---|
| 1867–1884 | Department of Marine and Fisheries — Fisheries Branch |
| 1884–1892 | Department of Fisheries |
| 1892–1914 | Department of Marine and Fisheries — Fisheries Branch |
| 1914–1920 | Department of Naval Services — Fisheries Branch |
| 1920–1930 | Department of Marine and Fisheries — Fisheries Branch |
| 1930–1969 | Department of Fisheries |
| 1969–1971 | Department of Fisheries and Forestry |
| 1971–1976 | Department of Environment — Fisheries and Marine Service |
| 1974–1975 | Department of Environment (Minister of State — Fisheries) |
| 1976–1979 | Department of Fisheries and the Environment — Fisheries and Marine Service |
| 1979– | Department of Fisheries and Oceans (DFO) |

## II. The Post-War Period (1945–76)

### A. OVERVIEW

#### 1. Atlantic

*Early Days*

In the aftermath of the Second World War, many argued that Government programs were needed to help adjustment to peacetime and to address the huge demands that would be placed on the economy. With respect to the fishing industry, the arguments for adjustment assistance were particularly cogent.

Canada's trading patterns for seafood products shifted substantially during and after the Second World War (DFO, Annual Reports, various years). The United States became more important and Commonwealth and European markets relatively less so. A severe dollar shortage affected some markets. Canned fish production which had been stimulated during the war had to be reduced in volume and sold in new markets.

By 1948, food surpluses appeared in North America. The salted fish business required support and fish oil markets were weak. The situation was such that the Minister of Fisheries in 1949 announced a program for the development of Canada's commercial fisheries designed to expand markets and to provide assistance to the fishing industry.

By the early 1950's, the industry had increased its efficiency and reduced dependence on low income markets. Increased world demand for fish products provided the rationale for industry to argue the need for more capacity. Since fishing fleets were mainly pre-war vintage, new vessels were required. The Government adjusted regulations restricting the import of trawlers, and fishing fleets on both coasts expanded considerably throughout the 1950's. In addition, evolution in vessel, gear and processing technology increased productivity (and capacity).

From 1947 to 1952, North American consumption of groundfish fillets doubled. This allowed for a doubling of Canadian production to 45 000 tonnes. Groundfish filleting and freezing capacity expanded substantially in the early 1950's, and exports of groundfish "blocks" to the United States in particular increased. The increased volume meant that industry had to devote more attention to marketing. With the increase in frozen production there was a corresponding reduction in salted fish production (a trend that has continued to present).

By 1957, Canadian exports of fish products had achieved a new reasonably stable pattern which persists to this day. Approximately 70% of annual production was being exported, with the United States accounting for about three quarters of total exports. Markets for salted fish had continued to decline and per capita consumption of fish was constant (it edged up during the 1980's). The market for fish meal was depressed due to large increases in production in South America (during the 1980's there was increased competition from the southern hemisphere with respect to a number of species and product forms).

### *1960's and 1970's*

In Atlantic Canada, the early 1960's were characterized by strong demand for salted fish and good conditions in the fresh and frozen sector. This was the first time since the war that salted fish markets had shown some strength; this resulted from low supplies in Europe. Nonetheless, only 61% of Newfoundland cod was used for salted fish in 1962 compared to 84% in 1953. In the early 1960's, "new" groundfish species such as redfish and flatfish were increasingly used for frozen fillets and blocks. Scallop and herring harvests increased substantially.

By the early 1960's, it had become apparent that one of the problems facing the Atlantic fishing industry was increased pressure on fish resources from the expanded fishing fleets of other nations. Competition caused Canadian fisheries managers to argue for action to protect the resource. There was also a growing concern about issues such as productive capacity, efficiency and limited entry. A marked expansion of the fishing fleet occurred in the early 1960's. One measure was the number of licensed Atlantic Coast trawlers and draggers which increased 50% from 1963 to 1967. In 1964 alone, the number increased from 385 to 452 vessels.

By the mid 1960's, the Atlantic fishing industry was optimistic. Fish landings had attained record levels in 1965 and 1966. Canada ranked second only to Japan in terms of exported value of fish products. Many argued that groundfish landings could increase 100% in the following decade with expansionary programs by government and industry. The industry was attracting foreign investment from Britain and other European countries, while prices for most fish products rose continuously from 1961 to 1966.

In late 1966, however, prices for frozen groundfish products in the United States began to decline. The 1967 output of frozen groundfish fillets and blocks dropped to 100 000 tonnes from the previous year's record of 112 000 tonnes. During that year, salted cod production increased by more than 20%, as many fishermen in Newfoundland returned to their traditional practice of salting.

Frozen groundfish fillets and block output rose to a record high of 117 000 tonnes in 1968. Output in other major producing countries also reached high levels. To exacerbate matters, the Pope rescinded the decree which had compelled Roman Catholics to eat fish each Friday. As a result of all of these factors, the United States market for groundfish weakened considerably. Notwithstanding these market problems, many new processing facilities were constructed in the late 1960's and early 1970's, often with government assistance.

While groundfish marketing problems continued during the late 1960's, most other sectors of the fishing industry remained profitable. Atlantic herring and crab fisheries continued to expand. In 1969, the Department of Fisheries and Forestry introduced major new policies including input restrictions, marketing agencies and the establishment of a 12 mile exclusive fishing zone. In the following year, fisheries "closing lines" were imposed to put major bodies of water (such as the Gulf of St. Lawrence and Dixon Entrance on the Pacific) exclusively under Canadian management (DFO Annual Reports 1969, 1970).

There were also some administrative changes at this time (DFO Annual Reports, various years). The Fisheries and Marine Service (established in 1973) assumed new responsibilities for **small craft harbours** and a service was initiated to deal with international matters. Prior to 1973, fishing harbours were the responsibility of the Department of Public Works and administrative control rested with the Ministry of Transport. As well, considerable attention was being focused on the preparation of Canada's position for the 1974 United Nations Law of the Sea Conference. Many were anticipating that substantial changes would come about with the extension of fishing jurisdiction (that eventually occurred in 1977).

Favourable market conditions resulted in high prices for the main fishery products in 1971 and 1972. Production of salted fish continued its long-term decline during these years, with 52 000 tonnes put to salt in 1972 compared to 70 000 tonnes in 1970. In the early 1970's, Atlantic Coast herring landings declined from record levels in the 1960's. As landings declined more herring was used for food products instead of fishmeal. The year 1973 was the best year to date for the fishing industry due largely to buoyant markets — but this prosperity soon vanished.

By 1974, the fishing industry was experiencing declining catches, rising costs of production and sluggish market conditions. On the Atlantic Coast the drop in landings was attributed to over-exploitation of some fish resources as well as severe ice conditions and a 6-week strike by Newfoundland trawl fishermen. Markets were sluggish due to poor economic conditions and competition from beef and poultry. To address these crises in the industry, a new fisheries policy was developed in 1975, followed quickly by new programs. The crisis in the industry at this time as well as the Fisheries Rehabilitation Program are discussed later in this paper.

Markets for most fish products began to improve in the latter part of 1975, but resource scarcity persisted, especially for many Atlantic groundfish species. Canadian fisheries jurisdiction was extended to 200 miles in 1977, resulting in major changes in Canadian fisheries management. This subject is covered later in this paper.

### 2. Pacific

The fishing industry in B.C. has traditionally been based on salmon, the single most lucrative fishery in all of Canada. Canned salmon was the most important product throughout the early post war period and Canada still produces nearly 25% of reported world canned salmon production (Crowley and Brown 1989). However, fresh and frozen salmon products have continuously gained in importance, as have herring, halibut and groundfish products.

After 1945, the B.C. fishing industry provided good economic returns to participants. Markets were strong and until 1948, when European fisheries resumed significant production, the United Kingdom purchased large quantities of both canned salmon and canned herring.

Market conditions weakened in 1949 and B.C. canning companies launched a campaign in 1950 to increase sales of canned salmon in Canada. The success of this campaign resulted in many similar campaigns in ensuing decades. When the industry was faced with excess inventories, it was usually able to market them in Canada with the aid of promotional campaigns (Commission on Pacific Fisheries Policy 1982).

During the 1950's, the West Coast fishing industry attracted many new vessels. Economic conditions varied with prices, the size of salmon runs, and collective bargaining disputes. The year 1958 was significant because of a very large run of Adams River sockeye salmon. Landings were the highest in 50 years and created great prosperity in the industry.

During the 1960's, the Pacific groundfish fishery expanded; the herring reduction fishery was terminated due to over-fishing of the herring stocks; and the Salmon Licence Control Program was implemented. Economic conditions in the industry continued to fluctuate from year to year, however, largely reflecting salmon abundance and the degree of over expansion of the fleet.

In the early 1970's, the herring roe fishery was developed to meet the expanding demand for "kazunoko" in the Japanese market. This fishery usually lasts only 1 month during the spring of the year, but it has added greatly to the incomes of fishermen and processors. It has more than compensated for the decline in halibut landings which began just prior to the termination of reciprocal fishing privileges coinciding with the 1977 extension of fishing jurisdictions by Canada and the United States. During the mid 1970's, there were some lean years for the industry due to depressed groundfish markets, as well as low salmon landings, but both of these subsequently rebounded as they had in the past.

## B. Marketing Assistance

In this section, we examine marketing assistance programs implemented during the post war period, including those that have carried on (in whole or part) to the present day. In later sections, we examine other forms of assistance: assistance to fishermen, assistance (largely to processors) for facilities and other institutional assistance/developments. These divisions are simply for organizing the discussion; some overlapping is unavoidable. (For an overview see OCED 1980).

## 1. Fisheries Prices Support Board (FPSB)

The *Fisheries Prices Support Act* was passed in 1944 (Fisheries Prices Support Board Annual Reports, various years). It provided machinery to address price decreases such as those that occurred for fish products after the First World War. Under the Act, a Board was created to advise the Minister of Fisheries on applications for assistance from the fishing industry. Subject to approval by the Governor-in Council, the Fisheries Prices Support Board (FPSB) was empowered to implement two types of programs:

> *Deficiency payments* or payments representing the difference between a prescribed price and the average market price. In prescribing prices the Act instructed the FPSB to ensure adequate and stable returns for fishermen.
>
> *Inventory financing* or payments to remove quantities of products temporarily from the market. Such products are sold back to original suppliers at cost, when market conditions have improved.

The FPSB at first was provided with a working fund of $25 million and authorized to hire staff for research, marketing and administration. All of this was to be complementary with resources from the Department of Fisheries, and the Board was to make use of departmental services, in particular its economic research divisions. In July 1947, the government appointed the first five board members and several months later established regional advisory committees of prominent industry members.

Almost immediately, the FPSB undertook studies of fishermen's incomes and domestic fish consumption. By 1949 the FPSB employed 20 persons including nine economists. During the period 1951 to 1953, the staff economists were transferred to the Markets and Economics Service of DFO with the proviso that the Department would carry out economic research on behalf of the FPSB. The Chairman and Executive Director of the FPSB were given responsibility in 1954 for administering the Fishermen's Indemnity Plan (later named the Fishing Vessel Insurance Plan) and, later, for organizing the 1959 community stage program in Newfoundland (discussed below).

The first full-fledged FPSB support program took place in 1948. An order-in-council authorized the Board to purchase east coast mackerel, cod and canned herring as a means of stabilizing prices. The next major support program in 1949 was for Manitoba Lakes frozen fish. Factors in recommending this program were large inventories, the decline in prices in the United States and the fact that surplus fish was threatened with spoilage unless it were moved into cold storage. Prescribed prices were about 80% of then-going prices for frozen fish.

Between 1950 and 1953, the FPSB initiated several price support programs for salted fish. As well, a build-up in frozen whitefish stocks in Manitoba and Saskatchewan led to price support in 1953. Deficiency payments for pickled mackerel fillets were also implemented in 1953. During the following year, a buy program for herring bloaters (whole smoked herring) was put in place as was another program for salted fish. In 1955, a special program was instituted to pay salted fish producers a proportion of their costs for salt. This "Fisheries Salt Assistance Program" involved an annual cost of about $500 000 and was in place until the late 1960's when the Canadian Saltfish Corporation took over the program. The Board also established a price support program for salted fish during 1955–56.

In 1954, B.C. fishermen requested assistance to destroy dogfish. Dogfish have been a nuisance to fishermen on both coasts since commercial fishing began. Prior to World War I the Government built three fish meal plants for dogfish on the Atlantic Coast but these proved not to be viable due to the high costs of reducing dogfish and the low price of fish meal. During the period from 1941 to 1948, high prices for Vitamin A obtained from dogfish livers led to heavy fishing and substantially reduced dogfish populations on the Pacific Coast. The Atlantic species was said to lack the high Vitamin A content. In 1949, low-priced synthetic Vitamin A became available. The dogfish fishery in B.C. terminated, and stocks again became super abundant.

The 1954 request for assistance was based on three arguments: (1) dogfish reduced incomes from normal fishing, (2) dogfish destroyed habitat and gear, and (3) dogfish were predators of food fishes. The Board did not recommend assistance, although assistance was later given in 1955 and again in 1959. The Board was subsequently given responsibility for administering the "Pacific Dogfish Destruction Program", even though the program was not established under the Act.

In 1958, cod landings in Newfoundland declined sharply (by about 30%) from the previous year. This affected inshore fishermen in particular. In assessing the request for assistance, the Board found the problem unrelated to price changes and therefore did not recommend assistance. However, among its recommendations the Board suggested that community processing facilities be created to encourage greater uniformity of product and to improve quality. The Government subsequently embarked on a program to build community stages in Newfoundland in order to provide facilities for the salted fish trade. (Community stages are large docks with a building for processing and storage.)

In the late 1950's, Board activities were therefore restricted mainly to saltfish and Pacific dogfish. This situation changed somewhat during the 1960's. In 1961, the Board approved assistance to develop new markets for frozen turbot in view of the fact that traditional pickled turbot markets had been declining for some years. In the following year, a special program was implemented for pickled herring, mackerel and bloaters. This program continued into 1964 during which time bloaters were purchased and deficiency payments were made on pickled products. In addition, during 1964 the Board was authorized to purchase frozen sauger (from the Prairies) for which there had been marketing problems resulting from several cases of botulism in the United States. Staff of the Board were also called upon to administer a program of income supplementation for inland fishermen affected by catch failures in 1964.

In 1965, the Board implemented a buy program for salted pollock, hake, cusk and cod due to marketing problems caused by political disturbances in the Dominican Republic. Most of the product that was bought by the Board was sold to the External Aid Office of the Department of External Affairs for donation to the Dominican Republic. Some of the salted cod products were eventually shipped to Africa under the World Food Program.

In 1966, the government authorized the Board to stabilize the price of perch originating from the Great Lakes. This stability was thought necessary because in most years prices declined drastically during the 2-month fishery (in April and May). Prices declined because there was no freezing capacity and firms were unable to pay storage for the 6–8 months that were required to use the supply. In a rather straightforward intervention, the Board purchased perch products and resold them later in the year, recovering nearly all costs. The program was extended through 1969, during which time losses of $663 000 occurred on the purchase and resale of products. After record landed values for perch in 1968, the Board decided that future programs should not result in losses to Government.

For the first time in 1966, the government also authorized the Board to purchase fish for the World Food Program. A total of 150 tons of canned mackerel were purchased from packers in New Brunswick and Prince Edward Island. Similar purchases have continued to the present time.

The program which provided one-half of the costs of salt to salted fish producers was abandoned in 1969 (but subsequently continued by the Canadian Saltfish Corporation). Due to continuing weaknesses in frozen groundfish markets, deficiency payments (of $4.2 million to processors) were implemented in May 1968 to maintain returns from cod, flounder and redfish. Processors were required to maintain prices to fishermen at or above the levels which had prevailed during the previous year. The FPSB also purchased 13.1 million pounds of dried salted fish during the year, half of which was supplied directly to African countries as part of a food relief program and half to the Canadian International Development Agency and CARE Canada for their programs.

A frozen groundfish stabilization program was implemented in 1969, designed to raise market prices and to prevent distress selling. This program was undertaken with the agreement of Scandinavian countries. The Board purchased 17.3 million pounds of cod blocks (26% of total production) as well as 240 000 pounds of fillets. By March 1970, the entire purchase had been returned, at cost, to the industry. This program was well received both by Canadian industry and by other countries because it was thought to have stabilized markets. A salted cod program was put in place during the year and was continued for two subsequent years to assist fishermen with price declines caused by oversupply and currency devaluation in some traditional markets.

## Organizational Changes

In 1970, a separate **Marketing Services Branch** was established within the Department of Fisheries and Forestry. This Branch (which became a directorate in 1981 and was subsequently dissolved in 1986) was created following the groundfish crises of 1968 and 1969 and the subsequent market recovery through market intervention. Its function was to work with the FPSB to analyze market trends, and to provide marketing services, including promotion of fish products in Canada and abroad. In 1986, market analysis became part of the economic and commercial analysis function of the Department. Promotion and government marketing assistance abroad has been assigned to the Department of External Affairs.

## Groundfish Assistance During the 1970's

In 1972, the FPSB implemented a freshwater fish price stabilization program. Nearly all fish acquired by the FPSB under this program were resold to original suppliers, with few costs accruing to Government. During the following year, the Board initiated a buy program for fish meal. The federal government had previously imposed an embargo on fish meal exports because of a world protein shortage and domestic needs. This had an obvious effect on fish meal prices in Canada, but the Board had acquired only 960 tons of product before the embargo was lifted.

The year 1974 witnessed one of the many crises in the Atlantic groundfish industry. Low catches combined with poor markets caused a severe cost price squeeze and cash flow problems for processors. A number of measures were funded under both the *Fisheries Prices Support Act* and the *Fisheries Development Act*.

The Board implemented an inventory financing assistance program (buy program) to assure markets at stable prices and to help industry with marketing. Payments under the program totalled over $7.3 million for fiscal 1974–75. A further $5.5 million was authorized to provide credit for inventory financing of other fishery products. Moreover, $1.5 million was authorized in October 1974, and $3.0 million in December 1974, to purchase canned ocean perch and other canned products for use in international food aid programs. The strategy was to divert production into canned products to help reduce frozen inventories. Also in that year the Government approved $3 million under a working capital loans program to assist the reopening of seasonal plants in northern Newfoundland, Labrador, and the Quebec lower north shore. Many of those firms had experienced losses because of ice conditions.

In January 1975, the FPSB initiated a program involving $15 million for frozen groundfish. The FPSB paid processors of frozen fillets and blocks to maintain July 1974 prices for fishermen. A Newfoundland trawler strike restricted actual payments and only $4.4 million was paid before the program ended in April, 1975.

**The Groundfish Bridging Program** was introduced in April 1975 to succeed the short term assistance measures that had been in effect from January, 1975. The program was a response to recommendations of a task force studying the problems in the industry. The task force was headed by Mr. Fern Doucet, Chairman of the Freshwater Fish Marketing Corporation. It made a number of recommendations to strengthen industry, including better stock management, reduced fishing effort (both national and international), entry control, increased efficiency, and improved marketing. The **Groundfish Bridging Program** involved $50 million assistance to groundfish fishermen and processors on both Atlantic and Pacific coasts. The program was intended to provide assistance until fish catches had returned to normal levels and industry was on a sound footing. Payments to processors were conditional upon maintaining July 1974 prices to fishermen. Expenditures under the program for fiscal 1975–76 amounted to $20 million for fishermen and $16 million for processors.

During fiscal 1975–76, the Groundfish Bridging Program and other programs provided financial assistance of $44 million to various fisheries. Of these programs, eight were funded under legislative authority of the FPSB ($5 million) and six were funded under the *Fisheries Development Act* (DFO and FPSB Annual Reports 1976). Under authority of the FPSB, deficiency payments were made for crab, smelt, mullet, and whitefish and a buy program was implemented for canned lobster. Deficiency payments were also made to the Canadian Saltfish Corporation (which had been established in 1970) for salted product produced from small sized cod. In 1976–77 the FPSB approved and implemented five programs, involving expenditures of $2.5 million. Programs providing deficiency payments for smelt, herring, whitefish and salted cod were implemented. The Groundfish Bridging Program continued until March, 1978. Funding during the latter 2 years was authorized at about $100 million per year.

During the 1980's, intervention was severely limited, reflecting better economic conditions following extension of fisheries jurisdiction in 1977. As noted above, the FPSB has continued to purchase canned fish (mainly mackerel) for the Canadian International Development Agency (CIDA), thereby easing cash flow problems. This program during the 1980's involved about $3.5 million per year (Fig. 1), but all monies spent have been recovered through sales (except for minor administrative costs).

Throughout the postwar period, and subsequently, the FPSB helped to stabilize some of the effects of cycles cited at the beginning of this paper. It has never been extensively used, and as a result has helped stabilize the industry without incurring large costs.

## 2. Consumer Services

In 1950, the Department of Fisheries established a consumer service as part of the group responsible for inspection. This new service was called the Inspection and Consumer Service Branch. Its main purpose was to encourage consumption of fish in homes, restaurants and institutions. An experimental kitchen was set up in Ottawa, where home economists tested and developed fish recipes. Consumer educational material in both English and French was provided to the public, the industry, and wholesale and retail distributors of fish. Material covered subjects such as proper storage and handling procedures.

To carry out this work, the consumer group initially established district offices in Vancouver, Winnipeg,

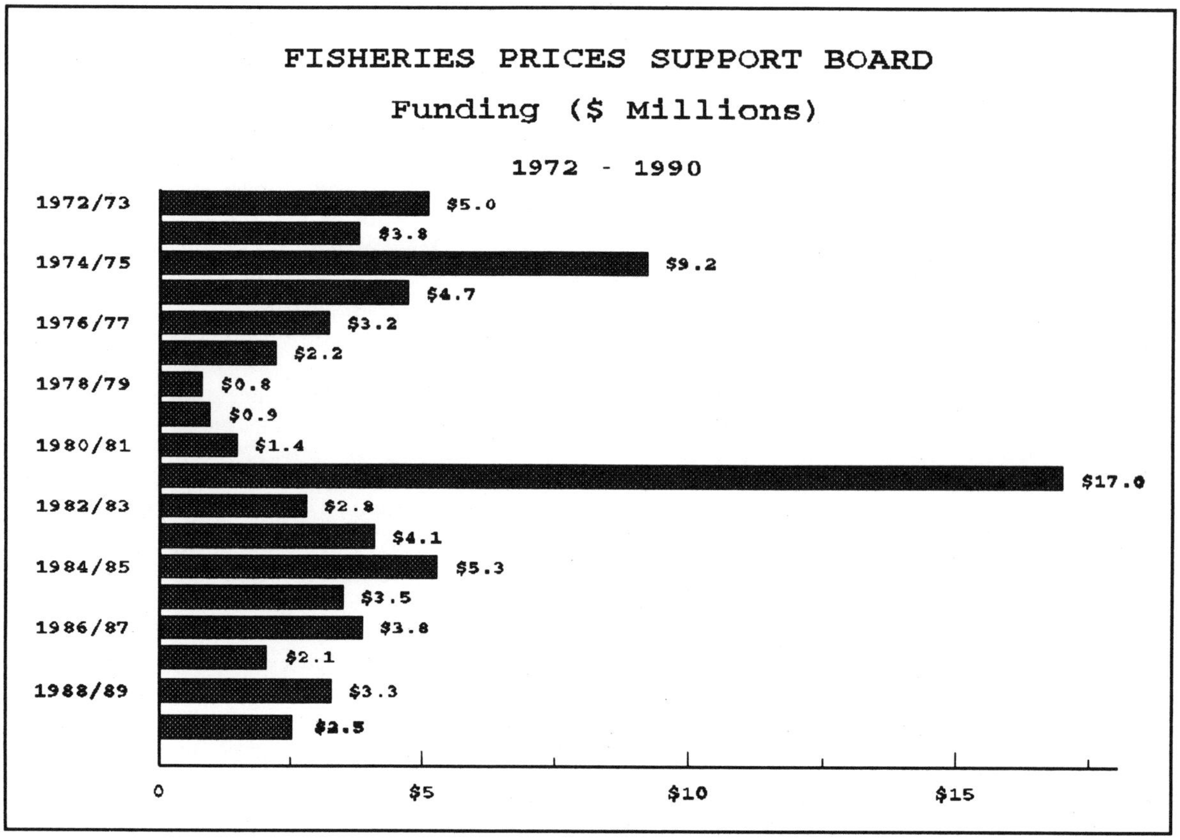


FIG. 1. Fisheries Prices Support Board, Funding, 1972–1989.

Toronto and Montreal. In January 1957 an additional office was opened in Halifax, and in the following year another in Edmonton. (By 1970, the offices in Toronto, Montreal and Edmonton had been closed). In March 1954, there were eight home economists employed in this service, but subsequently staff and financial resources were considerably reduced. A high point of the service occurred during a Federal-Provincial conference in 1955 when the test kitchen in Ottawa provided a fish luncheon for the Prime Minister (the Right Honourable Louis St. Laurent) and Premiers of the provinces. A national fish week was also initiated in that year, sponsored by the Fisheries Council of Canada.

The consumer services function was incorporated into DFO's Information Service in 1961, in a move to bring together all of the Department's public information activities. It was moved to the Marketing Services Branch in 1972. An industry review of the function in 1974 resulted in an up-dated approach. Promotion at major food shows and conventions continued. However, increased attention was given to underutilized species, especially to mackerel as part of the Atlantic Mackerel Development Program. (In 1986, the promotion and consumer services were terminated by the Department so that industry could assume its responsibilities in that regard — see below).

### 3. Freshwater Fish Marketing Corporation (FFMC)

The Federal-Provincial Conference of 1964 resulted in an inter-governmental group to investigate issues related to the marketing of freshwater fish (FFMC Annual Report 1971). This group subsequently recommended that an export-sales authority be established by federal legislation and that each provincial and territorial government provide for the appropriate reorganization of the fishing industry within its jurisdiction. Instead, the Government of Canada appointed a one man Commission of Inquiry to examine this and other options. Mr. C. McIvor, former Chairman of the Canadian Wheat Board, was given a wide ranging mandate to inquire into the marketing problems of the freshwater fishing industry in the provinces of Ontario, Manitoba, Saskatchewan, Alberta and the Northwest Territories (McIvor 1966).

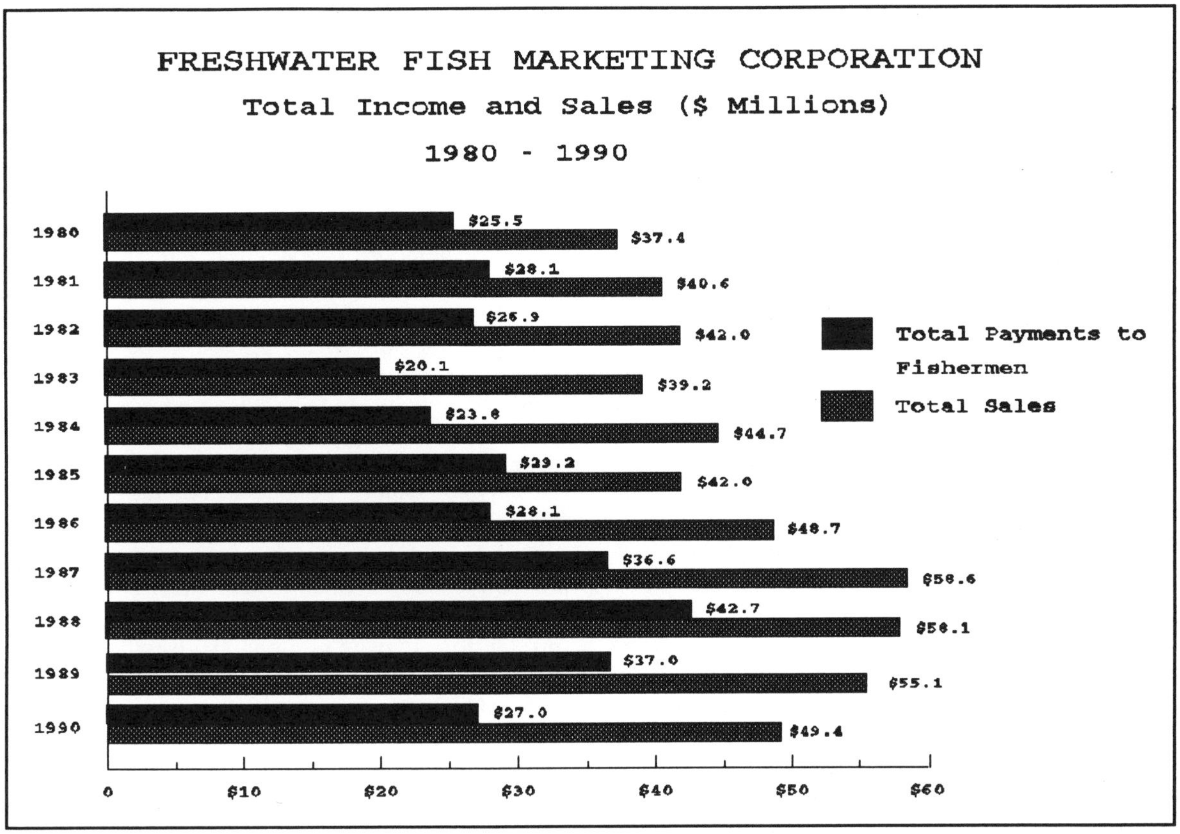


FIG. 2. Freshwater Fish Marketing Corporation, Income and Sales 1980–1988.

The Commission documented low prices obtained by numerous small Canadian producers and their poor bargaining position relative to larger United States importers. It cited a lack of quality control, product standardization and marketing coordination. It also noted that fishermen were often dependent upon buyers who supplied fishing gear and vessels (Judson 1961). Fishermen would receive a cheque at year end for any returns after their debt obligations to buyers had been satisfied. The Commission argued that this resulted in fishermen becoming indentured labourers since many received little or no compensation for their efforts. It recommended, as a result, that a fish marketing board be established under federal legislation with powers as the sole purchaser/seller of freshwater fish in the designated area.

The Freshwater Fish Marketing Corporation was established in 1969. Its objective was to improve and stabilize earnings of Canada's inland fishermen by controlling production, transportation, processing and marketing of freshwater fish.

All commercial fisheries in the provinces of Saskatchewan, Manitoba, Alberta and Ontario (excluding the Great Lakes) and the Northwest Territories were included. The Corporation received a statutory grant of $100 000 to meet initial operating and establishment expenses. Agents were appointed and assigned responsibilities for financing fishermen, receiving and grading their catch and making initial payments to the fishermen.

A system of initial and final payments was instituted. Initial payments were set at levels close to the commercial value of the product and final payments (net receipts in excess of initial payment) were distributed to fishermen at year end.

During its first operating year, the Corporation involved 35 agents and 11 processing plants. Earnings improved but it soon became apparent that centralized facilities for receiving, sorting and shipping fresh fish would improve earnings even more. This was subsequently accomplished at the Corporation's new facilities in Wascana, Manitoba.

During the 1980's, the Corporation increased its payments to fishermen virtually every year. Payments have ranged from $20 million in 1983 to $43 million in

1988 on total sales of from $37 to $59 million dollars (Fig. 2).

Today, the Corporation provides services to approximately 3,500 fishermen through 90 delivery points. High export prices and increased volumes to export markets have played a major role in increasing recent profitability of the Corporation. To date the Corporation has had the support of the federal government, participating provinces as well as the majority of fishermen. Maintaining this support while conducting its operations in a self-supporting manner are obviously necessary for the viability of future operations.

## 4. Canadian Saltfish Corporation (CSC)

At the turn of the century, the saltfish industry was located throughout Canada's Atlantic region (and there was still some saltfish activity in the northeastern United States). In the absence of refrigeration, fish and other proteins were most commonly preserved by salting and drying.

The provinces of Newfoundland and Labrador, Nova Scotia and Quebec produced the majority of saltfish. Changing social and economic conditions during the 1950's and 1960's, coupled with the development of the fresh and frozen fish industries, caused a decline in the production of saltfish products. Within Newfoundland, the southern coastal area moved to frozen fish, thus concentrating the saltfish industry in the northeastern part of the Province and in Labrador. Rising costs for fishing together with wide variations in landings from year to year, an inability to market the products as needed, and the loss of established markets in Greece and Brazil all contributed to deteriorating conditions.

In response to this situation, governments looked at ways to improve the situation. The Newfoundland Associated Fish Exporters Limited (NAFEL) was created in 1947 as the exclusive agency for the export of salted and dried cod fish. NAFEL had a monopoly in export trade and acted as a single desk sales organization until 1959 when it became a voluntary cartel (until 1970). Cooperation had begun to deteriorate by 1964, when at the request of the Government of Newfoundland and Labrador, the federal government created a Royal Commission to study the feasibility of a saltfish marketing board. The Commission rejected the notion of a marketing board, but determined that there were many marketing problems.

By 1970, persistent marketing difficulties, distress among primary producers and the emergence of the FFMC provided impetus for the creation of the Canadian Saltfish Corporation (CSC). *The Saltfish Act*, establishing the Canadian Saltfish Corporation, was passed by the House of Commons on February 25, 1970. Parliament gave the Corporation the exclusive right to purchase, process, package and market saltfish. CSC operations were required to be self-sustaining. Purchase of fish at competitive prices and marketing of processed fish were to be undertaken without government assistance. The objective was to improve and stabilize the earnings of primary producers of cured fish and to improve methods of production. All Atlantic provinces were invited to participate, but only Newfoundland and the Province of Quebec (with respect to the North Shore of that province) agreed to participate in the operations of the Corporation. Both provinces passed legislation parallel to federal legislation. Other provinces declined to participate, allowing their saltfish industries to fend for themselves in the market.

The CSC recorded sales of $43-$74 million annually from 1980 to 1988. Saltfish sales only accounted for $24–$45 million during the same period (Fig. 3). The 10 000–15 000 tonnes of saltfish produced annually accounts for one quarter to one half of Canada's salt cod production. Private companies elsewhere in Atlantic Canada account for the remaining production.

The CSC operated until 1991 primarily by contractual arrangement with ten individual plant proprietors acting as agents of the CSC. Fish was collected from fishermen throughout the CSC's jurisdiction and transported to agents' drying plants on the island of Newfoundland. The fish was either salted by fishermen or sold directly to agents as fresh fish for subsequent salting. Once in an agent's plant, the fish was either dried or shipped in the semi-processed wet saltbulk form, depending on market requirements with respect to moisture content, size and salt content (heavy or light salted). Many communities, especially in Labrador, still rely substantially on the CSC. In 1988, the CSC purchased 43 million pounds of its total yearly purchase of 56 million pounds from the communities located in northern regions. The CSC and its agents employed 1,180 plant workers of which 850 are in northern communities. But as electricity spreads and transportation improves, the population depending exclusively on the CSC will continue to shrink. One estimate is that no more than 200 fishermen are now totally dependent on the CSC, i.e., have no option other than selling to the CSC.

In recent years, the Corporation has not been successful in maintaining itself as a financially self-sustaining operation. This has reflected declining and changing saltfish markets, the increasing preponderance of small fish (not usually well suited for saltfish markets) and some poor management decisions. For example, for many years, dried fish has commanded a lower price than saltbulk yet the CSC continued to encourage

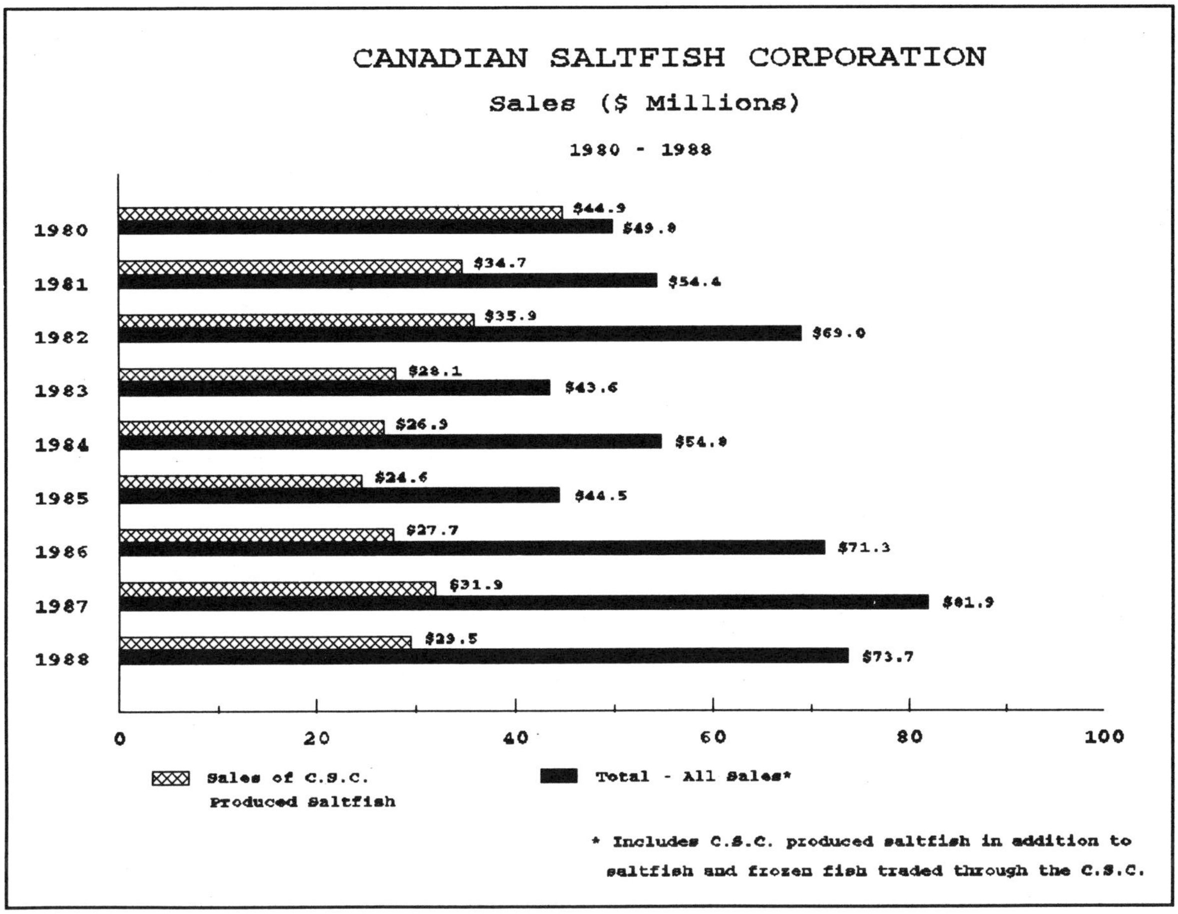


FIG. 3. Canadian Saltfish Corporation, Sales 1980–1988.

drying with its associated costs. The 1983–84 fiscal year was the last year the Corporation operated at a "profit". The reported accumulated deficit has risen to more than $40 million with about half of that being losses sustained in 1989 and 1990 fiscal years alone.

To keep the Corporation operating, the federal government made a $15 million, ex gratia payment during the summer of 1990 and the Corporation's debt was written off in 1991. Significant changes were made to operations in recent years. Full time staff was reduced to 13 persons from 45. Money losing contracts were cancelled and redundant assets disposed. The CSC monopoly on the island of Newfoundland was relaxed and licences issued to any registered processor for exporting saltfish. In 1992–93, the CSC was poised to return to a break-even position but with the moratorium on northern cod, production declined significantly resulting in continuing losses. However the CSC will continue to operate in the near future, albeit in a scaled back form.

## C. Assistance to Fishermen

### 1. Fishing Vessel Assistance Program (FVAP)

The Department of Fisheries introduced the Fishing Vessel Assistance Program (FVAP) in 1942. The objective of the program was to assist Canadian fishermen in the purchase of fishing vessels (this was amended in 1976 to include modification or conversion of fishing vessels). In doing so, it was expected to help modernize the fleet and improve its fishing capacity. Originally the program had its greatest take-up in the Atlantic fisheries, although all Canadian fishermen were eligible. The program was finally discontinued in 1986 because it had contributed to significant overcapacity and was exacerbating problems in the fisheries.

Under the original program, the Government paid provincial loan boards in the Atlantic provinces up to $165 per gross ton towards the cost of wooden vessels between 55 and 60 feet in length. This payment was also available for longer boats when they were owned by four or more fishermen. New regulations in 1961 increased subsidies: vessels greater than 40 feet in length and less than 100 tons were eligible for a subsidy of $250 per ton.

Before 1956, actual expenditures under the program were small. However, by the 1956–57 fiscal year expenditures had risen to $390 000. In the following years the number of subsidized vessels and the amount of subsidy continued to increase. By 1962–63, 49 vessels qualified for $500 000.

In addition, a subsidy was in effect in the 1960's to encourage diversification of inshore fisheries. The subsidy provided 25% of the approved cost of constructing fishing vessels between 35 and 45 feet in length. In 1969–70, the Department of Fisheries and Forestry assisted the construction of 44 vessels for a total outlay of $565 000.

After 1970, assistance was available for all vessels, not just wooden fishing vessels in the Atlantic Region, and terms of the program were expanded. With the expansion of the program, it was expected that Canada's fishing fleets would be able to compete better with other fishing nations. Under the expanded program, 35% of approved costs of vessels 45 to 75 feet in length were also paid by the Department of Fisheries and Forestry. This assistance was provided directly to vessel owners. In contrast, for vessels greater than 75 feet in length, the Department of Industry Trade and Commerce provided assistance to shipyards (see discussion of the Shipbuilding Industries Assistance Program below). During the first year of operations in 1970, 245 fishing vessels in the 45–75 foot range and 18 larger vessels were subsidized. During the following year, $4.2 million was paid in subsidies for 286 vessels in the former category and $5 million for 13 large fishing vessels.

By 1975, the Department had restricted applications to full-time fishermen, not engaged in limited-entry fisheries. Vessel subsidies of $4 million dollars were provided during the year. To prevent further expansion of the Pacific fleet, at the request of the fishing industry, subsidies were suspended for British Columbia fishermen. So by 1975 the program provided assistance only to the Atlantic (and freshwater) fisheries.

From 1975 to 1985, assistance to fishermen ranged from $4 million to $11 million annually (Figs. 4 and 5). Throughout this period the minimum vessel length eligible for assistance, decreased. In 1983 assistance was provided to vessels as small as 16 feet in length. The average subsidy per vessel amounted to $13 000. Approximately 95% of the subsidies were for vessels in the Atlantic fisheries, with the balance in the freshwater fisheries.

The FVAP was discontinued in March 1986. The program had been introduced at a time when fishermen had few sources of capital. Given the overcapacity of the Canadian fishing industry, the program was incompatible with other efforts to reduce the number of licenses and to limit entry.

#### (i) Shipbuilding Industry Assistance Program (SIAP)

The Shipbuilding Industry Assistance Program (SIAP) was introduced in 1975 to provide direct subsidies for the construction of ships. It was one in a succession of shipbuilding industry assistance programs instituted after World War II (de Silva 1961). Prior to this beginning in 1961, assistance had been provided directly to shipyards. This assistance had helped to offset high costs of production and to enable Canada's shipyards to compete with subsidized shipyards in other countries.

SIAP had two components. The first was the system of Performance Improvement Grants (PIG) and the second was direct production assistance. The latter was discontinued in 1985. Performance Improvement Grants provided assistance for shipyards to modernize their plant and equipment. Shipyards earned credits under PIG by investing in capital equipment (such as special welding equipment and cranes) and were able to save these credits to offset up to 50% of approved productivity improvement investments. Each credit was equal to 3% of the vessel's cost.

The amount of direct production assistance varied considerably over the years. From 1961 to 1968, shipyards could obtain up to 50% subsidies for large Atlantic fishing trawlers. In 1968 this was reduced to 35%. When SIAP was introduced in 1975, the initial rate was 14% for vessels greater than 75 feet in length (SIAP did not apply to smaller vessels). It was intended that this rate would drop by 1% per year until 1981 at which time it would be 8%. This however did not occur. In 1977 the rate was increased to 20% and was maintained at this level until 1980. The rate was reduced to 9% in 1980 and an annual program spending limit of $74 million was established for each of the fiscal years 1981 through to 1984. Total assistance provided between 1980 and 1984 was $433 million with a yearly average at the program spending limit. Of this amount, approximately 3 million dollars annually was claimed for fishing vessel construction.

SIAP was established, of course, to assist the shipbuilding industry not the fishing industry and most assistance went to Quebec, Ontario and British Columbia (See Fig. 6). Assistance for fishing vessel construction

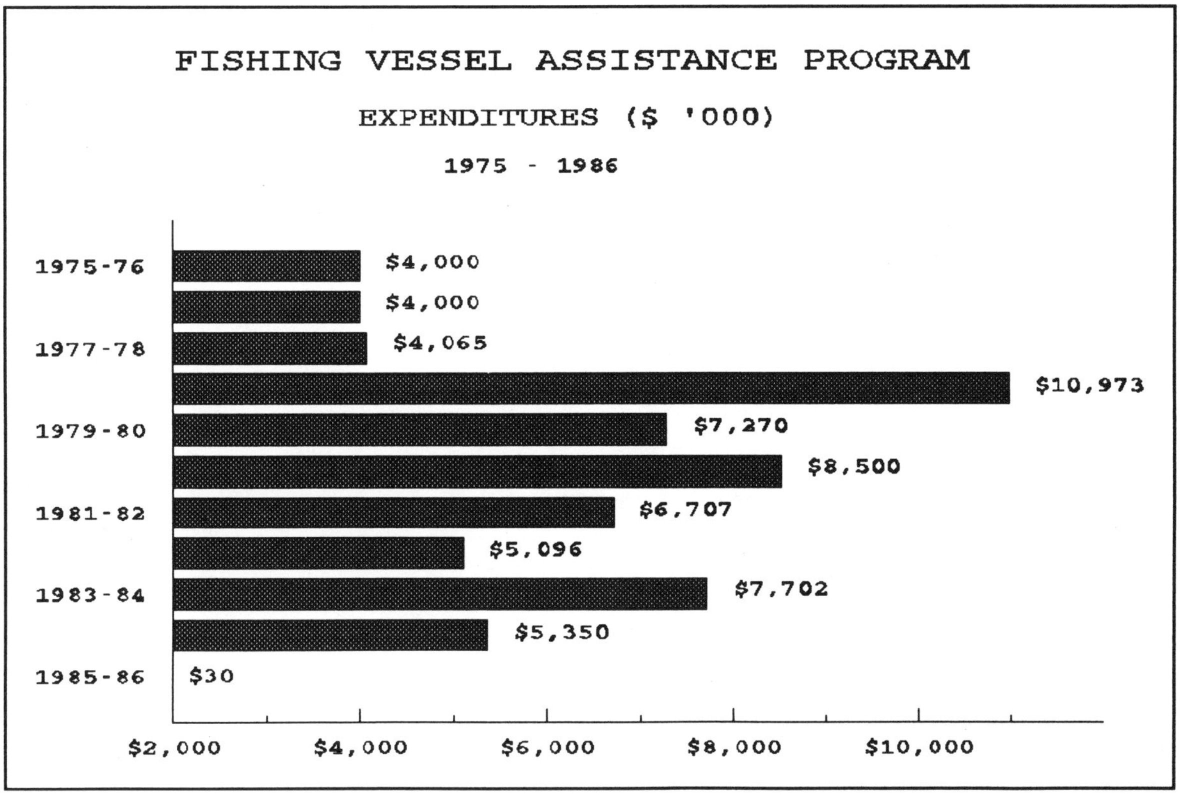


FIG. 4. Fishing Vessel Assistance Program, Expenditures 1975–1985.

amounted to less than 5% of total annual program expenditures. SIAP production assistance was discontinued for vessels delivered after June, 1985.

Beginning in the early 1980's, special shipbuilding assistance was also provided by the Department of Regional Industrial Expansion (DRIE), in conjunction with provinces. For example, in 1986 a DRIE grant of $10.5 million, matched by the province, was provided for the construction of seven vessels at Marystown Shipyards in Newfoundland. DRIE also provided assistance of $2.0 million, matched by the province, for the construction of three vessels at the Georgetown Shipyards in Prince Edward Island also in 1986.

## 2. Fishing Vessel Insurance Plan (FVIP)

In the winter of 1949, Atlantic fishermen experienced heavy storms causing severe losses of boats and gear. As a result, a Department of Fisheries Task Force examined insurance for fishermen to cover the special risks faced by fishermen. It found that commercial insurance was generally unavailable to many small boat fishermen and where it was, premiums were high. The dispersion of small vessels in many hundreds of fishing settlements made it difficult for commercial marine insurance companies to provide such coverage for fishermen (FVIP Annual Reports, various years).

The Government authorized a Fishermen's Indemnity Plan in 1953 to provide fishermen with insurance for vessels and gear. Administration of the plan was assigned to the Fisheries Prices Support Board, and an annual premium of 1% of the appraised value of vessels was set. The Plan provided fishermen with coverage for serious loss or damage to vessels and was enhanced in 1970 with the addition of third party collision coverage. Having recovered from an accumulated deficit in the early 1980's, the program has maintained profitability since 1985 (Fig. 7).

This insurance was initially available to all coastal fishermen operating vessels valued from $250 to $7,500. Maximum insured value increased to $10 000 in 1957, $12,500 in 1963, and $25 000 in 1968. The plan initially indemnified 60% of appraised value in the event of total loss and 85% of repair costs in excess of 15% of the appraised value. By March 1955, 2 360 vessels were covered under the plan and by 1960 the total was 5,560 vessels. The initial period of operation showed that premiums were close to covering costs of claims.

In 1970, the Department of Fisheries and Forestry assumed responsibility for administrating the FVIP.

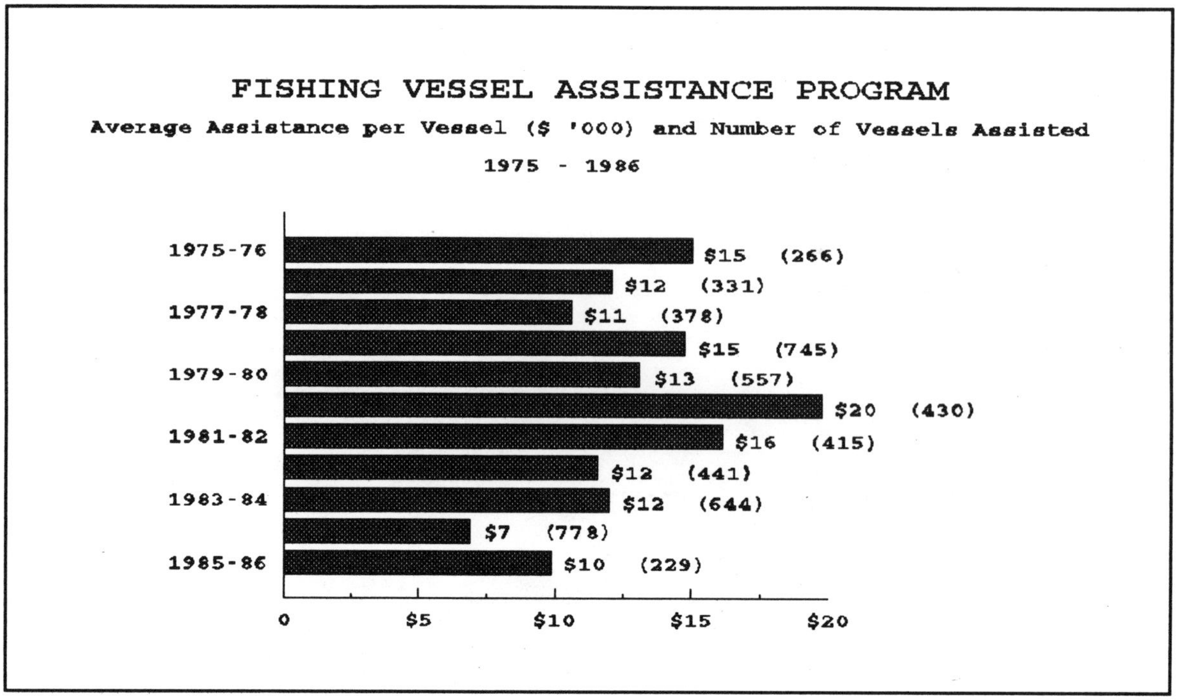


FIG. 5. Fishing Vessel Assistance Program, Assistance per vessel and Number of Vessels Assisted, 1975–1986.

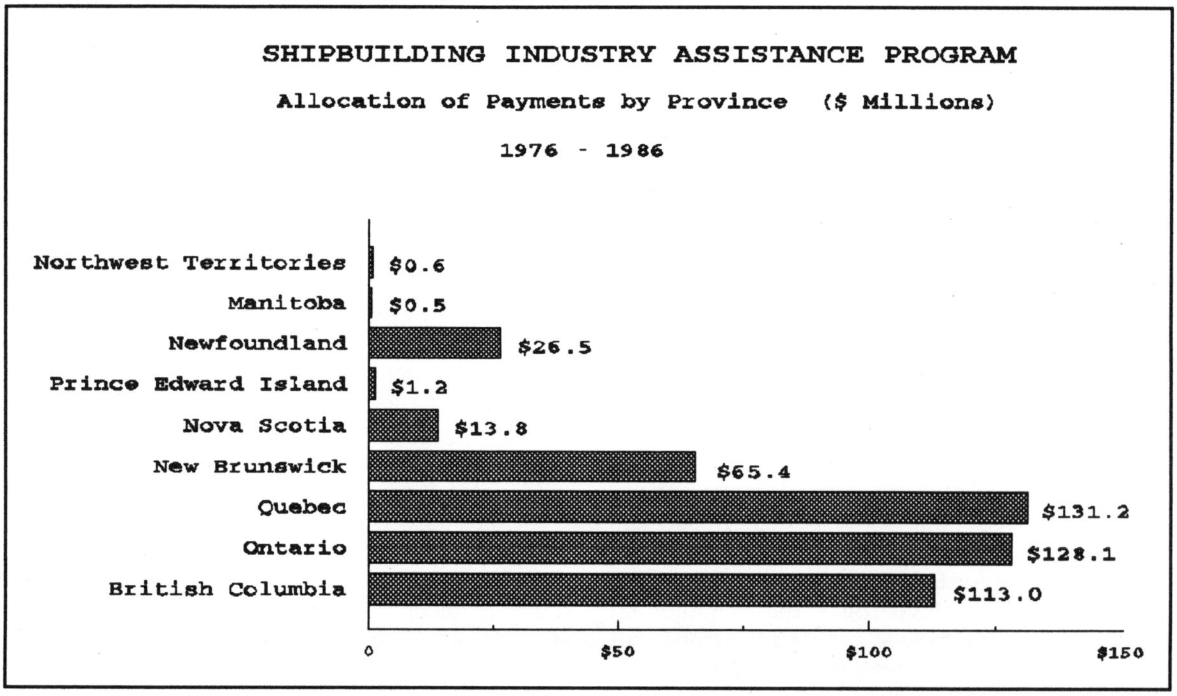


FIG. 6. Shipbuilding Industry Assistance Program, Payments by Province, 1976–1986.

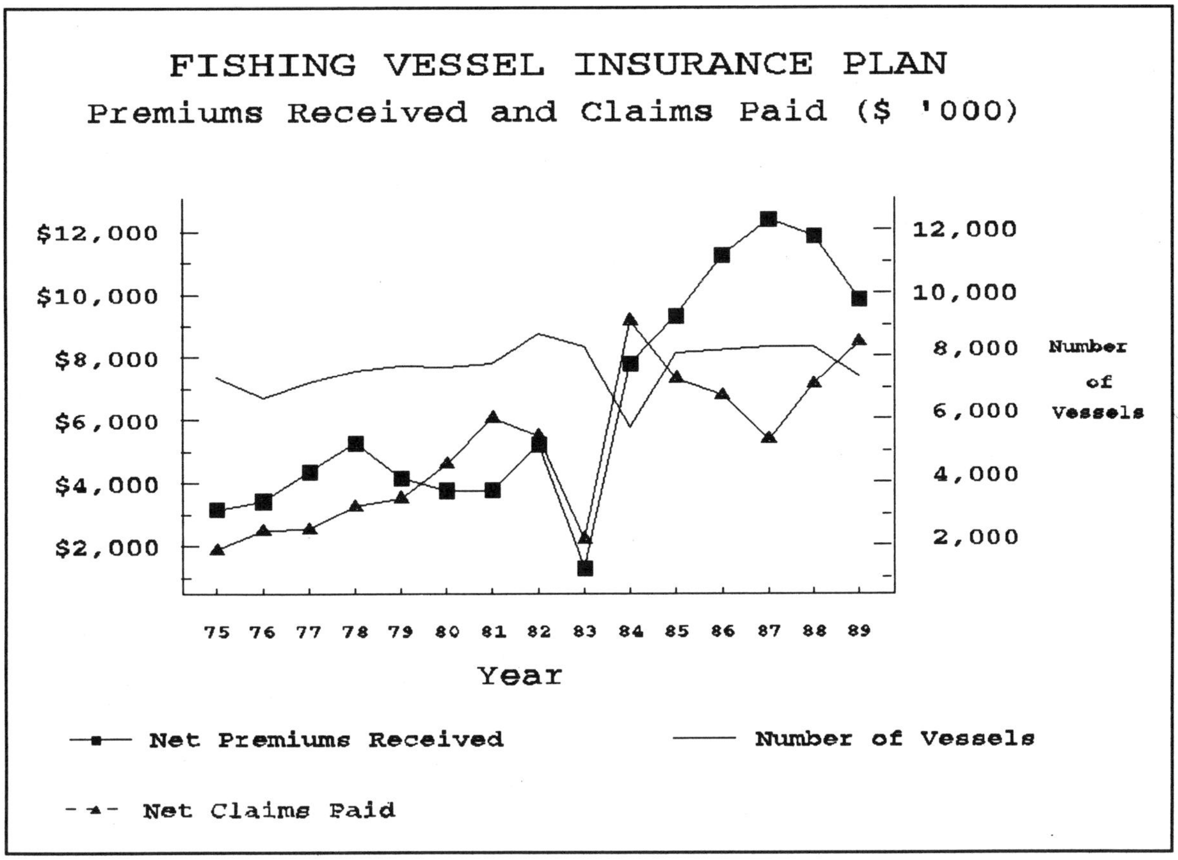


FIG. 7. Fishing Vessel Insurance Plan, Premiums Received and Claims Paid, 1975–1988.

Premiums were increased to 1 1/2 % for vessels valued up to $15 000 with a 2% rate on vessels valued at more than $15 000. Provision was made to include some large vessels in the plan. In 1968–69, a surplus of $155 000 was earned by the plan and in 1970–71 the minimum insured value for vessels was increased to $500.

The Plan initially provided protection to lobster fishermen for loss of traps. Premium rates varied, depending on season length and insured value. In 1966, coverage was extended to fixed fishing gear, shore installations and equipment used in all types of fishing. This insurance was discontinued in 1968 because it was costly and few fishermen subscribed to it. Acceptance of the lobster trap insurance, for the most part, had been limited to areas of heavy losses.

The Fishing Vessel Insurance regulations were amended in 1975. The significant change was more flexible premium rates to provide incentives for vessel owners to operate vessels safely and to maintain their vessels in good condition. Provision was also made to compensate reasonable expenses incurred in attempting to save vessels, or to minimize damage.

The years 1976 and 1977 proved to be good for the Plan. The number of insured vessels increased and, after claims and operating expenses, there was a surplus of revenues. Growth of the surplus continued until 1981, when the fishing industry experienced losses six times greater than in 1978–79. Premiums were subsequently increased and coverage was refused to fishermen with particularly poor claims histories. This continued until 1985 when the Minister announced the Plan would henceforth be strictly on a cost recovery basis. Since then, the Plan has been in a surplus financial position (Fig. 7). In recent years, premiums have been 50% greater than claims. Of course, this does not take account of administrative expenses (which would increase costs), or interest on premiums (which would increase revenues), both of which would be considered in a private sector plan.

Privatizing the FVIP has been discussed on a number of occasions. With private and mutual insurance companies now operating in the fishing vessel insurance sector, it can be argued that there is no longer a need for a government insurance program. In 1978 the government announced its intention to privatize the FVIP. Approval was sought from insured fishermen to introduce a form of mutual insurance. According to a survey undertaken by the FVIP, 44% of fishermen wanted a mutual plan, 42% wished to retain the FVIP, 6% wanted commercial insurance and 8% were undecided. The results did not suggest an overwhelming desire for change and no changes ensued.

Nevertheless the idea was considered again in 1984. An announcement to discontinue the plan was made in late 1984 but due to concerns raised by the fishing industry, the Minister rescinded his decision in 1985. In 1989, the idea was once again considered but abandoned.

### 3. Fisheries Improvement Loans Act (FILA)

The Fisheries Improvement Loans Act (FILA) was legislated to assist fishermen in accessing intermediate and short-term credit to improve their fishing businesses. The Act was passed in 1955 for an initial period of 3 years. FILA empowered the Minister to guarantee private sector lending institutions against loss on any loans made to fishermen for the improvement of fishing businesses. Amendments through the years kept FILA active until January 1987 at which time it reached its statutory guarantee limit. Loans outstanding reached a peak of $28 million in 1979, declining to $17.5 million in 1987 (Fig. 8). In June 1987 the FILA was allowed to expire, and Parliament amended the Small Business Loans Act (SBLA) to include fishermen.

A wide range of loans was available under this program: purchase, construction, major repair or overhaul of fishing vessels; construction, repair or alteration of any buildings required by a primary fishing enterprise; and "general works" related to the development or improvement of a primary fishing enterprise.

Prior to 1978, terms of repayment were a maximum 3 years, but in 1978 this was increased to 15 years. Loans had to be secured with equity from a fisherman's own resources. Before 1978, interest rates were set every 6 months based on government bond yields plus 1%. After 1978 the maximum rate of interest was based on the average prime lending rate of the six largest chartered banks plus 1%. The limit set by the government grew over the years. During its last year of operation, FILA could guarantee a total of $150 000 per borrower. The average loan ranged from $1 099 in 1957 to $23,172 in 1987. During the period from 1955 to March 31, 1990, a total of $15.7 million in claims was paid to banks for loans not repaid by borrowers, of which $766,142 was later recovered.

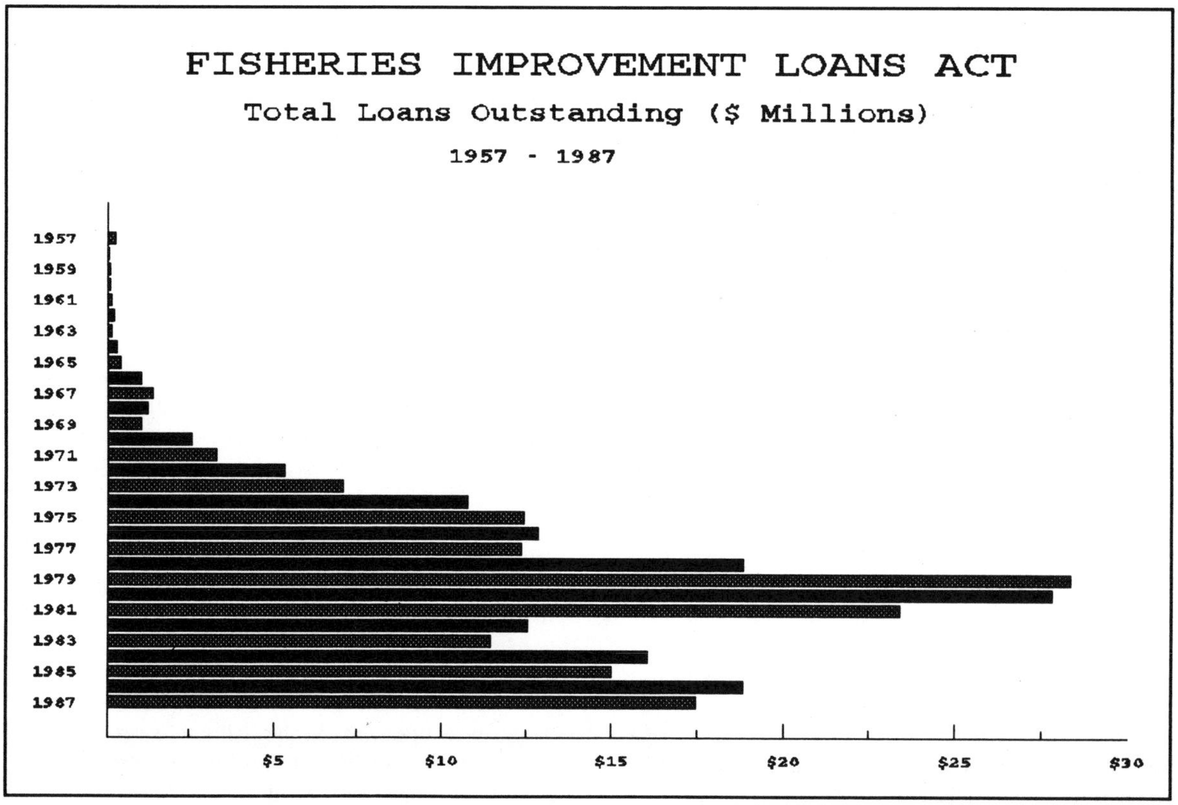


FIG. 8. Fisheries Improvement Loans Act, Total Loans Outstanding, 1957–1987.

### 4. Indian Fishermen's Assistance

From 1968 to 1986, four initiatives were implemented in B.C. to arrest the decline in Indian fishermen participation in commercial fishing (Native Brotherhood 1989). The funding for these programs ($42.7 million) came from both the Department of Indian and Northern Affairs and the Department of Fisheries and Oceans.

#### *(i) Indian Fishermen's Assistance Program*

The Indian Fishermen's Assistance Program (IFAP) provided grants and loans to native fishermen to upgrade their vessels. The intent was to increase the efficiency and mobility of the native fishing fleet so that it could compete with the non-native fishing fleet. The program was established in 1968 and discontinued in 1979. Approximately $16 million was spent under the program.

#### *(ii) Indian Fishermen's Emergency Assistance Program*

The Indian Fishermen's Emergency Assistance Program provided $4 million in financial assistance to native commercial fishermen in 1981. Economic conditions in the early 1980's (high interest rates, fuel prices and vessels prices) resulted in financial hardship for many native commercial fishermen. The program was designed to assist fishermen through this period.

#### *(iii) Northern Native Fishing Corporation*

The Northern Native Fishing Corporation was established in 1982. In the early 1980's, the large west coast fishing companies discontinued renting out their vessels. Since many native fishermen had rented vessels from these companies, it was necessary for them to purchase vessels to continue fishing. The Corporation assisted three northern B.C. tribal councils to purchase 243 gillnet vessels and 253 licenses from BC Packers Limited. Of the $11.7 million cost, 3 million dollars was applied to vessel improvement and operating costs with the remainder for direct purchase of vessels.

#### *(iv) Native Fishing Association Program*

The Native Fishing Association Program, established in 1986, has provided business management training and assistance to native commercial fishermen experiencing financial difficulties. The federal government provided the program's initial funding of $11 million which was used to endow a revolving loan fund.

### 5. Tax Provisions

The Department of Trade and Industry introduced accelerated capital cost allowances for ships after World War II with the objective of encouraging growth in Canadian shipbuilding. This provision permitted vessel owners to write off capital costs of a vessel over 3 years (changed to 4 years in 1982) instead of the previous 15 years. Recaptured depreciation was taxable when a vessel was sold, but the tax could be avoided if the gain were reinvested in the construction of a new vessel. This rapid depreciation provision no doubt contributed to vessel construction activity during the postwar period and continues to do so today.

Investment tax credits are also available under the Income Tax Act. Investors can deduct a varying fraction of new investment from tax otherwise payable in the year of acquisition. (Rates vary among regions.) Finally, fishermen qualify for excise tax rebates on fuel and equipment purchases.

### 6. Pacific Salmon Fishery Buy Back Program

This program was introduced in 1971, suspended in 1974 and re-activated for 1980–81 only. This program was part of the salmon licence control scheme designed to reduce the fleet by purchasing and retiring licensed vessels. Licence fees were doubled to fund the program. During the initial period in 1971, the program retired 362 vessels at a cost of $6 million. During the final period in 1980 81, $2.9 million was spent to retire 36 vessels. The program probably retired the least efficient vessels in the fleet and therefore had little long-term impact on capacity.

## D. Assistance for Facilities

Assistance for fish processors has been largely associated with industrial development. Such assistance was minimal until the early 1960's when government increased its emphasis on regional development.

### 1. Fish Chilling Assistance

Assistance for freezing and chilling fish was available from 1949 to 1980 and again during the late 1980's. Current assistance is available under Economic and Regional Development Agreements (Fisheries Subsidiary Agreements) between the federal and provincial governments.

In 1949, the government adopted regulations to encourage freezing and the development of bait storage facilities. Assistance to a maximum of 75% of cost, but not exceeding $10 000 to any one applicant, was available for maintaining and operating bait freezing and storage facilities for fishermen.

In 1972, the Department assessed the need for new ice-making and fish-chilling facilities. As a result, the **Fish Chilling Assistance Program** was initiated under the Fisheries Development Act in late 1973. It terminated in 1980.

The Program provided financial assistance for ice making and storage facilities; vessel chilling equipment and insulation; projects designed to improve on-board handling equipment for small inshore vessels; and promotion of the use of mechanical and sun drying methods for squid. Federal and provincial governments shared equally the cost of eligible projects with a maximum federal contribution for each facility of $25 000. In the first 3 years of the program, a total of $4.2 million was provided for equipment in plants and on vessels. In 1977, the maximum payment per facility was increased to $50 000. In the next two fiscal years, $4.0 million was spent.

An **Inshore Fish Handling System Program** was implemented in 1977, ending in 1981. This program provided assistance of $13 million to the Newfoundland inshore fishery for equipment such as hoists, graders, and insulated containers.

### 2. Newfoundland Bait Service

By the terms under which Newfoundland became Canada's tenth Province, the Department of Fisheries assumed responsibility for maintaining the Newfoundland Bait Service. This program was started in 1934 by the Newfoundland Commission of Government and has continued to the present day.

The program involves acquisition, storage, distribution and retailing of bait products from a network of facilities. Bait is purchased by the Government and then sold at lower prices to fishermen. The net cost of operating this program has normally ranged from 1 to 2 million dollars per year.

### 3. Department of Regional Economic Expansion

Federal assistance to fish processing plants in the Atlantic Region began in the early 1960's under the aegis of the **Atlantic Development Board (ADB)**. This board was created in 1962, and funded in 1963 with $100 million (later increased to $150 million). The objective of the ADB was to foster economic growth and development in Atlantic Canada. There were a number of other federal programs at the time to address regional economic disparities (such as the **Cape Breton Development Corporation**), but they did not provide much support specifically to the fishing industry. The main ADB program affecting the fishing industry was funding to improve water supply for industrial operations. Some infrastructure for the fishing industry was also funded under other ADB programs. In the late 1960's, ADB programs were extended to include export promotion (ADB Annual Reports, various years).

When the **Department of Regional Economic Expansion (DREE)** was created in 1969, ADB responsibilities as well as many other regional development programs were transferred to the new Department. DREE was created because regional development was to be given a top priority and an overall focus was thought to be necessary for planning and coordinating programs. Other programs absorbed into DREE were the **Area Development Agency (ADA)**, the **Fund for Rural Economic Development (FRED)**, the **Agriculture and Rural Development Act (ARDA)** and the **Enterprise Development Program (EDP).**

DREE programs for federal-provincial development agreements started in 1974. Ten-year **General Development Agreements (GDAs)** were signed with all provinces except Prince Edward Island where a 15 year Comprehensive Development Plan had been signed in 1969. The GDAs encompassed specific joint programs through **subsidiary agreements**. As of April 1981, 117 subsidiary agreements had been signed involving Federal commitments of $3 billion. The **Special ARDA Program** agreements, also under DREE, complemented the GDAs. Those agreements involved assistance for disadvantaged people, particularly those of Indian and Inuit ancestry.

The **Ministry of State for Economic and Regional Development (MSERD)** was established in 1979 as a new central agency (Crowley 1982). Under its authority, **Economic and Regional Development Agreements (ERDAs)** were signed with the Provinces and territories to replace the GDA's. When MSERD was disbanded in mid 1984, the Minister of **Regional Industrial Expansion** was made responsible for these agreements. Under the ERDAs, **Fisheries Subsidiary Agreements** were signed with each Atlantic Province. Management committees with equal federal and provincial representation were established to administer and manage these agreements. Fishery programs included general resource development, training, productivity improvements, energy efficiency, infrastructure for common use facilities and quality improvement.

The main DREE programs to assist the fishing industry in the 1970's were the **Regional Development Incentives Program (RDIP)**, the **PEI Comprehensive Development Plan** and the **GDA for Newfoundland**. Some assistance for construction of fish processing facilities was also provided during the 1970's under the **Special ARDA** agreement with B.C. In 1982, as a result of the growing concern with excess processing capacity in the Atlantic fisheries, DREE terminated assistance for the construction of new plants and plant expansions; funding continued, however, for plant conversions, upgradings and for facilities to produce "new" products.

In 1983, the **Department of Regional Industrial Expansion** was created through the merger of the **Department of Industry Trade and Commerce (ITC)** and DREE. In the early 1980's the **Industrial and Regional Development Program (IRDP)** supplanted previous programs such as EDP and RDIP. Assistance to the fishing industry under all of these programs for the years from 1974 to 1986 attained a peak in 1985 of $20.9 million (Fig. 9).

## E. Other Initiatives

Two other federal initiatives undertaken during the period which impacted significantly on the fishing industry were the Industrial Development Service (of DFO) and Unemployment Insurance. Any overview which did not consider these activities would be remiss, especially since they set the stage for programs which continue to the present day.

### 1. Industrial Development Service (DFO)

Parliament established an industrial development vote in 1950 to provide development programs for fishermen and the fishing industry in general. From 1950 to 1955, the vote was administered directly from the Deputy Minister's office in the Department of Fisheries. Projects such as exploration of new stocks of fish and development of fishing and processing gear and equipment were initiated.

A separate Industrial Development Service (IDS) with its own director was created in 1955. The main functions of the service were to apply scientific findings and to encourage the adoption of new catching, processing and distribution techniques. The service was organised with an engineering section, a vessels and gear section and a special services section. Fisheries development from the 1950's into the 1970's aimed to improve Canada's position among major fishing nations and to increase economic returns to the Canadian fishing industry.

During the first year (1955–56), 15 projects were financed for a total expenditure of about $430 000. The following year expenditures increased to over $640 000. In 1956–57, a committee was established to

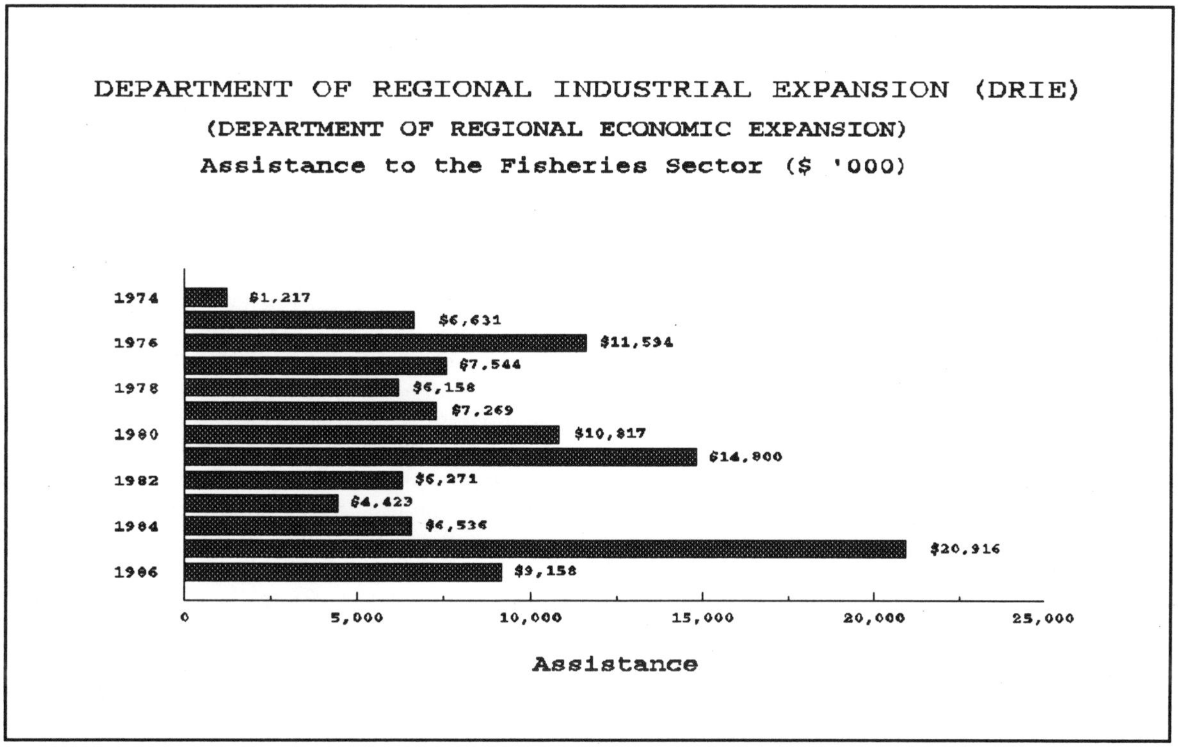


FIG. 9. Department of Regional Industrial Expansion, Assistance to the Fisheries Sector, 1974–1986.

coordinate work related to industrial development projects of the Department of Fisheries and the Fisheries Research Board (a research body which had not yet been integrated into the Department).

During the late 1950's, in addition to many gear and vessel technology projects, the Industrial Development Service conducted exploratory fishing projects and sponsored federal-provincial conferences to examine ways to modernize Atlantic fishing fleets. It also operated a processing plant at Valleyfield, Newfoundland to explore new methods for processing salted fish. Projects were initiated relating to new products, packaging and transporting fish. A major gear development at that time was the mid-water trawl which was first used in British Columbia and then introduced to Atlantic fisheries. Refrigerated seawater systems were also provided at that time for preserving fish while being held and transported. The Service administered the Fishing Vessel Assistance Regulations and the Frozen Bait Storage Assistance Regulations and was responsible for advising on marine works and aids to navigation.

The first Federal-Provincial Ministers Conference on Fishery Development was held in Ottawa in January, 1964. This was the first of several federal provincial conferences with sponsorship shared between the Department of Fisheries and fisheries authorities in Atlantic provinces. This conference concluded that Canadian fisheries resources could support long-term expansion of the fishing industry. Accordingly, Ministers mapped out a program for improving catching and processing techniques, expanding markets and reducing price uncertainties. Principles were similar to those being applied in agriculture.

In 1966, the Industrial Development Service sponsored two federal provincial conferences. An Atlantic herring fishery conference created great interest, even among B.C. fishermen who were facing a declining herring resource. A conference on offshore fishing vessels focused attention on concepts of new vessel design, safety and efficiency. From 1967 to 1970, additional conferences were held on fish protein, fishing vessel construction materials, Atlantic crab and automation in the fishing industry. In total, the Service sponsored nine conferences during the decade. These conferences set the stage for considerably enhanced consultation during the 1970's and 1980's.

By the mid 1960's, the Industrial Development Service was involved in over 100 different projects, applying new technology to fish finding and fish catching and to many other facets of the industry. The Service was active in expanding and developing the Atlantic herring and crab industries in the late 1960's and in assisting administration of the **Newfoundland Fisheries Household Resettlement Program**. The latter was a federal-provincial program designed to facilitate the movement of households from small fishing settlements to larger communities. From 1965 to 1970 a total of $7.4 million was spent moving 3 240 families; the federal share was $5 million. In 1970, responsibility for this program was shifted to the Department of Regional Economic Expansion.

The legislative authority for programs of the Industrial Development Service eventually became the *Fisheries Development Act* once it was passed in 1966. Prior to 1966, funding was obtained through a process called "vote netting" under various Appropriations Acts.

In 1973, exploratory fishing activities located sizeable stocks of capelin on the Grand Banks and mackerel off Newfoundland's north-east coast. In the following year, a major $1.3 million program was launched jointly with the Department of Regional Economic Expansion to establish a capelin fishery. A program to convert traditional side-trawlers to more efficient stern-trawlers, which had been established in 1972, continued.

From 1965 to 1975 the Service carried out 830 projects. Of these 226 dealt with fishing methods and equipment, 145 involved exploration of fish stocks, 80 were for vessel design, materials and propulsion, 174 for processingand product development, 104 for general development, 34 for aquaculture, 26 for development of electronic equipment, and 40 had to do with training and education of fishermen and plant workers. In 1975, the Industrial Development Service was disbanded. Development programs under the *Fisheries Development Act* have continued as part of regular departmental programming. (Fisheries Development Act Annual Reports, various years).

### 2. Unemployment Insurance

The objective of the Unemployment Insurance Act, as passed in 1940, was to provide income support during temporary interruptions in employment. The period during which benefits could be received was limited, and only involuntary unemployment was insured. The 1940 program was not a universal income support program; no support was provided for the self-employed, such as fishermen, on the grounds that they could manipulate their periods of unemployment to there advantage.

Unemployment insurance coverage was extended to most seasonal workers in 1950, but fishermen remained ineligible for benefits until April, 1957 (Dingledine 1981). After 7 years of study, increasing demands by fishermen for "equal treatment", and the active role of the powerful J.W. Pickersgill, Parliament passed unemployment insurance fishing benefit regulations in 1957 in the dying days of the Government.

Fishermen were covered on the basis that fish buyers would be treated as their employers. Benefits were payable to fishermen only during the seasonal benefit period of January 1 to April 15. (It is important to remember that fish plant workers are covered under provisions of the regular UI program).

The Unemployment Insurance Act in 1970 was introduced to ensure that unemployment insurance benefits were compatible with other social security programs and to eliminate existing gaps and overlaps. As well, it aimed to ensure that workers would be better able to deal with frictional unemployment (time between jobs), and re-enter the job market. The White Paper, on which the Act was based, recommended that self-employed individuals not be covered since the program was designed to provide assistance between jobs. However, self-employed fishermen continued to be insurable. It was stated at the time that their unemployment insurance coverage would continue for a 3-year period during which the government would develop a new program, unrelated to the Unemployment Insurance Act, to protect them from the uncertainties of the fishing industry.

In 1971, the Government authorized the development of a new program to replace the unemployment insurance program for fishermen. From mid 1971 to the start of 1974, however, work on developing a new program was suspended. In 1974 work resumed in conjunction with the Social Security Review.

Self-employed fishermen continued to argue that existing regulations were discriminatory, regressive and not suited to industry needs. In response to these criticisms from the Atlantic groundfish fishery, UI regulations were amended in Fall, 1976. The new regulations allowed benefits to start earlier (the week of November 1 rather than December 1) and authorized payment of regional extended benefits in the November 1 to May 15 season (additional weeks of benefits available based on the area's unemployment rate). Plans to develop an alternative program to unemployment insurance for fishermen were once again suspended.

In November 1978, the Government rejected a proposed amendment to the Unemployment Insurance Act to remove coverage for fishermen. The Unemployment Insurance Commission investigated the possibility of bringing premiums more in line with fishing benefits, which far exceeded contributions from fishermen and processors, but nothing came of this.

Two major changes occurred in 1980. First, both fishermen and their spouses became separately eligible. Previously, if a husband and wife were working together their earnings had been pooled, so they were not able to file separate unemployment insurance claims. Many complaints lodged with the Canadian Human Rights Commission prompted the change allowing separate treatment of wives and husbands. The additional unemployment insurance costs were estimated at $7.3 million in 1981, but this was undoubtedly an underestimate.

The second important change was the requirement that the unemployment insurance program be self-financing, with the cost of operating the program covered from employee and employer contributions. However, the federal government continued to make substantial payments for self-employed fishermen, payments relating to regionally extended benefits, and payments to those who took approved training or participated in job creation projects. All of these were excluded from the self-financing provision.

Under the current Unemployment Insurance program, fishermen are eligible for benefits depending upon their fishing activities (Employment and Immigration 1987). Fishermen who work as employees are eligible for regular U.I. benefits (as are fish processing plant employees). Self-employed fishermen are eligible for benefits under the (separate) Fishermen's provisions, either as "year-round" or "seasonal" fishermen. Year-round fishermen must have worked 20 weeks (fishing and other employment) out of the last 52 on a vessel designated by the Commission as a year-round fishing vessel. Seasonal fishermen require a minimum of ten weeks of insurable employment in order to qualify under the Fishermen's provisions (a minimum of 6 weeks must be weeks spent fishing). Unlike year-round fishermen who are eligible throughout the year, seasonal fishermen receive benefits only during the off-season (summer fishermen are eligible for benefits from November 1 to May 15 and winter fishermen are eligible for benefits from May 15 to November 1).

As with recipients of regular unemployment insurance, fishermen are also eligible for regionally extended benefits. Extra weeks of benefits depend on an area's unemployment rate. Those eligible for benefits in areas of high unemployment could receive an additional 32 weeks of benefits. Seasonal fishermen, however, are generally limited to a total of 29 weeks of benefits.

The majority of fishermen qualify for unemployment insurance with only a minimum number of insurable work weeks. Hence, only a small portion of benefits are actually funded by beneficiaries. In 1987, for example, the premium contribution from fishermen and processors (their "employers") was $17 million; benefits paid to fishermen were $223 million. (Fig. 10). In addition fish plant workers received $226 million in benefits in 1988.

Some additional changes to the unemployment insurance program came into effect in January, 1990. These do not affect the fishermen's component of the program

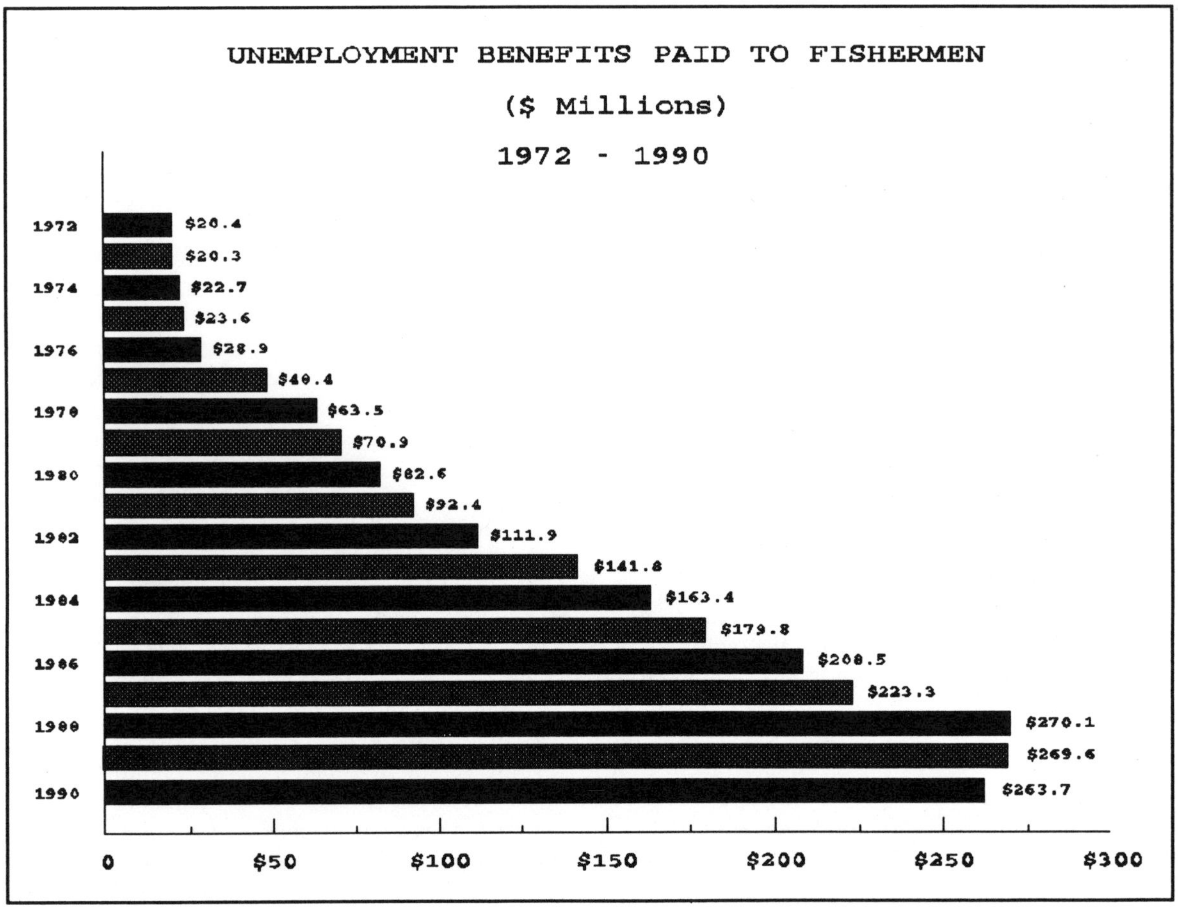


FIG. 10. Unemployment Benefits Paid to Fishermen, 1972–1988.

Source: Unemployment Insurance Statistics, Annual supplement to 73-001 monthly, 1988, Catalogue 73-202s Annual Supplement, Statistics Canada, p. 24.

except that payments to fishermen must in future be covered under the overall self-financing provisions of the program. In the long run, this may be significant because deficits in the fisherman's program will be funded from the unemployment insurance account in future. Other Canadians will be paying higher UI premiums to finance the fishermen provisions of the program.

Unemployment insurance has provided important social support but many critics claim that it has also distorted fisheries management and undermined healthy attitudes to work (Grady and Maclean 1989). There can be no doubt that support on such a grand scale has changed forever the Newfoundland economy and probably contributed to some endemic problems in the future. In effect, unemployment insurance and related programs have transformed a "labour surplus" situation to a labour scarce situation, with all that that implies. Governments have been reluctant to tackle this issue head-on but most observers believe it is a "time bomb" that has already had profound influence in Atlantic Canada and will continue to contribute to decreasing the competitiveness of the Atlantic economy.

### *Special Assistance*

As noted at the beginning of this paper, the fishing industry is cyclical and subject to uncertain biological and weather conditions. Moreover many fishermen earn low incomes from fishing. Following poor fishing seasons, fishermen may not have enough insurable weeks of employment to qualify for unemployment insurance benefits. This situation arises in localized areas virtually every year in Atlantic Canada. When eligibility for unemployment insurance is lacking, the federal government often has provided special employment projects to ensure eligibility.

In 1984 and 1985, spring ice delayed fishing activities in Newfoundland, Labrador and along the Quebec North Shore. In response, the federal government provided financial assistance worth $3.8 million in 1984 and $5.3 million in 1985. Similar requests are made virtually every year (again in 1991 such payments were made).

## III. Period Following Extension of Fishing Jurisdiction, 1977–1983

### A. Overview

Establishment of the 200 mile fishing limit resulted in major changes in the management of Canada's fisheries (Crowley and Palsson 1991). Strict conservation measures, and comprehensive fishery management plans and international agreements were put in place to increase fish harvesting in Canada. By the 1980's, the earlier expansionist philosophy of the 1970's had changed to resource conservation, domestication and economic efficiency.

Demand for seafood was strong through the mid 1970's and exports expanded. By 1981, however, the Atlantic fishing industry was once again experiencing financial difficulties. Fishing fleets which had expanded considerably in the late 1970's were now facing excess capacity and rapidly rising costs. Processing companies were undercapitalized and experiencing high interest rates and poor markets.

In the early 1980's, the B.C. salmon industry was affected adversely by a botulism scare. In Belgium, one person died and others became ill after eating contaminated canned pink salmon from Alaska. Another case of botulism poisoning was reported in 1982 in the United States. These incidents caused health alerts in many countries, as well as product recalls and, subsequently, a long period of slow sales. Salmon markets were also affected during the early 1980's by excess world supplies caused by record production in Alaska.

As a result of depressed industry conditions, the Government initiated reviews of the industries on both coasts — the "Pearse Commission" on the West Coast and the "Kirby Task Force" on the East Coast. The following section highlights the findings of these reviews and the assistance programs initiated during the period.

### B. (Kirby) Task Force on Atlantic Fisheries

The Task Force on Atlantic Fisheries (1982) is also known as the "Kirby Task Force", after its Chairman (now Senator) Michael Kirby. The Task Force had the mandate to develop measures for the long-term health of the industry as well as measures to deal with specific needs of the large companies. After 1 year of study and extensive consultations, the Task Force released its report in 1982. It contained 57 recommendations covering international issues, harvesting, resource-short plants, northern fisheries, quality control, port markets, productivity, profitability, marketing, fishermen's incomes, financial assistance, herring fisheries, and industry consultations. All of the recommendations were subsequently acted on (though some were not accepted in their entirety). (Parsons 1993a).

The Task Force recommended that government restructure fish processing companies and take an equity position which could later be recovered. This approach was in contrast to direct financial assistance provided during previous crises, and was additional to the 57 recommendations included in the report of the Task Force. The Task Force argued that equity participation was preferable because it was less likely to distort forces, it would minimize the risk of countervail action by the United States, and it would maximize the likelihood for Government to recover its money. Based on these arguments, the Government injected hundreds of millions of dollars into the major processing companies.

With respect to productivity, the Task Force recommended that regional development assistance be shifted from plant construction towards measures for increasing efficiency. This recommendation subsequently became part of DREE guidelines for the Industrial and Research Development Program (IRDP). The Task Force also recommended that research and development be given priority in government assistance and that agencies such as the National Research Council be involved.

The Task Force further recommended that exporters be licensed, a marketing commission and marketing councils be established, and that there be a major campaign for generic fish promotion in the United States. Some promotional was undertaken but funding for generic promotion ran afoul of budget cutbacks in 1986.

A major new system was recommended to stabilize fishermen`s incomes with funding to come from fishermen's contributions and the federal government. This proposal was rejected by federal and provincial fisheries ministers (at a 1985 First Ministers Conference). Although they concluded that fisheries incomes were low and often inadequate, they found that average year to year volatility was insufficient to justify such programs.

The Task Force recommended that there be no special assistance for vessel replacement but that fishermen be encouraged to purchase vessels from the most economical source. Subsequently, the Fishing Vessel Assistance Program was terminated and the import tariff was removed for vessels 30.5 meters and over in length. There still (1991) remains, however, a tariff of 25% on imported vessels under 30.5 meters in length.

## C. (Pearse) Commission on Pacific Fisheries

In January 1981, Dr. Peter Pearse of the University of British Columbia was appointed Commissioner to report upon fisheries policy on Canada's Pacific coast. The terms of reference for his inquiry were broad. They included condition of stocks, conservation and development of resources, fisheries management, fleet licensing, and license fees. The one man Commission conducted two rounds of hearings and received 193 written submissions from organizations and interested parties. The Commission's final report in September, 1982 contained numerous recommendations involving fundamental changes (Commission 1982).

Recommendations were made with respect to fish habitat management, stock management, the harvesting sector, indian fisheries, policy mechanisms as well as arrangements between the federal government and the Province of British Columbia. One of the central recommendations involved a new licensing system encompassing transferable fixed term licenses and a system of competitive bidding for fishing privileges. Due to the extreme fluctuations in year to year stock sizes of salmon and herring, an individual quota system could not be applied in these fisheries. Therefore, it recommended special measures to strengthen existing limited entry programs and to reduce the fleets in those fisheries by 50% within 10 years. In other smaller fisheries, it recommended a system of transferable quotas. The United Fishermen and Allied Workers Union launched a vigorous campaign against the proposed system arguing it would lead to competition for licenses, and the devastation of fishing communities. The Government did not act on the recommendations. However, they are currently being examined again, due in part to the success of individual transferable quotas in New Zealand.

In areas of fisheries assistance, the Commission found that some programs had encouraged expansion of fishing capacity and they concluded that they were a waste of taxpayer money. The Commission referred specifically to subsidies for the construction of fishing vessels such as the Fishing Vessel Assistance Program, accelerated depreciation for new vessels and the Fisheries Improvement Loans Program. (The first and third have since been eliminated, and the second has subject to recurrent scrutiny).

For an analysis of the relative impact of the Kirby and Pearse Reports see Parsons (1993a).

## D. Atlantic Fisheries Restructuring

Parliament passed the Atlantic Fisheries Restructuring Act in November 1983. This Act responded to recommendations of the Kirby Task Force to provide financial assistance for the large fishing companies which were then in financial difficulty.

Fisheries Products International Ltd. (FPI) was formed on December 31, 1984 through the amalgamation of 12 companies which had been involved in the fishing industry in Atlantic Canada: Fishery Products International Limited, Fishery Products (St. Anthony) Limited, Fishery Products (Harbour Breton) Limited, The Lake Group Limited, John Penny and Sons Limited, North Atlantic Fisheries Limited, Great Harbour Deep Seafoods Limited, New Fishco Limited, T.J. Hardy Company Limited, Marlake Limited, Caribou Fisheries Limited, and Cariboo International Limited.

FPI was conceived to stabilize the Newfoundland fishery by merging and restructuring fishing companies which had been unable to survive independently. FPI was initially controlled by a holding company owned 63% by the Government of Canada, 26% by the Government of Newfoundland and Labrador, and 11% by the Bank of Nova Scotia. In early 1987, FPI management made a public offering of common shares, and FPI returned to the private sector. Overall the restructuring of FPI cost the federal government $63 million plus an additional $168 million prior to privatization. The company that emerged had essentially no debt, so that it would be in a position to counter crises for many years to come.

National Sea Products Ltd. (NSP) was restructured in February, 1984 following the intervention of minority private sector shareholders (Kimber 1989). This restructuring involved direct funding of $130 million: $20 million from the private sector, $10 million from the federal government, $25 million from the province of Nova Scotia, and $75 million from the Toronto Dominion Bank. In addition, to retire debt owed to the Bank of Nova Scotia on a number of assets (formerly owned by H.B. Nickerson) the federal government committed an additional $80.5 million to NSHOLDCO Ltd. in exchange for common shares. NSHOLDCO Ltd. was a holding company created by the federal government and the Bank of Nova Scotia. This holding company was wound up and incorporated into NSP in March, 1988, resulting in a increase in the federal government's holdings in NSP.

As of April, 1988, shareholders of NSP were Scotia Investments Ltd. (26%), the Government of Canada (20%), Halifax Developments Ltd. (19%), and the Bank of Nova Scotia (14%). The remaining shares were held by smaller shareholders and the general public.

During fiscal 1983–84, the Government of Canada also acquired Pecheries Canada Inc., a holding company for Quebec-based Pêcheries Cartier Inc. (at a cost of about $31.5 million). This was later privatized in 1985 at a substantial loss.

## E. Fisheries Development

In 1978, Treasury Board requested that DFO prepare a plan for transition from previous "recovery" programs to "developmental" programs. The resulting document argued for fisheries rehabilitation and enhanced technology transfer. Following this, programs implemented during the 1980's emphasized "development" rather than the expansionist philosophy of the 1960's and 70's. Programs concentrated on competitiveness, efficiency, productivity and adding value domestically — all concerns that remain vitally important to the long-term health of the fishing industry.

Expenditures for technology transfer between 1978 and 1981 amounted to $12.3 million. During 1981–82, $6.5 million was spent on a wide range of development projects and $2.3 million on more general Economic and Regional Development Agreements (ERDAs). These projects covered quality enhancement, productivity improvement, fuel consumption reduction, processing, and methods of resource assessments.

At this time, there were three 5-year ERDA fisheries subsidiary agreements that were jointly administered by DFO and the Department of Regional Economic Expansion. The Coastal Labrador Program involved funding of $13.5 million; the South-East New Brunswick Program $2.7 million; and the Prince Edward Island Fisheries Development Program $7.0 million. Assistance was provided for fishing gear, processing equipment, cold storage facilities, community stages and aquaculture. In addition to fisheries subsidiary agreements, specific development assistance funding from DFO amounted to over $7.0 million per year over the period 1982 to 1985.

In 1983, DFO presented a plan for Quebec fisheries. This plan resulted from the repatriation of federal responsibility for Quebec fisheries (as recommended by the Kirby Commission). It involved funding of $135.2 million over 5 years (from 1983 to 1988). The plan covered resource development, fleet improvement, new products and northern Quebec economic development. A Quebec Region was created within the Department of Fisheries and Oceans.

In 1983–84, the Government introduced a program for "recovery capital", aimed at facilitating economic recovery. The fisheries component was $43.9 million, with funding as follows: Nova Scotia, $6.9 million for ice making and storage, industrial water and power; Quebec, $29.7 million for ice making and fish unloading, bait storage and salting facilities; Newfoundland, $3.0 million for two marine service centres; New Brunswick, $2.2 million for fish unloading; and $2.1 million for planning, implementation and covered co-ordination.

In 1984, the federal government signed ERDA fisheries subsidiary agreements with New Brunswick and PEI involving funding, respectively, of $45 million and of $10.5 million over 5 years. In 1988, Canada and Quebec agreed to a 3-year, $17.5 million fisheries subsidiary agreement under the Quebec ERDA. A 5-year inshore fisheries subsidiary agreement was also signed with Newfoundland ($60 million). In 1989, Canada signed new agreements with New Brunswick ($20 million) and P.E.I. ($16 million) to replace expiring fisheries subsidiary agreements. These agreements involved programs to enhance quality, improve efficiency of the fishing and fish processing industry, adapt new harvesting technology, broaden the resource base, and increase production of underutilized species.

# IV. Recent Initiatives

## A. Overview

In 1984, the newly elected federal Government emphasized reducing the deficit and the size of Government. At the time, good salmon runs on the West coast and the effects of industry restructuring on the East coast were providing good returns for industry. By the late 1980's, however, endemic problems began to emerge. The large increase in world salmon production from natural and farmed sources caused decreased prices and lower returns on investment on the Pacific coast. On the Atlantic, a resource based "crisis" developed with considerably reduced landings and employment. This brief section traces assistance programs from 1984 to 1990.

## B. (Nielsen) Task Force on Program Review

The new government established a Task Force on Program Review ("Nielsen Task Force") in 1984 to review expenditures throughout the government. As in other sectors, the fisheries component involved both industry representatives and government officials.

With respect to the fishing industry, the Task Force noted that "a labyrinth of government assistance programs and regulations [have] arrested the ability of the Canadian fisheries to adapt to market forces and [has] undermined its competitiveness" (Task Force on Program Review, p. 27). It stated that after extension of fishing jurisdiction to 200 miles, the Federal Government was spending some 1 billion dollars per year on various programs in order to sustain an industry that generated landings of about the same amount and ultimate sales of only 2 billion dollars. The Task Force also cited the possibility of countervailing duties being applied by the United States. It concluded that the fishing industry could not be profitable if it continued to be burdened with welfare and employment responsibilities.

The Task Force also stated that DFO should be reduced in size and its mandate narrowed to focus on conservation and basic services for maximizing economic returns. Subsequent to the report, DFO staff and budgets were reduced, first at headquarters and then in the regions. The domestic marketing programs of DFO were transferred to the private sector; the Fisheries Food Center was terminated; and some functions (such as market information) were taken over by the Program for Export Market Development (External Affairs).

### C. Program for Export Market Development

The Program for Export Market Development (PEMD) was created in 1971 to encourage Canadian firms to sell their products and services in foreign markets. It is administered by External Affairs and International Trade Canada in conjunction with Industry, Science and Technology Canada (formerly DRIE). Financial assistance provided by the program is designed to reduce risks involved in entering new foreign markets and is available for various promotional activities. Assistance is provided for firms to participate in foreign trade fairs, to explore potential foreign markets, to finance visits of foreign buyers to Canada, and to undertake certain special activities.

The Program for Export Market Development of Fish (PEMD FISH) was created in 1983 in response to recommendations made by the Task Force on Atlantic Fisheries. The objective of the program was to encourage Canadian firms to initiate or expand sales of groundfish and herring in foreign markets. Financial risk was shared between individual firms and government, with government covering up to 50% of costs. This assistance was to be repaid if a firm were successful in developing new markets. Funds allocated to the program were to develop satisfactory promotional and marketing programs by the industry.

The program was implemented in Atlantic Canada in November of 1983 and in British Columbia a year later. The program was shortlived due to criticism from the (Neilson) Task Force on Program Review. In the fall of 1985, PEMD FISH was terminated. This did not exclude the fishing industry from PEMD financing, however, as funding became available through other PEMD programs.

For numerous reasons, response to the program had in fact been modest. Many companies were not eligible under the FISH component of the program but were eligible under PEMD FOOD. A highly priced Canadian dollar and declining international demand for Canadian fish contributed to decreased demand for program funds. Many smaller companies were deterred by the requirement for detailed financial and sales information. Often, it was easier to obtain provincial grants.

### D. Industry, Science and Technology Canada (ISTC), Atlantic Canada Opportunities Agency (ACOA), and Western Diversification (WD)

Industry, Science and Technology Canada (ISTC) was formed in 1989 by integrating the Department of Regional Industrial Expansion (DRIE) and the Ministry of State of Science and Technology. ISTC's mandate focuses on building competitiveness in Canadian industry and promoting scientific excellence. ISTC initiated a Fisheries Products Sector Campaign in 1989 designed to improve the long-run international competitiveness of the fishing industry. The program is still operating in 1991.

In 1989, as well, responsibility for regional economic development in Atlantic and Western Canada was transferred from DRIE to the Atlantic Canada Opportunities Agency (ACOA) and the Department of Western Diversification (WD).

The Atlantic Canada Opportunities Agency (ACOA) was established in June, 1987 as a special 5-year economic development program for Atlantic Canada. Some $1 050 million over 5 years was allocated to the program. ACOA has two major activities: the Action Program and Economic and Regional Development Agreements.

The **Action Program** has provided loan insurance, interest buy-downs, technology innovations and plant and equipment modernizations. It has funded fishing industry projects with the stated intention that they should not expand capacity to process fish. The ACOA Action Program was a successor to two previous programs: the Industrial Regional Development Program and the Atlantic Enterprise Program. The ACOA Action Program was originally to run from 1987 to 1992 but has been extended to 1994. No additional funding, however, was allocated at the time this extension was announced.

The other major activity involves the five fisheries subsidiary agreements that exist for Quebec and the Atlantic provinces (see the previous section on fisheries development). There subsidiary agreements contained programs for upgrading fleets, enhancing quality, exploiting underutilized species and saving energy. All are administered jointly with DFO.

Assistance to Atlantic fish processing has increased sharply since 1986. Under the ACOA Action Program nearly $30 million was committed to various fisheries projects in the Atlantic Provinces in 1988 and 1989 alone notwithstanding ACOA's stated policy. Some of this has undoubtedly increased processing capacity and thereby exacerbated current problems in the fisheries.

Western Diversification (WD) was created in 1987 to enhance Western Canadian economic development. This agency administers the 5-year Western Diversification Fund and provides assistance to enhance productivity of Western Canadian industry.

WD has provided assistance to the fishing industry by allocating $53 million for the Salmon Enhancement Program in British Columbia, assisting development of small craft harbours (e.g. the **South Moresby Island Agreement)** and supporting fisheries related industries to develop new products. WD has also provided assistance to the aquaculture industry through the **Canada-B.C. Economic and Regional Development Agreement on Small Business Incentives**.

Finally, there have been special programs of assistance during difficult times in certain sectors of the industry. In the case of mussels (1989) and lobster (1990) this assistance amounted to several million dollars each and covered promotional marketing and, in the case of mussels, a special program of loans through the Federal Business Development Bank.

## V. Synopsis

Federal assistance to Canada's fisheries increased significantly after World War II and then declined, over a long period, with the rate of decline accelerating during the early 1980's. In the late 1980's, new programs of assistance were initiated to facilitate marketing, meet specific needs of fishermen, provide facilities for handling and processing of fish, as well as encourage industrial regional development and the restructuring of companies.

Objectives of assistance changed over time. Immediately after the war, programs were designed to assist the transition to peace time. During the next period (the 1950's), new programs were targeted on industrial development. It was thought that fisheries could be the basis for expanded output and employment. In the 1960's and 1970's programs were targeted on rural economic development, reflecting the Government's general concern with regional disparities. In the 1980's the development philosophy changed to increasing efficiency, enhancing productivity and technological innovation. Notwithstanding this, however, fishing and processing capacity in the industry increased substantially throughout the period under review.

During the past half century, the volatility of the industry provided the rationale for most government assistance. By contributing to the expansion of fishing and processing capacity, assistance programs probably exacerbated cycles and enhanced dependence on the public treasury.

## VI. Epilogue

By early 1990, it was apparent that yet another "crisis" had developed in the Atlantic fisheries. This was one more in the long succession of crisis noted at the beginning of this paper. No similar "crisis" currently exists on the Pacific coast, but no doubt the endemic problems there will result in a full fledged crisis in due course.

Current problems on the Atlantic are resource rather than market based, but with a broader incidence than many crises in the past. That is, the current situation reflects a dramatic decline in the northern cod stock, overfishing of several species on the Scotian Shelf and the Grand Banks, and several localized problems. This has been coupled with mainly market problems in the case of shellfish/crustacean species such as crab and lobster.

The previous major "crisis" in 1980 — the one that led to the Kirby Commission and the massive restructuring of the Atlantic industry — was essentially financial (Nicholson 1988). The offshore companies had been built on a foundation of sand: little equity and ever increasing debt. When interest rates increased substantially in 1981, companies simply were unable to cope.

Danger signs with respect to the current resource situation appeared initially in 1987. Catches stabilized or even decreased. But these decreases were camouflaged by high prices and boom conditions, and consequently ignored. In 1989 the total allowable catch (TAC) for Northern Cod was reduced from 266 000 tonnes to 235 000 tonnes; in 1990 a further reduction brought the TAC to 197 000 tonnes and in 1991 to 185 000 tonnes. Quotas also declined for haddock, redfish and some flatfish species. These reductions led to operating losses and corresponding plant closures by the two large offshore companies in 1990. In addition resource and market problems for other species caused financial and related problems throughout the Atlantic Provinces.

The federal government established a spate of task forces in response. In Newfoundland, the 1989 Independent Review of the State of the Northern Cod Stock was chaired by Dr. L. Harris, President of Memorial University (Harris 1990), and this was followed by a Task Force chaired in 1990 by Mr. Eric Dunne, Regional Director-General, DFO (Dunne 1990). In Nova Scotia, a Groundfish Task Force was chaired in 1989 by Mr. Jean-Eudes Hache, Regional Director-General, DFO (Haché 1989). As well, the Northern Cod Task Force in 1989 was led by the former Associate Deputy Minister of DFO, Mr. Ken Stein.

All of this finally resulted in the Atlantic Fisheries Adjustment Program (AFAP) in May 1990. This program provided $584 million over 5 years to four main funding areas: curbing foreign overfishing; rebuilding fish stocks; adjusting to the realities of a smaller industry and diversifying provincial economies. Initiatives under AFAP include the plant workers Adjustment Program (to assist workers and trawlermen who have been laid off), fishing licensing policy such as individual transferable quotas and initiatives to reduce capacity. Diversification measures include proposals involving underutilized species and other innovative approaches to existing harvesting and processing.

Under AFAP, the Atlantic Canada Opportunities Agency set up a program to create permanent employment for displaced workers and to diversify the economies of coastal communities. This program (the Fisheries Alternative Program) is directed to economic development outside of the fishing industry. The Fisheries Sector Campaign of the Department of Industry, Science and Technology, is designed to improve the long run competitiveness of the Canadian fishing industry. A special "Program d'essai experimentation", a 5-year federal initiative designed to maximize the economic viability and stability of fish harvesting, processing and aquaculture, has been set up for Quebec.

Then in 1992 there was a major collapse of the northern cod stock. In July 1992 the government implemented a 2-year moratorium on fishing for northern cod. This was accompanied by a major program of assistance to compensate fishermen during the moratorium and to provide some downsizing of the industry. For details, see Lear and Parsons (1993) and Parsons (1993b). It is too early to evaluate the impact of this program. Further assistance was provided in 1993 to deal with TAC reductions and closures of other cod fisheries on the Atlantic.

## References


Atlantic Development Board. 1963–1968. Annual Reports. Department of Regional Economic Expansion, Ottawa, Ont.

Canadian Saltfish Corporation. 1971–1988. Annual Reports. Department of Fisheries and Oceans, Ottawa, Ont.

Commission on Pacific Fisheries Policy ("Pearse"). 1982. Turning the tide — A new policy for Canada's Pacific fisheries. Minister of Supply and Services, Ottawa, Ont.

Crowley, R.W. 1982. The new power focus in Ottawa: MSERD. Optimum 13(2): 5–16.

Crowley, R.W., and W. Brown. 1989. Canada's Pacific fisheries — Issues in business development. Department of Fisheries and Ocean, Econ. Commer. Anal. Rep. 15: 18 p.

Crowley, R.W., and H. Palsson. 1991. Rights based fisheries management in Canada. Mar. Resources (in press).

Department of External Affairs. 1988. Program for Export Market Development (PEMD). Handbook: A Financial Assistance Program for Canadian Exporters. Department of External Affairs, Ottawa, Ont.

Department of Fisheries and Oceans. 1930–1988. Annual Reports. Ministry of Supply and Services, Ottawa, Ont.

1972–1987. Fisheries Development Act: Annual Reports, Minister of Supply and Services, Ottawa, Ont.

1947–1988. Fisheries Improvement Loans Act: Annual Reports, Minister of Supply and Services, Ottawa, Ont.

Department of Regional Industrial Expansion. 1981. Summaries of federal-provincial general development agreements and currently active subsidiary agreements. Minister of Supply and Services, Ottawa, Ont.

de Silva, K.E.A. 1981. An economic analysis of the Shipbuilding Industry Assistance Program. Discussion Paper No. 351, Economic Council of Canada, Ottawa, Ont.

Dingledine, G. 1981. A chronology of response. The evolution of Unemployment Insurance from 1940 to 1980. Minister of Supply and Services for Employment and Immigration Canada, Ottawa, Ont.

Dunne, E. 1990. Report on the implementation of the Task Force on Northern Cod. Department of Fisheries and Oceans, Ottawa, Ont.

Employment and Immigration Canada. 1987. Summary of U.I. Benefits. Minister of Supply and Services, Ottawa, Ont.

Fisheries Prices Support Board. 1947–1988. Annual Reports. Minister of Supply and Services, Ottawa, Ont.

Fishing Vessel Insurance Plan. 1953–1988. Annual Reports and Financial Statements, Minister of Supply and Services, Ottawa, Ont.

Freshwater Fish Marketing Corporation. 1970–1988. Annual Reports. Queen's Printer, Ottawa, Ont.

Grady, P., and D. MacLean. 1989. An income stabilization program for fishermen: Its feasibility and cost. Unpublished report for the Economic and Commercial Analysis Directorate, Department of Fisheries and Oceans, Ottawa, Ont.

Hache, J.E. 1989. Report of the Scotia Fundy Groundfish Task Force. Minister of Supply and Services, Ottawa, Ont.

HARRIS, L. 1990. Independent review of the state of the northern cod stock. Minister of Supply and Services, Ottawa, Ont.

JUDSON, T.A. 1961. The freshwater commercial fishing industry of Western Canada. Ph.D. dissertation, University of Toronto, Toronto, Ont.

KIMBER, S. 1989. Net profits: The story of National Sea. Nimbus Publishing Co., Halifax, N.S.

LEAR, W.H., AND L.S. PARSONS. 1993. History and management of the fishery for northern cod in NAFO Divisions 2J, 3K and 3L, p. 55–89. *In* L.S. Parsons and W.H. Lear [ed.] Perspectives on Canadian marine fisheries management. Can. Bull. Fish. Aquat. Sci. 226.

MCIVOR, GEORGE, H. 1966. Report of Commission of inquiry into freshwater fish marketing. Queen's Printer, Ottawa, Ont.

NATIVE BROTHERHOOD OF BRITISH COLUMBIA. 1989. A comprehensive plan to increase native participation in the British Columbia fishing industry. The Native Fishing Association, Vancouver, B.C.

NICHOLSON, P.J. 1988. A retrospective on Atlantic fisheries restructuring, 1979–87. Unpublished notes for a seminar at Fisheries and Oceans. March, 1988.

ORGANIZATION FOR ECONOMIC CO-OPERATION AND DEVELOPMENT. 1980. Financial support to the fishing industry. Paris.

PARSONS, L.S. 1993a. Shaping fisheries policy: The Kirby and Pearse Inquiries — Process, Prescription and Impact. p. 385–409. *In* L.S. Parsons and W.H. Lear (eds.) Perspectives on Canadian marine fisheries management. Can. Bull. Fish. Aquat. Sci. 226.

PARSONS, L.S. 1993b. Management of marine fisheries in Canada. Can. Bull. Fish. Aquat. Sci. 225: 784 p.

TASK FORCE ON ATLANTIC FISHERIES. 1982. Navigating troubled waters: A new policy for the Atlantic fisheries. Minister of Supply and Services, Ottawa, Ont.

TASK FORCE ON PROGRAM REVIEW. 1986. Natural resources programs: From crisis to opportunity. Minister of Supply and Services, Ottawa, Ont.

# Reliability of Scientific Advice on Fishery Management Measures

W.G. Doubleday

*Deparment of Fisheries and Oceans*
*200 Kent Street, Ottawa, Ontario, Canada K1A 0E6*

Doubleday, W. G. 1993. Reliability of scientific advice on fishery management measures, p. 369–383. *In* L. S. Parsons and W. H. Lear [eds.] Perspectives on Canadian marine fisheries management. Can. Bull. Fish. Aquat. Sci. 226.

Reliability of scientific advice for the management of fisheries is defined in terms of regularly meeting the expectations of fishermen and fishery managers who receive it. The possibility of defining objective criteria for reliability is examined in three situations: establishing closed areas, regulating mesh size, and setting catch quotas. Factors which influence the achievement of reliabilty are examined. Reliability of historical advice is considered for examples from Northwest Atlantic finfish fisheries. Suggestions are given to improve the reliability of advice in practice. Consistency of scientific estimates and advice over time is found to be as important as accuracy in some common situations.

---

La valeur des avis scientifiques en matière de gestion des pêches se mesure à la capacité de satisfaire régulièrement les attentes des pêcheurs et des gestionnaires des pêches auxquels ces recommandations s'adressent. La possibilité d'établir des critères objectifs d'évaluation est étudiée dans trois situations : la désignation de zones fermées à la pêche, la réglementation de la grandeur des mailles et l'établissement de contingents par rapport aux prises. Les facteurs qui influent sur le niveau de fiabilité sont examinés. La fiabilité des recommandations historiques est, par exemple, considérée dans des activités de pêche liées à la récolte des poissons de l'Atlantique Nord-Ouest. Des suggestions sont apportées afin d'améliorer, dans la pratique, la fiabilité des recommandations. La constance, au cours des ans, des prévisions et des avis scientifiques, est perçu comme étant aussi importante que l'exactitude dans les situations courantes et ponctuelles.

## Introduction

Major fish stocks are common property resources managed by governments. The public management of fisheries requires policies to mediate conflicts between groups of fishermen, to ensure that the productive capacity of the fish stocks is not depleted by overharvest, and to achieve economic objectives. Governments have supported scientific research to improve the management of fisheries since the nineteenth century (T.D. Smith in Gulland [ed.] 1988, p. 2). The Biological Board of Canada, for example, was empowered by a 1912 Act of the Canadian parliament for "conduct and control of investigations of practical and economic problems associated with marine and freshwater fisheries, flora and fauna" (Johnstone 1977). Today, scientific information and advice in Canada regarding the biological consequences of fishery management policies is provided mainly by scientists employed by federal and provincial governments. National governments around the world employ scientists to advise them on the need for and consequences of alternative fishery management measures.

Administrators commonly seek a rational basis for policies and fishing plans. Fishermen, administrators and government Ministers normally do not consider themselves experts in ichthyology and population dynamics of exploited fish stocks. They expect scientists to provide sound information and advice which can confidently be used as the basis for managing fisheries. Issues such as determining how much fish can be taken without compromising the continued productivity of stocks can be addressed by scientists who develop models based on the biology of the fish and the effects of fishing on the stocks. The technical details and uncertainties associated with estimating the abundance and trends of wild fish stocks are assumed to be within the competence of the scientists concerned.

This study considers what is involved in providing scientific advice that can be confidently used by fishermen, administrators and Ministers of the Crown. The discussion is focussed on the concept of reliability. Scientific advice, like a tool, can be confidently used if it regularly meets expectations for performance. Thus reliability is part expectation and part performance.

Both the expectation and the performance depend on specific circumstances of the fishery being managed and of the policy being followed. In this chapter reliability of scientific advice is examined in three general situations:

- closure of an area to fishing for some period of time to avoid capture of young fish,
- setting a minimum mesh size for otter trawls to allow the growth potential of young fish to be realized before harvest, and
- setting a catch quota to take a target fraction of the fishable stock.

Examples are primarily chosen from Canadian Atlantic fisheries and are not intended to illustrate the full range of the world's fisheries. The setting of catch quotas is the most demanding of the three examples in terms of expectations and the most challenging in terms of providing accurate estimates.

## Meaning of Reliability

Reliability can be defined in many ways. Concepts closely linked to reliability are soundness, dependability, credibility, and consistency. A reliable person can be depended on. Reliable information can be taken at face value and believed. A reliable machine performs as expected.

It is not enough to perform occasionally as expected. A car which fails to start two days out of three but runs perfectly otherwise is considered unreliable. Reliable machines or systems routinely and with very few exceptions meet expectations for performance.

Reliability is not absolute. Even the rotary dial telephones which have worked flawlessly for many years in most people's experience do occasionally break down. They are highly reliable but not perfectly reliable.

Reliability has been studied and quantified by engineers designing complex products which require many components functioning properly for the product to perform. The analysis recognizes that one can never be certain of expected performance and looks at the number of failures during a long period of use or the time between failures, particularly the average or mean time between failures of a system whose components have a known probability of failing after working properly for a given period of time. Mean time between failures is a convenient yardstick to compare reliability of alternative products designed to serve the same purpose.

Reliability theory translates a reliability performance target into corresponding requirements for components, taking into account whether some components can fail without the entire machine or system failing. Much of the early theory was associated with making aircraft more reliable by using more than one engine. Use of reliability theory in manufacturing has had some outstanding successes. Modern electronic devices such as computers often have mean times between failure of tens of thousands of hours of use.

The expectation on which reliability is judged need not be a high standard of performance. The old, rotary dial telephones were not expected to call forward or do speed dialling, but they almost invariably allowed a connection to be made without incident. Newer, computerized telephones are capable of much more but may drop calls or fail in other ways frustratingly often. The expectations of performance for the two are different and the key question is whether expectations are met often enough for a telephone system to be confidently used. Readers interested in more information on reliability theory could consult the book Reliability Technology by Green and Bourne (1972).

Examining the reliability of scientific advice on the management of fisheries requires consideration of what is expected and how often expectations fail to be met. In fisheries science as in business, it is the expectations of the customers that matter most. Scientific advice on fisheries issues is typically provided to governments. Thus the direct customers are officials. Administrators and elected officials are, however, sensitive to the views of fishermen and the fishing and fish processing industry. Consequently, the expectations of fishermen underpin those of administrators. In the following analysis, the expectations of fishermen and the fishing industry and the effects of scientific advice on these people is the basis for judging reliability.

## Objectives and Tools of Fishery Management

Fishery management serves many purposes. Specific measures may be taken, among other reasons, to allocate catch or access to fish resources, to regulate conflicts between groups of fishermen, or to conserve the fish resources. Conservation in this context is taken to mean maintaining the fish resources at a productive level indefinitely. Biological advice is often of little or no importance in questions of allocation of access and regulation of conflict between fishermen since the effect of a fish removed from the stock is much the same no matter who catches it, provided all fishing gears are regulated to avoid capture of significant amounts of young, rapidly-growing fish. Scientific advice is of central importance in choosing between conservation measures because it is directly relevant to the question at hand and because it is seen to be objective.

Conservation of fish resources involves controlling the exploitation of more or less biologically self contained management units called stocks. In many instances these stocks are distinct or nearly distinct spawning populations of a single species which inhabit a single geographic area throughout the year, while, in others, several intermingling spawning populations of a single species or several similar species may be managed together in a geographical area.

Controlling the exploitation of fish stocks usually involves controlling rates of fishing mortality. A fishing mortality rate can be compared to the interest rate applied to a daily interest savings account. The total amount of interest earned over a year is obtained by multiplying each daily balance by the interest rate and averaging these numbers over the year. If this interest, together with a steady flow of payments is taken from the account while money is added only at the first of the year, both the account balance and the accumulated interest would follow time trends similar to temperate marine fish stocks and catches. In general, the number of fish of a given age caught in a year is equal to the fishing mortality rate for that age times each daily number of fish in the stock, averaged over the year.

Typically fishing gears and fishing practices do not subject all age groups of a stock to the same fishing mortality. Young age groups are normally not caught at all and the fishing mortality rate normally increases with age until it reaches a maximum at which age the fish are called fully recruited. The fishable stock is defined as the total over all age groups of the number of fish alive at that age times the fishing mortality rate for that age divided by the fishing mortality rate for fully recruited fish (Fig. 1). Setting catch quotas to harvest a target proportion of the fishable stock can be an effective conservation measure.

Conservation, in the context of managing a fishery, means not impairing the productivity of the fished resource and not wasting its potential harvest. In turn this means not catching too many fish in general and not catching too many young fish or too many spawning fish in particular. Catching too many fish reduces stock sizes to low levels where the fishery becomes dependent on each year's new group of young fish reaching harvestable size. Catching too many young fish wastes the potential of growth to maturity which could give larger catches. Reducing the spawning stock to low levels risks the continued production of young fish.

Relevant management measures include restrictions on fishing gears and methods to avoid catching young fish, prohibitions on fishing in some areas and seasons to avoid catching young or spawning fish or to avoid exerting too much fishing pressure when fish are concentrated, and limitations on the amount of fish that may be caught to achieve a target harvesting rate and avoid reducing stocks by over-harvesting. Another means of controlling harvesting rates is to limit the fishing effort by restricting entry, controlling vessel and gear characteristics to limit efficiency and applying short open seasons.

## Scientific Advice in Support of Fishery Management

Administrators normally seek analysis of the consequences of proposed management measures before implementing them. For fishery management, key questions are the effects on fish stocks and the effects on fisheries. Scientists can shed light on both these questions through modelling of the dynamics of fished stocks and of the interaction between fishing vessels and gear and the fished stocks. Economists use biological impacts such as changes in annual catches and in the quantity of fish caught per hour or per day fished as a starting point for estimation of economic consequences of new policies.

Biologists combine general knowledge about the species in question with specific observations collected by research vessels or by observing the commercial fishery. The spatial and temporal coverage of sampling is coarse and the biological and technological processes involved change over time. It is impractical to determine with great accuracy the state of stocks and the implications of regulatory changes. Nevertheless, estimates and approximations can be obtained in most cases which are useful in choosing between alternatives and adequate to conserve fish stocks.

Some management measures require advice or estimation of a target which is revised infrequently, such as delineation of an area and season to define a closure or setting a minimum mesh size for trawls. Other measures such as catch quotas require frequent revisions. In many cases scientific estimates of fish abundance and fishing mortality are revised annually in a cycle tied to the development of fishing plans.

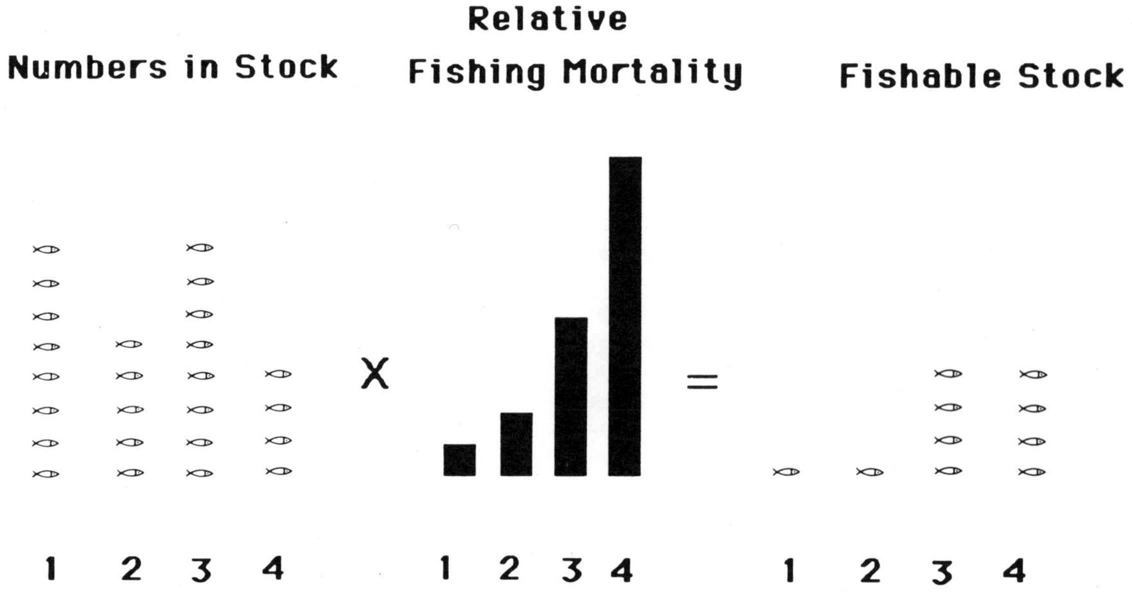


FIG. 1. The fishable stock is the sum of a fraction of each age group in the total stock. The fractions are the ratio of the fishing mortality for each age group to the fishing mortality for fully recruited ages.

In much of the world, including North America and Europe, fishery management follows an annual cycle of scientific analysis of the previous year's fishery followed by publication of scientific advice on the state of stocks, the development of fishing plans and regulations and their administration during the following fishing season. Scientists are often asked to estimate how far the management of the past year's fishery deviated from stated policy goals (what was last year's fishing mortality rate?) and what changes are needed to ensure that policy targets are met in the coming year. For example, Canadian Atlantic fisheries policy for marine groundfish stocks involves a target harvest of a fixed percentage of the fishable ages. The target harvest rate is called $F_{0.1}$ (Gulland and Boerema, 1973). This target was adopted in the Northwest Atlantic by the International Commision for the Northwest Atlantic Fisheries (ICNAF) following the September 1975 Special Meeting of its Standing Committee on Research and Statistics which found the earlier target of maximizing the catch from each year class of fish to be inadequate to conserve stocks (ICNAF Redbook 1976). In part, lower target harvest rates were advocated because of weaknesses in the overall management regime in place at the time such as a tendency for the Commission to adopt higher catch levels than scientists considered prudent and the lack of effective enforcement of agreed management measures. $F_{0.1}$ is a theoretical reference level for fishing mortality, intended to aid conservation and maintain good commercial catch rates. Scientists estimate the harvest rate (called fishing mortality or F) in the most recent year of fishing and project the catch corresponding to the target rate $F_{0.1}$ in the coming year. Other options are also provided.

Management plans often establish Total Allowable Catches (TACs) and sub-allocate these to individual enterprises or to groups of fishermen. A TAC is the maximum amount of fish which may be caught from a stock or other management unit, usually in one year. In the case of a sub-allocated TAC, fishing for each component continues until its limit is reached and the fishery as a whole is closed when all sub-allocations have been taken. This approach requires accurate and timely monitoring of catches during the fishing season.

### Criteria for Performance Measurement

The performance of scientific estimates can be quantified in the statistical language of accuracy, precision and, for time series, consistency. Accuracy is defined relative to a theoretical "true value". Accurate estimation procedures give answers close to the "true value" while precise estimation procedures show little variation when applied to many samples from the same population but may have a substantial bias leading to a set of estimates close to each other but systematically high or low relative to the "true value" (Fig. 2). A series of estimates of a parameter which varies over time can fluctuate more or less widely and rapidly than the "true value" being tracked (Fig. 3). A consistent estimation procedure gives a series of estimates for the same parameter which does not change sharply from one time period to the next.

Estimating the abundance of fish stocks and forecasting the consequences of new regulatory measures depends on the information available. Some of the most important information comes from monitoring the fishery itself and is subject to distortions from incomplete or inaccurate reporting of catches and unreported discards of small or undesired fish. Some of the information comes from sampling by research vessels and is subject to uncertainties due to the small number of samples taken and the imperfections of nets as sampling instruments. No scientist can remove the limitations inherent in the data he must use. Consequently, scientific advice and estimates for fish stocks are inherently imprecise and the true state of nature can only be approximated.

In the study of wild fish stocks, the "true value" is rarely known, even in retrospect. Calculations of the consequences of systematic over or under estimation of fishing mortalities and stock sizes must therefore rely on theoretical considerations. Year over year changes in estimates of abundance are visible to fishermen and administrators while theory-based measures of precision and accuracy often are not.

In order to determine how much precision, accuracy and consistency is needed, it is useful to look at the consequences of basing action on estimates that are not completely accurate, precise or consistent from year to year.

## Consequences of Error

The consequences of implementing a harvesting regime based on erroneous scientific estimates can be considered on several levels. Most serious is adoption of measures which allow the resource to be extirpated or reduced to such a low level that its productivity is severely impaired for many years (recruitment overfishing). Another consequence of erroneous estimates can be the loss of yield potential due to overharvesting (growth overfishing) or harvesting too lightly. It is also appropriate to consider losses in economic efficiency and potential employment resulting from estimates which are too high or too low. Finally, there are the impacts of changes in harvest level and harvest capacity and the economic dislocations associated with unnecessary changes in the management regime which result from errors in estimation of fish abundance or other parameters.

In the context of reliability, it is appropriate to concentrate on large losses which represent a visible failure of the harvest regime rather than small, theoretical losses in yield or efficiency. If a fish stock is depressed due to basing fishery management on optimistic scientific estimates (not due to setting catch levels above those advised) to the point that no harvest can be permitted, the estimates were not sufficient to conserve the stock. Such a failure must be extremely rare if the estimates are to be accepted as reliable by fishery managers and fishermen. If the yield might have been a few percent higher had the estimates been accurate but there were no other implications, no failure to perform would be perceived. A small, theoretical loss which is not felt by fishermen could occur constantly, with the estimates nevertheless considered reliable.

### *Strategic Goals vs Present Status*

Scientific estimates are often needed both to establish policy and to apply it routinely. This is particularly evident in situations where estimates of the current status of a fish resource are used to develop management plans aimed at achieving target harvest rates or target spawning escapements.

Catch quotas are based on applying a fishing mortality rate to an estimate of the fishable stock. In Atlantic Canada, the main reference point for projecting catch levels is $F_{0.1}$, as mentioned previously. The technicalities of estimating this reference point are beyond the scope of this chapter. The reasons for adopting this reference point are discussed in ICNAF (1976). In the case of Atlantic cod, $F_{0.1}$ is approximately 20%.

The consequences of errors of estimation in setting the target harvest rate are normally not serious. More significant is the choice of a reference point at or below the rate which leads to the theoretical maximum yield. The $F_{0.1}$ reference point allows a margin for error in estimating projected catches. During the late 1970's the Canadian Atlantic Fisheries Scientific Advisory Committee updated the estimated $F_{0.1}$ reference each year to allow for changes in the size of fish and the pattern of fishing over ages. This practice was soon discontinued when it became clear that the changes in the reference point for projections were introducing changes in fish quotas with no obvious benefits.

Projecting catches at a target fishing mortality rate also requires an estimate of the abundance and growth characteristics of the stock. These estimates are normally updated annually and errors in estimating abundance translate nearly proportionately into changes in the overall catch quota.

The way a fish resource would respond to management measures is never fully known. Management regimes which aim to stabilize stock size or spawning escapements reduce the insights which could be gained from experiencing a wider range of states of the fished

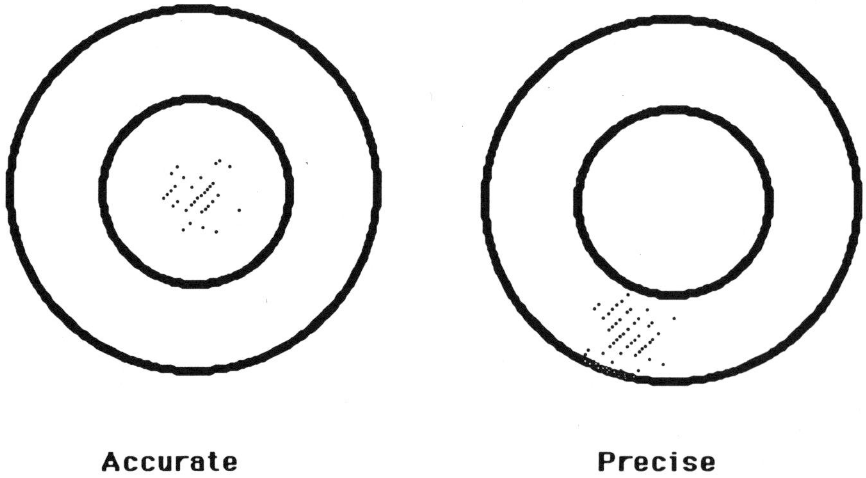


FIG. 2. The left bull's-eye shows accurate estimates which cluster around the centre of the target. The right bull's-eye shows precise estimates which are clustered in a small part of the bull's-eye but which miss the centre of the target. The situation on the right is more common for abundance estimates of fish stocks than is the situation on the left.

resource. Carl Walters and colleagues have developed an approach called Adaptive Management which aims to improve knowledge of how the productivity of natural resources relates to the way they are managed while minimizing short-term losses during the learning process. This approach is very relevant to the question of accuracy in setting policy goals, but is of lesser relevance to the question of reliability of scientific advice. The book, Adaptive Management of Renewable Resources, by Carl Walters (1986), examines the subject in considerable detail.

**Recruitment Overfishing**

Looking more closely at the possibility of inadvertently driving a resource to extirpation or commercial extinction, it is immediately clear that this means reducing the abundance to a very low level. Many years of monitoring stock abundance and reproductive success of commercially important marine fish stocks has revealed an almost universal pattern. There is generally a very broad range of spawning stock sizes within which the number of young fish produced has a weak or undetectable relationship with the spawning stock size. Sometimes the largest spawning stock sizes are associated with reduced production of young fish, but spawning stock size ten percent or less of the observed maximum is almost always associated with production of very low numbers of young fish.

For the spawning stock to be reduced suddenly to less than ten percent of maximum when it is thought to be much higher, e.g. twenty or thirty percent, the catch must be a large proportion of the available stock. This can result from a small error in setting a quota which is intended to be a large fraction of the fishable stock or from a large error in setting a quota intended to be a small fraction of the fishable stock. The requirement for accuracy increases with the target harvest rate. If the target is in the range of 15–25% as is the case with most marine finfish fished at $F_{0.1}$, overestimating by a factor of two still leaves 50–70% of the fishable stock in the water. If the target is 50% and the quota is overestimated by a factor of two there would be nothing left if the quota were taken.

Errors of this magnitude were observed for some major stocks of herring in the 1970's when it was found that the estimates of abundance based on catch rates of mobile gear such as purse seines remained high despite dramatic declines in stock size. Spawning stocks declined rapidly over a few years, approaching zero in some cases before fishing was stopped. Since then, more emphasis has been placed on other indices of abundance such as acoustic and egg and larval surveys, but the potential for rapid, undetected decline of schooling pelagic species remains high, particularly if the exploitation rate is high. Demersal fish are typically less aggregated than the pelagics and are less likely to be greatly overestimated. Nevertheless, when an index of abundance such as trawler catch rates is distorted by undetected increases in efficiency, such as improvements in acoustic fish-finding technology, stock declines can be masked here too.

When two or more valid indices of abundance are used, spawning stock is unlikely to become depressed

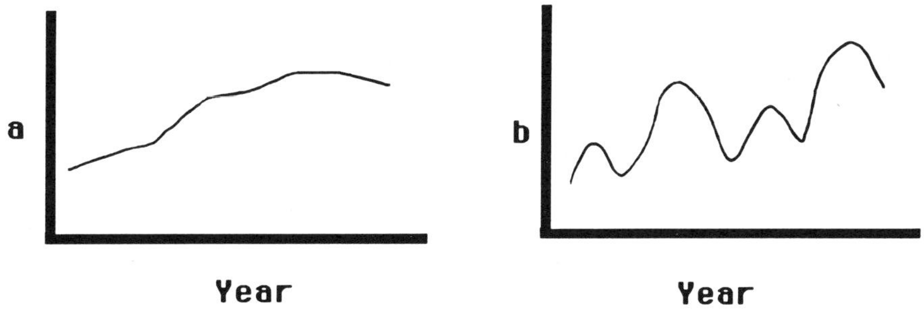


FIG. 3. The left illustration (a) shows a series of consistent estimates which increase gradually over several years. The right illustration (b) shows a much less consistent series of estimates which also increase gradually over several years but which fluctuate to a much greater extent from year to year. Scientific estimates of fish stock abundance are typically more like the series on the right.

without clear warning signs. With adequate monitoring of commercial fisheries and an independent research vessel-based index of abundance, errors of estimation are unlikely to lead to recruitment overfishing. From this perspective, provided the intended harvest rate is in the 15–25% range of fishable biomass, accuracy of plus or minus 50% nine years out of ten can be considered an adequate standard of performance since the spawning stock will be within a factor of two of the estimate nine years out of ten . If the target harvest rate were in the 40–45% range, overestimating abundance by a factor of two has more serious consequences and the corresponding performance standard might be plus or minus 25% for nine years out of ten.

Overfishing a stock to the point where subsequent recruitment is impaired is primarily an issue for catch quota management. Closure of areas does not directly control the quantity of fish removed but may reduce a fleet's ability to catch fish when they are aggregated which would reduce fishing pressure provided fishing effort did not increase outside the closed area. In practice, catch quotas are more effective than area closures in limiting fishing mortality rates since there is sufficient harvesting capacity in most instances to catch the available quotas in a fraction of the fishing season. Recruitment overfishing is even less relevant to the regulation of minimum mesh sizes for trawls, although the unexploited young age groups form a reserve which could aid recovery of a depleted adult stock.

### Wasting the Growth Potential

Young fish grow rapidly. Typically, growth in weight exceeds mortality due to natural causes at least to the age of maturation. If large numbers of young fish are caught, their potential to grow is wasted and, over the long term, catches from a stock are lower than they could be. Large numbers of young fish may be caught when the fishing gear does not let them escape after entering a net or when the rate of fishing is so high that many are killed even if most small fish encountering the gear escape. If small fish are caught and discarded rather than kept and sold, the economic losses are even larger.

Normally the characteristics of the fishing gear such as the size of the meshes of nets are regulated to achieve most of the growth potential of the fish. The size selectivity of otter trawls was the focus of study in the 1950's when international management of marine fisheries aimed to avoid overfishing through mesh size regulations. The theory presented in Beverton and Holt's 1957 book remains relevant today.

Small fish pass through the meshes of a trawl. Large fish are retained in the cod end. There is a range of intermediate sizes for which an increasing proportion of fish entering the net are retained. The fraction of each length group encountering the net which is retained is called the selectivity of the net for that length group. The age at which half of the fish encountering the trawl are caught and half pass through is called the age of 50% retention. This age is a key parameter in the yield models developed by Beverton and Holt (1957). During the 1950's and 1960's scientific effort was focussed on the estimation of selectivity patterns for trawls. ICNAF Special Publication No. 5, The Selectivity of Fishing Gear provides more information on selectivity studies.

A mesh size may be chosen to maximize the potential yield from a year class of young fish entering the fishery. This yield or catch in weight is smoothly related to the mesh size in the analytical models used. The curve becomes flat near the maximum of the relation between yield at a given harvest rate and age at first capture so that errors in estimating the location of the maximum have little effect on the yield achieved (Fig. 4) unless the species is subject to a high rate of natural mortality, which is the case for caplin, for example.

The calculations leading to estimates of optimal mesh size rarely take into account possible changes in growth rates of fish resulting from an increased abundance of larger fish due to postponement of exploitation.Because trawlers catch a range of groundfish species over wide areas, minimum mesh sizes are often applied to several species and stocks at once. The growth characteristics of important groundfish species such as cod, haddock and pollock are such that a mesh size suitable for one species is acceptable for other species as well. Such is not the case for smaller groundfish species such as redfishes or silver hake. Growth characteristics of cod vary considerably from the Scotian Shelf to Labrador and differing minimum mesh sizes have been applied to parts of this area.

John Gulland (1988, p 397) concluded that lost yield due to catching small fish could be as high as half the potential total in extreme cases, but is generally about 10–20% of the potential maximum. Thus, the long-term theoretical consequences of inaccurately estimating an optimal mesh size are small. In practice it would be difficult to demonstrate whether a theoretical gain had been achieved given the large variations in abundance which complicate any analysis.

The relationship between mesh size and yield from a year class may be flat near its maximum, but such is not the case for catch rates, particularly when intense fishing pressure rapidly reduces the abundance of exploited age groups. Under these circumstances most of the catch consists of two or three age groups. An increase in mesh size which effectively increases the age when fish are first subject to heavy fishing pressure may reduce catch rates substantially by sharply reducing the catch of one of these age groups. This is a transition phenomenon which is greatest immediately after a mesh size change. There is a convergence to a new steady state, often close to earlier catch rates, over a few years. Estimates of the short-term drop in catch rates can be compared directly with observed catch rates after a mesh change, but here too, the comparison is complicated by changes in abundance of the various age groups from one year to the next and changes in fishing practice and the degree of concentration of fish. Year to year comparisons of monthly catch rates can show changes of 25% with no mesh change and no change in assessed abundance.

**Dislocating the Fishery**

In addition to the question whether the target harvest rate is set too high or too low and whether the estimated current harvest rate is too high or too low, there is the possibility of estimates of abundance or harvest rate changing substantially from year to year without corresponding changes in the fished stock.

The effects of a change in catch quotas can be significantly magnified for some fleet components. In the case of Northern cod, for example, almost the entire reduction in total allowable catch in 1990 was borne by the offshore companies. In percentage terms, the reduction in their enterprise allocations was twice the reduction in the overall TAC. Furthermore, fishermen may have high fixed costs associated with loans for their boats and equipment so that small changes in yearly catch have a large impact on incomes and profits and large changes in yearly catch may trigger bankruptcies.

Another example of painful adjustment was the revision of advice on Southern Gulf of St. Lawrence herring in 1978. Steady declines in the resource had been masked by continuing high catch rates of purse seiners. When scientists concluded that these catch rates did not reflect abundance, their estimates dropped dramatically. Many of the purse seiners left the fishery entirely, while in subsequent years, the remainder fished much smaller quotas than had previously been available to them.

Dislocating fishermen and plant workers is anathema to elected officials and administrators. Rapid adjustments in coastal economies due to changes in estimated fish stock abundance which do not correspond to changes in the resource itself are a clear signal that the estimates are not reliable.

**Performance of Scientific Estimates**

Until recently, the accuracy of scientific estimates related to fisheries was rarely studied. This was due, at least in part, to technical difficulties in analyzing their performance. The technical difficulties have largely been overcome by the adoption of estimation procedures which remove subjective judgement as a factor and by improvements in computer calculation capabilities.

***Closed Areas***

Reliability of scientific estimates seems not to be an issue in establishing areas closed to fishing.

Closure of part of the Scotian shelf, off the Atlantic coast of Nova Scotia, to haddock fishing during part of its spawning period has occurred since 1970 (Halliday 1989). Much of the pressure from fishermen to implement the closure was due to a belief that spawning success would be improved. The original delimitation of closed areas in ICNAF Divisions 4X (the Southwestern Scotian Shelf) and Subarea 5 (Georges Bank) was based on analysis of the distribution of ripe, spawning, and spent fish in commercial and research vessel catches, planktonic egg distributions, and seasonal egg production curves. Despite the use of these data to define the closed areas, biologists never claimed that the closure would improve the production of young fish because they considered that reducing the overall fishing mortality was a prerequisite to increasing the numbers of spawning fish and the production of more young haddock. The Canadian Atlantic Fisheries Scientific Advisory Committee concluded in 1981 (CAFSAC 1981) that there was no scientific evidence that spawning success of Div. 4X haddock had been increased due to the closure. Nevertheless, catch rates were very high during the spawning period and Halliday concluded that the closure was an effective way of reducing fishing pressure during this period of the year. Since there was more than enough capacity to take the total allowable catch outside the closed period, the closure was not sufficient by itself to restrain the fishing mortality rate.

There was a recovery of the haddock stocks in the late 1970's. Reliability of the scientific basis of the closure seems not to have been an issue, even though haddock catches declined to very low levels in the 1980's. The boundaries of the closed areas were modified from time to time to reduce restrictions on fisheries for other species such as silver hake and argentine.

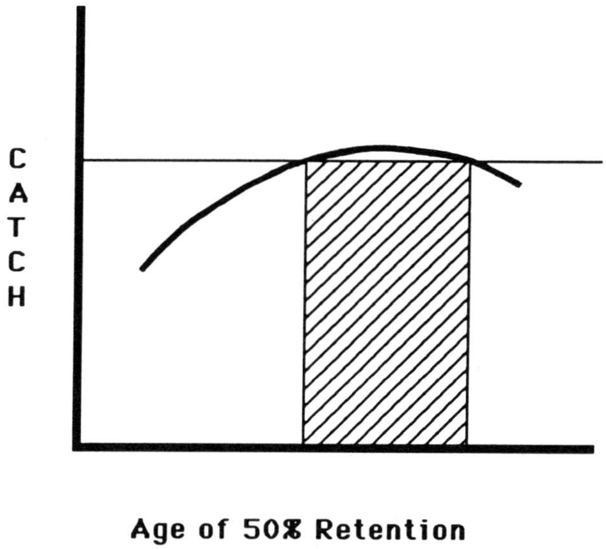


FIG. 4. Over a wide range of ages near the optimum, there is little change in the theoretical long-term catch associated with different mesh sizes corresponding to different ages of 50% retention in the fishery.

### Mesh Size

There are few published evaluations of the results of mesh size increases. J.R. Clark (in ICNAF 1963) reported the results of a comparison of catches by two groups of otter trawlers fishing for George's Bank haddock with mesh sizes of 2 7/8 in. and 4 1/2 in. during 1954–56. The introduction of the 4 1/2 in mesh was the first management measure adopted by ICNAF. It came into effect in 1953. This comparison was facilitated by licensing one group of trawlers to continue fishing with the pre-regulation mesh while another fished with the larger mesh. Such a direct comparison is rare.

Clark found that the selectivity of the new regulation mesh was as estimated by Graham prior to adoption of the measure. He also found that the larger mesh nets were 8% more efficient at catching large haddock. This was not forecast and meant that the short-term loss in catch per hour associated with the mesh size increase was less than expected. This error on the side of overestimating the short-term losses seems not to have called the reliability of the scientific calculations into doubt.

### Catch Quotas

Recent attention has focused on the estimation of catch quotas. Quota management is ubiquitous in marine finfish fisheries and errors in quota estimates are more likely to have serious biological and economic consequences than errors in estimates of other management measures like optimal mesh sizes or reference fishing mortality rates.

The performance of quota estimates has been examined by the Canadian Atlantic Fisheries Scientific Advisory Committee (CAFSAC) and by scientists associated with the International Council for the Exploration of the Sea (ICES).

The CAFSAC retrospective analysis of 1989 (Sinclair et al, MS 1990) found a systematic tendency to overestimate stock size in five of seven cod and haddock stocks considered and frequently found large differences between estimates of abundance obtained using information up to 1989 and those estimated using the same currently accepted methods with data available several years earlier. Because uncertainty in the stock abundance due to imperfections in abundance indices becomes rapidly smaller for years increasingly farther in the past, this approach gives an indication of biases and variance associated with the estimation techniques. Although the precision of catch projections was not dealt with, the report noted that coefficients of variation of 35% or more for population abundance could be expected, based on simulations of realistic errors in input data.

"Status-quo" TACs aim to maintain fishing mortality at its current rate rather than to achieve an absolute target like $F_{0.1}$. An analysis of "status quo" TACs for some north-east Atlantic stocks by Brander (1987) found errors of 14% for one year projections and 21% for two year projections. These errors are much smaller than the CAFSAC values not because abundance estimates are more accurate in the North Sea but rather because estimates of the fishing mortality rate err in the opposite direction to errors in estimating stock abundance. These errors largely cancel each other when the TAC for the coming year is calculated by (effectively) multiplying stock size by fishing mortality rate.

John Pope (in Gulland [ed.] 1988, p. 78–80) considered that current estimates of projected catch for North sea cod were subject to coefficients of variation of about 27% to about 33% depending on the choice of biological reference point. These values are similar to those implied by the CAFSAC retrospective analysis.

There have been substantial revisions in scientific assessments of many groundfish and pelagic stocks on both sides of the Atlantic in recent years, particularly for cod and haddock. When Canada extended fisheries jurisdiction in 1977, scientists projected that catches of the Northern cod stock off northeast Newfoundland would increase steadily to about 400,000 t. These projections assumed that the productivity of the stock complex would return to the high levels observed in the 1960's. Despite caveats which emphasized that future recruitment was unknown and could differ substantially from the historical average, these projections influenced expectations of the fishing industry. During the 1980's, these expectations were clearly not met on at least two occasions. In 1984, scientists reduced their long-term yield projections from about 400,000 t to about 300,000 t recognizing that the number of young fish entering the fishery since the early 1970's was less than historical numbers and that the growth rates of individual fish had declined. A 33% reduction in estimated stock abundance between 1987 and 1989 caused a crisis in the Canadian Atlantic fishery.

Abundance estimates can be no more accurate than the abundance indices on which they are based. Even an extensive trawl survey is usually subject to errors of half or double at least one year out of twenty. When two or more independent valid indices of abundance are available this uncertainty can be reduced. The sequential population analysis methods used in North America and Europe have evolved over the past decade to combine information from several sources, particularly research surveys and commercial catch rates.

Within the framework of a moderate target harvest rate consistency is far the most important aspect of reliability for catch forecasts. It is also easier to achieve consistency than accuracy. When abundance estimates are based on ever-lengthening time series of the same observation, such as the catch rate of a class of trawlers or of a standard research vessel, successive years' estimates share much of the same data and accordingly tend to err in the same direction. The use of a new index of abundance is frequently a major source of inconsistency. Another source can be the introduction of new mathematical estimation procedures.

Both these factors contributed to the major change in estimated abundance of Northern cod in 1989 — a time series of research vessel catches was given equal weight to commercial trawler catch rates for the first time and the estimation method was changed to reflect theoretical advances of the previous few years. A thorough review by the Harris panel left no doubts that the new estimates were closer to the truth than the earlier ones, but it was obvious to everyone that the scientific estimates provided since fisheries jurisdiction was extended in 1977 had not performed to expectations. The inconsistency had demonstrated unreliability. The northern Cod assessment of 1989 was judged by the independent Harris review panel (Harris, 1990) to be subject to uncertainty of plus or minus 25%.

### Setting Realistic Expectations

A 10% change in total allowable catch is considered large by fishermen and fishing companies and they are not satisfied with scientific estimates which can easily change by twenty percent from one year to the next with no corresponding change in the stock. Such changes in scientific estimates are caused by imperfections in measures of abundance which cannot be eliminated by sophisticated analysis. In this context fishermen are unlikely to consider scientific performance satisfactory unless the random fluctuations in catch quotas are small relative to other uncontrolled factors in their work, such as the year to year fluctuations in fish prices. Such a high standard of accuracy is not practically achievable at this time, although it may be an achievable target for consistency for long lived species subject to moderate rates of fishing mortality.

In the author's experience, fishermen and the fishing industry, together with many administrators, expect well managed fish stocks to be very stable with catches nearly constant or slowly increasing. This ideal world does not correspond to reality. Even with perfect management, the abundance and productivity of fish resources would be highly variable over long periods.

The mathematics of fish population dynamics can easily give a false impression of stability and predictability. This is most evident in multi-year projections of abundance and catches. These are calculated by adding the contributions to catch and the live weight of the stock of each of the year classes present in a given year. The projections are mechanistic and deterministic, using single estimates of the initial size of each year class and for the weight of each age group in the stock and in the catch. When no estimate is available for a year class that is young or not yet hatched at the beginning of the projection, an average value based on historical observation is normally used. The net result of this approach is that projections of catch and stock size for a given pattern and rate of fishing mortality always show a steady convergence to a constant "average" level. This level is effectively reached when the first year class of assumed average size has passed through the fishery. Such projections do not reflect the uncertainty of estimates nor the variability of future year class sizes and growth rates.

It is easy to be misled by the stable and predictable behavior of catch projections to conclude that the fish stocks themselves are inherently stable and predictable. Although some stocks, particularly flatfishes, have shown remarkable stability over many years, this is the exception rather than the rule. At the opposite extreme are the haddock stocks where year classes vary in abundance by a factor of one hundred or more. The variation of year class sizes leads to variation in the fishable stock which is the sum of year classes of exploitable ages. When stocks are heavily fished, year classes of fishable age are rapidly reduced to low numbers. Under these circumstances, the variability of the fishable biomass approaches the variability of individual year classes. If fishing is light, several year classes contribute significantly to the fishable stock, reducing the effect of variation in year class size. Projections inevitably diverge from reality as errors in estimation of the initial parameters are compounded over time and as more and more of the stock and catch are made up of year classes about which little or nothing is known and which are not of average size.

Scientists are very conscious of the dependence of the projections on initial estimates and assumptions and take pains to point out caveats applicable to their use, but these words of caution are often overlooked by the reader who then feels cheated when the future does not unfold as projected. Scientists are increasingly reflecting uncertainties in year class size and other factors by calculating a number of projections with parameters such as year class sizes chosen from statistical distributions. The range of these projections gives some understanding of the uncertainties associated with projecting future abundance and yield.

There is a public expectation that with proper management, nature can be controlled within narrow limits, if only the scientific basis were sufficiently reliable. This is not a valid expectation for fisheries. The productivity of fish stocks is highly variable due to changes in the numbers of young fish produced and the weight of fish of various ages. The fishery manager can only influence these factors indirectly by ensuring that spawning stocks are not allowed to decline to very low levels and by limiting exploitation of key prey species. The only factor fishery management can control is the catch, either directly through quotas or indirectly through limited entry or other capacity restrictions and controls on fishing gear, areas and seasons. These adjustments in catch can moderate fluctuations in the productivity of stocks but cannot remove them and stability of catches is antithetical to stability of stocks. In practice, stable catches can only be expected in the long run when the annual catch is a small fraction of the fishable stock.

A fish stock is part of a highly variable natural marine ecosystem and cannot be tightly controlled by man. Fluctuations in the numbers of young fish produced by fish stocks will continue to show a factor of ten to a factor of one hundred between good and bad years no matter how the fishery is managed. The growth rates of fish will continue to vary no matter how well the fishery is managed. The occurrence of prolonged periods of higher or lower than average productivity should be expected. Fishery management can only moderate the impact of these natural variations on the fish stocks and on the fisheries and can not tame them completely.

The tendency of fishermen and the public to expect too much of fishery management and scientific advice can be addressed by informing the public of the nature and limitations of management of fish stocks and fisheries. One initiative of the Canadian Atlantic Fisheries Adjustment Program of 1990 is the establishment of an enhanced dialogue between scientists and fishermen. Scientists discuss their estimates informally with groups of fishermen before finalizing and formalizing their advice. Public criticism of scientific estimates in Atlantic Canada has been less since the introduction of this program.

If the accepted aim of fishery conservation is to maintain stock sizes above minimum levels and scientific estimates are good enough to permit this indefinitely, then they are sufficiently reliable for conservation, even if their accuracy is low. If the accepted aim of fishery management is to maintain high constant catches unless reductions are essential to preserve stocks, scientific estimates must perform much better to be considered reliable.

Thus, reliability expressed in terms of conservation needs requires a lower standard of performance than reliability expressed in terms of the needs and expectations of fishermen. Also, consistency is as important as accuracy since it is changes which are perceived as unjustified which often make reliability of scientific estimates an issue.

## Reducing the Need for High Levels of Performance

It is clear that the performance of scientific estimation of fish stocks is not sufficient to meet the high expectations which can arise in a demanding quota management regime. This is somewhat analogous to the early days of aviation when engines could not be depended on for long flights. Despite these limitations, ways were found to achieve sufficiently reliable intercontinental flights. Aircraft were designed with four engines but needing only two operational to continue flying. Routes were planned to ensure the distance from the nearest airport was limited. Similar care in designing fishery management regimes can reduce the need for accuracy, precision, and consistency of scientific estimates to achievable levels.

Any regime with frequent adjustment in the level of a controlled parameter is vulnerable to unreliable estimates of that parameter. The regulation of the Canadian Atlantic groundfish fishery involves many different types of management measures with greater or lesser conservation value and greater or lesser needs for scientific accuracy and consistency. The regulations used include closure of areas to fishing for part of the year, restrictions on the size of meshes of nets, limits on the amount of fish that can be caught from a management unit in one trip or in one year by a vessel or class of vessels and limits on entry into the fishery and on replacement of fishing vessels with new vessels.

### Choosing the Management Regime

One way to overcome the limitations of abundance estimates is to adopt a fishery management regime which does not require them. Management measures such as closed areas and seasons and control on the amount and characteristics of fishing gears do not require frequent revision and are less likely to lead to questioning of the reliability of their scientific basis than catch quotas. Control of the dimensions and numbers of traps, together with limited numbers of licences and lack of fishing offshore, has been associated with continuing high lobster catches in Atlantic Canada for many years. Unfortunately, such indirect restrictions on the fishing mortality were found to be insufficient to conserve marine finfish stocks during the period of international management (ICNAF 1976). Such measures may, nevertheless, contribute to a conservation regime which also limits fish catches.

The core of the Atlantic groundfish management policy is the limitation of the rate of fishing mortality to the $F_{0.1}$ target through a series of catch quotas. The use of low target fishing mortality rates greatly reduces the need for accuracy in setting catch quotas. For species like cod and haddock, the margin for error in setting quotas for conservation purposes at $F_{0.1}$ is about twice the margin at the $F_{max}$ level. $F_{max}$ maximizes the theoretical yield from each year class entering the fishery. This approach provides greatest benefits for long lived species where the theoretical loss in long-term average catch is modest and there is a substantial gain in the catch per hour fished. In the case of the Northern cod stock, overestimation of abundance for several years was associated with a leveling off of growth, and not a decline. Had the target fishing mortality rate been in the 40–45% range of $F_{max}$, such an overestimation would have led to rapid decline in the stock during the late 1980's. If the demand for accuracy is modest, scientists can put more emphasis on achieving consistency from year to year.

The application of the $F_{0.1}$ target has been mediated by the "50% rule" since 1987. This rule reduces large adjustments in catch quotas by setting them at a fishing mortality rate half way between the current rate (status quo) and the reference rate ($F_{0.1}$). Estimates of the status quo catch are much more consistent than estimates of the catch at a fixed fishing mortality rate because overestimates of stock abundance are largely cancelled by underestimates of the current fishing mortality rate. The 50% rule nearly halves the random changes in catch quotas due to random changes from year to year in scientific estimates of stock abundance.

The 1990 NAFO Scientific Council Special Session on Management under Uncertainties found that compound management strategies which interpolate between classical strategies ($F_{max}$ or $F_{0.1}$) and status quo management to be superior according to various criteria and to have lower variability of catch and effort than the or strategies, strictly applied. The 50% rule is one of an infinity of such compound strategies. There are, however, limits to the extent to which a compound management strategy can be expected to converge to a target fishing mortality rate. In the mid-1970's ICES followed a policy of advising what catch corresponded to reducing a high fishing mortality rate by 10% of the difference between the estimated current rate and $F_{max}$. An analysis by this author (Doubleday 1979) found that uncertainties in estimating stock sizes were sufficiently large for some major stocks, including North Sea cod and North-East Arctic cod, that such a policy was as likely to lead to an increase in fishing mortality as to the intended decrease.

Besides adoption of a compound strategy the most promising way to achieve acceptable reliability is to aim for a low exploitation rate. For many stocks a harvest rate of 10% of the fishable stock can be approximated by constant catches over many years. This achieves stability of catch at the cost of substantial loss of potential yield which may make sense in the context of weak markets. This is the history of most of the northwest Atlantic redfish stocks.

Recently, multi-year fishing plans have been introduced in Atlantic Canada. These plans provide fishermen with an expected trend in catch levels over three years. The planned catch levels may be revised if new evidence shows that revision is necessary. There has also been discussion of basing advice on a description of the risks associated with maintaining the current catch level. Avoiding yearly changes to catch quotas will result in periods of high consistency which may correspond to high perceived reliability of the scientific information. If corrections are only made when estimates have deviated far from the initial values, the changes to TAC's may be very large when they do occur. It remains to be seen in practice whether a series of small changes to quota levels will be seen as more or less reliably based than occasional large ones.

There is an inherent tendency for harvesting capacity of fishing fleets to increase. If corresponding increases in fishing mortality rates are ignored until they accumulate to a large change such as 50%, which can happen in five years or less (e.g. the periods 1973–76 and 1980–84 for northern cod — Baird et al. (1991)), the adjustment necessary to restore a balance between the fleet and the resource will be very large and painful. The reduction in fishing capacity would likely come at a time when a natural downturn in stock productivity made changes particularly difficult.

### The Economic Structure of the Fishery

The economic structure of the fishery is an important factor influencing the ability of fishermen to tolerate changes in catch levels, and the level of consistency needed for reliability in their eyes. This is a complex subject, but some examples can give an idea of the possibilities which exist. The fluctuations of catch quotas have less impact on net revenues when fishermen have low fixed costs than when they have high fixed costs. This is typically the case for inshore small boat fisheries. Proportionate sharing of increases and reductions by all participants in a fishery avoids amplifying the impact on some groups. If a fisherman or company holds shares of quota for many species and stocks, the effects of fluctuations in quotas are diversified and consequently reduced. Catch reductions are particularly unwelcome when other factors like low fish prices have lowered profitability.

### Formulating Scientific Advice

For long-lived species subject to a conservative management regime, consistency of scientific estimates is more important than accuracy. Use of several measures of stock abundance and fishing effort reduces the impact of a random change in one index on catch forecasts. Direct measures of changes in fishing effort could complement indices in stock abundance and help to stabilize catch forecasts. Adoption of a new analytical method or reliance on a new index of stock abundance often leads to a large change in scientific estimates. If the resource is not at a dangerously low level, this inconsistency can be reduced by phasing in new methods and new abundance indices over two or more years. The deferral of improvements to scientific accuracy would be compensated by substantially increased consistency from year to year. Improvements in methods of estimating stock abundance, will, over a period of years, also lead to improved consistency from year to year by reducing random changes in abundance indices.

Beginning in 1990, Canadian scientists assessing Atlantic marine fish stocks have discussed their estimates and projections with groups of fishermen before formalizing their advice. Greater involvement of fishermen in the scientific advisory and quota setting process increases their willingness to accept the limitations of scientific estimates.

## Conclusions

To be accepted by fishermen and to be confidently used by fisheries Ministers and administrators, scientific advice and information must regularly meet their expectations. Objective standards of performance can be arrived at by considering the consequences of errors in estimation with respect to achieving biological or economic goals of fishery management. These standards can be more or less demanding of high levels of precision, accuracy and consistency depending on the fishery management regime. Expectations of fishermen and the general public may be much higher than the requirements for conservation.

Management measures such as closure of an area for some period to avoid catching spawning fish or young fish require little in the way of scientific information. They are generally set in place based on all the relevant information available at the time and are infrequently revised. Since little or no fishing occurs in the closed area, little new information is collected to challenge the scientific base for the closure. If everyone acknowledges that the closure will not, in itself, ensure maintenance of an abundant stock but will only avoid the capture of young or spawning fish previously occurring within it, expectations are virtually certain to be met. Since the measure itself is rarely revised and the offering of scientific advice is correspondingly infrequent there is very little risk of inconsistency or perceived unreliability. Unfortunately, this type of measure is not sufficient to meet the conservation and economic goals of management of most fisheries.

Setting a minimum mesh size for trawls or gillnets requires a higher standard of scientific information. Historically, mesh sizes were increased in order to increase future catches by allowing young fish to achieve their growth potential. Since mesh sizes are usually increased, catches and catch per day fished usually decrease immediately after introducing the new mesh size. Subsequently catch per day and total catch is expected to increase. Fishermen can easily compare estimated short-term losses to their own experience and eventually it will be obvious whether catches have increased in the long term. The long-term changes due to increased mesh size are, however, confounded with long-term changes in the productivity of the fish stocks. As in the case of closed areas, mesh sizes are infrequently revised so there is no problem with year to year consistency of scientific advice. If public expectations for increased future catches are not set too high, the major reliability risks come from discrepancies between the forecasted short-term losses and fishermen's experience soon after the change is made. In most marine fisheries, mesh sizes are large enough to achieve almost all the growth potential of young fish. In most recent cases, changes in mesh size no longer are associated with claims that catches will be increased, but are motivated by other considerations such as increasing the mean size and value of fish caught.

Setting annual catch quotas to achieve a target fishing mortality rate is very demanding of scientific advice. It is also the most direct and effective way to prevent catching too large a fraction of the fishable stock. The need for accuracy increases with the target fishing mortality rate. If one quarter or less of the fishable stock is to be caught each year, errors of 50% in catch forecasts every few years will not cause stock collapse, while a target fishing rate of 50% of the fishable stock requires estimates twice as accurate. More significant than the accuracy of annual catch forecasts is their consistency from year to year. Changes in catch quotas dislocate fishermen. Depending on the way in which increases and decreases are shared among participants, and on the extent of fixed costs such as mortgages on vessels, small changes in catch quotas may translate into large changes in incomes and profits of fishermen and fishing companies. A high degree of consistency from year to year is needed if catch forecasts are to be considered reliable by fishermen. At the very least, chance changes in quotas should be no greater than the year to year changes in other key factors of the fishery such as fish prices. Since there is less tolerance for error in government-imposed measures than for changes in market prices, the scientific advice should very rarely be seen to produce unnecessary reductions in the fishery.

A number of factors can reduce the need for high standards of consistency and accuracy of catch forecasts:

- A moderate target harvest rate does not require accurate catch forecasts.
- Use of compound harvest rate strategies such as the 50% rule reduce changes in Total Allowable Catches resulting from changes in scientific estimates, moderating the need for consistency.
- Equal sharing of quota changes among fishermen on a per cent basis avoids magnifying changes for some groups of fishermen.
- Gradually phasing in new abundance series and new methods of analysis in stock assessments reduces abrupt changes in scientific advice
- Use of several abundance indices reduces the risk of a stray high or low value in one index producing a large change in catch projections.
- Direct measures of changes in fishing fleet harvesting capacity could provide a direct indicator of any growing imbalance between the fleet and the resource harvested.
- Scientists should communicate the limitations of their estimates to fishermen so that they are well informed when they form their expectations.

Widespread use of catch quotas to manage marine fisheries is only some twenty years old. There has been a rapid evolution of scientific methods of estimating abundance and of fisheries management policy during those twenty years. Experience in Atlantic Canada over the past decade indicates that consistency of scientific advice is as important as accuracy in meeting expectations of fishermen and administrators. While striving to make technical improvements in their methodology, scientists should also foster realistic expectations among their clients.

## References


BAIRD, J.W., C.A. BISHOP, AND E.F. MURPHY. 1991. An assessment of the cod stock in NAFO Divisions 2J3KL. Can. Atl. Fish. Sci. Adv. Comm. Res. Doc 91/53: 84 p.

BEVERTON, R.J.H., AND S.J. HOLT. 1957. On the dynamics of exploited fish populations. Fishery

Investigations Series II, Volume XIX, Her Majesty's Stationery Office, London.

BRANDER, K. 1987. How well do working groups predict catches? J. Cons. int Explor. Mer 43: 245–252.

CANADIAN ATLANTIC FISHERIES SCIENTIFIC ADVISORY COMMITTEE. 1981. Advise on 4X Haddock Closed Area, CAFSAC Advisory Document 81/9, Dept. of Fisheries and Oceans, Dartmouth, N.S.

DOUBLEDAY, W.G. 1979. A sensitivity analysis of selected catch projections. ICES CM 1979/G:53, 7 pp + tables (mimeo.)

GREEN, A.E., AND A.J. BOURNE. 1972. Reliability technology. Wiley-Interscience, John Wiley and Sons, New York, NY.

GULLAND, J.A. [ED.]. 1988. Fish population dynamics. John Wiley & Sons, New York, NY.

GULLAND, J.A., AND L.K. BOEREMA. 1973. Scientific advice on catch levels. Fish. Bull. (U.S.) 71: 325–335.

HALLIDAY, R.G. 1988. Use of seasonal spawning area closures in the management of haddock fisheries in the Northwest Atlantic. NAFO Sci. Coun. Stud. 12: 27–36.

HARRIS, L. (CHAIRMAN). 1990. Independent review of the state of the northern cod stock. Department of Fisheries and Oceans, Ottawa.

ICNAF. International Commission for the Northwest Atlantic Fisheries. 1963. The selectivity of fishing gear. Int. Comm. Northwest Atl. Fish. Spec. Publ. No. 5. Dartmouth, N.S.

1976. International Commission for the Northwest Atlantic Fisheries Redbook, 1976, Int. Comm. Northwest. Atl. Fish., Dartmouth, N.S.

JOHNSTONE, K. 1977. The aquatic explorers: a history of the Fisheries Research Board of Canada. University of Toronto Press, Toronto, Ont.

NAFO. 1990. Scientific Council Report, Annual Meeting, September 1990. NAFO, SCS Doc 90/25

SINCLAIR, A., D. GASCON, R. O'BOYLE, D. RIVARD, AND S. GAVARIS. 1990. Consistency of some Northwest Atlantic groundfish stock assessments, NAFO SCR Doc. 90/96.

WALTERS, C. 1986. Adaptive management of renewable resources. MacMillan Publishing Company, New York, NY. 374 p.

# Shaping Fisheries Policy: The Kirby and Pearse Inquiries — Process, Prescription and Impact

L.S. Parsons

*Department of Fisheries and Oceans*
*Biological Sciences Directorate*
*200 Kent St., Ottawa, Ontario, Canada K1A 0E6*

Parsons, L. S. 1993. Shaping fisheries policy: the Kirby and Pearse inquiries — process, prescription and impact, p. 385–409. *In* L. S. Parsons and W. H. Lear [eds.] Perspectives on Canadian marine fisheries management. Can. Bull. Fish. Aquat. Sci. 226.

There have been over 100 reports of task forces, special inquiries and royal commissions investigating Canada's marine fisheries over the past century. This chapter examines the impact of two of the most significant recent reports — the Kirby and Pearse reports of the early 1980's — on Canadian marine fisheries policy. The recommendations of the Kirby Task Force, while they did not revolutionize the Atlantic fishery, provided a blueprint for a potentially more viable industry in the long term. The majority of its recommendations were accepted and implemented by government. The Pearse Report had less impact on commercial fisheries policy. Pearse's major recommendations for restructuring the commercial licensing system were rejected by industry and government. Pearse's major influence was on the development of subsequent policy for habitat management, native fisheries issues and recreational fisheries. Much of the difference in impact on public policy of the two reports can be traced to the differing composition, nature and scope of these inquiries and the process followed by each.

---

Au cours du dernier siècle, plus de 100 rapports de groupes de travail, d'enquêtes spéciales et de commissions d'enquête parlementaire ont été publiés sur les pêches canadiennes en mer. Le présent chapitre examine l'impact des rapports Kirby et Pearse publiés dans les années 1980, deux des plus importants rapports sur la politique des pêches canadiennes en mer. Les recommandations présentées par le Groupe de travail Kirby, bien qu'elles n'aient pas transformées radicalement la pêche dans l'Atlantique, ont jeté les bases d'une industrie potentiellement plus viable à long terme. Le gouvernement a accepté la plupart des recommandations formulées et les a mises en application. Par contre, le rapport Pearse a eu moins d'impact sur la politique des pêches commerciales. Le gouvernement et l'industrie ont rejeté les principales recommandations sur la restructuration du système de délivrance de permis de pêche commerciale. La plus grande influence de ce rapport se situe au niveau de l'élaboration ultérieure d'une politique de gestion de l'habitat, des pêches autochtones et de la pêche récréative. L'impact différent de ces deux rapports sur la politique officielle peut être lié à la nature et la portée de ces enquêtes, et au processus appliqué.

"Fishing as you know is Canada's oldest industry. That brings us to Canada's oldest headline: 'Fishery in Crisis'. And the second oldest headline (with variations): 'Task Force formed on fishery crisis'"

A. W. May
*Speech to the Fisheries Council of Canada*
*St. John's, Newfoundland, May 2, 1983.*

## Introduction

There have been over 100 reports of task forces, special inquiries and royal commissions investigating Canada's marine fisheries over the past century. These show that the industry has a long tradition of difficulty.

The Kirby and Pearse Reports were produced about the same time in 1982–83. They are the most comprehensive studies to date of the Canadian east and west coast fisheries. These special studies were undertaken in the context of the 1980–83 fisheries crises on both coasts. The structure and process of the inquiries differed significantly. This had a significant influence on their recommendations and the manner in which these were addressed by the federal government.

In this chapter, I examine the major recommendations of these two reports and describe the government, industry and public response to each. I then examine the reasons for their comparative impact on subsequent fisheries policy.

## The Two Inquiries' Differing Approaches

The Task Force on Atlantic Fisheries, headed by then-senior civil servant Michael Kirby, was a multi-person effort. It included:

— individuals from the fishing industry (Peter Nicholson, former Vice-President of H.B. Nickerson and Sons Ltd., and Father Desmond McGrath, one of the co-founders of the Newfoundland Fishermen's Union)
— representatives from various federal government departments (Arthur May, Richard Roberts and Paul Sutherland from the federal Department of Fisheries and Oceans, Roy Atkinson from the Department of Finance, David Patterson from the Department of Regional Industrial Expansion, Bruce Little from Kirby's staff at the Federal-Provincial Relations office, David Mann, a Halifax lawyer, and Victor Rabinovitch from the Canadian Labour Congress).

The Task Force conducted extensive public consultations and received written briefs. The process was, however, less formal than the west coast study.

The Pearse inquiry was a one-man Commission of Inquiry into Pacific Fisheries Policy, officially established under the Inquiries Act. Pearse, a professor of economics at the University of British Columbia, conducted formal public hearings and received formal written submissions from interested groups. He produced a report which represented the conclusions of one individual, the Commissioner. The Report of the Atlantic Task Force, in contrast, while it bore the imprint of Chairman Michael Kirby, was more of a joint report representing a blend of differing viewpoints of the Task Force members.

Another significant difference was that the Task Force on Atlantic Fisheries attempted to examine the entire Atlantic fisheries system from harvesting through processing to marketing. The Pearse report concentrated on the harvesting sector. In addition, it examined habitat management, the native fishery and salmon enhancement issues.

Despite these differences, the two reports taken together provided the most in-depth examination of Canada's marine fisheries to that date. They attracted considerable public attention on their respective coasts and helped shape fisheries policy over the next decade. Fisheries interest groups' reactions to the two reports differed significantly. This affected the extent to which the federal government accepted their recommendations.

## The Report of the Task Force on Atlantic Fisheries

The Prime Minister announced the establishment of the Task Force on Atlantic Fisheries (TFAF) on January 8, 1982. Its report was released by Fisheries and Oceans Minister Pierre De Bané the following February.

The Task Force's mandate was to advise the government on long-term policies for a healthy fishery. Key players on the Task Force, particularly Chairman Kirby and members Nicholson and Mann, continued to be heavily involved over the following year in the negotiations leading to the restructuring of certain offshore Atlantic fishing companies. In this chapter I will confine my remarks to the official Report of the Task Force titled "Navigating Troubled Waters — A New Policy for the Atlantic Fisheries" (Kirby 1982). For details on the Atlantic Fisheries Restructuring initiative, see Parsons (1993).

## Recommendations

### *Policy Objectives*

The crux of the Task Force Report was its articulation of three objectives for Atlantic fisheries policy and 57 specific recommendations on particular fisheries policy issues.

The objectives for Atlantic fisheries policy were:

1. The Atlantic fishing industry should be economically viable on an ongoing basis, where to be viable implies an ability to survive downturns with only a normal business failure rate and without government assistance.
2. Employment in the Atlantic fishing industry should be maximized subject to the constraint that those employed receive a reasonable income as a result of fishery-related activities, including fishery-related income transfer payments.
3. Fish within the 200 mile Canadian zone should be harvested and processed by Canadians in firms owned by Canadians wherever this is consistent with Objectives 1 and 2 and with Canada's international treaty obligations.

Conspicuous by its absence was a reference to the conservation objective which had been at the forefront of Canadian fisheries policy for more than a century. See Parsons (1993). It appears that the conservation objective really was taken for granted with the focus being on sorting out the social and economic objectives (Dr. A.W. May, St.John's, Newfoundland, pers. comm.).

The Task Force asserted that these objectives, ranked in order of priority, would provide a clearer framework for fisheries policy. However, it recognized that these did not represent operational objectives for fisheries managers.

### *International Issues (Recommendations 1–5)*

The international issues addressed by the TFAF are examined in detail in Parsons (1993). The Task Force Report recommended the following measures for government to improve international efforts in fisheries management and increase Canadian access to foreign markets:

1. Allocate non-surplus resources to foreigners as part of agreements for reciprocal fishing rights.
2. Allocate resources surplus to Canadian harvesting capacity and a fixed amount of 'non-surplus' resources preferentially to countries maintaining a satisfactory fisheries relationship with Canada.
3. Pursue reductions in tariff and non-tariff barriers affecting trade in Canadian fishery products through General Agreement on Tariffs and Trade negotiations.
4. Permit joint venture arrangements with foreign interests where appropriate.
5. Permit direct sales to foreign vessels 'over-the-side' and 'over-the-wharf' only in well-defined circumstances.

### *The Harvesting Sector (Recommendations 6–9)*

#### *Quota Licences*

Kirby proposed a system of quasi-property rights to deal with the common property problems of the fishery. The key recommendation was a set of new licensing principles under which a licence to catch fish commercially would, in effect, allocate to each licensed fisherman an individual quota of fish. Commercial fishing licences would become quota licences. Licences could then be sold or traded by fishermen wishing to increase or decrease their involvement in the fishery. By assuring each fisherman a right to a specified amount of fish, Kirby hoped such a system would end the destructive effects of the annual 'race for the fish'. He anticipated that over time this new licensing system would lead to the best possible match of harvesting capacity to the resource.

#### *Atlantic Fisheries License Review Board*

The TFAF proposed establishment (R8) of "a quasi-judicial Atlantic Fisheries Licence Review Board that would act in a review and appeal capacity for the current licensing system, as well as for the system of enterprise allocations and quota licences."

#### *Resumption of Federal Jurisdiction over Marine Fisheries in Québec*

Kirby proposed consolidating federal management of the fisheries in the Gulf of St. Lawrence through resumption of full federal responsibility for licensing and other aspects of marine fisheries management in Québec. A 1922 Canada-Quebec agreement delegating certain administrative responsibilities to Québec had led to confusion because of the division of responsibility for administration of the harvesting sector. For details, see Parsons (1993).

### *Resource-Short Plants (Recommendations 10–11)*

The Task Force identified the highly peaked, seasonal nature of inshore catches of groundfish, squid and pelagic species, particularly on the east coast of Newfoundland, as a major problem. This required a large processing capacity which was under-utilized in the off-season. These seasonal, under-utilized facilities are commonly known as 'resource-short plants'.

To address the problem of supplying resource-short plants with fish in the off-season and improve their degree of utilization and their chances of economic viability, the Task Force proposed establishing specific allocations of fish for delivery to these plants. It recommended that this fish be delivered by a self-financing fishing company or consortium, with the only government contribution being an allocation of fish. Only Canadian vessels would be permitted to catch these allocations by a target date of 1987.

### *Utilization of the Northern Cod Stock (Recommendations 12 and 13)*

The Task Force only addressed one major resource allocation issue: how to share the anticipated growth in the northern cod stock. For its scenarios, the Task Force assumed that the TAC of northern cod would increase to 400 000 tons by 1987, with a Canadian share of 380 000 tons. This was a highly speculative number based on one scientific projection, which assumed rapid stock rebuilding. In fact, the TAC increased only from 230 000 tons in 1982 to 266 000 tons in 1984, remaining at that level until 1988, then declining to 197 000 tons in 1990. Following a sudden, drastic and unexpected decline in the northern cod stock during 1991, a two year moratorium on fishing for northern cod was imposed, commencing in July 1992. For details, see Parsons (1993) and Lear and Parsons (1993).

Eager to distribute the anticipated growth, the Task Force presented the following allocation scenario (R12):

| | 1982 | 1987 |
|---|---|---|
| 1. Inshore allowance | 120 000 t | 145 000 t |
| 2. Existing trawler fleet | 87 250 | 145 000 |
| 3. Resource-short plants | 5 250 | 50 000 |
| 4. Other fixed and mobile gear | 2 500 | 40 000 |
| | 215 000 t | 380 000 t |

This would have provided equal shares for the inshore allowance and the existing trawler fleet, 50 000 tons for the proposed Resource Short Plant program (see above) and 40 000 tons for a new fleet of Scandinavian type longliners. Equal sharing between the inshore and offshore sectors was a major change from the policy adopted following the 1979 Northern Cod Seminar. It gave the offshore fleet a greater share of the growth. The allocation for resource short plants addressed the seasonality issue.

The Task Force also concluded that a new fleet of Scandinavian-type longliners was a promising alternative to greatly increased catches by the existing mix of inshore gear types. The Task Force thought this offered a technological option that would not exacerbate the seasonality of the inshore fishery but would nonetheless be compatible with a smaller plant and with ownership by small groups of fishermen. The basic concept was that the longliners would produce fish of top quality and could fish nearly year round. This technology, plus the RSPP, would help spread the seasonal peak of northern cod landings.

In retrospect, the attention given by the Task Force to allocation of future growth for this one cod stock was a mistake. It legitimatized a highly speculative, optimistic projection and raised expectations which were to be shattered 7 years later.

### *The Northern Fisheries (Recommendations 14 and 15)*

The Task Force addressed the problem of chronic under-development and instability of the fisheries economy in the northern part of the Canadian Atlantic region, defined roughly as the area north of 50° latitude. The Task Force proposed that:

1. The governments of Canada, Québec, Newfoundland and Labrador attempt joint economic and social development initiatives in this area; and
2. The Saltfish Act be amended to allow the Canadian Saltfish Corporation to buy, process and market fish and fish products in addition to saltfish in the area.

### *Quality (Recommendations 16–21)*

At the time the Task Force on Atlantic Fisheries undertook its work, the Department of Fisheries and Oceans had undertaken a major effort to improve fish quality. The Task Force reinforced the need for an initiative of this nature.

The major policy issue was whether quality improvement could be best achieved by government regulation and incentives or should be left to individual processors and fishermen to achieve voluntarily. The Task Force opted for the multi-faceted government-led approach which DFO had already initiated.

The key elements of the Task Force's recommendations were (1) dockside grading, (2) mandatory bleeding, gutting and icing of fish at sea, and (3) final product grading.

### *The Port Market (Recommendations 22–24)*

The Task Force briefly examined the port market, a term used to refer to the "approximately 1200 landing sites throughout Atlantic Canada where transactions to buy and sell fish took place between some 300 processing companies and 20 000 fishing enterprises."

The Task Force considered several options for intervention at the port market level, concluding that:

1. Processors should establish price differentials for landed quality, and cases where adequate differentials do not exist should be publicized.
2. Provincial governments that had not adopted collective bargaining legislation for inshore fishermen should do so. The federal government should support such collective bargaining when introducing new competition legislation.
3. The federal and provincial governments should study the various types of institutions that may bring greater order and efficiency to the port market.

### *Improving Fish Plant Profitability (Recommendations 25–31)*

The Task Force concluded that the crisis in the Atlantic fishery in the early 1980's was rooted in the low operating profit which, at an average of 12% of sales in 1981, could not support the processing industry. It suggested that industry required an operating profit (gross margin) of at least 18% of sales averaged over good and bad years. It concluded that the industry must reduce its costs and produce more high-value products. These cost reductions could not be achieved at the expense of reasonable earnings for fishermen and plant workers. The emphasis should be on increasing operating efficiency in the fish plants.

Despite these constraints, the Task Force envisaged a constructive role for government. It recommended that governments give high priority to upgrading the skills of fish plant workers and supervisors (R25). It recommended that development assistance target on improving efficiency (R26). The Task Force also proposed continuing annual compilation and analysis of industry financial statements (R31).

### *Marketing (Recommendations 32–39)*

The Kirby Task Force undertook an extensive analysis of markets and marketing (see Chapter 6 of the TFAF Report). It concluded that providing profitable outlets for the anticipated growth in Canada's Atlantic groundfish stocks required a broad range of market development activity by government and industry.

For the industry to become primarily market-driven rather than supply-driven, the Task Force suggested that marketing become the key management function that synchronized the various activities of the business and coordinated supply with demand.

The Task Force discussed at length various options for consolidating marketing activities, ranging from single-desk marketing to a state-owned marketing company, to a government umbrella organization under which consolidation of marketing by smaller processors would take place. There was a wide division of opinion within the Task Force on the preferred option. In the end, the Chairman, Michael Kirby, had to arbitrate between the conflicting views. He recommended that the Government:

1. License all processors of Atlantic groundfish and herring and exporters of whole or dressed fish, as a condition of selling their products internationally and interprovincially. This licensing was to be linked to final product grade standards.
2. Establish an Atlantic Fisheries Marketing Commission. This Commission would in turn create, initially, three product marketing councils for (a) fresh and frozen groundfish, (b) salted and dried groundfish, and (c) herring. The commission would advise the Minister of Fisheries and Oceans on marketing policies and coordinate the Council's activities. The councils would prepare marketing strategies and address product grade standards, export licensing and enforcement procedures.
3. Commit new federal funding for generic promotion of Atlantic groundfish and herring products in North America and Europe.

### *Fishermen's Incomes (Recommendations 40–44)*

The Task Force undertook a detailed survey of the incomes of Atlantic-coast fishermen in 1981. The survey documented consistently low incomes for fishermen and their households. The Task Force noted: "There is a considerable element of instability in pursuing a living from fishing. Fishing is a hunting occupation, and the 'beast' that is chased may outwit its pursuers or may not appear at all. Difficult weather and ice conditions or unpredictable species behaviour will play havoc with the revenues of even the most skilful fishermen."

The Task Force recommended a series of measures for the government to help stabilize and supplement fishermen's incomes. These included:

1. A 'production bonus' system to supplement incomes by rewarding desirable fishing practices.
2. A gross income stabilization plan to smooth out the high and low points in individual gross revenues over a rolling five-year period. This would be funded jointly by fishermen and the federal government.
3. Phasing out the unemployment insurance program for self-employed fishermen following establishment of the production bonus and income stabilization programs.

### *Financial Assistance (Recommendations 45–47)*

The Task Force considered but rejected the need for special new programs of financial assistance to fishermen or processing companies. It recommended that the government:

1. Not provide direct special assistance for vessel acquisition or replacement, but ensure that vessels could be purchased from the most economical source, unhindered by tariff or other barriers.
2. Initiate consultations between the federal and provincial governments on developing a uniform set of financial assistance policies and a possible new form of lending institution for working capital.

### *The Herring Seiner Fleet (Recommendations 48–51)*

Peripheral to its main mission, the Task Force had undertaken a study of the Atlantic herring purse seine fleet. It concluded that the herring resource on the Atlantic coast could no longer support the purse seine fleet at its existing size (66 licences). It suggested reducing the total Atlantic purse seine fleet by at least some 20 to 25 vessels. Among the measures recommended to facilitate this reduction was for the government to assign transferable vessel quotas to seiners, with the initial allocation distributed based on relative catches in the past three years. A licence buy-back program funded by industry levies on domestic purchases and over-the-side sales was also proposed.

### *Consultation and Decision-Making Processes (Recommendations 52–57)*

The Task Force recommended a number of measures to increase and improve consultation with industry and the public. These included:

1. Encouraging fishermen to organize;
2. Creating an Atlantic Fisheries Consultative Group to advise the Minister and senior officials on major strategic issues;
3. Pursuing more effective communication of scientific and policy information to industry and the public.

## Response to Kirby's Report

With certain exceptions, the Report of the Task Force on Atlantic Fisheries was relatively well received by the major sectors of the fishing industry.

Not all of the reaction was positive, however. The Newfoundland Fishermen's Union contended that the Task Force proposed nothing to help fishermen at a time when government was perceived to be positioning itself to bail out the large companies at considerable public expense. There was also division within the industry on the Task Force's marketing organization proposals. The Newfoundland Fishermen, Food and Allied Workers' Union (NFFAWU) felt that the TFAF proposal for an Atlantic Fisheries Marketing Commission would merely represent a rejigged status quo. Processors, on the other hand, through the Fisheries Council of Canada (FCC), supported Kirby's marketing recommendations.

Union President Richard Cashin and Secretary-Treasurer Earle McCurdy called the report a failure. The union expected that the Task Force would address "three problems of overriding importance: the economic plight of inshore fishermen, restructuring the offshore industry, and what to do about marketing fish products." The union levelled most of its blast at the lack of attention to the incomes of fishermen. McCurdy said:

> "This is all the more incredible in the light of the extensive survey of fishermen's incomes conducted by the Task Force and which concluded that the annual income of fishermen on the east coast remains below the poverty line.... One would have looked to Kirby to dispel the myth of public assistance to fishermen. He (the fisherman) is the victim of a monopoly or near monopoly in the selling of his fish; he is denied the advantage of an open marketplace; and he is unable, given today's prices, to make

any return on his investment. But Kirby ignored all this and recommended no meaningful changes for the inshore sector" (*The Evening Telegram*, St. John's, Newfoundland, February 18, 1983).

Barrett and Davis (1984), from an academic perspective, also criticized the Task Force Report:

> "The Task Force has failed to come to grips with the dominant problems in the Atlantic fishing industry. Their analysis of the harvesting sector is superficial and misinformed.... The Task Force's basic view, on the one hand, emphasizes overcapacity in the harvesting sector as a consequence of the common property character of the resource while, on the other hand, isolating the seasonal glut of resource, a major reason for overcapacity in the processing sector. Systemically linked in the Task Force's analysis, their solution is to rationalize the harvesting sector and, concurrently to reorganize the debt burden of the large processors.... In short, the Task Force's recommendations are intent on providing a brighter, more certain future for the corporate sector — all at a direct cost to independent fishermen, small fish processors, and many fisheries dependent coastal communities."

When the Atlantic report was released, the government at the same time announced that it had accepted 51 of the 57 recommendations. Two recommendations were rejected (a quasi-judicial licensing agency and a sunset provision for the fishermen's unemployment insurance program). Minister De Bané announced that, instead of the quasi-judicial board, the government would establish a consultative advisory committee on licence decisions and appeals, and also "establish a procedure whereby Members of Parliament can have input into licence decisions" (DFO 1983a). An Atlantic Fisheries Licence Appeal Board was finally established in July 1986. The Board acts as a last level of appeal for fishermen. The Task Force's call for a UI sunset provision was linked to its proposals for a production bonus system and a gross income stabilization plan. However, these were never implemented even though the government accepted them in principle at the time the Task Force Report was released.

The government had reserved its position on the four recommendations pertaining to the establishment of an Atlantic Fisheries Marketing Commission. Since these latter four recommendations were never accepted, the score card remained the same, 51 out of 57 recommendations accepted. In terms of normal government responses to reports such as these, this represented a very positive government endorsement of the Task Force's proposals for policy reform.

In announcing the government's response, Fisheries Minister De Bané stated:

> "Some of the recommendations call for extensive consultations with fishermen, processors and provincial governments on how they should be implemented.... If, in the course of these consultations, a consensus is reached on ways in which the recommendations can be improved, it will be accepted, provided that it is consistent with the fundamental thrust of the Report as reflected in the objectives" (DFO 1983a).

Following the Task Force Report and announcement of the government's response, there was a concerted attempt by DFO and other concerned agencies to implement the agreed recommendations. As a result of subsequent consultations with provincial governments and the fishing industry, not all of the recommendations that were accepted initially were implemented. Another eight recommendations were not implemented because of lack of agreement of the parties concerned. Examples include the recommendation to amend the Saltfish Act with respect to Northern Fisheries, the recommendation for export licensing, and the continued regular monitoring of the financial state of the industry. Another half dozen recommendations were only partially carried out or implemented in a modified form. In this category I would place the recommendations regarding quality improvement.

The government accepted the Task Force's policy objectives in the order of priority set out in the report. In general, these objectives were accepted as appropriate by provincial governments and most major groups in the Atlantic fishing industry. One major exception was the Newfoundland Fishermen, Food and Allied Workers Union which perceived this ranking of objectives as inimical to the interests of fishermen and plant workers.

The Task Force recommendations regarding international issues were accepted by the federal government and were implemented. The general recommendation that resources surplus to Canadian harvesting capacity and a fixed amount of "non-surplus" resources be allocated "after the fact" to countries that maintained a satisfactory fisheries relationship with Canada (including fisheries trade and conservation) guided government policy up to the mid-1980's. In June 1986 this policy was modified such that:

a) Canada no longer granted non-surplus allocations (except for an existing treaty commitment to France);
b) Surplus allocations were granted only to countries which cooperate with Canada in conservation efforts both inside and outside the 200-mile zone; and
c) Commercial cooperation became a factor only in arrangements with Eastern Bloc countries and even there was given secondary emphasis to the primary conservation concern.

In announcing the government's acceptance of Kirby's recommended quota licensing system, Minister De Bané indicated that the system would be phased in over the following three years, beginning with the offshore fleet (DFO 1983a). Implementation would require extensive consultations with those affected, with the introduction of any such system dependent on its acceptance by a substantial number of the parties affected.

Given the caveat about acceptability to the affected parties, remarkable progress was made during the 1980's on individual quotas for various sectors of the Atlantic fishery. The first major examples were enterprise allocations in the offshore groundfish fishery which had already been introduced by DFO on a trial basis in 1982 in advance of the Task Force Report and individual transferable boat quotas in the Bay of Fundy herring purse seine fishery. For details, see Parsons (1993).

One important aspect of the Kirby recommendations, open transferability of quota licences, was not generally accepted either by fishermen or by the federal and provincial governments.

As of 1989, the only transferable quota system in effect was in the Bay of Fundy purse seine fishery. A similar provision was later adopted for the Gulf of St. Lawrence small dragger fleet fishing groundfish and at the beginning of the 1990's was being extended to other fleets.

The federal government initially reserved judgement on the proposal to resume jurisdiction over marine fisheries in Québec. It had already begun negotiations with Québec on this issue. Attempts to reach a negotiated settlement with Québec failed, and on April 1, 1984, the federal government resumed administration of the marine fisheries. This coupled with major expenditure initiatives on fisheries development fulfilled the TFAF recommendation.

The government heeded Kirby's call for a resource-short plant program which was implemented from 1983 to 1988. Progress was slow on the Canadianization objective. By the late 1980's, however, considerable progress had been achieved with the formation of RSPP consortia and the acquisition by the Newfoundland consortium of its own vessel. However, reductions in the TAC for northern cod led to the elimination of the cod allocation for the RSPP in the multi-year Groundfish Plan for 1991–1993. This cast doubt upon the continued viability of the RSPP.

The government accepted the Task Force scenario to guide future allocations of the northern cod stock. It made a specific allocation to the RSPP, allocated more of the growth that did occur to the offshore fleet than to the inshore, and allocated some fish for experiments with Scandinavian longliner technology. With the reductions in the northern cod TAC between 1989 and 1991, this allocation scenario changed dramatically. The brunt of the significant reductions in the TAC between 1988 and 1990 was borne by the offshore fleet, the Resource Short Plant Program and the Scandinavian longliner allocation. In July 1992 a 2-year moratorium on fishing for northern cod was introduced (Lear and Parsons 1993). The sharing arrangements for northern cod are certain to be re-examined when the moratorium ends.

In some respects, the northern cod chapter of the Task Force Report fuelled a euphoria similar to that resulting from the 200-mile limit in 1977. It also undermined its own attempts to rationalize the processing sector through closing some plants. It was impossible for the federal Team in the restructuring negotiations to convince the Newfoundland government that it was necessary to close certain fish plants when the Task Force Report was projecting significant growth in the northern cod stock. The net effect was to postpone for nearly a decade a necessary rationalization of processing and harvesting capacity. In hindsight, the chapter on Northern Cod was the Achilles Heel of the Task Force Report.

Responding to the Task Force recommendation on Northern Fisheries, Fisheries Minister De Bané noted that the issue here was more a problem of economic development than of the fishery *per se*. He stated that the federal government had accepted the recommendation to work with Newfoundland and Québec on an economic and social development package for this area. Federal-provincial fisheries agreements subsequently brought money to address some of the needs of fishermen and communities in both northern Québec and northern Newfoundland and Labrador. The federal and Newfoundland governments explored options for a Northern Fisheries Development Corporation but could not agree on any of the options developed. In the end the concept withered on the vine.

The federal government accepted all of the Task Force recommendations dealing with quality. Minister De Bané, however, felt that the Task Force recommendations did not go far enough in one area: "The most effective way to improve the quality of groundfish is for buyers on the dock to pay higher prices for higher quality fish landed by fishermen. The present financial condition of the industry, however, makes it very difficult to pay such premiums for better quality fish. I have asked my officials to examine other options for providing, on a transitional basis, price incentives for quality" (DFO 1983a).

One of the first steps DFO took to deal with raw material quality was to amend the Fish Inspection Regulations dealing with construction and equipment requirements for fishing vessels and with chilling of fish on board and during transportation. Other regulations for unloading, handling and transporting fish took effect in 1983. Progress was also made in providing infrastructure, containerization and fish chilling facilities.

The Task Force recommendations regarding the port market had no impact on subsequent Atlantic fishery management initiatives. The federal government supported the call for collective bargaining for fishermen. However, collective bargaining was already well established in Newfoundland. New Brunswick had passed legislation prior to the Task Force Report. There was no action by other provincial governments.

Progress was made in the 1980's on some of Kirby's measures to improve fish plant profitability. There are currently two colleges in Atlantic Canada training plant managers and supervisory staff. Training covers the spectrum of plant management skills with a special emphasis on quality control. Industrial and Regional Development Program guidelines established by the Department of Regional Industrial Expansion in 1984 emphasized productivity improvements and modernization as a condition for financial assistance. Probably the most significant recommendation in this package, namely that the financial surveys and analyses initiated by the Task Force be continued annually, was not implemented. Once the financial situation of the industry began to improve in the mid-1980's, the industry lost interest in such surveys.

The federal Cabinet was divided on the Task Force's marketing recommendations. In his February, 1983, announcement, Minister De Bané accepted recommendations 32 and 33 dealing with the licensing of exporters. The government endorsed the proposal for a generic promotion program and committed $28.3 million for this purpose over five years. The government also earmarked more than $20 million for Atlantic groundfish and herring exporters under the Program for Export Market Development over five years.

As stated earlier, the government reserved its position on recommendations 34 to 37, which proposed the establishment of an Atlantic Fisheries Marketing Commission, its structure and functions. Minister De Bané and certain of his colleagues felt that the recommendations did not go far enough and, in some respects, would give industry a free ride with the generic promotion and other funding, with insufficient government involvement.

Minister De Bané would not support an industry-dominated Atlantic Fisheries Marketing Commission. In a June, 1984, statement detailing progress to date on the Task Force recommendations, he noted that the financial restructuring led by Mr. Kirby had given the newly created companies "great marketing power":

> "Given this consolidation and existing industry groups such as the Canadian Association of Fish Exporters, the setting up of a marketing council appeared mainly a reshuffling of the status quo. Therefore, the government has reserved judgement on marketing arrangements, pending further review" (DFO 1984a).

Unable to secure sufficient support for a more interventionist option, he preferred to let this set of recommendations die.

Accepting the Task Force's call for a major generic promotion campaign, DFO launched in January, 1984, a five year $7 million advertising campaign. The program emphasized the dietary and culinary advantages of fish, as well as their nutritional value. Advertisements featuring the theme "Today's dish is fish" appeared on television and radio and in consumer and trade magazines. This federal program started in 1983–84 was wound-up in 1985–86 with a total expenditure of $5.5 million.

DFO terminated fish promotional activities in 1986. It intended to privatize most of these activities. To assist privatization, DFO entered into a five-year transition program with the newly-established Canadian Seafood Advisory Council (CSAC) and the Fisheries Council of British Columbia. The government provided funding during 1986 to 1990 with the industry expected to assume all costs during the fiscal year 1990–91. From 1986 to 1990 the Seafood Advisory Council received $893 000 and the Fisheries Council of British Columbia got $583 000.

The government funding was to assist these two groups in taking over promotional activities that DFO had previously carried out. As one example, the Seafood Advisory Council continued the November Fish and Seafood Month, a Canadian promotion program carried out by DFO during the previous eight years. It also operated a Seafood Information Centre similar to the Food Centre previously operated by DFO.

In the mid-1980's extensive federal-provincial discussions took place on Quality Improvement. The federal and provincial Ministers of Fisheries agreed on the need for improved quality and recommended the following modified approach to the November 1986 meeting of First Ministers:

1. Encouraging of voluntary gutting and/or bleeding of groundfish at sea;
2. Promoting voluntary raw material (dockside/point-of-sale) grading by industry;
3. Consulting with industry about making approved in-plant quality control programs a condition of registration for a processing plant; and
4. Once in-plant quality control programs were established and evaluated, governments would assess, in consultation with industry, the feasibility of end product grading programs (DFO 1986).

The November 1986 First Ministers Conference agreed to consultations with industry about recommendation number 3. DFO and representatives from the fish processing industry subsequently proposed a Quality Management Program to require that companies have a formal system that would ensure existing minimum regulatory requirements would be met.

The resulting QMP was accepted by the Seafood Inspection Policy Advisory Committee (SIPAC) in December 1988. In October 1989, SIPAC recommended that DFO proceed with introduction of the initiative to industry at large.

QMP is a program which sets out means for an establishment to ensure products are processed, stored, graded, packaged, marked, transported, and inspected safely. Companies must have an acceptable QMP to maintain federal registration for their processing establishment. However, a company's QMP will be evaluated according to existing minimum requirements. The role of DFO inspectors shifted from inspection of products and plants to monitoring and auditing a company's ability to meet its obligations to comply with existing regulations.

The QMP was introduced voluntarily for seafood processors in 1990 and became a requirement for all Canadian fish and seafood plants in 1991.

Thus from 1986 onward the emphasis shifted from the earlier mandatory grading approach to federal monitoring of in-plant quality management programs. Overall, the Task Force's recommendations had a marginal effect on Quality Improvement measures.

The proposed campaign of generic promotion in North America and Europe by the proposed Product Marketing Councils was derailed by the decision not to establish an Atlantic Fisheries Marketing Commission. The overall intent of the Task Force recommendations on marketing was therefore never fulfilled.

The government accepted the recommendations not to establish a new general program of financial assistance for either fishermen or processors (R45) and to initiate consultations with provincial governments on a uniform set of financial assistance policies and a possible new form of lending institution for working capital (R47). However, it said it would consult fishermen's and processor's organizations before responding to the recommendation against direct special assistance for vessel acquisition or replacement (R46). The government later terminated the Fishing Vessel Assistance Program. See Crowley et al. (1993). Recommendation 45 was implemented in that no new general program of financial assistance was established. Regarding Recommendation 47, the federal government could not reach agreement with the provinces about financial assistance policies.

The government followed the general thrust of the Task Force recommendations regarding the herring seiner fleet. The Management Plan for the Herring Fishery implemented in 1983 gave purse seine licence holders a quota licence for a portion of the annual TAC of herring in the Scotia Fundy Region. This licence was permanently transferable by sale and purchase. See Parsons (1993) for details.

A buy-back program was also established in Northern New Brunswick funded under the Canada — New Brunswick Fisheries Subsidiary Agreement. Two seiners were retired from the fishery and their quotas allocated to the remaining operators.

The government generally accepted the recommendations about consultation and decision-making processes, and these were implemented over time. Membership on all advisory committees was reviewed and, where necessary, revised for equitable and broadly-based representation of major user groups. Various subcommittees were established to deal with specific issues, such as the introduction of EAs in various fleet sectors.

DFO continued to encourage fishermen to organize. One initiative provided start-up funds for L'Alliance des Pêcheurs Commerciaux du Québec.

In August, 1985, Minister Fraser established an Atlantic Regional Council to advise him on major policy issues affecting the Atlantic fishing industry.

Overall, about 65% of the Task Force recommendations were followed as proposed, another 11% implemented partially or in a modified form, for a combined total of 76%. Given the complexity of the issues and the diversity of views, this was an impressive record for any public inquiry or commission of this nature.

## The Pearse Report

The Commission on Pacific Fisheries Policy, headed by Dr. Peter Pearse, originated in the 1980 crisis in the

Pacific salmon fishery when a weak salmon run coincided with a weak market and a coincidental downturn in the roe herring fishery. An overcapitalized fleet appeared to face bankruptcy.

On January 13, 1981, Fisheries Minister LeBlanc announced the appointment of Dr. Peter Pearse as a Commissioner under the Inquiries Act to examine the state of the industry and to recommend strategies and policies for the longer term (DFO 1981). Pearse was mandated to inquire into and report upon:

(a) the condition of the stocks of fish within Canada's jurisdiction off the Pacific coast, current levels of utilization and their relationship to optimum rates of use;
(b) the provisions for conservation, management, protection and development of the fish resources, including the protection of their tidal and non-tidal habitat and the enhancement of salmonid stocks;
(c) the structure and size of the commercial fishing fleet and the relationship between the capacity of the fleet to harvest fish and the optimum rate of harvesting the stocks;
(d) the policies and procedures for licensing commercial fishing, and for regulating the size and structure of the fishing fleet, including the charges to be levied by the Crown for fishing privileges; and
(e) the nature and amount of non-commercial fishing in tidal waters for salmonid species, its impact on the stocks and the commercial fishery, and the policies and procedures for regulating non-commercial fishing.

Although not as comprehensive as the Report of the Task Force on Atlantic Fisheries, which examined fish harvesting, processing and marketing, the scope of the Pearse Inquiry was nonetheless wide, including such matters as fisheries resource management and conservation, industrial regulation, sportfishing policy, native rights, environmental protection, problems in the Yukon, intergovernmental arrangements, administration, enforcement and research. Specifically excluded from his mandate was the matter of "arrangements between Canada and foreign nations governing fishing rights and conservation of stocks."

Pearse produced a preliminary report in the fall of 1981, and his final report was released the following September (Pearse 1982). The latter was in many respects much more detailed than the TFAF Report. It made 292 recommendations compared with 57 in the TFAF Report. While the TFAF recommendations were generally focused on policy, leaving the details of implementation to government, Pearse's recommendations were much more detailed and prescribed in many instances how policies should be implemented, as well as what should be done. It is not feasible to examine all of Pearse's recommendations here. I will address the general thrust in particular subject areas, and the government and industry responses.

## Recommendations

### *Policy Objectives*

Pearse concluded that Canada's Pacific fisheries were in crisis. The economic circumstances of the commercial fisheries were bleak. In addition, there was growing concern about the precarious condition of many of the fish stocks. Natives were increasingly anxious about their traditional fishing rights, and sport fishermen were worried about recreational opportunities. Pearse observed that "the economic problems and other concerns are rooted in fundamental deficiencies in fisheries policy."

He contended that basic changes were needed. Thus he proposed the following framework of policy objectives for the Pacific fisheries:

*Resource Conservation* — Fisheries policy must first and foremost ensure that the resource is properly protected and, wherever advantageous, enhanced.

*Maximizing the Benefits of Resource Use* — This meant ensuring that the resources available for harvesting are allocated to those who will make the best use of them. Despite this, policy "must also take account of prevailing concepts of fairness in resource allocation."

*Economic Development and Growth*: This involved improving incomes in the fisheries and development of economic opportunities for coastal communities and Native people.

*Social and Cultural Development*: In addition to the needs of coastal communities, Pearse saw a need to consider the special economic problems of Natives and their unique dependence on fish for nutritional needs and cultural activities. He also saw the need to preserve recreational opportunities and protect the lifestyle of commercial fishermen.

*Returns to the Public*: Fair and reasonable returns should accrue to fishermen, but the public should receive some of the gains of improved economic performance.

*Flexibility*: Policies must be resilient and durable despite continuing shifts in the external environment, such as changes in resource abundance, in markets and in fishing technology.

*Administrative simplicity*: Policies must be practical and enforceable.

### *Habitat Management*

Pearse made 23 recommendations pertaining to habitat management. The general thrust of these recommendations was as follows.

The department should establish explicit fish production targets as long-term management objectives. These targets should be based upon full utilization of the existing natural habitat and additional production resulting from enhancement.

Pearse observed that DFO's objective for habitat management was too general. He proposed that:

> "The policy of the Department should be to ensure that the total fish production capacity in the region will not be diminished as a result of industrial and other activities that impinge upon fish habitat. Identifiable and measurable harm to fish habitat should be tolerated for any particular development only if the damage is fully compensated through expanded fish-production capacity elsewhere."

Pearse proposed that:

> "The Department should adopt an explicit policy for assessing proposed developments that threaten fish habitat and for determining compensation where required...."

Cash compensation was to be paid into a proposed Pacific Fisheries Conservation Fund, to be administered by DFO and spent only on habitat improvement and other fish production measures.

Pearse further proposed that:

1. The Minister be given the authority to convene public hearings concerning any proposed development;
2. A joint DFO/B.C. program be developed to monitor all operations with significant potential to damage habitat;
3. DFO be made responsible for all habitat protection legislation with appropriate transfer of resources.

### *Fisheries Management*

Pearse observed that the various systems of stock management on the Pacific coast had worked with "mixed success." He concluded that the most conspicuous management weaknesses were in the most valuable fisheries — salmon and roe-herring.

He proposed that the Department formulate and publish long-term plans and objectives for managing each of the major species. These should be supported by publishing information on the state of the stocks, a review of emerging problems, alternative proposals for improved management, and a review of the implications of proposed fisheries management techniques. Quantified targets for production by species and management regions should also be provided.

Pearse recommended that the management plans for each fishery be formulated through annual review and consultation.

### *Salmonid Enhancement*

In 1977, Fisheries Minister LeBlanc announced a Salmonid Enhancement Program (SEP) to double the stocks of salmon and anadromous trout, restoring them to their former levels of abundance. The program was to take advantage of the return of anadromous fish to their natal streams making them potentially easy to harvest. Such fish stocks are amenable to enhancement through improving spawning beds and constructing facilities that increase spawning. Larkin (1973) described the rationale for SEP.

The goals were:

1. To augment national and provincial income.
2. To create employment opportunities.
3. To improve economic opportunities for Native peoples.
4. To foster development of economically disadvantaged communities and regions.
5. To increase recreational opportunities.

The program was to have two phases. Phase I was originally planned as a five-year program beginning in 1977 but was subsequently extended to 1984. For this phase, the federal government committed $150 million and the province $7.5 million.

Pearse recommended that the projects planned to the end of 1984 proceed and the results of projects already in place be thoroughly evaluated. This evaluation should determine longer term priorities for future enhancement.

Pearse suggested that artificial enhancement projects be considered only if equivalent net gains could not be achieved through improving fisheries management or reducing fishing pressure.

Pearse recommended that governments recover the costs of the program. He suggested a second-phase enhancement program beginning at an annual level of at least $12 million plus a provincial contribution. Factoring in the operating costs of completed projects,

Pearse' proposals would have cost approximately the same as the existing enhancement effort.

### *Research and Information*

Although a great deal of high quality research had been done to support fisheries management, enhancement and habitat protection, Pearse concluded that "our understanding of the biology of fish and their relationship to the environment remains weak; this is a serious impediment to improved protection and utilization of the resources." He said that requirements for effective resource management should guide the Department's research program.

He recommended improving the quality and completeness of statistical information on catches, reinstating coastwide sampling of salmon catches and expanding programs for determining the racial composition of salmon and herring catches. Also, the programs of collecting and collating information on salmon and herring escapements and spawning needed to be strengthened. He recommended a substantial expansion of scientific research on fish habitat.

### *Licensing and Fleet Development Policy*

Pearse concluded that the central economic problem of the commercial fishery was the chronic overcapacity of the fleets. He concluded that licensing policy on the Pacific "lacks coherence and consistency, particularly in the provisions relating to such matters as the appurtenancy of licenses, the fee structure, transferability and criteria for renewability and replacement." Moreover, in the more important fisheries, he observed, "the system had obviously failed in its primary purpose of controlling fleet expansion."

Pearse identified the shortcomings of limited entry licensing and proposed these systems "be replaced by more effective quota licences in those fisheries where it is feasible to do so." To him, this included all of the significant commercial fisheries other than salmon and roe-herring. For the complicated salmon and roe-herring fisheries, he concluded that "quota licensing is not feasible, at least at present." He proposed improvements to the limited-entry licensing system to strengthen control of fishing capacity and to dampen incentives to invest in expansion as well as a program to reduce the fleets.

### *A Framework for Commercial Licensing*

Pearse made 39 specific recommendations on revising the framework for commercial licensing. He proposed licensing by species and by area, with the Pacific coast divided into three broad zones for commercial licensing purposes: north, south and west. He suggested that limited-entry licences specify the type of gear to be used, but quota licenses should not. Pearse proposed that all commercial licenses be issued to persons or companies rather than to vessels.

Pearse proposed that limited-entry licences and quota licences have terms of 10 years. For each limited-entry fishery, the government should determine and periodically revise the fleet capacity required in each gear category for each licensing area. One-tenth of that capacity by gear category should be available for allocation each year. For quota-licence fisheries, the government should calculate and periodically revise the total allowable catch; one-tenth of the total allowable catch should be allocated under new quota licences each year.

Pearse also proposed adopting competitive bidding procedures to allocate the total capacity for limited entry fisheries and the total allowable catch for quota fisheries. During the first 10 years, only holders of licences already active in the fisheries should be eligible for new licences. Thereafter anyone would be eligible.

Pearse's rationale for proposing this overall system was as follows. First, existing fishermen would enjoy greater security and certainty about their fishing privileges under long-term licences. Second, after a transitional period, newcomers could enter the fisheries by dealing directly with the government instead of having to purchase a licence from someone else. Third, the Department could adjust access to the fisheries as needs change. Fourth, he considered competitive bidding to be an objective means to determine who would have access to scarce and valuable public resources and to ensure that, over time, licences would be utilized by those who can make the best use of them.

Pearse recommended that, subject to specific limits, all limited-entry and quota licences be freely transferable from person to person. However, Pearse also recommended ceilings on the proportion of licences which could be held by a person or private corporation, 5% for major fisheries and 15% for other fisheries.

Within this broad framework, Pearse made proposals for transitional arrangements, specific policies for licensing individual fisheries and rationalizing those fisheries where overcapacity was a major problem. I will briefly summarize these proposals and then deal with the public and government response to the overall licensing package.

### *Rationalizing the Salmon and Roe-Herring Fisheries*

Pearse concluded that the existing licensing system had failed to control excessive expansion of fishing capacity in the salmon and roe-herring fisheries. He proposed that:

1. All existing ordinary and Indian salmon licenses and roe-herring licences should be replaced by new 10-year licences.
2. The three-area licensing system for the roe-herring fishery should be continued with certain modifications, and a similar system introduced for salmon.
3. A program to reduce the two fleets by half of their existing licensed capacity over the next decade should be introduced.
4. Competitive bidding for licenses should be introduced (one-tenth each year).

This would provide for a reduced fleet after a decade. Pearse proposed to supplement these arrangements with a means of withdrawing licences during this transitional period. He argued that government must facilitate fleet reduction. He recommended that the proposed Pacific Fisheries Licensing Board reduce the excess licensing capacity in the salmon and roe-herring fleets by offering compensation to licensees for voluntarily relinquishing their licences. He suggested that industry and government together finance this compensation program.

### *Royalties*

Pearse proposed a system of royalties on quota licences, or in the case of limited-entry licences, royalties on landings. He suggested a schedule of fees and royalties for vessel licences and quotas. He called for a long-term policy of maintaining royalty rates between 5 and 10% of the gross value of the landed fish.

### *Pacific Fisheries Licensing Board*

Pearse proposed a Pacific Fisheries Licensing Board be created as a Crown Corporation. This Board would administer commercial fishing licences within the general policy set out in the Fisheries Act and regulations. It would also conduct competitions for new licences, carry out a licence retirement program, maintain a public record of licence holdings, decide licence appeals and advise the Department on changes in licensing policy.

### *Indians in the Commercial Fisheries*

Pearse concluded that large numbers of Indians had been displaced from the commercial fisheries in recent decades. He observed: "This displacement has generated serious economic and social distress in Indian communities, many of which offer no alternative employment. The relative immobility of Indian people has left them heavily dependent on unemployment insurance and welfare payments."

Despite this, it was obvious that Indians could adapt and perform well in the commercial fisheries. Pearse concluded that the fisheries were an obvious base for policies aimed at Indian social and economic development. He recommended that:

1. The federal government quickly proceed to implement the Indian Fishermen's Economic Development Program, a proposal by the Native Brotherhood of British Columbia;
2. Licences held by Indian fishing corporations not be transferable to non-Indians and licensing policies be developed to enable such licences to be leased to individual Indians; and
3. The Department of Indian and Northern Affairs provide Indians and Indian corporations with financial assistance to compete successfully in the proposed periodic reissuing of licenses by competition (the 'auction' proposal discussed elsewhere).

### *Subsidies for Vessel Construction*

Pearse criticized the government for providing financial incentives to build new fishing vessels despite the overriding problem of too much fishing power. In addition to the Fishing Vessel Assistance Program, other Departments provided assistance in various forms. Pearse recommended that general subsidies through tax credits, accelerated depreciation allowances, subsidies to shipbuilders and loan guarantees be immediately terminated for fishing vessels on the Pacific coast.

### *Unemployment Insurance*

Pearse recognized that the insurance scheme was national and raised issues of social policy well beyond the scope of his inquiry. After noting a number of complaints at Commission hearings, he recommended that the Department, in consultation with the proposed Pacific Fisheries Council, investigate the desirability and feasibility of catch insurance for fishermen engaged in Pacific fisheries.

### *The Processing Industry and Its Regulations*

Pearse touched on this topic only briefly. He recommended that the government not become directly involved in marketing fish products produced on the Pacific coast.

### *Product Quality*

Pearse devoted only one page to this topic. He recommended that the Department, in close consultation with the fishing industry, explore grades for fish landed, with special attention to salmon. He also recommended that the Department consider product quality standards for food-herring products.

### *Export Regulation*

Pearse expressed reservations about the Department's "attempt to use its regulations to restrict export opportunities in the interest of promoting local processing." He considered this objective to be quite separate from the Department's responsibilities for product quality standards and in conflict with the objective of enhancing export opportunities. Pearse urged the Department to continue to develop its program of quality certification for exported fish products to ensure adherence to product standards but to refrain from "using quality controls as a means of restricting export trade."

Less than a decade later this issue would mushroom into a major bilateral trade issue between Canada and the United States.

### *The Indian Fishery*

Pearse observed that the existing policy governing the Indian fishery was not only unsatisfactory to many Indians, but caused management and enforcement problems for the Department. Concluding that "Indian fisheries policy cries out for reform" he proposed major changes including:

1. The Department should allocate a specific quantity of fish to be available annually to each Indian band involved in the Indian fishery.
2. The catch allocated to bands should have priority over commercial and sports fisheries.
3. No royalties should be levied on fish harvested by Indians under the proposed allocations.
4. Indian fishing permits should be issued annually to individual fishermen through band councils. Permits should authorize Indians to take fish for food and ceremonial purposes only. They should specify the quantity and composition of the authorized catch, and the location, time and method of fishing as required for management purposes.
5. The Department should enter into 10 year Indian Fishery Agreements with bands. These agreements should specify the band's allocation of fish, provide for an annual fishing plan, and where appropriate, authorize the band to engage in fisheries enhancement on or near their reserves.
6. Where they are willing and able to do so, band councils should be encouraged to take management responsibility for Indian fisheries.

### *The Sport Fishery*

Pearse focused on the sport fishery as a major component of the fishery for Pacific salmon. He estimated that spending related to saltwater sportfishing on the Pacific approached $100 million annually. Despite increasingly stringent regulations on the sport fishery, there was no sound basis to identify the impact of these regulatory measures.

Pearse observed, correctly, that "until relatively recently, sportfishing was of little consequence to resource managers." He felt that the Department's approach to sportfishing was poorly integrated with overall fisheries policy. To this end, he recommended that "the government's policy should explicitly recognize sportfishing as a legitimate, valuable and significant use of fish resources, and this should be reflected in a commitment of staff and budget."

Rather than spreading the available catch among more fishermen, thus enabling a greater number to participate but reducing the quality of the experience for each, Pearse proposed a five-year program to constrain the growth of sportfishing pressure on the resouıce while maintaining high quality sportfishing opportunities.

Pearse also recommended:

1. Developing a comprehensive data and information system for the sport fishery;
2. Developing a rapid data processing system designed to integrate sportfishing information into general salmon management planning; and
3. Research on the value of sportfishing opportunities on the Pacific coast and the effect regulations have on these values.

### *Enforcement*

Pearse examined the difficulties of enforcing regulations. He made 14 recommendations to improve enforcement procedures, strengthen the personnel and organization of the enforcement function and improve the prosecution process.

Pearse concluded that the existing scheme of financial penalties under the Fisheries Act was "replete with ambiguities, inconsistencies and anachronisms." He recommended substantially increased fines for offenses under the Fisheries Act and more vigorous use of license suspension and cancellation. He called for an aggressive policy in seizing offenders' vessels and equipment. Illegal catches and equipment should be forfeited to the Crown.

### *Consultation*

Pearse recognized the importance of consultation, and identified a number of deficiencies in existing processes. His proposed changes to improve consultation included a new Pacific Fisheries Council.

### *Federal Arrangements with British Columbia*

Pearse observed that the absence of a formal working relationship between the two governments was a serious deficiency in the existing policy framework for the Pacific fisheries. Accordingly, he recommended that the Government of Canada invite British Columbia to join in a comprehensive intergovernmental agreement on fisheries matters. He also proposed a new Canada-British Columbia Fisheries Committee at the Deputy Minister level.

Pearse also made a series of recommendations aimed at strengthening the administration of the Pacific Region.

### *Policy Implementation and Review*

Pearse concluded that his proposals would require changes to federal legislation, related regulations, fishing licences and administrative policies and procedures. He recommended that the Fisheries Act be replaced by "a modern, lucid statute containing the main principles of fisheries policy for Canada." The new Act should:

(i) Include a clear statement of national fisheries policy objectives;

(ii) Set out the Department's management responsibilities and planning procedures. The scope of these should be broad, leaving no doubt about the Department's mandate to effectively manage fisheries and fleet development;

(iii) Commit the Department to integrated resource management and planning, and set out arrangements for dealing with projects and developments that affect fish habitat;

(iv) Devote a separate part to Pacific fisheries, consistent with the national policy framework;

(v) Set out the legal authority and procedures to be followed in allocating the sport, commercial and Indian fishing rights recommended in his report;

(vi) Provide for the appointment of the proposed Pacific Fisheries Council and Pacific Fisheries Licensing Board;

(vii) Formally delegate decision making authority to the licensing board and, where applicable to regional officials of the Department; and

(viii) Include a clear and consistent structure of penalties.

To expedite these changes, Pearse proposed appointing a temporary Minister of State for Pacific fisheries, junior to the Minister of Fisheries, to implement reforms in Pacific fisheries policy.

## Response to Pearse's Report

Given the myriad of recommendations in the Pearse report on a multitude of subjects, it is difficult to see the forest for the trees. Many of the administrative proposals were implemented. However, the heart of his report consisted of proposals to restructure the commercial licensing system. Pearse proposed radical reform, replacing the existing system by 10-year licences, a competitive bidding or auction system for one-tenth of the licences each year, and transferable quota licences for all fisheries other than salmon and roe-herring.

Nearly half the recommendations dealt with the commercial licensing system. The crux of his proposed reforms for the Pacific fisheries rested on these licensing changes. It is thus not surprising that the public focused on this aspect of the report. Pearse's proposals clearly were comprehensive, bold and imaginative. But they touched off a firestorm of protest from commercial fishermen's associations. Following Pearse's interim report, halibut fishermen appeared somewhat receptive to his proposal for individual quotas. A detailed proposal for introduction of individual quotas in the halibut fishery was developed by DFO in consultation with industry. In August,1982, Minister LeBlanc gave his approval to proceed with the program for the 1983 fishery.

In September, Mr. Don Cruikshank, the Chairman of the Fleet Rationalization committee established by the Minister following Pearse's interim report, informed Minister LeBlanc of the Committee's great concern regarding the proposed implementation of the quota licence system in the halibut fishery (Cruikshank 1982).

Pearse's final report, released in September 1982, generated major concern in the industry over his licensing proposals. A backlash against the implementation of any Pearse recommendation began to develop. Various fishermen's groups voted in public meetings throughout British Columbia to reject the major recommendations. A November, 1982, Conference co-sponsored by the Western Fishermen's Federation and the University of Victoria called for new consultative arrangements. The Conference was unanimous in:

1. opposition to 10-year licences;
2. opposition to separating the licence from the vessel and auctioning renewable licences;
3. agreement that excess fishing capacity is a problem and that solutions should include both voluntary buy back and methods of increasing the amount of catch (Anon. 1982).

Also in November, 1982, the Fleet Rationalization Committee submitted its report to the Minister. It discussed individual quotas under the category of *Unacceptable Methods of Rationalization*. It stated:

> "We are opposed to the application of personal quotas on any fisherman."

Cruikshank, in a letter to Minister De Bané, dated January 10, 1983, reiterated opposition to individual quotas:

> "Before you commit this industry to an unworkable system, please recognize that the implementation of a quota system is virtually irreversible. To put this theory in place, to test it, is like a woman getting a little bit pregnant to see if she will suffer from morning sickness.
>
> "If you are truly intent on determining the viability of quota systems, scrutinize our present Abalone fishery. There are only 26 licences to police in this fishery, yet your staff members charged with this responsibility admit privately that they have no control" (Cruikshank 1983).

By now the industry was concerned that an individual quota system for the halibut fishery would automatically lead to a new licensing system which it generally opposed. On January 21, 1983, the Minister's Advisory Committee (MAC) requested that halibut quotas be delayed until MAC could suggest alternatives.

On February 18, 1983, MAC advised the Minister to reject all of the Pearse recommendations that would have fishermen bid for their licences in an open auction. They also suggested rejecting the proposal for a Crown Corporation to administer licensing and appeals.

At a meeting with the Minister in Vancouver on February 18, 1983, MAC presented to the Minister their advice based on a detailed examination of the Pearse Report. The Council recommended that 43 of the Pearse Report recommendations be implemented essentially as formulated by the Commission. These recommendations dealt with fisheries management, data systems, habitat management, protection, consultation and administration.

The Council recommended that a further 9 recommendations be implemented with some modifications. These recommendations dealt with enforcement capability in the Pacific Region, reform of the policy level consultative process, and administration of licensing in the Pacific Region of the department.

The Council recommended action on 23 recommendations whose implementation would depend on the cooperation of other agencies or governments.

Finally, the Council recommended rejecting 12 of Pearse's recommendations. These were the recommendations that would have fishermen bid for their licences in an open auction and the recommendations to establish a Crown Corporation for licensing and appeals.

The Minister reacted swiftly to this advice. At a Press Conference following his meeting with MAC, Minister De Bané rejected the idea that established fishermen must compete in an auction for the right to remain in the industry. He also rejected the proposal for a Crown Corporation to administer the licensing of fishermen: "I am not prepared to delegate to an outside body the decision-making authority entrusted to Parliament" (*The Vancouver Sun*, February 19, 1983).

Minister De Bané was seen as having rejected the major proposals of the Pearse Report. At the same time he indicated that the 76 technical recommendations on fisheries management, habitat management, enforcement and departmental administration would be adopted. He also said that the Department was developing national policies on fish habitat management and on Indian fisheries issues.

Meanwhile the problem of overcapitalization in the major Pacific fisheries remained. MAC was given until November 30 to produce proposals for fleet reduction, licensing and allocation of fishing rights among seiners, gillnetters and trollers.

*The Vancouver Sun* editorialized (February 20, 1983):

> "For all the good he accomplished in 20 months of hearings and work on behalf of the West Coast fishery, economist Peter Pearse might just as well have spent the time in Austria skiing....
>
> "Mr. Pearse was in Austria...when Fisheries Minister Pierre De Bané took a fish knife to Mr. Pearse's blueprint, 'Turning the Tide,' and gutted its major recommendations.
>
> "Cut away was the idea of competitive bidding among fishermen for licences, government

> financing for a buy-back of excessive fishboats, and a Crown Corporation to administer licensing of fishermen. About the only things saved were 76 purely technical recommendations — 76 out of a total of 300 put forward by Mr. Pearse as essential to the survival of the Pacific fishery.
>
> "What is left of the Pearse report is little more than a skeleton, an odd assortment of bones that Mr. De Bané hopes will somehow fall into place by themselves."

Peter Pearse indicated that he was not upset that some of his recommendations had been rejected but he would be if his opponents did not come up with some solutions of their own. "We all agree on the problem," he said, "but if you don't agree with my recommendations then you have to find a better solution." (*The Province*, February 22, 1983).

In a letter to the Editor of *The Vancouver Sun* (February 27, 1983) taking issue with a subsequent story headlined "Pearse Miffed at Probe Rejection," Pearse stated:

> "I am not 'miffed' because the Minister of Fisheries and Oceans rejected my proposals for reducing the excessive capacity of the fishing fleet; as far as I know he did not reject them. He did reject any plan that would depend on large amounts of government funding, and he did so in the context of some alternative proposals advocated by a fishermen's advisory group that called on the federal government to put up some $100 million to finance reduction of the fleet....
>
> "The only recommendations of mine that were rejected relate to the allocation of new fishing privileges through competitive bidding.... But at the same time, the Minister announced his acceptance of 76 of my commission's recommendations, and I think that deserves more recognition than you have given it. The changes recommended promise significant improvements in managing fish stocks, habitat protection, salmon enhancement, enforcement, and administration, and we should be pleased that they have been adopted so quickly."

In a speech to the May, 1983, Annual Meeting of the FCC, Fisheries Minister De Bané stated:

> "I support the Pearse goal of fleet reduction. We have to outwit that persistent pattern of over-investment that bleeds away profits. The Pearse report makes the goal clear. I particularly agree with the post-Pearse process, in which industry people themselves — fishermen and processors — are devising the method to reduce the fleet and give fishermen a reliable living" (DFO 1983b).

On September 28, 1983, MAC presented the Minister with a general discussion paper entitled *A First Step Towards Rationalizing the West Coast Fisheries*. That paper stated:

> "In the case of fleet rationalization, industry was united in its rejection of Dr. Pearse's proposed solutions. For the most part, the rejection stemmed from the fact that the Pearse recommendations did not take into account the social nature of the fishery and participants. In addition, where fishermen were looking for some stability and certainty with regards to their future, the ten-year licence bidding system provided just the opposite, with nobody really knowing what their future in the industry would be. However, the one point in the fleet rationalization recommendations which has received industry support is that which suggested a Buy-Back program....
>
> "After careful and exhaustive deliberations, the Minister's Advisory Council has determined that a Buy-Back program is not *only* viable, but it is the only option that will provide some stability in the industry while allowing us to deal with the issues of licensing and allocation, which are part of the long term solutions necessary for a stable and profitable industry to develop on the west coast" (MAC 1983).

On June 18, 1984, Minister De Bané and Senator Jack Austin announced in Vancouver a new Policy for the Pacific Fisheries with two stated objectives:

1. Eliminating the overcapacity of the fleet and strengthening its financial condition; and
2. A more rational approach to harvesting.

The proposals aimed at reducing the size of the fleet by as much as 45%, through a Buy-Back of at least $100 million, accompanied by a management system "that puts a final end to the 'race to the fish'." Mr. De Bané announced the government's intention to implement area and gear licensing systems and individual fishing allocations (DFO 1984b).

The proposals included a Buy-Back Corporation, with funding by government and financial institutions. Although a total amount of $100 million was mentioned, the details were vague "pending completion of negotiations."

The Minister indicated that individual fishing allocations would be introduced starting in 1985 on a demonstration basis with selected areas in the troll and net fisheries by assigning certain percentages of the total allowable catch to individual fishermen. He predicted that a full system of individual fishing allocations would be in place for all salmon fisheries by 1987. These allocations within each area and for each gear type would be transferable.

To provide the framework for these measures, the government tabled in the House of Commons draft legislation to amend the Fisheries Act and to create a new Pacific Fisheries Restructuring Act. On July 9, 1984, the new Prime Minister, John Turner, dissolved parliament and called an election for September 1984.

The British Columbia fishing industry reacted angrily to the restructuring proposals. Jack Nichol, President of the United Fishermen and Allied Workers Union, said the scheme was "throwing fishermen an anchor to sink them instead of a life raft to keep them afloat in hard times" (*The Globe and Mail*, June 13, 1984). Al Meadows, President of the Pacific Trollers Association said that "the legislation had received near-universal condemnation. Its compulsive nature and dictatorial intent we find particularly galling, especially in light of the considerable consultation that took place between the Minister and his advisory council." Throughout the process, Meadows added, Minister De Bané and his staff were aware that the industry would never accept regulatory measures such as area licensing, individual quotas and single-gear licensing (*The Vancouver Sun*, June 27, 1984). David Elliott of the PTA said that Minister De Bané was told repeatedly by his industry committee that quotas and area licensing were repugnant to fishermen. A British Columbia Fishermen's Coalition demanded the "retention of common property rights for all fishermen" (*The Vancouver Sun*, June 19, 1984).

New Fisheries Minister Herb Breau met with MAC on August 10, 1984. He agreed that there should be no short term action on a comprehensive system of individual allocations in the salmon fishery. However, he requested the Department and the industry to examine the feasibility of testing individual quotas in a mutually agreed salmon fishery.

In September, 1984, the Liberal Government was defeated and a majority Conservative Government swept to power. The previous government's proposals to restructure the Pacific fisheries disappeared with their fall from power. The new Government launched its own consultations with the industry on how to address the Pacific fishery. The government considered funding a voluntary buy-back program but this never materialized. The concept of individual fishing allocations was dropped because of opposition by the major sectors of the industry.

Demands for fleet rationalization and a Buy-Back Program diminished with the resurgence of the Pacific salmon fishery during the mid to late 1980's. The catch and landed value of the salmon fishery in 1986 was the highest since 1950. The catch at more than 100 000 tons was up from lows of 50 000 and 55 000 tons in 1984 and 1985 respectively. The landed value more than doubled between 1983 and 1986. Landings remained high through the latter half of the 1980's. This relative prosperity masked the underlying chronic problem of overcapacity.

Meanwhile the major thrust of the Pearse Report, namely, his proposals for restructuring the commercial licensing system, had sunk in a sea of opposition. Government and industry had failed to grasp the nettle of opportunity and to agree on viable alternatives to reduce an overcapitalized fleet. The opposition to Pearse's licensing proposals tended to overwhelm any positive reaction to other aspects of his report.

Being the result of a Commission of Inquiry, the Pearse report was not considered by Cabinet prior to its release. Thus the government response process was considerably different from that of the Kirby report. The various players in the industry staked out their positions on Pearse's recommendations and lobbied government before any decisions were taken or announced. The result was that the only recommendations which were formally accepted by government were those announced by Minister De Bané on February 18, 1983, following his meeting with MAC.

In announcing the government's response to the Kirby report, Minister De Bané explained the differences:

> "The process being followed in the case of the Pearse report is, of course, different from that being followed in the case of the Kirby report. The government did not see the recommendations of Dr. Pearse until his report was made public. This is the usual process followed in the case of a Commission of Inquiry. In the case of the Kirby report, because the Task Force was established to report directly to a committee of Ministers on an ongoing basis, Ministers had a chance to learn about the results of the Task Force work as it was

being done. This is the reason why we have been able to respond more quickly to the Kirby report than to the Pearse report" (DFO 1983a).

### Overall Impact of the Kirby and Pearse Reports

In the early 1980's the Kirby and Pearse reports were seen as blueprints for the reform of fisheries policy on the Atlantic and Pacific coasts. There were, however, considerable differences in the nature and scope of the investigations, the nature of the reports and their recommendations, the industry and government response to each, and their overall impact on fisheries policy.

### Nature and Scope of the Investigations

The Task Force on Atlantic Fisheries, under Michael Kirby, was a multidisciplinary, multi-sector group, with members drawn from the central agencies of the federal government, the line department responsible for fisheries management (DFO), a former Canadian Labour Congress Official, a former Vice-President of one of the insolvent offshore companies and a senior official from the Newfoundland Fishermen's Union. While the Chairman was a powerful individual well-connected at the political and bureaucratic levels of the federal government, the Task Force Report was a collective effort and generally represented a consensus of sometimes divergent views.

The Task Force took a multidisciplinary, integrated, systems view of the fishery from the fish in the water to food on the plate of the consumer. It dealt extensively with a wide range of harvesting, processing and marketing issues. Its primary focus was on the Atlantic groundfish fishery but it dealt peripherally with the Atlantic herring purse seine fishery.

Peter Pearse, on the other hand, headed a one-man Commission of Inquiry into the Pacific fishery. Although he conducted public hearings and received numerous formal submissions, his report was really one person's view of the problems, opportunities and changes required in the Pacific fisheries. While his examination of the Pacific fishery dealt with a broad range of issues, it focused almost exclusively on the harvesting of fish. It touched only in a perfunctory manner on processing and marketing issues. On the other hand, Pearse dealt with policy issues of major significance on the Pacific coast which were not even mentioned in the Report of the Task Force on Atlantic Fisheries. These included such matters as habitat management, salmonid enhancement and the Native fisheries.

Another significant difference between the work of the Task Force on Atlantic Fisheries and the Pearse Inquiry was that the Task Force did not attempt to examine the administration of the fishery in terms of the effectiveness of the Department of Fisheries and Oceans and its bureaucratic structure. Pearse, on the other hand, spent almost as much time and paper on administrative and technical matters as he did on the major policy issues.

Both Kirby and Pearse consulted extensively during the course of their investigations. The consultations by the Task Force were semi-formal while Pearse conducted a formal, quasi-judicial Commission of Inquiry. Both Chairmen produced interim documents for public feedback. The TFAF document was a summary of issues and options, with no indication of the Task Force's preferences. Pearse produced a full-fledged Interim Report with interim recommendations and the rationale for these.

### The Reports and Their Recommendations

The TFAF commissioned a number of studies, the results of which are presented in its report. The information presented on the economic condition of processors, the financial situation of fishermen and the nature of and prospects in Canada's Atlantic groundfish markets was thorough and of considerable benefit to later investigators and commentators. This established a factual base from which the TFAF proceeded to identify key policy issues. Each issue was examined, options for future policy identified, the pros and cons of these options discussed and then a recommendation made as to which policy option the government should pursue. These options were examined in the context of the three objectives for the Atlantic fishery proposed by the Task Force.

Pearse presented some factual background on particular subjects in his report. His approach was not as comprehensive as that taken by the Task Force. He identified policy issues and recommended approaches but did not really present alternative options with a discussion of the pros and cons of each. While he proposed a set of objectives for Pacific fisheries policy, he did not relate his recommendations specifically to those objectives.

The Task Force confined itself to 57 recommendations on major policy issues concerning the Atlantic fishery. These recommendations were couched in terms of policy direction and did not attempt, except in rare instances, to prescribe how these recommendations should be implemented. It was relatively easy for the fishing industry and for government to focus on distinct policy issues and clear-cut recommendations for future policy direction.

Pearse produced a flurry of 292 recommendations, ranging from major policy changes to details of how many staff DFO should have for a particular function in a particular area. Major policy and minor administrative and procedural recommendations were mixed together throughout the report. Thus it was very difficult for the Department, the fishing industry, and the political process to grasp Pearse's recommendations in a comprehensive fashion. In many respects, Pearse's report failed to address issues at a consistent level throughout. It was difficult to cut through the underbrush to see the thrust of Pearse's thinking and approach.

There was some similarity in the conclusions and recommendations. Both reports recommended major changes in fisheries policy. Both Kirby and Pearse concluded that the common property aspect of the fishery was a fundamental problem leading to overinvestment and overcapacity. Both proposed quasi-property individual quotas as the preferred solution to the overcapacity in the harvesting sector. The TFAF confined its recommendations to the broad policy direction, leaving the details to DFO to implement. Pearse not only prescribed individual quotas for all fisheries other than salmon and roe-herring, he also spelled out in considerable detail exactly how the individual quota approach should be implemented, fishery by fishery.

Both Kirby and Pearse called for quasi-judicial licensing agencies separate from DFO and improved appeal processes. They also called for improved consultative processes, including broad fisheries policy consultative groups on each coast. Kirby dealt extensively with the need for quality improvement and made very specific proposals to achieve this. Pearse endorsed quality grading but did not give much attention to this matter.

Kirby and Pearse agreed in opposing general programs of financial assistance to the fishing industry. Kirby, however, became involved, as Chairman of the Federal Negotiating Team on restructuring, in promoting major government involvement in the Atlantic fishery through equity in the restructured offshore companies. This ran counter to the major thrust of Pearse's report. On the other hand, Kirby opposed government financial involvement in buy-back of excess fishing capacity. While he proposed reducing capacity in the herring purse seine fleet through transferable individual quotas and a buy-back, the buy-back was to be entirely industry financed. Pearse, in contrast, concluded that the government had contributed to the problem of overcapacity by issuing too many licences and hence had a responsibility to foot at least part of the bill for fleet reduction. This would be achieved through a voluntary buy-back program in the salmon and roe-herring fisheries.

Kirby dealt extensively with the question of marketing. While rejecting a role for government in the purchase and sale of fish, he proposed an Atlantic Fisheries Marketing Commission, with some government participation. Through this, the industry could better coordinate its marketing efforts. He also proposed a major generic promotion campaign, largely financed by government, to increase per capita consumption of fish in North America. Pearse, on the other hand, had little to say about marketing except that government should not get involved.

Pearse proposed major new directions for policy on habitat management, recreational fisheries, and Indian fisheries. Kirby was silent on these matters. These differences reflected the perceived relative importance of these issues on the Atlantic and Pacific coasts in the early 1980's.

Pearse proposed a variety of changes in administration of Pacific fisheries, including the organizational structure of the Department and personnel issues. Kirby was virtually silent on such matters.

The Kirby report had a significant impact on Atlantic fisheries policy in the 1980's. The Task Force objectives were generally accepted as guidelines to evaluate specific policies. The emphasis on an economically viable industry as a first priority helped to focus attention on the inherent problems of the fishery as the employer of last resort. The economic versus social debate re-emerged in the late 1980's — early 1990's as companies decided to close some fish plants to better match processing capacity with the available resource. At the same time the social dimension loomed large as the affected communities struggled to save essentially one-industry towns. The federal government initially set the Total Allowable Catch for some stocks higher than it otherwise might have in order to minimize the social and economic impact of quota reductions upon the industry and coastal communities.

Kirby's proposals to revise international fisheries policy set the scene for new policies to reflect the emerging realities of declining surpluses and foreign overfishing outside 200 miles.

The major impact of the Kirby report was probably the momentum it generated to introduce individual quotas in many fisheries during the 1980's. For details, see Parsons (1993). The theoretical precepts were not followed in many instances. Nonetheless, these initiatives set the stage for potential rationalization of harvesting capacity in the 1990's.

Kirby's recommendations on a resource short plant program and his proposal for allocation of the growth of northern cod in the short-term improved the economics of certain seasonal plants and helped moderate the previously strident debate over fish allocations.

His recommendations on quality improvement were only partially implemented. At the beginning of the 1990's, the jury was still out on whether the modified approach of in-plant quality management programs would ensure consistent, high quality products.

Kirby's recommendations on the port market and improving fish plant profitability appear to have had only a marginal impact on industry operations. The government's rejection of his marketing recommendations in effect maintained the status quo.

Regarding fishermen's incomes, his recommendations to amend the Fishermen's UI program were adopted. These removed some of the productivity disincentives. His proposals for a gross income stabilization plan were investigated more thoroughly. Analysts in 1986 suggested such a program was unnecessary since the problem with fishermen's incomes was income inadequacy rather than instability. In 1992 Fisheries Minister John Crosbie established a Task Force on Incomes and Adjustment to develop proposals to address income inadequacy and instability issues in the Atlantic fisheries.

Kirby's proposal to reduce the herring purse seine fleet by introducing transferable individual quotas was implemented in modified form. At the end of the first five-year years it appeared that the impact was beneficial. The program was continued as originally planned (MacFarlane 1989).

His proposals to change the consultative process were carried out.

Two major deficiencies are worth noting. First, the Task Force report, apart from the UI amendments, did little to address the problem of inadequate incomes of fishermen. Second, by crystallizing projections that the TAC of northern cod would grow to 350 000 to 400 000 tons by 1987, the Task Force raised expectations which were shattered when this growth failed to materialize. This set the scene for another major crisis in the Atlantic groundfish industry beginning in 1989.

Overall, the recommendations of the Task Force, while they did not revolutionize the Atlantic fishery, assisted in its evolution to a potentially more viable industry in the long term. A market upturn in 1985 to 1988 rekindled the old euphoria. Overexpansion occurred again in some sectors. But the progress made in introducing individual quotas could assist ultimately with a better match between harvesting capacity and the resource provided the problem of cheating can be addressed effectively. Progress has been slower than the Task Force envisaged or government desired. Unless the federal and provincial governments take steps to encourage a reduction in harvesting and processing capacity we can expect crises similar to those of the past. The fault will lie not with the Task Force recommendations but rather with government and industry actions in the years following the Task Force Report.

The impact of the Pearse report in shaping subsequent fisheries policy is less clear. The 76 recommendations which were followed do not appear to have profoundly affected management of the Pacific fisheries. Industry and government rejected the major thrust of the report — restructuring of the commercial licensing system.

One of Pearse's colleagues, Dr. Patricia Marchak, Director of the Fish and Ships Research Project at UBC, summarized the situation in a letter to the Editor of *The Vancouver Sun* (February 26, 1983):

> "The Pearse report has been shelved for the excellent reason that it proposes a competitive bidding system that was absolutely unacceptable to the fishermen on this coast. As well, it was strongly opposed by a good many employees of the department of fisheries and oceans. No federal minister would have tried to implement a report so firmly opposed by the majority of affected groups."

The radical reforms Pearse proposed may have delayed progress towards individual quota systems in some fisheries such as halibut. Unlike the Atlantic coast, where innovative approaches to individual quotas were introduced in several fisheries during the 1980's, individual quotas on the Pacific were introduced only in three small-scale fisheries during the 1980's. Not until 1989 was there a move to extend this to other fisheries such as halibut and sablefish. This is ironic since Pearse recommended introducing freely transferable individual quotas in virtually all of the Pacific fisheries, except salmon and roe herring, whereas Kirby had suggested a more cautious approach.

The most significant impact of the Pearse report occurred in habitat management, recreational fisheries and native fisheries. Pearse influenced the development and content of new national habitat management and recreational fisheries policies. His ideas on the native fisheries were progressive. While not implemented at the time, they influenced subsequent government thinking as it struggled to develop fairer policies for the native fisheries. Many of his ideas are reflected in the new Aboriginal Fisheries Strategy announced in June 1992 (Parsons 1993).

Both Kirby and Pearse proposed the establishment of quasi-judicial licensing allocation agencies. While this idea was initially rejected, in October 1991 Fisheries Minister John Crosbie proposed the establishment of two agencies or Boards separate from DFO, responsible for licensing and allocation matters. On this matter, Kirby and Pearse were a decade ahead of public policy.

**Conclusions**

Why was the Kirby Report largely accepted by government and the Pearse Report not? Why did the Kirby recommendations play a significant role in shaping Atlantic fisheries policy in the 1980's and the Pearse Report only a minor one in shaping Pacific fisheries policy? Many answers have been suggested. I believe that this outcome resulted from the following combination of factors:

1. Kirby was a senior federal civil servant, Deputy Clerk of the Privy Council, who had the kind of access to Ministers enjoyed by few at any level in the Public Service of Canada. Although experienced in resource economics and author of a successful report that more or less set the policy for B.C.'s forest industry, Pearse was an outsider in the federal system.
2. Kirby drew around him a multidisciplinary team of senior government and industry officials. The interplay of ideas was encouraged. Trade-offs occurred between the theoretically desirable and what was considered to be feasible and workable. Pearse, on the other hand, worked in isolation. Although he drew upon other expertise, his report was that of one individual, one who already held very strong views on how the industry should be reformed. Pearse lacked the kind of feedback provided by the varied composition of the Atlantic Task Force.
3. Kirby promoted policy reform but he wanted this reform to be implemented, not to be shelved. The recommendations of the Atlantic Task Force called for policy change and specified the direction of change but did not attempt to spell out how this should be implemented. The Task Force recognized the need for further consultation. In most instances, it left the details of implementation for DFO or other departments. Pearse also called for policy change but he was perceived as imposing on the industry a set of changes which it opposed. Pearse also specified in great detail how these changes were to be implemented. During the process of developing these recommendations, there were no countervailing forces to balance the theoretical approach of the resource economist with more pragmatic considerations.
4. Kirby was an experienced power broker with influence at all bureaucratic and political levels, including the Prime Minister, by virtue of his position as Deputy Clerk of the Privy Council. He reported regularly to a committee of Ministers. His report was considered by Ministers and approved, with only one significant reservation, before its release.
   Pearse had no such expertise in influencing the Ottawa approval process. He cast his report upon the waters of public opinion where it sank in a sea of opposition.
5. While there was opposition to some parts of the Kirby Report in the Atlantic, the east coast industry was more scattered and not united in opposition to any specific proposals. On the Pacific coast, the industry was articulate and united in opposition to the major thrust of the Pearse Report.
6. Pearse, as an outside expert, could only recommend changes to government. He had nobody on his team that could actually work inside government to see that proposed changes were implemented. Kirby, on the other hand, brought a number of government people into the Task Force, and he himself brought the Task Force's views to the central agencies and to Cabinet.

Once the Kirby Report was released, many of the Task Force returned to their government jobs charged with implementing the report. One of the members, Dr. Arthur May, became Deputy Minister of Fisheries and Oceans and hence was in a pivotal position to see that the Task Force recommendations were carried out. Government officials' participation in the Task Force made them a part of the process. As Arthur May put it:

> "Cliff Levelton, a retired fisheries manager..., has a saying that almost amounts to Levelton's Law: 'Do not separate the analyzers from the implementers.' The Kirby process followed that rule. For a leader we had an outside observer, disinterested in the best sense, to hold up the mirror and show us what needed changing and in the group, we had people who could help carry changes through" (DFO 1983c).

Pearse did not follow that process. At the end of the day his report foundered on the shoals of industry opposition.

Minister De Bané, addressing the same FCC meeting on May 3, 1983, commented:

> "Even if these reports are the best ever, some of the hundreds written before must have

been good. Yet more problems kept arising, and more reports kept coming out, for over a century now. Why didn't they solve the problems? Why has much of the East Coast industry gone from crisis to subsidy to crisis, and the West Coast gone from boom to battleground? "My second basic question (is): why are opinions so split and conflicts so common? Why do I hear from individual companies different views that I hear from the Fisheries Council as a group? I am getting more messages than Marconi ever thought of when he stood on Signal Hill" (DFO 1983b).

Having questioned the built-in hostility in the industry, Minister De Bané queried: "Could it have been (created by) lack of sufficient shared information? In the end, the best value of the Kirby and Pearse reports may be educational. They set forth the information, and they prove that getting progress and profits out of common property requires a common understanding of the way we want to go."

That common understanding of "the way we want to go" has proven to be an elusive goal. The fishing industry has been characterized by a "we versus they" attitude for generations. Conflict and hostility have been dominant forces. Kirby managed to overcome it, to some extent, and hence his report had a significant influence on Atlantic fisheries policy in the 1980's. Pearse was less fortunate. Conflict and hostility to his major recommendations killed the major thrust of his report before it had a chance to make an impact.

The book is not yet finally closed on either of these reports. New ideas in the fishery can easily take a decade or more to percolate through the system. This is exemplified by the move to "arms length" licensing and allocation boards in the early 1990's, a decade after Kirby and Pearse proposed such an approach.

The Kirby and Pearse reports were not the first to examine fisheries policy and propose changes; neither were they the last. But in terms of scope they were among the most significant. While their impact on fisheries policy differed, the information they provided and the lessons learned from the two contrasting processes may be their most lasting contribution to the ongoing evolution of fisheries policy in Canada.

**Acknowledgements**

I would like to thank Greg Sheehy for his numerous constructive comments on various drafts of this manuscript. Dr. Arthur May also made some helpful suggestions.

**References**

ANON. 1982. Concluding Statement of the Conference: Turning the Tide: a discussion of a new policy for Canada's Pacific Commercial Fisheries. University of Victoria, Victoria, B.C. November 14, 1982.

BARRETT, L.G. AND A. DAVIS. 1984. Floundering in troubled waters: the political economy of the Atlantic fishery and the Task Force on Atlantic fisheries. J. Can. Stud. 19: 125–137.

CROWLEY, R.W., B. MACEACHERN AND R. JASPERE. 1993. A review of federal assistance to the Canadian fishing industry 1945–1990, p. 339–367. *In* L. S. Parsons and W. H. Lear [eds.] Perspectives on Canadian marine fisheries management. Can. Bull. Fish. Aquat. Sci. 226.

CRUIKSHANK, D. 1982. Letter dated September 20, 1982 to the Hon. Roméo LeBlanc, Minister, Department of Fisheries and Oceans.

CRUIKSHANK, D. 1983. Letter dated January 10, 1983 to Hon. Pierre DeBané, Minister, Department of Fisheries and Oceans.

DFO. 1981. Inquiry into Pacific Coast fishing industry announced. NR-HQ-081-001E. January 13, 1981.

1983a. Response of the Government of Canada to the Report of the Task Force Study on the Atlantic Fishery. News Release. Ottawa, February 17, 1983.

1983b. Notes for a speech by the Hon. Pierre DeBané, Minister of Fisheries and Oceans, to the Annual Conference of the Fisheries Council of Canada. St. John's, Newfoundland, May 3, 1983.

1983c. Notes for an address by Dr. A.W. May, Deputy Minister of Fisheries and Oceans, to the Annual Conference of the Fisheries Council of Canada. St. John's, Newfoundland, May 2, 1983.

1984a. Progress Report on Reforms in the Atlantic Fisheries. Department of Fisheries and Oceans, Ottawa. 13 p.

1984b. Joint Statement on the New Policy for the Pacific Fisheries by the Hon. Pierre DeBané Minister of Fisheries and Oceans, and the Hon. Jack Austin Minister of State for Social Development. Vancouver, B.C. SP-HQ- 84-014E. June 18, 1984.

1986. Challenges Facing the Fishery Sector: A Report to the Annual Conference of First Ministers, Technical Appendices. 156 p.

KIRBY, M.J.L. [CHAIRMAN]. 1982. Navigating troubled waters — a new policy for the Atlantic fisheries. Report of the Task Force on Atlantic Fisheries. Supply and Services, Ottawa. 379 p.

LARKIN, P.A. 1973. Play it again Sam — an essay on pacific salmon enhancement. J. Fish. Res. Board Can. 31(8): 1434–1456.

LEAR, W.H. AND L.S. PARSONS. 1993. History and management of the fishery for northern cod in NAFO Divisions 2J, 3K and 3L, p. 55–89. *In* L.S. Parsons and W.H. Lear [eds.] Perspectives on Canadian marine fisheries management. Can. Bull. Fish. Aquat. Sci. 226.

MAC. 1983. A First Step Towards Rationalizing the West Coast Fisheries. Discussion paper prepared by the Minister's Advisory Council. September 28, 1983.

MACFARLANE, D.A. 1989. An economic assessment of the herring purse-seine fleet restructuring plan of 1983: the ten year management plan at mid-term. DFO Econ. Commer. Anal. Rep. 31: 49 p.

PARSONS, L.S. 1993. Management of marine fisheries in Canada. Can. Bull. Fish. Aquat. Sci. 225. 784 p.

PEARSE, P.H. [COMMISSIONER]. 1982. Turning the tide: A New Policy for Canada's Pacific Fisheries. The Commission on Pacific Fisheries Policy. Final Report. 292 p.

# Aquaculture in Canada

D.E. Aiken

*Department of Fisheries and Oceans*
*Biological Sciences Branch, Biological Station*
*St. Andrews, New Brunswick, Canada E0G 2X0*

Aiken, D. E. 1993. Aquaculture in Canada, p. 411–446. *In* L. S. Parsons and W. H. Lear [eds.] Perspectives on Canadian marine fisheries management. Can. Bull. Fish. Aquat. Sci. 226.

During the 1980's, Canadian aquaculture changed from a hobby activity to a producer of high value seafoods capable of generating more than a hundred million dollars for the Canadian economy. That dramatic growth was stimulated by the success of the east and west coast salmon industries — which expanded production a hundred-fold to a record 22 500 tonnes in 1990 — and by the blue mussel culture industry of the east coast. Led by Prince Edward Island, production of blue mussels from east coast farms increased 5-fold during the 1980's to more than 3000 tonnes. The balance of Canadian aquaculture production at the end of the 1980's consisted of Atlantic and Pacific oyster (roughly 2300 and 3900 tonnes, respectively in 1989), and trout (just over 3000 tonnes from freshwater and marine culture).

The Canadian aquaculture industry is maturing. The early preoccupation with financing, biology and growout technology has changed to a focus on production efficiency, product development and marketing. The long-term economic health and global competitiveness of the Canadian industry will depend on how successfully these issues are addressed.

---

Au cours des années 1980, l'aquiculture canadienne s'est transformée d'un passe-temps en une source de produits marins de grande qualité qui infuse plus de 100 millions de dollars dans l'industrie du pays. Cette croissance phénoménale a été stimulée par le succès des industries salmonicoles des côtes est et ouest du Canada, dont la production a centuplé pour atteindre 22 500 tonnes en 1990, et l'industrie mytilicole de la côte est. Avec l'Île-du-Prince-Édouard en tête, la production de moule bleue a quintuplé pendant les années 1980 pour atteindre 3 000 tonnes, la production aquicole canadienne se complétait de l'huître américaine et de l'huître du Pacifique (environ 2 300 tonnes et 3 900 tonnes, respectivement en 1989) et de la truite (un peu plus de 3 000 tonnes produites en eau douce et en eau salée).

L'industrie aquicole canadienne mûrit. Les préoccupations du début liées au financement, aux facteurs biologiques et aux méthodes d'engraissement et d'affinage ont fait place à l'efficacité de la production, la mise au point de nouveaux produits et la commercialisation. La santé économique et la compétitivité globale à long terme de l'industrie canadienne dépendront de la résolution efficace de ces questions.

## Introduction

Aquaculture, simply defined, is the husbandry of aquatic animals and plants. In general use it can include rearing to produce food for human consumption (comestible aquaculture), to enhance wild stocks (enhancement aquaculture), and to stock ponds for fee-fishing (recreational aquaculture). It can also include production of aquatic organisms for personal use (hobby aquaculture), including the aquarium or ornamental fish trade, and even the rearing of aquatic plants and animals to obtain chemicals used in pharmacology and pearls and other items that are used in the manufacture of jewelry.

Not everyone accepts this liberal interpretation of the term aquaculture. Extremists reject the many forms of aquatic husbandry that produce something other than a product for the table, including the cultivation of pearls and aquarium fish, and FAO requires that growth or quality be enhanced to some degree through human intervention in the rearing process, thereby excluding the storage and shipment of lobsters and other live aquatic organisms. FAO also requires individual or corporate ownership of the cultivated stock (thereby excluding enhancement aquaculture). These varying opinions of what constitutes "aquaculture" should be borne in mind when interpreting aquaculture statistics.

Aquaculture activities are also loosely classified according to the intensity of husbandry that is practiced. *Extensive* aquaculture involves a minimum of human intervention: no water treatment, no feeding, little or no confinement. Sea ranching of salmon, which involves release of hatchery-reared juveniles to the natural environment, is extensive aquaculture. At the other extreme is *intensive* aquaculture, where the animal's environment and well-being are controlled by the farmer. The rearing of salmon in sea cages is intensive aquaculture. *Semi-intensive* denotes something between these two extremes, although the lines of demarcation are blurred.

Most aquaculture systems developed for new culture species pass through three stages on the way to commercial viability: a research and development stage involving heavy investment in the development of new culture systems and strategies; a pilot scale stage during which new systems are tested on a commercial scale; and a commercial stage in which standardized systems are used and most major biological and technical problems are controlled

Commercialization of the industry also passes through three stages. Hunter (1989) suggested the terms *pioneer*, *entrepreneur* and *farmer* to describe these. In the pioneer phase the intent is to develop systems and strategies for culturing the species, so there is heavy investment with no appreciable return. However, that investment can pay off handsomely if the species proves to have commercial potential because during the succeeding (entrepreneur) phase the market is wide open, competition is minimal and there is a large profit margin. The lure of significant profits attracts competition and production begins to rise, eventually putting downward pressure on prices, reducing the profit margin and forcing growers to become increasingly more efficient in order to compete. At this point the industry is in what Hunter called the farming phase, characterized by reduced risk, increased stability and a slim profit margin. Hunter's terminology is useful for rapidly expanding industries such as those for salmon and blue mussel in Canada, but less applicable to those like the Canadian oyster industries in which development has been more erratic.

Early Canadians, blessed with productive marine and freshwater environments and suffering no shortage of either aquatic or terrestrial animal protein, had little incentive to grow their aquatic food. It was there to be harvested, and both the commercial and recreational fisheries developed to exploit the wild bounty. Ironically, it was through this harvest of wild species that Canadians first developed expertise in rearing salmonids (salmon, trout and charr species). Hatcheries were built to augment wild stocks of salmonids, and the resulting technology proved invaluable when private sector salmonid culture emerged many years later.

Brook trout and Atlantic salmon were reared in Québec hatcheries as early as 1857, and a trans-Canada network of federal hatcheries was subsequently established. Most of these were for rearing salmonids, but there were notable exceptions. The Dildo hatchery on Newfoundland's Trinity Bay was said to be the largest codfish hatchery in the world. It had a remarkable 9-year history starting in 1889. In addition to 200 million cod fry, it also produced salmon, capelin, eels, carp and lobster. In 1893 that one hatchery stocked 500 million lobster larvae as part of a 4-year effort that contributed nearly two billion young lobsters to Newfoundland waters. Unfortunately, as hatchery proponents eventually acknowledged, the success of an enhancement project depends less on the numbers stocked than on the numbers that survive, and there were indications that few of the billions released by some of these hatcheries survived to harvestable size.

## Global Perspective

The aquaculture industry makes an important contribution to the Canadian economy, but total production is small by world standards. According to the latest official FAO statistics (Nash and Kensler 1990), 20 countries dominate world aquaculture production, and 12 of these are Asian. Canada does not produce even 1% of the total world harvest in any of the major species groups (finfish, crustaceans, molluscs and seaweeds), and it is not among the top 16 countries in value of aquaculture commodities, even though Canadian aquaculture concentrates on high value species. Canada probably ranks about fiftieth in total aquaculture production, with roughly 18 000 tonnes worth US$109 million (Anon. 1990), of the US$20-25 billion value of world production (Fig. 1).

In short, compared with other industrialized nations with comparable climatic conditions, Canada's aquaculture industry started late and developed slowly, primarily because of the historical abundance of animal protein from both terrestrial and aquatic sources. Canadians have benefited from some of the most productive recreational and capture fisheries in the world, and the capture fishery, in spite of increasing problems, has allowed Canada to be one of the few net exporters of fisheries products in the West.

Finally, as Boghen (1989) pointed out, aquatic animals and plants are a relatively unimportant part of the Canadian diet. Annual per capita consumption of these products by Canadians (7 kg) is only a fifth of that in Japan, a fact that is reflected in an annual per capita aquaculture production rate (200 g) that is less than a third of the world average. The combined influence of these factors on aquaculture development was neatly summarized by Boghen (1989):

> For most nations where aquaculture is highly developed there is a significant direct correlation between consumption and production and very often an inverse relationship between the availability of arable land and the dependency of the population on fish products as part of its natural diet.

But this situation is changing in Canada. Awareness of the potential social and economic benefits from aquaculture has prompted increasing involvement at the various levels of government, and awareness of the financial rewards associated with a productive aquaculture industry has captured the attention of private investors. In a decade, the Canadian salmon farming industry has become the fourth largest in the world.

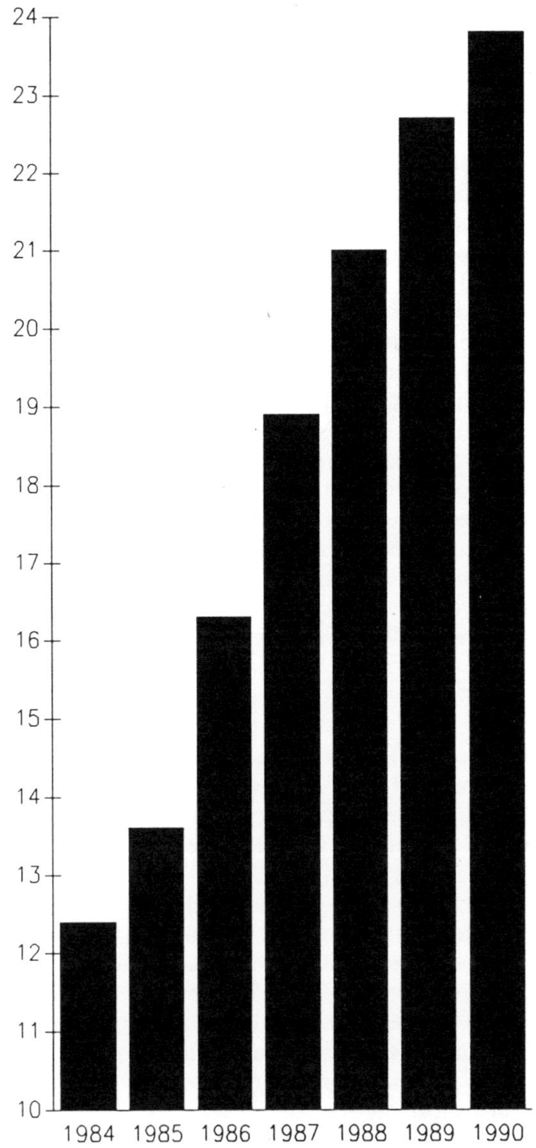


FIG. 1. Value of world aquaculture production, in billions of US dollars (from Nash and Kensler 1990; values for 1988–90 are projections).

## History

According to MacCrimmon et al. (1974), freshwater aquaculture started in Canada in 1857 when Richard Nettle — first Superintendent of Fisheries for Lower Canada — incubated and hatched brook trout and Atlantic salmon in Québec City (Dunfield 1985). Nine years later, Samuel Wilmot hatched 15 thousand salmon in Ontario, launching Canada's long history in enhancement aquaculture. Production from the Newcastle Hatchery in Ontario was described in a Special Report for 1868 as being adequate to "...convince the most skeptical of the reality of fish culture." Private aquaculture began about the same time. John Holliday established the first private hatchery in Québec in 1859 — a salmon spawning bed on the Moisie River — and by 1870 David Brown of Galt, Ontario, was working with a broodstock of some 10 thousand brook trout (Lafleur 1987; MacCrimmon et al. 1974).

This early interest in enhancement aquaculture generated a network of federal and provincial hatcheries that eventually spanned the country, and by 1950 these government hatcheries were producing approximately 757 million freshwater and anadromous fish annually. By 1972, annual hatchery production had dropped to 370 million fish, mostly salmonids (Boghen 1989), as a result of hatchery closures and implementation of a strategy to rear juveniles to larger sizes before release, thereby increasing survival rates at sea.

One of the interesting results of those years of enhancement aquaculture was the introduction of a large number of exotics. Government efforts to enhance local and regional fisheries resulted in the introduction of alien species to all the provinces of central Canada. In Ontario alone 10 species were introduced, starting with Pacific salmon in 1874. Today smallmouth bass and carp are locally abundant in the four central provinces but the primary exotics are salmonids, especially the rainbow trout (MacCrimmon 1984).

## The Modern Era

In the early 1980's Canadian private sector aquaculture changed from a novelty to a vocation, from a hobby to a revenue producing activity. Ken Lucas (1984) attributed this to five related factors:

- Advances in science and technology;
- A steady rise in world demand for fish products;
- Increasing population pressure on world protein supplies;
- The approach of a maximum sustainable yield from the world's capture fisheries; and
- The adoption of a 200 mile zone of national fisheries jurisdiction.

Lucas felt that four factors enabled Canadian aquaculture to develop rapidly during this period:

a) an abundance of clean fresh and salt water;
b) scientific expertise in fisheries management, animal husbandry and fisheries economics — disciplines that are important in aquaculture;
c) long-standing interest in sport fishing and its associated aquaculture activities; and
d) abundant domestic supplies of the raw materials required by aquaculturists for fish feed.

For practical purposes the "modern era" of Canadian aquaculture began with the National Aquaculture Conference held in St. Andrews, New Brunswick, in July of 1983. Jointly sponsored by the federal Department of Fisheries and Oceans and the Science Council of Canada, the conference was convened because it was apparent that aquaculture was coming of age in Canada, that it would soon establish an identity separate from the recreational and harvest fisheries that had dominated it for a century, and that it had the potential to become a dynamic growth sector in its own right.

The switch from aquaculture as a novelty to aquaculture as an industry was more complicated than might be supposed. The existing social and legal fabric had developed around the capture and recreational fisheries, and as a result aquaculturists in one province found they had no statutory rights to the fish they grew in the ocean, while those in another province were prohibited from selling the "game fish" they grew in their raceways and ponds. In the marine environment the federal government controlled the water column and the provincial government controlled key areas of the sea bed, creating overlapping jurisdictions and raising perplexing questions concerning lease rights and property ownership in the marine environment. Years would pass before the historical impediments to aquaculture development were removed by a combination of legislative action and memoranda of understanding (MOUs) between the federal and provincial governments. By then a new series of impediments had arisen under the banner of "multiuser conflict," the competition between aquaculture operations and other interest groups that use water resources.

Beyond the legal complexities, about which books have been written (see Wildsmith 1982), the emergence of aquaculture as a commercial industry required a change in attitude, an awareness that revenue could be generated directly by producing a marketable product rather than indirectly by producing fingerlings to enhance recreational or capture fisheries. The seeds of that awareness had been around for some time: in 1969 the now infamous Sea Pool Fisheries established a land-based salmonid culture facility near Clam Bay on the eastern shore of Nova Scotia for the purpose of producing luxury seafood. Unfortunately, the company was soon bankrupt, a victim of bad luck, poor planning and inept management.

Failures such as this are inevitable in a new industry, which must work its way through a succession of problems on the way to maturity and stability. The problems in the aquaculture industry were predominantly biological and technical in the beginning, but later they became more economically, socially and politically oriented. Canadian producers of salmonids, mussels and oysters have successfully dealt with the most serious of

FIG. 2. Most salmon farmed in Canada are grown in marine cages, but in 1988, this land-based facility on Quebec's Baie des Chaleurs produced 35 tonnes of Atlantic salmon. (Photo: Bill Boudreau).

their technical and biological problems and are now addressing the social, economic and political issues.

### The Salmonid Industry

The principal salmonids cultivated in Canada are Atlantic salmon (*Salmo salar*), Pacific salmon — primarily chinook (*Oncorhynchus tshawytscha*) and coho (*O. kisutch*) — rainbow trout (*O. mykiss*),[1] brook trout (*Salvelinus fontinalis*) and Arctic charr (*S. alpinus*). Adult salmon are grown to market size primarily in sea cages, although a land-based salmon farm on Québec's Baie des Chaleurs produced 35 tonnes in 1988 (Saunders 1989), and has the capacity to produce 500 tonnes (Fig. 2).

Trout are grown in both the marine and freshwater environments. Rainbow trout is the most common species, being grown from coast to coast in everything from net pens and raceways to potholes and dugout ponds. Atlantic and Pacific salmon have been introduced to the opposing coasts over the years, but Pacific salmon have never been widely cultivated on the Atlantic coast of Canada. Atlantics, on the other hand, comprise an increasingly important part of the farmed salmon production in British Columbia.

Sea cage culture of Atlantic salmon was attempted off the coast of Nova Scotia in the 1970's, but it failed when the fish froze to death in late winter. A similar fate befell initial attempts to overwinter Atlantic salmon in cages in the Bay of Fundy, and it appeared that salmonid cage culture would not be feasible on the Atlantic coast. On the Pacific coast similar attempts with pan-size coho and chinook also failed due to a variety of technical and marketing difficulties. The early failures did little to convince the conservative Canadian financial establishment to fund the development of marine finfish culture in Canada.

Some members of the scientific community, on the other hand, were convinced that commercial marine salmonid aquaculture was not only possible but inevitable in Canada. Roly Brett, a federal research scientist at the Pacific Biological Station in Nanaimo, B.C., extolled the virtues of salmon farming to a dubious public. "It is my conviction," he wrote in 1974, "that Canada has a great potential for marine aquaculture, particularly salmon culture...." Within 15 years, Canadian salmon farmers were producing enough salmon to rank fourth in the world (Fig. 3,4).

On the Atlantic coast another federal research scientist, Arnold Sutterlin, brought federal, provincial and private interests together in a demonstration project to prove his contention that salmon could be overwintered in the marine environment if discretion was used in the selection of sites. In business, according to conventional wisdom, three things are essential for success: *location*, *location* and *location*. Sutterlin insisted the same was true for marine salmon farming in the western North Atlantic. Canada's first crop of cage-reared

[1]Rainbow trout is now classified in the same genus as Pacific salmon (*Oncorhynchus* spp.), but in this chapter it is grouped with the other trout species.

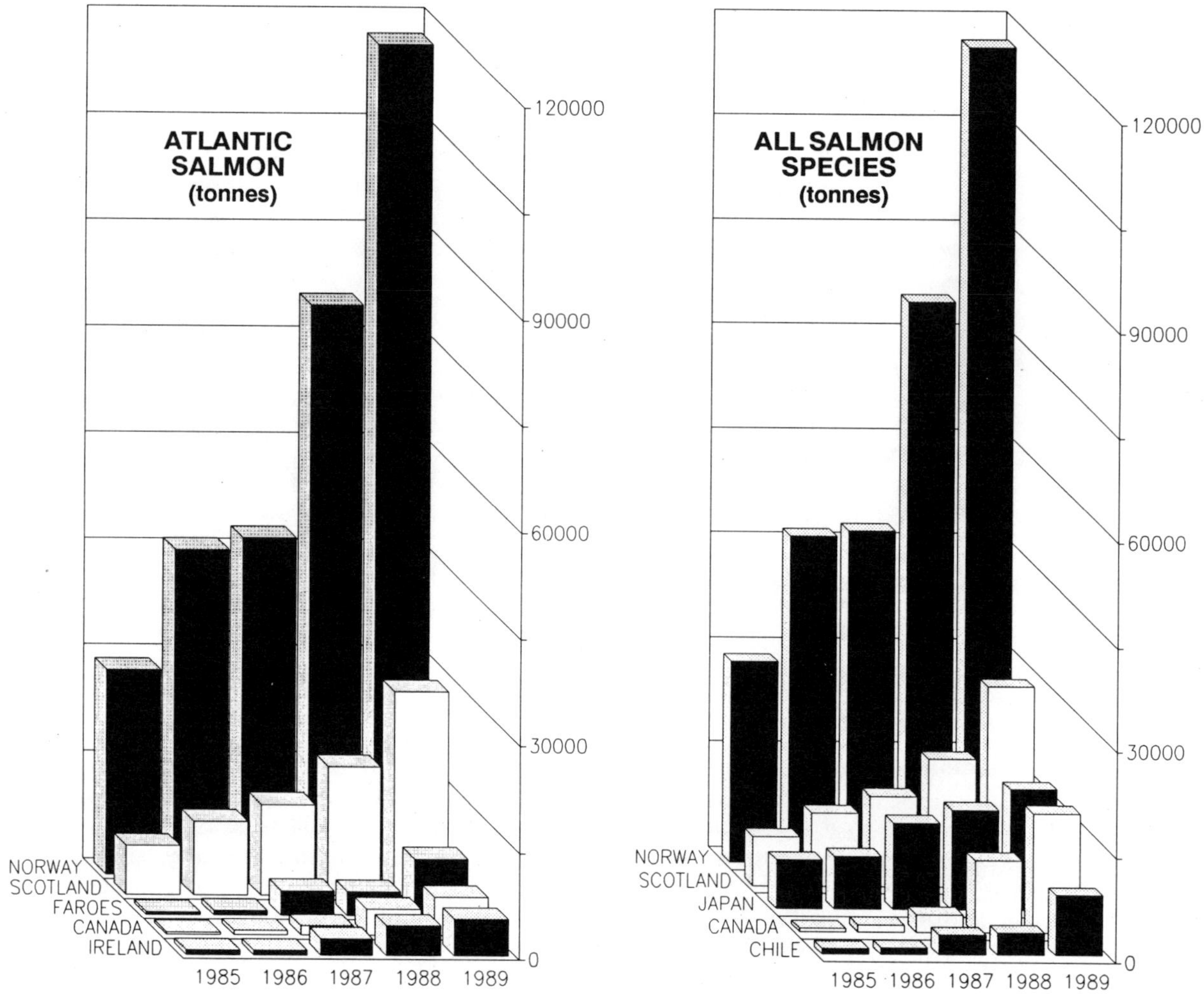


FIG. 3. Production of farmed salmon between 1985 and 1989 by the top five producer nations (B.C. Salmon Farmers Association).

Atlantic salmon was harvested in 1979 in the southwestern Bay of Fundy, and production from that relatively small corner of Canada has increased dramatically every year since.

### Trout

Trout — primarily the rainbow trout, brook trout and Arctic charr — are grown across Canada. Rainbow trout are grown in sea cages on both coasts because their shorter time to harvest (one sea year) generates cash flow and facilitates a more rapid response to changing markets. Marine rainbow are also a premium product in their own right, although this fact has not always assured success (Saunders 1989). Canadian production of trout in fresh water remained relatively stable at 500–700 tonnes per year until 1985, but has since expanded considerably (Fig. 5).

Canadian production of rainbow trout in 1990 was approximately 3150 tonnes, roughly equivalent to 12% of U.S. trout production and only 1% of the world total. Unlike salmon, which have considerable scope for expansion in western Canada, the rainbow trout industry seems unlikely to grow to any significant degree (Needham 1990).

Trout culture began in Prince Edward Island in the 1970's but has only matured in the last 5–7 years. Seasonal culture in marine cages is a recent innovation (winter temperatures in the southern Gulf of St. Lawrence are lethal for salmonids). Other recent innovations such as large high-tech land-based facilities have been less successful.

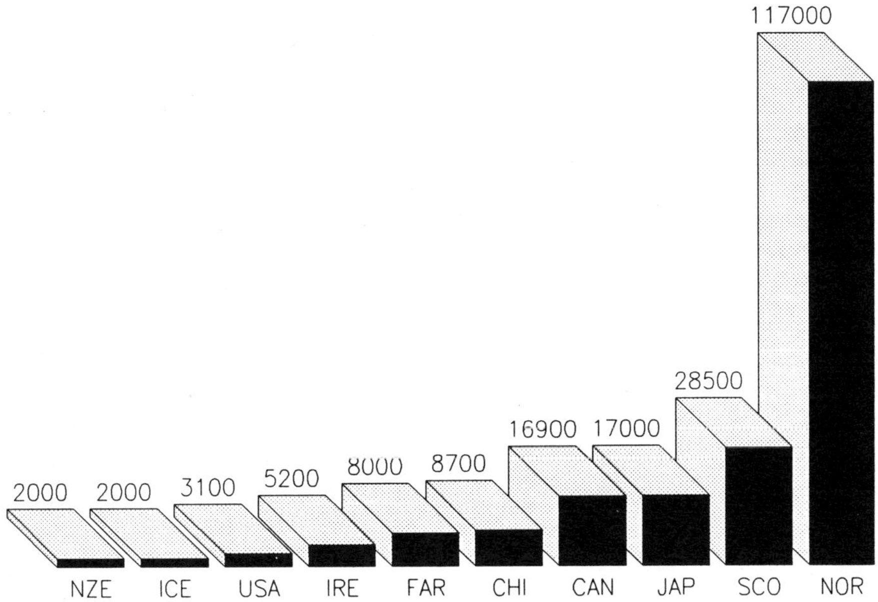


FIG. 4. Countries that produced at least 2000 t of salmon in 1989. In 1985, only Norway, Scotland and Japan exceeded 2000 t. New Zealand (NZE); Iceland (ICE); United States (USA); Ireland (IRE); Faroes (FAR); Chile (CHI); Canada (CAN); Japan (JAP); Scotland (SCO); Norway (NOR). Source: B.C. Salmon Farmers Association.

Trout are grown in Nova Scotia and New Brunswick in freshwater ponds and marine cages, and are produced for the comestible and recreational markets. As noted above, rainbows are often cultured in marine cages in conjunction with salmon, especially in southeastern New Brunswick. Marine cage operations in Nova Scotia are concentrated along the eastern shore and in Cape Breton. In both New Brunswick and Nova Scotia, unlike many areas, it is more lucrative to rear food for human consumption than to rear fish for fee fishing enterprises. In Nova Scotia in 1987 the recreational market was worth only C$21 thousand, whereas the comestible market had a farm gate value greater than C$1.2 million (Anon. 1988a).

Significant industries have also developed around trout culture in Québec and Ontario, and to a lesser extent in the Prairie Provinces. Brook (speckled) trout has been the dominant species cultured in Québec (80% of the 1350 tonnes produced in 1990), but most of this is targeted for recreation and enhancement. The remaining 20%, mostly rainbow trout, is cultured for the food market (200 tonnes in 1987). The latter is expected to be the growth segment of the industry over the next decade, and the Québec government feels that commercial trout production in the province could reach 1000 tonnes by 1995 (Anon. 1988a; Price Waterhouse 1990).

Nearly 80% of the trout produced in Canada in 1990 came from Ontario farms (Fig. 6). Most of the fresh product is sold within Ontario, but approximately 20% is exported to Québec and the northern United States (Anon. 1988a).

Ontario farmers have been raising fish commercially for over a quarter of a century, but the industry's most rapid growth has occurred since 1985. In 1987, when the 1500 tonne harvest was worth more than $7.5 million, roughly half of the 200 licensed operations in Ontario qualified as serious commercial enterprises. These ranged in size from small family farms to corporate entities with an annual production capacity of 250 000 fish. Another recent innovation is net pen culture of trout in Lake Huron. Although this is relatively common in other areas of the world, it has not been widely practiced in Canada.

The rainbow trout has long been the primary species cultivated in the Prairie Provinces. "Prairie pothole" culture began in Manitoba in 1968 in the natural pothole lakes and man-made "borrow-pits" that are common across the prairies (Fig. 7). Fingerlings were stocked in early spring and harvested in late autumn as 250–500 gram fish, most of which were retained for personal consumption by farmers who considered the fish a supplement to their winter diet. Extensive cultivation techniques were employed; supplementary feed was

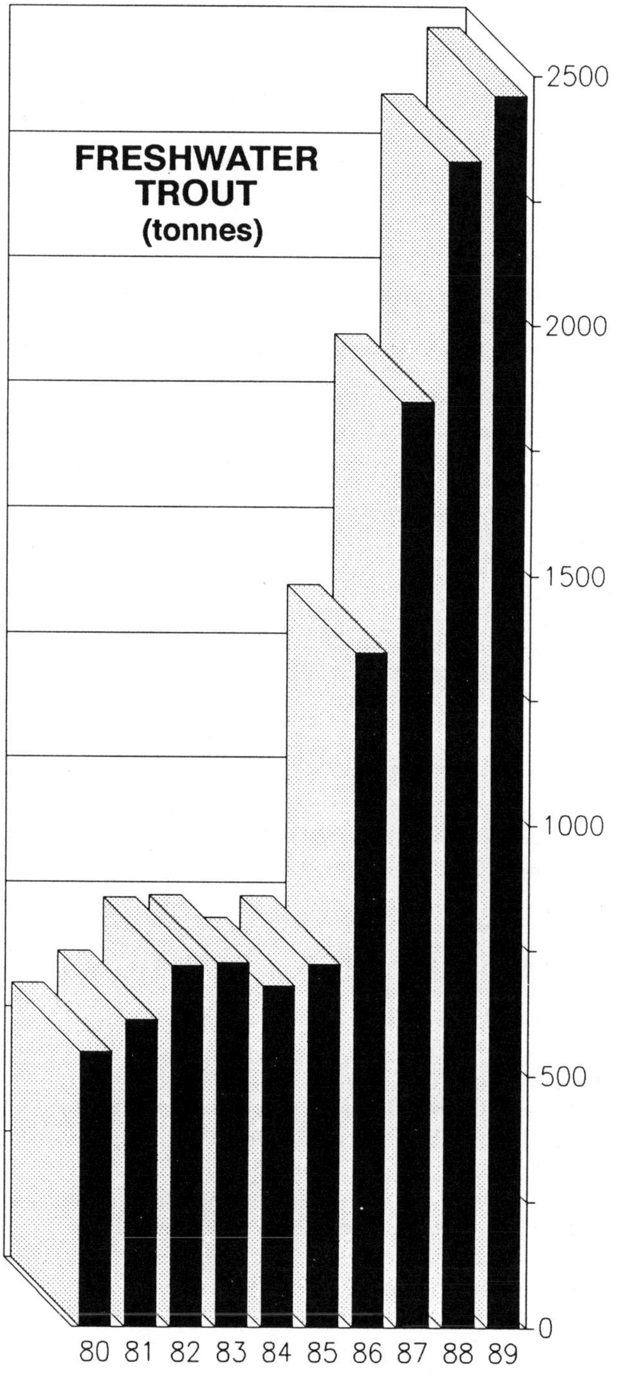


FIG. 5. Canadian production of trout in fresh water from 1980 through 1989. Production from the marine environment added another 584 tonnes in 1989 (Price Waterhouse 1990).

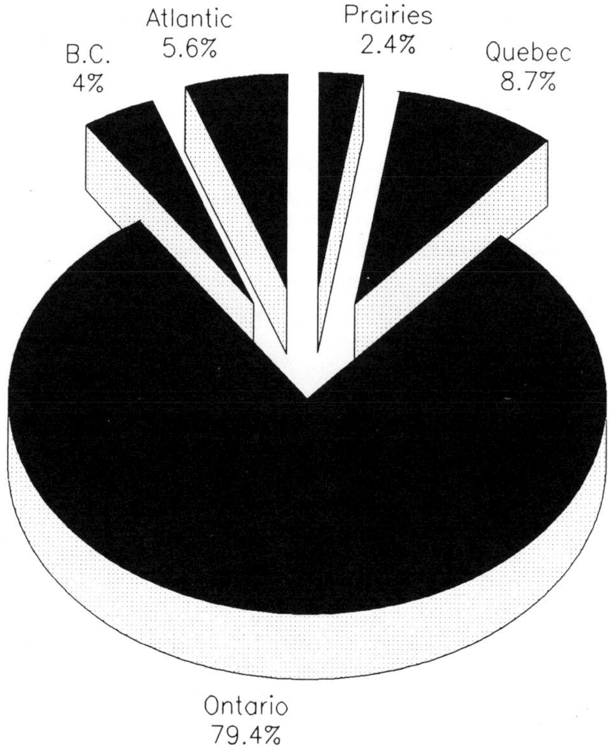


FIG. 6. Relative contribution of the various regions of Canada to the 3150 tonnes of trout produced in 1990 (Northern Aquaculture, Vol. 7 No. 3).

seldom provided and the trout were expected to subsist on the abundant natural invertebrate population. Pothole culture was an inexpensive but risky undertaking. Algal blooms are relatively common in these potholes and the oxygen depletion that accompanies their collapse may kill the trout. In addition, predatory birds take their toll and bacterial and algal growth can impart an undesirable flavor to the flesh.

In 1987 the Manitoba industry consisted of about 300 hobby farmers (cultivation for personal consumption) plus another 30 commercial operations. In 1989 the industry in Manitoba included one of Canada's first commercial producers of Arctic charr, a fish that is now being commercially cultured across Canada (18 farms produced roughly 100 tonnes in 1991). Apparently there is considerable variation in the performance of different charr stocks under culture conditions, and the shortage of broodstock of proven performance has delayed development of the charr industry (Painter 1990).

The industry in Saskatchewan and Alberta has been dominated by hobby farmers and private game fish growers, but in Saskatchewan there has been increasing interest in commercial production of fingerlings and pan-size rainbow trout. In 1987 only about 5% of the 2400 licenses issued in Saskatchewan went to commercial operations, and production from these farms varied from as few as 1000 to as many as 80 000 fish. Commercial growers in Saskatchewan have also begun experimenting with Arctic charr.

By any standards, aquaculture is a small industry on the Prairies. Hobby farming provides little employment in the area. Most commercial ventures produce fingerlings, operate seasonally and provide minimal direct employment. Commercial pond growout provides more economic benefits, both direct and indirect, because of the need for equipment service and repair as well as boats, motors and other equipment to work the ponds, feed and veterinary supplies to maintain the fish, and gill nets and other materials for harvesting, processing and shipping. As the commercial growout sector expands, so will the economic benefits to the area.

### Pacific Salmon

Commercial salmon farming began in British Columbia in the early 1970's with pan-sized coho and some chinook salmon, but most of those efforts eventually failed because of problems with broodstock, site selection, diets, and disease. Production grew rapidly from 1979 to 1982, when it peaked at 273 tonnes, but the subsequent failure of Apex Bio-Resources Ltd — a major producer of pan sized coho — reduced industry production to slightly over 100 tonnes the following year (Aiken 1986).

It turned out that important aspects of the biology of both coho and chinook were too poorly understood to support commercial production, and a long period of research and development was needed. By 1986 the fruits of some of this effort were becoming apparent, and between 1986 and 1989, production from B.C. salmonid farms increased from roughly 400 tonnes to more than 12 thousand tonnes (Aiken 1989a; Egan and Kenney 1990; Fig. 8), due to improvements in chinook husbandry, the introduction of Atlantic salmon (Fig. 9), and increased governmental and financial support. Coho production, on the other hand, remained relatively stable throughout this period (Egan & Kenney 1990).

Unfortunately for Canadian farmers, world production was also expanding dramatically during that period. In 1988 the top 10 producer countries — led by Norway — put close to 150 000 tonnes on the world market. An even greater production the following year caused the market to soften, triggering panic selling by farmers with a large financial exposure. This in turn caused prices to tumble to unprecedented lows, forcing many new farms into receivership.

In Canada the impact of this price collapse was most severe on the west coast. The Atlantic industry, having grown in conservative fashion, was able to withstand a brief period of instability. The British Columbia industry, on the other hand, had a less secure market position and greater financial exposure, a combination that resulted in numerous receiverships and a major restructuring in 1989 and 1990.

FIG. 7. In the "Prairie pothole" trout culture practiced in the late 1960's and early 1970's, fingerlings were stocked in small potholes (*top*) and "borrow pits" (*middle*) across the prairie provinces. Most production was for personal consumption, but some was marketed (*bottom*). (Photos: Province of Alberta).

### Atlantic Salmon

In October of 1986, representatives from 16 nations gathered in Biarritz, France and endorsed a resolution

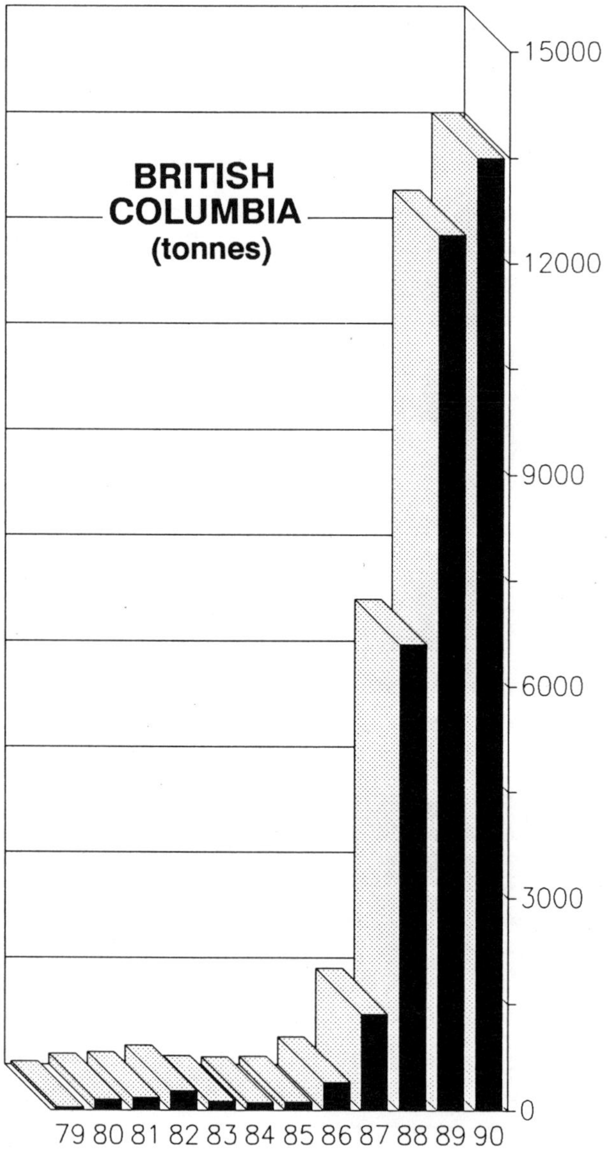


FIG. 8. Production of Pacific salmon by the B.C. salmon farming industry, 1979–1990 (Price Waterhouse 1990 and Northern Aquaculture Vol. 7 No. 3).

to reduce commercial fishing of Atlantic salmon. Canada had already banned the commercial fishery in the three Maritime Provinces and the Gaspé region of Québec. It had also shortened the season and revoked the licenses of all part-time salmon fishermen in Newfoundland. In 1987, $7.3 million was spent in a federal-provincial buyback program that removed nearly 80% of the 438 active commercial salmon licenses in the Maritime Provinces. This was a major step in the preservation of wild stocks of Atlantic salmon, a step which some people feel was made possible by salmon farming. In pre-farming days, the reasoning went, salmon were available to the general public only through angling and commercial fishing, so the commercial fishery was necessary in order to make this common property resource widely available to the public. However, the ready availability of farmed salmon eliminated the need for a commercial salmon fishery (Anderson 1988).

Commercial salmon farms are now scattered throughout the Atlantic Provinces wherever conditions are suitable, but the largest concentration and the focus of the industry is in the southwestern Bay of Fundy, in what has been termed the "Fundy Isles" region that extends from St. George to Grand Manan, New Brunswick. Here the bays, coves and channels created by the archipelago protect salmon cages from the ferocity of winter storms while providing access to the warmer water from the Bay of Fundy (Fig. 10).

Low winter water temperature is a major limiting factor in cage rearing of salmon in Atlantic Canada. In late winter the water in nearshore areas of the northwest Atlantic can cool to well below zero degrees Celsius. If the water temperature drops to roughly minus 0.7°C, the flesh of a salmon will freeze, causing instant death for the salmon and economic loss for the farmer. The deep water of the Bay of Fundy, on the other hand, seldom cools below 0°C, and the turbulent 8 metre tides in the area mix and warm the incoming water twice a day as it plunges through the Fundy Isles archipelago. By utilizing carefully selected sites within this narrow zone, salmon farmers can gain protection from winter gales and take advantage of favorable winter water temperature to produce some of the finest Atlantic salmon available in the world.

In 1979, the first harvest from Bay of Fundy salmon farms yielded 6.3 tonnes valued at US$39 thousand. Within 5 years the production had jumped to over 200 tonnes valued at over US$2 million (Fig. 11). More important, the price per kilogram was rising rapidly, an indication that the industry had passed from the pioneer phase into the entrepreneur phase, that period of rapidly expanding production, increasing market demand, and soaring profits.

Official fishery statistics for 1988 indicate that 62% of the total value from fishery landings in the Bay of Fundy came from salmonid aquaculture, and that proportion is expected to increase. In 1988 — the year those statistics were compiled — the landed value from salmon farming was roughly C$35 million. In 1990, the 8182 tonnes produced by the industry was worth some C$70 million.

FIG. 9. Canada's two major species of farmed salmon — the Atlantic salmon (*top*) and the chinook. The Atlantic — the more domesticated of the two — is now grown on both coasts, whereas the chinook is exclusively a Pacific species. (Photos: William Pennell [*top*]; B.C. Salmon Farmers Assoc.).

FIG. 10. The east coast salmon farming industry is concentrated in the southwestern Bay of Fundy, where high tides, deep water and numerous protected coves provide ideal conditions for marine sea cages. (Photos: Dave Aiken)

### *Economics*

Net-pen salmon farming is an expensive undertaking that requires significant amounts of working capital. Once the cages have been purchased, outfitted with nets and moorings and stocked with smolts, the new farmer faces 18 to 20 months of expenditures for fish feed, labor, insurance and disease management before the product is ready for market. New salmon farmers in the Bay of Fundy require upwards of three-quarters of a million dollars in working capital to operate through the second year.

An excellent economic assessment of the Bay of Fundy salmonid industry in 1987 was prepared by Fiander-Good Associates under contract to the Province of New Brunswick and the federal government. Their analysis indicated that a new salmon farmer starting with eight cages each stocked with 2500 smolts and expanding to 24 cages within 3 years could expect to lose more than C$315 000 the first year. However, under market conditions that existed in 1987–88, substantial earnings would accrue at the end of the second year and annual after-tax profits could be expected to stabilize around C$300 thousand by the fifth year.

These figures would, of course, be altered by subsequent changes in market conditions, and the impact of any change would reflect the sensitivity of different components in the culture system. For example, the economic viability of salmon farming in the Bay of Fundy is considered to be highly sensitive to changes in the selling price and mortality rate of fish, moderately sensitive to changes in feed costs, and relatively insensitive to changes in smolt cost. The global market squeeze precipitated by the Norwegians in 1989 affected selling price, one of the most sensitive elements of the culture system.

The economic benefits from a major industry extend well beyond the direct employment, capital investment and operating costs of the farms themselves. The salmon farming industry requires a wide range of products and services, and these are provided by individuals and small companies in the local communities. The Fundy salmon industry, for example, has fostered the development of three major satellite industries: smolt production, feed manufacture and equipment manufacture. Smolt supply was a severe restraint in the early days, but private hatcheries responded and in 1987 more than 900 thousand smolt were available from eight private, one non-commercial and two federal hatcheries (Fig. 12). Sea Farm Canada, with three hatcheries, employed more than 40 persons and paid more than C$750 thousand in wages. In the feed industry, 40 person-years of employment and a half-million dollars in wages were attributed to the salmon industry in 1987. The fabrication of cages and miscellaneous equipment (moorings, flotation, packing boxes) and the provision of marketing services added another 74 person-years and C$550 thousand in wages. In 1987 the ancillary industries and services employed 114 persons and paid nearly C$2 million in wages. By 1990 the number of full-time jobs attributable to the salmon aquaculture industry in the southwestern Bay of Fundy had risen to 500. In rural maritime areas that depend heavily on the troubled capture fisheries, salmon aquaculture is proving to be an important alternative.

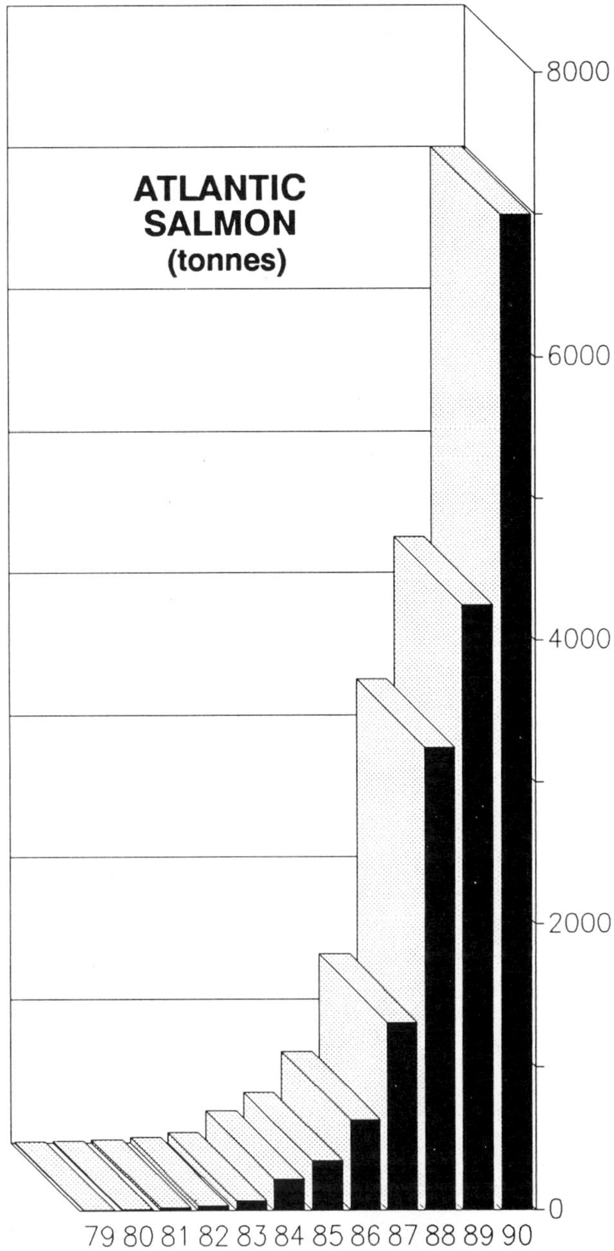


FIG. 11. Production of Atlantic salmon by the New Brunswick industry, 1979–1989 (Price Waterhouse 1990). Data for 1990 are preliminary.

## Future Prospects

Salmon culture is now a mature industry in Canada. Throughout most of the the 1980's the industry enjoyed

FIG. 12. The economic benefits of salmon farming extend well beyond the sea cages. Satellite industries and additional employment develop for feed and equipment manufacture, processing and shipping (above) and smolt production at hatcheries such as this one on Digdeguash Lake in New Brunswick (lower). Five hundred full-time jobs were attributed to the Bay of Fundy salmon farming industry in 1990. (Photo: P.W.G. McMullon, DFO).

the benefits of being in the *entrepreneur* stage of development (open markets, expanding production, large profit margin), but that changed abruptly in 1989 when a record 209 500 tonnes of farmed salmon hit the world market (Fig. 13), sending world prices tumbling. The impact in Canada was especially severe on the more heavily leveraged west coast industry, and the number of salmon companies declined nearly 40% during 1989 (Egan and Kenney 1990). However, some observers feel that the rationalization will be beneficial in the long term, producing a leaner, more competitive west coast industry.

The future of the B.C. industry is difficult to predict. Physical attributes give the region tremendous potential for expansion, yet the industry has suffered from the ravages of plankton blooms — especially in the Sechelt area where the industry initially developed (Fig. 14) — and the focus of the industry has gradually shifted to the north end and west coast of Vancouver Island where more favorable conditions exist. In addition, the preferred Pacific species (chinook and coho salmon) have been less tolerant of domestication than Atlantic salmon. Diseases — especially bacterial kidney disease (BKD) and vibriosis — and early maturation of male Pacific salmon ("jacking") have caused problems for the industry. One response to the problems encountered with Pacific salmon has been the introduction of Atlantic salmon to west coast farms, and in 1990 Atlantic salmon constituted about 15% of the 13 500 tonne harvest. Future imports of Atlantic salmon ova for production purposes have been constrained (Egan and Kenney 1990), but local broodstock should allow a continuing increase in the proportion of Atlantic salmon in the B.C. harvest, perhaps to as much as 50% by the end of the century (Needham 1990).

Although the disease and domestication problems have had a serious impact on the industry, British Columbia growers have suffered as much from disjointed marketing, a fact that became painfully clear when chinook prices dropped well below those for Atlantic salmon when the market softened in 1989. Fortunately, this deficiency has been recognized and steps are being taken to improve downstream marketing and develop product forms that better match the consumer's needs.

The impact of the global price collapse on the east coast industry was much less severe because of that industry's slower growth and consequently more secure financial position. Today there are over a thousand sea cages with a combined volume of 1.1 million cubic metres in the southwestern Bay of Fundy. In 1989 the industry produced 4250 tonnes worth C$45 million, and the 3.1 million salmon that were in the cages in 1989 yielded 8182 tonnes worth roughly C$70 million in 1990. The maximum number of salmon approved for net pens in the southwestern Bay of Fundy is 4.4 million fish, so unless there is a major change in technology, the industry will be limited to an increase of roughly 42% over current size.

Total Canadian production of farmed salmon in 1990 was expected to be close to 20 thousand tonnes, worth nearly C$145 million. This would give Canada 8% of the world production of 250 thousand tonnes and the number three ranking, ahead of Japan but trailing Scotland and Norway (Fig. 15), a position Canada is expected to retain through at least 1993.

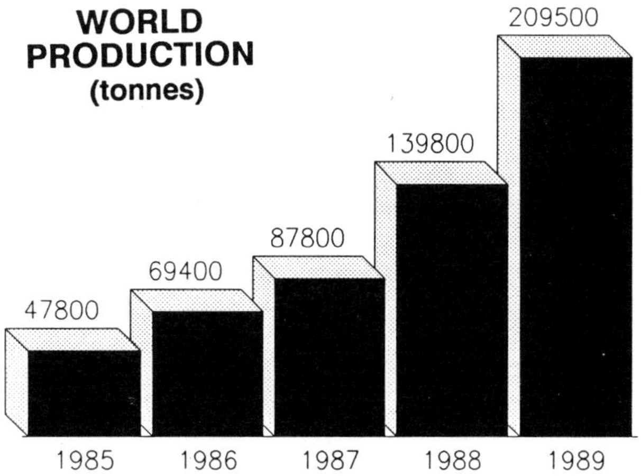

FIG. 13. Growth in world production of farmed salmon from 1985 through 1989 (B.C. Salmon Farmers Association, 1990).

Future growth in the Canadian farmed salmon industry will probably be determined more by marketing than by any other single factor. The major producer regions are focusing intently on marketing, searching for advantage in a highly competitive industry. Canada has the huge advantage of proximity to the United States market, the largest in the world, but Chile has other advantages that could make it a serious competitor. Right now Canadian producers benefit from the U.S. consumer's perception that Canada is the primary supplier of salmon (Egan and Kenney 1990), although this is not the case. The United States is, however, the major market for salmon produced on both Canadian coasts (Fig. 16), so the continuing growth of the Canadian industry will require that Canada adopt an aggressive marketing posture in the U.S.

In the United States, consumption of salmon is highest on the U.S. West Coast, where the preferred species is the chinook salmon. This combination of proximity, perception and preference provides the Canadian industry — especially the B.C. component — with a base on which to build a dominant market presence. The challenge to the Canadian industry is to do exactly that.

## The Shellfish Industry

Commercial shellfish culture in Canada is currently limited to three species: blue mussel (*Mytilus edulis*), American oyster (*Crassostrea virginica*) and Pacific

FIG. 14. Plankton blooms have caused heavy loss of fish for the salmon culture industry in some areas of British Columbia, and disposal of the fish is a difficult task. Here the accumulated "morts" are being pumped by a commercial firm — Coastal Seatrucking Ltd — for delivery to a composting operation. (Photo: Jim and Brian Peers, Aquarius Seafarms Ltd.).

oyster (*Crassostrea gigas*). The mussel culture industry went through the standard development stages and evolved much like the salmonid culture industries on the east and west coasts. However, the oyster industries have developed in an atypical manner on both coasts. The Atlantic oyster industry, in particular, has been a source

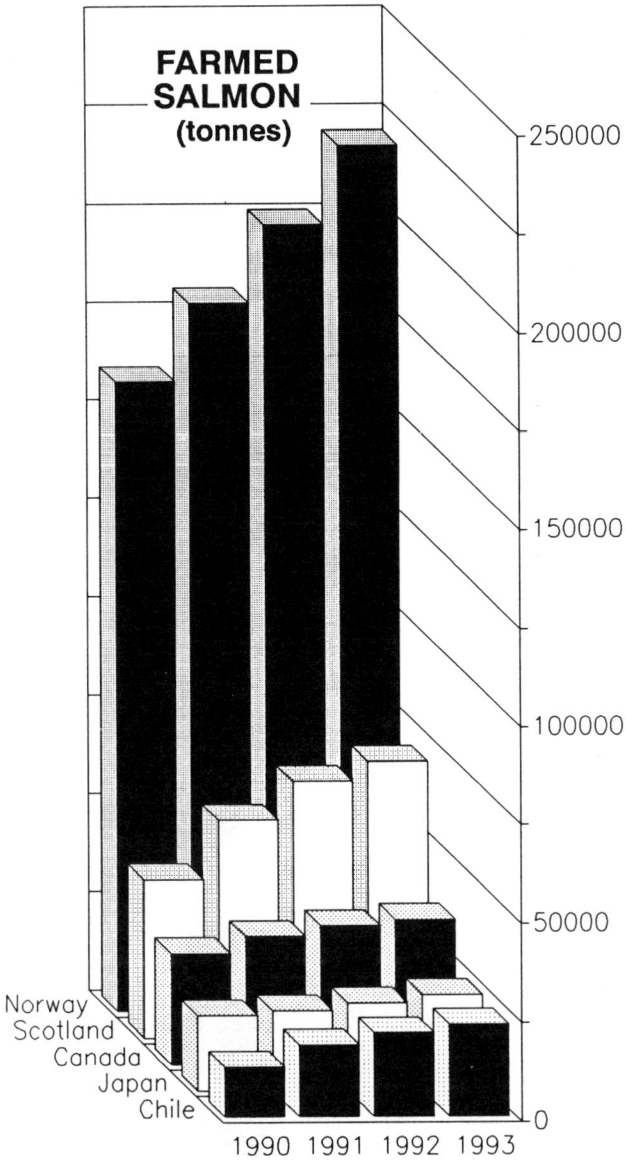


FIG. 15. Projected production of farmed salmon of all species by major producer nations between 1990 and 1993 (B.C. Salmon Farmers Association 1990).

of frustration for government specialists who feel that regional potential has always exceeded production by a considerable margin.

## Blue Mussel

Although the blue mussel is an excellent product, its recognition and acceptance in North America is relatively new. According to Lutz (1980), the harvesting of mussels began with the pickling and canning of the wild

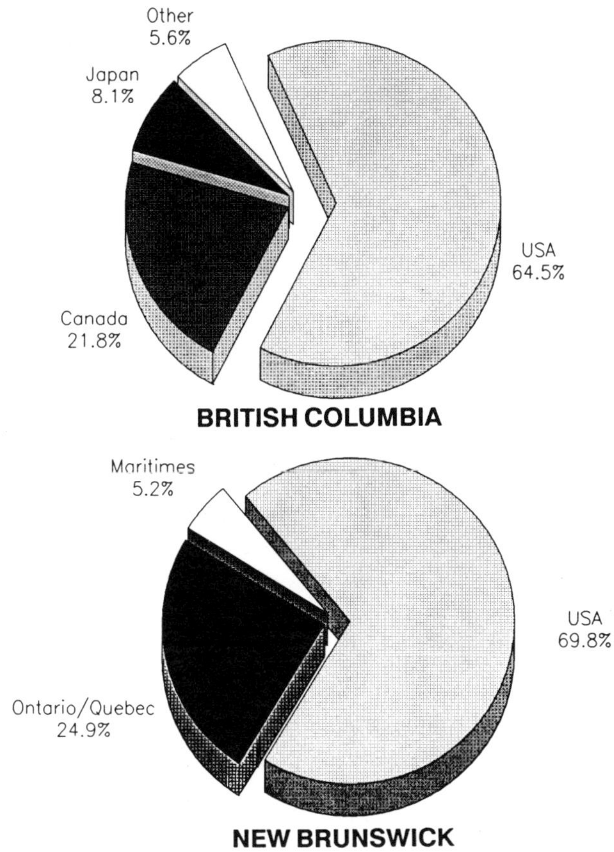


FIG. 16. Major markets for British Columbia and New Brunswick farmed salmon in 1989 (B.C. Salmon Farmers Association 1990).

crop during the two world wars, and Newfoundland dominated that early production with 1000 tonnes of canned mussels annually (Sutterlin et al. 1981). That industry eventually waned, but the emergence of a successful mussel culture industry in Europe in the early 1970's rekindled Canadian enthusiasm. The problem at that time was the absence of an established North American market. Except for an ethnic market in major urban centers there was no widespread demand for, or appreciation of, the cultivated product. Even in coastal areas where the mussel was at least familiar to consumers its reputation was sullied by the generally poor quality of slow-growing and gritty wild mussels harvested from the intertidal zone.

Most aquaculture industries develop by producing a product to fill an established demand. In the case of products destined for human consumption, that demand is usually created when the capture fishery cannot meet the demand. If the cultivated product is better in quality or more consistent in supply, consumer preference can be switched to the farmed product. The

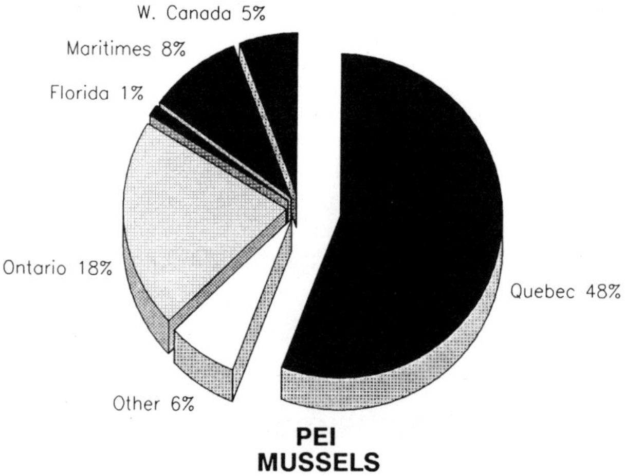


FIG. 17. Markets for mussels produced by the industry in Prince Edward Island in 1989 (Price Waterhouse 1990).

mussel industry in Canada was an anomaly in that no significant demand for the fresh product had ever been created by the wild harvest, so those contemplating mussel culture had to develop consumer awareness and demand for the product as well as the culture technology to produce it.

Suspension culture technology was developed in pilot-scale trials conducted by the Nova Scotia Department of Fisheries between 1968 and 1972. That led to a slow but steady growth of commercial cultivation in Nova Scotia over the next few years, but it was in Prince Edward Island that the industry blossomed. In 1978 the Island's first commercial mussel farm was established by Joe Van Den Bremt, a former Belgian tobacco farmer who was inordinately fond of mussels (Aiken 1987). By emphasizing product quality and marketing the industry was able to create and expand markets across Canada (Fig. 17). In 1989 the Prince Edward Island industry employed 250–300 persons and placed 2700 tonnes of their "Island Blue" mussels on the market (Fig. 18) — six times the production of "Scotian Pride" mussels by rival Nova Scotia — and returned C$4.3 million to the Island economy. According to current projections, the Prince Edward Island industry will continue to expand until it reaches roughly 4000–5000 tonnes annually, at which time all available lease sites will be developed (Muise 1990).

There are several reasons for the industry's success in developing a market where none existed, but the most important factor has been the product itself. Mussels grown in suspension culture are an excellent product, significantly better than their brethren grown on the bottom (Fig. 19,20). Mussels on the seabed grow more slowly, allocating a higher proportion of their

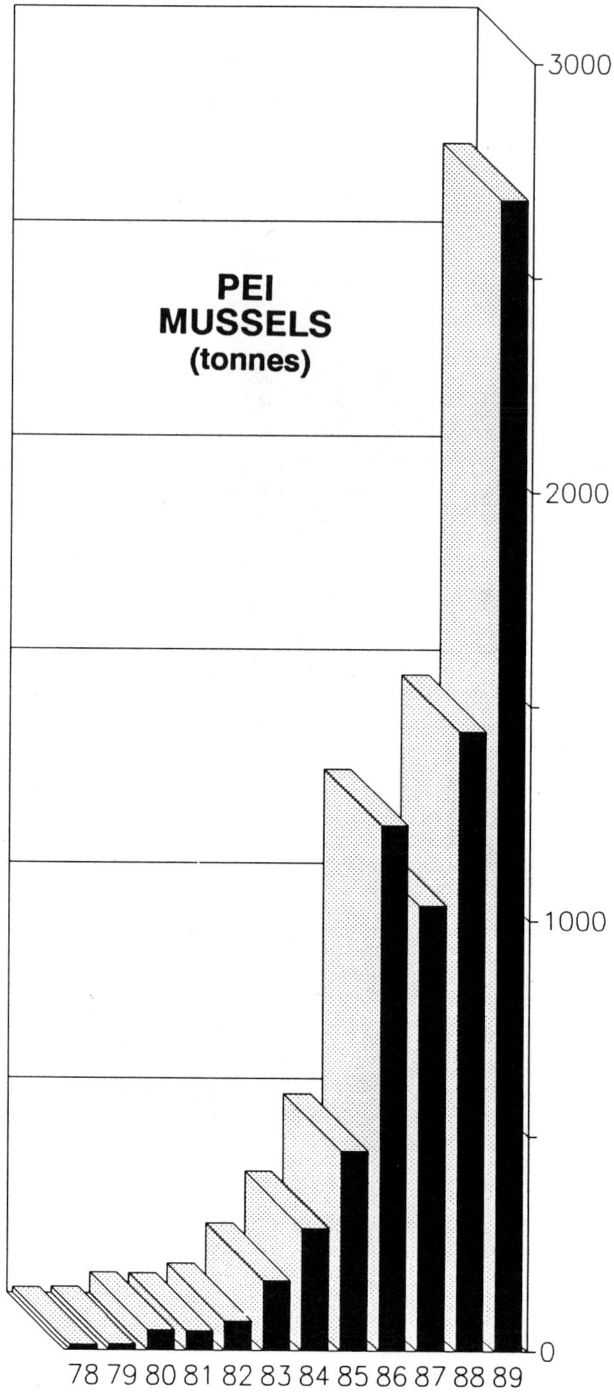


FIG. 18. Production of blue mussels by the industry in Prince Edward Island, 1978–1989 (Price Waterhouse 1990).

weight to shell instead of meat, and are more likely to be "gritty." In one analysis where a kilogram of mussels from the seabed was compared to an equivalent weight

FIG. 19. Much of the success of the mussel culture industry in Atlantic Canada is due to the quality of the product, and those grown on ropes suspended above the bottom are among the best. (Photo: Wayne Barrett Photography).

grown in suspension, the suspension mussels had 36% less weight in their shell and 57% more weight in their meat (Mallet 1989). For the consumer that means mussels from suspension culture will yield more meat of higher quality.

Nova Scotia, Québec, New Brunswick and Newfoundland all have productive mussel culture industries, although none approaches Prince Edward Island in total output. In fact, the combined production of those four provinces is consistently less than is produced by the P.E.I. industry alone (Fig. 21).

Nova Scotia's mussel culture industry employed 154 persons and produced 400 tonnes in 1989, 60% of which was exported from the province. Most of the

FIG. 20. Mussel culture in the Gulf of St. Lawrence presents special challenges during the winter. To harvest after the Gulf has frozen, growers must use divers to retrieve their culture lines from beneath the ice (*upper and middle*). Special airboats (*lower*) facilitate travel and harvesting during the spring when ice conditions can be treacherous. (Photo: Wayne Somers).

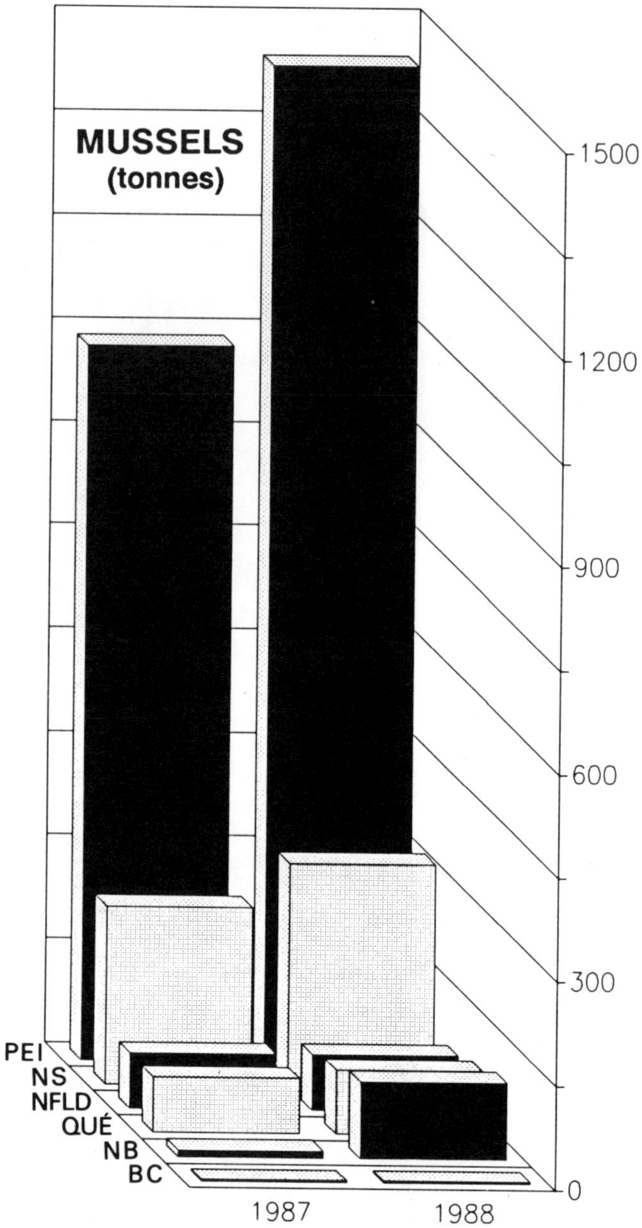


FIG. 21. Production of blue mussels by the various provinces in 1987 and 1988 (Price Waterhouse 1990).

production came from only three of the 55 growers in the province. Nova Scotia's marketing strategy also differs from the approach adopted in Prince Edward Island, where most of the production is sold through two major processors. In Nova Scotia the growers have tended to act independently, selling directly to wholesalers, retailers and local restaurants, but as competition intensifies in the marketplace the benefits of a cooperative marketing strategy are becoming more apparent (Muise 1990).

Québec's 1989 production of 172 tonnes came primarily from the Îles de la Madeleine in the Gulf of

St. Lawrence. Their market was the smaller cities and towns in Québec (the major centers having been preempted by the P.E.I. industry), but the relative isolation of the island growout sites makes marketing difficult. Air freight from the Îles de la Madeleine is prohibitively expensive at current levels of production, and surface transport is unreliable in winter due to ice conditions in the Gulf.

In 1987, New Brunswick was the smallest producer of mussels in the Maritimes region. Depending on the authority you consult, they either expanded dramatically thereafter (Price Waterhouse 1990) or remained relatively stagnant (Muise 1990). In any event, production in that province is centered in the northeast because the potentially productive Bay of Fundy coastline to the south suffers annual incursions of paralytic shellfish poisoning (PSP). New Brunswick growers market their product in the towns and smaller cities of Québec, thereby competing with their counterparts in the Îles de la Madeleine.

Newfoundland growers produced 70 tonnes of mussels in 1989 from two full-time and 60 part-time operations (Muise 1990). Like the growers in the Îles de la Madeleine, Newfoundland farmers face transportation costs that make it difficult to be competitive in the lucrative but distant markets. In 1989 all of their product was consumed within the province. However, growers along the east coast of Newfoundland are trying to turn adversity to advantage. Their cooler water temperatures delay spawning to late summer, leaving their shellstock in peak condition during mid-summer when demand is high and supplies from other segments of the industry are minimal (Muise 1990).

Interest in mussel farming has increased throughout the Atlantic region in recent years and there are now roughly 8000 hectares under lease to 250 independently owned enterprises. Most of these produce fewer than 10 tonnes annually and are typically operated by the owner part-time. Mallet (1989) found that 90% of the mussels produced in the region come from just 10 farms that occupy less than 2% of the leased area, primarily on the eastern end of Prince Edward Island and the eastern shore of Nova Scotia.

Mussel farmers in Canada have not yet encountered serious disease problems, but there have been setbacks due to blooms of toxic phytoplankton and to public opposition to the visual impact of mussel longlines (what critics refer to as "visual pollution"). The most serious shellfish toxicity incident recorded in Canada struck the P.E.I. industry in 1987, when a bloom of the diatom *Nitzschia pungens* contaminated the mussel crop with domoic acid, a potent neurotoxin not previously recorded from shellfish (Bird and Wright 1989). That incident cost the P.E.I. industry $300 000 and a year of production (Shumway 1989), but the industry emerged from it better equipped to detect and deal with toxic blooms.

Although virtually all of the commercial production of blue mussels in Canada is from the Atlantic side, the blue mussel grows on the Pacific coast as well. Pilot scale studies several years ago indicated it would be possible to culture mussels in British Columbia, and several companies eventually attempted to do so (Pennell 1987). However, three major problems have prevented the development of a commercial industry on the West Coast: fouling of culture longlines, predation by sea ducks in winter, and mortality of mussels in summer just before market size is reached (Bourne and Brett 1984).

The mussel industry in the five provinces of eastern Canada produced 3137 tonnes worth C$5.5 million in 1989, but the possibilities for expansion are narrowing in Prince Edward Island and Nova Scotia, the two major producers. The estimated maximum production from all available lease sites around Prince Edward Island is roughly double the 1989 production of 2700 tonnes (Muise 1990), and Nova Scotia growers are encountering increasing public opposition to expansion of their industry. Newfoundland seems to have the most potential for expansion with 60 part-time growers contemplating commercialization (Muise 1990). Their constraint will be the market. The total Newfoundland production of 70 tonnes was sold within the province in 1989. Any significant expansion will require the development of markets off-island plus innovative methods for overcoming the higher shipping costs to mainland centers.

## Oysters

The oyster culture industries of Canada have been a source of frustration for proponents of the aquaculture industry (Fig. 22). Culture of the American oyster *Crassostrea virginica* (also known as the Atlantic or Virginia oyster), began in Canada in 1865 when the government of Prince Edward Island allowed the leasing of designated areas for oyster culture (Mathieson 1912). However, the industry never approached its potential. In 1970 the gourmet market for this species in Europe and the United States was estimated at more than 3600 tonnes, but Canadian production was seldom greater than a third of this, primarily because the industry was "small, technically and managerially deficient, poorly organized..." and dominated by part-time producers (MacCrimmon et al. 1974).

In 1969 the European flat oyster, *Ostrea edulis*, was successfully imported to Prince Edward Island and subsequently transferred to Nova Scotia for culture in locations where water temperature was too low to

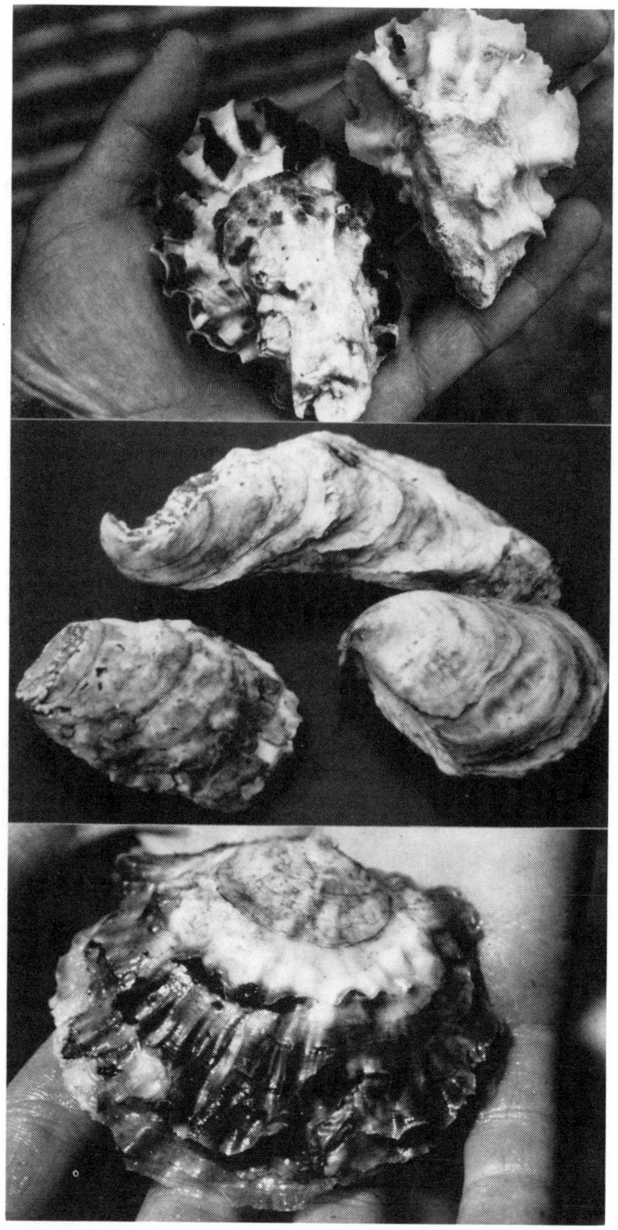

FIG. 22. The three species of oyster being cultivated in Canada: Pacific or Japanese oyster (*top*), American oyster (*middle*), and European oyster (*bottom*). The European oyster has been favorably received in the market, but the industry has had difficulty overcoming technical problems. (Photos: William Pennell [*top*]; Dave Aiken [*middle, bottom*]).

support the American oyster (Newkirk 1989). However, in spite of the export potential represented by markets in the United States and Europe, growers have had difficulty overcoming the lack of natural seedstock and the high cost of growout.

On the Pacific coast the early history of oyster culture is similar. The native oyster, *Ostrea lurida*, enjoyed high value and a good reputation with oyster gourmets, but because it is small and slow-growing, it was never cultured (Pennell 1987). The American oyster, brought to British Columbia around the turn of the century, survived but never prospered. A small residual population remains in Boundary Bay (Bourne et al. 1989). Finally, the Pacific oyster (*Crassostrea gigas*) was imported from Japan over a 35-year period beginning in the early 1900's (Bourne and Brett 1984; Bourne et al. 1989; MacCrimmon et al. 1974). Culture began in the 1920's and expanded in the 1950's and 1960's (Pennell 1987) but — like its counterpart in the Canadian Maritimes — languished thereafter, falling far short of its potential and the existing market demand.

Thus the history of oyster culture in Canada has been one of chronic underproduction relative to potential and market demand. The two species with the most potential and the fewest problems — the American oyster and especially the Pacific oyster — are generating renewed interest that may, finally, reverse that trend.

### *American Oyster*

The American oyster is abundant from Massachusetts to the Gulf of Mexico and into the Caribbean; in Atlantic Canada it is found primarily in the southern Gulf of St. Lawrence along the shores of New Brunswick, Prince Edward island and Nova Scotia, where water temperatures are sufficiently warm to support reproduction. In these areas the American oyster is grown on private leases, of which there are approximately 2000 varying in size from 0.4 to 40 hectares (Lavoie 1982). These occupy 3314 hectares of an estimated 6200 hectares of good oyster ground, which could theoretically yield almost 50 000 tonnes annually (Lavoie 1989), and yet in 1987 only 2200 tonnes were harvested from the leases in the three Atlantic provinces. Production from P.E.I. leases increased through 1989, but declining yields from Nova Scotia and New Brunswick offset most of the increase (Fig. 23).

The chronic underutilization of oyster grounds in the Atlantic provinces is a result of historical, administrative, social, economic and political factors, and these are elegantly summarized by Lavoie (1989). The important question is whether this hidebound industry can cast off the baggage of history and respond to the increasing national and international demand for top quality American oysters.

In Atlantic Canada the American oyster is grown either in suspension or on the bottom, but because of the capital investment and labor costs associated with suspension culture, plus the conflicts that arise with other users of the coastal environment, few growers in Atlantic Canada raise American oysters to market size using suspension culture. Instead, the seed — obtained in a variety of ways (see Lavoie 1989) — are transferred to a suitable cultivation site, or *lease*. Seed can actually be of any size, and one of the more common practices involves *re-laying*, in which fairly large oysters are transferred from contaminated natural beds to lease sites for depuration. Since improvement in quality and some growth usually occur on the lease, re-laying is generally considered to be a form of culture.

In commercial aquaculture, market is paramount. The paradox in Atlantic Canada is that the mussel culture industry, with no established market, has grown dramatically while the oyster culture industry, blessed with an entrenched and unfulfilled market, has languished. The market for American oysters is one of the oldest in Canada — oysters from the Maritimes were consumed in Québec City in the 1600's — and it is also a luxury halfshell market in which the rewards for quality are considerable. The price difference per tonne between fancy and commercial grade was more than a thousand dollars in 1986, and culture techniques for improving the quality of oysters are well known.

Thus the American oyster is a species that commands a reasonable price in an established and apparently unfulfilled market, a species whose culture requirements are well known and which suffers few serious biological problems; in short, an excellent candidate on which to build a growing industry. As Lavoie (1989) aptly phrased it: "The fresh oyster has a shelf life superior to that of any other shellfish presently offered in our markets, and foreign markets have yet to be tested. It is up to producers and marketers to recognize these opportunities and to take the necessary steps to capitalize on them."

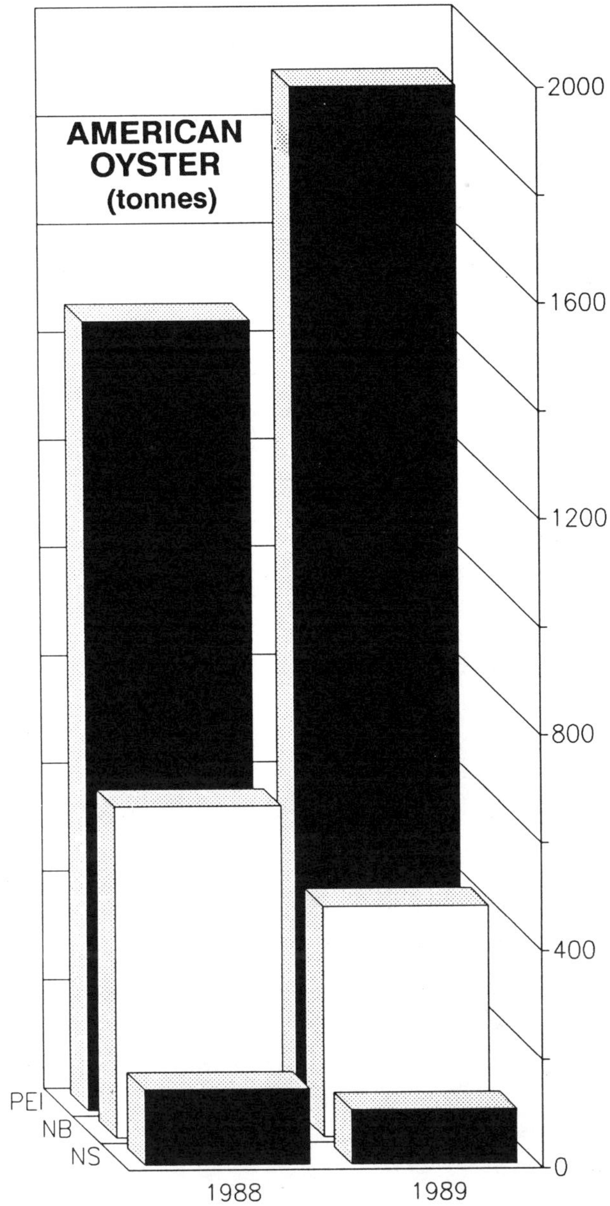


FIG. 23. Oyster production in the Atlantic provinces, 1988–1989 (Price Waterhouse 1990).

### *Pacific Oyster*

In 1948 a natural breeding area for Pacific oysters was discovered in Pendrell Sound, British Columbia, the result of repeated introductions of seed from Japan spanning more than 20 years. This ended the region's dependence on Japan as a source of seed for the culture industry, and stimulated local interest in oyster culture.

During the 1950's, numerous family oyster farms began operating on the intertidal grounds of Georgia Strait (Pennell 1987). Production of Pacific oysters reached a peak of 5900 tonnes in 1963 and then, in spite of increasing demand and rising prices for oysters, began a decline that continued until the 1980's. Meanwhile, a thriving oyster culture industry developed in Washington State using the same species and the same culture techniques. In 1970 the Canadian industry supplied less than a third of the table oysters consumed in the country, and the equivalent of nearly 14 000 tonnes of live oysters had to be imported to meet the demand (MacCrimmon et al. 1974).

During the 1970's Pacific oysters were produced primarily by family enterprises utilizing bottom culture techniques, but because of the limited potential of the intertidal grounds, suspension culture was recommended

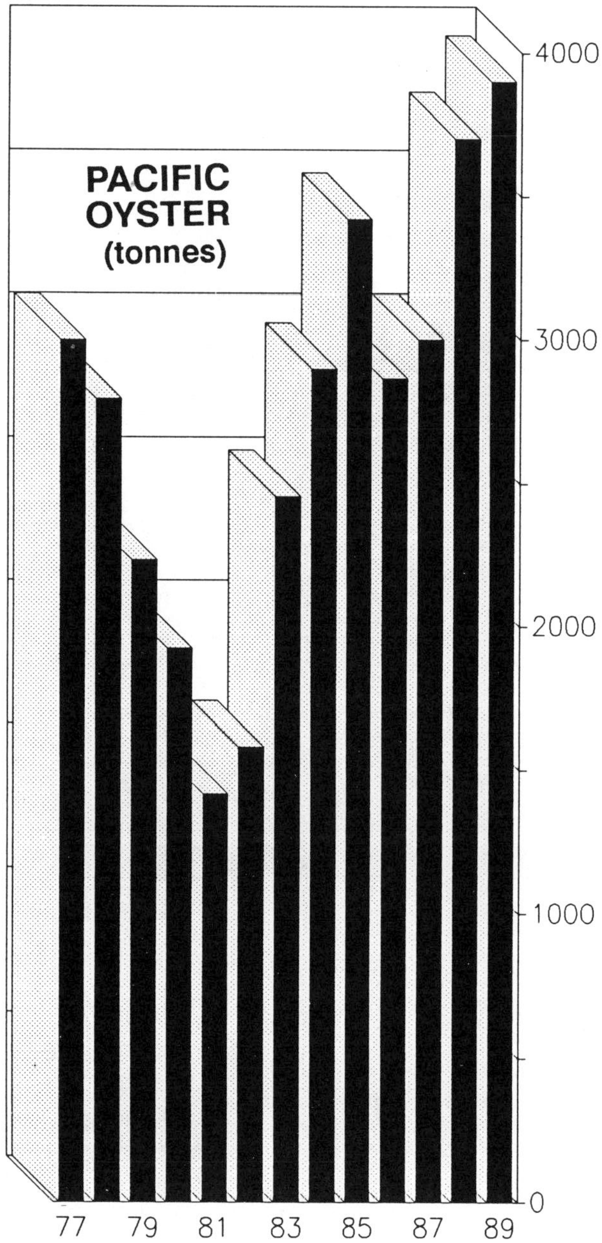


FIG. 24. Production by the B.C. oyster culture industry since 1977 (Price Waterhouse 1990).

(MacCrimmon et al. 1974). Some growers successfully shifted to suspension culture, and these became the nucleus for a revitalized oyster industry in British Columbia starting about 1981 (Fig. 24). In 1987 roughly 2800 tonnes were harvested and another 530 000 dozen were produced for the half-shell trade for a total value of C$2.6 million (Bourne et al. 1989).

The British Columbia oyster industry is concentrated in the Strait of Georgia, 70% of the total production coming from the Baynes Sound region between Denman and Vancouver Islands. The industry uses three methods of culture: intertidal bottom, near-bottom, and suspension culture. Most oysters are currently produced in intertidal culture on firm sand-gravel bottom (Fig. 25). Where the bottom is soft the oysters can be held above the substrate by stakes, plastic umbrellas and racks (near-bottom culture). In suspension culture the oysters are grown on ropes suspended from rafts or longlines, or in trays suspended from rafts. With suspension culture the oysters reach market size sooner than in intertidal bottom culture — generally two growing seasons instead of three — and the yield per hectare is more than 25 times greater (Bourne et al. 1989).

Most new growers in British Columbia have concentrated on the half-shell product grown in trays for export to the United States. Half-shell oysters are a fairly new product in British Columbia, and the industry is still working on an effective marketing technique. The market softened considerably toward the end of 1989 because the sizes being produced were not in demand. The East Coast market requires a smaller oyster and the B.C. industry may be able to adjust production to meet this demand (Talley 1990).

The only constraints on the industry, aside from the need to develop and service good markets, are those of seedstock supply and pollution. Wild seedstock supplies have been unpredictable in recent years, and local hatcheries have been unable to meet the demand for eyed larvae (Pennell 1987). Sewage pollution has also become a concern. It was first noted in 1963, when some oyster beds were closed, and pollution of beach leases has since ruined some established companies (Bourne et al. 1989).

The Canadian oyster industry will produce about 4000 tonnes in 1990 and, according to Needham (1990), this is not likely to increase much beyond 10 thousand tonnes by the turn of the century because the technology that offers the greatest opportunity for expanded production — suspension culture — is so expensive. Another possibility for expansion is improved utilization of oyster grounds in the Maritime provinces. However, a complex of historical, administrative, social and political factors make this unlikely (Lavoie 1989).

### Other Candidates

The established commercial aquaculture industry in Canada has a narrow base: essentially four salmonids, two oysters and the blue mussel. In 1987 this industry produced $109 million worth of product, with almost half coming from salmon (Anon. 1990). The value in this narrow focus has been a concentration of effort in the development of appropriate technology. The hazard now that the industry is established is the vulnerability that attends specialization in any product line. This fact was illuminated by the domoic acid-PEI mussel disaster of 1987 and emphasized by the salmon price collapse of 1989. Diversification is the key to long-term survival in any business because of the reduced probability that a disaster will strike two or more products simultaneously.

An impressive number of *potential* aquaculture species exist in Canada, but few are expected to make a significant contribution during the 20th century. The possibility of culture is determined by biology and technology, but economic factors determine whether culture of a particular species will be commercially viable.

## Finfish

Thirteen species of finfish are usually mentioned when alternative species are discussed in Canada. Eight are either marine or spend some part of their life cycle in the marine environment. The remaining five are considered freshwater species.

The Arctic charr is considered by many to have the greatest immediate potential as an alternative to the rainbow trout. In the past it has suffered from a reputation as a niche product, and an expanded role will require broader market acceptance. Charr are more tolerant of cold water and high rearing densities, and are said to be more efficient than salmon at converting feed to flesh, which is an economic advantage. However, there appear to be large strain-specific variations in growth rate as well as ability to handle the marine environment. To date the industry has been held back by a shortage of broodstock for those strains that possess desirable traits. In 1989 there was only one commercial producer of Arctic charr in Canada, but a second facility was anticipating a commercial harvest of some 4 tonnes. By 1991 there were a couple of dozen farms with an anticipated total production of some 35 tonnes. Initially most of the interest was in the west, but the majority of farms are now in Atlantic Canada (Anon. 1991).

In the freshwater environment the lake whitefish (*Coregonus clupeaformis*), yellow perch (*Perca flavescens*), walleye (*Stizostedion vitreum vitreum*), sturgeon (*Acipenser* spp.) and possibly the smallmouth bass (*Micropterus dolomieui*) are considered to have potential, although each has problems that make commercial comestible culture unlikely for a few years.

In the marine environment there are several candidates but nothing that holds immediate commercial promise.

FIG. 25. In British Columbia, Pacific oyster seed are encouraged to settle on cultch (bags of oyster shell, *top*) in remote setting tanks (*middle*), after which they are grown out on intertidal beaches (*bottom*) or suspended from floats or rafts. (Photos: Dave Aiken [*top and middle*]; William Pennell [*bottom*]).

The Atlantic cod (*Gadus morhua*) is being "farmed" in sea cages in Newfoundland, but there are technical problems as well as questions about economic viability. On the west coast the sablefish (*Anoplopoma fimbria*) or black cod (not a real cod) is in demand, but significant

Fig. 26. The Atlantic halibut, under evaluation at DFO's research laboratory at St. Andrews, New Brunswick, as a potential culture species. (Photo: P.W.G. McMullon, DFO).

technical problems stand in the way of commercial culture. Halibut (*Hippoglossus* spp.) is another species on which a great deal of research money has been spent (Fig. 26). Five countries, including Canada, have research programs on the culture of halibut, but reasonable survival from hatching through to the juvenile stage remains an elusive goal, and the economics of growout are still questionable. Haddock (*Melanogrammus aeglefinus*) may actually have more potential than halibut. It is a high-value species — at times more expensive than salmon — that appears to present fewer biological obstacles to culture. Striped bass (*Roccus saxatilis*) are also frequently mentioned as a culture prospect in Canada because they have market appeal and are adaptable to culture, but there are problems with broodstock domestication and some question about economic viability due to low growout temperatures in many regions. Finally, there is the chubby little lumpfish (*Cyclopterus lumpus*), prized not for its flesh but its eggs. Wild stocks of lumpfish have difficulty withstanding the depredations of roe hunters, and the combination of supply, demand and price makes culture worth pursuing.

## Molluscs

Of the marine animals currently being investigated as aquaculture candidates in Canada the scallops — the Japanese scallop (*Patinopecten yessoensis*) on the west coast and the sea scallop (*Placopecten magellanicus*) and especially the bay scallop (*Argopecten irradians*) on the east coast — appear to have the most potential (Couturier 1990) (Fig. 27). The blue mussel, an established industry in the Maritimes, has not been commercially profitable on the west coast due to heavy fouling of gear, predation by sea ducks and the mussel's distressing tendency to die off just before reaching commercial size. If these problems could be solved the blue mussel could become an alternative crop at selected locations on the B.C. coast.

The Belon, flat or European oyster has been cultivated along the Nova Scotia coast for several years. The species grows well in suspended culture and has been popular in test marketing, but two problems make commercial culture risky: seed is available only from hatcheries, and suspension culture is expensive. One species of abalone, the pinto abalone (*Haliotis* sp.), was evaluated in British Columbia for several years. The culture technology is intensive and expensive, and the economic viability of the cultured product remains to be demonstrated. The only clam under consideration in Canada is the Manila clam (*Tapes philippinarum*), an exotic that jumped from Japan to British Columbia with shipments of Pacific oyster seed. Recent changes to the regulations in British Columbia have stimulated commercial interest in this species.

### Crustaceans

Only two crustaceans have potential in Canada: the lobster (*Homarus* spp.) and the signal crayfish (*Pacifastacus leniusculus*). The potential of crayfish is limited (Thompson 1990). Work began on the American lobster in the 1970's, and at this point a great deal is known about the biology of this animal; more, in fact, than is known about many species that are in commercial culture. The culture technology is available for lobster, domesticated strains have been developed and progress has been made on distinctive products, but the problem is economics. Lobster culture is intensive and expensive, and with landings from the capture fishery at historic highs there is no immediate prospect that lobster culture can be competitive in Canada.

### Marine Macroalgae

Three species of seaweed are being investigated for commercial culture in Canada: kombu (*Laminaria* spp.) and nori (*Porphyra* spp.) in British Columbia, and Irish moss (*Chondrus crispus*) in Nova Scotia. Nori and kombu appear to have significant potential on the west coast. On the east coast, Irish moss has been the object of collaborative research by the National Research Council of Canada and a private company in Nova Scotia, and in 1989 the collaboration resulted in commercial production from a land-based facility at Charlesville (Isaacs 1990).

FIG. 27. The three species of scallop considered to have commercial potential in Canada: the Japanese scallop (*top*), sea scallop (*middle*) and bay scallop (bottom). (Photos: Christine Hodgson [*top*]; P.W.G. McMullon [*middle*]; P. Smith [*bottom*]).

## Human Health Issues

Mussels and oysters are filter feeders; that is, they obtain their food by filtering impressive quantities of water and extracting the nutrients contained therein. This is convenient for shellfish growers because supplementary feeding is not required. The obverse is that the grower has no control over what the shellstock consume.

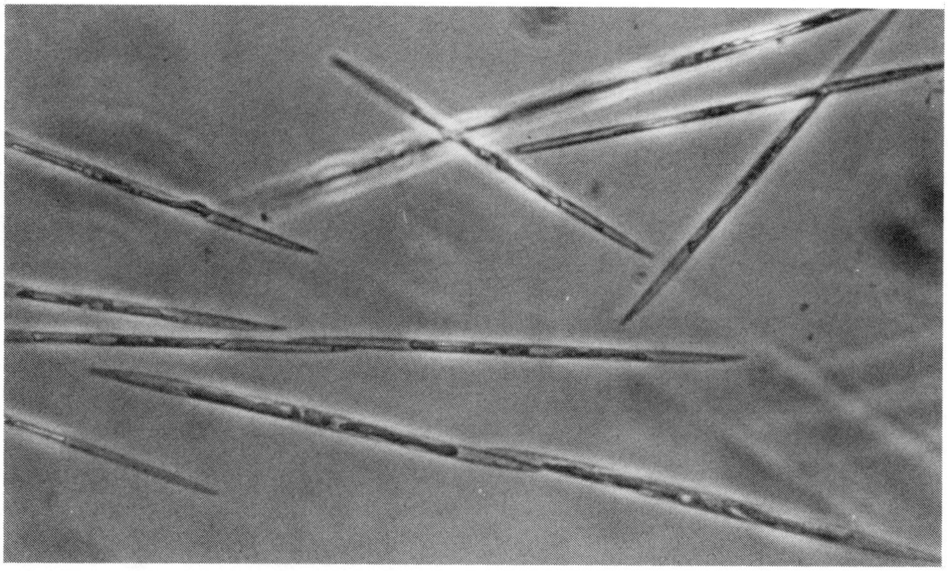

FIG. 28. The diatom *Nitzschia pungens*, producer of domoic acid, which can cause Amnesic Shellfish Poisoning (ASP). In 1987 this diatom bloomed off the eastern end of Prince Edward Island, causing economic hardship for the Island's mussel growers. (Photo: Dave Aiken).

This makes the farmer vulnerable to two potentially troublesome situations: if the water contains insufficient nutrients there is no convenient way to supplement the diet; and if the water contains toxins or contaminants there is no convenient way to prevent their uptake.

The nutrient dilemma has not been a serious problem, although it could become one for mussel growers as the intensity of mussel farming increases in key estuaries. The toxin and contaminant dilemma, on the other hand, is already a serious global problem. An expanding plague of toxic algae is responsible for paralysis (paralytic shellfish poisoning, or PSP), gastrointestinal distress (diarrhetic shellfish poisoning, or DSP), neurological problems (neurotoxic shellfish poisoning, or NSP) and amnesia (amnesic shellfish poisoning, or ASP) from Australia to Scandanavia. Throughout the world in areas of dense human habitation the coastal environment is being inundated with human and industrial waste, and public health warnings regarding shellfish concentration of viral, bacterial and chemical threats are carried in the world press on a regular basis, much to the distress of shellfish farmers everywhere.

The Canadian shellfish industry is not insulated from these problems. In the winter of 1987 the mussel growers of Prince Edward Island experienced the worst kind of incident — contamination of their shellstocks with a potentially lethal, previously unknown and totally unanticipated toxin of unknown origin. The lethal aspect made it newsworthy and the unknown nature of both the toxin and the source gave it an air of mystery that assured maximum coverage in the press and maximum damage to the P.E.I. industry. While scientists scrambled to identify the toxin and government spokespersons struggled to explain the inexplicable, panic spread, briefly encompassing all shellfish. Reason eventually prevailed and both the toxin (domoic acid) and its source (the diatom *Nitzschia pungens*) were identified (Fig. 28). However, the P.E.I. industry lost a harvest year while measures were being developed to prevent human health problems should the organism recur. The solution — for this and similar problems elsewhere — is an efficient, broad based inspection and monitoring program that can detect the presence of toxins in shellfish and shellfish growing waters before they become a hazard to human health (Shumway 1989). Efficient monitoring has long been in place in Canadian waters for paralytic shellfish toxins (PSP), but the program was broadened considerably in the wake of the domoic acid incident. Public confidence has been restored and the P.E.I. mussel industry has recovered, but government and industry now realize how quickly disaster can strike.

## Multiuse Conflicts

Canadian aquaculture is encountering increasing public opposition. The increase in the number of salmon sea cages and the longlines of suspension mussel and oyster culture in the coastal environment has generated a widespread public resistance to further expansion (Fig. 29). Applications for coastal leases in many areas have been facing aggressive opposition from upland owners, traditional fishermen, real estate developers and recreational interests. Some of this is being reflected in government policy, not always to the benefit of aquaculture. In an effort to reduce the backlash, shellfish growers in some areas have started submerging most of their longline flotation, reducing the visual impact and leaving more of the surface available for other uses.

Salmon farms have come under scrutiny in the Maritimes for three perceived transgressions: altering the sea bed in the vicinity of sea cages (buildup of fecal material and uneaten feed); littering nearby shorefront property with residue and refuse from day to day operations; and negatively affecting established local fisheries. In this arena charges are easy to make and difficult to refute. Public officials tend to assume there is a problem until research provides evidence to the contrary, but obtaining the necessary data requires time and money that has not been available. In some cases sea cage operators have been careless, lending credence to a broader spectrum of unsubstantiated charges. In addition there is the vested interest or NIMBY (**Not In My Back Yard**) phenomenon in which individuals are opposed to local aquaculture activities for purely personal reasons.

Research is slowly defining the local and downstream environmental impacts of sea cage operation, and farm operators are becoming more conscious of the rights of coastal landowners and more responsible in their operating procedures. However, aquaculture impact on local fisheries remains a difficult issue to resolve. Salmon farms can preempt large areas of the seabed and exclude trawling activites. Has this exclusion measurably affected the catch of local scallop draggers? Salmon farms can also alter the seabed immediately beneath their cages; is this portion of the seabed an important breeding or nursery ground for lobster or some other commercially valuable species? Will its loss adversely affect local fisheries several years later? In coping with these questions regulators have tended to either establish aquaculture zones and accept whatever fisheries impacts might be involved, or insist on a detailed evaluation of each proposed site, an approach that is both expensive and time-consuming.

The southwestern Bay of Fundy is also the focus of a passive weir fishery for herring, an industry that in recent years has annually landed about C$6.5 million worth of fish and is estimated to employ 1800 people and contribute C$70 million a year to the local economy. The conflict between this traditional harvest fishery and

FIG. 29. Multiuse conflict in the marine environment. As the aquaculture industry grows it comes increasingly into conflict with other users. Prominent among those are traditional industry (such as the herring weir fishery in the Bay of Fundy, *top*), recreational users, who object to encroachment on their fishing and boating space (*middle*), and upland owners, who complain about perceived environmental insults, including the visual impact of floating aquaculture gear (*bottom*). (Photos: Dave Aiken [*top*, *middle*] and J. Sylvester [*bottom*]).

the expanding aquaculture industry involves some perplexing issues. In the words of Stephenson (1990):

> "We know that weirs rely on subtle movements of herring near shore and that herring

are sensitive to things such as water currents, temperature, water quality, noise and light. We can predict potential effects from cage sites that block fishways, generate noise and operate bright lights, and we can speculate on the detrimental effects of metabolites and excess feed on herring movements in the area. We know also that impressive quantities of organic compounds are deposited around cage sites and these may be carried considerable distances by the spectacular tidal currents in the Bay of Fundy. However, the impact of these exudates is hard to quantify. While herring can be sensitive, we know also that at times. . .they will go most anywhere, and anecdotal evidence since the development of aquaculture has been conflicting. We have seen early juvenile 'brit' herring around salmon cages and have even heard of herring inside salmon cages. On the other hand, we have yet to see a weir perform well in close proximity to a major cage site."

Providing the research data required to resolve all issues involved in the conflict will take years and may not even be possible. In the short term a practical solution is for regulators to take a conservative approach to the major issues, for the holders of aquaculture leases to operate their sites in a responsible manner, and for participants at all levels to promote a continuing dialogue of education and accommodation.

## The Infrastructure

An aquaculture enterprise requires that private property rights be secured, financing obtained, rearing systems purchased and seedstock procured. Once the system is operating there are additional requirements for research and development, disease control, environmental manipulation, and marketing. Many of these activities affect public and private interests and conflict with other uses, especially in coastal marine and public inland waters, and this makes regulation inevitable. Support is also required to assist the industry and prevent problems from stifling its development. This combination of support and regulation is provided by government as a part of the industry "infrastructure," and the effectiveness of this infrastructure is usually reflected in the rate of development and ultimate success of the industry.

The rapid growth of aquaculture in Canada (production increased more than 500% in the 1980's) suggests there was a reasonably effective infrastructure in place during the development period, and yet few in the industry have warm feelings about the level of support they received from governments in the early days. Obtaining official permission to grow fish in the marine environment was a nightmare made more complex by overlapping jurisdictions at the federal and provincial levels. The federal government was responsible for fisheries, while the provinces were responsible for property rights. As a result, an individual wishing to own and grow fish was faced with two or more levels of bureaucracy, often with overlapping jurisdiction and conflicting priorities.

In British Columbia as recently as 1987, for example, a prospective aquaculturist was required to deal with three levels of government plus other agencies and associations in order to acquire the necessary licenses, permits, approvals and registrations. Once a business license and zoning compliance were obtained from municipal officials, the aquaculturist would move to the *provincial* level to register their company, submit a business plan and aquaculture management plan, and obtain an aquatic land lease and a pollution control permit. If salmon farming was contemplated a water license and a commercial fish culture permit were also required. If oysters were on the plan, it was necessary to register with the B.C. Oyster Board as well as the B.C. Ministry of Agriculture and Fisheries. Once all this was done, the *federal* government became involved. A salmon grower with net pens in navigable waters would need an aquaculture permit from DFO plus clearance from Transport Canada. Furthermore, if seedstock were obtained from a hatchery that was more than 50 km from the growout site, transplant approval had to be obtained from a federal-provincial commmittee (Anon. 1988a). The result, in addition to confusion and aggravation, was a considerable delay in the processing of applications. Fortunately, much of this has changed. Under new regulations effective at the end of 1989, a single aquaculture license is issued for registration of an aquaculture enterprise as well as operation of the facility (Price Waterhouse 1991).

The infrastructure that currently supports the aquaculture industry in Canada falls into three broad categories: (a) research and education, including the training and extension programs operated by government and universities; (b) financial and employment assistance programs; and (c) the regulatory structure maintained by the various levels of government to mannage aquaculture development and ensure harmony with competing public and private interests.

### Research and Education

Modern aquaculture relies heavily on science and technology. With its long history in fisheries research and an established technological capability, Canada already had many of the resources it would need to support the developing aquaculture industry. The federal hatchery system, with its years of experience in salmonid enhancement, was able to provide technical assistance and an interim supply of seedstock to help the industry get started. Research on reproductive physiology, genetics, disease and nutrition, already underway in the national network of 12 federal research intitutions and satellite laboratories, quickly adjusted to include the research needs of the aquaculture industry.

Today, aquaculture research is an integral part of the federal fisheries program. The Department of Fisheries and Oceans (DFO) currently allocates 200 person-years of effort and spends C$16 million on aquaculture research and development, and DFO laboratories on both coasts conduct research in support of the salmonid farming industries and do biological and development work on other species with potential. In addition, the broad-based multidisciplinary capability of the federal research system is occasionally mobilized in response to crisis situations that affect the industry. An example was the "domoic acid incident" that struck the Prince Edward Island cultured mussel industry in 1987, making their esteemed "Island Blue" product suddenly toxic and unmarketable. The toxin could not be identified with any of the standard shellfish toxicity tests, and the causative agent was not immediately apparent. However, in a surprisingly brief period a multidisciplinary research effort identified both the toxin (domoic acid, not previously known as a shellfish poison) and the diatom that produced it (*Nitzschia pungens*, not previously known to produce toxins). The rapid identification by the research community allowed monitoring and quality control measures to be put in place in time to quell public concern and avoid irreparable long-term damage to the P.E.I. industry.

Education and training capability are also essential for aquaculture development. Salmonid culture is intensive and technically demanding, and because of its rapid expansion the industry had an immediate need for large numbers of personnel trained in the variety of skills demanded on an aquatic farm. This requirement was addressed through the community college system on both coasts. Eventually, degree courses in aquaculture were offered by several universities across Canada, providing the industry with skills and knowledge ranging from the practical to the esoteric.

If there has been a weakness in the education and training made available to the Canadian industry it has been in the area of extension services. No formal mechanism was developed to bridge the void between research results in the lab and their application on the farm. The Atlantic Salmon Demonstration and Development Farm, established under the Canada/New Brunswick Economic and Regional Development Agreement (ERDA) in 1985 to serve the Atlantic salmon industry in the Bay of Fundy, was an excellent example of aquaculture extension. It provided practical information to help growers improve the efficiency of their farming efforts, and supported practical research on equipment and rearing techniques. Government funding for the service was terminated in 1989, and the operation was turned over to the New Brunswick Salmon Growers Association (Aiken 1989b). However, DFO is committed to continued involvement with this project and to the establishment of a similar industry-managed development farm in British Columbia (Anon. 1990).

### Assistance Programs

The Department of Fisheries and Oceans (DFO), as the federal agency with primary responsibility for aquaculture development in Canada, maintains a variety of programs to assist in that development. In addition to the research and hatchery capability noted above, DFO handles some aquaculture licensing, enforces fish health regulations that govern movement of salmonids across provincial and national borders, conducts market studies and provides disease diagnostic services in support of fish health regulations. An example of the latter is the Fish Health Unit in Halifax. It conducts the bacteriological, virological and parasitological inspections required for health certification of fish hatcheries and facilities producing fish for sale nationally and internationally, assists federal hatcheries in the diagnosis and treatment of disease outbreaks, and inspects shellfish being transferred to hatcheries for commercial culture. DFO also maintains regional programs in support of aquaculture, and has established Aquaculture Coordinators in the regions to act as conduits of information between industry and government.

Other federal departments are active as well. The now defunct Unsolicited Proposals Program (UPP) maintained by the Department of Supply and Services once funded basic and applied research by industry in cooperation with federal science and technology, and the Industrial Research Assistance Program (IRAP) has six components that are accessible by the aquaculture industry and provide everything from advice or technological assistance to the funding of major research and development projects (Anon. 1988b). The Natural Sciences and Engineering Research Council (NSERC) also offers programs of value to aquaculture, and regional agencies such as ACOA (Atlantic Canada Opportunities Agency) and WED (Western Economic

Diversification program) have played a role in local development.

Financing is often the most critical need perceived by the industry. Salmon farming requires significant amounts of working capital, and in the development years that capital was not available from commercial sources. In general, both commercial and government sources have been more readily accessible to the aquaculture industry in the east than in the west (Price Waterhouse 1991), and this has forced the British Columbia salmon farming industry to obtain its funding from offshore banks, primarily Norwegian. The Federal Business Development Bank (FBDB) now has a suite of programs that benefit aquaculture, and some have been established in response to specific events. The Mollusc Growers' Loan Program, for example, was established to alleviate the cash flow problems of Atlantic mussel growers as a result of the domoic acid incident of 1987. The FBDB also provides venture capital, term loans to acquire fixed assets, loan guarantees for companies dealing with a chartered bank, and counselling to improve management capability (Anon. 1988b).

## The Regulatory Structure

Under the "Constitution Act," the federal government has jurisdiction over the sea coast and inland fisheries, public harbours and navigation in marine and inland waters, while the provincial government is responsible for property rights and civil rights (Anon. 1988a,b). However, over the years the federal government has delegated administrative responsibility for fisheries to provincial agencies in Nova Scotia (trout fisheries), Québec (freshwater fisheries), Ontario, Manitoba, Saskatchewan, Alberta and British Columbia (freshwater fisheries).

The aquaculture industry created regulatory problems because the activities almost always impacted on the jurisdiction of more than one level of government. Technically, fish in floating cages were a public resource under federal jurisdiction, while the seabed on which the floating cages were moored was under provincial jurisdiction. To address this problem of overlapping jurisdictions, Memoranda of Understanding (MOUs) were developed between the federal and provincial governments starting in 1986. The objective was to establish two things: a single office for aquaculture licensing and leasing; and a coordinated approach to government support for aquaculture. To date the federal government has signed MOUs with the Northwest Territories, Nova Scotia, Québec, Prince Edward Island, Newfoundland, New Brunswick, and British Columbia (Anon. 1990).

Under the Fish Inspection Act, the federal government is responsible for inspecting fish and fish products that move to or from Canada or from one province to another. The federal government is also charged with ensuring the quality and safety (as food) of molluscan shellfish. This latter requires coordinated effort by three federal departments: DFO monitors for paralytic shellfish toxins, for which Health and Welfare provide analytical support, and Environment Canada monitors the coastline for bacterial and chemical contamination.

One contentious regulatory issue is the Fish Health Protection Regulations, the intent of which is to prevent the introduction and dissemination of infectious salmonid diseases by regulating the movement of fish into Canada and across provincial boundaries. Industry representatives have challenged the content and enforceability of these regulations and insist they unfairly restrict the activity of Canadian fish farmers. Proponents point out that Canada is still free of some infectious salmonid diseases in spite of the tremendous trade in this group of fish, and credit rigorous application of the fish health regulations for this fact. Effective fish disease control regulations also help attract investment for aquaculture. The Fish Health Protection Regulations are being revised to include new concepts (e.g., zoning) and diagnostic procedures. They will also be applied to all finfish (not just salmonids) and to molluscs and crustaceans as well.

## National Aquaculture Strategy

In 1988 the government of Canada announced its intention to develop a long-term strategy for Canadian aquaculture development. Two years later that strategy was described in a document titled *Cultivating the Future: An Aquaculture Strategy for the 90's* (Anon. 1990). The Department of Fisheries and Oceans, as lead agency for aquaculture, developed its strategy around five goals, which can be paraphrased as follows: Science and Technology; Environmental Protection; Quality Control; Market Analysis; and Advocacy.

1. ***Science and Technology***. Recognizing the current and future needs of the aquaculture industry, DFO intends to support research in three key areas: new candidate species for aquaculture; genetics and broodstock development; and nutrition and physiology. Lobster, sablefish, cod, halibut, walleye, Arctic char, abalone and scallops are noted in the strategy document as species on which research will be supported directly, and perch, sturgeon, clams, quahaugs and marine

plants are considered to be candidates for cooperative involvement with the private sector.

A second component of this science and technology goal is fish health, and the government plans to foster this through tighter controls on the introduction and transfer of aquatic organisms (a national policy will be developed), and by influencing aquaculturists to utilize private, provincial and university laboratories for their disease diagnostic services, freeing DFO scientists to expand their surveys of diseases of aquatic organisms.

The third component of this goal will be the enhancement of extension services. DFO plans to achieve this by assisting in the development of industry-managed demonstration farms on the east (Atlantic Salmon Technology Centre) and west coasts, and exploring the potential for additional units for shellfish, marine plants and freshwater species.

2. ***Environmental Protection***. The Department of Fisheries and Oceans is responsible for maintaining a healthy aquatic environment for all users. Aquaculture is a new element in that environment, and little is known about the long-term consequences for the aquatic habitat or the wild stocks of plants and animals that share that habitat. DFO therefore plans to undertake research on several important areas related to the impact of aquaculture, including:

   - Genetic interactions between wild and cultured stocks;
   - Exchange of disease agents between wild and cultured stocks;
   - Potential impact of fish farms on seabed communities and fish habitat;
   - Impact of industrial pollutants on aquaculture operations; and
   - The accumulation of toxins in cultured organisms.

   In addition, DFO plans to formulate a national set of basic habitat considerations for aquaculture site selection and produce regional guidelines for aquaculture facility development and operation.

3. ***Quality Control***. To promote consumer confidence in the safety and quality of farmed fish, DFO intends to supplement its existing Fish Inspection Program with a new inspection system that will monitor cultured finfish for drugs, chemicals and additives, to ensure that only the highest quality cultivated products reach the consumer.

4. ***Market Analysis***. DFO already provides market and commercial analysis, research and intelligence for use by industry in making marketing decisions. To address the problem of gathering accurate and comprehensive production statistics, DFO intends to work with the provinces to develop a national aquaculture statistics system that will provide current data on number of farms, employment, purchases and sales by species.

5. ***Advocacy***. Finally, DFO plans to expand its role as industry advocate and facilitator, developing initiatives that are sensitive to both industrial and provincial needs. A corporate division for aquaculture will be established to implement the national Aquaculture Strategy, and Memoranda of Understanding (MOUs) will be negotiated with those provinces not yet included — currently Ontario, Saskatchewan, Alberta, Manitoba and the Yukon.

"The national Aquaculture Strategy is full of good intentions," an industry representative recently said, "but good intentions won't ensure the future of this industry." True, but there are indications that DFO is taking its Aquaculture Strategy seriously. One year after the Strategy was announced, DFO published a status report (Anon. 1991) documenting its progress on the five goals. To those in the industry accustomed to seeing government initiatives end with a statement of intent, that was a positive sign.

One thing is clear — aquaculture will succeed or fail in Canada on the strength of private sector vision, capability and adaptability, not government advocacy. Government support can assist in making a good thing better, but it cannot compensate for insular thinking or lack of good business sense on the part of the industry. Canada's success in the international aquaculture market will depend less on government support than on how astutely the industry reads and responds to the requirements of the future.

## Summary

The Canadian aquaculture industry — a mosaic of small family enterprises, publicly owned and listed companies, and subsidiaries of large corporations involved in international seafood markets and foreign owned enterprises — has come through an exciting

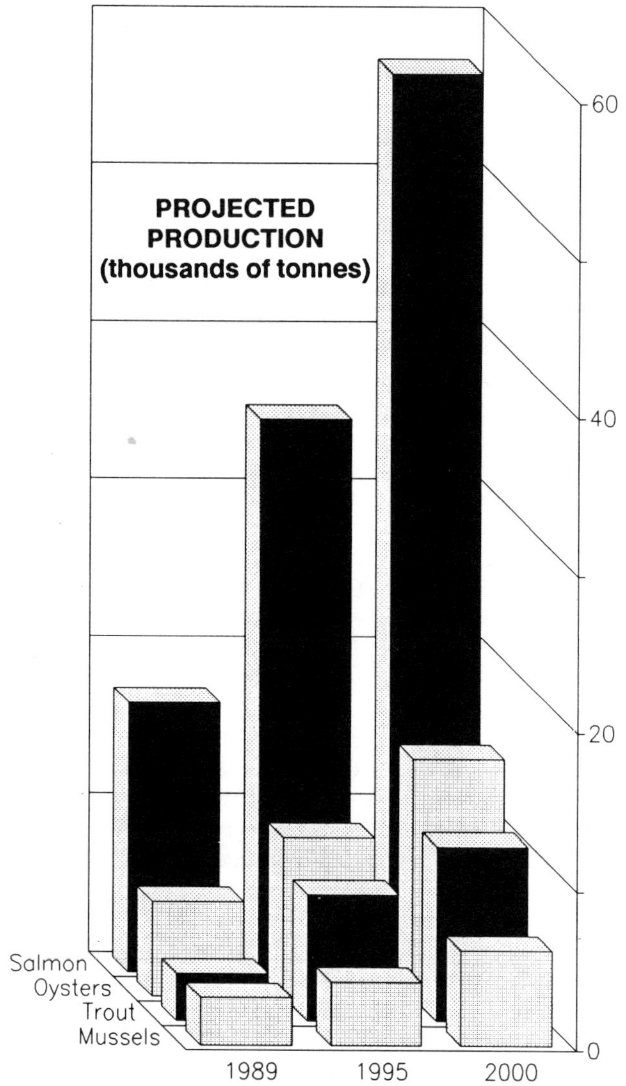


FIG. 30. Projected growth of Canada's four major aquaculture products through the year 2000, assuming a moderate growth performance (Price Waterhouse 1990).

decade of commercial development. Many of the problems created by overlapping governmental jurisdictions have been addressed in federal-provincial Memoranda of Understanding (MOUs), private property rights are now being granted to aquaculturists, and the financial sector has become more understanding and supportive.

On the political side, the aquaculture industry has finally been recognized as a significant socioeconomic force, one that is capable of rejuvenating economies and providing employment in areas hard hit by declines in traditional industries. In 1984 the annual value of the industry was C$7 million. By 1987 the value had increased to C$62 million. During 1988 the value expanded by 76% to C$109 million, and current projections indicate that by the year 2000 the Canadian aquaculture industry will have an annual value of C$500 million to C$1 billion and provide direct employment for 5 thousand persons (Anon. 1990). In addition, the aquaculture industry is expanding relative to the traditional capture fisheries. In 1987 the yield from aquaculture was only 3% of the landed value of Canadian fisheries. During 1988 that figure doubled to 6%. By the year 2000 the portion of landed value attributable to aquaculture will probably be 25% or more.

Between 1982 and 1990 several sectors of the Canadian industry matured, and are expected to continue growing for the rest of the decade (Fig. 30). Salmonid culture — especially sea cage culture of Atlantic and Pacific salmon — moved through the commercialization phase and into the farming phase to take its place as part of a global industry. The mussel industry has also matured, although its markets are still primarily domestic.

Even the phlegmatic oyster culture industry has showed signs of rallying, at least on the west coast. Finally, the aquaculture industry is showing interest in diversification, one of the surest signs of maturity. Farmers who to this point have concentrated on a single species and a single product are now looking for additional species and innovative products. For the Canadian industry of the 1990's, four species and four products will not be enough.

In less than a decade the Canadian aquaculture industry has grown from a novelty to an important sector of the Canadian economy. In so doing it has become an important part of the socioeconomic fabric of many communities. However, the price of rapid growth in many areas has been a broad-based public resistance to further expansion, especially in the heavily utilized coastal environment. Established industries resent the intrusion of a new and competing force, landowners and developers complain that aquaculture development devalues coastal property, and recreational interests object to interference with their use of public water. These are important issues that must be taken seriously by the aquaculture industry and by all levels of government. They are issues that must be handled with sensitivity and skill for the future benefit of all Canadians.

## References

AIKEN, D.E. 1986. The B.C. salmon culture industry. Bull. Aquacult. Assoc. Can. 86-3: 4–13.

1987. Island Blues, the Prince Edward Island mussel culture industry. Bull. Aquacult. Assoc. Can. 87-1: 10–20.

1989a. Salmon farming in Canada: the Pacific coast. World Aquacult. 20(2): 11–18.

1989b. The economics of salmon farming in the Bay of Fundy. World Aquacult. 20(3): 11–19.

ANDERSON, J.M. 1988. Hunting and farming salmon. Bull. Aquacult. Assoc. Can. 88-1: 22–29.

ANONYMOUS. 1988a. Commercial aquaculture in Canada. DFO/4151, Cat. No. Fs 23-139/1988E. Communications Directorate, Department of Fisheries and Oceans, Ottawa, 30 p.

1988b. Aquaculture development in Canada: a guide to federal government programs. DFO/4173, Cat. No. Fs 23-137/1988E. Communications Directorate, Department of Fisheries and Oceans, Ottawa, 83 p.

1990. Cultivating the future, an aquaculture strategy for the 90's. Communications Directorate, Department of Fisheries and Oceans, Ottawa. DFO/4243, 39 p.

1991a. Cultivating the future. Aquaculture status report. Communications Directorate, Department of Fisheries and Oceans, Ottawa. DFO/4566, 15 p.

1991b. Arctic charr wins fans at Canadian trout farms. Northern Aquacult., May/June 1991, p. 17.

B.C. SALMON FARMERS ASSOCIATION. 1990. Current developments in world salmon markets: implications for the Canadian salmon farming industry. Econ. Commer. Anal. Rep. 46: 115 p.

BIRD, C.J., AND J.L.C. WRIGHT. 1989. The shellfish toxin, domoic acid. World Aquacult. 20(1): 4041.

BOGHEN, A.D. 1989. Introduction: the state of aquaculture in Atlantic Canada, p. 3–26. *In* A. D. Boghen [ed.] Cold-water aquaculture in Atlantic Canada Canadian Institute for Research on Regional Development, Moncton, N.B.

BOURNE, N., AND J.R. BRETT. 1984. Aquaculture in British Columbia, p. 25–41. *In* G. I. Pritchard [ed.] Proceedings of the National Aquaculture Conference — strategies for aquaculture development in Canada. Can. Spec. Publ. Fish. Aquat. Sci. 75.

BOURNE, N., L. CLAYTON, AND C. HODGSON. 1989. West Coast oysters. Bull. Aquacult. Assoc. Can. 89-1: 5–12.

COOK, R.H. 1990. Salmon farming in the Bay of Fundy — the challenge of the future. World Aquacult. 21(2): 46–53.

COUTURIER, C. 1990. Scallop aquaculture in Canada: fact or fantasy? World Aquacult. 21(2): 54–62.

DUNFIELD, R.W. 1985. The Atlantic salmon in the history of North America. Can. Spec. Publ. Fish. Aquat. Sci. 80: 181 p.

EGAN, D., AND A. KENNEY. 1990. Salmon farming in British Columbia — an industry in transition. World Aquacult. 21(2): 6-11, 24–29.

HUNTER, G.A. 1989. Innovation in the 1990's: Lessons from the B.C. salmon farming industry. Bull. Aquacult. Assoc. Can. 89-4: 7–16.

ISAACS, F. 1990. Irish moss aquaculture moves from lab to marketplace. World Aquacult. 21(2): 95–97.

LAFLEUR, P.-E. 1987. Quebec trout farming. Bull. Aquacult. Assoc. Can. 87-3: 26–33.

LAVOIE, R.E. 1982. The oyster industry in eastern Canada: present status and recent developments, p. 261–265. *In* K.K. Chew [ed.] Proceedings of the North American oyster workshop, World Mariculture Society Special Publication No. 1, Baton Rouge, LA.

1989. Culture of the American oyster, *Crassostrea virginica*, p. 125–155. *In* A. D. Boghen [ed.] Cold-water aquaculture in Atlantic Canada. Canadian Institute for Research on Regional Development, Moncton, N.B.

LUCAS, K.C. 1984. Aquaculture in Canada — getting our act together, p. 1–5. *In* G. I. Pritchard [ed.] Proceedings of the National Aquaculture Conference — strategies for aquaculture development in Canada. Can Spec. Publ. Fish. Aquat. Sci. 75.

LUTZ, R.A. 1980. Introduction: mussel culture and harvest in North America, p. 1–17. *In* R. A. Lutz [ed.] Mussel culture and harvest: a North American perspective. Developments in Aquaculture and Fisheries Sciences No. 7. Elsevier, NY.

MACCRIMMON, H.R., J.E. STEWART, AND J.R. BRETT. 1974. Aquaculture in Canada — the practice and the promise. Bull. Fish. Res. Board Can. 188: 84 p.

MACCRIMMON, H. R. 1984. An overview of aquaculture in central Canada, p. 42–55. *In* G. I. Pritchard [ed.] Proceedings of the National Aquaculture Conference — strategies for aquaculture development in Canada. Can. Spec. Publ. Fish. Aquat. Sci. 75.

MACLEOD, L., M. HILL, C. ENRIGHT, AND K. LIPSETT. 1989. Aquaculture in Nova Scotia. Bull. Aquacult. Assoc. Can. 89-2:5–14.

MALLET, A.L. 1989. Culture of the mussel *Mytilus edulis*, p. 179–207. *In* A. D. Boghen [ed.] Cold-water aquaculture in Atlantic Canada. Canadian Institute for Research on Regional Development, Moncton, N.B.

MATHIESON, J.A. 1912. Oyster fishery of Prince Edward Island. Sea-fisheries of eastern Canada. Commission of Conservation, Mortimer Co., Ottawa, Ont.

MOJSEJ, J. 1991. Salmon market outlook, January 1991. Econ. Commer. Anal. Rep. 81: 91 p.

Muise, B. 1990. Mussel culture in eastern Canada. World Aquacult. 21(2): 12–23.

Nash, C.E., and C.B. Kensler. 1990. A global overview of aquaculture production in 1987. World Aquacult. 21(2): 104–112.

Needham, T. 1990. Canadian aquaculture — let's farm the oceans. World Aquacult. 21(2): 76–80.

Newkirk, G.F. 1989. Culture of the Belon oyster, *Ostrea edulis*, in Nova Scotia. *In* A.D. Boghen [ed.], Cold-water aquaculture in Atlantic Canada. Canadian Institute for Research on Regional Development, Moncton, N.B.

Painter, R. 1990. Warming up to Arctic charr. Can. Aquacult. 6(1): 32–35,45.

Pennell, W. 1987. B.C. shellfish culture. Bull. Aquacult. Assoc. Can. 87-4: 5–15.

Price Waterhouse Management Consultants. 1990. Long term production outlook for the Canadian aquaculture industry (1990 edition). Econ. Commer. Anal. Rep. 67: 100 p.

1991. The financial structure and performance of the salmon farming industry in Canada. Econ. Commer. Anal. Rep. 94: 144 p.

Saunders, R.L. 1989. Salmonid aquaculture: present status and prospects for the future, p. 31–75. *In* A.D. Boghen [ed.] Cold-water aquaculture in Atlantic Canada. Canadian Institute for Research on Regional Development, Moncton, N.B.

Shumway, S. 1989. Toxic algae, a serious threat to shellfish aquaculture. World Aquacult. 20(4): 65–74.

Stephenson, R.L. 1990. Multiuse conflict — aquaculture collides with traditional fisheries in Canada's Bay of Fundy. World Aquacult. 21(2): 34–45.

Sutterlin, A., D. Aggett, C. Couturier, R. Scaplen, and D. Idler. 1981. Mussel culture in Newfoundland. Mem. Univ. Nfld., Mar. Sci. Res. Lab. Tech. Rep. 23: 82 p.

Talley, K. 1990. Industry excess continues to plague marketplace. Can. Aquacult. 6(1): 31.

Thompson, A.G. 1990. Crayfish — a candidate for culture in British Columbia? World Aquacult. 21(2): 92–94.

Wildsmith, B. 1982. Aquaculture: the legal framework. Emond-Montgomery Ltd., Toronto, Ont. 313 p.